Volume 09

Make:

technology on your time™

Shock and Awe
Journey to the outer limits of belief in our special fringe technology section.

ON THE COVER: Kirlian Photo of a Coin
Jasper Nance, who took the cover photo, writes that "the light you see during a Kirlian discharge comes from an electric field ionizing the air and parts of your Kirlian subject. The charged air gives off intense heat and ultraviolet radiation as well as visible light." *(See page 66.)*

Special Fringe Section

Columns

KIRCHER'S CLOCK
As if by magic, a floating fish points at the current hour on a rotating globe of the Earth.

Vol. 09, March 2007. MAKE (ISSN 1556-2336) is published quarterly by O'Reilly Media, Inc. in the months of March, May, August, and November. O'Reilly Media is located at 1005 Gravenstein Hwy. North, Sebastopol, CA 95472, (707) 827-7000. SUBSCRIPTIONS: Send all subscription requests to MAKE, P.O. Box 17046, North Hollywood, CA 91615-9588 or subscribe online at makezine.com/offer or via phone at (866) 289-8847 (U.S. and Canada), all other countries call (818) 487-2037. Subscriptions are available for $34.95 for 1 year (4 quarterly issues) in the United States; in Canada: $39.95 USD; all other countries: $49.95 USD. Periodicals Postage Paid at Sebastopol, CA, and at additional mailing offices. POSTMASTER: Send address changes to MAKE, P.O. Box 17046, North Hollywood, CA 91615-9588.

Make: Projects

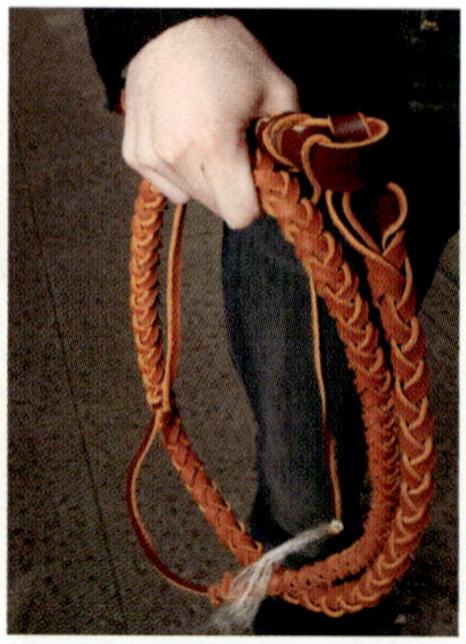

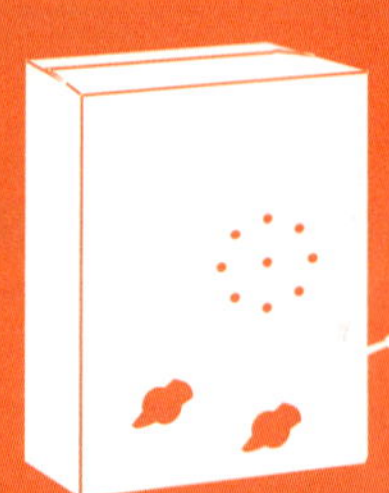

Volume 09

Make:

technology on your time™

Maker

READ ME: Always check the URL associated with each project before you get started. There may be important updates or corrections.

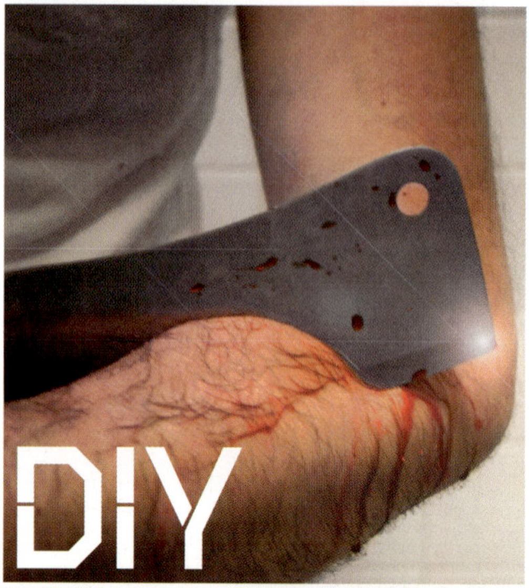

115

77" of your "thing" capacity
30" of your "thing" capacity
39.5" of your "thing" capacity
You have "a thing." And no. It's not your "issues." Or your "stuff." It's the one thing you love doing. And maybe even being. Maybe you feel it when you're plugging in a guitar. Maybe you feel it when you're sitting in the middle of the ocean, alone. Waiting. Maybe you feel it when you're taking a load of Little Leaguers to Districts, two states away. Whatever it is, it's your thing. And it's always been your thing. Okay, sure. People may share the same thing. But deep deep, deep down, you know your thing is a little different. (Okay...a lot different.)
The Passat Wagon is a thing for people who have a thing, not just things.
So fill it up. Load it up. All 73.77 cubic ft.* of it. And do your thing.

Make:
technology on your time™

EDITOR AND PUBLISHER
Dale Dougherty
dale@oreilly.com

EDITOR-IN-CHIEF
Mark Frauenfelder
markf@oreilly.com

MANAGING EDITOR
Shawn Connally
shawn@oreilly.com

ASSOCIATE MANAGING EDITOR
Goli Mohammadi

SENIOR EDITOR
Phillip Torrone
pt@makezine.com

PROJECTS EDITOR
Paul Spinrad
pspinrad@makezine.com

STAFF EDITOR
Arwen O'Reilly

TECHNICAL EDITOR
Brian Jepson

COPY CHIEF
Keith Hammond

COPY EDITORS/RESEARCH
Rachel Monaghan
Marlowe Shaeffer
Barbara Talley

EDITOR AT LARGE
David Pescovitz

CREATIVE DIRECTOR
Daniel Carter
dcarter@oreilly.com

DESIGNER
Katie Wilson

PRODUCTION DESIGNER
Gerry Arrington

PHOTO EDITOR
Sam Murphy
smurphy@oreilly.com

ONLINE MANAGER
Terrie Miller

ASSOCIATE PUBLISHER
Dan Woods
dan@oreilly.com

DIRECTOR OF EVENTS
Sherry Huss

CIRCULATION DIRECTOR
Heather Harmon

SALES & MARKETING ASSOCIATE
Katie Dougherty

MARKETING & EVENTS COORDINATOR
Rob Bullington

MAKE TECHNICAL ADVISORY BOARD
Gareth Branwyn, Limor Fried, Joe Grand, Saul Griffith, William Gurstelle, Bunnie Huang, Tom Igoe, Mister Jalopy, Steve Lodefink, Erica Sadun

PUBLISHED BY O'REILLY MEDIA, INC.
Tim O'Reilly, CEO
Laura Baldwin, COO

Visit us online at makezine.com
Comments may be sent to editor@makezine.com

For advertising and sponsorship inquiries, contact:
Dan Woods, 707-827-7068, dan@oreilly.com

Customer Service cs@readerservices.makezine.com
Manage your account online, including change of address at:
makezine.com/account
866-289-8847 toll-free in U.S. and Canada
818-487-2037, 5a.m.–5p.m., PST

MAKE is printed on recycled paper with 10% post-consumer waste and is acid-free. Subscriber copies of MAKE, Volume 09 were shipped in recyclable plastic bags.

Please Note: Technology, the laws, and limitations imposed by manufacturers and content owners are constantly changing. Thus, some of the projects described may not work, may be inconsistent with current laws or user agreements, or may damage or adversely affect some equipment.

Your safety is your own responsibility, including proper use of equipment and safety gear, and determining whether you have adequate skill and experience. Power tools, electricity, and other resources used for these projects are dangerous, unless used properly and with adequate precautions, including safety gear. Some illustrative photos do not depict safety precautions or equipment, in order to show the project steps more clearly. These projects are not intended for use by children.

Use of the instructions and suggestions in MAKE is at your own risk. O'Reilly Media, Inc., disclaims all responsibility for any resulting damage, injury, or expense. It is your responsibility to make sure that your activities comply with applicable laws, including copyright.

Contributing Editors: William Gurstelle, Mister Jalopy

Contributing Artists: Doug Adesko, Meiko Arquillos, Scott Barry, Howard Cao, Roy Doty, Jason Forman, Marcia Friedman, Karen Hansen, Dustin Amery Hostetler, Timmy Kucynda, Erika Larsen, Tim Lillis, Ross Orr, Charles Platt, Hal Robins, Nik Schulz, Damien Scogin, Robyn Twomey, Izu Watanabe

Contributing Writers: Tim Anderson, Thomas Arey, Sunny Armas, Andrew Michael Baron, Dennison Bertram, Blind Lightnin' Pete, Shawn Carlson, Larry Cotton, Cory Doctorow, George Dyson, Michael Flynn, Michael John Gorman, Saul Griffith, William Gurstelle, Kaden Harris, John Iovine, Tom Jennings, Alek Komarnitsky, Heidi Kumao, Todd Lappin, Andy Lee, Mister Jalopy, Steve Nalepa, Nicole Oncina, Tim O'Reilly, Ross Orr, Tom Owad, Bob Parks, Jonah Peretti, David Pescovitz, Charles Platt, Michael H. Pryor, Douglas Repetto, Donald E. Simanek, Sparkle Labs, Bruce Sterling, Zack Stern, Brian Stucki, Cy Tymony, Ed Vogel, John Wanberg, Audra Wolfman, Tom Zimmerman, Lee Zlotoff

Interns: Matthew Dalton (engr.), Adrienne Foreman (web), Jake McKenzie (engr.)

Founding Creative Agency: Albertson Design

Contributors

Jake MacKenzie (MAKE engineering intern) started taking stuff apart as soon as he discovered it was possible, and can get most of it back together. He enjoys bike riding, table tennis, video games, and foreign or classic films (boo Hollywood). An engineering student, he likes working on vintage amps, DIY radio gear, synths, and bikes. His current project is restoring a Tektronix oscilloscope from the 50s ("They don't make 'em like they used to — this thing is truly a boat anchor"). He will always eat udon, phô, pizza, or waffles.

Donald Simanek (*Reinventing the Square Wheel*) is a retired professor of physics with much experience dealing with things that don't work as they should. He has long been concerned with how students learn (and fail to learn) basic concepts of physics, and with the continuing failures of institutionalized education. The culmination of these concerns is his website (www.lhup.edu/~dsimanek/home.htm), which includes the occasionally acclaimed "Museum of Unworkable Devices." His other interests include the history of science, the science/pseudoscience boundary, photography, and classical music. He lives in Pennsylvania with his wife, Connie, a mathematician, a field where everything works, by definition.

Robyn Twomey (*Workshop* photographer) is "nonstop, curious, obsessive, resourceful with material things, and indulgent with expensive food." Her shoot with Kevin Binkert was a great success: "At first he was too humble to ham it up for the camera, but I couldn't get him to stop posing by the end of it." She is currently working on a photo essay about people with Tupac tattoos, but for fun she likes to "drink pinot noir, drive down the coast, read 'The Ethicist,' and go dumb." She lives in the Mission District of San Francisco, loves the idea that "30 is the new 20," and will always eat Delfina's flatiron steak and her mom's homegrown artichokes.

John Iovine (*Digital Kirlian Photography*), a tinker and science experimenter, is the author of many books, including *Kirlian Photography* and *Holography for Photographers*. He lives in Staten Island with his children, Anna Rose and James, his wife, (the "one and only") Ellen, dog Chansey, and "a number of neighborhood strays." What interesting projects is he working on now? "That could fill a book." He likes to read and write science fiction ("writing is hard, fiction is harder"), and explore the areas of science that "mainstream science doesn't want to look at." His favorite food is steak, barbecued.

Michael John Gorman (*Kircher's Clock limited edition booklet*) is the author of several books and various articles about the history of science and technology in journals including *Nature*, *Science*, and *Leonardo*. He is a founder of SEED (seedartscience.blogspot.com), a Dublin-based group dedicated to the dialogue between art and science. Michael John lived for some years in Rome, where he was cofounder and director of the Athanasius Kircher Correspondence Project, before returning (via Stanford) to his native Dublin, where he lives with his wife, Caroline Bouguereau, and their 2-year-old daughter Maia. He is currently director of the Science Gallery at Trinity College.

Sam Murphy (MAKE photo editor) lives in the Mission District of San Francisco with her cinematographer boyfriend, Mike, and their two kitties, Presley and Priscilla (who they did not name). Together, Mike and Sam make nonfiction films for nonprofits (cyclepathproductions.com), and annoy people with their photo/film-geek talk. Sam is a slightly angry, smart-ass, vegetarian bicyclist and bluegrass lover. When she is not watching a documentary film, she can be found playing soccer or drinking a soy chai. Her photographs can be seen in this issue opening each major project, and Mike is the star rocking out with the cracker box amp (and cracking the bullwhip).

FRINGE BENEFITS

By Mark Frauenfelder

STEWART BRAND, FOUNDER OF THE seminal *Whole Earth Review*, once famously said, "You've got to explore the edges to see where the middle is going." It's good advice because interesting changes often occur where cultures, ideas, and disciplines merge and mutate.

The special section in this volume of MAKE explores ideas on the edge. For example, Bob Parks' article on the Global Consciousness Project in Princeton, N.J., looks at the heretical notion that random number generators might be able to predict catastrophic events hours before they happen. We also show you how to take Kirlian photographs, which reveal the "auras" surrounding objects and living things. The principle behind Kirlian photography is well understood, and there's nothing supernatural about it, but the history of the process is interestingly laced with pseudoscientific speculation, and the resulting photographs are beautiful.

In this issue we also explore ideas that go way beyond the edge of traditionally accepted notions into the domain of the inexplicable, the unreasonable, and the impossible. Why should we be interested in this outer fringe? Is there anything of value being made there? More often than not, the answer is no. But that doesn't mean it's not worth making the occasional foray into the murky areas where it is difficult to distinguish between quackery and the genuine-but-inexplicable.

The most obvious reason for exploring the outer fringe is to discover the limits of the natural world. Take perpetual motion, for instance. To this day, a curious subculture of cockeyed optimists are tinkering away in self-funded laboratories, trying to make perpetual motion machines that defy the cardinal rules of physics (which say you can't get something for nothing). Occasionally, some of these inventors even get funding and public attention for their outlandish contraptions. The most recent free-energy company to shine under a surprisingly uncritical media spotlight is the Dublin-based Steorn, which claims to have developed a technology that uses magnetism to produce more energy than it consumes. If Steorn's claims prove true, it would mean the end of the energy crisis, and the potential to raise the standard of living of every person on the planet to the kind enjoyed by Halliburton's board of directors.

Anyone who thinks Steorn might be onto something would do well to read Donald Simanek's article on the history of perpetual motion machines in this volume (*page 70*). Simanek explains why making a perpetual motion machine is a worthwhile experience — not because it will work, but because you will gain firsthand knowledge learning why it does not work.

Rest assured, this issue contains plenty of projects that *do* work, like the amazing Hilsch vortex tube, which uses compressed air to push hot air out one side of the tube and cold air out the other side, and a pinhole camera that doesn't require your own darkroom to develop the photos. (We leave it up to you to come up with a way to take Kirlian photos with a pinhole camera.) Our nifty, easy-to-make Curie heat engine looks like something that could be found merrily swinging back and forth in the corner of a perpetual motionist's laboratory, but its operation depends on the energy of a candle to heat up a nickel-iron alloy wire so that it temporarily loses its attraction to a magnet. You might be able to use it as a metronome when you play your cigar box guitar (*from MAKE, Volume 04*) through the cracker box amplifier we'll show you how to make in this issue (*page 104*).

Finally, I'd like to remind you to mark your calendars for this year's Maker Faire, which will be held at the San Mateo County Fairgrounds in Northern California on the weekend of May 19–20, 2007. Last year over 20,000 people came to participate in demonstrations, projects, entertainment, and learning. If you're a maker and have something you'd like to show at the faire, visit makerfaire.com for information. I hope I see you there!

Mark Frauenfelder (markf@oreilly.com) is editor-in-chief of MAKE.

DOOHICKIE CENTRAL
Recycling & Remarketing High Technology
WEIRDSTUFF®
WAREHOUSE
21 WEIRD YEARS of the Best Surplus ~ Deals ~
Software • Computers • Electronics • Equipment • Doohickies
Sound Blaster Live! 32-bit Sound Card $10
Jetway V2MDMP Socket A Motherboard $18
Seagate Barracuda 40GB IDE Hard Drive $25
Seagate Barracuda 80GB IDE Hard Drive $35
Seagate Barracuda 160GB IDE Hard Drive $45
GeForce 4 Ti4200 64MB AGP Video Card $15
200 Watt Ablecom 1U ATX Power Supply $5
Compaq 7000 Series 19" 4-Post Rack $300
HP 4M HPIB/GPIB Cable 10833C $20
6-Way Adjustable Office Chair In Store Only $40
CyberHome 5.1 Surround Headphones $35
USB 2.0 Cardbus Adapter for Laptops $20
48" Banana Plug to Minigrabber Test Clip $15
3M Workstation Static Monitor System $35
Agilent Laser Diode Sensor Head E1711A $50
PRI Automation Robot Controller TTR-300 $50
Tygon 50 Ft. General Lab Tubing $35
Motor - 230V 60HZ 200 RPM $8
Hard to Find Products, High Tech Stuff, Closeout Items, Obsolete Stuff, & Much More!
The majority of our products are surplus (new and used) and are available in limited quantity. First come first serve - while supplies last! Don't delay!
WEEKLY SEALED BID SALE AUCTION
Hi-Tech Items, Electronics, Test Equipment and more! New Auction every other Friday
WE BUY AND SELL EXCESS INVENTORIES FOR CASH!
Visit Our Sunnyvale Store
384 W. Caribbean Dr.
Sunnyvale, CA 94089
Mon-Sat: 9:30-6:00 Sun: 11:00-5:00
(408)743-5650 Store x324
GIANT AS-IS SECTION!
10,000 Sq. Ft. of Computers, Electronics, Software, Doo-Hickies, Cables, and More!
FREE COMPUTER RECYCLING
WeirdStuff will Recycle most Computers, Monitors, and Electronic Equipment.
Sign up for our SPAM FREE newsletter!
WEIRDSTUFF® Web Cam! view online!
WWW.WEIRDSTUFF.COM
* = for the latest, lowest prices, please call our store or visit www.weirdstuff.com
© 2007 WeirdStuff, Inc.

Use these:
SUN SPOT JAVA DEVELOPMENT KIT
Experimental Technology from Sun Labs
Java™
To build these:
http://www.sunspotworld.com
Sun microsystems

PERVASIVE PUBLISHING

By Tim O'Reilly

PETER BRANTLEY, OF THE CALIFORNIA Digital Library, has written a fascinating meditation on the future of publishing (xrl.us/brantley):

Books are artifacts — loved, endearing, effective in their form. They will endure. But they and other media will be consumed in a wide range of environments, more or less virtual, more or less complex, more or less multimedia. It is all of a degree.

It is not, therefore, whether the reading experience will be textual or richly media immersive. All the world will be a skein of media, linked and interacting, flowing and increasingly seamless. There will be text, but text embedded within other environments, available to and linking to a rich cornucopia of information, amusement, and education.

Brantley nominated Second Life as a candidate for both the platform, and the channel for media in the future. Much as I like SL, I'm not sure that's right. Second Life as currently constituted is a pale shadow of what is to come. What I took away from Peter's post was the idea of Second Life as a metaphor for pervasive publishing. We already live in Second Life; we just haven't recognized it yet.

As the poet Wallace Stevens said, "It is not in the premise that reality is a solid." Reality is always a dialogue between the world and the mind, what we see and what we make of it. Stored ideas and images are used to filter and enrich the stuff of everyday experience. A baby spends its first years building a map of the world; an active, thoughtful reader continues that process while a passive media recipient lets others do the mapping. Ideas and images are tools we think with. Mastery of these tools makes the difference between creativity and mere consumption.

Books (and in fact, all media, including commercial media like advertising) have always provided "a context in which other people can think," as Edwin Schlossberg famously said, but what is different today is the pervasiveness of that media.

Think back to the days before Google and Wikipedia, when a gap in your knowledge was not so easily filled, and now roll forward, to imagine a world where the ability to check facts, to add background, and to share the experience with others, is truly ubiquitous. We can look forward to a future in which information retrieval about the things we encounter in our daily lives, and contextual advertising to go with it, is not limited to text on a screen, but becomes a component of audio and video overlays on our perceptual field.

In the past, the dialogue between mind and matter was carried out in the imagination. But increasingly, it will happen as part of our daily experience, mediated by technology. We will need to find new ways of engaging, new ways of being active rather than passive, or our thinking will be done for us.

And that brings us back to Second Life. What is important about Second Life is not just its immersive quality. It's the fact that its members are active participants in creating it. They are makers, not consumers. That's what is also important about the World Wide Web, Wikipedia, YouTube, and Flickr.

Stevens talked about this idea, too. He wrote about reality as "the supreme fiction," and foresaw a future in which we sought not "truth" but rather aesthetic visions that would form a new shared reality.

The future of media is the future of thinking, the future of the shared imaginings that we call culture, and science, and religion. How we communicate, and what we communicate, are inseparable. What does not change is the imperative to engage, to imagine, and to remake what we are given into something new that we can share.

Check makezine.com/09/nff for links to related stories.

Tim O'Reilly (tim.oreilly.com) is founder and CEO of O'Reilly Media, Inc. See what's on the O'Reilly Radar at radar.oreilly.com.

MAKER'S CORNER

By Dan Woods, Associate Publisher

Advice and news for MAKE readers.

Maker Faire 2007: Save the Date!

One of the more popular emails we receive is, "Are you guys doing another Maker Faire this year?" Hell yes, we are! In fact we're planning two Maker Faires this year:

- SF Bay Area Maker Faire: May 19–20 at the San Mateo County Fairgrounds
- Austin, Texas Maker Faire: October 20–21 at the Travis County Fairgrounds

And yes, we are also actively exploring other potential Maker Faire venues across the planet. If you'd like to be alerted as to upcoming MAKE events, including future Maker Faires, visit makezine.com/faire and sign up for the MAKE newsletter.

So save the dates, be a good maker and spread the word, and sign up for the free MAKE newsletter (makezine.com/newsletter) for further details.

What Is a Premier Maker?

Answer: Someone who finds the least expensive way to subscribe to MAKE (and more).

Become a Premier Subscriber, and each year when your subscription is up for renewal, we'll send you a single email notification letting you know that we're about to renew your MAKE subscription and charge your card. You'll always receive a subscription price at least $5 off the standard subscription price; you'll receive just one email versus multiple emails and snail mail notifications; and for being a Premier Subscriber, you also receive full access to the entire MAKE archive when you access the MAKE: Digital Edition, a complimentary service available exclusively to MAKE subscribers.

Upgrade to Premier Maker at makezine.com/premier.

MAKE: The Next Year Boxed Collector's Set Now Available

It's finally here: MAKE: The Next Year! The original boxed collector's set, MAKE: The First Year, proved so popular with our readers (three reprints!) that we decided to put together a special collector's edition of our entire second year. Hence, MAKE Volumes 05 through 08 are now packaged up in a sturdy (and handsome) collector's box. Our price: $60. That's the same price as buying the individual copies on the newsstand, plus we throw in the box. Hard to think of a better gift for a maker (or maybe yourself). Order yours today at store.makezine.com.

MAKE Has a Brand New Store

If you weren't among the record number of makers who visited the Maker Store (store.makezine.com) last December, you haven't yet seen our brand new storefront or our growing list of cool MAKE: It tools, project kits, books, apparel, and toys. Just a small sample of our new products:

- MAKE Warranty Voider tool
- Limor Fried's MintyBoost USB Charger Kit for your MP3 player, camera, cellphone, etc.
- MAKE Controller Kit
- MAKE Open Source MP3 Player Kit
- Andre Lamothe's XGameStation Pico Edition 2.0
- Building Small Form Factor PCs (downloadable book)

Stop by and check out the latest merch at store.makezine.com.

CRAFT, Volume 02 Is Out!

When the first box of CRAFT, Volume 02 arrived in the MAKE offices, it may have set a Maker Media record for vanishing acts. I stashed a copy for some perusing at home that night. No better way to say it than Frickin' Awesome! Check out the TOC and sample articles at craftzine.com/magazine. And if the spirit moves you, consider giving CRAFT as a gift for any and all occasions. (Maybe you forgot Valentine's Day?)

Dan Woods is associate publisher of MAKE and CRAFT magazines. When he's not working on circulation and marketing or finding cool new stuff for the Maker Store, he likes to hack and build barbecues, smokers, and outdoor grills.

I AGREE

IN ORDER TO "PROTECT INTELLECTUAL PROPERTY," THESE BANDITS ARE PERFECTLY WILLING TO TAKE AWAY YOUR RIGHTS TO YOUR PHYSICAL, PERSONAL PROPERTY.

By Cory Doctorow

BY READING THIS SENTENCE, YOU indicate your agreement to the following terms and conditions:

1. *You agree that you will buy two copies of every issue of MAKE until further notice.*
2. *You agree that you will not lend either of your MAKE magazines, or allow them to be lent by a third party.*
3. *You agree that you will not sell your MAKE magazines.*
4. *You agree that MAKE is provided to you under license, and that it remains the exclusive property of O'Reilly Media, Inc.*
5. *You agree that you will be subject to these terms and conditions and other terms and conditions that O'Reilly Media might, from time to time, publish.*
6. *You agree that you have made this agreement freely and of your own volition, that if any term of it is found to be unenforceable that the remainder will remain in force, that the jurisdiction for any litigation arising from this agreement is the State of California, and that you will pay any legal fees accruing to O'Reilly Media in the case of such litigation.*

☐ I Agree (Check this box to proceed.)

It's amazing how little it takes to form an "agreement" these days. I once heard an MPAA executive explain that you could "agree" to waive your fair use rights to record a TV show by changing the channel. My ThinkPad arrived from Lenovo with an eight-page "acquisition agreement" that explained all the things I'd just "agreed" to by sending them thousands of dollars. Amazon Unbox makes you agree to install spyware on your computer in order to download their videos. From the dry-cleaner's ticket to the hotel room to the small print on the back of your cellular bill, these gotcha-greements lurk everywhere, waiting to leap out and clobber you with abusive terms and services. Woe betide the fool who installs Vista (you "agree" not to solve your own problems without authorization) — and don't let's get started on all the things you "agree" to before you get on a plane.

Civilization begins with an agreement: the social contract. The idea that we can meet as relative equals and hammer out a compromise practically defines what it means to be human. Can you have a social contract if you reduce the idea of a contract to "By standing there, shaking your head, and screaming, 'No no no no no no! I don't agree,' you agree to let me come over to your house, wear your underwear, make some long distance phone calls, and clean out your fridge. Run away screaming to indicate your assent"?

These "agreements" set out the terms under which you can use your own property. They waive fair use, prohibit lending, resale, or reverse engineering — sometimes even swear you to secrecy!

If you're a maker, your vocation is voiding warranties and creating fire hazards. These "agreements" set out two classes of people: lordly manufacturers who own and control everything, and lowly tenant farmers who merely borrow their goods from the lord.

Being a maker is about owning and Owning — having access to the lowest-level guts of the machine, putting Tab A into any slot that takes your fancy. It's the right to break your stuff and fix it again. It's all about real, non-metaphorical property.

Start a count of the "agreements" in your life. How many times a day do you "agree," and what is it you "agree" to? Start pushing back. Don't take it lying down.

Be disagreeable.

Cory Doctorow (craphound.com) is a science fiction novelist, blogger, and technology activist. He is co-editor of the popular weblog Boing Boing (boingboing.net), and a contributor to *Wired*, *Popular Science*, and *The New York Times*.

MADEONEARTH

Report from the world of backyard technology

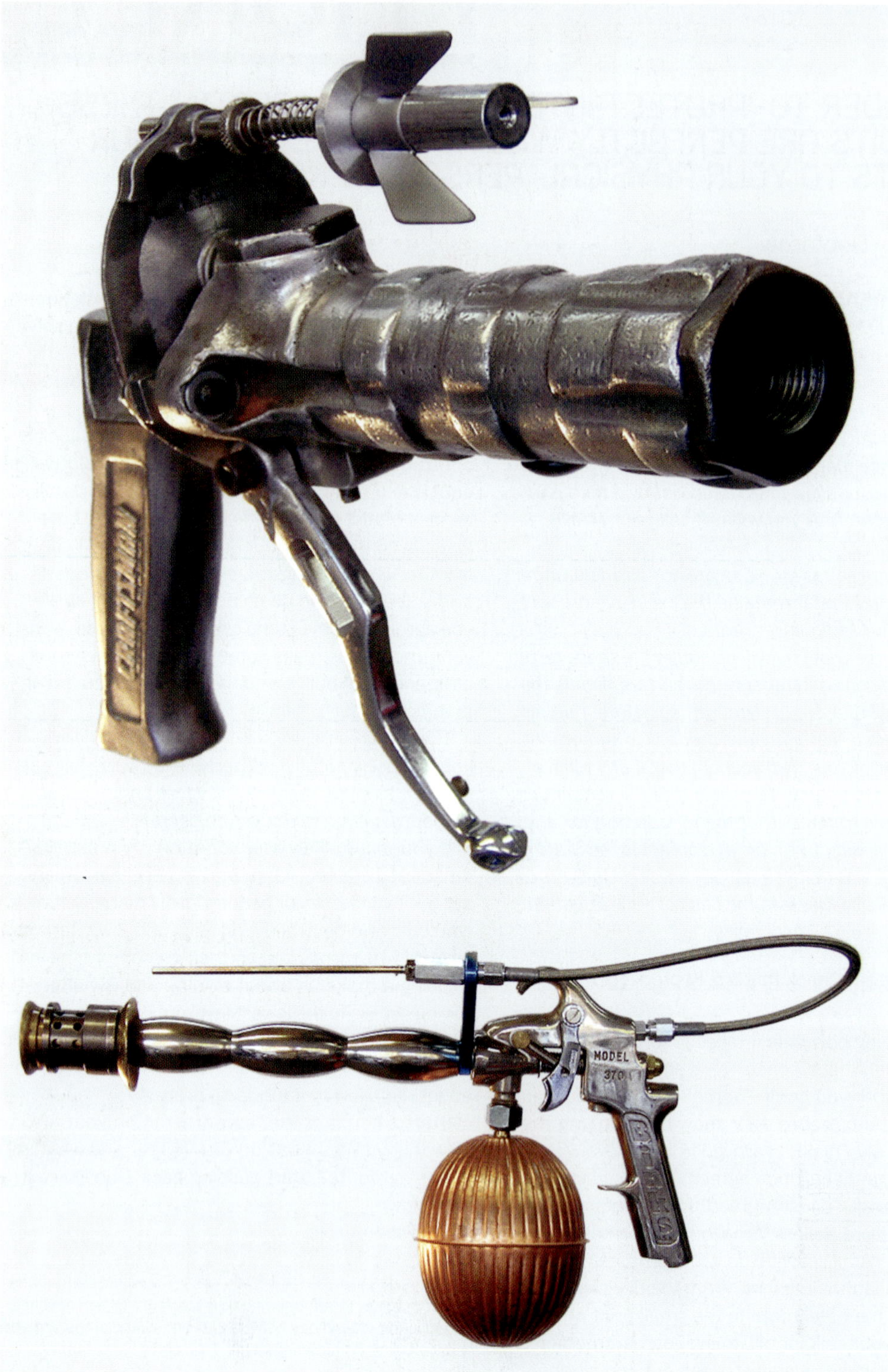

NO POSTAGE
NECESSARY
IF MAILED
IN THE
UNITED STATES

BUSINESS REPLY MAIL

FIRST-CLASS MAIL PERMIT NO 865 NORTH HOLLYWOOD CA

POSTAGE WILL BE PAID BY ADDRESSEE

O'REILLY

Make:

PO BOX 17046
NORTH HOLLYWOOD CA 91615-9588

NO POSTAGE
NECESSARY
IF MAILED
IN THE
UNITED STATES

BUSINESS REPLY MAIL

FIRST-CLASS MAIL PERMIT NO 865 NORTH HOLLYWOOD CA

POSTAGE WILL BE PAID BY ADDRESSEE

O'REILLY

Make:

PO BOX 17046
NORTH HOLLYWOOD CA 91615-9588

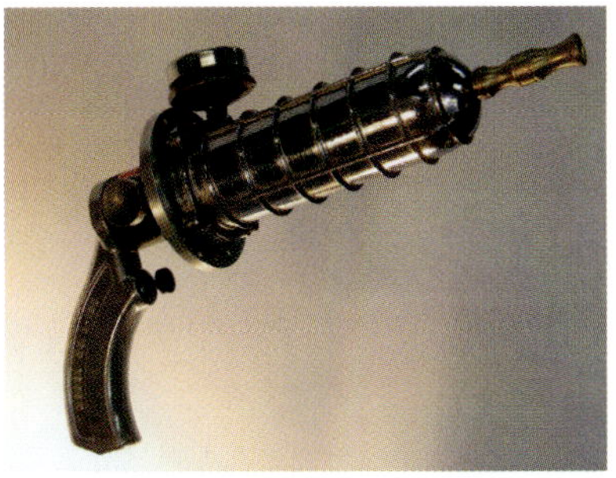
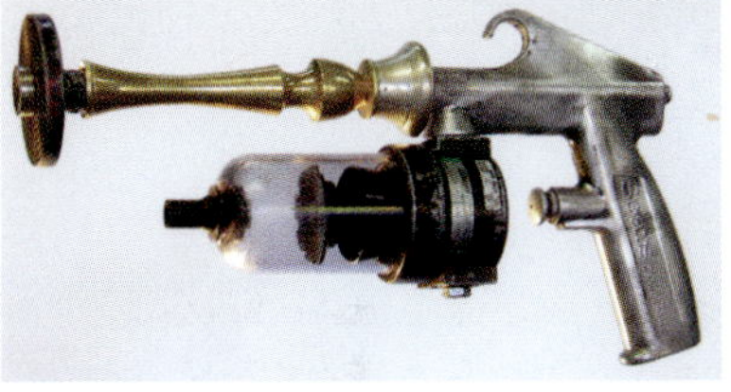
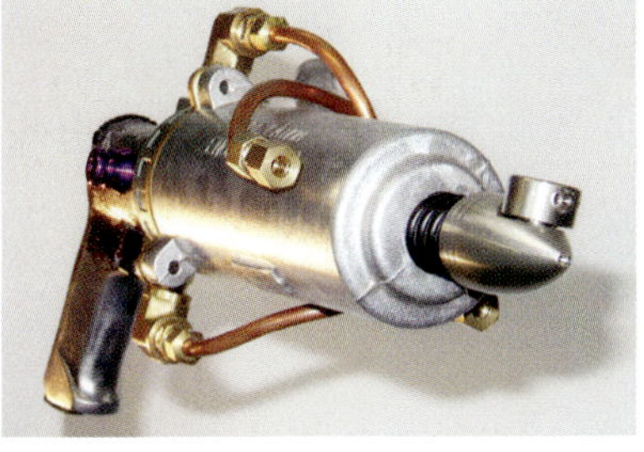

Photography by Paul Loughridge; photograph far right by Cindy Loughridge

Raygun Reverie

"You know, guys never really grow up," says **Paul Loughridge**. His hand-built, retro raygun models have the authentic menace of *Star Wars* and the wacky form factor of *Men in Black*. But their mid-century, all-metal look is straight out of his childhood favorites: *Flash Gordon*, *Amazing Stories*, *Lost in Space*.

Collectors think he's on to something. His work has a satisfying heft from a time before plastics and semiconductors ruled the Earth, when science fiction was built to last with rivets and bolts, copper flanges and chrome fins. Who wouldn't love to fry a Martian with the Aluminizer, the Shrink Ray, or the Double Barreled Intergalactic Self-Defense Mechanism?

Loughridge, 51, is a Silicon Valley graphic designer, trade-show director, and compulsive collector whose itch for vintage toys took him down the DIY path. Maybe it was his wife banishing his collection ("rayguns, rocketships, and robots ... all pre-*Star Wars*") to a storage unit. Maybe it was the raygun that got away.

"I'd just got back from a toy show," he recalls, "and must have seen some bitchin' raygun I couldn't afford. I decided to make my own out of metal. Used an old drill handle that had belonged to my dad." He Googled "raygun," and master maker Clayton Bailey became his inspiration. Dozens of rayguns followed.

In a garage outfitted with a drill press and grinder, Loughridge bolts together flea-market junk. Pistol grips come from old air tools and hacksaws. Barrels and muzzles might be a mean-looking orange juicer, BMX foot peg, or — just once — a perfectly good flower vase. ("Sorry, honey!") Finishing touches include cocktail shakers, brake hoses, and copper toilet floats.

"Rayguns are definitely a guy thing ... it's a Tim Allen grunt kinda thing. Boys want to touch them." But he's broadening his audience. "I made a couple robot dogs and female robots, and women are buying them."

Now when he's not tending his bonsai tree collection (he's got 70 in the backyard), you might find Loughridge in the garage, creating toy robots, rocketships, and rayguns. "I just love old retro stuff. It all has to do with being a kid." —*Keith Hammond*

» **Rayguns:** lockwasherdesign.com/index_4.htm
claytonbailey.com/galleryrayguns.htm

Illegal Soapbox Derby Races

There are no permits from City Hall, no advance promotions, no schedules, and only one ironclad rule: every car must have a beer can holder. Yet even that edict often falls by the wayside because no one knows who's in charge of enforcement. And besides, at the Bernal Heights Illegal Soapbox Derby races — held each autumn in San Francisco —the point isn't to follow the rules, or even to go fast.

Instead, it's about figuring out how to build a gravity-powered car out of whatever materials you can find, then taking your homebuilt wheels on a white-knuckle ride down a twisty mountain road that overlooks the sprawling city below.

"Some guys want to win more than others, but pretty much it's all just for fun," says **Scott Strebel**, 29, co-creator of a car unofficially known as the Keg Racer. Built over the course of a week by combining the front end of a Quarter Midget go-cart, a few wheels purchased at Orchard Supply, and three old beer kegs he found lying around his garage, Strebel's vehicle looks like a Homer Simpson-inspired cross between Luke Skywalker's Landspeeder and Anakin Skywalker's Pod Racer — though he insists the similarity is purely coincidental.

"We just arranged all the materials on the floor, then figured out what we could build with them," he says. Other cars on the hill in 2006 included a bullet-shaped streamliner with a fully enclosed cockpit, a four-wheeled surfboard, an unmodified 1960s pedal-car called the Dude Wagon, and a rolling coffin emblazoned with Dale Earnhardt's signature "3" logo.

Soapbox derby races have been taking place atop Bernal Hill since the 1970s, but it's not entirely clear what makes the event illegal — some say the outlaw tag refers to the unauthorized appropriation of a normally serene city park, while others claim the race is a defiant antidote to the more uptight (read: rulebook) American Soap Box Derby.

Of course, the AASBD doesn't require each car to have a beer can holder. But then again, at the Bernal Heights Illegal Soapbox Derby Races, no one really does either.

—Todd Lappin

Photograph by Telstar Logistics

Photograph courtesy of San Francisco Maritime National Historic Park

History Junkies

As an undergrad, **John C. Muir** came across 19th century photos of Chinese junks sailing the San Francisco Bay. Curious about what these foreign vessels were doing in California, Muir dove into maritime research and was soon building a replica himself.

Muir learned that several immigrant Chinese shrimping communities settled in the San Francisco and San Pablo Bays between 1860 and 1910. Little remains of these villages, but the well-preserved remains of two redwood junks were discovered during low tide in the mudflats of China Camp State Park in Marin County.

Muir, now a curator of small craft at San Francisco Maritime National Historical Park, decided to recreate a traditional California-Chinese shrimp junk from the ground up. With Park Service approval he snowshoed through the mud to take photos and measurements, and enlisted the help of third-generation China Camp resident **Frank Quan** and a crew of mostly amateur boat-builders, who called themselves "Junkies."

To build the 42-foot replica, they used traditional techniques Muir learned from trips to boatyards in the Guangdong province of China. The Junkies bent the redwood of the keel and planks over a fire, shaping them around a fulcrum while clamping or weighing the ends down with buckets of rocks, and constantly spraying the lumber with water, to prevent burning.

"At first fire-bending seemed counterintuitive," says Junkie **Inka Petersen**, "but it works instantaneously. Plus we got to have a bonfire on the beach every day."

When it was time to piece the vessel together, volunteer blacksmiths taught the Junkies to forge headless iron nails based on samples Muir brought from China. Headless nails are integral to the art of edge-nailing, which joins the planks to each other as well as to the main structure.

Six months later the junk was christened the *Grace Quan* after Frank's mother. The crew raised the hand-sewn and tanbark-dyed cotton canvas sail, perhaps the vessel's most stunning feature, and the *Grace Quan* set sail on San Francisco Bay.

—Audra Wolfmann

» **Shrimp Junk Project:** nps.gov/archive/safr/junk.html

Purrfect Relaxation

Feeling stressed? You need a visit to the studio of **Duncan Laurie**, a three-story structure perched on a gorgeous spit of land that overlooks Rhode Island's Narragansett Bay. Laurie built the structure from mostly salvaged parts, and if the salt air and crashing waves outside don't ease your mind, the studio's Purr Generator surely will.

The Purr Generator is based on Radionics, an alternative technology founded by Dr. Albert Abrams in the early 1900s, on the observation that all matter emits radiation. Radionics surmises that a healthy person attains energetic equilibrium, while unhealthy people are essentially out of tune. The Purr Generator aims to restore balance with healing frequencies that neutralize energy blockage.

The Purr Generator attempts to replicate and amplify the sense of relaxation people experience while holding a purring cat. The device produces its main signal wave at a happy-feline frequency of approximately 25Hz, and directs it into the user as sound, physical vibrations, and "radionic intent," bathing the body in good vibes. A second channel generates a signal that's close to the first but offset by a user-adjustable +/- 2Hz, which adds a dramatic spatial effect and a throbbing beat frequency.

To use the device, you lay on a bed suspended inside a cube-octahedron structure. A coneless magnetic coil under the pillow generates a tuned magnetic field, two Buttkicker-like transducers make the bed vibrate physically, and speakers attached to the geometric shell above and below the bed produce sound. A controller box at your right lets you tweak and mix the various waveforms, insert a radionic command, and add in white noise.

Laurie's studio is filled with other fascinating equipment: a brainwave-to-MIDI converter, Faraday cages, and ultrasound microphones. But the Purr Generator is a popular favorite. Laurie has successfully used it to encourage bone healing for his own hip replacement, and rumor has it the Purr-cure also works for ailments ranging from back pain to depression. Purrfect!

—Steve Nalepa

» **Duncan Laurie:** duncanlaurie.com

Photograph by Todd Thille

Photograph by Charles Platt

Throwing to the Dogs

When **Ken Schroeder** was repairing appliances for a living, he decided that a spring-loaded switch from a dishwasher would be ideal as a trigger for some kind of device that his dog, Bender, could activate. Two years later, while studying industrial design at the Art Institute of Pittsburgh, Ken mounted the switch behind the sensor plate of the world's first automated dog-biscuit thrower.

Bender places his paw on the sensor plate. An electric can opener turns gears that feed a biscuit from a magazine. Bender waits expectantly. Tension builds. The motor from a hand-held kitchen mixer starts whirring, driving eggbeaters coated in silicone caulking. The biscuit hits the beaters, which kick it out of an ejection port. While Bender chases the biscuit, the machine resets itself, ready for the next cycle.

The bone thrower satisfied three goals for Schroeder. "I had to make a project that involved gears and electrical," he recalls. "Also I made a video about teaching an old dog new tricks, for a psychology class. And, Bender and the bone thrower were attractive to potential girlfriends." He pauses. "Can you express that in the nicest possible way?"

Schroeder has a long history of building gadgets. "When I was a kid," he says, "I used to take Legos and add motors and paddlewheels, and play with them in the pond. Also I used an angle grinder to cut notches in the rims of bicycle wheels, so that I could ride on ice. Unfortunately, that didn't work very well."

Currently Schroeder lives in Florida, where he and his brother hope to start a business selling furniture fabricated from unusual materials, such as string soaked in resin. What motivates him in his design projects?

"It's fun to be creative and make things," he says with a shrug, as if the answer should be obvious.

—Charles Platt

» **Dog Biscuit Thrower:** ktschroeder.com/Products.php

Photograph by Jeffrey Traer Bernstein

The Sound of Music Blocks

The Tangible Music Sequencer lets people of all musical abilities feel like Julie Andrews conducting an orchestra of robotic von Trapp children.

The Sequencer is a collection of candy-colored, palm-sized boxes that semi-autonomously "play together" by making sounds (and light) and then triggering adjacent blocks to follow. Simple software lets you drag-and-drop any sounds, such as drumbeats, words, or even whole songs, onto icons representing the blocks. You then place the physical blocks in any configuration — next to each other, one after the other, or in forks — and press their Play buttons to hear the sequence of sounds you've created. If you press multiple Play buttons, you can create polyphony or other overlapping sequences of simultaneous sounds. You can also move the blocks back and forth to make them repeat.

"I'm taking advantage of the knowledge people have of interacting with their surroundings, and giving them a simple, yet expressive way of using it," says **Jeffrey Traer Bernstein**, a Ph.D. student at Princeton's Sound Lab, and the maker of the Sequencer.

Each box is controlled by a Freescale 8-bit microcontroller, sends signals via infrared, and uses Panasonic low-power RF modules to communicate with a hub on the host computer. The host doesn't just upload sounds to the boxes; it also calculates what each sound "looks like" in blinking light, and uploads this info to the blocks for great visual effect.

"I wanted to make the simplest musical instrument possible that allows for expression and yet makes immediately recognizable music," says Bernstein. "The idea is not to make John Cage."

For Bernstein, the success of his Sequencer will be determined by the surprising interactions between users and the instrument. He's working on manufacturing his Sequencer, and ultimately getting it into the hands of everyone from DJs to kids to orchestras.

"People have this idea that what they make is theirs, that they have exclusive control over how it'll be used, but I think that's counterproductive."

—Nicole Oncina

>> **Princeton Sound Lab:** soundlab.cs.princeton.edu

Hacker of Yesteryear

When **Jake von Slatt** was 14, he was the only kid in his neighborhood (maybe the only kid in *any* neighborhood) who owned his own brazing torch. At 16, he lost part of a finger in a "hovercraft incident" (now that's an emergency room visit you don't see every day). Jake von Slatt's not even this colorful character's real name, but a persona he uses online, his "brand," as he calls it.

By day, he's a Linux sysadmin for an aerospace firm outside Boston, but in his spare time, von Slatt likes tinkering with and modding castoff tech of the past. He has several websites chronicling his progress in everything from converting a 1989 Thomas Saf-T-Liner MVP bus into a gorgeous motor home to his Steampunk Workshop, where he mods kerosene lamps and experiments in brass etching. And then there's his page where he enthusiastically logs his town dump and dumpster-diving finds.

Given Herr von Slatt's interest in tech resuscitation, it's no wonder that one day he looked at a Western Electric Bell Model 500 rotary phone, that icon of 20th century telephony, and saw its future as a 21st century mobile.

Unlike other retro handsets he'd seen tethered to a mobile phone, von Slatt wanted his creation to be totally portable, sans cord, so he got out his Dremel tool and grafted his mobile LG phone's belt clip onto the back of the Model 500's earpiece. He upgraded the electret in the Model 500 with an element from a computer mic, and used a mini-plug out of the Western Electric so he can still unplug the mobile phone and use it without the handset.

Some wire soldering, J-B Welding, and black Krylon painting later, a comfortable, hands-free mobile Model 500 emerged that fits perfectly into the brassy von Slatt lifestyle. Picture him tooling down the street talking on his Western Electric cellie in a steam-powered car (a project he's considering tackling next).

—Gareth Branwyn

» **Jake von Slatt:** vonslatt.com

Photograph by Jake von Statt

MARK TWAIN: TECHNOLOGIST

HOW THE AUTHOR OF *HUCKLEBERRY FINN* COULDN'T MAKE IT AS A VENTURE ANGEL.

By Bruce Sterling

AS A TEEN, HE EARNED HIS LIVING AS A printer in St. Louis, Philadelphia, New York, and Cincinnati. He roamed the Mississippi as a steamboat pilot. He was a soldier for a couple of weeks, and a Nevada silver miner for a spell. He was a roughin' it, hands-on, jack-of-all-trades.

Then his literary genius began to tell on him. Soon it was clear that he was much better at telling stories than he would ever be at making things.

Still, Samuel Clemens never shook off his romance with technology and invention. It's in his books; *A Connecticut Yankee in King Arthur's Court* is a time-travel yarn about a can-do techie who destroys medieval England through his ability to "make anything in the world." Clemens himself was an ardent inventor: he created a perpetual-calendar watch-fob, a self-adjusting elastic strap to anchor his pants, and "Mark Twain's Self-Pasting Scrap Book."

The scrapbook made some money because any book with Mark Twain's name on it would sell. His other hobbyhorses perished through public indifference. This hurt Clemens' pride a bit, but at least he could write about invention: he was ever the stout public defender of the lone inventive genius.

Clemens realized early on that inventors were mostly put-upon, solitary types, rarely properly rewarded. He also knew full well that the profit from Yankee ingenuity went mostly to investors and capitalists. He was an idealist, but he'd been around.

As one of the best-known celebrities of his era, Clemens had money to invest. He badly needed to do this. Like most best-selling writers, Clemens had an impressive income, but it was dangerously sporadic. Furthermore, Clemens was very much living the high life in New York in a grand mansion, built to his specs with all modern conveniences: six servants, private tutors for his daughters, and a needy host of guests, builders, plumbers, doctors, Tiffany decorators, and similar colorful encumbrances.

Even when Clemens was in top creative form, he was forced to hustle and make do, working the treadmill as a lone artist at the mercy of Gilded Age publishers. His lifelong dream was financial independence — a stable way to sit back, breathe easy, and thrive off investments.

Clemens was an ardent inventor: he created a perpetual-calendar watch-fob, a self-adjusting elastic strap to anchor his pants, and "Mark Twain's Self-Pasting Scrap Book."

One obvious scheme was to start his own publishing house. He did this, and it was a quick success — not through his own writings, but from the best-selling deathbed memoirs of Ulysses S. Grant. This business success emboldened him, but Clemens soon found that the hassles of small publishing were even more repulsive than the hassles of big publishing.

So his attention returned to his first craft: printing. Clemens was fascinated by the technical underpinnings of the printing industry. He was willing to

The "Compositor" of James Paige, the Victorian robotic equivalent of a nimble-fingered, teenage Sam Clemens. The Charles Babbage Difference engine, also a conspicuous failure, was a "mere commonplace" compared to this "awful mechanical miracle."

Photograph courtesy of The Mark Twain House & Museum

take a flutter on many other forms of tech — food additives, bicycles, steam-pulleys, carpet-making machines, a cash register, a steam generator (the list did go on) — but he truly understood printing. He sensed, correctly, that the medium of print was long overdue for a technical revolution. Maybe he, Sam Clemens, could do to printing what Eli Whitney had done to cotton-picking.

Clemens was a classic early-adopter, power-user type. He was a keen student of communication technologies. Patent fountain pens attracted his authorly interest. He was a pioneer user of the typewriter, and he used an early Edison cylinder as a dictation machine. Clemens' mansion had one of the world's first private telephones. Clemens even had electronic music piped in by telephone, through that legendary example of a dead medium, the Cahill Teleharmonium. Toward the end of his life, he was to champion a primitive European television.

Still, he'd grown up inside printing, and he'd been compositor, editor, publisher, journalist, author — he knew that business up and down. If he was ever to strike it rich as a venture capitalist, it made sense that printing would be his field. And in 1889, Clemens found his lone inventive genius: James Paige.

Since the days of Gutenberg, workaday printers had set up lead type by assembling pages one metal letter at a time. The Paige Compositor was a new machine that could hand-set type rapidly and automatically. This ingenious device had been patented in 1877, but the fat-headed printing industry had been too lazy to build it — too sluggish to make the greatest leap in typesetting in four centuries!

It remained to procure the funds to break this device from its visionary blueprints and actually build it, and Clemens, who'd caught wind of it through his publishing house, burned to see it take form.

The Compositor was a stunner. As Clemens wrote: "All the other inventions of the human brain sink pretty nearly into commonplaces contrasted with this awful mechanical miracle. Telephones, telegraphs, locomotives, cotton gins, sewing machines, Babbage calculators, Jacquard looms, perfecting presses, Arkwright's frames — all mere toys, simplicities! The Paige Compositor marches alone and far in the lead of human inventions."

In 1889, Clemens signed a contract, gallantly giving Paige the ownership of his grand invention, but granting himself a hefty 500-dollar royalty for every such machine sold. Clemens struggled hard with the business model for his oncoming tech revolution, but no matter how he tried to figure it, it seemed impossible for him to avoid making millions. There were thousands of printing presses in the world. Every such machine with hand-assembled type was bound to become obsolete overnight. He had a huge, disruptive innovation that was striking at the root of the industry.

And Clemens was right: the hand-assembled press was doomed. And Paige was right, too: his amazingly elaborate invention, elegantly mimicking human movements with its 18,000 parts, could set type ten times faster than any human being could. The Paige Compositor was a kind of robot teenage Sam Clemens; it did just what Sam himself had once done, but on an industrial speed and scale.

Then, however, came the human element. James Paige was brilliant, a great talker, a mechanical genius, and a hacker perfectionist. Obsessed with his own brilliance, Paige couldn't budge his machine out of the start-up garage and into the hurly-burly of commerce. With 18,000 different parts, there was always some nifty upgrade to be made to the Compositor. Then there was the allure of Paige's moonlighting side projects, such as electrical generators. Paige couldn't be bothered to field-test his machine under real-world conditions. His Compositor was always in beta and never quite ready to ship.

In the meantime, the Mergenthaler Linotype appeared on the publishing scene. The Linotype was a rugged, stupid, IBM-PC of a beast. The Linotype was 60 percent slower than the elegant Paige Compositor, but it was also the first to market. Furthermore, since the Linotype wasn't quite so saturated with technical genius, it was easier to maintain, repair, and improve.

Ottmar Mergenthaler had never bothered to mimic any human movements. Trained as a watchmaker, not a printer, he'd invented an entirely new way to line up type mechanically. So, in the race toward a printing revolution, the Paige Compositor never left the starting gate.

Clemens had happily trifled with many tech toys over the years, but the Compositor was his demon. He sank $200,000 of his own wealth into his grand dream of reinventing print. But his steel darling was obsolete before it could hit the streets, and Clemens hit a cash-flow crunch that he could not escape.

He finally wrote: "I've shook the machine, and never wish to see it or hear it mentioned again. It is superb, it is perfect, it can do ten men's work. It is worth billions, and when the pig-headed lunatic, its inventor, dies, it will instantly be capitalized and make the Clemens children rich."

The Compositor was indeed superb, it was perfect, and it could do ten men's work, but as an investment, the Paige Compositor was poison. It never made anyone rich. Superb perfection and the work of ten men were not at issue. Ease of maintenance, ruggedness, mass production, cheapness of operation, room for improvement — a machine that worked like a real machine — that was what the industry required. The Paige Compositor was as rare a thing as Clemens himself.

Crushed by debt, Clemens shut down his mansion, abandoned his crumbling publishing interests, gave up all hope for a settled, bourgeois existence, and fled with his family to Europe, for what turned out to be nine years of globe-trotting exile. Within four years, scraping frantically, he'd managed to pay off his creditors. Still, he never again wrote in the easy, funny, chatty way that he'd written before trying and failing to become a tech mogul.

Clemens sank $200,000 into his grand dream of reinventing print.

Great wealth would always be denied him. Great fame would fall on him in heaps, and that would bury every other aspect of his efforts. He would never become Samuel Clemens, Venture Angel. To this day, he's Mark Twain, Famous Author.

As for James Paige, "a most great and genuine poet, whose sublime creations are written in steel" — he is entirely forgotten, except as the man who bankrupted Mark Twain.

The last models of the Paige Compositor were bought by the Mergenthaler company, picked up as curiosities. In 1964 — while their Linotype was still a going business — they donated the last surviving Compositor to Mark Twain's house and museum. There, the genius machine still stands today, admired by tourists: gorgeous, unworked, and unworkable.

Bruce Sterling (bruce@well.com) is a science fiction writer and part-time design professor.

1+2+3

Aircraft Band Receiver

By Cy Tymony

Modify an AM/FM radio to pick up air traffic control communications.

You will need: AM/FM radio, small Phillips screwdriver, small flathead screwdriver

Extending the Range of Your Radio

The aircraft band, 108 to 138MHz, is directly above the FM band. But aircraft signals are broadcast in an AM format. Amazingly, it's possible to modify a typical AM/FM radio to receive aircraft signals in the proximity of an airport, without removing or adding any parts! Here's how.

1. Identify the radio parts.

Use a battery-powered, inexpensive, analog radio. Remove the back cover and locate the main tuning capacitor (A). It's easy to find — just turn the tuning dial, and you'll see its parts move through its clear case.

Near the main tuning capacitor you should see one or two small coils of copper wire (B) mounted on the PC board. These coils are used to limit the frequency range of the radio.

Next, locate the tuning transformers (C). They look like square, metallic boxes with tuning slots in the top. One of the tuning transformers may have a couple of small diodes near it mounted on the PC board. This is the tuning transformer that you'll adjust. Its function is to filter out AM noise.

2. Modify the radio.

Tune the radio to an FM station at the upper end of the FM band. Notice where the dial is positioned.

Spread apart the small coils near the main tuning capacitor using a small flathead screwdriver. When you finish, tune the dial and you'll notice that the broadcast stations have moved down the dial. The radio is now able to receive stations well above 108MHz.

Tune the radio between stations so you can hear a slight hiss. Notice the position of the slot on the top of the tuning transformer that is near the main tuning capacitor (nearest to the small diodes). Slowly turn its screw until the hiss sound is at its maximum level. Note exactly how many turns and in which direction you turn the screw, for easy repositioning later. The radio is now able to receive AM signals in its newly expanded FM band.

3. Test the modified radio.

Replace the radio's cover. Take it — and the screwdrivers — near an airport during a peak air traffic period. Turn the radio on the FM band with the volume up and slowly adjust the dial. You should be able to hear air-to-tower transmissions. If needed, make adjustments to the tuning coils and the tuning transformer.

Illustration by Timmy Kucynda

Cy Tymony is the author of the *Sneaky Uses for Everyday Things* book series.

What Rough Beast Slouches Toward Arcata?

MAKE goes to the World Championship Kinetic Sculpture Race. By William Gurstelle

WHILE IT COULD BE NAVIGATED FASTER, it usually takes at least eight hours to drive up Highway 101 from San Francisco International Airport to Arcata, Calif. If you go, it's worth taking your time, because this road takes in a lot of interesting society and scenery changes along the way.

At the start, you cross the heart of downtown San Francisco, slowly cruising up Van Ness Avenue northbound across the Golden Gate Bridge toward the congested freeways traversing the Bay Area's northern suburbs. Soon, though, the milieu changes for the better as you pass through bucolic Sonoma County; the vineyards of California's wine country are wonderful viewing even at freeway speeds. Farther north, the Japanese and German luxury cars of wine country become less dominant, their space on the highway usurped by pickup trucks and far, far older Fords and Chryslers. Winding on, the 101 slows down, speed limits changing in inverse relationship to the height of the redwood trees flanking the road. As it morphs into the main drag of the small towns of Laytonville and Willits, the 101 slows to 25 miles per hour. Finally, the highway meanders north and west through Rio Dell, Fortuna, and Eureka before reaching the tranquil burg of Arcata, the starting point of the World Championship Kinetic Sculpture Race.

Photography by Karen Hansen

As twisty, turny, and motion-sickness-inducing as the 101 can be in redwood country, kinetic sculpture racers will tell you that it's nothing compared to the road they traverse from Arcata to Ferndale. In fact, it's not a road at all. It's a 42-mile ordeal of land, sand, muds, and suds. The race starts and ends on pavement, but in between, the course encompasses dirt tracks, 30-foot-high sand dunes, open water, and gooey, sloppy mixtures of mud and muck.

Every May in the redwood-friendly moistness of the Northern California coast, a wonderful exhibition of true maker determination and creativity takes place. Approximately 30 racing teams and thousands of spectators descend on eccentric and unconventional little Arcata to attend what might be the most interesting sporting event in the world, at least to the eccentric and unconventional kind of people who read this magazine.

> Below Dead Man's Drop, emergency medical technicians wait, ready to deal with breaks, blood, and twig-pierced eyebrows.

The size and time invested in such an outré event defy easy understanding. On the morning of race day, two-and-a-half dozen human-powered sculptures assemble in the town square. Watching and waiting, the great throng of onlookers stands six deep along the streets.

The elaborate mechanicals and their maniacal makers wait for the blast of the town's noon whistle. At the sound of the klaxon horn, all begin to circle the town square once, twice, and then a third time until the exit gate is flung open and the racers begin the marathon quest for glory. The crowd cheers them, then heads off to its favorite viewing points along the route. For the next three days, the racers pedal their vehicles up hills and down sand dunes, across the waters of the Eel River and Humboldt Bay, and through the towns of Manila, Eureka, and Loleta.

It's grueling work, but the harsh conditions are offset by post-race basking in the personal triumph that comes from merely finishing, and in the wry sense of humor that permeates the entire event. That has to be enough, for there's absolutely nothing to motivate the racers except the glory of participating and crossing the finish line in Ferndale three days later. Luckily, there's plenty of glory to go around.

Polymaths, All

What sets this event apart from other competitions with a homebuilt component is that it celebrates the polymath instead of the specialist. To understand this event, and to excel at it, one can't be only a sculptor, or a mechanical engineer, or a cyclist. One needs to be all three, for the final judgment depends on the sum total of the builder's artistry, engineering, and physical fitness.

This is an event for polymaths, of the sort that Leonardo Da Vinci, Blaise Pascal, and Buckminster Fuller would enjoy. It's not just about having overdeveloped leg muscles, and pedaling fast and long. The best, most glorious participants must be clever technologists, designing gear trains and vehicle superstructures that can finish the marathon-and-a-half-long course of diverse terrain.

Some of the racing machines weigh more than a ton fully loaded, and a fair amount of engineering is required to make such weightiness sufficiently mobile to be pedaled across the many conditions encountered. Typically, this involves designs incorporating swing-down pontoons, differential gear boxes, variable-angle recumbent seats, and most important of all, massive drive trains, with more gear ratios than you can shake a Cannondale at. The pilots of kinetic racers often have more than 600 different gear ratios from which to choose.

And even that isn't all. To be competitive, a racer has to be beautifully sculpted as well. Builders must have artistic talent, for skill and brawn mean little without artfulness.

The Racers

It's the artistry, whimsy, and imagination that the designers and builders incorporate into their machines that make the experience what it is. The best machines are true works of art, evocative of the sculpture of Marcel Duchamp, Alexander Calder, and Jean Tinguely.

June Moxon's racer is called *Skaredy Kat*, but she

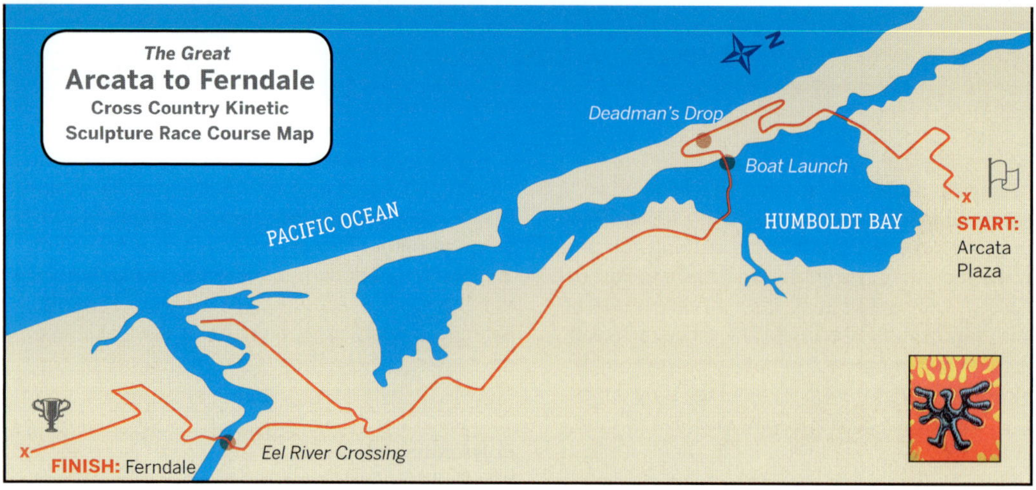

A map of the 42-mile course, which traverses pavement, sand dunes, mud, and water. For a more detailed account of the course, visit kineticsculpturerace.org/map.shtml. Image shown on previous image: *Flash Gourd'n*.

most certainly is not one herself. The year 2006 marked her 24th year of competing. During that time, she's powered herself across a lot of territory. In fact, she and fellow racer Ken Beidleman once pedaled a racer across the United States.

"In 1989, Ken and I left Ferndale on our racer with $200 in our pockets, and we survived by donations and doing odd jobs," she says. "Each day on the road was different: heat, snowstorms, steep hills, rain." Eventually, they pedaled all the way to St. Augustine, Fla.

Skaredy Kat is, as June describes it, "a ginormous, black and white, spooked-out, kinetic kruising kitty kat." More specifically, it's a 500-pound tandem tricycle with an immense gear ratio and a sophisticated system of pull-cords that allows June and her co-rider to move the puppet parts of *Skaredy*'s head, eyes, whiskers, and tail while riding.

June's boyfriend, Ken Beidleman, is equally enamored of kinetic sculpture racing. A metal sculptor from Redding, Calif., Ken first became interested in the activity back in 1987 when he volunteered to work with the event's founder, Hobart Brown, on his racer. Beidleman has participated more or less continuously since then.

"Every year I try to come up with something new," he says. "For instance, I'll say, let's do something contemporary, a racer [with a theme based] on a current event or movie."

In the past, that line of thinking produced the machines he named *Watermelon World*, *Hog Wild*, and *Nightmare of the Iguana*. The 2006 entry was *Flash Gourd'n*, an upgrade of an earlier year's *Gourd of the Rings*, but *Flash Gourd'n* has better steering, fewer squeaks, and reduced weight.

The night before the race, builders work into the wee hours making final adjustments. At the Arcata Kinetic Lab, some pause to chat with the curious. Others are furtively focused behind drawn curtains. In the back corner, sculptor Duane Flatmo is hard at work, putting the final touches on a comically scary, bug-eyed, four-person tricycle.

Extreme Makeover is a heavy beast, weighing 1,200 pounds when fully staffed. A good portion of that weight is invested in teeth and eyeballs. Once the race starts, it's the scores of gear ratio choices that allow the pilot to take it over water, turf, asphalt, and even sand dunes.

Like fairy tale creatures, the racers lumber along the beach, making the best time they can before attempting the more difficult parts of the race. The hardest obstacle is a high, steep, and perilous sand dune called Dead Man's Drop.

Atop the dangerous dune, sunburned, mosquito-plagued spectators call out encouragement as racers make the tough climb at tortoise speed. Below Dead Man's Drop, emergency medical technicians wait, ready to deal with breaks, blood, and twig-pierced eyebrows. As the racers plunge over the edge of the dune, the crowd follows in their wake, cheering each guts-and-glory descent.

For what it matters, the 2006 grand champion was *East Shark*, a two-person land submarine designed by high school students from Eureka. They finished the race in under eight hours of actual pedaling time. *Flash Gourd'n* and *Extreme Makeover* finished respectably in the middle of the pack, taking about 20 hours. And *Skaredy Kat*? It, too, finished. Eventually and gloriously.

MAKE Contributing Editor William Gurstelle wrote the rubber-band ornithopter project (*MAKE, Volume 08, page 90*).

Clockwise from top left: *Skaredy Kat*, June Moxon in *Skaredy*, *Stag Party*, *Brain Power*, *Pear Country Chopper*, *East Shark*, *Mardi Gras*, and *Extreme Makeover*.

ART WORK

The Joy of YDKEWYGUYGI

By Douglas Repetto

LIKE MOST TECHNOLOGIES, MANY EARLY mark-making machines were purely practical, usually addressing the pre-printing-press need for quick and easy copying. The automaton craze of the 18th century produced the first completely mechanical drawing and writing machines, as well as some of the first machines with mechanisms that were as interesting as their outputs. (And sometimes more so!)

Swiss inventor Henri Maillardet made a machine that could produce four different drawings and three poems encoded on a large set of complex cams. It signed its own work by writing "Ecrit par L'Automate de Maillardet" ("Written by Maillardet's Automaton") in the border of the final poem — perhaps evidence that even the earliest mark-making machine makers were playing with questions of authorship and intention. Maillardet's automaton is on display at the Franklin Institute. It still works!

You might think that advances in printing technology would have meant the end of homemade mark-making machine mania, but they haven't. Although inkjet and laser printers do a great job of creating clean copies of full-color text and images, they're not so great at doing some of the things that make manual mark-making so interesting.

Sure, you can simulate brush strokes in your digital painting program, but making real brush strokes is more fun. While untold thousands of engineering hours have gone into making sure that WYSIWYG when you send a job to your printer, there's still a lot of interest in systems where you don't know exactly what you'll get until you've got it (YDKEWYGUYGI).

Jean Tinguely, a 20th-century Swiss artist, is best known for his large kinetic sculptures made from scavenged industrial debris. Starting in the late 1950s, Tinguely made a series of "Méta-matic" drawing machines, complex jumbles of gears, springs, and weights. The unpredictability of the mechanism coupled with the choices made by the user (selecting different kinds of papers and brushes or making adjustments to the mechanism itself) assures that no two runs of the machine are alike. This is a theme that, ironically, appears often in machine-made art: non-repeatability and the potential (and pitfalls) of infinite variety. If you're ever in Basel, make sure to check out the Tinguely Museum — there's even a working Méta-matic that you can play with for €1.

In the late 1970s, Croatian-American artist Anton Perich built a large, light-controlled painting machine from cobbled-together parts of sewing machines and industrial debris he found on Canal Street in Manhattan. There's a telling quote in *The Andy Warhol Diaries* about Perich's machine (they were neighbors): "Anton was home with his painting machine and I was so jealous. My dream. To have a machine that could paint while you're away. But they said he had to be there while it painted because it clogs up."

"Anton was home with his painting machine and I was so jealous. My dream. To have a machine that could paint while you're away." —*Andy Warhol*

Warhol was being wry here, but he brought up a common misconception; the motivations of mark-making machine makers are generally more complex and interesting than "I want a machine to do my work for me." For some artists, the output of the machine isn't even the main point; the process or method of making marks can sometimes be as, or even more, interesting than the marks themselves. As for Perich, 30+ years later he's still making big, complex paintings with his original machine.

Sabrina Raaf's "Grower" and Fernando Orellana's "Drawing Machine 3.1415926 v.2" are two machines that make marks that would be impossible for humans to make on their own, and both make nods toward early analog scientific recording equipment.

Raaf's machine is a robot that draws green marks (evoking blades of grass) on a wall, the heights of which correspond to the amount of CO_2 in the air. Over the days and weeks that the robot is working, it builds up a field of data-driven "grass" as it tirelessly treks back and forth along the wall.

Orellana's machine is a lot like a mobile with a pen attached to it. But rather than just twirling in the wind, the pen's position on the canvas is controlled by the local ambient sound, similar to the way a seismograph pen records shock waves moving through the Earth. Both works create a long-term record of their local environment, but in a highly stylized form meant to leave an impression, rather than an exact accounting of some environmental variable.

Hektor, a "graffiti output device" by Swiss artists Jürg Lehni and Uli Franke, uses a deceptively simple and elegant mechanism in a small, portable package. Two motors are attached to a wall with a belt hanging between them. A spray paint can and activator hang from the belt, and as the motors turn they pull the spray can around the canvas. Because the mechanism can bring the spray can to any point, the content of the paintings is separate from the machine's mechanism, making Hektor the equivalent of an enormous general purpose plotter (it's even driven by Adobe Illustrator). The catch is that despite its elegant design and vector-based data, Hektor's fuzzy, dripping output is far from pristine. It's another take on the tension between computer-based precision and real-world grunge.

Some people make machines with outputs that are simply expressions of their own construction (Jonah Brucker-Cohen's simple plastic-cup-and-motor "drawbots" are a good example of this). Others focus on environmental inputs, or algorithmic control, or AI, or ... well, the list goes on. Finally, some people just want to make a machine that makes pretty pictures. There's room for everyone in the world of mark-making art machines. What's yours going to do?

Photography by (clockwise from top): Jürg Lehni (Hektor); Sabrina Raaf (Grower); Anton Perich (machine at bottom); Fernando Orellana (machine at middle left)

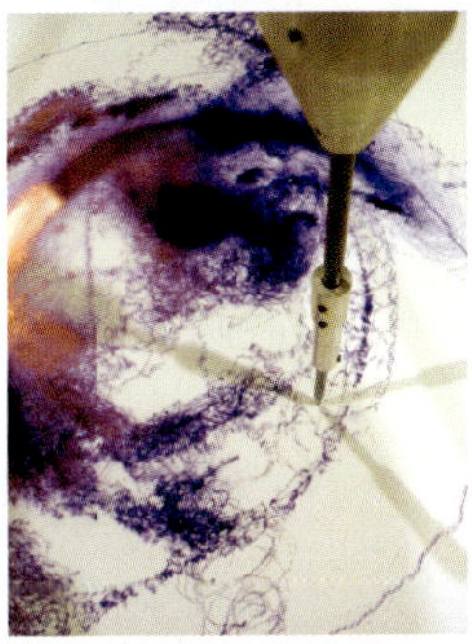

Top: Hektor enjoys painting both landscapes and portraits. Middle: Orellana's Drawing Machine and Raaf's Grower sense the world and then leave their marks. Bottom: Perich's steampunk painter is half airbrush, half sewing machine.

Douglas Irving Repetto is an artist and teacher involved in a number of art/community groups including dorkbot, ArtBots, organizm, and music-dsp.

P.O.K.E.R. Night

Or, Let's Build a Vortex Tube Party!

By Mister Jalopy

IN 1867, SCOTTISH PHYSICIST ROBERT Clerk Maxwell proposed a "thought experiment," suggesting that it may be possible to break the second law of thermodynamics, if you could find a molecularly minded demon that was willing to cooperate.

Start with two hypothetical containers of equal temperature with a connecting trap door controlled by a gatekeeper to sort hot from cold molecules. When the demon sees a fast/hot molecule racing toward the door, he opens the trap to let the hotter molecule speed into container A. Similarly, if a slow-moving molecule is headed toward container B, the demon would allow it to pass until the two containers are at different temperatures. Assuming that the demon would have modest demands for his efforts, you would break the cornerstone of physics laws — something for nothing.

The Hilsch Vortex Tube

After an impressive 72-year run, the "Amateur Scientist" column fell victim to a 2001 *Scientific American* modernization effort. A terrible loss, the "Amateur Scientist" was an inspirational resource that dedicated considerable effort to explaining the hard science behind the projects.

Whether working on "Cloud Chambers to Detect Nuclear Events" (*see page 156 for MAKE's version of the cloud chamber*) or "An Inexpensive X-Ray Machine," you really felt like you were in the midst

Photography by Mark Frauenfelder and Mister Jalopy

of serious scientific pursuit and the feeling of exploration was very high — especially with the scant direction.

In November 1958, *Scientific American* published one of my favorite projects — the Hilsch Vortex Tube. Invented in 1933 by a French physicist named George Ranque, and improved by a German physicist named Rudolf Hilsch in 1955, it uses nothing but compressed air. The design of the device created a vortex that acts as a demon to "split" the hot molecules from the cold by spinning air to very high speeds. How fast? Half a million rpm fast.

Spinning air is enough to generate a temperature change of a couple hundred degrees? Yes, but at the time of the original 1958 article, they weren't exactly sure how it worked. As a very amateur scientist, I will try to explain the process. Compressed air is shot into a vortex that creates a hollow cyclone of fast, hot air that travels down the interior of the hot pipe. If it were visible, it would look like a tube of hot air that lines the inside of the pipe, and then a smidge of the hot air is exhausted through the almost-closed stopcock.

The remainder of the air is reflected off the nearly blocked stopcock through the center of the hot-air cyclone — in effect, a column of air shooting through the hot-air tube. As this happens, this column of air "gives up" its heat to the surrounding cyclone as it passes through the center. When the cold column reaches the vortex, it shoots through the center and exits the cold side. Whoa.

Cold-Cut-and-Soda-Pop Engineering

Anticipating a lukewarm response from my buddies for a "Let's Build a Vortex Tube Party," I invited my amateur scientist friends for a poker night (little did they know that "poker" was an acronym for Proof Of Koncept/Experimentation Revue).

Fearing a rebellion when they discovered there were no poker chips or one-eyed jacks, I made certain to lay out an impressive spread of top-notch cold cuts, provolone, olives, marinated pepperoncinis, and plonk chianti. I would have served a better wine, but I figured these wily characters would have realized they had been duped by my dodgy scheme, checked that their wallets were intact, and hit the door.

Original Mister Jalopy Proposal

With a deep breath and steely resolve, I told the fellas that instead of a freewheeling night of gam-

MAXWELL'S DEMON THOUGHT EXPERIMENT CIRCA 1867

1. **Chambers A and B contain gas at equal temperature. The demon has the door closed but is ready to move quickly to let the molecules through.**
2. **As a fast/hot molecule approaches from chamber B, the demon throws open the door to admit the molecule to the soon-to-be-hotter chamber A.**
3. **Much door opening and closing ensues until the demon has tirelessly isolated all the cold from hot molecules.**

Demon illustration by Mark Frauenfelder

bling and tasteless jokes, we would be attempting to channel Maxwell's demon. They went for it — though they insisted on keeping the tasteless jokes. Using segments of iron plumbing pipe as the hot and cold sides, I suggested that we epoxy together a vortex assembly with three ½" slices of a Delrin plastic 3" rod.

Revised Approach

The guys are not scientists or machinists but they are talented, creative folks that are deeply interested in how things work. Hoping that would be enough, the group reduced the vortex assembly design to use a single Delrin disk. In my original plan, the washer was replaced with a ½"-thick slice of Delrin, but the revised plan brought back the washer. We didn't know if the thicker Delrin disk would affect performance, but since we had no idea what we were doing, we decided the fewer changes, the better.

The finished vortex tube looks like a champion! From appearances, it would seem to instantly freeze tissue samples for plague research or, perhaps, cut 1"-thick steel plate with an awesome 1,000-degree cutting flame! Real life offered slightly less spectacular results, but we did measure a temperature spread of 8 degrees between the hot and cold sides. We raised the flag and declared victory!

If I were a real scientist, I would continue to refine the design using our modest success as a starting point, but as an amateur, I am happy with the Proof Of Koncept/Experimentation Revue results. Until an engineer shows up with some imported spicy salami, the vortex tube will sit on the shelf as a memento of a fun late night with good friends.

Mister Jalopy referenced C.L. Stong's collected columns from the out-of-print The Scientific American Book of Projects for The Amateur Scientist, *but the complete, 72-year archive is available on CD-ROM (*Scientific American's The Amateur Scientist, Science Fair Edition, *$30, brightscience.com).*

AiRTX International (airtxinternational.com) sells commercial vortex tubes capable of generating temperatures from -40°F to 250°F. We thank them for allowing us to use their superb illustration.

Mister Jalopy is an earnest but *amateur* scientist. Know more about vortex tubes than he does? We wouldn't be surprised. Join us for vortex discussions at makezine.com/09/vortex_tube, as we have lots of questions.

Despite being hoodwinked, my friends met the vortex tube challenge with enthusiasm and a determination to succeed, provided the beer did not run out.

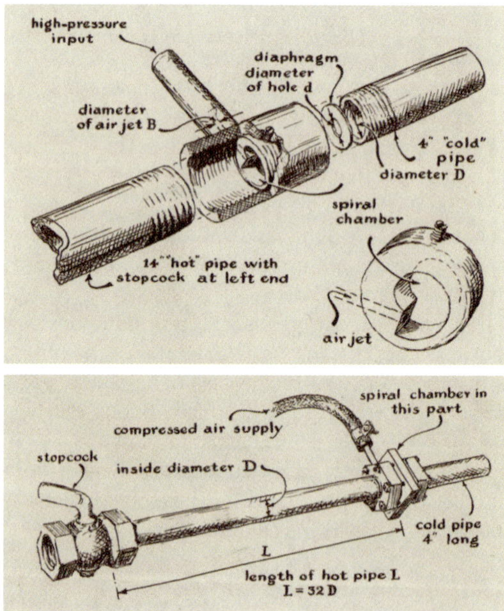

In the original "Amateur Scientist" design, adjustments are made via the stopcock on the hot side and the interchangeable washer on the cold. (Source: *The Scientific American Book of Projects for The Amateur Scientist*, C.L. Stong, 1960)

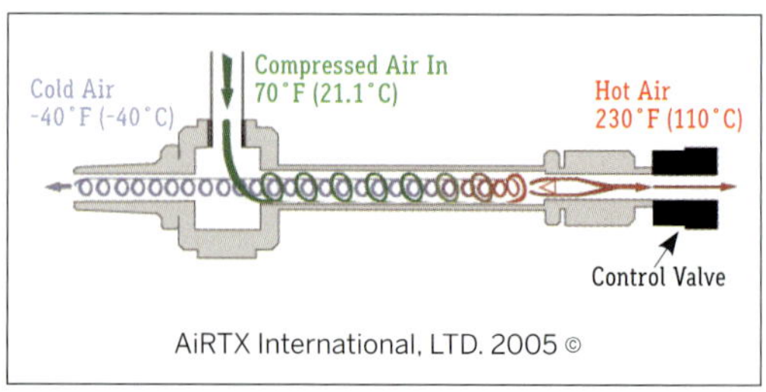

AiRTX International's commercial vortex tube ($109-$149) produces frigid air from room-temperature compressed air.

THE P.O.K.E.R. TUBE, VERSION 1

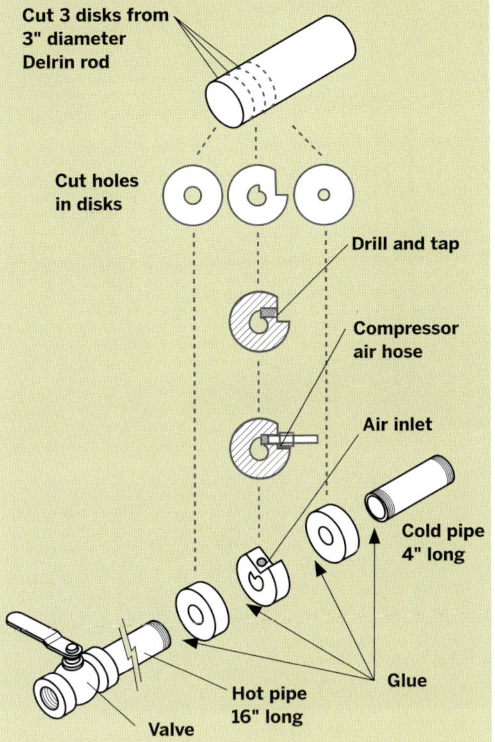

VERSION 2

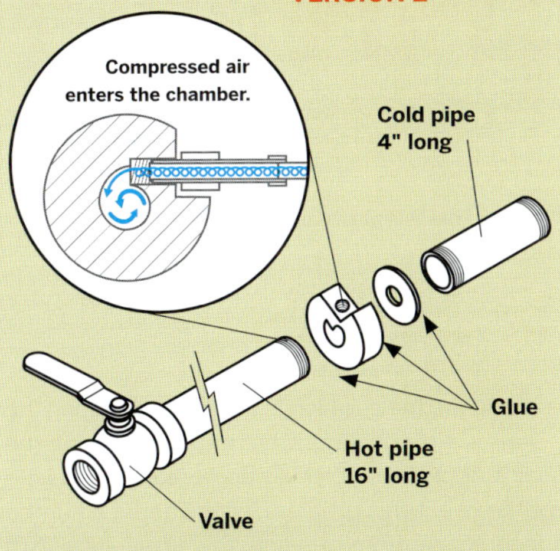

In just a few hours, we managed to cobble together a proof-of-concept Hilsch vortex tube that produced two streams of air with small, but measurably different temperatures. You're welcome to try to improve on our efforts. If you succeed (or even if you fail) let us know at makezine.com/09/vortex_tube.

Illustration by Gerry Arrington

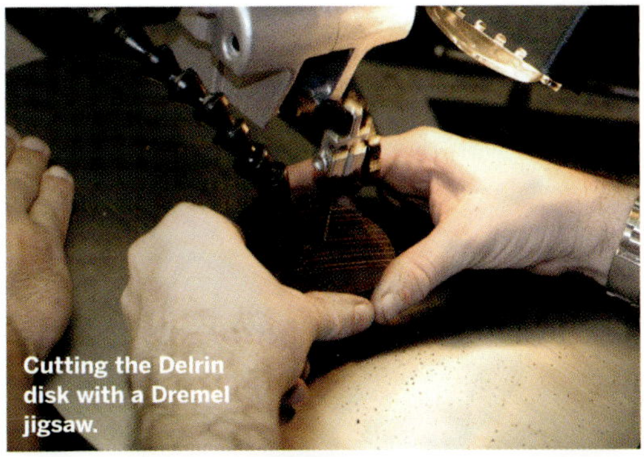

Cutting the Delrin disk with a Dremel jigsaw.

The actual vortex chamber is very small, as seen here.

The glue only held for seconds; epoxy putty was much stronger.

Our Hilsch tube is not as pretty as SciAm's, but it worked (sort of).

DISAPPEARING APPLE

Remove the logo from an iBook.

By Tom Owad

"I am writing a book/TV documentary for Canongate and the BBC called *Bonfire of the Brands*. The short story is that I am attempting to overcome my addiction to brands by burning all my branded stuff on a great bonfire on 17th September 2006. Afterwards, I will be attempting to live a life brand-free."

This was the introduction to an email I received from Neil Boorman last August. It continued, "I have allowed myself one luxury which will be spared from the fire, which is my iBook. But to keep it, I must de-brand the thing, which is why I'm contacting you."

There's a particular problem with removing the iBook's Apple logo: it leaves an Apple-shaped hole. To circumvent this problem, I considered a number of techniques, from molding an entirely new lid, to simply covering the Apple with a big white sticker. Ultimately, I decided the best way was to cut out the logo on a milling machine, then mill a replacement piece out of another lid. I called PreOwned Electronics and ordered a stack of grade B iBook lids to experiment on, then headed down to the

Photography by Tom Owad

garage. My mill is a Homier Mini-Mill (and currently halfway through a homebrew CNC conversion). I mounted the lid on the mill, outside down, using a clamping kit. The Mini-Mill was just big enough for the job — half an inch less travel on the Y-axis would have made the job extremely difficult.

I loaded a ⅛" end mill and cut out a rectangle around the Apple logo. Because end mills are round, the mill can't cut inside right angles, so I then removed the lid from the clamps and filed the corners down (this only took about two strokes with the file). To make the inset, I took another iBook lid and hacksawed it down to a bit bigger than I needed; I then clamped it into the mill and milled it down to the exact size.

Next I filed off all the burrs and snapped the piece into the lid. A little super glue on the inside of the case held everything snugly in place, but still left a visible crack around the perimeter of the rectangle. I filled this with Bondo Glazing and Spot Putty, then sanded it down. Finally, I spray-painted the lid with Gloss White Krylon Fusion for Plastic. To prevent light from shining through where the logo used to be, I merely taped on a piece of aluminum foil.

Removing the logos from the operating system (Mac OS X 10.3) was much easier, thanks to a shareware app by Geekspiff called ThemePark (geekspiff.com/software/themepark) and CodeServant's open source application ThemeChanger (codeservant.com/themechanger.php). ThemePark presents a window with the graphical elements in the user interface and allows the user to just drag and drop replacement images over them. I replaced each Apple logo with Neil's initials. Once I was finished with my changes, I saved the theme and installed it with ThemeChanger.

The boot panel — the window that appears during startup with the Apple logo and progress bar — can be modified just by replacing the PDF stored at */System/Library/CoreServices/SystemStarter/QuartzDisplay.bundle/Resources/BootPanel.pdf*. This leaves a single Apple logo: the dark grey apple that first appears when you turn on your Mac. With Ryan Schmidt's Startup Syringe (ryandesign.com/jagboot), replacing it is just drag and drop.

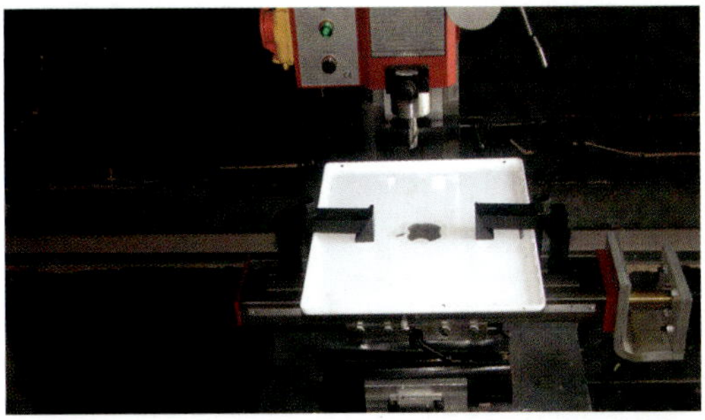

Tom Owad (owad@applefritter.com) is a Macintosh consultant in York, Pa., and editor of Applefritter (applefritter.com). He is the author of *Apple I Replica Creation* (Syngress, 2005).

INSTANT HAMMOCK

By Tim Anderson

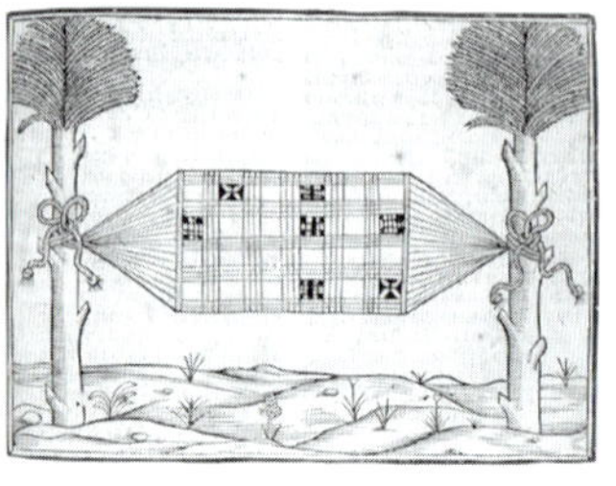

Make your own hammock in less time than it takes to read this.

Hammock. Barbecue. Canoe. Hurricane. All are important elements of life in the Southeast and Caribbean. And it turns out these are all words from the Arawak Indian language, dating back to before Columbus arrived in the "new world."

Some things just don't change, and it's a good thing. The hammock is one of those things. There are lots of types of hammocks. The ones we've all seen are fancy, complicated ones that are good to sell to rich, industrialized people like us.

There are other types that are much easier to make. Ever wanted a hammock and not had one? You probably weren't more than a few feet away from the materials needed to make one, and it wouldn't have taken you more than a couple of minutes to make it.

Here's how to make your own hammock in less than a minute:

1. Get a large piece of cloth such as a bed sheet. I didn't have one to spare so I bought two roll ends of cotton cloth for a dollar a yard at a discount store. Use any kind of cloth that appeals to you. Most types of cloth will be strong enough when new.

SAFETY WARNING: When it gets old, eventually it will rip and you'll fall onto the ground, just like the fancy commercial hammocks you've fallen out of in the past.

2. I'm making a big hammock, so I sewed the pieces of fabric side by side using a "flat felled" seam. Skip this step if you're in a hurry or if you're starting out with a sheet that's big enough.

A few hammocks and tying techniques:

A. Varieties of hammocks are found all over the New World. These Nicaraguan students are sleeping while waiting for nesting turtles to come ashore in their study area. These very small hammocks are one of the traditional styles and are surprisingly comfortable. The ones shown are made of uncoated polypropylene and are sold in all the markets for next to nothing. Stacked on a shelf they look like shopping bags, and you wouldn't know they were hammocks.

B. Detail of how this hammock is hung: The rope is passed through a large hem at the end of the hammock, bunching the cloth together. Then that end of the rope is tied with a bowline knot to make a loop. The other end of the rope is tied to a pillar or tree.

C. A fancier style of large hammock, this denim model is in a hotel on Isla Ometepe, Nicaragua. It was sewn just the same as the smaller hammocks seen above. Then a dozen or so slits were cut in the hem. The fingers of cloth thus created are tied with bowline knots to the ends of 5-foot chunks of colorful cord, forming loops to make the bridle.

D. Other end of the bridle: The loops of cord are gathered together into a single loop. This is whipped with a covering of cord knotted with a series of half hitches, one at each turn. The covering keeps the bridle lines from slipping and prevents the suspension rope from chafing through them. The suspension rope is tied to the loop with a lark's head knot.

3. Bunch up both ends of your cloth. If your cloth is long enough and you want to be extra secure, tie an overhand knot in the ends of your sheet to keep the rope from sliding off.

OPTIONAL: If your cloth is long enough and you want to be extra secure, tie an overhand knot in the ends of your sheet to keep the rope from sliding off.

4. Tie a chunk of rope to each end of your sheet using a lark's head knot (*see MAKE, Volume 01, page 77*).

5. Hang your new hammock between two stationary objects, get in, and relax. A nice feature of these cloth hammocks is that you can pull the sides closed over yourself and be very cozy inside. If the hammock is large, you can lie diagonally in it and have your spine totally straight. That's how the Mayans do it.

Photography by Tim Anderson

Tim Anderson, founder of Z Corp., has a home at mit.edu/robot.

Turn On, Tune In, Bliss Out

Dr. James Hardt hopes his DIY neurofeedback tech will fire up the global brain.

By David Pescovitz
Photography by Howard Cao

BLUE. THE COLOR OVERWHELMS ME AS I STEP inside a nondescript office park building across the street from the San Jose airport. There's blue everywhere. Blue carpet. Blue walls. Blue fabric draped here and there. Even the wall art, cosmic fantasy framed prints that are the New Age equivalent of black light posters, is heavy on the azure.

The cracked geodes decorating the conference table are bluish-purple. This is the Biocybernaut Institute, home to neurofeedback technology that the inventor believes could someday be used to awaken the global brain. Enter the mind-machine maker himself, Dr. James Hardt, sporting what else but a blue fleece jacket. "You apparently like blue?" I ask him.

"It's not about liking blue," Hardt says. "It's about the fact the blue enhances the brain's alpha state."

Hardt should know. Over three decades he's trained hundreds of people to use a neurofeedback system of his own design to control their own brain waves and spend more time in an alpha state.

For those who may not have lived in California during the 1970s, the alpha state is a brainwave pattern with a frequency of 8-13Hz that's commonly associated with relaxation and meditation. Increased alpha wave activity has also been shown to help alleviate ADHD and depression, boost creativity and alertness, and facilitate out-of-body experiences and other psychic phenomena, depending on who you ask.

Hardt claims that one week of alpha training can yield changes in brain waves comparable to 21 to 40 years of Zen meditation training. All for the low, low price of $15,000 per person. That's how much a weeklong training session of roughly 12-hour days at the Biocybernaut Institute runs. Bliss doesn't come cheap.

"Alpha power, when it's high, is like being in heaven," Hardt says. "You're filled with joy, enthusiasm, and a sense of wonder."

Neurofeedback was born during the psychedelic 1960s out of earlier experiments to detect the current in our heads. The core technology is the electroencephalograph (EEG), a device used to record electrical activity in the brain by sticking electrodes to the scalp. The field began in 1908 when Hans Berger wired his son up to a galvanometer and recorded his brain waves. Berger named the signals alpha waves because they were the first ones discovered.

But the real biofeedback breakthrough occurred in the 1960s, when UC San Francisco psychologist Joe Kamiya closed the loop by providing subjects with real-time data of their brainwave activity in the form of audio tones. Kamiya hooked up all kinds of people to his machine, from Zen monks who exhibited heightened alpha states, to Summer of Love seekers eager for a new trip, to a curious physics graduate named James Hardt.

One day, a lab technician left for lunch, forgetting that Hardt was in the chamber. Over the next several hours, Hardt says he went on a transcendental "adventure" in the alpha state, complete with "ego dissolution," that hooked him for life. Learning to control your mind from the inside out, he believes, is the first step to manifesting a human superconsciousness.

"If you've got a mess in your mind, you're not going to be as good a candidate for a linkup," he explains. "You usually take a shower before making

"In physical fitness, the goals are strength, flexibility, and endurance," Hardt says. "In mind fitness, the goals are the same. You want to learn to increase and decrease the strength of your alpha waves and then build up the endurance to do that for long periods of time. It's literally pumping neurons."

love, right? This is the mental equivalent of that."

From 1971 to 1977, Hardt worked closely with Kamiya to see how neurofeedback might push the limits of human potential. Meanwhile, he rolled his own belief system from a smorgasbord of spiritual traditions and philosophers.

Two decades later, the Biocybernaut Institute apparently has no shortage of students. Hardt says he's trained everyone from Silicon Valley big shots to pro football players to U.S. Army Green Berets. The latter group signed up during the 1980s at the bequest of Maj. Gen. Albert Stubblebine, onetime head of the United States Army Intelligence and Security Command, better known for his involvement in recently declassified psychic spying programs.

Every weeklong training at the Biocybernaut Institute begins in the conference center, where Hardt gives a PowerPoint-powered rap to orient the participants. Next, his staff applies a half-dozen scalp electrodes to each trainee's head. After the trainees are wired up, each moves into one of four soundproof chambers (blue, of course) where the neurofeedback session actually takes place. Every chamber contains a surround sound system, a computer on a desk, and a task chair. (If you're doing theta training, you get to kick back in a La-Z-Boy. Theta brain waves are associated with deep relaxation and light sleep.) Baseline EEG measurements are taken, and then a symphony of electronic tones reminiscent of a minimalist techno album signals that the long, strange trip is about to begin.

"When you go to the theater, you have to practice suspension of disbelief," he says. "To participate in this training, you need to have a suspension of belief and go in only with what you discover. With that mindset, you have the openness of a baby and of an advanced mystic."

Rip off its New Age wrapper and the real magic of the Biocybernaut training lies in the technology behind the blue curtain. An array of amplifiers and filters extract the brain's electrical signals collected by the scalp electrodes. That data is then analyzed by PCs that generate the feedback tones heard in the chambers. All of the neurofeedback gear — except two hulking polygraphs that print out the raw EEG data — is Hardt's own design. No commercial product was ever good enough, Hardt says, so he assembled and patented his own. One former student posting online describes the Biocybernaut facility as "having an 'I built a rocket ship in my garage' kind of feel." Indeed, a fair bit of, er, unconventional engineering went into the system.

"This advanced technology allows scientists to talk the language of mystics, and mystics to talk the language of scientists. I look at what we're doing as creating the science of spirituality."

Early on, Hardt explains, he had a quality control problem with components on the filter boards. Some of the resistors and capacitors behaved erratically at the boards' operating temperature of 104 degrees. Hardt's solution was to heat a room to 104 degrees and test each component individually.

"Once the room was hot enough, my staff and I went in wearing bathing suits and metered each part," Hardt recalls. "We put them into labeled bins so we had known values."

Next, Hardt wrote software to specify how those resistors and capacitors with slightly different values could be combined into circuits that would operate within his strict tolerances. Only then could his circuit board assembly line begin.

"This advanced technology allows scientists to talk the language of mystics, and mystics to talk the language of scientists," Hardt says. "I look at what we're doing as creating the science of spirituality."

Back in the chamber, the trainees hear various synthesized tones that change according to how their alpha states increase or decrease. The tones,

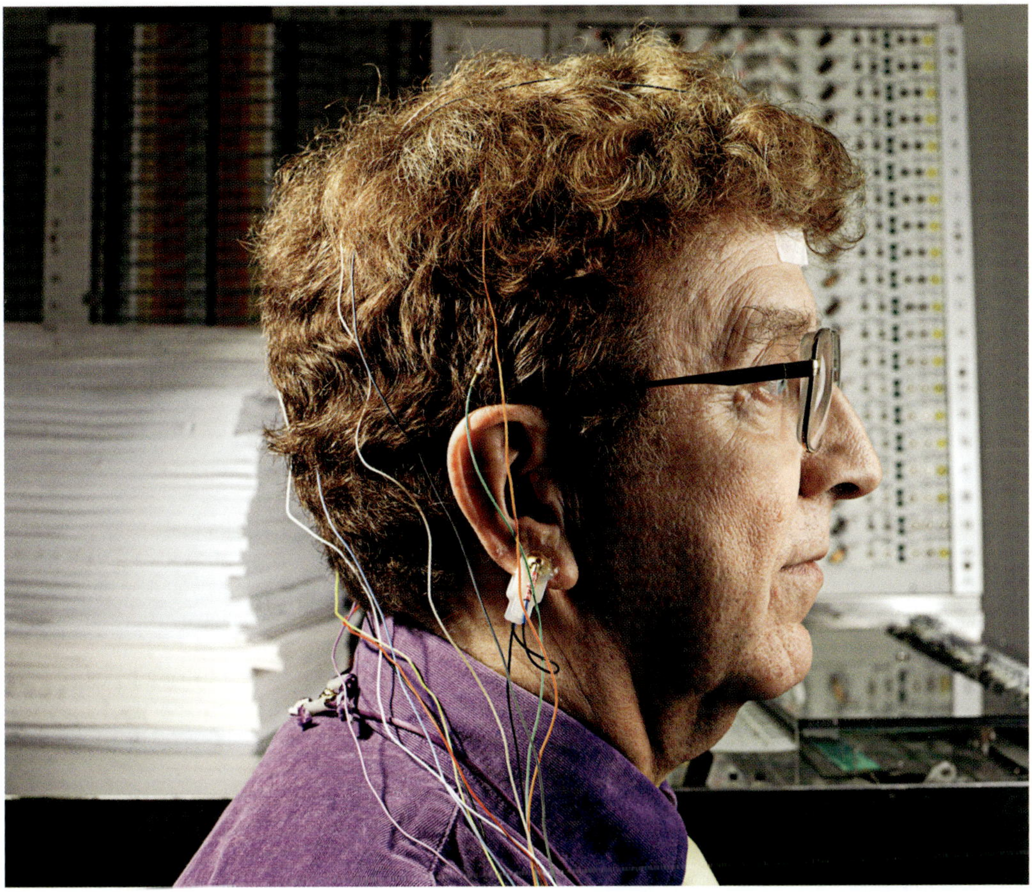

he explains, "are like an auditory mirror for your mind." The trainee is taken through a series of survey questions and alpha suppression and enhancement exercises. Numbered scores appear on the computer display, providing more feedback to the trainee. Hardt won't tell me how much "objective time" each trainee spends in the chamber. Wristwatches are confiscated at the door.

After each session in the chamber, the trainees retire to the "canopy room," a lounge that reminds me of a cheaply decorated chill-out room at a rave. Here, Hardt or his staff debrief with the trainees about their time in the chamber.

The cycle repeats every day for a week per training, with various levels of training offered. (Prepay for 12 trainings and you get a 30% discount!) It's not until the more advanced trainings, though, that Hardt's vision for a global brain comes into play. Shared alpha biofeedback trainings involve two students pairing up in one chamber to synch their own alpha states and "jam together," Hardt explains, "with one playing the (psychic) oboe and the other the violin."

Those shared sessions are practice for the

Hardt sits, wired with eight scalp electrodes, in one of the Biocybernaut Institutes's four brain wave training chambers. Multiple speakers deliver separate feedback tones in response to activity in four parts of the brain.

mental orchestra Hardt hopes to help conduct in the future. He believes that with 5 to 10 billion Biocybernaut students all linked up, a self-reflexive human superconsciousness may emerge. When I ask him how that could possibly work at a cost of $15,000 a head just for the basic training, he explains that economies of scale could cut the cost in half. As a step in that direction, he's planning to open new centers in Canada and two in Austria this year. Then it would just be a matter of getting, say, the U.S. military to subsidize the training cost with a percentage of the annual defense budget.

"After we were done, there'd be so much mind power that they wouldn't need a defense budget."

MAKE Editor-at-Large David Pescovitz is co-editor of boingboing.net and a research affiliate of the Institute for the Future.

WHO WANTS TO BE AN INVENTOR?

DON'T WORRY, NO ONE IS GOING TO STEAL YOUR SECRETS.

By Saul Griffith

FAR, FAR OUT THERE ON THE FRINGE OF society's imagination is the backyard inventor who strikes it rich with some cunning little gizmo that improves life. You've seen this person portrayed by Wayne Szalinsky in *Honey, I Shrunk the Kids*, Doc Brown in *Back to the Future*, Bruce Wayne in *Batman*, and countless other fictional characters, including my favorite, Wallace, and the real brains of the operation, Gromit.

Every time Squid Labs is in the news somewhere, we receive a new batch of emails from home inventors: "Dear Sirs, I've invented an energy device. I need help in patenting it and prototyping it. It will revolutionize the world and save the children of Africa from starvation."

Against my better judgment, I often send a reply: "Can you tell me a little more about your device so I can figure out if we can help?" Generally, I get the paranoid response: "If I tell you any more, you will know what the invention is!" To which I have to say, "Without more information I can't help you. Good luck." And in the back of my mind I imagine another perpetual motion machine.

This phenomenon reminds me of Tim O'Reilly's writing regarding authors: "Lesson 1: Obscurity is a far greater threat to authors and creative artists than piracy." Which might just as easily be translated as: Obscurity by secrecy is a far greater threat to inventors than having your ideas stolen by corporations.

I know a lot of people who invent things, even for a living, but I can't say I know anyone who represents the mythical backyard inventor who retires to live a life of luxury on the royalties of their patent. It is a concept fueled by the American dream and in the media offerings that support that dream, but it is not a reality.

I suspect that what would be most helpful is to step back and understand what it is that "inventors" really want, and figure out what would be their best strategy to get it. Like musicians, artists, and the other obsessively creative types, most innovators really just want to earn a good living doing what they love. Inventors, deep down, would like to be paid for full-time tinkering. If they could retire with the proceeds and buy yachts, that would be ideal, but I suggest that having the resources to do what they love and a workshop to do it in is way more important than the pot of gold at the end of the rainbow.

I'll return to O'Reilly's words to string my own thoughts together: "For all of these creative artists, most laboring in obscurity, being well-enough known to be pirated would be a crowning achievement." Again, these words apply directly to inventors and the process of invention. You'll be better off in the long run being known for inventing something — even if you give it away — than hoarding the idea in a padlocked file cabinet in your basement.

I know a lot more people who publicized ideas that turned into careers or job offers than I do people who made it big with that mousetrap. I don't actually know anyone who has been "screwed" by "bastard corporations" or other "thieving inventors."

Of course I always console my friends when they don't think their licensing deal was good enough, but it's really the same counseling they get from me over thinking about exes: "Of course they were [insert epithet] but now you are out of it, and better off for it."

Let's briefly look at the other option, the "patent and sell" dream. A single patent through to prosecution will cost you anywhere from $20,000 to $100,000, depending on how much of the world

Famous inventors of fiction and reality. Top: Bruce Wayne and Wayne Szalinsky. Bottom: Thomas Edison and who knows? It could be you. Just don't expect to retire to a life of luxury after filing a single patent.

it covers. This is before you've gone through the expense of trying to license your patent to someone (before you make any money, you'll have to spend a lot letting interested parties know about the idea). Alternatively, you can wait until someone infringes your patent and then try to recover monies from their error. This is even more expensive. I rarely recommend this process to any individual inventor. The likelihood of a yacht retirement outcome is similar to the likelihood that you look like Bruce Wayne.

Patents and secrecy don't really serve the individual anymore. Unfortunately, they have become trading cards in a grand corporate shell game. I'd stand by the statement that the most likely road to the tinkering nirvana is to become known for what you do. That means doing it well and telling people about it, even telling them how to do it.

The reality is that the invention or the idea is a tiny fraction of the work to be done, and in all probability other people have already had the idea or patented the same thing. The real tradeable value is your capacity to implement the idea. I've hired people because they've built cool projects on instructables.com that demonstrated their skill, and I've seen people turn their kit-building websites into profitable small businesses. It all started with pretty much giving a few of their inventions away — loss leaders to support a lifetime of tinkering.

To read Tim O'Reilly's essay "Piracy is Progressive Taxation, and Other Thoughts on the Evolution of Online Distribution," go to makezine.com/go/piracy.

Saul Griffith thinks about open source hardware while working with the power nerds at Squid Labs (squid-labs.com).

Illustration by Dustin Hostetler

Photograph by Howard Cao

Patently Curious

One sunny morning in the environs of Washington, D.C., I'm standing in an overgrown suburban backyard, slapping mosquitoes while trying to get a photograph of a "free energy" model railroad car on a small folding table. Tom Valone lines up the car, lets it go, and an array of cunningly oriented magnets draws it forward. The little car accelerates, runs off the end of its track, and tumbles into the grass.

This is more impressive than I had expected. But what would happen if the track were circular? Would the magnets on either side of the track push the little car around and around indefinitely?

Valone isn't sure about that. "You should look at a different patent from the same inventor," he says. "It uses magnets in the form of archways. Two witnesses have told me that when they visited the inventor, they saw a model running in a circle continuously for two hours."

I always seem to have this kind of problem with free-energy demos. The real demo is someplace else, or it happened at some other time, or the people who saw it have disappeared, or the inventor refuses to talk to a journalist because he's afraid that his idea will be stolen. "Where, exactly, does the inventor live?" I ask.

"In Virginia."

Valone's own house is in Virginia. "So why haven't you visited him?"

"Oh, I'm too busy."

This is the odd thing about Tom Valone. He has a fevered interest in fringe phenomena, catalogs them with obsessive zeal, yet seems more interested in describing them than evaluating them. As he puts it, "I hope to create a ripple effect. If I make people aware of things, their perception of what they should do on their own may change."

Despite his unconventional interests, Valone has a solid background in conventional science. He received a Bachelor of Science in physics and a Bachelor of Science in electrical engineering at SUNY Buffalo in 1974, and remained at SUNY, teaching college classes in subjects such as engineering physics, technical physics, electronics, digital logic, microprocessors, and environmental science. After he received a master's in physics in 1999, he took a job as an examiner at the U.S. Patent Office, specializing in the utterly conventional subject of electrical testing and measurement.

He had an ulterior interest, however. "Academic scientists," he says, "tend to value journal publication exclusively. That's their benchmark for credibility. If they don't see it in an academic journal, they don't think it's true. But this is a very tunnel-vision approach. Patents provide an alternative view of innovations occurring in the world, from authors who usually don't publish journal articles."

And so in his spare time he gathers information — patents, especially — on every obscure invention he can find. At this point he is probably the most broadly informed authority on the kind of basement tinkering that ventures into gray areas beyond the conventional laws of physics. While most of us tend to dismiss such work, Valone is like an open-minded juror who regards a defendant as being innocent until proven guilty. In his world, claims are worthy of consideration until proven false.

His open-mindedness led him into big trouble in 1999, when he conceived a "Conference on Future Energy." He obtained permission to hold it in an auditorium at the State Department, taking advantage of an Open Forum Speakers Program that existed "to explore new and alternative views on vital policy issues of the day." Unfortunately, "new and alternative" didn't mean the same thing to them as it did to him.

Robert Park, a PR writer for The American Physical Society (APS), is notorious for denigrating, ridiculing, humiliating, and eviscerating any individual who propounds propositions that Park, in his wisdom, finds implausible. When he learned that Valone's

Photography by Charles Platt

Physicist Tom Valone looks at technologies most scientists consider junk. By Charles Platt

Tom Valone outside the main building of the Patent Office, where he is employed as a patent examiner.

conference would include a former Los Alamos scientist discussing cold fusion, he was appalled that federal real estate should host such heresy. He contacted Peter Zimmerman, who had just been hired as a State Department science advisor, and Zimmerman duly used his clout to have the conference evicted.

Worse was to come, as Valone found himself the target of a jihad waged by guardians of establishment science. At a meeting of APS, Zimmerman boasted about his efforts to have the conference shut down. Meanwhile, Valone was fired from his job at the Patent Office. He tried to get it back, but the process lasted more than six years, during which he furthered his education and received a doctorate. Finally an arbitrator reinstated him and awarded back pay for the entire period, less 30 days.

"My conference was just an open forum for people who had unusual inventions in the energy arena," Valone says, still sounding genuinely puzzled that it could have sparked such retaliation. "Some were deserving of critical review, but others were solidly acceptable."

The magnetically powered railroad car in his backyard does not seem solidly acceptable (to me, at least), but the local mosquito population is waging such a relentless jihad of its own that I decide I have given the railroad car as much critical review as it is going to get. We retreat into Valone's house, where two walls of the living room are covered with shelves carrying books on topics from yoga to UFOs. A back room crammed with disorderly stacks of papers serves as an interim office for the Integrity Research Institute, a small enterprise that he runs with his wife.

Since I seem skeptical about the railroad car, he digs out the relevant patents: #5,402,021 for the linear track and #4,877,983 for the one with archways, both issued to Howard R. Johnson. The magnets are mounted in such complex configurations that analysis of the aggregate force is almost impossible. Valone nods enthusiastically. From his point of view, this means the design cannot be refuted. From my point of view, it means it cannot be evaluated.

He shows me another patent, #4,215,330, issued in 1980 for a similar array of magnets that entices a steel ball up to the top of a ramp, where it drops down and apparently could roll back to its starting point, ready for another trip. This idea has been picked up by Jean-Louis Naudin, a French free-energy advocate who provides detailed construction blueprints at jnaudin.free.fr/html/s102jlnp.htm.

PERPETUAL MOTION DEMO: Magnets separated by diminishing intervals provide a net force to drag a steel ball up a wooden ramp. The ball falls from the apex of the ramp into a channel, guiding it around for the next ascent. Would it work? The best way to find out would be to build it.

THE KURE TEKKO MOTOR: Mutual magnetic repulsion forces the rotor around inside the spiral stator until it reaches a gap where an electromagnet encourages it to make the next revolution.

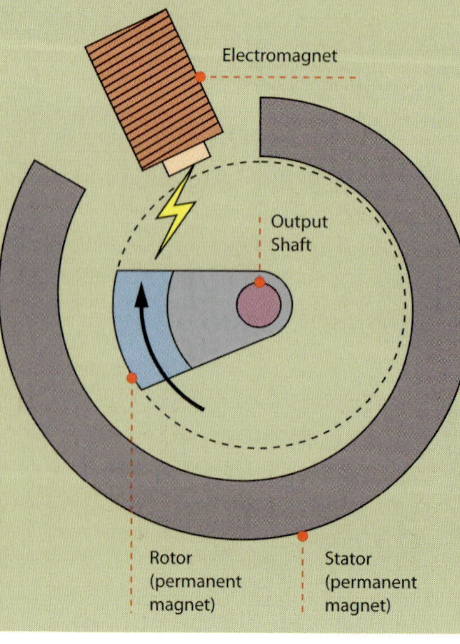

Valone is probably the most broadly informed authority on the kind of basement tinkering that ventures into gray areas beyond the conventional laws of physics.

Right: Valone in a shed he uses as a workshop in his backyard. Above: The magnetically powered model railroad car in the backyard of Valone's suburban home.

Valone unearths another magnetic propulsion concept, described in *Popular Science* in June 1979. This is the Kure Tekko magnetic motor, which interests Valone because it uses the under-exploited concept of a "magnetic gradient." A magnetic rotor turns inside a spiral-shaped stator that is also magnetic. Mutual repulsion accelerates the rotor as its distance from the inner surface of the spiral increases. At the end of the spiral is a gap that the rotor must cross before it overcomes the repulsion of the stator to begin its next cycle.

"The Japanese used a solenoid to take the rotor across the gap," says Valone. "Of course, this consumes electricity. But there may be a better way." He shows me a paper from the IEEE, entitled *Transactions on Magnets*, describing a piezoelectric laminate that may be capable of switching the force from a static permanent magnet in the same way that a transistor switches electricity. This indeed has fascinating implications — depending how much switching power it consumes.

When I ask Valone how he came to be so open-minded, he turns the question around. "I always wondered why other people were less open-minded," he says. "I promised myself that as I grew older, I would never lose my curiosity, and the ability to solve mysteries that others had overlooked. I felt that this was a youthful quality I needed to keep."

He agrees that his open-mindedness has the potential to lead him astray. "My ability to perceive reality is conditioned by my beliefs," he says. "If I believe in free energy, I'll start thinking I see it in all the wrong places. So many inventors have been self-deceived. Ninety percent of the people who claim they have free energy don't have it."

Still, he suggests that open-mindedness is precisely what the world needs right now. "Where are we going to get energy from in the future?" he asks. "None of conventional science can answer that question. Unfortunately federal dollars are tied into projects that aren't going anywhere, because politics have dominated science, and scientists all over the world are afraid to stick their necks out. We are losing our innovative ability in this country." He pauses to emphasize the point: "The only approaches that have promise today are unconventional."

+ Additional Sources:

Tom Valone's Integrity Research Institute (users.erols.com/iri).

The Basement Mechanic's Guide to Building Perpetual Motion Machines (makezine.com/go/simanek) includes actual photos of construction projects which, unfortunately, don't quite work. *(See Donald Simanek's article about perpetual motion, page 70.)*

Wikipedia article about the Simple Magnetic Overunity Toy (SMOT), which some claim produces more energy that it uses (wikipedia.org/wiki/SMOT).

Wikipedia article about Steorn, Ltd., a Dublin, Ireland-based company that claims to have invented a perpetual motion machine (wikipedia.org/wiki/steorn).

Charles Platt has been a senior writer for *Wired* magazine and has written science fiction novels such as the *Silicon Man*. He wrote "Electric Avenue" in MAKE, Volume 05.

5 Fringe Power Plays

Since Tom Valone maintains such a generously stocked clearinghouse of unconventional concepts, he's well-situated to suggest a Top Five of those that he believes could have the most far-reaching consequences. Here are his picks, in no special order:

THE CONCEPT	THE PROMISE	THE PROBLEM
1 ZERO-POINT ENERGY Quantum physics predicts that a minimum state of energy always exists, even in a vacuum with a temperature of absolute zero.	Inexhaustible power permeating every cubic centimeter of the cosmos.	Energy flows from areas of higher potential to lower potential. Since zero-point is, by definition, as low as you can go, energy would be more inclined to flow to it than from it.
2 BIOELECTROMAGNETICS Nikola Tesla advocated "electrotherapy" in 1898, and electromagnetic radiation enjoyed a brief vogue as a rejuvenation treatment. Recent research shows that near-infrared LEDs provide symptomatic relief from neuropathy while also accelerating cell growth (see makezine.com/go/nasa). Valone points out that light is a form of electromagnetic radiation, and suggests that since antioxidants inhibit the transfer of electrons, an electromagnetic field could encourage this beneficial effect.	Feel-good pick-me-ups, plus a possible cure for terminal cancer.	Evidence for far-reaching claims is anecdotal.
3 AIR-POWERED AUTOMOBILES In France, the MDI Aircar is powered by tanks of air compressed to more than 4,000 pounds per square inch. In Australia, a rotary air-powered engine has been developed by a startup named Engineair. Carbon-fiber air tanks may be cheaper, smaller, and lighter than other power storage systems such as batteries, and the electricity to pressurize them can be generated by non-oil sources such as hydro, wind, solar, nuclear, or coal.	Zero local pollution, reduced global pollution.	The first MDI prototype had a practical range of only about 5 miles. The manufacturer claims that a higher tank pressure and other enhancements may raise the range to at least 100 miles, but these claims have been questioned. Combustible fuels offer much more energy per pound than compressed gas.
4 ELECTROGRAVITICS Legendary American physicist Townsend Brown believed that an asymmetric attraction between high-voltage electrodes might partially nullify gravity. His British patent number 300,311 for "A Method of and an Apparatus or Machine for Producing Force or Motion" led him into research that may have hit a dead end — or may have been incorporated in secret military aircraft.	Low-powered levitation.	The work has never been fully replicated.
5 MAGNETIC MOTORS The seemingly mysterious nature of magnetism has always attracted perpetual-motion enthusiasts. Howard R. Johnson's classic paper is at makezine.com/go/johnson. It begins, "Using the Bohr model of the atom, and knowing that un-paired electron spins created a permanent magnet dipole, I kept wondering why we couldn't use these fields to drive something."	More power comes out than goes in.	Working models remain elusive.

CURRENT STATUS	OPTIMISTIC ASSESSMENTS	REALITY CHECK	OPPORTUNITIES FOR AMATEUR EXPERIMENTERS
Zero-point advocates suggest ingenious methods such as using molecular nanotechnology to access the energy. These ideas remain unproven.	Tom Valone has written his own book, *Practical Conversion of Zero-Point Energy*. Nick Cook, an editor at *Jane's Defence Weekly*, speculates on military applications in his book, *The Hunt for Zero Point*.	Try the 1997 article from *Scientific American*, archived at padrak.com/ine/ZPESCIAM.html.	Few at this time.
Early adopters can pick up a Healthlight LED infrared therapy pad for a mere $4,496, or a Curatron 2000HT pulsing electromagnetic field emitter for $2,170.	*Bioelectromagnetic Healing*, a book by Tom Valone.	"Magnetic and Electromagnetic Field Therapy" by Marko S. Markov and Agata P. Colbert in the *Journal of Back and Musculoskeletal Rehabilitation*, Volume 15, Number 1.	High. You can buy near-infrared LEDs in bulk, or install a Tesla coil under a plastic patio chair, but we don't recommend it.
Apparent lack of investment capital.	Automotive engineer and inventor Guy Negre describes his vision at theaircar.com. Engineair is at engineair.com.	en.wikipedia.org/wiki/air_car	Doubtful. Achieving and storing the very high necessary air pressure is challenging and potentially dangerous.
An endless source of speculation among UFOlogists and conspiracy theorists.	Try Paul Schatzkin's book, *Defying Gravity: The Parallel Universe of T. Townsend Brown*. Also try the official Townsend Brown website, soteria.com.	en.wikipedia.org/wiki/thomas_townsend_brown	Dozens of sites tell you how to make your own electrostatically energized "lifters" from aluminum foil. Don't expect them to carry a payload, though.
A backwater.	In his home workshop, Tom Bearden claims to have built his own Johnson motor and talks about "oceans of free energy" (cheniere.org/misc/oulist.htm).	The Philadelphia Association for Critical Thinking (PhACT) offers sobering advice regarding magnetic machinery at phact.org/e/z/freewire.htm.	Excellent, so long as you are willing to build things that don't quite work. If violating the second law of thermodynamics were easy, everyone would be doing it.

Illustrations by Scott Barry

FRINGE

In Search of Psi

At the tender age of 13, I embarked on a serious mission to tap the psychic waveband. While other kids were playing team sports or ogling girls, I was fabricating divining rods and a Hieronymus Machine.

My guide into the queasy world of quasi-science was John W. Campbell, Jr., the irascible editor of *Astounding Science Fiction* magazine. He sounded like one part engineer, one part wide-eyed visionary, and two parts drill sergeant. His enemy was unthinking orthodoxy. His precept was that you could crack any problem with the scientific method, so long as you didn't pay too much attention to "experts" warning you that what you wanted to do couldn't be done.

In 1956, Campbell published an informal position paper, asking his readers if they were ready for a serious exploration of "psi," by which he meant the entire subject of psychic phenomena. If psi existed, it should yield to a mix of logical deduction and Yankee ingenuity, just like anything else. So how about it? He wrote as if he were addressing a platoon of fresh-faced recruits. Were we man enough (surely, none of his readers was female) to tackle some really life-changing, head-bending ideas, no matter where they might lead us?

In his June 1956 issue, Campbell reported that his readers shared his zeal. They were gung-ho, ready to lock-and-load and follow him over the top into combat with prejudiced naysayers who dismissed psychic research as frivolous or wacky. So be it! He started publishing a series of revelatory articles that I read with utter fascination.

DIVINING RODS

I especially liked a feature about "divining rods" because they were so simple. All you needed were two pieces of rod and two lengths of tube. You bent each piece of rod at a right angle and hung it in a length of tube. You held the tubes, one in each hand, so that the rods pointed away from your body, like a pair of handguns. If you rotated your wrists slightly, the rods would swing from side to side.

Supposedly they would detect cables, conduits, water pipes, sewer pipes, aquifers, or anything else that ran in a straight line underground. You held the divining rods and started walking. As you approached the linear underground object, the divining rods started to turn. When you were directly above the object, the rods aligned themselves with it.

The rods were said to be standard equipment among hard-hat maintenance workers across Heartland America. The magazine published a picture of the Distribution Supervisor from a water company in Flint, Mich., holding a pair of the rods — or at least, so the caption claimed.

Campbell theorized that the human brain might have unknown powers to sense underground objects. The brain simply needed a means to express itself, perhaps via unconscious muscular movements of the type that old-fashioned water diviners had applied with a yew twig. The rods probably worked the same way.

I chopped a wire coat hanger into two pieces, hung each of them in a spool of thread that I filched from my mother's sewing kit, and stretched a piece of string across my bedroom floor. I closed my eyes and stepped forward. One pace, two, three. I opened my eyes, and sure enough, I was standing directly on the string, and the rods had turned (or, I had turned the rods) to align themselves with it.

I was thrilled, but of course I realized this was not a fair test, since I had known where the string was located, and I had known the result that I was hop-

---

Author's note: Since the inventor's patent was filed under the name "Hieronymus," I have used this spelling, even though Campbell, who was in contact with the inventor, spelled his name as "Hieronymous."

A boy and his Hieronymus Machine. By Charles Platt

In October 1958 this seemingly undistinguished science fiction magazine contained a revelatory story on "divining rods." (Original cover by Kelly Freas.)

The brain simply needed a means to express itself, perhaps via unconscious muscular movements of the type that old-fashioned water diviners had applied with a yew twig.

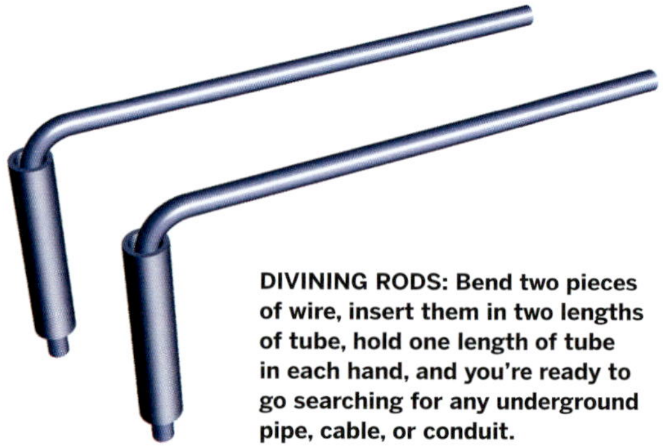

DIVINING RODS: Bend two pieces of wire, insert them in two lengths of tube, hold one length of tube in each hand, and you're ready to go searching for any underground pipe, cable, or conduit.

ing to achieve. I asked my father if he knew of any underground objects buried outside the house, but he said he didn't. So I rode my bicycle to the home of a friend whose father was an electronics engineer in the Air Force.

My friend and I took turns holding the rods while we walked to and fro across his backyard. The rods swung in our hands, and we started drawing a map of underground objects. The objects formed an X-shaped pattern, linking opposite corners of the yard. But what, exactly, were we mapping?

We turned to my friend's father, who had been watching us with interest. "Is there *something* under your lawn?" I asked.

He nodded very seriously, although he seemed to be having trouble repressing a smile. "There are two drains," he said. "They run from corner to corner."

My eyes widened. "In an X-shaped pattern!"

"Absolutely right," he said.

PSIONIC DETECTION DEVICE

Well, after that, John W. Campbell was more than just a guru to me. He acquired godlike status. In a previous issue he had published a piece about a "psionic detection device." This had seemed an overly ambitious project at the time, but I was willing to work a little harder, now, as a foot soldier in the resistance movement against blinkered orthodoxy.

The psionic device had been invented by a lone experimenter named Thomas G. Hieronymus who claimed that it could detect "eloptic radiation," emitted by every object in the universe. (He received U.S. patent number 2,482,773, and although he is no longer alive, his legacy lives online at sites such as rexresearch.com/hieronym/1hieron.htm.)

The Hieronymus Machine featured a tuning dial coupled to a prism that would refract the eloptic radiation, feeding it through an amplifier to an output coil. To use the device, you placed your hand on a "sensor plate" above the coil and rotated the tuning dial. When the machine resonated with matter in front of the prism, the sensor plate would feel sticky to the touch (although some people described it as hot, or cold, or oily).

Campbell built his own version of the machine and announced, with utter certainty, that it worked. Six months later he came up with a creative modification. If he was correct that psychic phenomena were inherent in the brain and merely needed an output device to express themselves, the choice of raw materials in a Hieronymus Machine was no more important than the type of metal you used to make a divining rod.

Why not reduce the electronic components to a symbolic level? If you merely printed them on paper, they should still work (if they worked at all), in the same way that other psychic phenomena worked symbolically. For instance, some people claimed you could kill pests in a field by dripping a chemical on a photograph of the field, while others believed you could induce pain in a person by sticking a pin in a doll. Supposedly the symbol enabled expression of a latent mental power.

Campbell duly published a circuit diagram, advised his readers to add a symbolic prism and sensor plate, and told them to let him know what happened.

The more I thought about it, the more I wanted a Hieronymus Machine. I craved it with the same intensity that I craved Lucilla Stanley, a golden-haired, freckle-faced goddess who sat opposite me

SYMBOLIC HIERONYMUS MACHINE

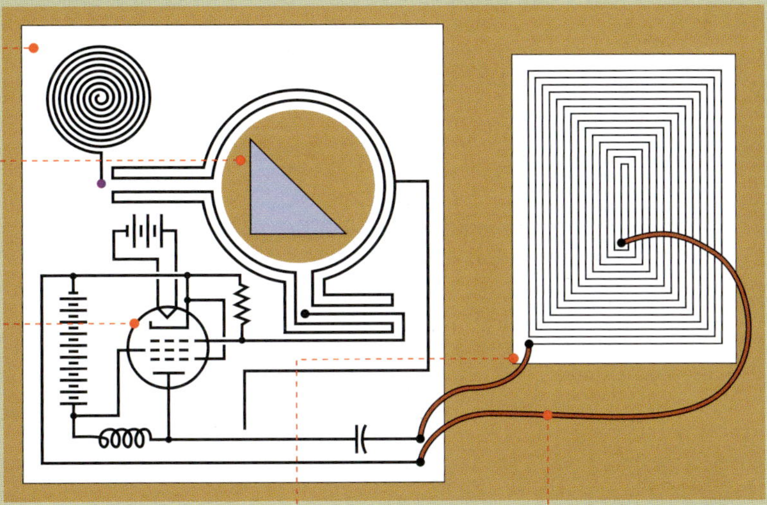

Draw the circuit on a piece of card.

Cut out a circular hole in which a triangle of clear plastic can rotate on the end of a shaft attached to a dial.

Symbolic vacuum tube amplifies the psionic signal, powered by a high-tension symbolic battery. A transistorized version might have a shorter warm-up time.

Draw the symbolic output coil on another piece of card, which you will glue under the sensor plate.

Attach the amplifier circuit to the coil with some pieces of thread glued at both ends.

From a concept by John W. Campbell, Jr.

FRONT VIEW

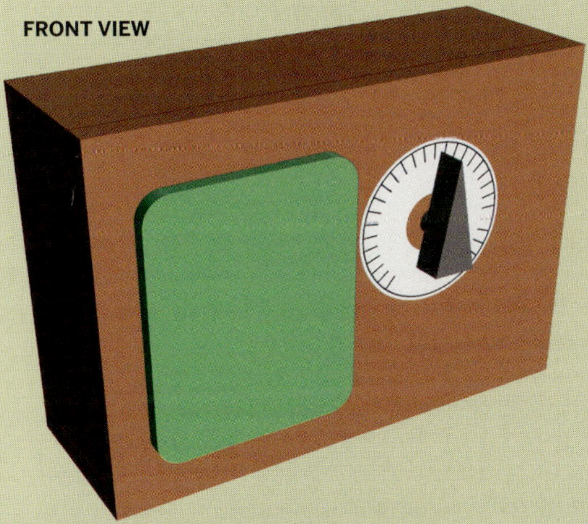

DETAIL OF REAR VIEW

REAR VIEW

These renderings suggest a simple construction project using plywood, a control knob on a shaft, and a plastic prism on the other end of the shaft, rotating in a hole cut out of a circuit board. No need for tiresome electrical wiring; just paste a printed copy of the circuit onto the board, and a printed coil under a "sensor plate" such as the lid of a lunch box. Connect the coil to the circuit by gluing the ends of sewing thread or string, and you're ready to search for eloptic radiation on the psionic waveband.

Illustrations by Charles Platt

in my French class in school, inducing adoration that made me powerless to speak. Realistically, I had to admit that the Hieronymus Machine would be more easily attainable than Lucilla, so I banished her from my mind while I made a laborious hand-drawn copy of Campbell's schematic circuit diagram (this was before the invention of Xerox machines). I used a triangle of Lucite as the prism, attached it to a wheel and a rod from my Erector Set, and contrived a sensor plate from the lid of my lunch box. Now I was properly equipped to boldly go into the unknown territory of psi!

During the next week I reduced my popularity among kids at my school to a new low as I pestered them to put their hands on the sensor plate while I twiddled the tuning dial — keeping its position hidden from them, naturally. Did the plate feel sticky? Or hot, maybe? Or different in any way at all?

I received a lot of positive responses. Definitely, the characteristics of the plate seemed to change.

The only problem was that these alleged changes did not coincide with any position of the dial. Positive results were scattered randomly across the entire 270 degrees of rotation. I was forced to face the fact that the only power I had detected here was the power of autosuggestion.

There was one last candidate for my test. His name was Leonard Fisk. Leonard was a gawky kid with a bad haircut and exceptionally large ears. He was not especially good academically, and not popular in team sports either. He was a bit — odd. I hesitated to add him to my list of experimental subjects because I wasn't sure he would cooperate.

I was wrong about that. He turned out to be very enthusiastic, and to my amazement, he detected a change in the sensor plate when the tuning dial was in the same position every time, within a margin of plus or minus 10 degrees.

I wondered what the difference was between him and all the other people (including myself) who had tried unsuccessfully to get results from the Hieronymus Machine. Maybe it had something to do with his oversized ears. I imagined them functioning like psychic antennas. It seemed unlikely, yet who was to say how these things worked or didn't work?

I went home and wrote a letter to John W. Campbell, Jr. describing my results, although I decided not to mention my theory about Leonard's ears. A few weeks later I was thrilled to receive a reply from Campbell, telling me that he had forwarded my mail to someone named Smith, who was working to collate results from serious experimenters such as myself, scattered across the entire globe.

After another two weeks I received a letter from Smith. Unlike the fancy publisher's stationery that Campbell had used, this letter was a ragged, plain piece of paper on which text had been badly typed with numerous handwritten corrections. Smith wanted me to conduct a whole series of new experiments. The trials should be double-blind and independently witnessed. I had to record the age, gender, body weight, and other details of my experimental subjects. They must all perform the tests in the same place, ideally at the same time of day, and in the same room temperature.

I didn't like the sound of this at all. I had been able to persuade the kids at school to stop for a couple of minutes on their way to lunch, but getting them to sit still in a "controlled environment" was never going to work, especially since my school's heating system was notoriously unreliable.

THE PROCESSES OF SCIENCE

Campbell had excited me with the idea that a few gifted amateurs in basement workshops could validate ideas that conventional scientists had been too prejudiced to take seriously. He made it seem as if psychic research could be a slam-dunk, a quick fix. But now Smith wanted meticulous data collection under controlled conditions, requiring weeks of repetitive work. There was no excitement in this scenario. It sounded really boring — like real science.

I have to admit that I lost interest in the Hieronymus Machine. John W. Campbell, Jr. issued more wake-up calls in his magazine, more rants insisting that "We Must Study Psi!" but he no longer captivated my attention. In June 1960 he ran a big feature on the Dean Drive, a system of rotating weights that was supposed to lift itself by taking advantage of an undiscovered nonlinear correction to Newtonian laws of motion. Forget about NASA's costly boondoggle, trying to use dangerous liquid-fueled rockets to put men on the Moon by the end of the decade. We could bolt a Dean Drive to a submarine that would serve as an ad-hoc airtight space vehicle, and move men to Mars within a year — if the Dean Drive worked the way it should.

Somehow I didn't quite buy it. I had learned a painful lesson about the processes of science. In the real world, science is not quick, not simple, and seldom

By January 1959, Campbell was still insisting, in an editorial, that "We Must Study Psi!" (psychic phenomena). But his table thumping was attracting fewer takers. (Original cover by Kelly Freas.)

In June 1960, Campbell proposed bolting a Dean Drive to a submarine to create an instant airtight spaceship, ready to put men on Mars within a year (assuming the Dean Drive actually worked). By this time he was changing the name of the magazine from *Astounding Science Fiction* to *Analog Science Fact & Fiction*, in a bid for respectability that was never entirely successful. (Original cover by John Schoenherr.)

exciting. I was now 16, the same age at which Einstein had suddenly wondered if he would be able to see himself in a mirror if he and the mirror were traveling at the speed of light. From this moment of intense conceptual excitement came the theory of relativity; but the details, of course, consumed the rest of his life, and he never did come up with a unified field theory. As Thomas Edison supposedly said, genius is one percent inspiration and 99 percent perspiration.

I never went into science, or even pseudoscience. Instead I became a science writer, which enabled me to enjoy all the excitement of discoveries vicariously, without the hard work of making them. In retrospect, I think John W. Campbell, Jr. was not so different. He was another scientist-wannabe looking for a shortcut around the monotonous, challenging terrain of real-life research.

This does not mean that he was necessarily wrong about everything. Discoveries may come from unexpected sources. Divining rods still may work (for some people, at least), and even Thomas G. Hieronymus could have been onto something.

In fact, maybe his machine still merits a couple hours of construction time. Your friends may be sufficiently intrigued to tolerate a few sessions at the sensor plate before they wander off to do something more important, such as browse YouTube or check email. Depending on your attention span, you may even gather some usable data.

I only ask one favor. If you find someone, as I did, who yields positive, repeatable results — please be sure to note the size of his ears.

Charles Platt is a frequent contributor to MAKE, has been a senior writer for *Wired*, and has written science-fiction novels, including *The Silicon Man*.

The Music of Chance

How random is this? I'm sitting in an empty classroom at Princeton, a few feet away from Roger Nelson, the founder and director of the Global Consciousness Project. The setting seems plucked right out of *A Beautiful Mind*, with a thin band of blackboard framing the room, broken by windows overlooking an immaculate campus. There are chalkboard leavings from the previous occupants: elaborate differential equations from a math class, Arabic conjugations from some linguists. And in the nearby engineering quad sits a server, quietly humming away, polling dozens of computer randomizers around the world for Nelson's grand experiment. Based on that data, he's written several carefully worded scientific papers that claim that human minds are somehow influencing the random numbers in subtle ways.

The project surprised me when I first heard about it. Nelson uses well-documented statistical methods to look at 10 million datapoints per day from 65 computers around the world between San Francisco and French Guiana. Nelson examines trends that he claims correspond to global events. Many of his best examples tend to be large catastrophic occurrences — embassy bombings, earthquakes, 9/11 — anything that "rends the social fabric and holds human attention." There's more: in a November 2006 paper published by the American Institute of Physics, he explores the idea that trends in random numbers begin hours *before* an event.

I wasn't sure I believed it, but Nelson impresses me as a straight-up guy in the first few minutes of our meeting. Herbal tea in hand, he walks around looking for a quiet place for us to talk, finally ducking into the empty classroom. He's in his mid-60s, with a neatly trimmed beard and striped broadcloth shirt, eminently professorial in appearance.

The Global Consciousness Project, or GCP, is unconnected with Princeton University, except that Nelson's friends in the engineering department allow him to keep a server, and Nelson himself worked as a researcher there for over 20 years, retiring in 2002. GCP is a bootstraps project with a tiny budget, mostly stemming from PayPal donations. Friends of the project designed one of the matchbox-sized gadgets that perform random-number generation. Nelson's adult son wrote the software to poll the devices, and 65 volunteers around the globe host the generators in their spare bedrooms, university labs, and home offices.

"They're like one-way modems, blasting out random bytes at 9600 baud," Nelson says, holding out one of the randomizing devices used in the project. He passes me the tiny gadget, and I involuntarily wonder if there's something fishy about the apparatus — the image comes to mind of a magician tapping the sides of a wooden cabinet before the lady climbs in. I know he wouldn't appreciate the comparison.

Nelson is diligent in his research. Over the last few years, he's been the focus of respectful, if cautious, articles in *The New York Times* and *Time*, the latter going so far as to point out that many of our greatest scientific breakthroughs were once ridiculed. The November paper came out of a symposium on "Retrocausation" — organized by the respected American Association for the Advancement of Science (AAAS) — exploring the theory that time flows symmetrically.

"The theoretical questions about the direction of time go back 150 years," says Daniel Sheehan, organizer of the AAAS symposium and a physics professor at the University of San Diego. "Nelson and other experimentalists are starting to put some experimental background to the idea that retro-causation may be a real phenomenon. The experimental evidence is not all in, but his results are tantalizing."

Photograph by Erika Larsen

Roger Nelson's Global Consciousness Project searches for meaning in random numbers. By Bob Parks

Nelson at his office in Princeton, N.J. "Sometimes those meaningful coincidences are so potent that it's really hard not to believe that there's something, some kind of connection there that is real and meaningful."

AT THE FOUNDATION OF GCP'S NETWORK are 65 random-number generators. Truly random numbers are important to hard science, and these common devices harvest the white noise from various physical sources, like the heat coming from a resistor, the time intervals between particles arising from radioactive material, or electrical fluctuations in analog components.

The device used by the GCP is called Orion ($550, randomnumbergenerator.nl) and uses two diodes to generate a chaotic signal. The Orion sends current the wrong way through a diode so that only a few miscellaneous electrons make it across the junction. A very weak, fluctuating current results. The circuit amplifies the signal so that a microcontroller can sample and compare it with a threshold value. Current above the line receives a "1" and current below, a "0." The circuit compares the bit streams against stored random data using an exclusive-or gate (XOR) to further eliminate mechanical drift. (The GCP network also uses a similar randomizing device with a field effect transistor as the entropy source.) Host computers on the GCP network sample 200 of these bits per second and add them together for sums of between 0 and 200. Since it's random data, the sums usually come in around 100, with a few outliers in the 80s and 120s.

A few minutes later, the data pours into the Princeton server. The ensuing analysis is a formal, deliberate process. Nelson emphasizes that he conducts research as a scientist, and follows widely accepted research methods. In general, to make a scientific claim, one must set up a formal hypothesis and a null hypothesis before looking at experimental data. Then, he'll fire up his Linux machine to feed the data through a statistical analysis package called XploRe.

As an example, let's pick a day in the life of the GCP network — a biggie, September 11, 2001. Nelson, who lives with his wife on a quiet Princeton street and conducts all research at home, was casually watching the network from the five computers he has running all the time. But there didn't seem to be anything going on in the world worth analyzing. Then an airplane hit.

"By the time the first building collapsed, I realized we had a global event maybe bigger than we'd ever looked at," he says. He started outlining a formal hypothesis with an expected result and the formal time period for the experiment — a window starting ten minutes before the first plane hit to four hours after. With those parameters fixed, he was then free to look at all the data surrounding the event for a more exploratory look. Nelson has since examined the 24-hour period starting at midnight of Sept. 10. Graphed against a pseudo-random line, the data shows a striking upward trend. The calculated probability of that trend is 1 in 100,000. What's more, the break from chance begins at 4 a.m., hours before impact.

As Nelson himself points out, the possibility of network precognition is clearly exploratory. "There's no way you can use that as evidence in the scientific sense," he says. Nelson frequently underscores that he wants to be critiqued principally on his formal hypotheses. But the results of the formal hypothesis are actually pretty boring. In more than 200 time parameters studied under hypotheses, only 21 show a significant deviation against a distribution of purely random events. For example, the formal September 11th events showed 2.8 percent odds of occurring against chance. Some larger odds come from other formal studies such as the May 2006 Indonesian earthquake (.2 percent); the September 2004 Chechen hostage standoff in Russia (1.2 percent); and the March 2000 Papal visit to Israel (.8 percent). Given the number of tests, however, it's possible that the GCP researchers are getting lucky by picking

HOW THE GCP WORKS

1. Random-number-generating devices (right) are placed in spots around the planet. They connect via serial cables to PCs in home offices, workplaces, and the basements of academic institutions.

2. The RNG gadget is the size of a matchbox. It works by sampling a fluctuating signal inside the device, generating a random stream of 1's and 0's. Each host computer grabs 200 samples per second, adds them together, and sends the sums off in batches every 10 minutes to a central computer in Princeton.

3. At the GCP headquarters, a Linux PC collects the data and analyzes it for deviation from normal statistical probability.

"Nelson and other experimentalists are starting to put some experimental background to the idea that retro-causation may be a real phenomenon. The experimental evidence is not all in, but his results are tantalizing."

small, anomalous windows in random data.

"Statisticians will say, 'Well, there are so many opportunities to find anomalies,'" says Nelson. "But sometimes those meaningful coincidences are so potent that it's really hard not to believe that there's something, some kind of connection there that is real and meaningful."

MANY UNIVERSITY ACADEMICS HAVE dismissed GCP on its face, but few have been willing to crunch the numbers themselves (the raw output is available on noosphere.princeton.edu). A common critique involves electromagnetic waves — doesn't the use of cellphones, television transmission trucks, and two-way radios increase during world tragedies? Wouldn't that addle the randomizing hardware? Nelson counters that the devices are carefully shielded. Besides, each device does generate random data during the experiments. It's the statistical trends among the 65 devices that are the subject of study.

There have only been two serious looks by scientists critical of the work. S. James P. Spottiswoode, chief statistician for entertainment-monitoring company Neilsen, reproduced the computer analyses of the formal 9/11 findings, and confirms Nelson's result. He notes, however, that any event with a 2.7 percent chance of happening shouldn't surprise anyone.

Nelson's other auditor, Jeffrey Scargle, a staff astronomer at the NASA Ames Research Center in California, contends that the math Nelson uses to plot graphs makes minor trends look more meaningful than they are. Scargle also questions the XOR operations inside the random-number generators. "They're scrambling the bit stream in order to get rid of drifts in the instrument, but by doing that they're getting rid of any signal that would possibly have a direct effect on the random-number generator," he told me. Nelson himself clearly seems interested in mind-quanta interaction. Why not let the bits fall as they may?

Nelson responds by saying that the XOR doesn't entirely remove the influence of the original quantum events; it merely smoothes out the data. "He thinks I'm throwing the baby out with the bathwater," he says. "But from laboratory tests, we know we still have some baby."

Nelson's critics also dismiss the work as carefully selecting data to guarantee a positive result. The exploratory tests boast larger departures from chance, but these sessions are often conducted by laying out all the statistical deviations on a timeline and zeroing in on the surprising ones. Nelson deflects this by saying the exploratory claims are clearly labeled as such. "It's unacceptable that they criticize the GCP's hypothesis testing on the basis of postfacto analyses," he says calmly. "These are independent projects. It would be stupid not to see what the data looks like."

While we talk, I try to drill down on the exact mechanism of the phenomenon. "During these global events, do you think the devices are not performing in a scientific manner?" I ask.

"That's a pretty logical kind of thought," he says affably. "It sounds as if you're saying human consciousness is changing the way these devices operate. But probably not." Nelson suggests the possibility that the numbers aren't changed at all while they are being generated, but don't become fixed on computer hard drives until they are analyzed. Then he stops talking and sort of shifts in his chair. "This is a lot of hand waving, for sure," he says.

Clearly, there are plenty of other people curious about global consciousness, too. Traffic to the GCP website nearly doubled to 60,000 last May during a busy month that included the Indonesian earthquake and Earth Day. And on Dec. 22, 2006, the pro-sexual liberation group Baring Witness held a "Synchronized Global Orgasm" and used the network data to explore the impact of mass nookie on GCP's spreadsheets. The GCP reported a "positive outcome, but it is a mild trend" from the experiment. "The picture has a feeling of promise, but it is not possible to know whether the trend is evidence of a signal or not."

People are encouraged to try harder next year.

Bob Parks is the the author of *Makers: All Kinds of People Making Amazing Things In Their Backyard, Basement or Garage*, published by O'Reilly Media.

FRINGE

Digital Kirlian Photography

Shoot "auras" without film. By John Iovine

Kirlian photography records a high-voltage corona discharge around objects. Some people call this discharge an "aura" and attribute metaphysical and paranormal factors to its varying parameters. Originally, Kirlian photography was a contact print process that used film. This article illustrates another method of shooting Kirlian photographs, using a digital camera.

The technique can be traced back to the late 1700s, when Georg Christoph Lichtenberg first created "electro-photographs" in dust using static electricity and sparks. Nikola Tesla photographed corona discharges using his famous Tesla coil in the 1880s and in the early 1900s. Others followed, and in 1939, the Russian husband-wife research team, Semyon and Valentina Kirlian, began their 30-year investigation into electro-photography techniques.

In 1970, the landmark fringe book, *Psychic Discoveries Behind the Iron Curtain* by Sheila Ostrander and Lynn Schroeder, popularized the Kirlians' work in the English-speaking world, and electro-photography has been known as Kirlian photography ever since.

Many claims have been made regarding Kirlian images, but most phenomena have conventional explanations. For example, changes in the "aura" of an honest individual who is lying result from the same stress-related increase in galvanic skin resistance that a polygraph lie detector measures. Changes in skin resistance can also be caused by illness, fatigue, drug or alcohol consumption, and other factors, which make these conditions observable using Kirlian photography with no paranormal "bio-plasma" explanation required. The Kirlians themselves believed that their photography could diagnose illnesses before noticeable symptoms manifested, an idea that generated interest, but was never verified by scientific investigation.

One famous Kirlian phenomenon, the "phantom leaf," cannot be explained by known physical laws — but it may be a fake. The experiment is easy to perform: you take a Kirlian photograph of a leaf after cutting off a small portion. Phantom leaf photos from Soviet-era proponents showed the removed portion of the leaf appearing as a ghostly apparition, suggesting an ethereal "bio-plasma body" persisting where the physical leaf no longer existed. Many researchers and experimenters have been unable to replicate this effect (myself included) although it is easy to fake with a simple double exposure: take a short exposure of the entire leaf, then cut off a piece and continue the exposure.

Although I have never observed any paranormal phenomena with Kirlian photography, I like Kirlian photographs, which are unique and often beautiful. I also like exploring, so I've continued to look for the phantom leaf effect over the years, taking sporadic excursions back into Kirlian photography when I think of new twists to incorporate. If the phantom leaf effect exists and someone finds an exact combination of voltage, frequency, pressure, ambient humidity, exposure time, and other experimental parameters that can reproduce it consistently, it will be the starting point of a new paradigm.

FILM AND DIGITAL TECHNIQUES

The process for traditional, film-based Kirlian photography is simple. In complete darkness, place sheet film on top of a metal plate, and then an object to photograph on top of the film. If the object is inanimate, ground it. Then apply high voltage to the plate momentarily. The corona discharge between the object and high voltage plate is recorded onto the film as a contact print. Develop, and you have a Kirlian photograph of the object.

To shoot Kirlian photographs with a digital camera requires a slightly different technique that uses a transparent discharge plate instead of a metal plate. This discharge plate has a coating of tin oxide on one side that is so thin it is visually transparent, but still electrically conductive.

Place the object on one side of the plate and the camera on the other. As with the film process, connect the object to an earth ground if it's inanimate.

Photography by John Iovine

With all the room lights turned off, open the camera's shutter and apply high-voltage power to the transparent discharge plate. The camera captures the corona discharge between the object and the transparent plate.

With film Kirlians, the electrical discharge has to travel through the film's 3 primary color dye layers and activate the silver crystals in each layer at different depths; whereas with digital, it all happens at one level on the plate. This makes Kirlian photographs with film more colorful around the edges, but less accurate than digital Kirlians.

KIRLIAN PHOTOGRAPHY WITH A DIGITAL CAMERA

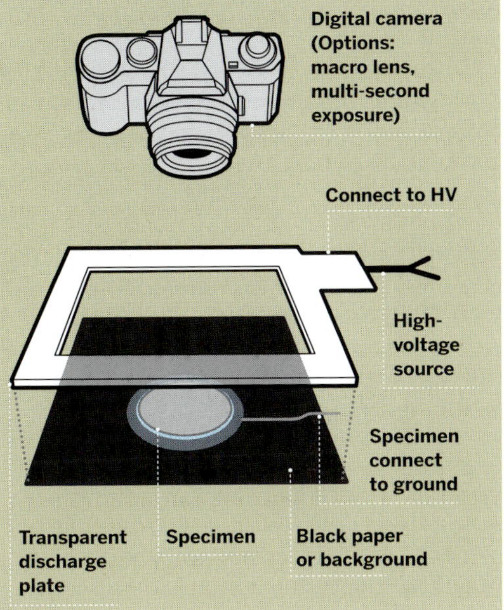

DANGER: HIGH VOLTAGE!
Do not attempt this project without expert assistance unless you are an adult experienced in working with extremely high-voltage power sources. This project is intended only for very experienced adults. Severe injury, death, or property damage may result from failure to use adequate safety gear and precautions.

MATERIALS

Digital camera with manual focus and a shutter capable of multi-second exposure times. It also helps to have a macro lens setting for close-ups.

High-voltage power supply You can use a Tesla coil, induction coil circuit, high-voltage flyback transformer circuit, or any other source that puts out 5,000+ volts, between 1-5 milliamps, at any frequency (or DC). Variable frequency lets you experiment. The PG13 supply I sell is great for this; see makezine.com/go/highvoltage.

Transparent discharge plate You can buy one of these, or make one out of conductive glass (tin oxide coated), clear plastic, thin copper plate, silver epoxy, and some HV wire. For sources and instructions, see makezine.com/09/kirlian.

Stand or tripod or another way of holding the camera still

Black paper or foamcore

High-voltage wire, Teflon-coated or other, sufficient for your HV power supply

READY, STEADY, SHOOT

Long exposures are required to capture the corona discharge, so you will need to keep your camera still and the object steady. I use a camera copy stand that holds my camera pointing downward, and I set up the object and plate horizontally underneath it. Once you have a steady arrangement, here's the procedure.

1. A black background works best, so put down a piece of black paper or foamcore (must be nonconductive) on your work surface. If you're shooting an inanimate object such as a leaf or coin, you need to

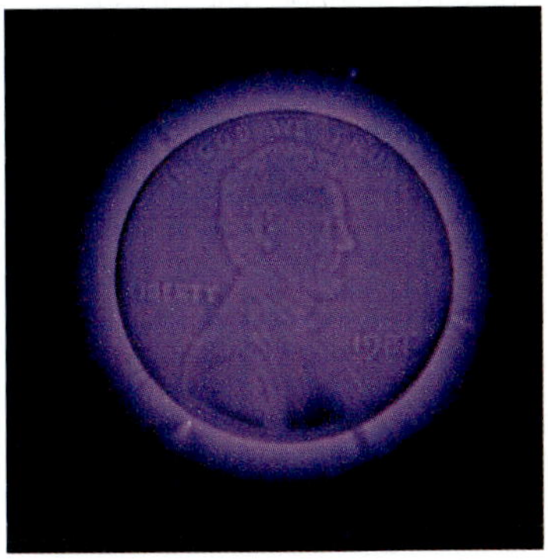

Illustrations by Damien Scogin

ground it. To do this, cut a small hole in the center of the black background, and attach a length of HV wire to an earth ground (a pipe will work). Then strip the other end of the wire, run it through the hole in the background, and place the object on top, making sure it has contact with the wire. You can also use a small grounded copper plate under the background instead of a wire, so long as the object you are shooting makes contact with the plate through the hole.

2. Attach the high voltage lead from the HV power source to the discharge plate's electrode, and place the plate on top of the object.

3. Position your camera to frame the object, and focus it manually. If you can set the camera's f-stop, open the aperture (reduce the f-stop) as much as possible. You will usually shoot a flat object, which doesn't require much depth of field. Set the shutter speed to 15 seconds or more.

4. Turn off all room lights. Turn on the high-voltage, and adjust its frequency to provide the brightest discharge. Hit the camera's shutter, and keep the power going for the full duration of the exposure.

5. Evaluate your picture, adjust, and reshoot. Raise the f-stop and reduce the exposure time if the image is washed out; do the reverse if it's too dim.

SHOOTING PEOPLE

Sooner or later you are going to want to shoot people. The best place to start is with a person's fingertip. To avoid electric shock, make sure that the subject doesn't touch ground. They also need to hold still, of course, and they shouldn't put too much skin surface in contact with the discharge plate. Touching an entire hand to the discharge plate can dissipate the corona discharge so much that it won't photograph, even with a long exposure.

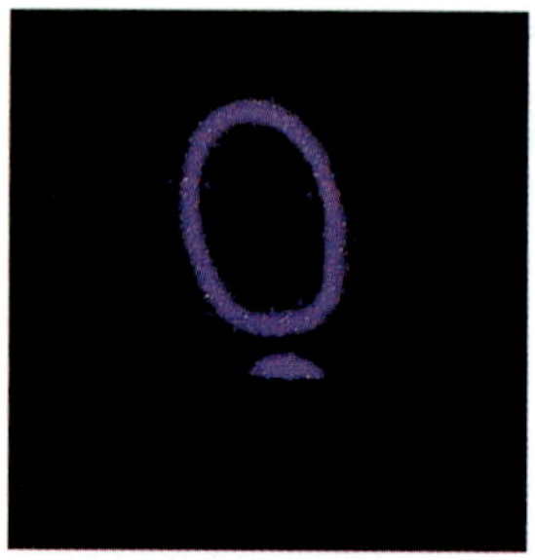

WARNING: When shooting people, make sure the subject doesn't touch ground.

DIGITAL MANIPULATION

My digital camera has a maximum exposure time of 16 seconds, during which the image accumulates some digital noise — mostly gray, red, and yellow pixels of various shades sprinkled throughout the resulting photograph. I remove this in Adobe Photoshop as follows:

1. Starting with the base image of the leaf, select Image » Adjust » Brightness/Contrast, and increase the contrast.

2. Zoom picture to 500% so you can clearly see the background pixel noise. Using the Magic Wand (set to 30 tolerance), click on a noise pixel. Then go to Select » Similar. Now all the pixels with that color in the photograph will be selected.

3. Set the foreground color to black and hit the Delete key to turn all the selected noise pixels black. Repeat this process by selecting and deleting various other noise pixels, but don't select any noise pixel that has a similar shade to the main image you are trying to enhance.

Kirlian photographs are mostly blue-violet with a little white. This is because the corona discharge in nitrogen (air is 78% nitrogen) generates light in

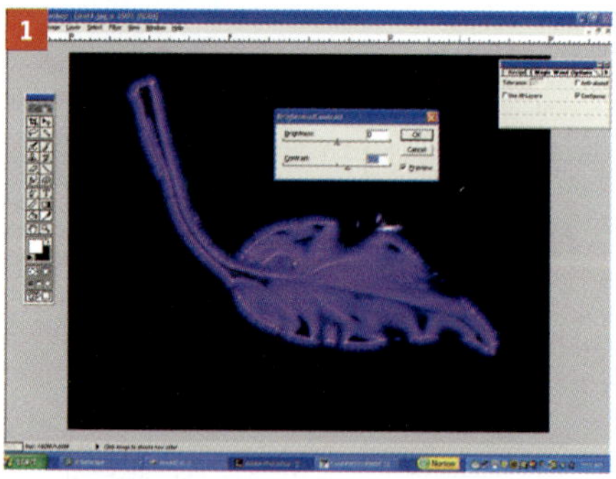

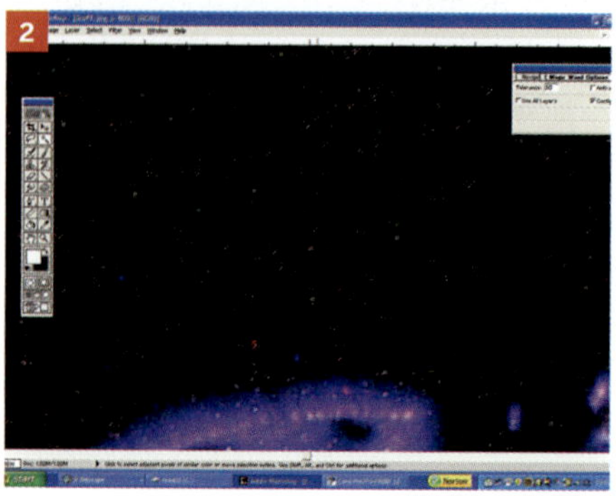

the blue-to-ultraviolet range. You can use Image » Adjust » Color Balance in Photoshop to change your image's color for some beautiful effects.

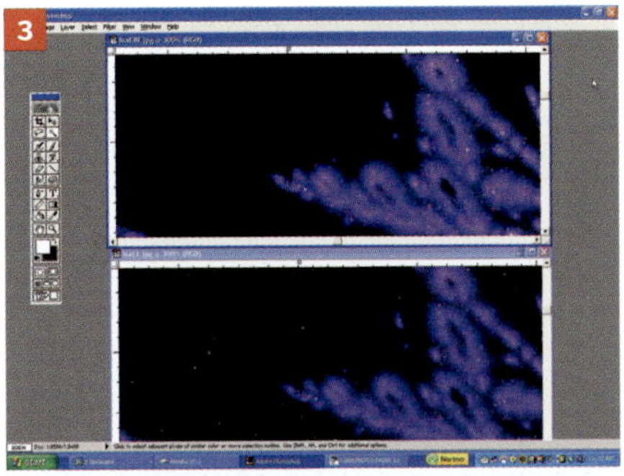

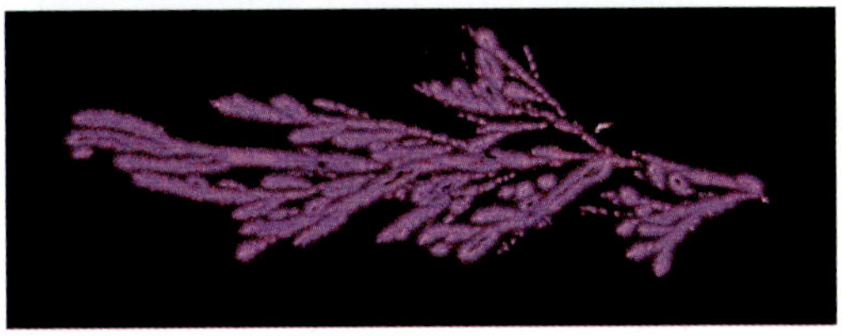

GOING FURTHER

Using digital cameras to capture Kirlian photographs is relatively new. Because it's a low-light process, we can borrow techniques from the astronomers who use digital cameras, as well as astronomy software such as AstroStack (astrostack.com), which stacks multiple images of the same object to create a brighter, more detailed image.

Shooting Kirlian photographs in atmospheres with other mixes of inert gases, such as helium, neon, and carbon dioxide, will create corona discharges in different colors. Just make sure the gases are inert.

To learn more about Kirlian photography using film, refer to my book, *Kirlian Photography: A Hands-On Guide* (Images Publishing, 2000).

John Iovine is a science and electronics tinkerer who has published a few books and articles and owns and operates Images SI Inc. (imagesco.com). He lives in Staten Island, N.Y.

Film-Based Kirlians

The light you see during a Kirlian discharge comes from an electric field ionizing the air and parts of your Kirlian subject. The charged air gives off intense heat and ultraviolet radiation as well as visible light. Photographic emulsions interact with all of these, plus the electrostatic field itself, while digital sensors mainly just capture visible light. As a result, traditional film-based Kirlian images require shorter exposure times than digital methods, and can show the lightning-fast discharge arcs in greater detail.

For example, my Kirlian photograph of the quarter on this issue's cover took only a ¼-second exposure. With faster photo papers and films it might take even less. In addition to improving detail, such short exposures also decrease the chances of vaporizing or burning your subject. And when photographing a human being, short exposure times are less painful!

Open-film Kirlian requires a fully darkened room, tens of thousands of volts, and expensive sheet film. All of this can be daunting and, well, dangerous — or at least expensive. But you can start out and get wonderful results using black and white photo paper instead of film, and working under a darkroom safelight or red LED flashlight to help you steer clear of the power supply. Later, you can graduate to a color image, and the digital method is also worth a try! —*Jasper Nance*

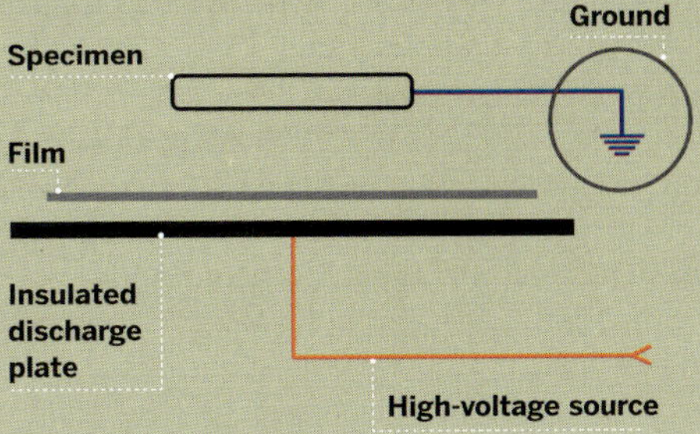

Photograph by Jasper Nance

FRINGE

Reinventing the Square Wheel

Why build an unworkable device? Every basement tinkerer has probably built at least one device that didn't work as expected. One can learn a lot from such failures, because they teach us the pitfalls of design, the limitations of materials, and the restrictions imposed by nature's laws. It's easy to superficially examine a drawing or blueprint of a device, and glibly say, "That will never work." It's quite another thing to reason out exactly why it doesn't work, and one can gain even deeper insight from hands-on manipulation of the recalcitrant mechanism.

PERPETUAL FUTILITY

When people say, "It'll never work," they may sometimes be right. Most new ideas turn out to be wrong and are swept under the rug of history. But even wrong ideas can be useful because they give us valuable information about what doesn't work, and narrow the field of things we might try that could work. Unfortunately, some people never give up on a discredited idea. Such is the case with the perennial search for perpetual motion. Even today some "perpetual motionists" firmly believe it is possible, devoting considerable time and money to their attempts to achieve it.

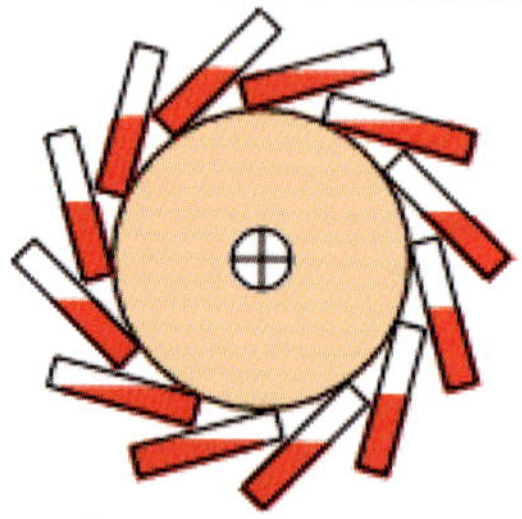

One of Bhaskara's wheels

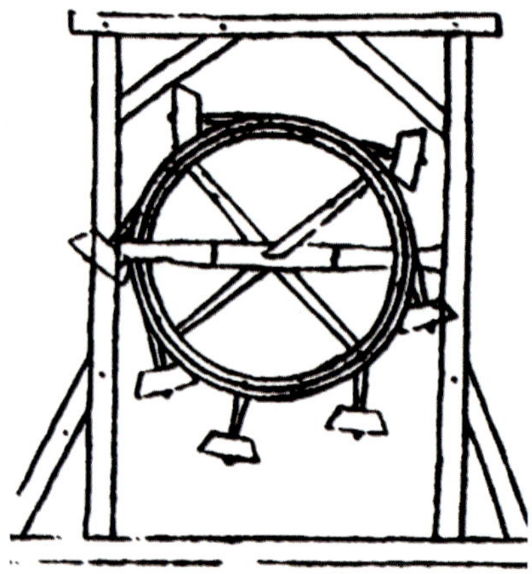

Villard de Honnecourt's wheel

Top illustration by Hans-Peter Gramatke

Early Attempts at Perpetual Motion

A perpetual-motion device is easier to define than to make. It is a wheel that continues its motion — undiminished — forever, without input or output of energy. Even better would be a machine that continues its motion indefinitely while continually putting out additional useful power (output power greater than input power). Hopeful inventors call this an over-unity machine, because it would have a power efficiency greater than 1. If some of its output were used to power the input, it would require no fuel, but would simply continue doing useful work. Needless to say, no such machines have ever been successful, or we would not have an energy crisis.

Perpetually turning wheels were described in Sanskrit treatises in the fifth century, and later by

The production of useful work is strictly limited by the laws of nature. The production of useless work seems to be unlimited. By Donald E. Simanek

the Indian mathematician Bhaskara (c. 1159). One such wheel had cylinders around its rim partly filled with mercury. As the wheel turned, the motion of mercury was supposed to keep the wheel heavier on one side of the axle, causing it to rotate perpetually. (If you build this model, use colored water. Mercury is an environmental hazard.)

This idea reappears in Arabic manuscripts, one of which contained six perpetual motion devices. These designs later reached Europe. A wheel described by French architect Villard de Honnecourt in 1245 is identical to a design from Arabia — an overbalanced wheel with hinged hammers or mallets equally spaced around its rim. De Honnecourt insisted on an odd number of hammers. He claimed his device would be useful for sawing wood and raising weights.

Villard de Honnecourt's Wheel

One reason these overbalanced wheels seem plausible to some people is because a static picture is used to represent a dynamic situation. People suppose that the falling hammer gives the wheel an extra impetus. But when the wheel moves clockwise (perhaps initiated by a push), the weight falling from A to B pushes the wheel backward (counterclockwise) during its fall, which probably isn't what the inventor wanted. Casual inspection of the diagram wouldn't lead you to expect this, but it's easy to observe once you build one and try it.

Of course some of the forward motion of the wheel is recovered when the weighted arm hits the peg at position B, but there's a net loss because all materials are somewhat elastic, converting kinetic energy to thermal energy when compressed. The hammers' fall doesn't generate additional energy but causes loss of energy. A simple flywheel without the hammers would work better.

All perpetual-motion machine designs, when carefully examined, can be proven unworkable for at least two reasons: (1) theoretical — they are based on incorrect assumptions about physics, or

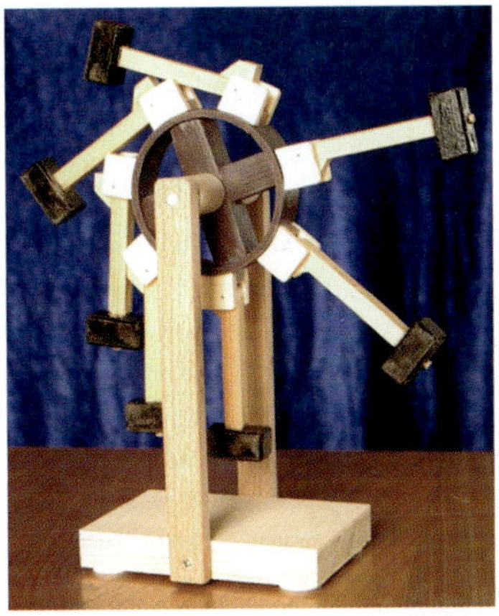

Wooden model of Villard's wheel

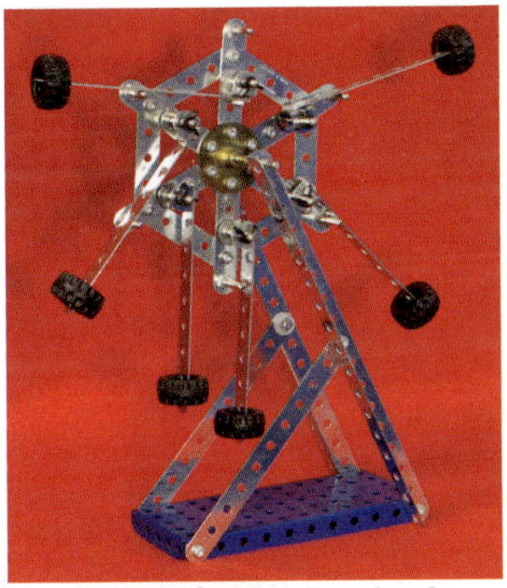

Model of Villard's wheel

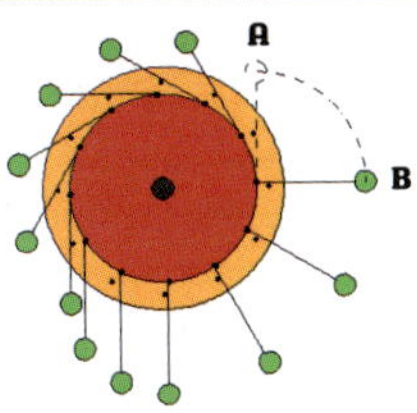

The action of the flipping weight

Photography by Donald Simanek; illustration by Hans-Peter Gramatke they apply physics incorrectly; and (2) experimental — if you build and test them, they don't work.

Simon Stevin's Ball-Ramp and Woodward's Wheel

Flemish mathematician and engineer Simon Stevin (or Stevinus, 1548-1629) analyzed a device consisting of a chain of balls on a frictionless double ramp. The chain was supposed to slide counterclockwise since there was more weight of chain to the left of the vertex. But Stevin's analysis showed that the system had no inclination to move at all, because it was always in static balance. These results led Stevin to formulate a fundamental principle of mechanics, known as Stevin's Principle of Virtual Work, which is found in textbooks even today.

Something about the notion of perpetual motion tantalizes those with restless minds, who think, "Perhaps there's some principle of physics, as yet unknown, that we might discover if we just alter the design a bit." So, through the centuries, these people tinkered with the designs, in a fantastic variety of ways, until the proposed machines became hopelessly complex. Such is the perversity of nature that none of them worked.

Some inventors were so confident of success that they included a brake in their designs to prevent the machine from turning so rapidly that it would tear itself apart. They needn't have worried.

Some drawings included an arrow showing which direction the inventor supposed that the wheel would turn. This is especially helpful; without it, we physicists wouldn't have the slightest idea which way the wheel should turn. If laws of physics allow the wheel to turn equally well in either direction, you can be quite sure the wheel won't initiate motion by itself and won't sustain motion forever when you give it a push.

By the 19th century, there was a refreshing return to simple basics. F. G. Woodward proposed a hoop wheel supported by two rollers. This is about as simple a design as you can imagine. Since the wheel always has more weight to the left of the rollers, that side should move down, as the arrow indicates. But the wheel stubbornly defies that "logic," refusing to budge.

PERPETUAL OPTIMISM

Inventors before the 17th century had no physical laws of energy and momentum that would

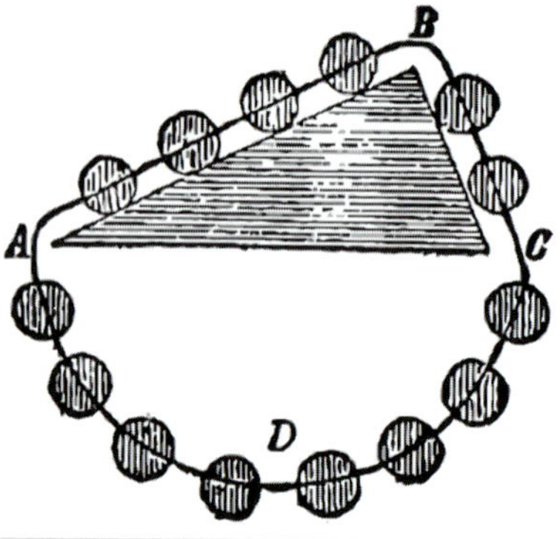

Simon Stevin's ball-ramp

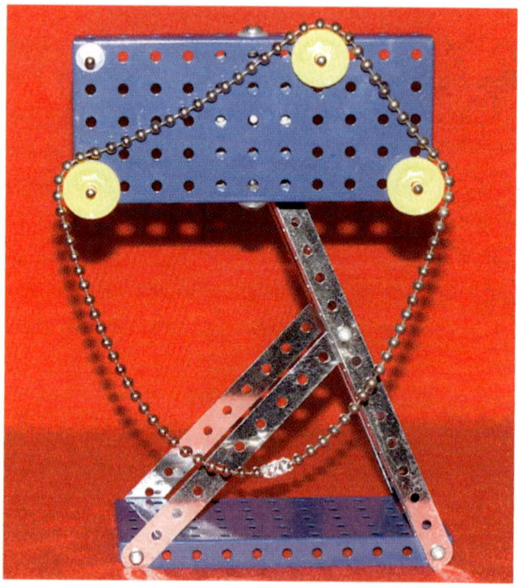

Model of Stevin's chain

Richard's overbalanced chain

cause them to doubt the possibility of perpetual motion. But today we have a body of well-tested and accepted laws of mechanics that clearly tell us that certain things are impossible in nature, and no experimental evidence suggests otherwise. So how can anyone still expect to invent a perpetually moving mechanism? Well, many do, undeterred by a long history of total failure, contemptuous of the laws of physics, and oblivious to scientists who declare, "It'll never work."

Hans-Peter Gramatke has identified 1,800 perpetual-motion patents in the United States, England, France, and Germany between 1860 and 2000, and a dozen patents from other countries. From 2000 on, an average of about 50 patents per year were issued worldwide. Most were in the four countries named above, although Japan, China, and Korea show recent increases. These are patents that were "obviously" intended to be perpetual motion. There may be many more unworkable device patents in other categories.

Patent examiners do look for serious flaws in patent applications, but their primary criterion for patentability is that a device be a new idea; it doesn't need to be workable or practical. That's why unworkable devices get patented. Today, inventors avoid using words that would signal to an examiner that their device is supposed to get something for nothing, or that it is violating any well-established laws of nature.

Has all of this effort been a waste of time? Not entirely. Failed experiments give us a better understanding of the inherent limitations in nature's operations. Science learns from its mistakes. Perpetual motionists' mistakes get perpetually repeated.

PIXIE DUST

Perpetual motionists are especially attracted to those parts of physics they don't understand. William Gilbert (1544-1603) summarized his experiments on magnetism in his book *De Magnete* (1600). The book was a sensation, widely read and widely misunderstood. It stimulated all sorts of incorrect, pseudoscientific notions about magnets, some of which still persist today. One fellow, heavily invested in a scam perpetual-motion device, wrote to me that every physicist should know that magnets contain infinite stored energy. As an example, he cited the lowly refrigerator magnet, which, he said, "can hold itself in place forever, working against the

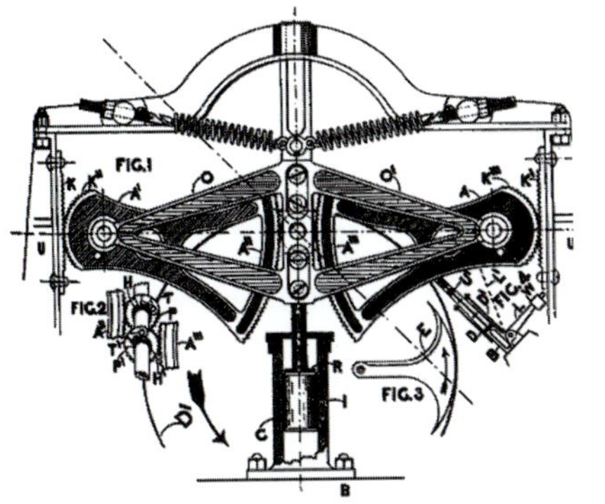

Gilbert's perpetual motion

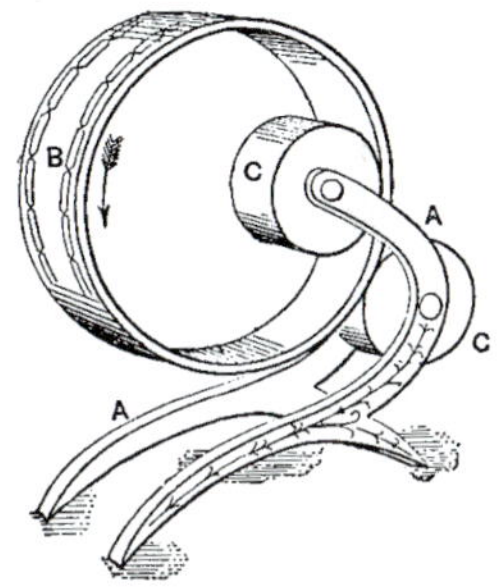

Woodward's wheel

pull of gravity, so it obviously has unlimited stored energy." Obvious to him, perhaps, but physicists know that work is the product of force and the distance that force moves something. A magnet does not move as it clings to the refrigerator, so it does no work and expends no energy.

Since simple mechanisms have been so well studied for so long, perpetual motionists today concentrate on elaborate and complex schemes, and then challenge skeptics to find the flaw. Science writer Bob Schadewald once observed, "A perpetual motionist typically concocts a scheme so complicated that he can't see why it won't work. He then assumes that it will work." Some perpetual motionists pin their hopes on cutting-edge speculations of theoretical physics. Zero-point energy and dark energy are popular now. Perpetual motionists use these as the stage magician uses "magic pixie dust" — to justify apparent miracles. Some postulate as-yet undiscovered laws or hidden energy sources in nature, give them names, and then assume they have just those properties required for the success of their perpetual motion machine. They forget that these properties exist only in their imaginations.

Zero-point energy is a useful concept that has been experimentally confirmed in many ways. But this energy is "locked up" by constraining laws of physics that prevent it from producing useful work. There is not a shred of evidence, from experiment or established theory, that suggests any way to utilize zero-point energy as the driving energy for a cyclic machine. Dark energy is still a speculative hypothesis whose usefulness in physics theory remains to be seen. It, too, seems to be unavailable for conversion to useful work.

HOW CAN WE SAY THAT SOME THINGS ARE IMPOSSIBLE?

Of course, science has not yet discovered all of nature's secrets. In the future we are sure to formulate new principles of physics. Even some of the laws we now know to be valid may need modification or reinterpretation as future discoveries are made. It is even possible that some basement inventor might stumble onto a useful device or even a new physical principle while trying to achieve perpetual motion.

Should we fund research into perpetual motion? I think not. A thing is not necessarily true just because we'd like it to be. It would be foolish to pursue this research because we haven't a single indication, no evidence, no theory — nothing — that even suggests that a perpetual-motion machine is possible, and nothing to suggest how to go about achieving it even if it were possible. Research into antigravity devices, vehicles that travel faster than light, gravity shields, or ways to travel back in time would be just as promising. We haven't a clue how to make them, either, and well-established laws predict that they aren't possible.

Every law of nature that tells us how nature works also tells us how nature doesn't work. The negative part is abhorrent to some people, who desire a magical universe where anything you can imagine is possible, if you tinker with things enough. Laws of nature express the constraints imposed by the geometry of the universe. You can imagine other geometries, but they are not recognized in our universe. For example, you may imagine a triangle in a perfectly flat plane that has exactly equal angles but very unequal sides. But that's not achievable in our universe. Geometry is also the fundamental limitation on the performance of machines.

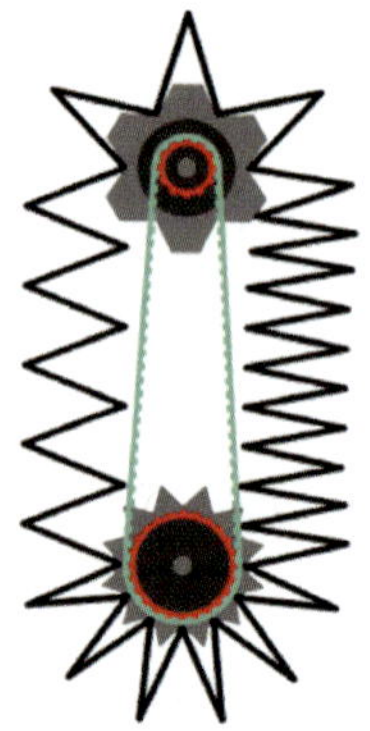

Simanek's silly spring device

Simanek's Silly Spring Device

Most perpetual-motion-machine designs that people send me are variations of old and discredited ideas. It gets boring after a while, so I decided to invent one of my own — to make a point. It uses a flexible, coiled spring looped over two pulleys. One side of the spring is kept compressed by a non-slipping belt running over smaller pulleys fixed to the larger ones. Clearly it is heavier on one side of the pulley axle, and this "unbalance" is maintained as you manually turn it. It can be rotated freely, in either direction. But it will not turn on its own, even if you could reduce friction to zero. Thanks to Hans-Peter Gramatke's animation, you can see it in motion on my website (see References, below). It demonstrates that computer simulations can model unreality just as well as reality.

References

At these two sites, you can find pictures of many perpetual-motion devices you might want to build yourself. They are guaranteed to be unworkable.

Donald Simanek's Museum of Unworkable Devices: www.lhup.edu/~dsimanek/museum/unwork.htm

Hans-Peter Gramatke's pages: www.hp-gramatke.net/index.htm

Donald E. Simanek is emeritus professor of physics at Lock Haven University of Pennsylvania. Visit his pages of science, pseudoscience, and humor: www.lhup.edu/~dsimanek.

Illustration by Hans-Peter Gramatke

Masonic Conspiracy Revealed!

When engineering culture detaches from reality.

By Paul Spinrad

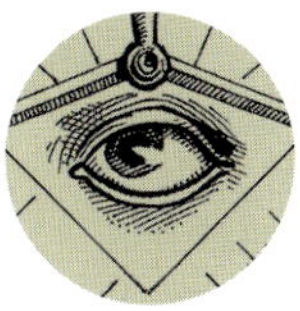

Generations of conspiracy-minded observers have nurtured theories about powerful secret societies such as the Masons. Here is my theory: Occult institutions evolve out of professional guilds after they stop caring about how to actually build things.

In every age, geeks gravitate to where the interesting action is, and in medieval Europe, this was the cathedral — the structure itself, the art, and the pipe organ inside. Like most people, I've always been amazed by the great gothic cathedrals, but it was French historian Jean Gimpel's book, *The Cathedral Builders*, that gave me the real nerd's-eye view.

Gimpel explains that starting in the last half of the 12th century, competitive cathedral builders broke the world's record for the highest interior vault five times within 62 years: from the completion of Notre-Dame de Paris in 1163 (32.8 meters high), through Chartres, Rheims, Amiens, and finally to Beauvais in 1225 (48 meters high, collapsed in 1284). This space-race happened during what's now called the Early Gothic era, when cathedral building was literally at its height.

Engineering culture flourished during this era, and the cathedral builders were the alpha geeks. They and their kind traveled throughout Europe and contracted independently with parishes and municipalities to build churches, bridges, and other public works. They scouted talent by hiring unskilled stonecutters from the local populations, and invited the most promising ones to apprentice. They formed an international maker subculture that was surrounded by an illiterate and innumerate population that relied on their work, but could never understand it. To outsiders (non-engineers), they must have seemed like wizards: they used strange implements and symbols, spoke their own terminology, and produced magical-seeming sights, sounds, and structures.

Meantime, these builders had a unique relationship to the Church, which monopolistically financed and controlled all culture — like today's Hollywood and Madison Avenue, combined. The Catholic Church needed the building expertise of "free masons," who didn't have to buy into any ideology or hierarchy, or even obey the local clergy. This gave the masons an unusually powerful position, which was inevitably eyed with suspicion.

Toward the end of the 13th century, as Europe became more nationalistic and political, Church funding for great cathedrals dried up. Design expertise was also tragically lost, and no one knew how to build as high anymore. This resulted in the Late Gothic style of cathedral, characterized by smaller structures that were built using old plans, but with more ornamentation added — lame.

Not surprisingly, the masons' professional culture also began to change during the Late Gothic period. Inside the guilds, nepotism and politics replaced ability and contribution as the path to master status. What had become known as Freemasonry was more secretive, perhaps out of defensiveness; the great cathedrals stood as prominent witness to all the knowledge that had died with previous generations. By the 15th century, masons in Germany adopted professional codes that forbade speaking publicly of lodge activities or divulging old methods for rendering drawings. Throughout Europe, the building trades carried an overlay of secrecy, antique reverence, and the promise of access to great lost wisdom.

From Operative to Speculative

Jumping ahead, Gimpel describes Scotland in the 16th century, where a boom in castle building took place. Job seekers answered the calls for work, and traveled from all over to converge at building sites. But the assembled workers didn't know each other and had no educational degrees to go on, so the structural-engineering geeks developed a secret system of hand signals to identify themselves to one another at job sites and distinguish themselves from the unskilled stonecutters who could only build walls. After recognizing the signs and gathering together, these masons could then approach the local authority and say, "We're the master builders here, and you need to put us in charge."

Around this same time, the masons in Scotland also formed a network of lodges where traveling members could stay and meet with other builder geeks. Conspiracists point to these Scottish lodges as the cradle of a mysterious occult society tied to the medieval Knights Templar, but as Gimpel writes:

> *Masonic historians have long thought that [the] secrets which the workmen were asked to keep were of an esoteric nature. They were nothing of the sort ... there is no reason to suppose that these secrets contained anything more esoteric than ... discussions in the lodge ... as well as technical secrets of the trade concerning, for example, the design of an arch.*

But then something happened that changed Masonry completely, and (I believe) did eventually turn it into an occult conspiracy. Upper-class men who weren't trained masons wanted to join Scotland's Masonic lodges. They were refused at first, but were later accepted under the provision that they paid double the dues of working masons. We can only guess at the mixture of motivations that led these fancy lads to seek companionship in a professional builders' society, but as a result, they gained access to a nationwide network of lodging and social contacts — an exclusive club, in the more modern sense.

This new membership policy gradually detached Masonry from its engineering roots, and within a couple of generations, more Masons were wealthy gentlemen than actual builders. But as the makers gave way to the schmoozers, the lodge system proved to be a resilient social construct. Masonry became political and social, rather than practical — a cabal of insiders, rather than a collective devoted to expertise and education. Expert trade knowledge and levels of technical proficiency evolved (devolved?) into an artificial system of signs, symbols, and access, whose function was to distinguish insiders from outsiders. But in the social context within which this knowledge operated, it was still powerful.

Here's how it works. Let's say you're an insider who knows some secret mumbo-jumbo. Promoting the belief to outsiders that your secrets have great power makes this belief self-fulfilling because it makes outsiders think you have something on them. The more mysterious and powerful these inside secrets seem, the more effective the hype becomes. Now, if you imagine running this secrets-as-status dynamic recursively, you get concentrically nested levels of power, a hierarchy that runs from complete outsider to innermost circle. Add the human imagination on both sides, and you spawn a thousand occult rituals and a thousand conspiracy theories.

Masonic historians refer to this great shift away from building as the shift from "operative" to "speculative" masonry, and the event that signified the completed transition is considered to be the establishment of the Grand Lodge in London, in 1717. Numerous books trace modern Masonry's lineage from the Grand Lodge to today's Masonic lodges, which combine, to varying degrees, the roles of community service organization, local chamber of commerce, and drinking club.

The Masonic blueprint has also informed other secret societies and organizations, ranging from the occultist Golden Dawn to college fraternities. However, I believe that the true heirs to the Masonic tradition are at makers' fairs, s33krit hackers' conferences, engineering departments, and other places where people share practical knowledge with other wizards.

So, what is to be done? Cities across America have beautiful Masonic lodge buildings sitting on prime downtown real estate. Lodge membership is graying, and many chapters have closed due to the lack of new members. I think this presents a grand opportunity. Let's start a new Masonic Conspiracy! Let's take Masonry back, and convert it from speculative back to operative! Let's get all the makers we know to become Masons, and turn all those gorgeous, unused lodge buildings into temples of geekdom! Who's with me?

Paul Spinrad is projects editor of MAKE.

Freemasonry in a nutshell: 16th century structural engineers invent secret system of hand signals to identify themselves at job sites. Later, non-engineers pay to join the secret society, turning Freemasonry from operative to speculative. Now, it's time for makers to add the geek factor back into Freemasonry.

Illustration by Hal Robins

FRINGE

The Essential Strangeness of Time

Reconstructing Athanasius Kircher's magnetic clock. By Jonah Peretti

Caroline Bouguereau, a French artist now living in Ireland, was commissioned by Stanford University Libraries in 2001 to reconstruct the magnetic clock described by German Jesuit polymath Athanasius Kircher (1602-1680). Jonah Peretti, emeritus director of R&D at Eyebeam in New York City, conducted the interview.

How did you decide which machine to build?
I liked the way the clock was a miniaturized model of the cosmos. It was a puzzle, an enigma ... a fish pointing mysteriously at the globe of the Earth, as it rotates slowly. The visual impact of the engraving appealed to me, and the magic of the machine: how were the globe and fish suspended? What cosmic force made the globe turn around? And, above all, why a fish? As an artist, I've worked with fish before. I also like curious scientific apparatus, machines that have an aesthetic appeal, or an unsettling, disorientating effect.

What were the biggest challenges in making the clock? Was it hard to make the clock work? What techniques did you have to learn?
The biggest challenge was to find a way to suspend the small sphere in the middle of the larger globe. In the reconstruction, I tried to imitate Kircher's methods as closely as possible to produce the Baroque aesthetic effect of his device. To allow the sphere to appear to "float" within the globe, Kircher suspended it at the junction of two immiscible liquids of different densities. My concerns were which liquids to use for this purpose, and how to make the ball stay at the same position, right at the middle of the globe. Most of the organic solvents that worked well had serious risks. Acetonitrile, in particular, suggested enthusiastically by a Stanford chemistry professor, was apt to produce hydrogen cyanide on contact with air — probably not a good thing for an art exhibit in a public indoor space.

Did the project feel contemporary, historical, or both?
I had to immerse myself in the culture of the 17th century for the design; I had some freedom of design but there were very strict rules. The map of the world painted on the globe is an example. I copied a 350-year-old map. To paint on copper, I looked at how people did this in the 17th century — they rubbed the copper with garlic first, and scratched it to make the paint stick. I think the clock also has a very contemporary "feel," though. I've always found time [to be] something very mysterious. We are so used to clock faces and digital representations of time that we have forgotten how strange it is. I think Kircher's machine restores the essential strangeness of time.

From an artist's perspective, what's important or interesting about Kircher?
Kircher is probably the most diverse and prolific kinetic artist of his time. Kircher is one person who really brought the worlds of mechanical ingenuity and arcane knowledge together in his Wunderkammer [curiosity cabinet]. With Kircher you never know what to take at face value.

See the artist at makezine.com/09/kircher.

Photography by Jason Forman

A floating fish always faces the current time (painted on a globe) in a reconstruction of Kircher's celebrated magnetic clock from the 17th century. An ordinary water clock "magically" turns the globe with hidden magnets.

Thank you, subscribers!

We're giving you a special, limited-edition booklet.

This miniature book, produced as a period-authentic chapbook popular in Europe at the time, tells a historically speculative tale of one of the most remarkably creative inventors in history, Athanasius Kircher. This is the first volume in our forthcoming series of Maker Tales books, and this limited first edition is anticipated to quickly sell out.

In celebration of the start of MAKE's third year, we're giving all subscribers this Maker Tales booklet, which should arrive with your copy of MAKE 09. If you're not a subscriber, you can still get a copy by purchasing one of a limited number available at makezine.com/store, or by subscribing at makezine.com/subscribe, using promotion code M7AK by May 15, 2007.

Make: Projects

Learn how to truly crack the whip by plaiting your own supersonic boomer. Or join the pinhole paparazzi and create unique panoramic images without touching a lens. If that doesn't focus your attention, get cracker-jacked with a simple amp built from a cardboard box.

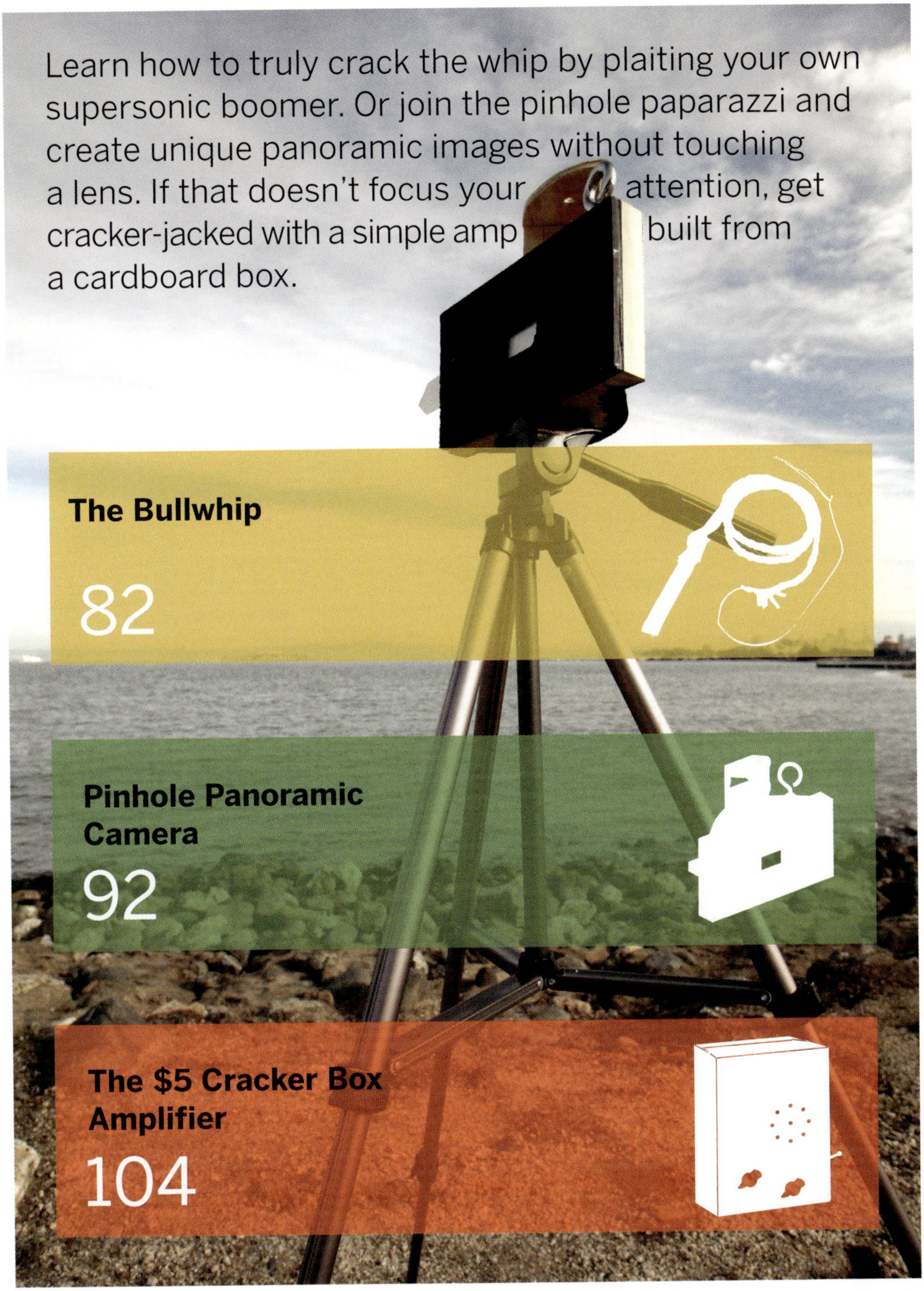

Photograph by Sam Murphy

THE BULLWHIP

By William Gurstelle

WHIP IT GOOD

You don't have to be Chuck Yeager to break the sound barrier. You just need a good bullwhip, which converts arm movement into supersonic speed and triggers the sonic boom we call a whip crack.

I love the scene in *The Blues Brothers* when Joliet Jake Blues (John Belushi) and his band are playing at a roadside dive. The tough crowd is throwing bottles, and the chicken-wire screen in front of the stage is doing only a mediocre job of protecting them. Jake spies a coiled bullwhip hanging on the wall, and a light bulb flashes in his head. He grabs the whip and gets the band to strike up the *Rawhide* theme. With each chorus, with each crack of the whip, the crowd's demeanor changes, and by the time the Blues Brothers finish the song, the crowd is won.

Everyone loves the sound of a whip. But music and culture aside, whips are interesting physical devices: self-contained, human-powered machines that exceed the speed of sound, which is roughly 770 mph. Bullwhips are easy to make at home, and their deep, explosive crack is far more affecting when you experience it live, rather than just hear it recorded.

Photography by Sam Murphy

Set up: p.85 **Make it:** p.86 **Use it:** p.90

William Gurstelle is a MAKE contributing editor. He wrote the ornithopter flying machine project in Volume 08. His fifth book, *Whoosh, Boom, Splat*, comes out this March.

WHIP SMARTS

A whip is no mere rope on a handle — it's a system of finely tuned interacting parts: the *handle*, the *thong*, the *fall*, the *belly*, and the *cracker*.

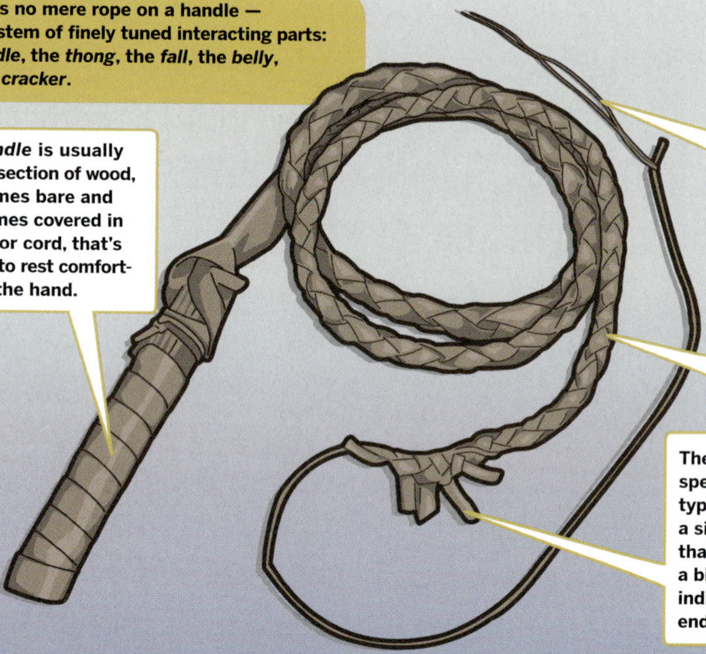

The *handle* is usually a short section of wood, sometimes bare and sometimes covered in leather or cord, that's shaped to rest comfortably in the hand.

The *cracker* (or *popper*) is a small, replaceable piece of string or nylon cord at the very end of the whip. Because it moves so fast, this piece is subject to much wear.

The *thong* is a section of leather braid wrapped around a flexible core that whip makers call the *belly*.

The *fall* is a series of specialized knots, typically made from a single piece of leather that's 1-2 feet long and a bit wider than the individual braids at the end of the thong.

Speed Over Mass:

A whip tapers gradually from base to end, to smoothly transfer momentum and energy from the thick plaits near the handle to the thin, flexible popper.

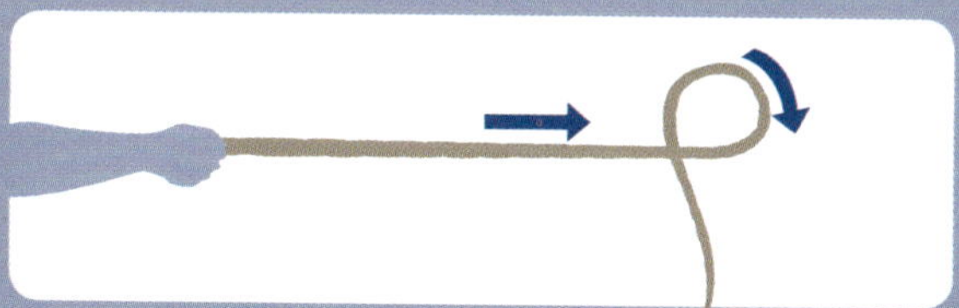

1. **The whip-throwing motion causes a loop to propagate down the whip's length.**

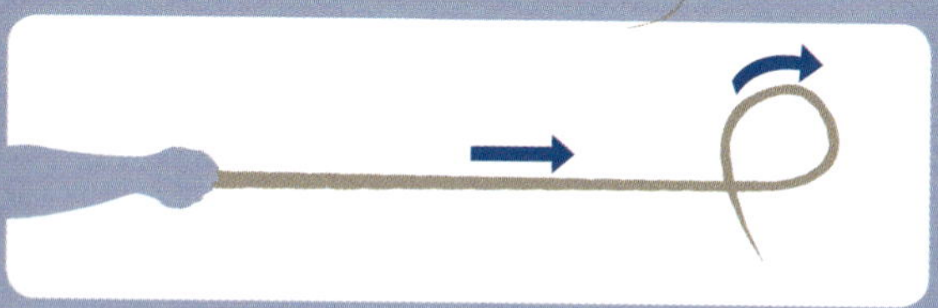

2. **As the loop continues down, its speed increases following the mathematical inverse of the thickness and weight of the whip's cross-section.**

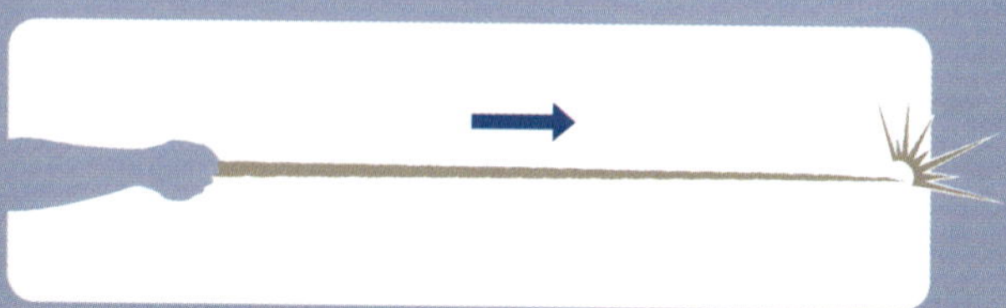

3. **By the time the roll approaches the end, the loop velocity has gone transonic, crashing through the sound barrier. *Crack!***

Illustration by Tim Lillis/narwhalcreativve.com

SET UP.

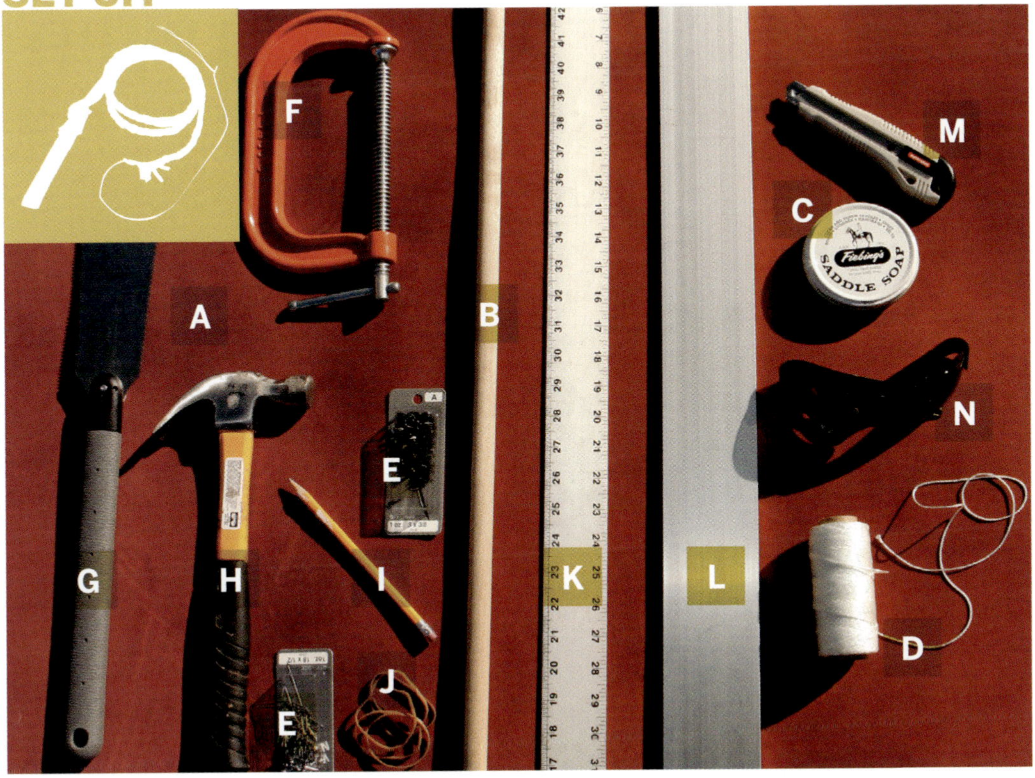

MATERIALS

[A] Leather hide, about 3⁄16" thick and at least 8' long Purchase a hide from a leather shop. I purchased mine at Tandy Leather, a large national chain, and found the staff extremely knowledgeable. The shop buys hides from a tannery and then sells its products by the side, hide, skin, split, or belly. The size, thickness, and texture of each hide vary, so you'll need to inspect several to find one that works. Try to get one that will let you cut straight 8'-long strips. Whip makers can cut curved strips from smaller hides, but this adds complexity.

[B] ¾" diameter smooth wooden dowel, about 7" long This will be the handle. You can make it longer or shorter according to your taste.

[C] Saddle soap

[D] Light cotton or nylon cord, at least 1' or horsehair, hemp, or any other thin, fibrous material

[E] Small nails or heavy staples

TOOLS

[F] Vise or C-clamp on a sturdy support

[G] Saw

[H] Hammer

[I] Soft-leaded pencil

[J] Rubber band

[K] Ruler

[L] Straightedge (or use the ruler), the longer the better

[M] Sharp utility knife with spare blades

[N] Eye protection

[NOT SHOWN]
Large, flat cutting surface

Sturdy support you can tie around

MAKE IT.

CRAFT YOUR BULLWHIP

START Time: **4–8 Hours** Complexity: **Medium Easy**

1. CUT THE LONG PIECES

1a. Cut the belly. Mark and cut a leather strip at least 8' long, that tapers evenly and symmetrically from 3⁄16" at one end to 1⁄8", as shown in the Belly Cutting Diagram (below). This will be the whip's belly, around which the thong plaits are laid. The plaits form a hollow tube through which the belly runs. If your hide is longer or shorter than 8', adjust your layout accordingly. With a longer hide, you can make a longer whip.

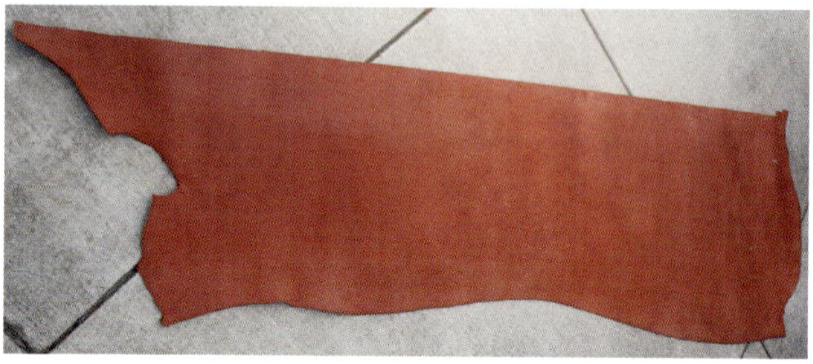

Belly Cutting Diagram (not to scale)

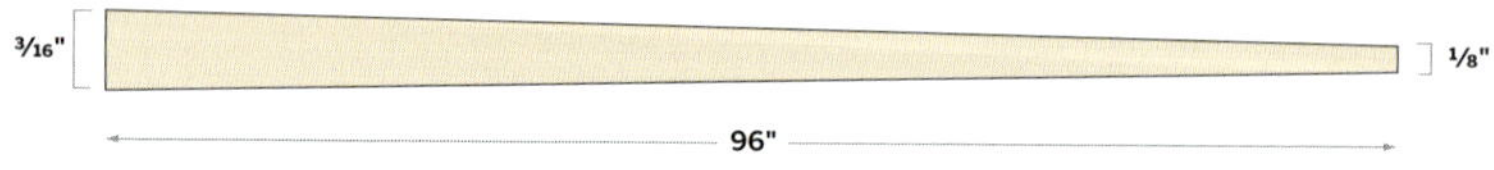

1b. Lay out the thong. The thong consists of connected strips of leather that are woven or plaited together to form the main body of the whip. Thong layout is the most critical portion of the entire whip-making procedure, so take time to do this accurately. Adjust the length of the thong to match the length of the belly if it's different from 8'.

Thong Cutting Diagram (not to scale)

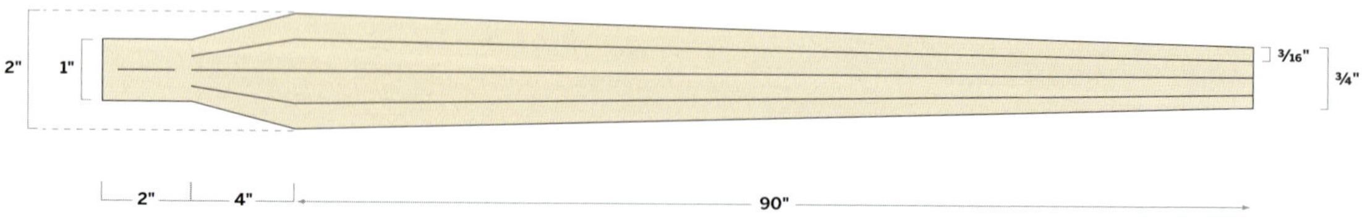

Use a soft-leaded but sharp pencil to draw the outline of the thong on the leather hide, 2" wide at the widest point, following the Thong Cutting Diagram at the bottom of the previous page. The main concern is to keep the width of all 4 strips equal to one another at every point along the thong's length.

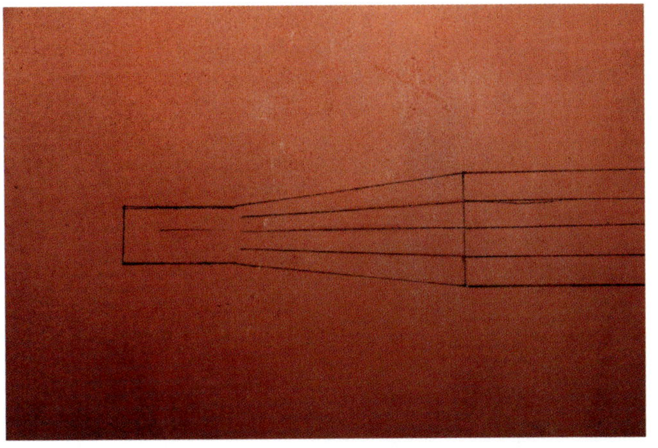

1c. Cut the thong. Use a very sharp knife and a straightedge to guide your hand as you cut through the leather, following the layout.

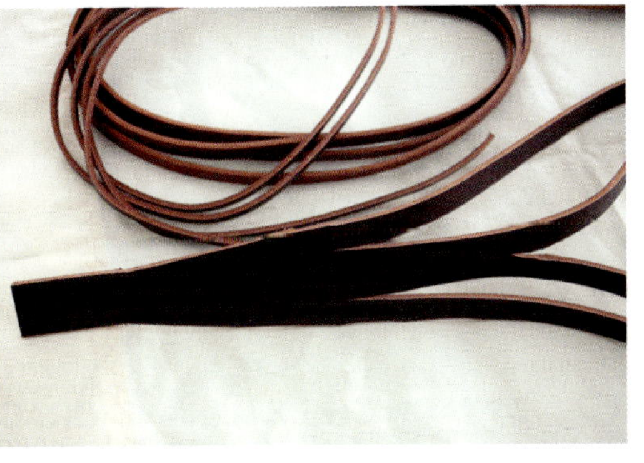

2. PLAIT THE THONG

This is the most time-consuming portion of the project. Study the plaiting sequence photos carefully. The first time you plait your thong into a whip, it might be somewhat confusing, but keep trying, and before long the process will be not only easy, but also quite a bit of fun.

2a. Hang the belly and the thong. Secure the wide end of both the belly and the thong together, with the belly centered on top, with a C-clamp, vise, or other sturdy support. Tie the other end of the belly to another support, so that it's stretched tight and will hold steady while you plait around it.

2b. Dress the leather. This prepares the thong to make it easier to pull tightly into good, even plaits. I apply saddle soap to the leather with my fingers and then work it into the leather, warming and softening it. Some serious whip makers use their own special dressings concocted from beef fat, beeswax, oils, and other stuff, although these will also make the whip more attractive for dogs to chew on.

2c. Do the actual plaiting. This takes a while to accomplish, but it's pretty easy once you get the hang of it. Every wrap goes around the belly, which is the thinner strap shown in the background. Repeat the process, pulling the thongs tight after every step, until there is about 4" of unplaited thong left at the bottom. The belly will be longer than the plaited whip, and will extend out of the bottom. Tie off the plaited thong temporarily with a rubber band to keep it from unraveling.

3. ATTACH THE FALL AND CRACKER

The fall is a thin strip of easily replaceable leather attached to the end of the whip. Since the whip tip moves at supersonic speed, the end gets worn fairly quickly. The replaceable fall protects the plaits in the whip.

3a. Cut a 2' strip from the hide, as shown in this Fall Cutting Diagram. Cut a 1" slit near the wide end.

Fall Cutting Diagram (not to scale)

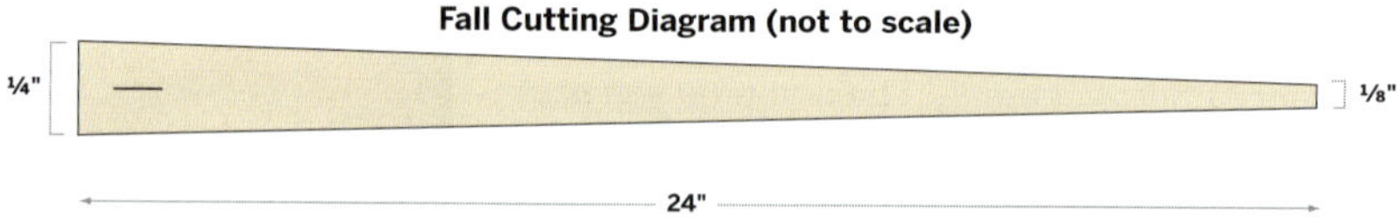

3b. Attach the fall to the whip by placing the whip's tip through the slit in the fall. Then tie 4 single overhand knots in each of the strands of the thong. Pull the knots tight and cut off the excess.

3c. Make the cracker by cutting an 8" length of light cotton or nylon cord (or horsehair) and tying it securely to the fall with a sheet bend knot (wikipedia.org/wiki/sheet_bend). Unwind a bit of the other end into individual fibers.

4. MAKE AND ATTACH THE HANDLE

4a. Cut the dowel down to handle size (I find 7" convenient), and cut a shallow groove in the handle 1" from one end.

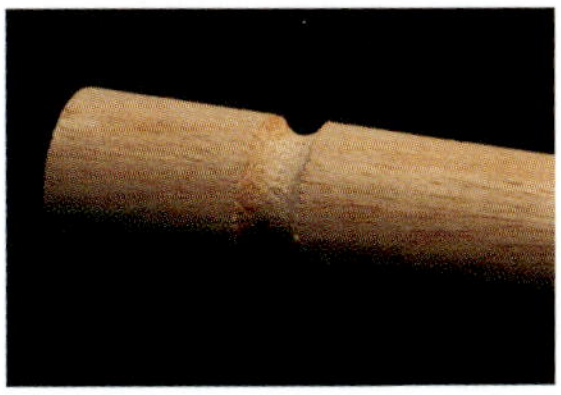

4b. Cut an even leather strip ½"–¾" wide by 2'–3' long, with gradually angled cuts at both ends. Use a nail or heavy staple to tack the strip to the top of the handle, opposite the groove. Wind the leather around the handle and tack again at the bottom end of the strip.

4c. Cut a 1"×7" leather strip from the hide. This will be the keeper, the piece that connects the whip plaits and belly to the handle.

4d. Cut a very thin leather strip, about ⅛"×15", from the hide. Bind the keeper to the handle with this thin strip, using the groove to anchor it firmly in place. First, tie the thin strip directly into the groove with a clove hitch (wikipedia.org/wiki/clove_hitch), then wind the leather strip several times around the keeper, and tie it off. Pull hard on the keeper to test it for strength. If necessary, retie or use a second thin strip to bind it on.

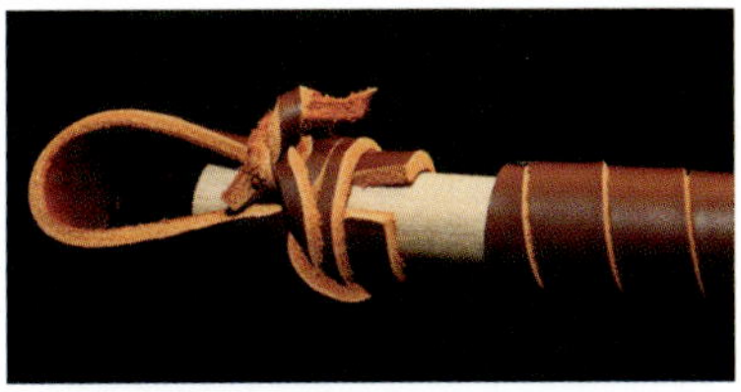

4e. Attach the plaited thong to the handle by first passing the handle through the slit in the thong, then pulling the thin end through the keeper. The whip is now ready for use!

FINISH X

NOW GO USE IT »

USE IT.

LET'S GET CRACKING

SAFETY

Until you get a feel for how to handle a whip, expect to endure a few accidental welts. Things to remember:

• Eye protection is absolutely required. A hat that covers your ears is also a smart precaution.

• Wearing gloves, a jacket, boots, and heavy pants will reduce your chances of getting a whip-lash.

• Keep people, animals, and things you don't want broken clear of your practice area.

PRACTICE

There are many methods of whip cracking. All require practice to obtain consistent results. Practice makes perfect, and builds arm strength too!

The Forward Crack Start with the whip laid out on the ground in front of you. Bring your arm up and back. Then with a sharp motion, bring your hand forward, similar to a baseball pitcher's arm motion. Done correctly with a well-made whip, this results in a nice, sharp crack. Try various arm speeds to see what works best for your whip.

The Overhead Crack Begin by twirling the whip overhead to gain velocity. When the fast-moving whip tip is just behind and to your right (assuming a right-handed user), bring your hand forward smartly, and quickly snap your wrist. Done correctly, this will reward you with a brisant crack.

CRACKER CARE

Cracking a whip subjects the popper end to a lot of punishment, and they don't generally last too long. I like to experiment with different cracker materials to see what gives the best report. To begin, try soft, thin cotton or nylon twine with one end unwound into its individual fibers. In my experience, different cracker constructions give widely different results. Experiment with various materials and lengths to see what works best with your particular whip.

Photograph above by Corbis; photograph below by Sam Murphy

RESOURCES

For more family-friendly whipping tips and resources see makezine.com/09/bullwhip.

LASH PROGRAMMING

Whips evolved as herding and cattle-driving tools. A good whip handler can produce a cracking sound that bursts from any location within the whip's reach. Cattle instinctively turn away from the loud sound's source, so the ability to precisely place these small sonic booms anywhere in midair lets cattle drivers steer their herds.

Crafting a whip that cracks well requires time-tested methods. A well-designed whip is a lesson in supersonic fluid mechanics. It must taper gradually from base to end in a specific fashion, and it must be made of materials that gracefully and easily transfer the energy and momentum of the moving internal coil along its length.

Alain Goriely, professor of mathematics at the University of Arizona, has published several research papers on the way whips work, and has come up with several fascinating conclusions about how waves propagate through these flexible devices. His research describes how a loop moving down the length of a whip accelerates as the whip tapers, and he developed a series of equations that accurately model the curvature, boundary conditions, tension, and speed of the whip. Goriely's computer analysis of whip dynamics confirms that the cracking sound results from hypersonic movement, and he has measured tip speeds in well-made whips that exceed Mach 3 — three times the speed of sound.

Goriely divides the world of whips into two categories: pain-making whips and noise-making whips. The pain-makers, such as the infamous cat-o'-nine-tails, are short, bulky, and made up of several strands that hang separately. They don't crack. Noise-making whips are long and tapered, consisting of a braided single strand. Despite their frightening appearance, the noise-making whips are not used as weapons or torture devices; their purpose is to produce and aim incredibly high tip velocities.

Photograph: Corbis

PINHOLE PANORAMIC CAMERA

By Ross Orr

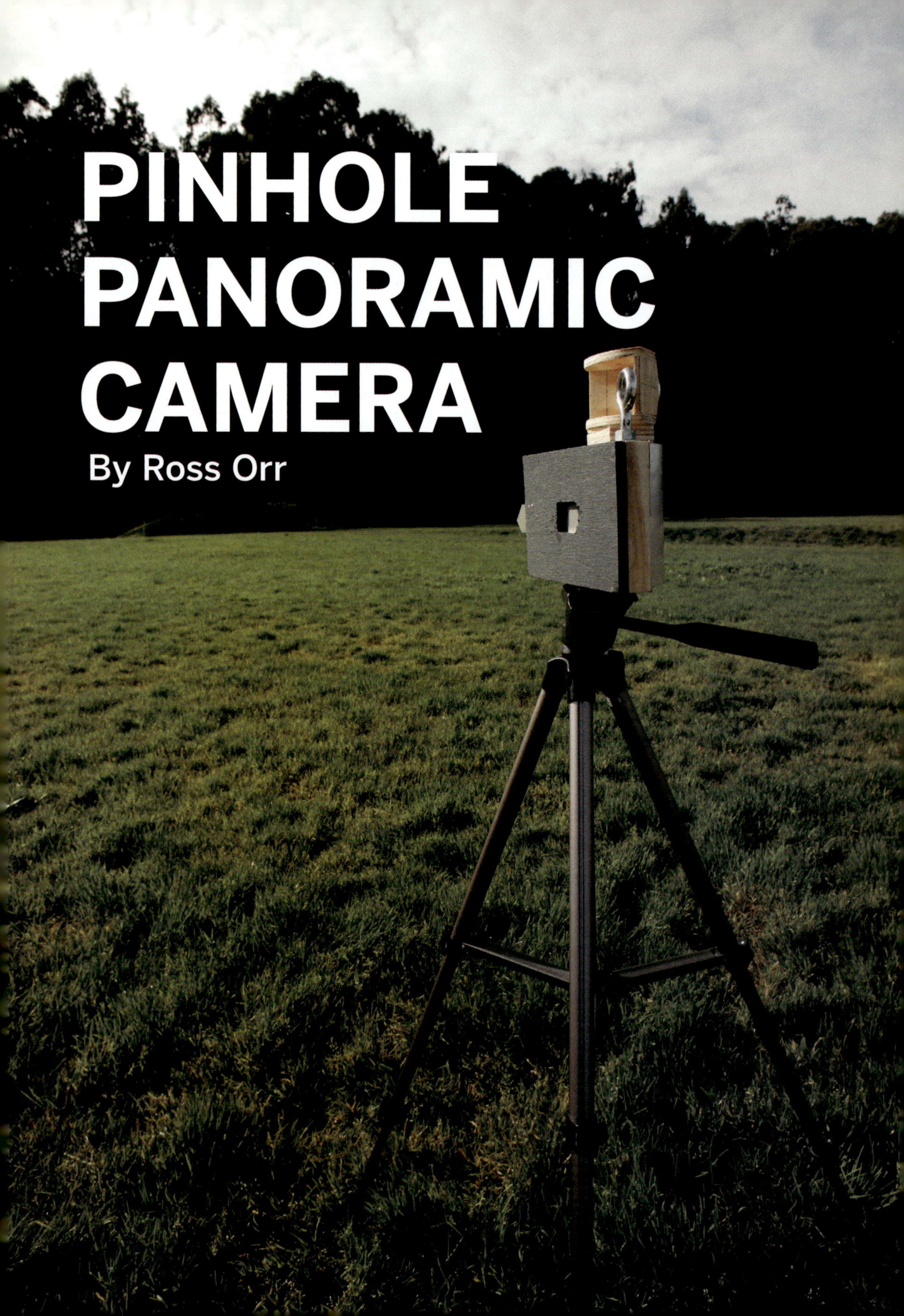

PIN-O-RAMA

Lensless and low-tech, pinhole cameras have always been maker-friendly. But forget the Quaker Oats carton, and go wide with this roll-film, panorama design.

I bought a new scanner recently, and soon found myself spelunking through drawers of old photos from my many misspent years in photography. Some of the most interesting shots were the pinhole camera experiments I had done as a teenager. With ghostly outlines from multi-minute exposures, and shapes warped into boomerangs by curved film, these otherworldly images got me dreaming about pinhole cameras again.

So I headed to the workshop to build a new one. And then another, and another. I eventually made more than a dozen, and this "Pin-o-rama" design is my favorite. Unlike simpler pinholes, it uses standard 120mm roll film, which means you don't have to open the camera and reload after each exposure, and you don't need a darkroom to process the results — just take the rolls to a photo lab. Also, it's built entirely from scratch, rather than hijacking the film-transport from an existing camera.

Photograph by Sam Murphy

Set up: p.95 **Make it: p.96** **Use it: p.101**

Ross Orr keeps the analog alive in Ann Arbor, Mich. A frequent contributor to MAKE, Ross hacks low-tech gadgets and invasive plants in his spare time.

THE HOLE PICTURE

A pinhole camera is a light-tight box with a piece of film (or photo paper) on one side and a tiny hole in the other. An image forms because each point on the film can only "see" the one patch of the outside world that's lined up with the pinhole, whether it's light, dark, blue, red, etc. Because the pinhole does not focus light like a lens, the film can be any shape and a flexible distance from the hole, and the enclosure can be made from any lightproof material.

With construction so forgiving, pinhole tinkerers have produced a riot of camera creations, from mint tins to airplane hangars — even animal skulls and hollowed-out vegetables. Pinholes are the most hackable camera type ever devised.

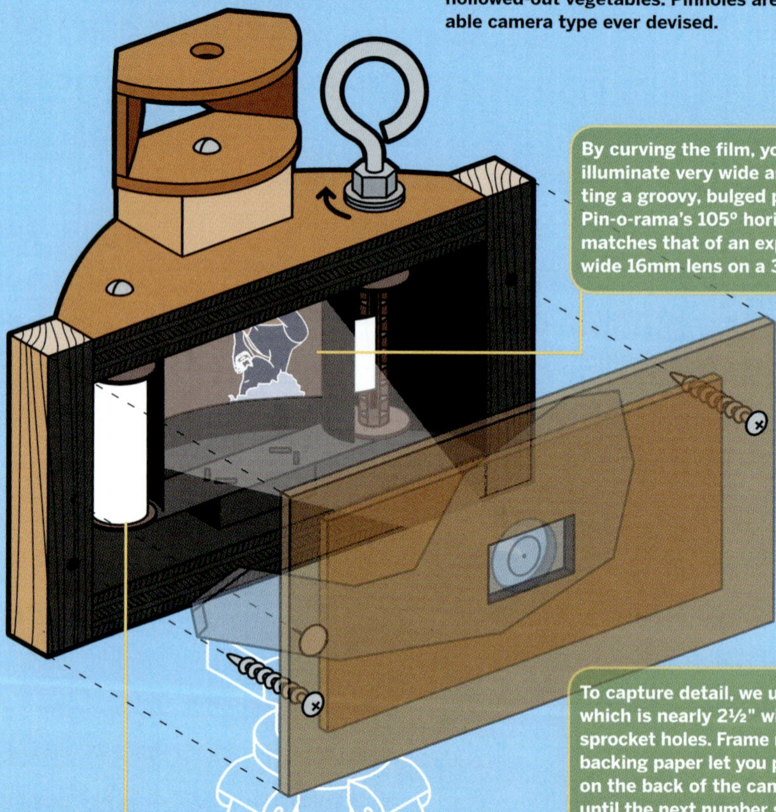

By curving the film, you can uniformly illuminate very wide angles, while getting a groovy, bulged perspective. The Pin-o-rama's 105° horizontal coverage matches that of an expensive, super-wide 16mm lens on a 35mm camera.

To capture detail, we use 120mm film, which is nearly 2½" wide and has no sprocket holes. Frame numbers on the backing paper let you put a peep sight on the back of the camera, and wind until the next number shows. To shoot panoramas, we count off every other number, yielding 6 shots per roll.

PROPER PINHOLE SIZE

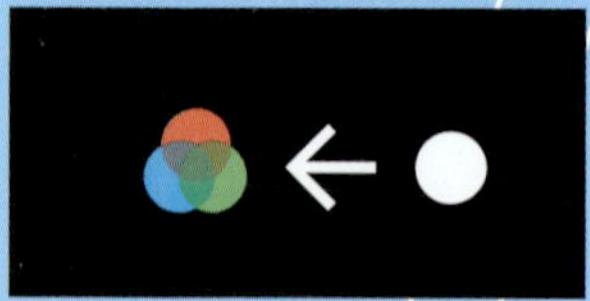

TOO LARGE: The pinhole's image comes from overlapping pinhole-sized blobs of light. A large pinhole loses all detail smaller than its own diameter.

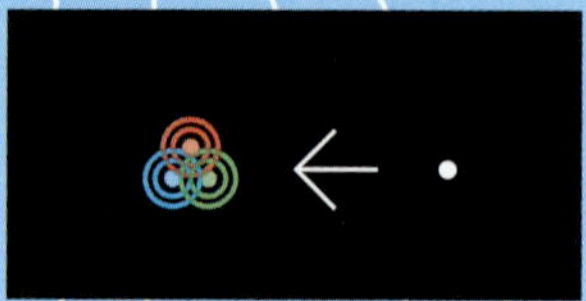

TOO SMALL: With a too-small pinhole, light diffraction smears the image. A smaller hole also decreases brightness, requiring longer exposures.

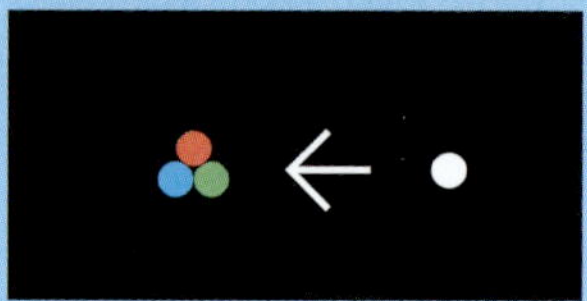

JUST RIGHT: Calculating proper pinhole size has absorbed much scholarly brainpower. For cameras with typical focal lengths, it's 0.2mm to 0.5mm.

Illustration by Nik Schulz

SET UP.

MATERIALS

[A] Scrap piece of ½" MDO plywood at least 6" square. Medium density overlay (MDO) has a smooth finish that looks nice.

[B] ½"×¾" pine strips (2) about 6" long each

[C] Aluminum sheet 0.01-0.02" thick (e.g. roof flashing), about 1' square

[D] Tinsnips

[E] Sewing needle

[F] Utility knife

[G] Jigsaw or scroll saw or band saw

[H] Drill and assorted bits

[I] Drawing compass

[J] English and metric ruler or calipers

[K] Flat black spray paint

[L] Roll of 120mm film and cheap, expired film rolls for testing (2) or one old roll and a spool

[M] Light meter or camera with a built-in meter (optional)

[N] Slide projector or slide scanner

[O] Tuna can or other source of springy steel

[P] Black silicone sealant

[Q] Pop rivet tool and pop rivets (rivets not pictured)

[R] Ball-peen hammer

[S] ⅜" threaded eye bolt (or thumbscrew) with matching nuts (2) and washers (2)

[T] #10×1" bolt with assorted matching nuts and washers

[U] ¼" #20 nut

[NOT SHOWN]
Epoxy or wood glue

Scrap piece of ¼" plywood at least 9" square

5/16" fender washer

¾" round-head wood screws (4) and 1½" round-head wood screws (4)

Cereal box cardboard

Staples or tacks

Electrical tape

320-grit sandpaper

Carpenter's square

Hacksaw

Metal file

Screwdriver any type

Block of scrap wood

Photography by Ross Orr

MAKE IT.

BUILD YOUR PINHOLE PANORAMA CAMERA

START **Time:** A Day or Two **Complexity:** Medium

1. MAKE THE BODY AND WINDER

1a. Draw a paper template for the top and bottom pieces. We're going to cut 2 identical D-shaped pieces of plywood; the curved side is a 105° chord of a circle with radius 65mm (about 2.56"), which continues along tangent flat planes to form 2 corners that extend ¼" past the circle's center. Download the template I used at makezine.com/09/pinhole.

1b. Use a jigsaw or band saw to cut the ½" plywood into 2 D-shaped pieces, following your template.

1c. Clamp the 2 Ds together, and sand until the cut perimeters are smooth and matching.

1d. Use the template to mark the spool centers on the top piece. Drill a ⅜" hole for the film take-up spool, which will be on the left as you face the camera's curved back. Drill a 3⁄16" hole for the supply spool on the right. In the center of the bottom piece, on the underside, drill a ½" hole partway through.

1e. Epoxy a ¼" #20 nut into the partial hole in the bottom piece, to make the camera's tripod mount.

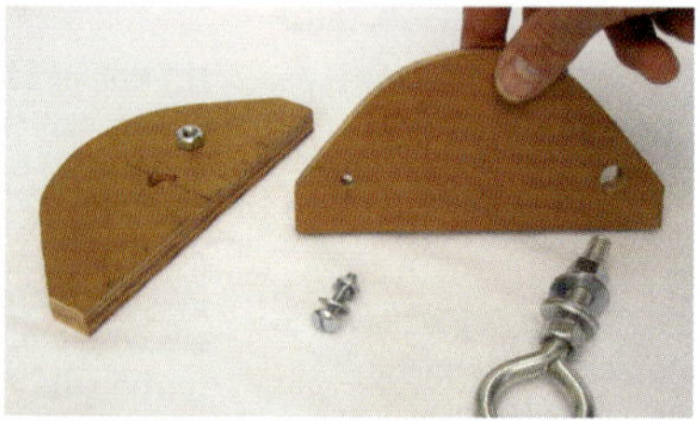

1f. Fit the ⅜" eye bolt through the top piece with nuts and washers on each side, then measure the distance that the nut opposite the winder eyelet will stick into the camera. Find a combination of washers and nuts that fits onto the #10 bolt (for the supply reel) and matches this height.

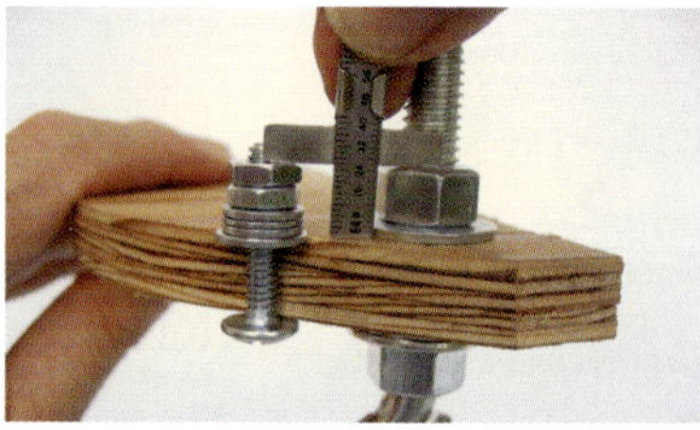

1g. Cut both bolt ends so each protrudes ⅛" beyond the nut face. Get an empty 120mm film spool from a friendly camera lab, or untape one from a cheap, expired roll of 120mm film; this will be our take-up spool. File the ends of each bolt flat on 2 sides, so that they engage the slot of the spool.

1h. Use thread-locking compound or mash the bolt's threads slightly to keep the nuts in place, so that the bolt assembly spins easily through the wood without loosening or tightening.

1i. Calculate the inside height of the camera by measuring the spool height and adding twice the height of the nut stacks; using this dimension will center the film vertically. Cut a scrap block of wood down to this dimension.

1j. Calculate the outside height by adding 1" (twice the plywood thickness) to the inside height, and cut 2 side rails to this length from the ½"×¾" wood. Clamp the top and bottom pieces around the scrap block, and glue on the side rails after making sure that all 4 pieces fit together evenly and are perfectly square.

1k. After the glue has set, cut a 1"×5" strip of springy steel — I used the sidewall of a tuna-fish can. Drill ⅛" holes that exactly match the spacing of the bolts on the top, and then attach pop rivets through these holes. The rivet nubs will catch the bottoms of the film spools.

1l. Cut a small wooden block the same thickness as the nut stacks, and staple or tack the spring to it, centered. My block was ⅜"×1"×1½". Apply glue to the block and position it on the floor of the camera. Load the film and take-up spools between the spring nubs and the bolts, and slide the spring block around until the film spools are exactly vertical. Clamp, and let the glue set.

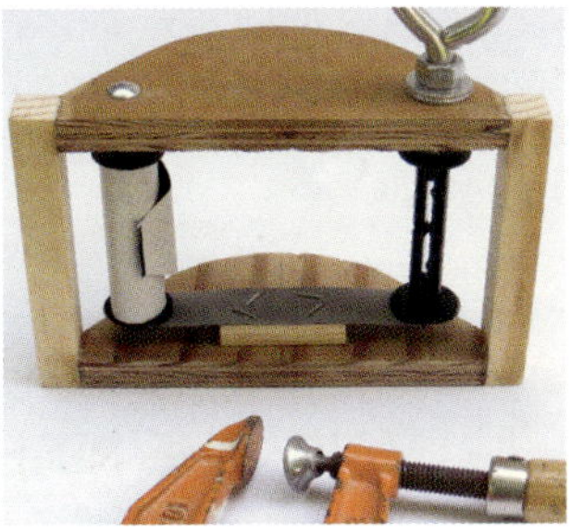

2. MAKE THE FILM GATE

2a. Cut aluminum flashing to the height of the camera body, and 8" wide. Draw a 2¼"×4¾" rectangle in the center. At each corner, draw lines that parallel the top and bottom edges, ½" in (to match the plywood thickness), and extending to the same longitudes as the sides of the rectangle. Measure the camera's curved plywood edge, and symmetrically mark the corners of the aluminum where the edge overlaps this distance. For example, my camera back measured 7¾" around, so I marked in ⅛" from each side.

2b. Use tinsnips to cut along the 4 lines that extend in from the sides, and then cut the excess-overlap corners off of each piece, as marked. Then score the center rectangle lines with a sharp utility knife, and flex along the score lines to snap through the aluminum.

2c. Fold the 2 center flaps inward from each side, using a screwdriver shaft as a brake to form smooth, 90-degree bends. Sand any rough edges, especially around the opening in the center.

2d. Fit the gate around the back of the camera. Cut notches in the flaps so that they clear the spring strip, and carefully curl them around so they fit around the film spools.

2e. Spread a thin bead of black silicone sealant along the plywood edges, position the film gate onto the back, and tape it into position until it cures.

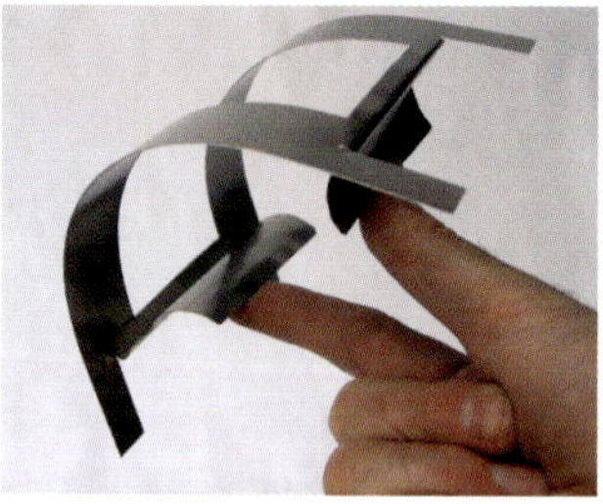

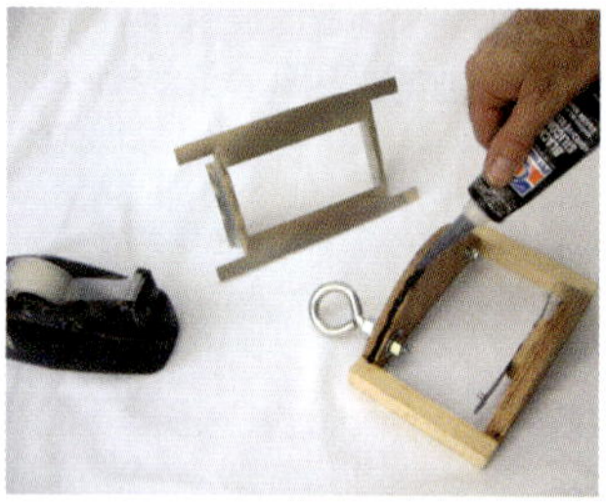

2f. Wrap your test film in position around the back of the gate (if it isn't centered, add or remove washers on the film-winder bolts). Mark the edges of the film on the gate.

2g. Cut 2 straight strips of cardboard from a cereal box, to use as guide rails for the film. Glue them in place along your marks on the gate with a thin layer of silicone. The film should easily slide between these guides with a little wiggle room.

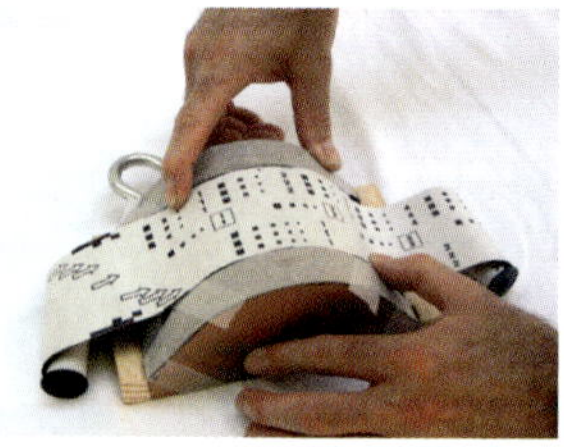

3. MAKE THE PINHOLE

3a. Cut several 1⅞" squares of sheet metal; we'll put a pinhole in each of these, and load them into a 35mm slide projector or slide scanner later to choose the best ones.

3b. For each square, use a ball-peen hammer to tap a small bump into the center of the metal.

3c. Thin the bump with #320 sandpaper. Stop sanding when light pressure with a needle telegraphs a tiny dent through the other side.

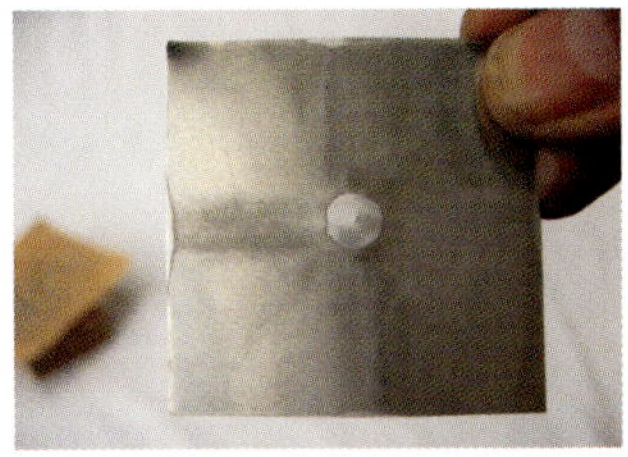

3d. Back the bump against a phone book, and lightly press the needle until a hole shows through. Sand away any raised burr, blow through the hole to remove gunk, and check the diameter.

3e. Use a slide scanner or projector to check each hole's diameter and roundness. For our camera's 65mm focal length, we want a pinhole diameter of 0.33mm, plus or minus 20%. Focal length divided by hole diameter yields the equivalent f-stop, and we're aiming for about f/200.

With a scanner: Set the scanner to its highest resolution, scan the hole, and read its size by setting your image software's units to millimeters.

With a projector: Set the projector up so that a full 35mm slide image measures 52"×35". Load the pinhole into the projector, and look for the light spot to measure about ½" in diameter.

4. MAKE THE FRONT AND SHUTTER

With long exposure times, pinhole shutters can be very low-tech — a piece of black gaffer's tape will work fine but I like this simple "cigar cutter" design. The lip that's formed by the inner and outer panel edges deflects any light that leaks through the sides, making it disappear before it gets a direct line to the film.

4a. Cut 2 rectangles of ¼" plywood, one to the camera front's inside dimensions and one to its outside dimensions. Mine measured about 4"×5¾" and 3½"×5¼".

4b. Drill a ⅞" diameter hole in the exact center of the smaller (inner) piece, and cut a 1"×1½" rectangle in the center of the larger (outer) piece. If you want to use a simple piece of tape as a shutter, glue these pieces together and skip ahead to step 4e.

4c. Use paper and a thumbtack to make templates for a pivoting shutter and the chamber it moves within. The pivot point will sit at the right edge of the front panel (if you're right-handed). pinhole by about 2"; this will be the shutter's inside edge. The chamber should limit the shutter's travel so that its hole swings between fully visible to fully tucked away. When you have shapes that work, cut the shutter out of aluminum flashing and the chamber out of cardboard.

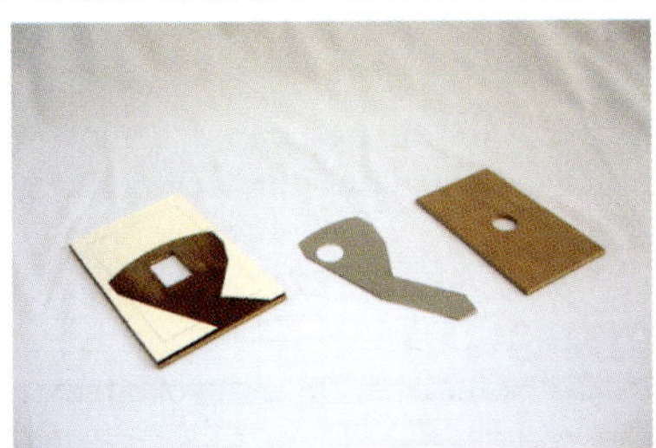

4d. Glue the cardboard chamber to the outer panel, thumbtack the metal shutter into place at the pivot point, and position the small panel in back. Test the shutter to make sure the alignment works.

4e. Disassemble the pieces, and spray-paint black both sides of the inner panel, and the back sides of the shutter and outer panel. Reattach the shutter and glue everything back together.

4f. Center the pinhole behind the rear opening and tape it in place. Hold the panel against the camera, point the pinhole toward some light, and sight along the edges of the film gate to check for any obstructions. The plywood shouldn't block any light from reaching the corners of the film.

4g. Drill clearance and pilot holes, and use two ¾" wood screws to attach the front panel to the side rails, in place.

5. MAKE THE BACK

5a. Cut a piece of flashing to match the height and width of the camera back, bending ½" flanges for the side rails. Drill a ½"-diameter peep sight for the frame counter, offset 1¼" left of center. Put a small square of black tape over this hole; you'll keep it there to cover the hole whenever you aren't winding the film.

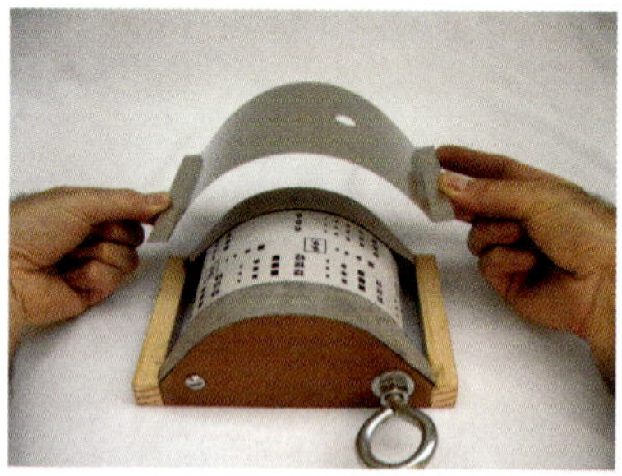

5b. Spray-paint black the inside of the back piece you just cut, and the interior of the camera itself. But don't paint the wooden side rails, which you'll be gluing, or the gate where it touches the film. Mask these areas with tape before spraying.

5c. Glue down the back with a sparing amount of silicone, to avoid squeeze-out into the film path. Tack and clamp the side flanges to the wood rails, and hold the back against the D's with tape until the silicone cures.

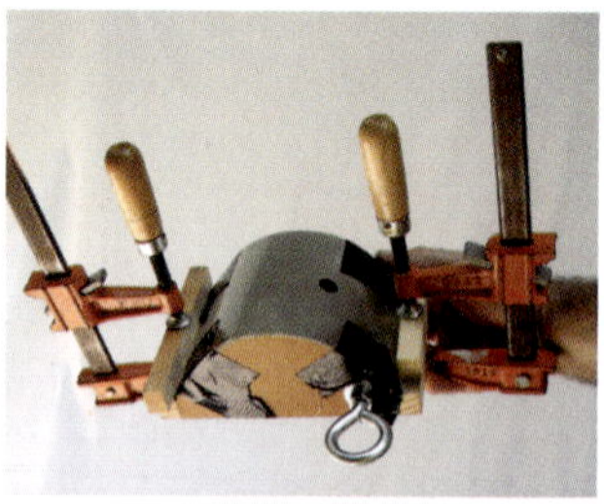

5d. That's it! Your camera is now ready to use.

NOTE: The finished camera shown above includes an optional viewfinder. To see the step-by-step process for making the viewfinder, please go to makezine.com/09/pinhole.

FINISH X

NOW GO USE IT »

USE IT.

JOIN THE PINHOLE PAPARAZZI

120MM FILM

Once you remove 120mm film from its foil pouch, the only thing that keeps ambient light from exposing it is the backing paper at each end of the roll, and the fact that it's wrapped so tightly around the spool. So be careful, and don't reload in the sun.

Choose 100- or 200-speed negative film. Anything faster makes sunny-day exposures too brief to time accurately. I recommend Fuji's Superia color and Acros black-and-white films.

LOADING THE CAMERA

When the roll is done, you don't rewind; instead, wind it all onto the take-up spool. The original inner spool becomes the take-up for your next roll.

1. Unscrew the front of the camera, unspool a few inches of backing paper, and use your (moistened) fingertips to fish the paper through the film gate.
2. Tape the paper onto the take-up spool, wind it one turn, and then close up the camera.
3. Wind until the "1" shows in the peep sight. You're on a roll, and ready to shoot! (After each winding, stick the black tape back down over the hole.)

TAKING PHOTOS

With the viewfinder, you will need to move your eye around to see all the edges of the frame.

After every shot, wind to the next odd number, so you don't accidentally double-expose. This yields 6 double-width panoramas per roll.

With 100-speed film, try 2-second exposures in sun, or 10 seconds in bright shade. At this slow shutter speed and camera height, you'll need a tripod to keep your images sharp. In a pinch, resting the camera on a flat surface will work, too, but be careful not to jiggle the camera while exposing the film. For other light conditions, use a light meter or another camera to get exposure times. Your Pin-o-rama has an f-stop of about f/200. Few light meters give readings for f-stops this high, but you can take the shutter speed indicated for f/16 and multiply the time by 150. Don't hesitate to double or triple the calculated time. Negative films are tolerant of overexposure. To time exposures without looking away from the subject, hold a watch to your ear and count ticks.

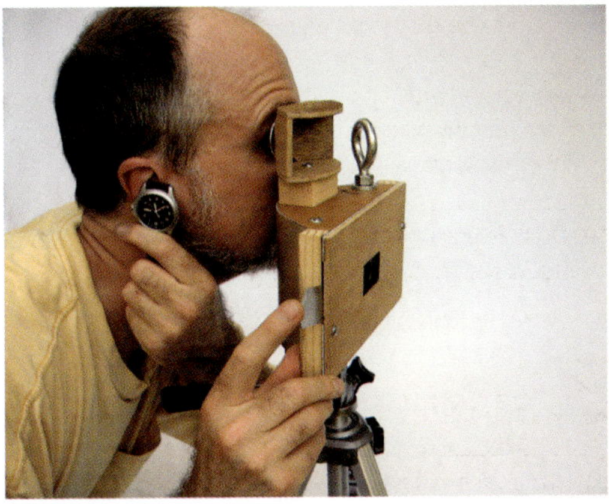

FINISHING A ROLL

After exposing shot #11 (the sixth and final shot), keep winding until the paper backing disappears from the peep sight. Open the camera in dim light, and use the adhesive band to tape the roll tight. (If you find the take-up reel has wound too loosely, add more spring pressure on the supply spool.)

Any lab or camera shop that caters to professionals can develop your 120mm negatives for about $5. The 6×12 format is nonstandard, so they should develop the negatives only, and return them uncut. Then you can scan them yourself with a flatbed film scanner. The generous negative size means that even a budget scanner will do a decent job — resolution is a non-issue given the softness of pinhole images. (Darkroom prints are possible too, using a 4×5 format enlarger.)

For pinhole photography resources and additional information go to makezine.com/09/pinhole.

PIN-O-RAMA PHOTO GALLERY

The Pin-o-rama camera design was a brainwave from early 2006. Since then, I've loved exploring how it transforms familiar scenes from my hometown into fresh and surprising images.

Big Boy The limitless depth of field of a pinhole lets you move as close as you want to your subject, while leaving distant details sharp — making all kinds of playful juxtapositions possible. The Pin-o-rama's curved perspective, so obvious when manmade straight lines are in the frame, is much less noticeable when they're absent. With this camera's wide, cinematic framing, I like compositions where the subject is strongly off-center.

Cafe Scene This one is a sentimental favorite from my first test roll with the Pin-o-rama prototype. During the 9-minute exposure, people got up and sat down, unaware that the strange object resting on my table was taking a photograph. The clock's minute hand also has blurred into nothingness. At the time, I had no idea what this camera's images would look like, but after pulling the film out of my developing tank, I was delighted.

PINHOLE PORTFOLIOS

See more of Ross Orr's pinhole photographs at flickr.com/photos/vox/tags/pinorama

Homemade Pinhole Flickr Group: flickr.com/groups/homemadepinhole

Liberty Plaza Tree Lights
I started this 15-minute exposure just as daylight was fading from the sky — slightly self-conscious about loitering in a darkened park with my unidentifiable apparatus. I was surprised that on the negative, it's quite noticeable that the pinpoints of light shatter into tiny ripples, from the diffraction of light waves passing through the pinhole. In the background, the taillights of cars creeping by in stop-and-go traffic add their own sparkle to the scene.

Liberty Block Darting into the street between passing cars, I set down my tripod to make this exposure of a nice old commercial block in downtown Ann Arbor, Mich. The couple waiting at the crosswalk stood motionless enough during the 3-second exposure to register on film, but another shopper walking through the scene dematerialized completely. The glint off the window was pure serendipity; pinhole photography seems to invite fun surprises and accidents.

Peaches Pinhole photos can evoke mysterious, even somber moods. But I enjoy working against that stereotype, seeking out subjects that are vibrant and colorful. Strong crossing light can provide extra punch and contrast. I also sometimes use Photoshop's Unsharp Mask filter in an unconventional way: with a very large radius and the amount set to about 12%, to lift the "fog" from pinhole images and regain the brilliance of the original scene.

Yellow Bakery Vertical I've always enjoyed how the afternoon sun lights up this brightly painted neighborhood bakery. This was shot using a Kodak pro film 3 years past its expiration date; any minor color shifts are easily corrected when scanning. Historically, the biggest users of 120 film were wedding photographers and other professionals. But these shooters migrated to digital so abruptly that camera stores often have excess out-of-date rolls for sale, at half price or lower.

THE $5 CRACKER BOX AMPLIFIER

By Ed Vogel & Blind Lightnin' Pete

BIG SOUNDS FROM A SMALL PACKAGE

In MAKE, Volume 04, I presented my version of the venerable cigar box guitar. The instructions for the project included adding an electric pickup so you could play the guitar through an amplifier.

People from around the world emailed me to tell me they'd built cigar box guitars based on my instructions. I struck up a conversation with one gentleman from Europe who goes by the moniker Blind Lightnin' Pete. He made a couple of beautiful cigar box guitars, including onc he calls the Vintage Blues Texas Rattlesnake Special model. He then went one step further, and built a cracker box guitar amplifier. This outstanding little amp cost all of $5 to build (depending on where you get the parts). Pete kindly allowed me to modify his design and present it as a project for you to build. (See page 111 for a word from Pete about the origins of the cracker box amp.)

My amp differs a little from Pete's because I wanted to make a workable little practice amp with parts and tools that could be purchased "one-stop shop" at RadioShack and built in an hour.

Photograph by Sam Murphy

Set up: p.107 **Make it:** p.108 **Use it:** p.111

Ed Vogel lives in Minneapolis and believes that nothing may just be the next big thing.
Blind Lightnin' Pete is the online pseudonym of Howlin' Mississippi Slim.

THIS BOX ROCKS

Hi-Carb Sound

The heart of this surprisingly loud, clear-sounding, battery-powered guitar amp is National Semiconductor's LM386 series low-voltage op-amp IC. Two potentiometers in the circuit control the gain and the volume. For the cleanest sound, turn down the gain knob all the way and turn up the volume knob to the maximum. Then slowly turn up the gain. For a raunchier, distorted sound, start with the volume knob all the way down, and the gain knob at maximum. Then crank up the volume. You can achieve lots of different sounds by playing with the knobs. Experiment!

Illustration by Timmy Kucynda

SET UP.

MATERIALS

[A] A box of some sort or another (cracker box shown)

[B] Toggle switch, single pole single throw

[C] 9V battery

[D] Battery connector

[E] 0.047μF capacitor

[F] 220μF capacitor (biggest)

[G] 0.01μf capacitor

[H] 100μf capacitor

[I] Hookup wire, 20 or 22 gauge AWG solid core is best.

[J] 5KΩ potentiometer (audio or log taper)

[K] 25-ohm (25Ω) rheostat

[L] LM386N audio amplifier

[M] 8-pin DIP IC socket

[N] Chicken head knobs (2)

[O] Prototyping PC board

[P] Soldering iron

[Q] Solder

[R] Speaker, 8Ω impedance

[S] 10Ω resistor

[T] ¼" mono phone jack

[NOT SHOWN]
Speaker grill (optional)
Glue gun

MAKE IT.

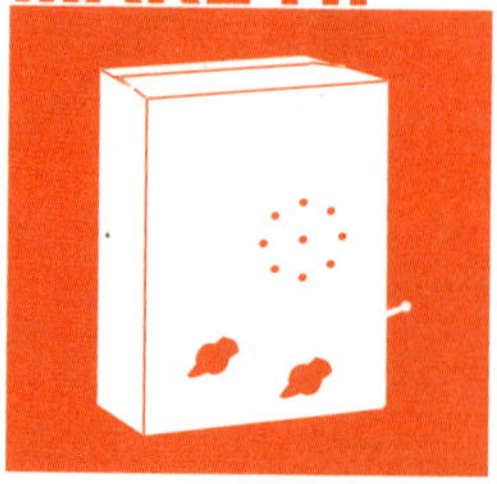

BUILDING THE CRACKER BOX AMP

START **Time:** **An Afternoon** **Complexity:** **Medium**

1. MAKE THE CIRCUIT

1a. Make a copy of this schematic, or download the PDF at makezine.com/09/crackerboxamp and print it out.

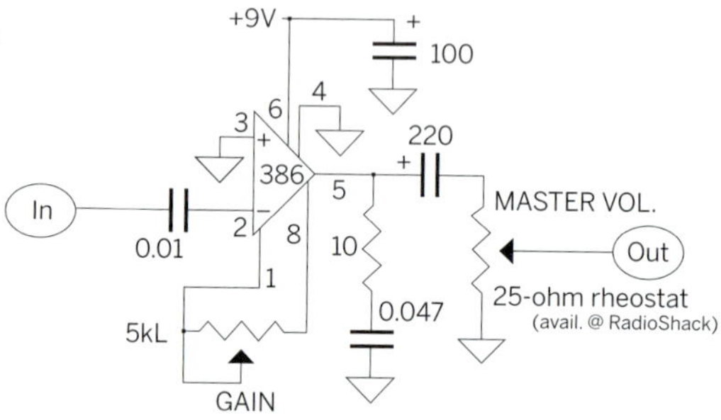

1b. Install the socket in the printed circuit board.

1c. Solder it down.

1d. Install the chip. I like having the chip in the printed circuit board while I build because there can be no doubt as to where pin 1 is. This is also why I install parts and make wire connections on the top of the printed circuit board.

1e. Install the 0.01µF capacitor so one leg connects to pin 2 of the chip and one leg is in a "proto row." Flip it over and solder it.

Photography by Ed Vogel

1f. Install the 9V battery clip and mark a plus sign for the red wire and a minus sign for the black wire.

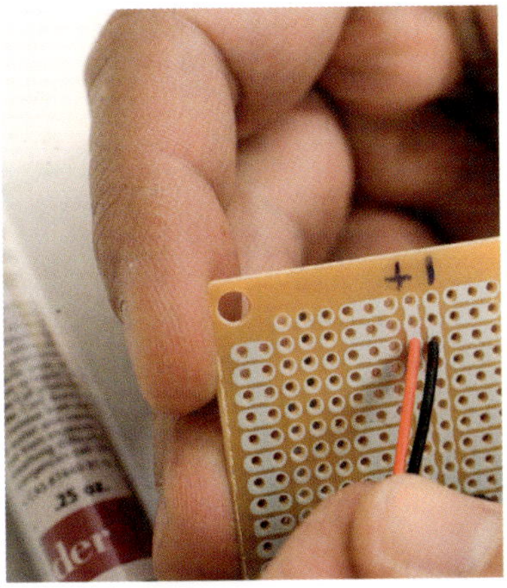

1g. Install the 10Ω resistor and the 0.047μF capacitor. Take advantage of the "proto rows" to make the connections:

- Chip pin 5 to one leg of the 10Ω resistor.
- The other leg of the 10Ω resistor to one leg of the 0.047μF capacitor.
- The other leg of the 0.047μF capacitor to "ground."

For our purposes "ground," which is shown on the schematic as a triangle with the point down, is the long "proto row" we marked with a minus sign.

1h. Use this same technique to install and make connections.

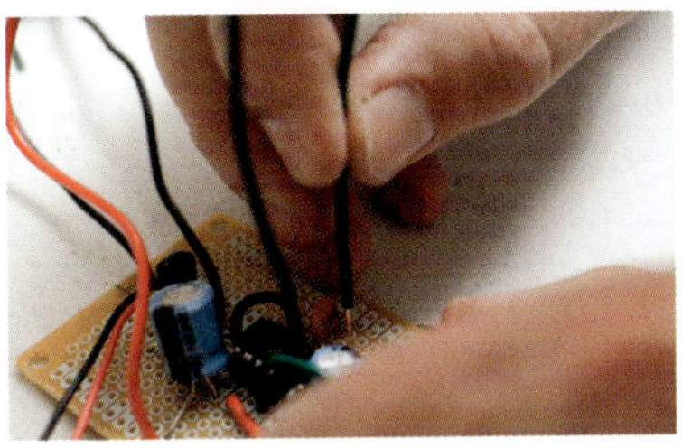

TIP: Every time you install a part or make a connection, mark it off on the schematic ("Little Gem" schematic courtesy of runoffgroove.com/littlegem.html).

1i. Solder the wires to the phone jack. Use green for signal and black for ground.

1j. Install the wired phone jack to the circuit. You should end up with something that looks like the photo to the far right here.

1k. Cut the red lead and install the switch.

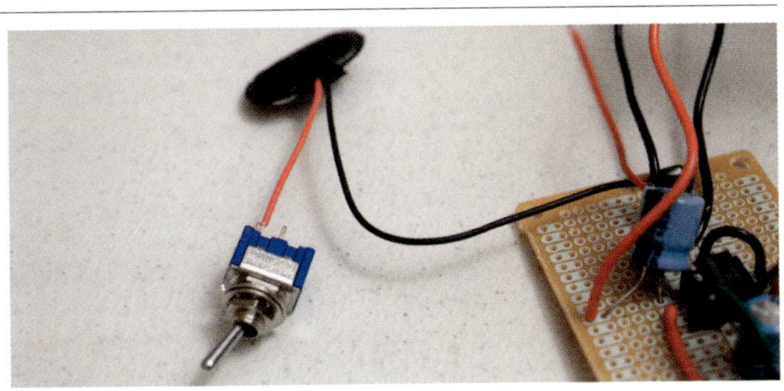

2. BUILD THE ENCLOSURE

2a. Make holes in the side of your box to fit the potentiometer, rheostat, and phone jack.

TIP: Why bother with drills and X-Acto knives when you can use your soldering iron to make holes?

2b. Make holes for the "speaker grill." You are going to find some hanging chads on the inside of the box. Reach in there with the soldering iron and burn them off.

2c. Make a hole for the switch.

2d. Pop your circuit into the box.

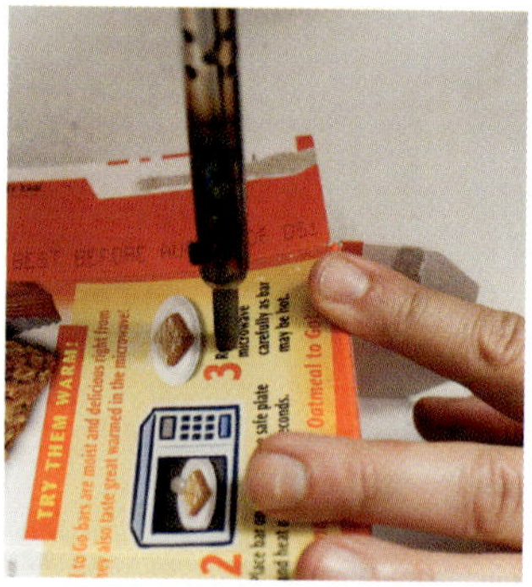

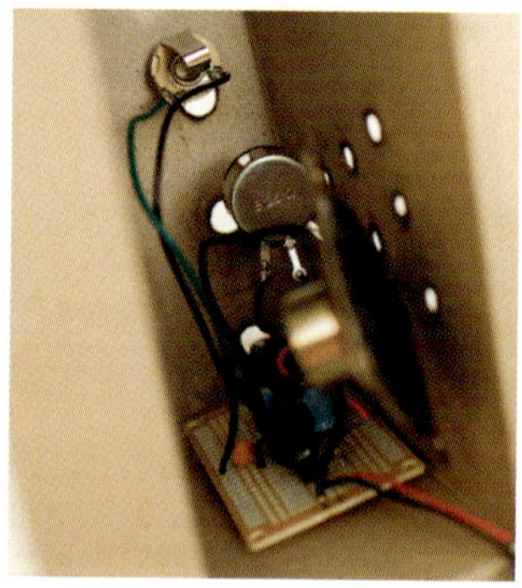

2e. Mount the speaker. Make some big glops of hot glue to act as "stand-offs" on the speaker.

2f. Mount the switch.

2g. Install the chicken head knobs.

NOTE: It's a proven fact that chicken head knobs greatly enhance the vintage sound of an amplifier. Use them liberally.

FINISH X

NOW INSTALL A BATTERY AND GO USE IT! »

USE IT.

THE ORIGINS OF THE CRACKER BOX AMP

Blind Lightnin' Pete's cracker box amp.

A FEW WORDS FROM BLIND LIGHTNIN' PETE

The cracker box amp I built cost $5. It uses an 8-pin National Semiconductor LM386 series low-voltage op-amp IC. The amp circuit unleashes the full potential of this beast and creates ¼ watt of arena-shaking power. Think of it as sort of a silicon shrunken head of the Marshall stack that Jimi Hendrix played at Monterey.

This integrated circuit has provided the basis for low-power solid-state amplifiers in recent years, including the famous Smokey Amp and a few of the designs at runoffgroove.com.

You can buy an LM386 for under a buck; it's a standard RadioShack item, the same one that was used in the MAKE project for turning your old computer mouse into a robot (*see MAKE, Volume 02, page 96, "Mousey the Junkbot"*). Our favorite hobbyist robot supply source, Solarbotics (solarbotics.com), sells them for 75¢ a piece.

I added a couple of capacitors, a couple of resistors, an LED, a ¼" jack, a potentiometer, and a $2 speaker, wrapped it all in a big blob of solder, crammed it in whatever empty box was laying around, and voilà!

The pot controls the gain, and it goes from California clean vintage Fender to Santana Mesa Boogie crunch to Hendrix Marshall. It runs off any combination of batteries — I usually use a 9V, but it's possible to get a cleaner tone with 12V (8 AA batteries in series). I have used it to drive a 4×12 Marshall cabinet, and it gets pretty loud. Not loud enough to compete with a rock drummer, but loud enough for me not to hear my wife screaming "turn it down," which I guess is enough for household use. If we had any neighbors, I could raise some complaints from it. Let's just say that even with a 2" speaker it's plenty loud enough for most apartment dwellers.

Interested in learning a bit more about the LM386? A great place to start is National Semiconductor's website (national.com) where you can download the data sheet. Even better, if you take the time to register on the site, they will generously send you a few samples for free!

Ask for the LM386N-4 series, as these are rated to handle up to 18V. Although any of the LM386 chips will work wonderfully for our hi-gain design, several experimenters and makers have found that cleaner tones with more headroom are achievable by running the circuit with a few extra volts.

OTHER OP-AMP 386 PROJECTS

Home-built bat detector:
bertrik.sikken.nl/bat/my_div.htm

Mini bench amp to test audio circuits:
makezine.com/go/minibench

Headphone amplifier:
radiowrench.com/sonic/so02144.html

See videos of Blind Lightnin' Pete playing his cigar box guitar through his cracker box amp at makezine.com/09/crackerboxamp.

Photography by Blind Lightnin' Pete

PROTOTYPE THAT

When Kevin Binkert moved into the old Standard Metal Products (SMP) building in San Francisco's South-of-Market neighborhood 15 years ago, all that remained of the 1920s metal foundry was a metal plaque. In homage, Binkert revived the name and built out an atelier that melds the most modern Computer Numeric Control machines with traditional hand tools.

As a maker-for-hire, Binkert has prototyped a handheld blaster for brain tumors, engineered hydrant valves for San Francisco's fire department, produced custom parts for the city's Italian streetcars, and restored two historic clock towers.

In between commercial jobs, he worked on the Spirit of America, a 45,000-horsepower jet vehicle that tore through the desert at an unofficial record speed of 675 mph; built The Spinner, a Ford V8-driven machine that whips braided cables to deafening supersonic speeds; and unleashed the Flame Tornado, a gas-powered sculpture that spews a 40-foot-tall vortex of fire.

These days, he's taking his talent to the small screen for *Prototype This!*, a forthcoming Discovery Channel TV show produced by *MythBusters* creators Beyond Productions. Binkert and MAKE contributor Joe Grand are among the five hosts.

Binkert's DIY career began in post-college jobs at movie special-effect houses. His defining maker moment came when he joined machine performance provocateurs Survival Research Laboratories (*see MAKE, Volume 07*) in 1989.

"I think that around retirement a lot of machinists start realizing they can make art with their tools," Binkert says. "I got into this the opposite way."

—David Pescovitz

1. Okuma three-axis milling machine, known for its reliability and accuracy. 2. Mori Seiki CNC lathe used by its previous owner to machine parts for high-speed centrifuges. 3. Marvel cut-off band saw. 4. Monarch Model EE lathe, arguably the best manual lathe ever. 5. Cat-40 tools for the Okuma. 6. Wheel balancer machined for an SF streetcar. 7. Chip conveyor catches the scraps from the CNC and drops them in the steel drum. Periodically, a man named Luther retrieves the material for recycling. 8. Dentures left by dentists who commissioned a custom denture-cutting tool. 9. Arbor press for installing bearings and pins.

Photograph by Robyn Twomey

+ More info: smpmachine.com

More images on makezine.com/09/workshop

1+2+3

Curie Engine

By John Iovine

Changing the temperature of nickel wire turns heat into motion.

You will need: A bit of low-Curie nickel alloy wire, 1" or larger neodymium "super magnet," copper wire, birthday candle, brass screw and nut, small steel plate, glue or wood screws, 4"×¾" wooden slat, 2"×1¼"×½" wood block. Nickel wire and magnet, or a full kit, are available for purchase at makezine.com/go/heatengine.

1. Put it together.

Glue or screw the steel plate and 4" slat to the wood block base as shown below. Drill a pilot hole, and screw the brass screw at the top of the slat. Drill a ¼" hole in the base about 7⁄16" from the steel plate end. For complete measurements, go to the website listed above.

Wrap about 1" of nickel alloy wire around a pencil to make a coil. Take 4" of copper wire, and twist the coil onto one end. Make a 90° bend in the copper wire 1½" down from the end with the coil, and cut the other end so there's 1" after the bend. Hook the wire bend onto the brass screw, with the coil on the side nearer the steel plate.

2. Test the wire and magnet.

Place the neodymium magnet onto the metal plate. The wire should swing up to meet the magnet, pulled by the nickel alloy coil. Push down and release the wire, and it should swing back up. Insert a birthday candle in the base, and adjust the magnet position and wire so the coil is suspended above the wick.

3. Light the candle.

Light the candle. The wire should swing down and up repeatedly, out of and back into the candle flame. Like ordinary iron or steel, the nickel-iron alloy is magnetic, but it has a much lower Curie point — the temperature above which it loses its magnetic properties. (The copper wire and brass screw are not magnetic.) When the coil heats up to this point, it is no longer attracted to the magnet, so it falls away and out of the flame. As it cools, it becomes magnetic again and swings back up, over and over again, until the flame burns too low to touch the coil.

Illustration by Dustin Hostetler

John Iovine wrote the Kirlian Photography article on page 66.

STRUNG OUT

Make a one-string electric guitar from plastic tubing. By Tom Zimmerman

This project will show you how to build an amazingly inexpensive and rad-looking one-string electric guitar out of pine wood and PVC pipe. A single string purposely keeps the design simple. (*For a 3-string cigar box model, see MAKE, Volume 04.*)

But first, we need to establish the names and functions of the various parts of a guitar. Please refer to the diagram and materials list on the following page before starting.

1. Get your parts together.

Do you want a lead or bass guitar? A lead guitar has a thinner string and is used for playing melody. A bass has a thicker string and plays lower notes. Get the cheapest tuning machine and metal low E string you can find. If the tuning machines come in a strip, cut them apart with a hacksaw, preserving a mounting hole for each tuning machine.

2. Make the body.

The sound of the vibrating string is picked up by a piezo element, so you can design the body any way you please. Look at pictures of guitars to get ideas, or trace a real guitar.

Draw the pattern on the wood and cut it out using a jigsaw, scroll saw, or band saw. It helps to first make release cuts (cuts going from the edge of the wood to the pattern) that fall away, providing turning room for the blade.

Next, sand and round the edges to give it a finished look, and then hold it like you are playing it. This is your last chance to modify any curves for comfort. When you're happy with your design, clean off the dust with a damp rag and paint the body. Several thin layers of paint are better than one thick layer. Refer to the photo at the bottom of the next page to see the process of making the body.

Photograph by Sam Murphy

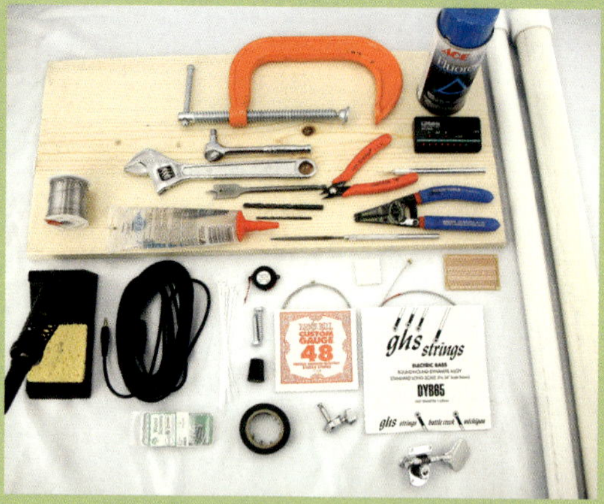

MATERIALS	TOOLS
Guitar (or bass) string	Jigsaw or band saw
Guitar (or bass) machine tuner	Drill
PVC pipe 4' long 1" dia. for guitar, or 1½" dia. for bass	Sander
PVC end caps (2) To fit pipe	¾" wood boring bit
Pine board 1"×12"×2' long	⅛" twist bit
Spray paint For coloring the body	¼" twist bit
Silicone sealer glue	Clamps (2)
⅛" extender cable (RadioShack #42-2562)	Thin flat file
Piezo element (RadioShack #273-073)	Razor knife
Rubber stopper	Wire stripper
Rubber feet	Wire cutter
Cable ties	Adjustable wrench
1½" hex lag screws (6)	¼" socket
Black tape	Soldering iron and solder

Guitar Parts and Their Functions

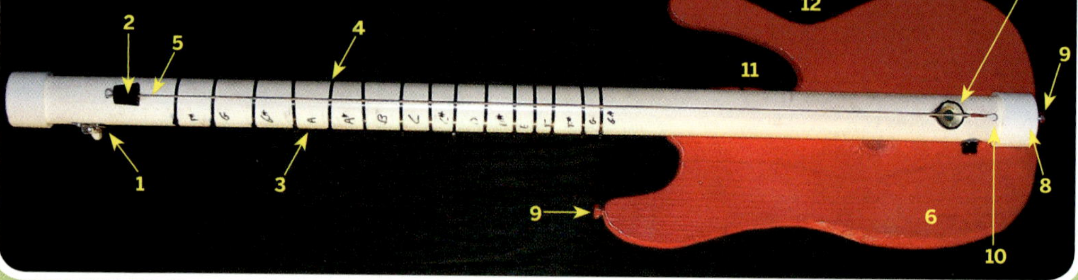

1. **Tuning machine** Pulls the string tight so it will produce sound.
2. **Nut (stopper)** Establishes the lowest note when the string is played open (no fingers).
3. **Neck** Supports the string and frets.
4. **Frets** Stop the string at precise lengths to produce notes.
5. **String** A metal wound wire that vibrates when plucked to produce sound.
6. **Body** Keeps the neck from rotating.
7. **Pickup** Converts string vibrations into electricity.
8. **Input jack** Connects pickup to amplifier cable.
9. **Strap buttons** Stubs to attach shoulder strap to the body.
10. **String anchor** Holds the end of the string.
11. **Neck cutout** Provides hand access to play high notes.
12. **Leg cutout** Contoured to fit the leg when playing in the sitting position.

You may notice a black knob sticking out of the PVC pipe below the pickup. I installed a potentiometer to adjust the volume, but I don't recommend it. It adds noise to the circuit. Instead, adjust the volume on the amplifier.

Make the Body

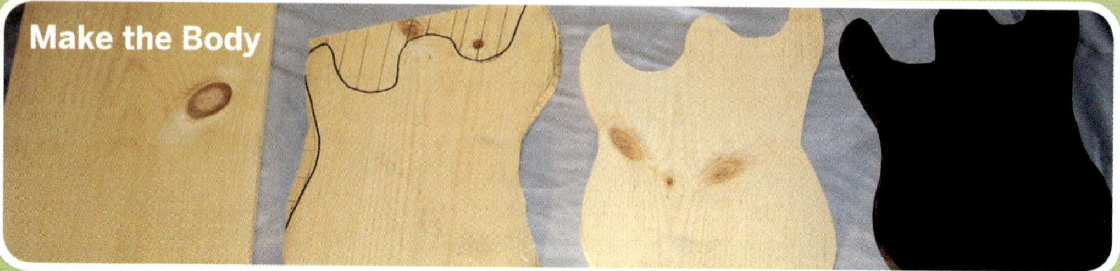

Photography by Tom Zimmerman

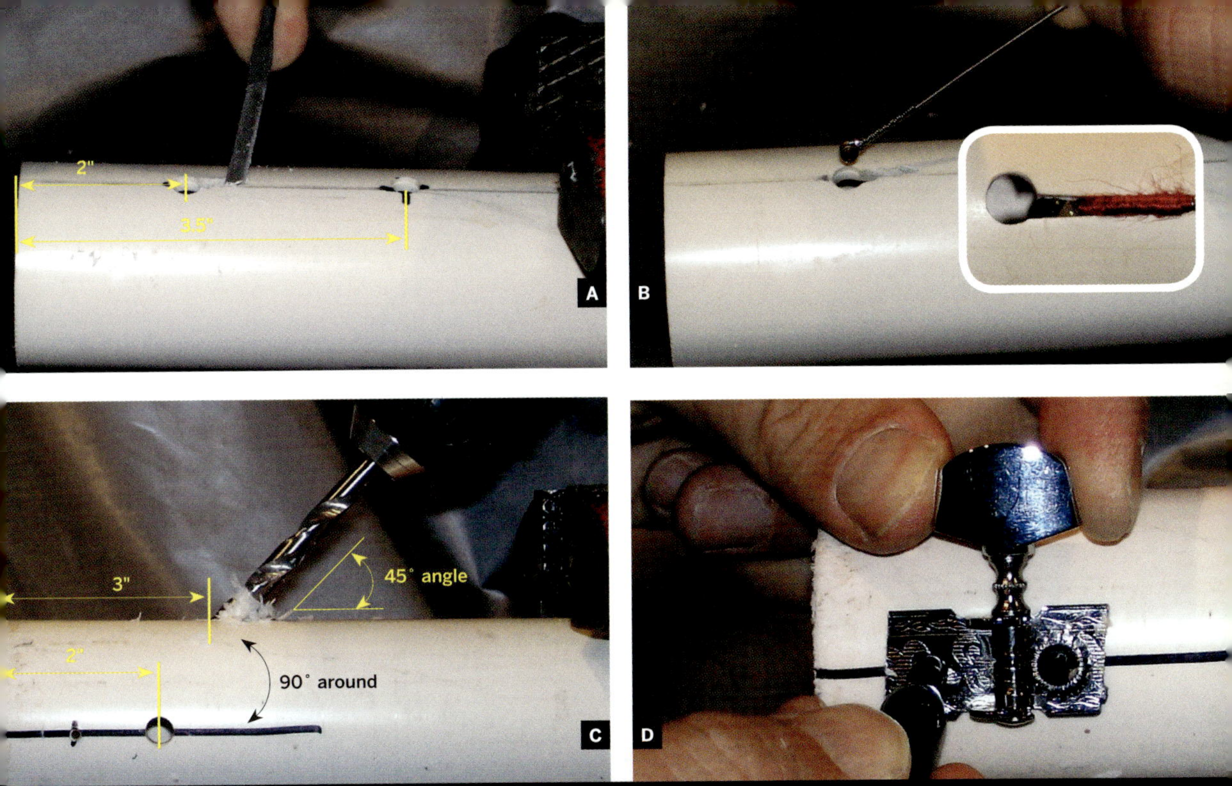

Fig. A: File the notch.
Fig. B: Insert the string ball (with close-up of the installed string).
Fig. C: Drill holes for the tuner.
Fig. D: Mark the tuner mounting screw hole.

3. Prepare the string anchor.

If you're making a lead guitar, cut the 1" PVC pipe to 39" using a hacksaw. For the bass guitar, cut the 1½" PVC pipe to 44". Clamp the string tightly to the PVC pipe and use it as a guide to draw a straight line down the pipe. Remove the string and drill two ¼" holes 2" and 3½" from the end of the PVC pipe along the line. In the hole closest to the pipe end, use a small flat file to carve a ½"-long notch toward the center of the pipe (Figure A). Test to see that the ball of the string fits in the hole and is grabbed by the notch when the string is pulled tight (Figure B).

4. Install the machine tuner.

Drill a ¼" hole 3" from the other end of the PVC pipe, on the line you drew. While the bit is turning, slowly tilt the drill toward the center of the pipe about 45 degrees (Figure C). This will create a slanted ledge for the string to rest on. Drill a ¼" hole 2" from the end of the pipe, 90 degrees around from the first hole.

Put the shaft of the machine tuner into the second hole and mark the location of the mounting screw (Figure D). Remove the machine tuner and drill a ¹⁄₁₆" pilot hole. (Careful: Keep the bit straight or it will snap). Put the machine tuner back into the hole and screw it in with the wood screw that came with the machine tuner.

5. Attach the neck to the body.

Draw a centerline down the back of the body and drill 4 evenly spaced ⅛" pilot holes. Countersink each hole halfway through the wood from the back with the ¾" wood bore bit.

Lay the PVC pipe on a table, string side down. Clamp the body on top of the PVC pipe along the centerline with the ends of the PVC pipe and body flush. Make sure the machine tuner is away from the body. Use the 4 pilot holes as a guide to drill ⅛" holes through the PVC pipe. From the back of the body, screw the lag bolts through the wood and into the PVC pipe with an adjustable wrench.

See Figure E on the following page for reference.

6. Install the strap buttons.

Hold the guitar against your body and find 2 comfortable anchor points for the strap. Drill ⅛" pilot holes and screw in the lag screws, leaving a ¼" gap between the screw head and the body for the strap to fit into.

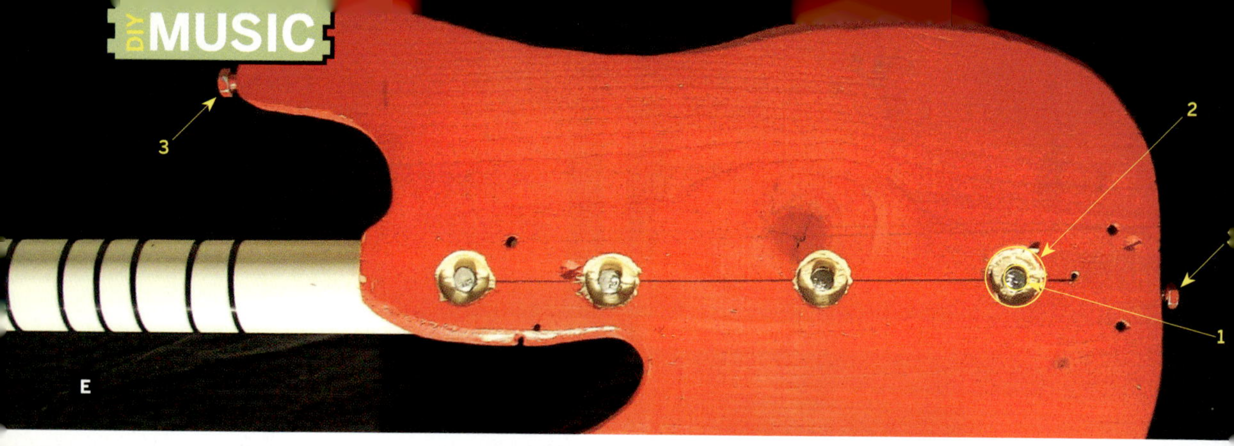

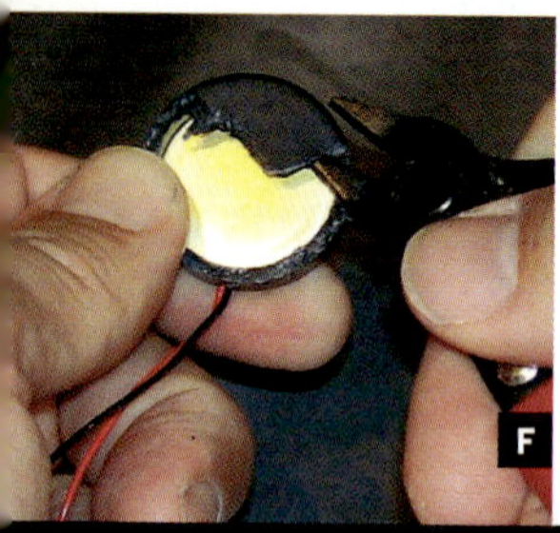

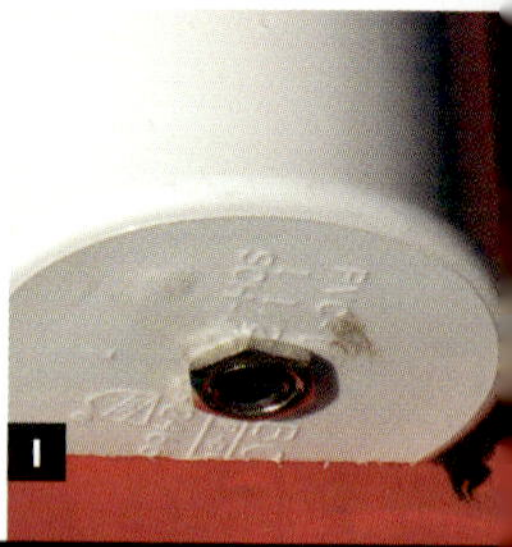

Fig. E: Attach the neck to the body. (1) Hex lag screws pass through wooden body into PVC pipe. (2) Countersink holes provide clearance for wrench and allow the screw heads to be flush with the body. (3) Lag screws are used as strap buttons. Fig. F: Prepare the pickup by cutting away the top of the plastic case. Fig. G: Attach a rubber foot in the disk center. Fig. H: Drill the cap. Fig. I: Sand one side flat.

7. Prepare the pickup.

Cut away the top of the piezo element's plastic case, starting at the hole in the center. Don't insert the cutters too far into the hole — you don't want to damage the piezo element. Cut away the plastic until you get to the side of the case, then start cutting around the side until the whole top comes off (Figure F).

Attach a ½" self-adhesive rubber foot onto the center of the piezo element (the brass shiny surface) or slice ¼" off the bottom of a rubber stopper and glue it to the piezo element with a dab of silicone sealer. The diameter of the rubber stopper must be small enough so it does not touch the plastic case (Figure G).

8. Install the pickup.

Drill a ⅛" hole in the center of one PVC end cap (Figure H) and sand one side flat so the pipe will rest against the body (Figure I). Cut the plug off the audio extender cable and feed it through the hole in the PVC cap. Tie a knot in the cable 1' from the end so it won't pull through the end cap. Feed a cable tie through the non-notched ¼" hole out the bottom of the pipe. Tape the cut end of the extender cable to the cable tie and pull it back through the non-notched hole (Figure J).

Cut and strip the cut end of the extender cable. The piezo has 2 wires: ground (black) and signal (red). The extender cable has 3 wires: ground (bare copper), left and right (both insulated). To confirm which is which, plug the extender cable into multimedia speakers and crank up the volume. When you touch the left channel wire, you'll hear a buzzing out of the left speaker. When you touch the right channel wire, you'll hear a buzzing out of the right speaker. When you touch the ground, you won't hear anything.

Solder the piezo ground (black) to the extender cable ground. Solder the right and left channels together to the piezo signal (red). Tape up the connections individually and gently pull them through the ¼" hole (Figure K).

9. Install the string.

Slice a wedge off the side of a single-hole rubber stopper. Push the ball end of the string through the notched hole and pull the string so the ball locks securely under the notch. While holding tension on the string, feed the other end through the wedged rubber stopper and into the hole of the tuning machine shaft (Figure L). Rotate the

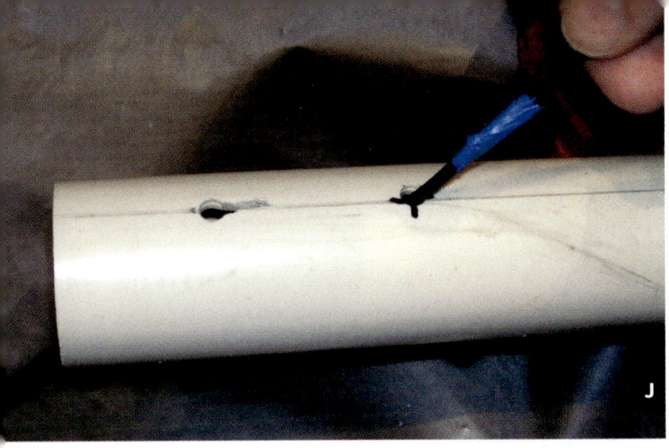

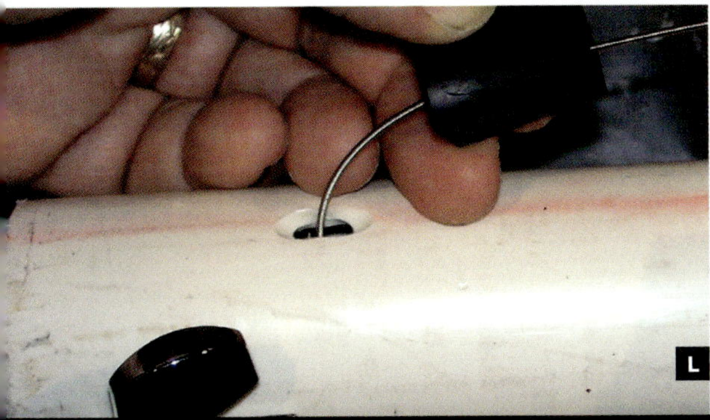

Fig. J: Feed the cable through the non-notched hole. Fig. K: Wire the pickup and gently pull the connecting wires through.

Fig. L: Install the wedged rubber-stopper nut. Fig. M: Make sure the guitar string rests on the center of the pickup's rubber foot.

tuning machine until the string is tight enough to stay in place, making sure the string rests on the center of the pickup rubber (Figure M).

Notice that we haven't glued the piezo element to the PVC pipe. The free-floating pickup guarantees that it is lined up with the string, preventing sideways forces that cause the pickup to distort. You now have a fretless guitar.

The gap between the string and PVC pipe should be uniform along the neck. If the string hits the PVC pipe when you pluck it, increase the size of the nut (wedged rubber stopper) and pickup rubber.

10. Add the frets.

Frets stop the string at fixed distances to produce discrete notes. If you want to add frets, tighten about 12 cable ties evenly spaced along the length of the neck. If you have a good musical ear, slide each of them to produce a chromatic scale (E, F, F#, G, etc.).

If you have no idea what this means, borrow a guitar tuner from a musician friend and turn the machine tuner so it produces an E note when the string is open (no fingers on the string). Press behind the first cable tie (closest to the machine tuner) and move the cable tie until an F note is produced. Move the next cable tie to produce an F# note, etc. Alternately, you could slide the frets to produce a major scale ("do, re, mi") or any scale you like, just like a sitar!

11. Learn to play.

Now that you have a guitar, it's time to play. Put on your favorite song and try and play notes that sound like the chords the band is playing.

Or get together with a musician friend and have him or her teach you scales and some basic riffs (series of notes). If nothing else, crank up the music, pluck away, and look cool. It works for music videos and it can work for you.

Tom Zimmerman is an inventor, educator, and researcher at the IBM Almaden Research Center who loves gadgets, LEDs, synthesizers, and hooking people up to computers.

TIME-BASED DEVICES

Transplant a heartbeat from cheap alarm clocks.

By Larry Cotton

Some projects need a nice, steady, low-voltage pulse at a constant rate. This is relatively easy to do with the ubiquitous 555 IC, but what's not so easy is keeping a 555-based circuit accurate to within a few seconds a month. For that, you're better off just gutting a cheap travel clock. This will give you an easy module that generates ultra-accurate pulses at 1Hz — something which, in my opinion, is more valuable than telling time.

Not just any clock will do, however: it should run on one battery (usually AA size) and have hands, not an LED readout. If it has a second hand, it must jerk forward every second (most do), rather than sweep. Clocks like this are easy to disassemble and cost less than $5 at some retailers. They all use the same basic mechanism: a 32,768Hz quartz-crystal oscillator, along with the electronic and mechanical parts that convert its oscillations to clock ticks.

Disassembly demands a bit of finesse. Although you're not tinkering with a fine Swiss watch, you are filleting some fairly delicate clock guts. What you're looking for is the electromagnet (solenoid), which turns the first gear in the gear train.

Your surgical procedure will depend on whether your project just needs the electronic pulse generator, or a clock with working hands that continues to tell time. If you just need the clock's ticker, you can disassemble without worrying about reassembly, and throw away the gears, hands, and even the electromagnet.

Here are the steps I followed with my clock. Yours may differ depending on its age. I've also successfully used wall clocks, which also come

Photography by Larry Cotton

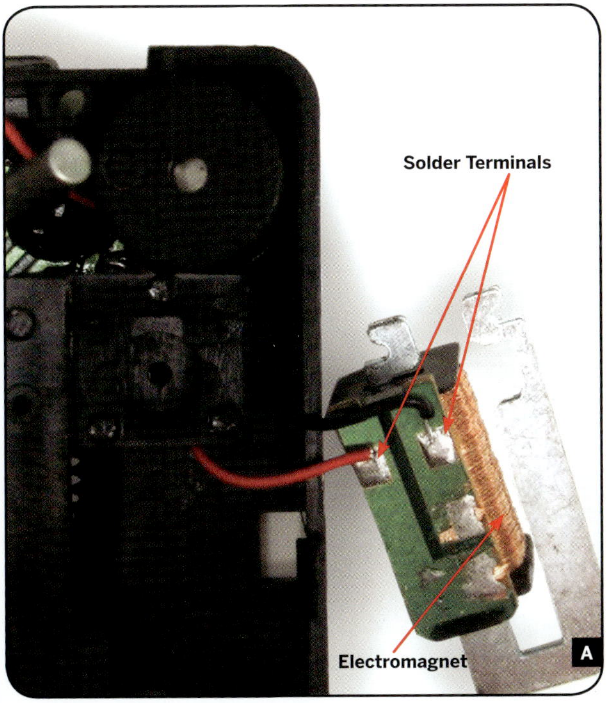

Fig. A: Lifting out the electromagnet reveals the two wires you need to splice into. Fig. B: Clearance hole cut through clock's front panel.

Fig. C: New wires to your own circuit connect to the electromagnet through the clearance hole in the now-bare clock face.

apart easily. The electromagnet in a wall clock may not be as readily removable, but it's a piece of cake to solder new wires to, and it will work fine with the electromagnet in place.

1. Remove the battery cover.
2. Remove the battery.
3. Remove the clear face cover.
4. Pull off the hands.
5. Remove a shaft nut and washer. This depends somewhat on the quality of the clock, and may not need to be done.
6. Pull off any setting knobs that attach to rear protrusions.
7. Remove clock face and mechanism. You may have to release the boxy little clock mechanism (roughly 2"×2"×½") from the main body of the clock; you may also need to remove a printed clock face in order to access parts, and make a wire-clearance hole. Read the remaining instructions to help you decide what to do.
8. Pry off the back. It's usually just snap-fitted together. Otherwise, carefully Dremel (is that a verb?) the case open. In some cases, the back piece holds tiny plastic gears in place, so they may fly out if you're not careful. But if you don't need the hands, don't sweat the flying gears.
9. Carefully lift out the little gear and the electromagnet. The electromagnet has fine copper wire wound on it, and the first gear in the train is surrounded by steel fingers. Two wires are attached to the magnet, usually red and black.
10. Discard the electromagnet. If you don't need the hands, you can do this by clipping or unsoldering its wires.
11. Pry the face off if you haven't already done so. Mine was attached with tape. You may have to route 2 new wires through the front of the clock.
12. With a Dremel and small bit, make a small hole in the plastic housing. You want to do this just below where the 2 solder terminals for the electromagnet's wires were. This will clear the new wires you will install. Alternately, if you have room, you may be able to route the wires around the circuit board and out the back.
13. Carefully solder 2 new thin, flexible, insulated wires in place. You can do this to either the electromagnet or just to the 2 wires. Polarity isn't important. You may have to mechanically attach your new wires with tiny loops before soldering.
14. Route the wires away from the clock. Go through the hole in the housing, if necessary. If you need the clock's hands, reinstall the first gear

D E

Fig. D: Moebius clock advances a single-sided paper loop with printed times. Fig. E: Dual-dial clock shows hours above and minutes below; hands swing back at the ends of their arcs.

Fig. F: Simple transistor circuit boosts clock pulse up to 5V, for driving a stepper motor or interfacing with other 5V-based electronics.

out, and reverse your disassembly.

The 2 wires you connected to the clock will put out a very short electrical pulse every second. Some clocks alternate a strong every-other-second pulse with a weak and less usable one. You can use a cheap NPN transistor such as a 2N2222 to amplify this pulse, to interface with the next stage of electronics. You can also add divider ICs, such as a 4017, to increase time between pulses to almost any length you wish.

I've transplanted tickers successfully into several alterna-clocks I've built. For my Moebius clock, I sent the amplified pulse into a stepper motor which advances a Moebius strip tape with times printed on it. For my dual-dial and arc clocks, I send the pulse to the motor through a BASIC Stamp microprocessor, which keeps track of the time. When it becomes 12:59:59, the code triggers a subroutine that quickly runs the hands (or arcs) backward and resets the time to 1:00 — which is always fun to wait for and watch.

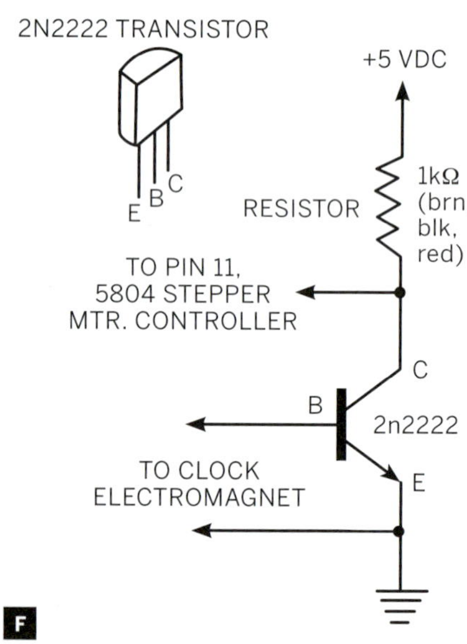

F

Larry Cotton is a retired power-tool engineer, musician, and part-time math teacher who just happens to have a bit of extra time on his hands in New Bern, N.C.

METAL ETCHING

Chemically carve detailed text or line art onto brass or copper. By Tom Jennings

I make forged, obsolete laboratory instruments as art pieces. To render intricate line art and text onto metal panels, I use the same chemical etching process that creates printed circuit boards. The etched areas are only a few thousandths of an inch deep, but they give a clear image that you can fill in with paint or enamel. If you're building the panels into an object, you should etch them before bending them or drilling any holes. This process only works on flat, solid metal surfaces.

Be warned: This is a highly variable process. I've done immaculately perfect panels with smooth 10-micron lines, but most often there are flaws in the materials, rough areas, loss of resolution, or other problems. Cleanliness and care get better results, but the trick is to go with the flow, and work with what you get.

Create the Transparency

You can create your artwork by hand or digitally, but it needs to be rendered as a mirror-image of your final image, in solid black on transparent film. Use black ink only, and print on the coated side, not the shiny side.

Illustrator, the GIMP, Photoshop, and other graphical utilities let you threshold any images into pure black-and-white, to make them suitable for etching. Then use Reverse to produce a printable mirror-image. You want the printed image to read correctly when you're looking at it with the ink-and-coating side down.

I do my artwork in Adobe Illustrator, and print it on HP transparency film in my Epson 777 inkjet. When my art is ready, I load the transparency film, hit Save, reverse the image, Print, then Undo.

Photography by Tom Jennings

MATERIALS

Artwork
Transparency film
Printer or copier to render your artwork onto the transparency film
Copper or brass plate Zinc will probably also work, but don't use iron or steel, and definitely not aluminum, which is dangerously incompatible with the etchant.
Lamp with 500W photoflood lamp bulb (3200° Kelvin) and reflector on improvised stand over work surface
Dishwashing liquid
Sandpaper and fine abrasives of varying grits up to 600 or finer You can also use red and white auto polish, cheap from any auto parts store.
Nitrile or other chemical-protective gloves
Protective eyewear
Pane of glass big enough to cover metal plate
Pyrex dishes (2) big enough to hold the metal plate
Piece of clean wood or other "carrier" for the metal plate so you can move it without touching it
A drawer or other clean, dry, and absolutely dark place to put the plate
Photo timer or stopwatch that's easy to read in the dark

CHEMICALS

All are available from GC Electronics (gcwaldom.com) or Minute Man Electronics (minute-man.com)
ER-71 negative acting photo resist liquid in a spray bottle
ER-8 photo etch developer
Ferric chloride etchant
Electroless tin plating solution (optional)

WARNING: The light-sensitive chemicals are flammable, and the ferric chloride etchant is extremely dangerous — it burns and stains skin, fabric, wood, and reacts violently with aluminum.

Polish the Metal Plate

To receive the photo-resist coating, the metal must be absolutely smooth and grease-free; a single fingerprint can ruin the surface.

1. Sand the surface with increasingly fine sandpaper, ending up with 600 or 1500 grit. Follow up with red and white automotive polishing compounds.
2. Wash the entire plate spotlessly clean with hand dishwashing soap, and dry with paper towels. Don't touch the surface!
3. Place the plate on a wood block or other carrier, bring it to a protected place, and let it dry. Be sure to handle the plate by the edges only. You can dry it in a clean, non-greasy oven at 120-150°F for 20 minutes, or leave it in a drawer for a few hours.

Set Up Your Work Area

The coating, exposure, and development steps need to be performed in a darkened room or under a red darkroom light, so it's important to find a good spot and set up ahead of time. And because solvents are involved, the area also needs decent ventilation. A kitchen sink by a back door at night would work.

1. In an area that's well-ventilated and can be made dark, set up the 3200° Kelvin photoflood lamp in the reflector, such that it shines straight down onto a flat tabletop, 18" from the table surface, plus or minus an inch. The lamp has a very short life (6 hours) and it generates a lot of heat. It exceeds the ratings of most cheap aluminum reflectors, but that was not a problem for me. It's not left on for long enough to do any harm that I noticed.
2. Put your photo timer or stopwatch someplace accessible and visible.
3. Clean and dry the sheet of window glass so that it's spotless on both sides.

Coat the Plate with Emulsion

This step coats the plate with a very thin light-sensitive film, and you need to do it in the dark. Be sure to keep the ER-71 liquid in its box when not in use.

1. Holding the prepared metal only by the carrier, tilt it, and pump-spray the surface with ER-71 photo resist liquid until it is soaking wet and excess liquid drips off one corner.

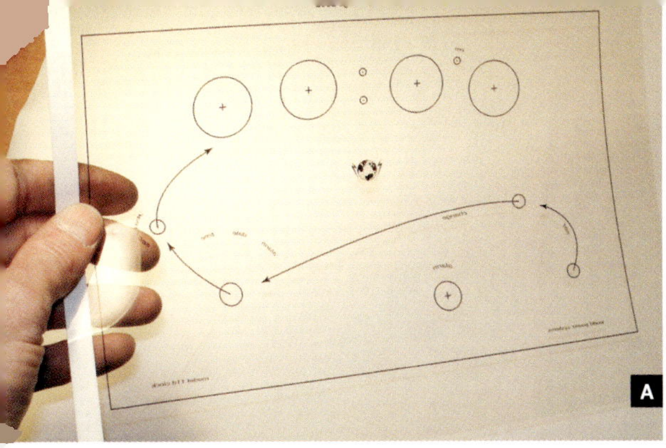

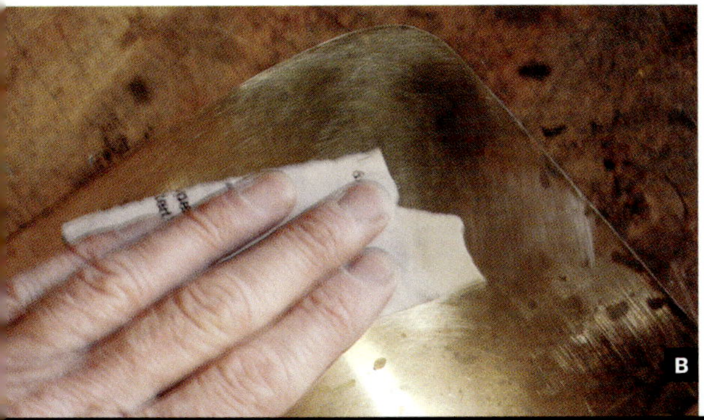

Fig. A: Print the etch image backwards onto a sheet of transparency film. Fig. B: Clean and polish the metal (copper or brass) surface. A single fingerprint can ruin the transfer.

Fig. C: Sensitize the metal by coating it with photo emulsion liquid. Keep the spray bottle out of the light when not in use.

2. Let the excess drip for 12 seconds, then hold it flat and look for 100% perfect wet coverage. If you see fisheyes, or the liquid pulls away from the metal, it wasn't clean enough. In this case, wipe off what you can, and finish cleaning with some ER-8 developer on a rag. Clean thoroughly and try again.

3. Put the plate in a dark, dry, flat, and level storage space, like a drawer. As the coating dries, it will level out and harden. Let it dry for 30 minutes.

4. A second coat isn't usually needed, but I've occasionally used one when the coating seemed too thin or etching didn't work the first time. Wait at least 1 hour before applying a second coat.

Expose the Plate to Light

This step also needs to be performed in the dark or under dim red light.

1. Place the metal plate on the table, photo-coating side up. If your carrier is very thick, raise the lamp to compensate, so that it's 18" above the object.

2. Place and carefully align your artwork onto the plate so that the image is righted. Because we printed the image reversed, the inked image now lies right up against the coated metal. That's good. If it were printed the other way, the transparency film would lie between the ink and the metal, which greatly lowers resolution.

3. Carefully place the cleaned sheet of glass over the transparency to hold it flat against the coated metal. Recheck image alignment.

4. Turn the 3200°K photoflood lamp on for 4 minutes, then switch it off. You may need to experiment with exposure time if you use a different lamp or height.

5. Keep the exposed plate in the dark until it's in the developer bath (next step).

Develop the Image

Developing the plate causes the parts that were exposed to the light to harden into a resist coating, and the parts that were directly under the black ink to wash away, leaving bare, exposed metal. You need to continue working in the dark until the plate is in the developer bath.

1. Place the exposed object in Pyrex dish #1, coated side up, and pour on enough ER-8 developer to cover it. Used ER-8 is fine for this step. Once the plate is covered, you can turn the lights on.

2. Rock the dish gently so that the ER-8 sloshes

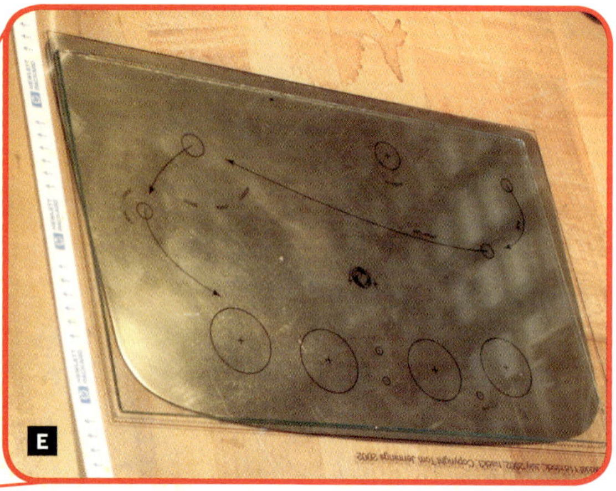

Fig. D: Expose the coated metal plate for 4 minutes under a 500W photoflood lamp, with the reflector positioned 18" above. Don't worry if the reflector isn't rated for a bulb that hot; it isn't on for long enough to matter.

Fig. E: During exposure, the transparency should be positioned with its inked side directly against the metal, pressed down with a plate of glass.

over the coated surface for 45 seconds.

3. Remove the object from the dish, draining developer back into the dish.

4. Place the object into Pyrex dish #2, and pour in enough new, fresh ER-8 to just cover the surface. The exposed coating is very soft and delicate at this point.

5. Very gently rock the dish for 10 seconds. Lift the object out of the dish and let it drain, handling it only by the edges.

6. The coating is now no longer light-sensitive, but it is invisible and extremely delicate until it's dry. Do not touch! Set the object aside to dry fully, for 1 hour.

NOTE: ER-8 developer is reusable an unknown number of times, probably a dozen or more. The second brief dip in fresh ER-8 just rinses off any crud that was in the used ER-8 from the main step. This stretches the life of the developer. I keep my used ER-8 in a clean glass jar.

Find and Fix Flaws in the Coating

If the exposure process failed, etching will just roughen the plate's surface and you'll have to sand it all off and start over. Here's how to make the coating pattern visible; then check it and correct any flaws before etching.

1. Fill a Pyrex dish with water and add just a few drops of etchant.

2. Immerse the plate in the dish; the coated areas will remain shiny while the open areas will faintly etch, making the coating pattern visible.

3. Compare the coating pattern to your original artwork and look for areas where the plate should be etched, but are still shiny (coated), and vice versa. For areas that should be etched but weren't, use a knife to carefully scrape the coating away. For areas that should be coated but are exposed, cover them using a paint pen or pieces of tape trimmed to shape. Correcting a few flaws this way is fine, but if large portions of your design didn't make it through, you're better off repolishing the plate and starting over.

4. You can discard this very weak etchant solution by pouring it down the drain and following it with copious amounts of water.

Etch the Metal

This step completes the process. The ferric chloride etchant eats away metal from the object anywhere that coating was not exposed to light.

Fig. F: After exposure, develop the plate in 2 baths of ER-8 developer. Used ER-8 liquid is fine for the first bath. Fig. G: Etch the plate in warmed ferric chloride solution, wearing gloves and eye protection at all times.

Fig. H: Monitor progress during etching by lifting the plate out of the etchant and inspecting the surface.

WARNING: FERRIC CHLORIDE ETCHANT IS VERY TOXIC AND CORROSIVE.

It eats people, cloth, and wood, as well as metal. It will kill your pets and your plants. You cannot dump it into any sewer system, since it eats through pipes. Store it in its original bottle, and when it's spent, take it to a chemical disposal site. When ferric chloride contacts aluminum, it makes heat, splatters everywhere, and ruins the metal.

These are not wimpy warnings. Etchant is dangerous stuff. Wear chemical gloves and protective eyewear, and have someone around in case of emergency.

1. Cover the bottom and sides of the plate with packing tape or spray paint, to protect it from being etched.
2. Warm up the etchant to about 120°F by putting the sealed container into a bucket of hot water for 30 minutes. Etching goes faster at elevated temperatures.
3. If the metal plate is heavy, warm it to about the same temperature. (If it's thin, don't bother.)
4. Put the plate into a Pyrex dish and pour undiluted etchant to cover it completely, as much as possible without overflowing.
5. Rock the dish slowly but constantly to etch away metal. I put a pencil under the Pyrex dish to make it easier.
6. The amount of time is highly variable, and depends on how much metal you want removed. With chemical gloves on, you can lift the object out to inspect it. Etching for a full hour might get you 0.02" of depth, which may not sound like much, but it's quite striking and effective. You can feel it, and it's plenty deep to hold paint.
7. When done, remove the object from the bath and let it drain completely into the Pyrex dish. Then rinse it off in a sink with copious flowing water, and wash with hand dishwashing soap.
8. Your item is done! The resist coating can be removed with a bit of lacquer thinner on a cotton ball, or with mild abrasives like the polishing compounds used earlier.

Plating and Painting

So far, the photographs in this article have been showing the brass plate from my Model 11d Nixie clock. Another example, probably the best etched

I

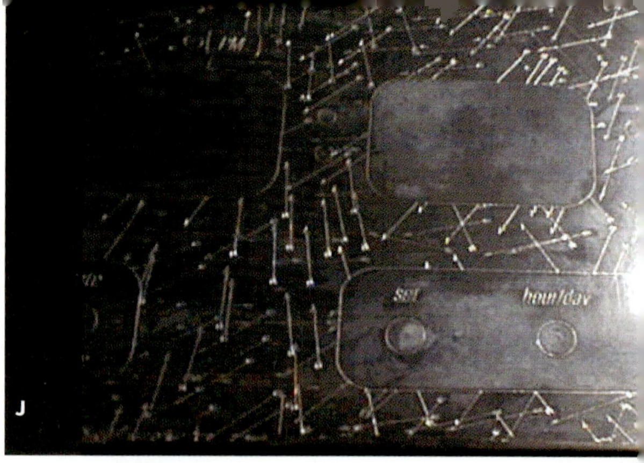

J

K

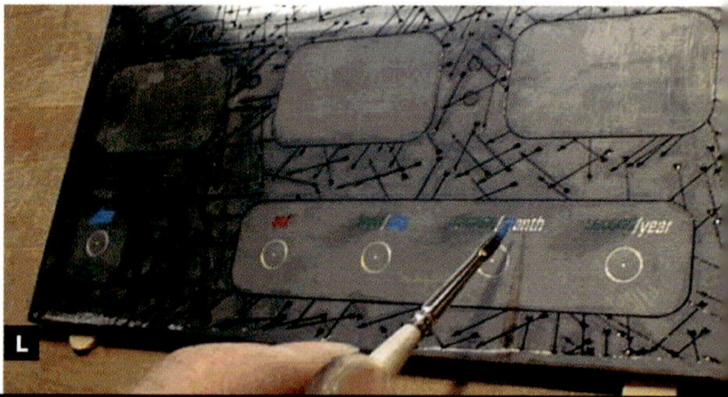

L

Fig. I: Another panel, copper this time, fresh out of the etchant bath and washed off. Fig. J: An electroless plating bath tin-plates the etched copper panel, giving it a more neutral color.

Fig. K: Filling in the etched arrow details with a fine paintbrush and black enamel paint. Fig. L: Filling in the color detail with more enamel.

panel I've done, and one of the more detailed, is from my Model 11b Nixie clock.

This was a copper plate, which I tin-plated after etching using electroless plating solution. After removing the plate from the tin bath, the tin plating was dull.

I polished it with dry cotton balls, then used a small paintbrush to fill the etched lines with enamel paint (One Shot brand). I used tiny scraps of paper to carefully wipe the enamel filling flush with the surface, then polished off the excess with small scraps of plain copy paper.

After the paint dried, I shaped the panel by drilling holes for the switch and buttons, and filing out the rectangular areas for the display.

I also used tin-plated copper, painted with black enamel, for my Model 13, shown at right. Instead of using Nixie tubes, this clock shows the time and date with an analog needle gauge.

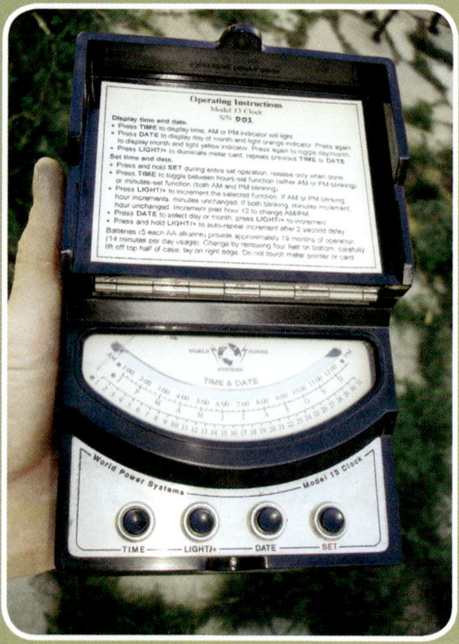

Tom Jennings (tomj@wps.com) is an artist with a background in technology.

ELEGANT PLYWOOD COFFEE TABLE

Making furniture with no nails, screws, or glue.

By Andy Lee

Photography by Sam Murphy

I've been interested in designing furniture for a long time. When I needed some of my own, I decided to build a set out of ¾" plywood. I chose to work with birch veneer plywood because it reminds me of Scandinavian furniture, which has a clean, elegant style that I love.

Having done a lot of moving in the last few years, I also wanted to make the furniture easy to assemble without any fasteners. To accomplish this, I designed each component piece with slots that mate with one another and hold the whole piece together. Not having fasteners makes the furniture easy to pack up and transport — each item breaks down into a stack of flat boards. Finally, I didn't want to create any waste, so each piece is cut exactly from a half sheet of plywood.

Here's how to make an elegant coffee table with a few simple cuts, using a drill, jigsaw or circular saw, clamps, and other familiar tools.

1. Get the wood.

Obtain a 4'×4' piece of ¾" plywood that is in good condition and free from warping and knots. You'll also need four 1"×¼" wooden pegs. Different grades of plywood vary in price, ranging from rough particleboard to finely veneered and pre-sanded. This project will work with any type. I chose Baltic birch veneered 5-ply from a local

Inset: Pattern and dimensions for table's 4 component pieces. Tabletop, brace, and 2 legs are all cut from a single 4'×4' sheet of plywood, with no waste.

hardware store. If your sheet does have a knot or two, plan your production so that they wind up on the underside, rather than facing out. After sanding and painting, the piece will look great.

SAFETY FIRST! Whenever working with or handling wood it's a good idea to use gloves to protect your hands from getting splinters. Wear safety goggles and hearing protection while using any tools, and always secure the material you're working with to a stationary object using clamps or a vice. It's best to work with a partner, and it also makes projects more fun.

2. Measure the wood.

Referring to the pattern above, measure and mark your plywood along lines 1, 2, and 3. These divide the sheet into 4 pieces: a tabletop, a brace, and 2 legs.

To ensure accuracy in measuring, the trick is to measure from 2 or 3 different directions and make sure that they all yield the same result. I often measure cuts I'm about to make from opposite ends of the plywood, and then use a square to ensure their proper angle against the edge of the material.

TIP: To make clean, straight cuts with a handheld jigsaw or circular saw, clamp down a fence at either end of the material and run it along the edge you want to cut. A fence is simply a straight, solid object that provides an edge for your tool to rest against.

3. Cut the wood into pieces.

These first cuts break the unwieldy 4'×4' sheet of plywood into more manageable pieces. I used a handheld jigsaw.

Cut 1 Halve the plywood sheet to produce two 4'×2' pieces. One piece will be the tabletop and the second one will become the legs and the brace that holds the other 3 parts together.

Cut 2 Make a cut parallel to the first that's 5"-10" in from the edge. This creates a narrow 48" piece that will serve as the table's brace. The wider you make this piece, the lower and more stable your table will be. I made a taller table by cutting a 5" brace piece.

Cut 3 Clamp the larger piece left over from Cut 2, and cut it evenly in half, perpendicular to the first 2 cuts. With my table's measurements, this turned a 19"×48" piece into two 19"×24" legs.

When finished, the pieces of the table should look like this. Disassembled as shown in the inset and stacked together, they're a cinch to store and transport.

To assemble the table, slide the slots in the legs into the slots in the underside of the brace, as shown here. Then pop on the tabletop.

4. Cut the slots.

Cut the 4 slots, following the measurements shown on the previous page. Note that although the plywood is labeled as ¾", its actual thickness may vary. So for each slot, you should first measure the thickness of the plywood it will fit around (calipers are handy for this), and make the slot a few hundredths of an inch larger.

There are several ways to cut slots with handheld tools. One method is to use a ¾" hole saw or paddle drill bit and drill a hole at the inside end of the "soon-to-be slot." Then cut 2 parallel lines, tangent to the hole, out to the edge of the board.

Another way is to drill holes at the inside corners of the slot that are just large enough to admit a jigsaw blade, then jigsaw out from those.

The third method is to cut a slot straight out using a router and a ¾" router bit. This is the best method, provided you have the equipment.

5. Test assemble, and set the pegs.

Slide the legs and brace together, to make sure they all fit. Along the top edge of each leg piece, measure 2" in from each end and mark a point centered on the edge of the plywood. Disassemble the pieces, and drill ¼" holes at each of these 4 points, ½" deep. On the underside of the tabletop, mark 4 corresponding points and drill ⅜" receiving holes for the pegs ½" deep. I marked locations by sliding pieces of pencil into the leg holes and centering the tabletop on top.

6. Finish the wood.

You have many creative choices when it comes to finishing your furniture. You can sand the edges, finish them with a router, or level them square. You also need to seal the wood, especially if you intend to use this as a coffee table where you will serve beverages. I used a water-based stain and 3 coats of high-gloss finish for the product, but there are other options. Ask the friendly people at your neighborhood paint store for advice.

Bonus: Now that you know how to make an awesome coffee table, you can use the same general method to make a bench or a desk or a dining room table or a stool. Go to town!

Andy Lee (andyleedesign.com) is a mechanical engineer and product designer. He specializes in consumer and household goods.

THE BYTELIGHT

Make a high-tech mood light from a fluorescent lamp and 54 obsolete SIMMs. By Ross Orr

Several years back, I had a volunteer gig with a local charity. My job was to take old Macs that had been donated, and jolt them back to life so they could be resold at our weekly rummage sale.

In those days, the low-end machines often came in with 1 or 2MB of memory — that M is no typo, you youngsters! — and I considered it a great victory if I could scrounge enough memory to upgrade them to 4 or 8MB. (I can also remember fitting System 6 on an 800K floppy — but Grandpa will save that story for bedtime another day.) Anyway, we routinely yanked out 256K SIMMs doing these upgrades, and I eventually accumulated a whole sack of them.

I knew these chips were utterly useless, but could never bring myself to throw them out. I kept dwelling on the thousands of dollars that their original purchasers had undoubtedly spent on them. Plus, these little wafers were kind of beautiful: each had its own subtly different color, delicate circuit traces, and cryptic labeling. They were almost like jewelry. I felt I needed to do something to honor their now-defunct preciousness, and thus the ByteLight was born.

The idea is simple, and can be used with any "interesting" piece of circuitry that has outlived its usefulness, as long as a bit of light shines through it. You just need to construct a shallow light box. Fronting it with diffusing plexiglass lets you glue down multiple circuit boards using silicone sealant. You can buy white plexi, but I just used clear scrap I had on hand. Sanding both sides with fine sandpaper on an orbital sander gave it a nicely frosted surface.

Photography by Ross Orr

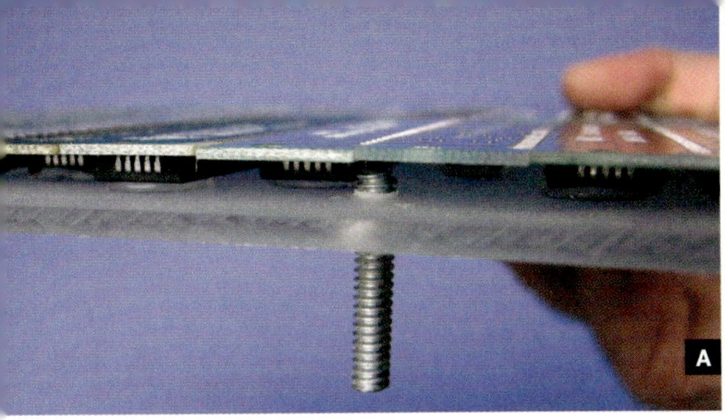

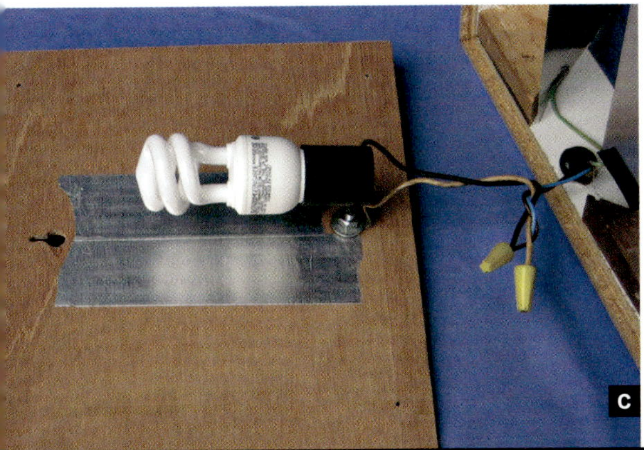

Fig. A: Silicone glue attaches wafers to plexiglass, which is held above the rim of the light box by a threaded rod. Fig. B: Wafers are glued chip side down to show tracery.

Fig. C: Removable light box back with lamp socket and compact fluorescent. Fig. D: The sheet aluminum lining stapled into the box reflects and diffuses light.

I made my box from ½" plastic-faced particleboard from an old sink cabinet. (A tip for neat mitered corners: after cutting the bevels, lay the boards flat, end to end, and run packing tape across each joint. Apply glue, and fold the corners into place. The tape holds the joints in position.)

The front edges of the box are cut away at an angle, so they won't cast shadows on the frosted panel. I added aluminum reflectors to even out the light, but that's optional. For easier access when it's time to change bulbs, I bolted the lamp socket to the removable back panel of the box. I also drilled a keyhole slot centered in the back panel, to make it easy to hang the finished box.

Use fluorescent lamps for your light source, to keep things cool. I used a 10W screw-base fluorescent, which puts out the same number of lumens as a 40W incandescent. Even so, I left a small gap between the plexiglass and the box to let heat escape. The box is sized to be 1" smaller than the height and width of the plexi, which eclipses the gap with a nice floating-plane look.

The trick to making the front float is to thread stubs of 10-24 rod into holes tapped in the plexiglass. Then drop these into holes in the box perimeter, drilled oversize and filled with silicone. I used kitchen matches to hold the gap spacing while waiting for the silicone to cure.

The power cable, appropriately enough, is also recycled from a computer. I snipped off its IEC connector, and wired it up to a screw-base lamp socket. The blue (or white) neutral wire in the cable should be connected to the shell of the socket. If your cord has a green or bare ground wire, you can connect that to any metal parts of the light box. Tie the cord into a knot inside the light box, so an accidental yank can't pull it out.

If you're laying out a grid of small boards like these SIMMs, it may turn out they aren't exactly identical in size. So test your layout beforehand, and sand down the edges of the oddballs as needed. I decided I preferred their look with the component side down, to emphasize the patterns of the circuit traces, leaving the RAM chips themselves as ghostly outlines.

The result is a softly diffused green light — a nice ambience for late-night relaxing. And I'm pretty sure I'm the only person on my block with a 13.5-megabyte night light.

Ross Orr hacks low-tech gadgets and invasive plants in Ann Arbor, Mich.

DIY CIRCUITS

Photograph by Sparkle Labs

DIGITAL CLOCK

Programming PIC microcontrollers, part 3.

By Sparkle Labs

In parts 1 and 2 of our microcontroller primer (*MAKE, Volume 04, page 158, "Microcontroller Programming," and Volume 07, page 149, "Hello, World"*), we got a PIC chip development system up and running, covered basic inputs and outputs, and left off by showing how to use a PIC chip by itself to control a simple pushbutton counter. In this article, we'll show how to use a PIC along with other chips to make a digital clock.

Our two peripheral chips are the DS1305 Real Time Clock (RTC), which keeps track of the time, and the MAX7221, an LED display driver. Both are manufactured by Maxim Integrated Products, with datasheets available at maxim-ic.com. We'll use the PDIP package versions, so they'll fit into a breadboard.

Just like our microcontroller, these special-purpose ICs have settable 8-byte registers that determine how they function. But peripheral chips don't store and run programs on their own, so we can't upload code to them and set their registers directly. Instead, we'll program our microcontroller to tell these chips what to do. The chips' datasheets explain how the microcontroller can communicate with them, as well as how they need to be connected into our circuit.

ICs exchange data over numerous communications protocols. Our clock will use Serial Peripheral Interface (SPI), a simple serial binary protocol that requires just four pins on the microcontroller, plus one more for each additional peripheral device. One pin sends data out to the peripheral(s), one pin receives data, a clock pin pulses high and low to mark each new send/receive cycle, and the latch (or load) pins, one for each peripheral, switch the peripherals on and off. These pins and their connections are collectively known as the data bus.

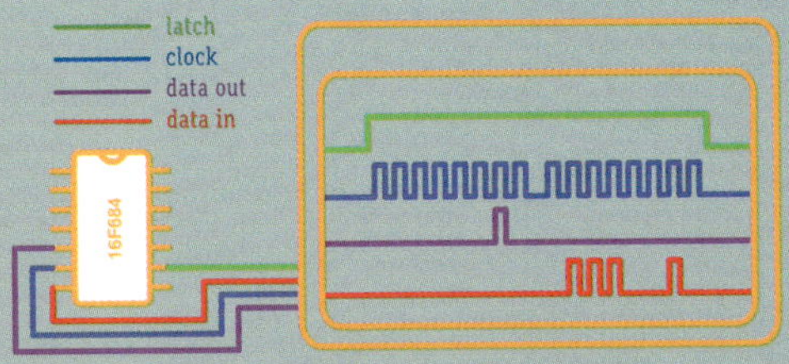

The LED Display

First, let's look at how to hook up our display driver to run the displays that will show our clock's time. The MAX7221 datasheet explains that this IC is capable of driving up to eight 7-segment digit displays (or up to 64 LEDs in an 8×8 matrix). Referring to the pin description on page 5 of the datasheet, we see that pins 1, 12, and 13 are the data-in, latch, and clock pins, respectively; we'll connect these to I/O pins on the PIC. Pin 18 connects to power through a resistor, the value of which limits the peak current that the chip supplies to the LEDs, to match the LED's power needs and desired brightness. Pin 24 is data out (which we are not using). Pin 19 connects to +5 volts, and pins 4 and 9 go to ground. Physically, pins are numbered on ICs counterclockwise from pin 1, just to the left of the notch on top.

The remaining 16 pins connect to the LED displays themselves. Eight of these pins are the digit drivers, each of which connects to the common cathode of a single-digit display and sinks the current that runs to the segments of the digit that are "on." The other 8 pins are the segment drivers, which supply current to all of the segments that share the same position on the displays.

To turn a digit on, you set the digit driver low for that digit. Then current will flow through from the segment drivers that are set high, and light the segments. Meanwhile, the driver scans through all the displays repeatedly, changing segment pins and refreshing digits fast enough that the LEDs look like they're always on.

Our clock needs 5 digits, 4 numbers and an AM/PM indicator, which we'll sink from the chip's digit driver pins 2, 3, 6, 7, and 11. We'll also use 2 individual LEDs in series for the colon character separating hours and minutes.

For our numeric displays, we'll use 5 Kingbright SA56-21GWA green LED packages. As their datasheet explains, these have 10 pins each. The middle pins on each side (pins 3 and 8) are the common cathode, which you connect to a digit driver on the 7221. The remaining pins connect with the identically positioned pins on other displays, and are collectively fed by the corresponding segment driver. The 2 LEDs for the colon character, meanwhile, are always on, connected between power and ground.

Testing the Display

With our PIC, the 7221, and the display all wired up and powered, we can write some code to talk to the driver and see if it works. Table 1 on page 6 of the 7221 datasheet shows us that the driver takes serial input in 16-bit chunks, and it ignores the first 4 bits of each. The next 4 bits specify which register we want to send the data, with the registers corresponding to the individual digit displays, plus special-purpose registers that control meta-states such as shutdown, display test, brightness/intensity, and decode modes. The last/least significant 8 bits of each 16-bit data chunk carry the data itself, as either individual on-off bits for each segment, or else (in decode mode) 4-bit binary numbers which the driver converts into readable numerals.

At the top of the code, we need to send some startup and initialization data. To send values to individual pins, for example to set the latch pin low or high in order to activate or inactivate the chip, we use the BASIC functions `low` and `high`. To send chunks of serial data to the driver's registers, we use the subroutine `shiftout`, which takes as arguments the data pin, the clock pin, the mode, and the value (the bits themselves). Our mode will always be 1, which specifies Most Significant Bits (MSB) first.

For example, the datasheet explains that after powering up, we need to set the shutdown register (address 1100) to normal operation (1). We do this with the call:

```
low s7221 ' make the latch pin low
shiftout dout, dclock, 1, [%00001100, %00000001 ]
high s7221
```

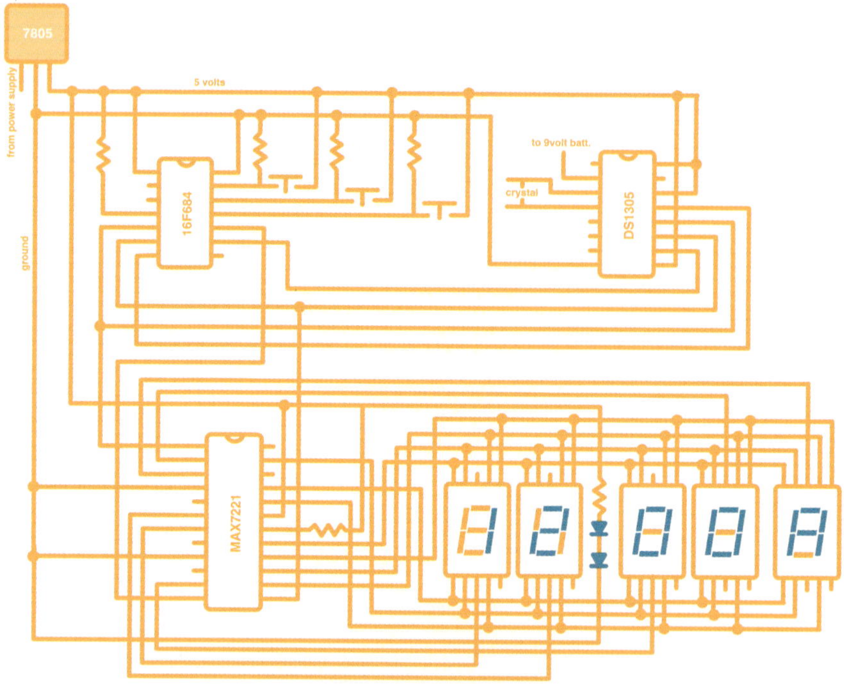

Digital clock wiring. ICs used are 16F684 microcontroller, MAX7221 display driver, and DS1305 real time clock. Circuit also includes a 32.768kHz watch crystal, 7-segment LED packages (5), and a 7805 voltage regulator. All resistors are 10KΩ, and the two single LEDs separating hour and minute displays are 200Ω. For fun we used tilt sensors along with regular buttons for the Set, Hour, and Minute switches, shown at right of the micro.

After our code makes a few more initializations, we can write information to the display digit-by-digit by calling the function shiftout with the register address of the digit you're writing to, and its new value. If the decode mode register is set to decode that digit, it will look at only the last 4 bits of the data and convert it to a numeral 0-9. Otherwise it interprets each bit as an on-off setting for each of the 8 individual segments.

We set our driver's Decode Mode register to decode just the first 4 digits (DIG 0-3), and our 5 displays are arranged physically in the order 3, 2, 0, 1, 4. At far right, DIG 4 serves as the AM/PM indicator, and because it isn't decoded, we can send it data segment by segment to make an "A" or a "P" character. For testing, we can set our clock to read "1:00" by calling shiftout to send register %0011 (DIG 2) the data %0001 and send registers %0001 and %0010 (DIG 0-1) the data %0000.

Now we have a clock that's correct twice a day. Let's add another chip to make it correct all day.

The Real Time Clock

The DS1305 Real Time Clock (RTC) keeps track of the year, date, day of the week, hour, minute, and second. It has registers that we can read the current time from, or write to, for resetting the time. It also has registers for 2 alarms, and even 96 bytes of RAM for data storage. To run, all it needs is power, ground, and a 32.768kHz watch crystal connected between pins 3 and 4.

To see how to connect the chip to our circuit, we run through the pin description on page 3 of the online datasheet. Pin 1 goes to ground, as we have no rechargeable battery. Pin 2 connects to a +9V battery. Pins 3 and 4 connect to the crystal. Pins 5-7 and 15 we won't use. Pin 8 goes to ground. Pin 9 connects to power, to tell the chip to use SPI protocol. Pins 10-13 are for the SPI bus, and connect to the microprocessor's latch, clock, output, and input pins, respectively. Pins 14 and 16 connect to power.

As with the 7221, you communicate with the DS1305 by sending the address you want to talk to, followed by the data. First, we initialize the chip by sending to the control register ($8F hex, or %10001111) the data %00000111. This sequence of bits starts the oscillator, write-enables the other registers, and enables both alarms, which we may want to use later.

Next, we need to think about how the user interacts with the clock. On power-up, the display shows 12:00 AM. Three buttons, Set, Hour, and Minute, work by switching 3 digital input pins on the microprocessor. Our code defines 2 modes: Display and Set. In Display mode, the PIC reads time data from the RTC, then converts it for display and transmits it to the display driver.

In Set mode, the two other buttons increment the time shown on the display, working like keyboard buttons — advance one with each press, and continuously if held down. As the PIC advances internal hour and minute variables, it updates the display with the new time, and when it leaves Set mode, it writes the new time out to the RTC. On our finished clock we used tilt switches for the Hour and Minute buttons, so you set the time by physically picking it up and steering it around.

The display driver reads digits differently from the way the RTC represents time, so our code needs to convert between the two. The display driver wants each digit in "binary coded decimal," where each digit is represented by a byte (even though it really only requires 4 bits). The RTC uses 4-bit binary coded decimal for separate hours, minutes, and seconds registers, and bits 5 and 6 of the hours register determine if the clock is in 12- or 24-hour mode, and in 12-hour mode, whether it's AM or PM. So, for example, the time 12:35 PM is represented in the RTC hours and minutes registers as:

```
%01110010 %00110101
```

But needs to be sent to the display driver as:

```
%00000001 %00000010 %00000011 %00000101
```

Below is a routine for converting the RTC time into display digits:

```
digit0 = clockmin & $0F
digit1 = clockmin >> 4
digit2 = clockhour & $0F
digit3 = (clockhour & $EF) >> 4
PMlight = clockhour.5
```

`digit0 - 3` are the variables we use for the display. `clockmin` and `clockhour` are the variables in which we store the data from the clock. Rather than using arithmetic operators, we can perform some binary-fu with the bitwise operator AND (&) and SHIFT (>>), which moves bits to the right and drops the lower ones out. By using these with the hexadecimal numbers `$0F` (15, or `%1111`) and `$EF` (239, or `%11101111`) as masks, we have all the hour and minute conversions we need. The last line of the routine looks at bit 5 of `clockhour`, which is high (1) for PM times in 12-hour mode.

Besides converting real-time clock values for the display driver, the other main thing our code needs to do is handle Set mode, which is easy once you can capture the button presses. You can see how that's done in the project code online.

Alarms and Other Enhancements

Moving forward with our clock, we will want to add an alarm or two. To do this, you need any device that can make noise, and a way for the user to set the alarm. We may want to show the date and make the alarm settable for specific days of the week, which the RTC will support. Perhaps we will also add a light detector, to dim the display at night. And we certainly need a nice case and some pretty buttons.

These are the basics of using a PIC with peripheral devices. There are thousands of such devices available, including accelerometers, temperature sensors, digital potentiometers, and storage devices. In this series of articles we have focused on PICs and PicBasicPro, but you can apply much of the information to other microcontrollers and compilers. Have fun!

Datasheets

MAX7219-MAX7221 display driver
makezine.com/go/driver

DS1305 Real Time Clock
datasheets.maxim-ic.com/en/ds/DS1305.pdf

SA56-21GWA/A single digit numeric display
makezine.com/go/display

Materials list, development environment, code, additional diagrams, and suggested reading online at makezine.com/09/diycircuits_clock.

Amy Parness and Ariel Churi work at Sparkle Labs. Amy is a digital artist and product designer who pets her cats while working with new technologies and materials. Ariel eats chapati and creates "hi-tech, hi-touch" toys and environments for all ages.

TV SET SALVAGE

Nothing good on TV? Well, there's plenty of good stuff *in* a TV. By Thomas Arey

During the late 1950s and early 1960s, electronics hobbyists found a wealth of parts for projects inside discarded television sets. It was considered a point of pride in amateur radio circles to homebrew a transmitter out of a broken "boob tube." Of course, those were the days of 6146B sweep tubes, big 1-watt resistors, and transformers that would break your foot if they fell off your workbench. Occasionally, if you were really lucky, that scrounged set would turn up a couple of those new transistor thingies.

Technology marches on, and the insides of more modern TVs turn up a very different parts complement, lacking in any "hollow state" components except the picture tube. As a radio hobbyist, I wondered: Is it still possible these days to take a discarded TV and salvage the stuff that radios (and many other projects) are made of?

Thanks to a 10-year-old General Electric portable I found on a dumpster dive, I happily found out that the answer is yes.

First, a few words of warning. Do not take a television apart unless you are comfortable handling both high voltages and potentially exploding glass. The power supply in a traditional TV presents 30,000 volts when energized. Even unplugged, the capacitors in the power circuit can hold a charge for months. Also, the picture tube is a large glass vacuum container. If it is dropped or cracked it can shatter into thousands of nasty shards. Wear eye protection and gloves when handling the tube. That said, if you are still ready to own your TV like a true maker, let's go!

Removing the back of my GE set revealed one large circuit board at the base and a smaller one on the back of the picture tube. A thick red high-

Photography by Thomas Arey

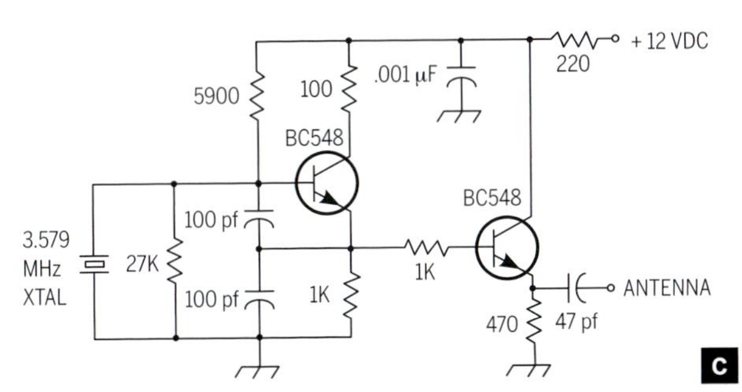

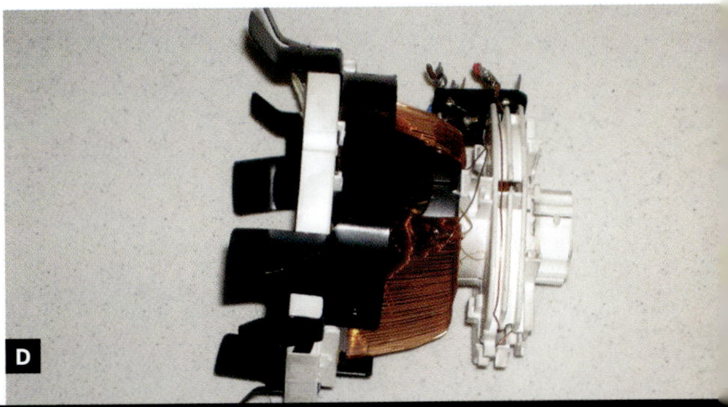

Fig. A: Opened TV with 2 circuit boards. Thick, red wire drives the tube from a high-voltage power supply, lower right. Fig. B: Circuit boards snapped out of the opened television without any unscrewing.

Fig. C: Colpitts oscillator circuit uses TV components to generate a sine wave you use as a broadcast carrier. Fig. D: The flyback transformer, which pulls the tube's beam along its scan pattern, has lots of winding wire.

voltage wire connected the power supply to the tube's anode cap. Again, be careful here. Cracking the case and liberating the PC boards required only a Phillips screwdriver. The boards themselves were "snap fit" into place without screws.

The PC boards yielded far more parts than I expected. There were only 3 proprietary integrated circuits; the rest of the circuitry was all discrete, standard components. At the time of this set's design, surface-mount technology had not yet kicked into high gear, so all the parts are easily usable for hobby projects, except fora few large SMT resistors.

I found 18 BC548 (2N2222) and BC558 (2N3906) transistors; both are commonly used in simple oscillators and transmitters. I also found 15 diodes and half a dozen semiconductor components of other types, plus an array of ¼-watt resistors, and ceramic disk and electrolytic capacitors. The RF chokes and slug-tuned inductors showed promise for future experiments, as did the infrared sensor for the remote control.

The power supply circuit included a mains-to-12V transformer and a pair of heavy-duty voltage regulators, and the flyback transformer on the back of the picture tube had lots of wire the perfect size for winding RF coils.

There was one great prize for any radio hobbyist, a "color burst" crystal with a frequency of 3.579MHz, which lies in the CW (continuous wave, used for Morse code) portion of the 80-meter ham band. There were also 4MHz and 8MHz crystals that have potential for other projects.

Using just the parts from this TV, there are many designs for radio transmitters you could build. One example is based on a Colpitts oscillator, which uses two BC548 (2N2222) transistors, a color burst crystal, and some resistors and capacitors. Add a Morse code key or other momentary switch at the oscillator's power point and a ¼-wavelength stretch of wire at the antenna point (65' 3" for a 3.579MHz signal), and you have a low-power radiotelegraphy station that will broadcast your dot-dash signals 25 miles or more, depending on conditions. Just remember to get your amateur radio license before you press the key down.

T.J. "Skip" Arey N2EI has been a freelance writer in the radio/electronics hobby world for over 25 years and is the author of *Radio Monitoring: A How To Guide*.

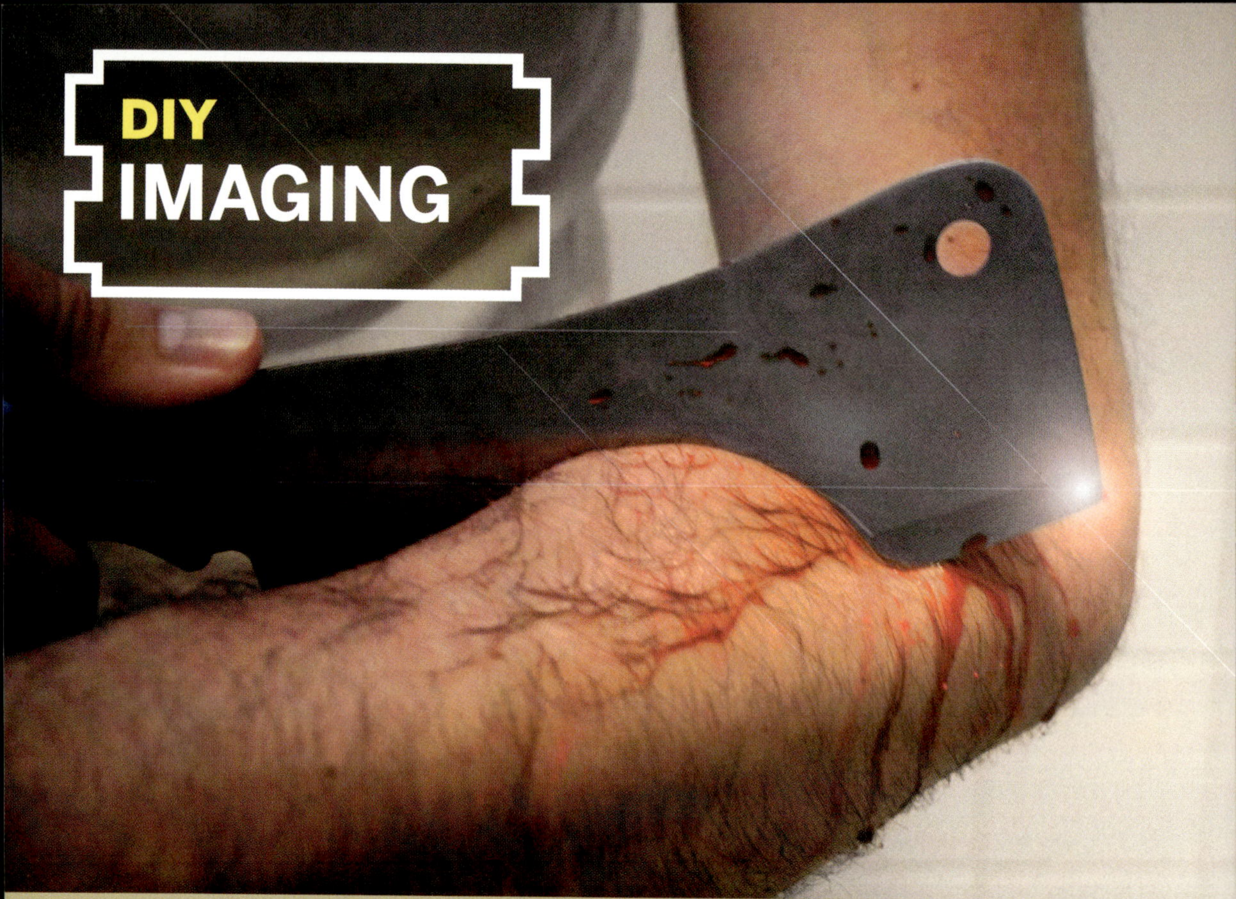

ACTION MOVIE EFFECTS

Shoot a fight scene with a blood-spurting knife wound and a head smashing through a window. By Zack Stern

The perfect action scene needs combat, gore, and at least one actor going through a window. Here's how I shot a scene with all these ingredients, from making the special effects to shooting and editing the footage.

Movie Glass

We wanted one actor to shove another's head through a kitchen window. Who wouldn't? We found a good location that had a single-pane window, 30"×26", which we could swap out for our own fake-glass concoction. We made a harmless shattering window pane by casting sugar glass in a wooden frame (recipe next page). If your window is much bigger and has several panes, you can add thin balsa divisions between the panes. These will hold everything together until it's shattering time.

First, glue together a rectangular frame to fit your window, using 3 layers of thin wood. Make all 3 layers flush around the window's outer perimeter, but around the inner edge, sandwich the front and back layers around a narrower middle layer. This leaves a gap for the molten sugar to flow into. Glue the frame together and press the corners under weights until it dries (Figure A).

Put the frame on top of a sheet of cardboard that's larger than the pane, and cut more cardboard to fit snugly in the frame and match the thickness of the bottom layer of wood. Cover the cardboard in the frame with aluminum foil (Figure B). Be careful to keep the foil flat; even small wrinkles in the foil will be cast into the sugar glass.

Cook the water, corn syrup, sugar, and coloring

Photography by Zack Stern

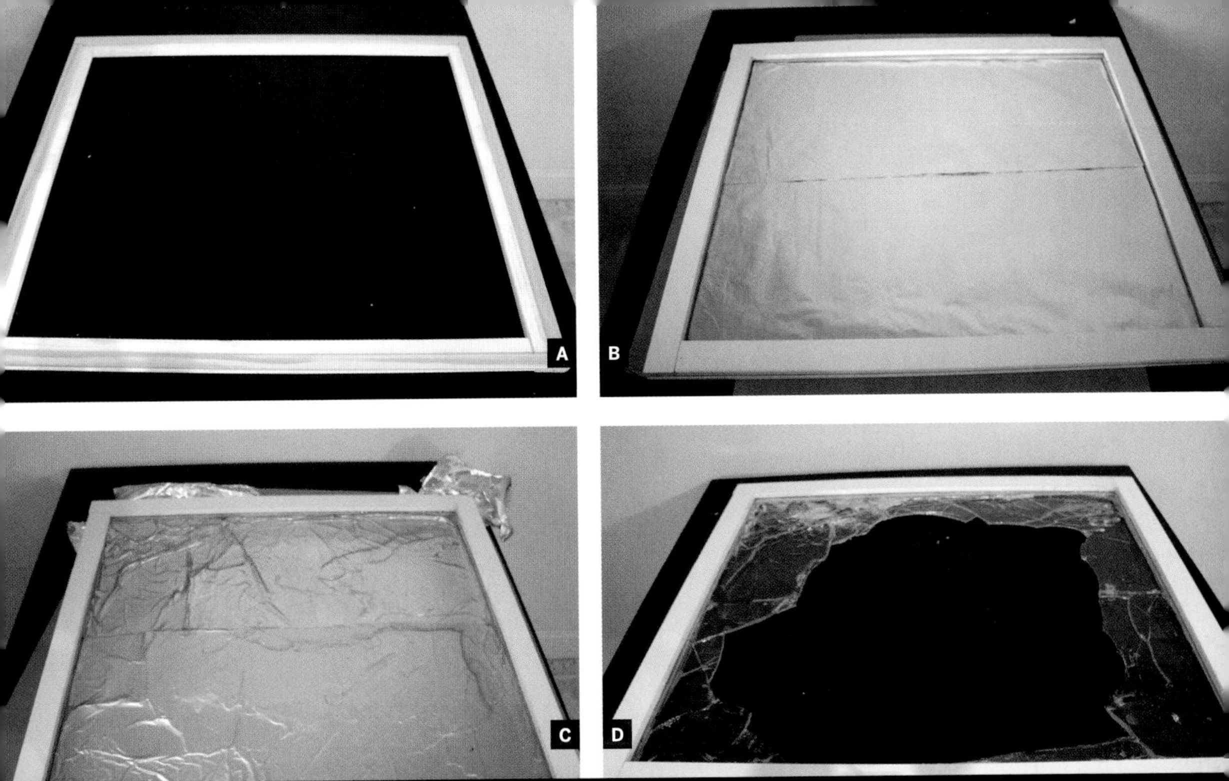

Fig. A: Making the frame.
Fig. B: Foil over cardboard supports the pane.
Fig. C: Sugar pane poured into the frame.
Fig. D: Finished pane, broken (by villain's head).

MATERIALS

- Thin strips of wood
- Wood glue
- Aluminum foil
- Sheet of cardboard
- Balsa wood strips **(optional)**
- 4 cups water
- 2 cups light corn syrup
- 7 cups sugar
- Food coloring **(optional)**
- Pot, stove, candy thermometer

in a pot on a stove. Use a candy thermometer to monitor the temperature, and when it approaches boiling (212°F), lower the flame and let it simmer for half an hour.

Turn the flame back up and heat the liquid to 302°F (about another half hour). Pour it into the frame (Figure C). Use a spatula to gently spread the mixture, but don't disturb the foil.

With each batch, pour an extra spoonful onto a piece of scrap foil; use this to gauge the composition, and convince your actors that they'll have fun smashing through it (Figure D). Aim for a thickness of about .75cm (or 7.5mm), otherwise the window will be either too easy or too hard to break. We needed 2 batches to fill our big window. If you use multiple batches, pour half at a time to avoid bumpy layers.

The sugar glass will take about 2 hours to fully cool and harden. After that, it'll slowly melt and otherwise decay, so you should plan to have the window cooled and ready somewhere between several hours and one day before the shoot. Be gentle moving the window, keep the foil on until shooting time, and peel it off carefully to avoid cracking.

Blood-spurting Knife Wound

Use a Dremel or other tool to cut an arc into the knife that's sized to fit around the victim actor's neck, arm, fingers, or other body part. Work on the back side of the knife (whichever side you don't want to show on-camera) to hide mistakes and gouges. Then use a Dremel and a metal file to dull all of the knife's sharp edges. Also dull the tip of the knife. Be sure to remove any metal filings. You don't want any actors getting hurt.

Cut a small length of tubing to run from the far edge of the arc to the knife's handle. If you have big hands and can palm a turkey baster, attach that to the tube, and glue the tube to the knife

MATERIALS

Cheap knives (2) Identical, with blades at least 1mm–2mm thick; ours cost $1.50 from a dollar store.
Turkey baster Or large medicine dropper or squeeze bulb
A few feet of thin plastic tubing From hardware store
Strong glue Or hot glue and glue gun
Red food coloring
Cornstarch
Water
Dremel Or cutting tool
Metal file

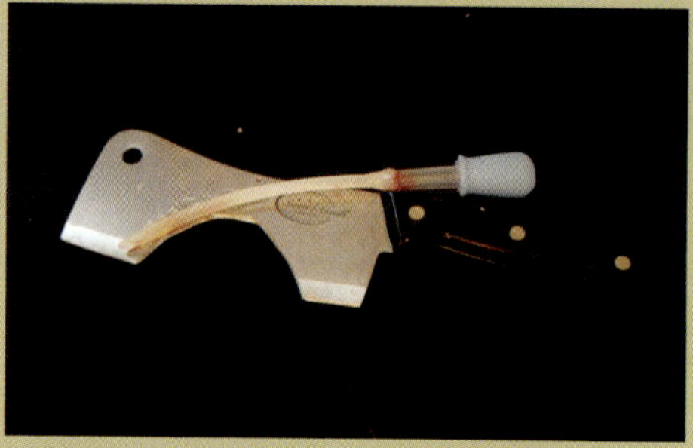

handle. Otherwise, use a medicine dropper. Or you can also run a longer tube through the knife-wielding actor's sleeve to his other hand, where he won't have to conceal the bulb.

Mix red food coloring, water, and a little cornstarch to make fake blood. Or use cheap red wine, or other on-hand substitutes. Pour it into a bowl and squeeze the bulb to load the knife up.

Roll Camera!

To get the most out of your special effects, make a shooting script that describes exactly what you need to include, shot by shot, and cross them off as you go. A storyboard can also help visualize each shot ahead of time, so that you don't miss anything important. For example:

SHOT 1. Wide, establishing shot of kitchen with HERO chopping a carrot.
SFX: CHOPPING

An establishing shot orients the audience to your location. Show the whole room, including the window.

SHOT 2. Close-up of chopping, a shadow falls over the hands.
SFX: CHOPPING
HERO: *I've been expecting you. I'm making us soup.*
ASSASSIN: *I've already eaten.*

Edit this to be just a few seconds. This shot will increase the pace of the sequence and establish the knife. Don't record the dialogue until the next shot.

SHOT 3. Wide shot from original, establishing perspective. Over-the-shoulder, ASSASSIN moves away from camera, toward HERO. HERO spins wildly, swinging the cleaver.

Record the action from both shots 2 and 3 from this angle, as one continuous shot. Mime the chopping, making no sound. When editing, begin the video of shot 3 a few frames after the Assassin begins moving. Let the dialogue lead the video, so you hear it begin over shot 2 and continue into shot 3.

Mark points on the floor where your actors can stand well apart during the knife swing. Be careful; you're using a real knife! Keep at least 1' of extra room between the blade and your actor. Because you're shooting the characters in-line, the audience won't be able to see the gap. When editing, cut this shot to end just when the knife would make contact.

SHOT 4. Cutaway, extreme close-up of HERO's eyes.
SFX: THUNK (knife drives into ASSASSIN)
ASSASSIN: *Aaagh!*

Cut to this 1-second (or less) shot when the knife would hit the Assassin. Later, slam a knife into a chicken to record this sound; it'll squish into the flesh and thud when it hits bone. When editing, layer that sound effect into the sequence. Don't record the Assassin's yell until the next shot.

SHOT 5. Medium, reverse shot of original perspective (over-the-shoulder from HERO to ASSASSIN). Knife is embedded in his arm, spurting blood. ASSASSIN grabs the knife.

HERO punches him in the face, knocking him down. He falls out of frame, pulling out the knife.
HERO: *That was for Lt. James Mitchell. He was*

one day from retirement.
ASSASSIN: (Groveling) *I just do what they —*
SFX: THUD
SFX: CLANK *(knife hits the floor)*

Use a waist-up medium shot to show the actors and the cleaver effect together. Use the prop knife, and for more blood spurting from under Assassin's shirt, hide extra blood-filled bulbs and tubes running out to production assistants (read: siblings, friends, or children) to squeeze. Assassin should yell from the knife impact to begin the shot; edit the beginning of that sound to overlap shot 4.

Punching is another trick of perspective. The instigator should be able to swing fully and not reach the other actor, while the recipient then jolts his head in response. The camera angle obscures the point where the contact would occur, so the audience believes what it hears. For the sound effect, punch a whole chicken or slab of beef.

SHOT 6. Cutaway, close-up of bloody knife.
HERO: *And this is ...*

This short cutaway should last 1 second or less. Bloody up the real knife and floor, and drop the knife into the frame. As with previous shots, let the sound of the knife begin in the preceding shot, and edit this shot to show the last 20 frames of the knife clattering on the ground. Also, don't record dialogue here; overlap it back from the following shot.

SHOT 7. Reverse shot, low perspective behind ASSASSIN toward HERO. HERO grabs the back of ASSASSIN's neck and hair, twisting out of frame toward the window.
HERO: *... for me.*

Shoot this with the full line, "And this is for me" so you can edit the audio to lead the video again. Add more blood running down the Assassin's arm and hand.

The audience thinks the Hero is pulling the Assassin, but for safety the Assassin is pushing the Hero's hand, controlling the action. The Hero should lightly hold the head and neck of the Assassin, but the Assassin should lead with his head and neck to trick the viewers.

SHOT 8. Wide shot of HERO driving ASSASSIN's arms and head through window.
SFX: CRASH

Use this break to make sure everything is set up properly for this, the one-take-only shot. Practice the choreography first with an open window.

As in the preceding shot, the weaker character leads all of the action, pushing through the fake window with their arms. While the sugar glass will crumble and shatter on impact, long sleeves can prevent scratches from the edge of the break. Have the actor tuck in his chin to protect his face.

For the sound effect, don some goggles and record a large, round bottle, like an apple juice jar, shattering on cement. Or search for a clip online.

SHOT 9. Wide, reverse shot of HERO pulling bloody ASSASSIN back in, then pushing him stumbling toward the door.
HERO: *You tell El Gato that I'm sick of running away. You tell El Gato I'm coming for him.*

Add blood to the Assassin's face before this shot, and he can stumble out of frame during the dialogue.

SHOT 10. Overhead crane shot starts tight, then pulls back. HERO drops to knees. The floor is filled with blood, glass, and the knife.
Cut to black.
HERO: (Yelling) *El Gatooooooooooooooo!*

To be your own crane, firmly attach the camera to a strong tripod and swing the tripod from a ladder.

For tips on how to record convincing effects and clear dialogue, visit makezine.com/09/diyimaging_effects.

Zack Stern (zackstern.com) is a San Francisco video producer and writer who has current or upcoming articles in CRAFT, *ReadyMade*, and other publications.

TILT-SHIFT PHOTOGRAPHY

Flexible lens makes scenes look miniature.

By Dennison Bertram

One of the fancier lenses in the world of SLR and DSLR photography is the tilt-shift lens. You might not know what these lenses look like, but you probably have seen their effects. Architectural photographers use tilt-shift lenses to eliminate the perspective distortions that sometimes give buildings the appearance of falling over. Aerial photographers use them to make large cities look like toy models. Art and portrait photographers use them to control exactly where the focus falls.

Tilt-shift lenses cost $1,000-plus, which is far beyond what most photographers will pay to experiment. Fortunately, building your own tilt-shift lens is easy, and doing so will open up a remarkable array of creative optical effects.

To build your own tilt-shift lens, you start with a spare lens that's built for a film format larger than that of the camera you'll use the lens on. For example, I used a 6×6 lens (designed for 6cm film) to make a tilt-shift lens for a 35mm camera body. With 35mm or APS format digital SLR cameras, you'll need a lens built for 6×6 film or larger. The oversized lens gives you extra room to move and distort the image that lands on the film or CCD, while still filling the frame. (You could use this hack to mount a 35mm lens on a 35mm camera, but it would only work with a macro lens, for very close objects.)

Assembly

Using a rotary tool or hobby knife, hollow out the middle of the camera body cap (Figure A), then grind or file it down smooth, so there are no rough spots or burrs (Figure B).

The plunger will act as a flexible camera bellows, allowing us to tilt and shift the lens to our heart's

Photography by Dennison Bertram

Fig. A: Hacked lens begins with a standard plastic body cap. Fig B: Body cap is hollowed out.

Fig. C: Humble plunger becomes camera bellows. Fig. D: Plunger bellows should fit tightly to lens.

MATERIALS

- **SLR (single-lens reflex) or DSLR (digital SLR) camera body** with interchangeable lens mount
- **Oversized lens** I used an old Carl Zeiss made for the now-obsolete Pentacon 6 camera format. Millions of these were manufactured, and eBay is full of good deals on them.
- **Rubber plunger with bellows design** Any will do, so long as it's flexible and not too large.
- **Stiff cardboard (non-corrugated)** Or stiff, black plastic
- **Plastic body cap to fit camera body** I use a lot of these in my work. They're the perfect way to attach your camera to your own hacked-lens creations, and they're also cheap.
- **Hot glue and glue gun or other adhesive** You need to bond the body cap to the cardboard/plastic and the cardboard/plastic to the plunger rubber. If you want a more robust setup, skip the glue and attach the pieces with small hobby nuts and bolts.

desire. Cut a hole in the top of the plunger, where the stick is (Figure C), making it just large enough to stretch around the base of your lens (or make it a bit smaller and enlarge it later).

Go ahead and stick your lens onto the plunger to see if it fits, and trim the rubber as necessary. Don't worry about gluing the lens down yet, but the hole should be tight enough so the lens fits snugly (Figure D). I even cut grooves in the rubber to let me screw the precious lens into place. Keep holding onto the lens, though; don't expect the rubber to hold so tight that you can let go.

If you're really enterprising; you could buy a bayonet adapter for the lens you're using, attach it to the plunger bellows, and then screw your lens onto that when needed.

Next you need to build the backing. To do this, I cut a ring out of cardboard with the inner circle the same circumference as the body cap and the outside matching the bottom of the plunger (Figure E). Although not shown here, it helps to paint one side of the cardboard black, to cut down light refraction inside the bellows.

Then I hot-glued the parts together, body cap into ring, and ring, black side in, onto bellows (Figure F). The hot glue is for expediency; if, after some experimenting, you think you'll use this lens setup often, I recommend finding something sturdier than hot glue and cardboard to hold it together.

You're all done! Attach your lens, and you're ready to shoot.

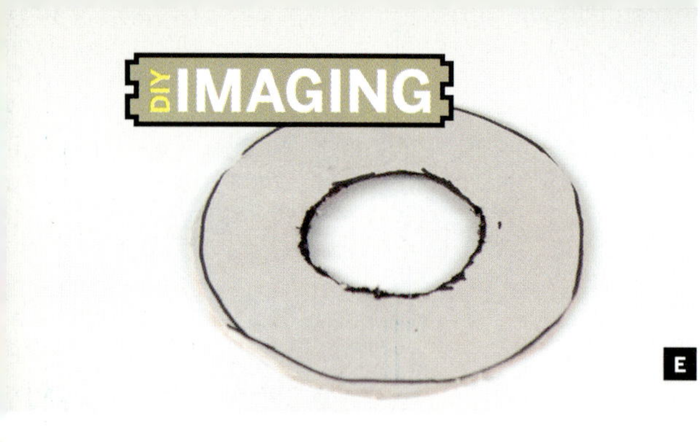

Fig. E: Cardboard backing ring.
Fig. F: Finished assembly.

Fig. G: Bendable lens makes Prague look miniature.

Shooting

This hack works surprisingly well. The image quality of the Zeiss is awesome, and I didn't get any optical interference between it and my Nikon DSLR's CCD chip. But there are a number of things to keep in mind while shooting. First, automatic exposure modes will not work with this lens, so you have to shoot manually. The apertures will still work, but in general you want to shoot with the lens wide open, or it will be far too dark to focus.

Needless to say, auto-focus also won't work. With this lens, you focus (or selectively un-focus) by squeezing the lens and plunger down, bending it, and twirling it around to get the cool effects you want. To give a "miniature" effect to a cityscape, tilt the lens forward or backward so that the only things in focus are in the middle ground, in mid-frame horizontally. The blurred foreground and background simulate the look produced by a macro lens taking a close-up of something small (Figure G).

If you want to use a lens like this in low light conditions, where you'll need to hold the lens in one position for a long time, you might augment this design with an adjustable mechanical frame that controls the lens' range of motion. If you build a tilt-shift lens with a frame, a photographic bellows will be more flexible than a plunger.

A note to digital camera users: dust is a common problem with most digital cameras. Projects such as this one can exacerbate the situation. Before using this lens, be sure to clean it out with some strong puffs of air, to get rid of any loose dust particles that might be inside.

Dennison Bertram is a fashion and beauty photographer who lives in the Czech Republic.

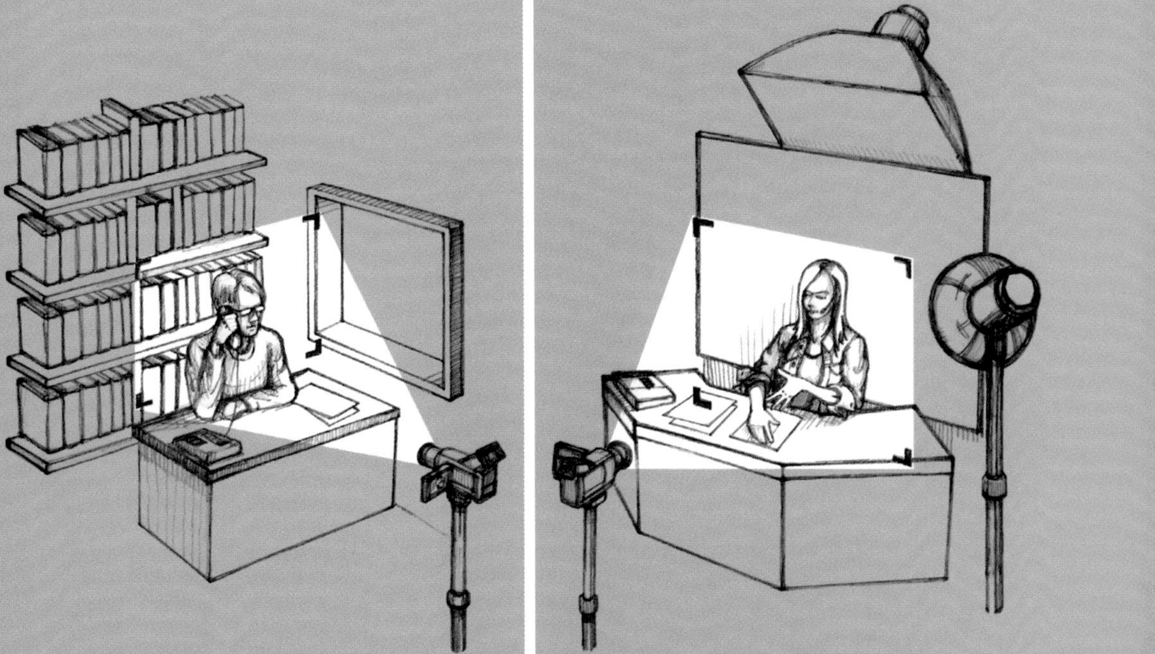

ONLINE VIDEO PRODUCTION

Tips and techniques for daily content.

By Andrew Michael Baron

As creator of *Rocketboom*, I've been producing a 3-minute daily video news report since 2004. Here's how I prevent the every day thing from becoming everyday.

Camera

Use a MiniDV camera that takes MiniDV tapes. We have a $350 Panasonic, an older $2,000 3-chip model, and a nice new Sony HDR-HC1 MiniDV/HDV camcorder, which at $1,400 is the best deal on the market right now. Tapeless cameras will also work, but their storage is too expensive. It's nice to keep all your source material archived at full resolution. A 60-minute full-res video will take up a whole hard drive, or you can capture it onto a tape that only costs a couple of bucks.

There's really only one feature I recommend for people investing in a new camera: make sure it has a microphone input or a line-in jack. If not, you'll be dependent on the camera's built-in microphone, which in many environments will prevent you from doing good interviews and getting clean audio.

Also consider whether the camera is a single-CCD or 3-CCD model. A 3-chip camera processes each color separately, producing a brighter and sharper picture. Most cameras under $500 are 1-chip; these are also fine, but may look a bit more grainy, especially in low light. Most people use 1-chip cameras, since 3-CCD models have only recently become affordable.

Illustrations by Damien Scogin

So, you want to produce an online news show year-round on a daily schedule, and keep it interesting? Use a good MiniDV camera with a tripod and a separate microphone. Pay attention to lighting and image compressibility. Always keep an eye out for good source material. Cook up techniques that work with what you have and are capable of.

In a pinch, however, anything will be fine. I often use footage from my little still camera that records video. If I happen to be somewhere without my camera, and someone else has one, I'll get their contact info and try to hook up with their footage later. Even if I do have my camera, it's nice to get more footage and a different perspective to work with. Pictures, video, audio: these are all assets that can be used.

We get source material from pretty much everything under the sun, so it's helpful to always look around and see what's available. Sometimes you don't need video at all; just an audio clip is fine, and you can build a story with images, website data, and other stuff surrounding the source.

Holding the Camera

Use a tripod whenever you can. Do it! Otherwise, hold the camera as still as possible. Too much jerkiness can ruin a scene. The "raw" quality of videoblogging can be charming, but not when the camera is jerky and moving around a lot.

Compression-wise Shooting

When a video is compressed, the file size can be twice as large if the frame contains a lot of fine detail and movement. If you are shooting trees and moving the camera around, your video won't compress easily. I never sacrifice any detail and movement that is important to a shot, but otherwise I try to keep these to a minimum.

Lighting

Most cameras auto-adjust for lighting, but the darker it gets, the more grainy the video will look. Too much darkness is not charming either. Use the same lighting sense you have from taking still pictures; for example, have the light coming from behind you, but don't stand in front of it. The brighter the light, the more brilliant the picture (until you go overboard and everything becomes overexposed and saturated).

Sound

Nicer handheld microphones use a microphone cable, and you will need an adapter like a Beachtek to convert their XLR plug to the ⅛" mini that you can plug into a camcorder. This is worth it and will make a big difference. Cheap microphones add a lot of noise to the signal.

If you're interviewing people on location in loud surroundings, it's best to have separate handheld

Every picture tells a story: Rocketboom production tricks have added visual appeal to phone interviews, made single-camera shoots look like multi-camera, and covered up creatively for missing footage and sound.

microphones for everyone. But if you don't, treat your camera like a microphone. When someone is talking, bring the camera way up close to the person, get a good audio signal, and don't worry about the video. Then, later, when they are finished, take B-roll footage of the environment, and try to capture the stuff that they were talking about. Then in post-production, take the audio track and lay B-roll over it where appropriate. The audio comes first, and you cut the video to match.

Situations

We've come up with a lot of tricks for navigating various constraints. Here are some examples, and you can watch the reports themselves at rocketboom.com/vlog/archives, where they are organized by broadcast date.

For our Nov. 30, 2005 webcast, we did a phone interview. To set up, we pointed a camera at Amanda Condit here in New York City, and our interviewee Joshua Davis pointed a camera at himself in California. Amanda called Joshua, and they did the interview with both people being taped. Then Joshua mailed us his tape, and we put the segment together with assets he gave us, website images, etc.

When you shoot a great musical performance, you can build an entire broadcast over one song. Shoot the entire song from one angle, keeping the mic the same distance from the performers throughout. Then shoot in and around the scene to get B-roll to add in. This is a favorite technique, and I used it for our June 17 and July 20, 2005 shows. Both of these look like 2-camera coverage, but if you are really savvy, you can see where I cheated.

Our April 6, 2006 show was a phone interview recorded as audio only, with no video. Our interviewee Erin Peterson didn't have to do anything but talk to us. We added visuals afterward using images from Flickr, and Amanda filmed an intro and outro of herself starting and ending the phone interview, to frame it.

For our episode on Aug. 10, 2005, Amanda went to cover a garden and did a lot of talking, but we screwed up and the audio was off the entire time. We didn't realize this until much later, and we decided to turn it into an "art piece" by adding music and effects to the silent footage.

Andrew Baron is the creator and producer of Rocketboom, a 3-minute online daily news show with a daily audience of more than 300,000.

Photography by Alek Komarnitsky

OUTDOOR WEBCAM ENCLOSURE

Capture winter scenes from hanging sewer pipes. By Alek Komarnitsky

I anteed up the big bucks for a wireless security webcam with motorized pan, tilt, and 10x optical zoom — specifically, the D-Link DCS-6620G. Nice webcam, but I wanted to put it outside so people everywhere could view my Halloween decorations and infamous Christmas lights.

The problem is that the webcam is rated only down to 32°F, and here in Colorado, temperatures can drop below zero. Suitable prefab outdoor enclosures cost about $500 and include a blower and heater, so I decided to build my own simpler webcam enclosure. It cost me a whopping $27, and it has successfully stood up to two full seasons of Rocky Mountain rain, cold, and snow.

I installed the webcam at my neighbors' house, hanging it from a 6" can light fixture under an eave that had good line-of-sight to our house. The basic idea was to attach 2 brackets hanging down from the inside of the can, and build an easily removable enclosure that would hang from a rod running horizontally through the brackets.

For the brackets, I straightened two 5" L-brackets in a vice, and extended their internal cut with a hacksaw. Then I used sheet metal screws to anchor the brackets to the inside walls of the recessed light fixture. I screwed in an adapter to convert the fixture's socket into a power plug for the webcam, which is the only physical connection the webcam needs.

I made the enclosure itself out of 6" inner diameter, foot-long ABS sewer pipe. (Yes, sewer pipe — no expense spared!) I cut a 1' length of

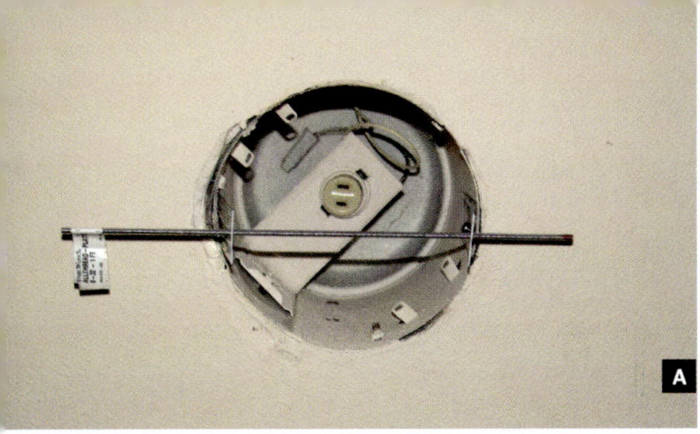

Fig. A: Metal rod "hanger" attaches to can light with metal brackets. Fig. B: D-Link DCS-6620G webcam with cover plate.

Fig. C: Internal view of webcam enclosure. Fig. D: Webcam assembly hangs from friendly neighbor's second floor balcony.

MATERIALS

- D-link DCS-6620G with threaded base and transformer
- 6" ABS plastic sewer pipe, 1' long cut down from 2' section sold at Home Depot
- 6" ABS end cap
- 5" round cover plate
- 5" metal L-brackets (2)
- Sheet metal screws (4)
- 1' threaded metal rods (3) with matching wing nuts (6)
- ¼-20 bolt with washers and nuts
- Light socket to 2-prong plug adapter
- Electrical tape
- Clear plastic CD case cover from free AOL install disc
- Outdoor recessed can light fixture already installed in good location for observation

pipe and drilled 2 holes at one end to hang it from. A threaded rod runs through the pipe and the hanging brackets inside, and 2 wing nuts hold the rod in place.

Before installing the webcam enclosure at my neighbors' house, I tested the basic concept on a recessed ceiling light in our upstairs hallway. The pipe hung securely, but this indoor can light was 6½" in diameter, which meant that the two brackets stayed visible on either side of the pipe. The light fixture at the neighbors' is just 6" wide, so the pipe fits around the brackets and conceals them, which looks nicer.

Inside the enclosure, the webcam itself hangs from 2 more rods that run through holes drilled side by side farther down the pipe. To suspend the webcam upside down, I ran a ¼-20 bolt through the center of a 5" cover plate and screwed it into the webcam's tripod mount. I added extra nuts onto the bolt to reduce the chance of stripping the webcam threads. Inside the pipe, the cover plate rests on the 2 lower rods, and the webcam hangs underneath.

For the enclosure's view port, I cut a hole in the pipe and made a window by taping on the clear plastic from a free AOL install CD case.

Fig. E: Portable webcam assembly can be hung under eaves or placed freestanding, weather permitting.

Fig. F: Webcam view at 2x optical zoom last fall. During the winter, the aspen leaves drop and open up the visibility. Fig. G. 10x webcam picture of Alek taking picture of it.

I put a flat ABS end cap at the bottom of the enclosure, which is where the webcam's 120VAC-12VDC transformer sits. This provides some warmth for low-temperature operation. I left the end cap vented (there's about a ¼" opening on the top) to prevent fogging in the viewport. The flat end cap also lets me unhook the webcam and set it down to operate anywhere, connected via extension cord, weather permitting.

You can optionally add a cheap wireless temperature sensor inside the PVC, and also one taped to the outside, in order to provide a "delta temperature" reading.

Actual installation was super-duper easy since the entire unit is self-contained, with only a power cord coming out of the top. Plug that in, run the threaded metal rod through the PVC-bracket-bracket-PVC, screw down the wing nut, and you are ready for action!

The installation has worked great, although I did crack some plaster around the light fixture — oops! The webcam works even with the temperature dipping below 0°F a few times. The only issue was that the wireless signal would sometimes drop out, so I installed a Pringles-can antenna at my house pointed toward the webcam, and have had no problems after that.

The webcam enclosure is an integral part of my Controllable Christmas Lights for Celiac Disease website, which lets people remotely view live images of thousands of Christmas lights, and also control them with a click of a mouse. Besides being fun for people around the world, the site has raised over $16,000 for charity.

(Before making my Christmas lights controllable through my website in 2005, I simulated this effect with canned photos and a CGI script. The ruse spread far and wide before I invited the *Wall Street Journal* to reveal it as a hoax. But that's another story.)

Alek Komarnitsky lives in the Republic of Boulder. When not spending time with his wonderful wife Wendy and two sons Dirk and Kyle, he enjoys tinkering with stuff. Read more at komar.org.

WINDUP CAR

Mod a cute convertible with a big rotating key.

By Sunny Armas (with Paul LeDuc)

When my pickup truck broke down, a friend lent me her Geo Metro convertible for 2 weeks. Well, I'm a good-sized guy, and at first I felt a bit silly driving this tiny car, but I quickly fell in love with it. It's easy to park, great on gas, and a blast to drive. I decided to buy my own, and I thought that if I'm willing to be seen driving this toy-sized car, why not go all the way and put a big windup key on the back?

I called my metal sculptor friend Paul LeDuc for his help and ideas, and we talked about the project. I knew I wanted a big key that would turn slowly, to make the car look as toy-like as possible. I also wanted the key to be easily removable and fit in the trunk to prevent people from trying to rip it off, and to keep the option of driving a little more low-profile. Also, I had just had the car painted, so I wanted to be careful that any mods would not damage the nice new finish.

We decided that the key should be about 16" across. For the flat part, we drew a template and used a plasma cutter to cut the shape out of ¼"-thick steel plate. The harder part was the center tube, which needed to mount neatly through the trunk of the car. For this, Paul strolled through the hardware store and found a nice chrome mounting ring for finishing bathroom sink drains. The threaded, 1½" inside-diameter ring had some height that would help guide the key, and included a rubber gasket, washer, and nut to secure it around both sides of an aperture.

Paul used a grinder to remove the strainer from the bottom of the sink mounting ring, then cut a hole in the center of the trunk deck (lid) with

Photograph by Sam Murphy

Fig. A: To turn the key, a windshield wiper motor is mounted under the trunk deck. Power comes from the dashboard cigarette lighter, via wires that run under the carpet and out the same hole as the brake light wires.

Fig. B: The key to making a cute little car even cuter.
Fig. C: Friction mount made from a foam bicycle handle grip makes the key easily removable, and lets you twist it and stop it without harming the motor.

a drill and a hole saw. He slid in the drain ring, and fitted the washer and nut on the back. Then he found some steel tubing in his shop that fit perfectly into the sink ring, cut a length of that to be the key's center tube, and welded it to the flat key shape.

To turn the key, we used a spare windshield wiper motor that I had. These run off a 12VDC car battery, of course, and they turn slowly with a lot of torque. To connect the key to the motor and make it removable, we made a friction mount out of a foam bicycle handle grip. We welded a 2½"×⅛" disk to the shaft of the motor, and welded that to a 5" steel rod that's the same diameter as bike handlebars (⅞"). The shaft protrudes from the middle of the hole in the trunk, and with the foam grip over it, the key slides over it snugly.

The great thing about the foam connection is that it lets the motor run freely if the key's movement is impaired, like if someone grabs it, and it also lets you turn the key when the motor is off. This means that little kids can turn the key before the car starts, and imagine that they're helping the car go!

To attach the motor, Paul removed the trunk deck, laid it upside down, and welded together a framework that would hold it securely and keep it aligned straight up. He made the framework out of 1"- and 2"-square tubing, welded to the deck's ribbing with minimal damage to the paint, and tapped screw holes in to serve as motor mounts.

The motor is powered from the car's cigarette lighter. To do this, I ran wires under the carpet, starting in front near the dash, between the seats, and threading them back into the trunk through the same hole that the brake light wires run out of.

To connect the wires to power, I cut the fan off an old cigarette-lighter-powered fan, and ran the circuit through an old VW bus light switch, which I mounted next to the car's gearshift. I positioned the switch so that when I shift between neutral and first, my thumb can easily turn the key motor off and on. That way, when I stop the car, the key stops, and when the car starts moving, the key starts up again — bringing smiles to all, or at least to most.

Sunny Armas lives in San Jose and enjoys bringing smiles to others. He has helped Paul LeDuc with many other art projects, including parade floats and metal waterfalls.

Photography by Paul Spinrad

YOU ARE A THINK TANK

FROM MAKER TO MAKER

Who doesn't appreciate a really good tip now and then? Especially the kind, as one reader put it, "that changes your life." Whether it's something as practical as looking good in party photos or something more creative like figuring out how to judge the amount of propane in a tank, we all rely on our friends and neighbors to tip us off to the new and the good. *—Arwen O'Reilly*

If the Shoe Fits

Gina Trapani from the Lifehacker blog has a great tip from a reader: buy a clear plastic shoe organizer and use it to store your gadgets. You'll be able to see everything at a glance, and each cable can have its own compartment. makezine.com/go/shoefits

Party Favor

I found this great lifestyle hack on Flickr, of all places: Meg Hourihan suggests using a cooling rack on top of a cookie sheet to dry wineglasses when there are too many to fit on a normal dish rack.
makezine.com/go/glasshack

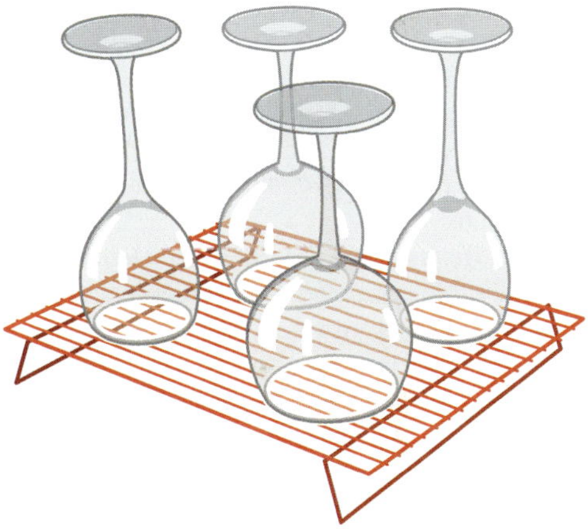

BB Q&A

Boing Boing reader Ross points out a great way to judge how well your barbeque will go: "There is an easy way to see the level of propane in a current tank. Just warm up some water and pour it on the side of the tank. Since the gas is liquid and cool, the water's heat will cause the tank to form condensation at the current level of the propane. I do this before any cookout so I can get propane before the guests arrive rather than the last minute dash."
makezine.com/go/propane

Don't Say Cheese!

Terrie Miller, MAKE's webmaster, has a couple of photography tips: "My brother-in-law relayed this great tip: when the photographer comes around taking candids at the next company party, or the waiter snaps a photo of your table at a restaurant, give him your most enthusiastic 'thumbs up' sign. You'll have a better chance of looking happy and engaged in what's happening, instead of looking awkward and embarrassed (if not actually sullen)."

She adds, "If you're the photographer, here's another good tip: tell your group to wave. I learned this from Derrick Story (see his blog, thedigitalstory.com/blog, for more good tips) when he took our favorite group photo at our wedding reception ... everyone was laughing and smiling, with no frozen poses in the bunch."

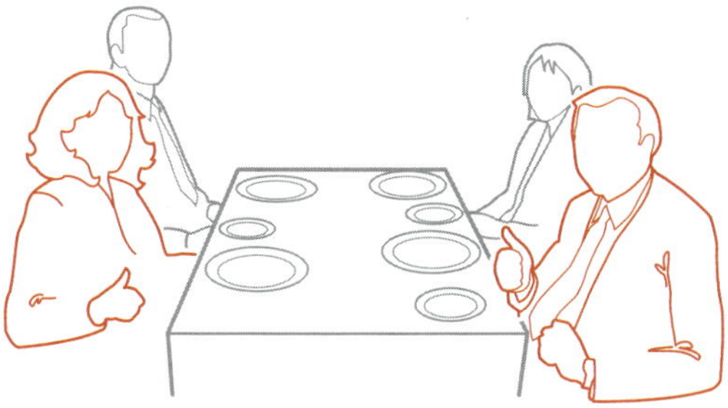

Illustrations by Damien Scogin

Have a tip for MAKE readers? Send it to tips@makezine.com.

Citizen Scientist

BUILD A CLOUD CHAMBER

CATCH THE GHOSTLY TRACKS OF COSMIC RAYS USING DRY ICE, ALCOHOL, AND A BASKETBALL DISPLAY CASE.

By Dr. Shawn

HOLD OUT YOUR PALM FACE UP. IN THE time it takes you to read this sentence, two high-energy subatomic particles will likely have pierced clean through your hand.

We live in a radioactive environment and there's no escape. At sea level, the flux is roughly one particle per palm size per second, which means San Franciscans sustain an astonishing 20 million hits through their bodies each day. Denver is higher in elevation, so its protective blanket of air overhead is thinner. As a result, Mile Highers get hit at roughly twice that rate.

These particles make their presence felt by ionization — that is, ripping electrons away from their atoms. Since atomic electrons are electrically charged, they interact with all other charged particles (e.g. other electrons, protons, helium nuclei, and so on), and so any of these, when sent screaming through space, becomes a type of ionizing radiation. In addition to charged particles, some photons, such as X-rays and gamma rays, also have enough oomph to whack electrons out of their atomic orbitals and so they, too, make the list.

Much of the radiation around us comes from nuclei in the soil and the air that have too much energy to be stable; they shed their excess by expelling elementary particles. But most of the damaging corpuscles have their origins deep in outer space where pulsars, supernovae, and even stranger cosmic players accelerate elementary particles to fantastic energies and hurl them out into the cosmos. After perhaps 100,000 years of silently streaking between the stars, some of these tiny wanderers end their existence when they strike the nucleus of an atom in the air high over our heads, and explode.

Since energy and mass are two sides of the same coin, some of the enormous kinetic energy these particles often carry instantly converts into wholly new particles, sometimes thousands of them, that shower down toward Earth. Most of these new particles interact with still more atomic nuclei and quickly range out — never reaching the ground. But one exotic species, the muon, does not feel the nuclear force. As a result, muons do not get absorbed in the air. Most survive to reach Earth's surface where they fall from the sky in all directions as a steady spray of subatomic machine gun bullets that we call "cosmic rays."

METHOD #1: THE 30-SECOND CLOUD CHAMBER

It turns out that these exotic particles, as well as the more mundane radiation produced by the decay of unstable nuclei near us, can be made visible inside a chamber that takes about 30 seconds to assemble if you've got some dry ice on hand. (Dry ice is easy to get from many ice warehouses or welder's supply stores. Visit dryicedirectory.com to search for dry ice dealers by area code.)

YOU WILL NEED:

- **Large jar with metal lid**
- **Rubbing alcohol**
- **Dry ice (1 cake)**
- **Towel**
- **Flashlight**

Simply take a large jar with a metal lid and warm the jar under a flow of hot water. Then thoroughly rinse the inside with rubbing alcohol and dump out the excess. Replace the lid. Center a cake of dry ice on top of a towel. Then place the jar, lid side down, in contact with the dry ice and wrap the

Inside the trophy case cloud chamber when chilled and illuminated by an LCD projector. Although they do not show up well in this photograph, the alcohol droplets and particle tracks are clearly visible to the naked eye.

remaining dry ice up in the towel to prevent it from "smoking." The alcohol will evaporate from the warm sides of the jar and condense near the frigid lid.

Do this in a darkened room. Shine a bright flashlight into the jar from the side, and you'll see pinprick-sized droplets of alcohol coalescing near the bottom. After a minute or two, when the dust inside the jar has settled out, you'll also see something extraordinary. About once a minute, just above where the droplets are forming, a ghostly line will suddenly appear and then disintegrate as falling rain of alcohol. These spectral emanations are caused by ionizing particles of radiation passing through your jar.

Muons fall from the sky in all directions as a steady spray of subatomic machine gun bullets that we call "cosmic rays."

These thin trails of vapor form because the ions that the passing particles leave in their wake attract the electrically neutral alcohol molecules, just like a balloon that has been electrically charged by rubbing on someone's hair readily attracts small bits of electrically neutral dander. The alcohol molecules just above where the cloud appears are almost, but not quite, cold enough to form droplets. But when a passing particle lays down a trail of ions, those ions can pull together enough alcohol molecules to entice droplets to form. These droplets coalesce along the track and essentially amplify its width over a trillionfold to make the particle's passage plainly visible. For obvious reasons, a radiation detector that uses tiny droplets to reveal its quarry like this is called a "cloud chamber."

While the cloud chamber in a jar couldn't be simpler to construct, it has three drawbacks. First, it's small, so you often have to wait a long time before another cosmic ray will happen to pass through at just the right spot. Second, the curved glass makes the tracks hard to see. And finally, the show only starts after any dust particles inside have settled, and it stops as soon as the alcohol has all condensed. That limits you to just a few minutes of good viewing.

A far better chamber would operate with a large enough volume for tracks to appear every few seconds, have flat sides for clear viewing, and would contain a reservoir of alcohol large enough to keep the show going for hours.

Photograph by Shawn Carlson

YOU WILL NEED:

Basketball display case
Thick black felt to cover the top and bottom of the chamber
Sheet metal a thin square measuring 26cm a side
Hardware cloth mesh same size as metal
Silicone aquarium cement one tube
Stapler
Rubbing alcohol
Dry ice
Small wood screws 4 to 6

METHOD #2: THE BASKETBALL CASE CLOUD CHAMBER

Here's how to make a superior cloud chamber.

The trick to creating cloud chambers is to find the right vessel. And nothing I've found so far trumps a basketball display case. I purchased a new one for $30 from a local craft store, but you may need to check online or peruse your local trophy stores. These cases are clear glass on the top and three sides. The back is lined with a mirror and the bottom is covered with thin particleboard. Each side is about 26cm long. When the chamber is operating, tracks form within at a height of about 4cm, which makes the active volume a whopping 2,700 cubic centimeters. That is far larger than any other homemade cloud chamber I've seen, and results in a particle track every few seconds.

First, to let the chill of the dry ice in, replace the thermally insulating particleboard on the bottom of the display case with a thin piece of conducting sheet metal. (Find a local supplier of sheet metal and have them cut it to size.) Gently push out the bottom and use the silicone cement and screws to seal your sheet metal in its place (Fig. A). Turn the assembly upside down and rest about 10 lbs. of books on top to insure a firm connection. Run a bead of silicone all around the inside joint to make sure the seal is airtight; air currents will destroy the tracks as fast as they form.

When the cement sets, cut a square of black felt to exactly fit inside the bottom of the case. This will soak up the excess alcohol used in the chamber and provide a good black contrasting background to view the tracks.

Adding the felt to the lid is trickier. First, cut a square just large enough to completely cover the exposed glass on the inside of the top. Turn

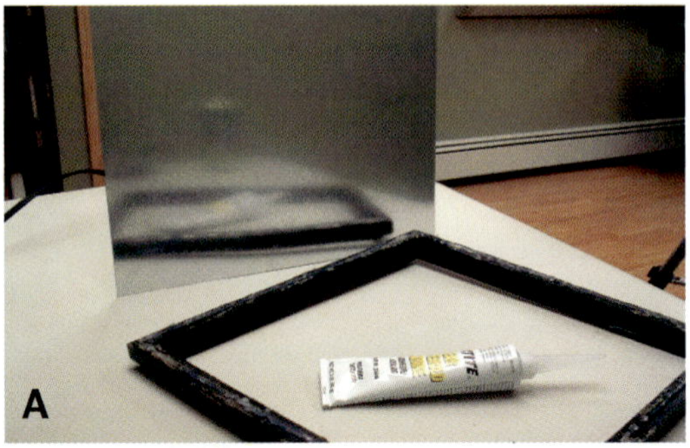

A

B

C

D

Photography by Shawn Carlson

the top of the case upside down and drop the felt inside. Now comes the tricky part — fastening the fabric in place. You can't simply use glue, cement, or epoxy because alcohol eventually dissolves virtually every adhesive.

A better solution is to pin the felt against the top using a wire screen. Cut the screen so that it is 2-3 cm (about 1") wider than the felt, and use a pair of pliers to bend down the four sides. Press the felt down, and then delicately staple the wire frame in place using ordinary office staples. (Figs. B and C)

Finally, to block out all of those track-destroying air currents, you need to seal all of the wood/glass joints with silicone. Run a bead of caulking along all of the joints inside and out, and then spread the silicone smooth with your finger. Let the silicone set up overnight before testing your chamber.

That's it. You're done.

ENTER THE SUBATOMIC UNIVERSE

Place the bottom of the chamber directly on top of a block of freshly cut dry ice (Fig. D). You want the entire bottom surface to come to the same temperature. (Any variation will cause air currents to flow inside the chamber and obliterate your tracks.) So if the chamber is larger than the block of dry ice, you'll need to create a tile of four blocks, and then center the chamber on that. As before, wrap exposed dry ice with a towel to prevent smoking.

NOTE: If your supplier gives you irregular hunks of dry ice, you'll need to create a dry ice and alcohol bath. First, wrap the dry ice chunks in a towel and pulverize with a hammer. Next, dump the pulverized pieces into a plastic kitchen trash bag. Lay the bag on top of a doubled towel to provide insulation, and pour in several cups of isopropyl (rubbing) alcohol. (You'll waste less dry ice if you chill the alcohol in your freezer first.) Then wet the top of the bag with alcohol to provide a conductive seal, and press the bottom of the chamber onto the bag. This will chill the chamber nicely and avoids the mess and alcohol fumes of a traditional alcohol and dry ice bath.

Next, fully charge the top pad of felt with alcohol and evenly moisten — do not saturate — the bottom pad. The alcohol on the bottom helps conduct heat out of the chamber and hastens the formation of the cloud. For this, you'll want the highest concentration of alcohol you can find. Go to your local auto supply store and purchase a bottle of Iso-Heat. This product is used to remove water from fuel lines and, as it turns out, is pure anhydrous isopropyl alcohol.

Now close up the chamber and wait. It will take a while for all the dust in the air inside to settle out and for the temperatures of the different parts of the chamber to equalize. The show will start in about 20 minutes and it will go on for hours.

To see it, turn out the room lights. Then shine a very bright column of light in from one side. A bright flashlight will work, but the batteries soon give out. I use the light from an LCD projector. The placement of the light is absolutely critical. The light must shine across the bottom of the chamber where the droplets are forming, and you must position your head low and at a steep angle relative to the light. Experiment a bit to find the positions that best illuminate the chamber and provide the most spectacular viewing.

After the chamber has reached equilibrium, droplets will form very fast near the bottom and less so as you move up. The top of the cloud will be about 4cm above the bottom. That's where you want to focus your attention, where the wispy telltale tracks you are looking for will suddenly spring into being. Their appearance, like tiny ghosts dashing through a fog, is something out of Harry Potter.

If you measure their rate and directions, you can learn a great deal. Consider adding a digital camera, shooting either through one side of the case, or through a hole cut in the felt and wire screen on the top. Take a steady march of images and keep only those that happen to capture a track. Long, straight, and slender tracks are most likely made by muons. Thick, stubby tracks that start and stop inside the chamber are created by alpha particles. Their rate will let you set an upper limit on large radioactive nuclei in your environment, like radon.

Once you've astonished yourself, make plans to show your chamber to every person you possibly can. No demonstration I know is better able to get people excited about science than this one.

Dr. Shawn (Shawn Carlson, Ph.D.) is a former columnist for *Scientific American*, and the only person ever to win a MacArthur Fellowship for science education. He is the creator of the Society for Amateur Scientists (sas.org), Super Science Fair Support Center (scifair.org), and Labrats (labrats.org).

HowToons.com

BY SAUL GRIFFITH & NICK DRAGOTTA '07

ver thousands of years sailors have learnt to estimate the speed of the wind just by looking about. This chnique matured into what we now call the Beaufort scale. The universe tells you everything you eed to know about it as long as you are prepared to watch, to listen, to smell, in short to observe.

ORCE	SPEED		
2	4-6 knots 4-7 mph 7-11 km/h	SEA	Small wavelets. Crests have a glassy appearance and do not break.
		LAND	Wind felt on face; leaves rustle; ordinary vane moved by wind.

"Light Breeze"

FORCE	SPEED		
3	7-10 knots 8-12 mph 12-19 km/h	SEA	Large wavelets. Crests begin to break. Foam of glassy appearance.
		LAND	Leaves and small twigs in constant motion; wind extends light flag.

"Gentle Breeze"

FORCE	SPEED		
6	22-27 knots 25-31 mph 40-50 km/h	SEA	Large waves begin to form; the white foam crests are more extensive everywhere.
		LAND	Large branches in motion; whistling heard in telegraph wires; umbrella use difficult.

"Strong Breeze"

FORCE	SPEED		
7	28-33 knots 32-38 mph 51-62 km/h	SEA	Sea heaps up and white foam from breaking waves starts to blow in streaks with wind.
		LAND	Whole trees in motion; umbrellas discarded; inconvenience felt when walking.

"Near Gale"

FORCE	SPEED		
10	48-55 knots 55-63 mph 88-102 km/h	SEA	Very high waves. Surface of the sea takes on a white appearance. Visibility affected.
		LAND	Seldom experienced inland; trees uprooted; considerable structural damage occurs.

"Storm"

FORCE	SPEED		
11	56-63 knots 64-72 mph 103-117 km/h	SEA	Exceptionally high waves. The sea is covered with long white patches of foam.
		LAND	Very rarely experienced on land; accompanied by widespread damage.

"Violent Storm"

"Hurricane"

MakeShift

By Lee D. Zlotoff

The Scenario: There was a time when high school, however trying and tedious it often seemed, was still a refuge from the realities of the big, bad world. Alas, those days have gone the way of the slide rule and mimeograph machine. The threats that once stalked only distant cities and bad horror movies are now just as likely to find their way onto a high school campus as anywhere else.

The Challenge: You're in metal shop one sunny morning when an announcement comes over the PA system that the school has been ordered into a lockdown. The teacher must lock the door to the room and let no one out for any reason. The announcement doesn't explain why (A riot in the cafeteria? A crazed gunman on the loose in the area?) or how long you may be stuck here. The lights and power are still on, but you quickly discover that the cellphone system has been shut down or is not working, and neither is the school's internal phone system. And if the school has a wi-fi network, it's also dead.

In order to find out what's going on, or let all your parents — who will no doubt be freaking out any time now — know that you're OK for the moment, or to alert someone to the fact that your buddy Chester here will need his insulin shot within the hour, the class concludes it needs to try and establish some kind of communication with folks outside the school. Thinking this will at least keep everyone in the class focused on something besides the crisis, the teacher gives her OK.

The metal shop itself is on the top floor of the school, with a row of windows that open and face out toward the surrounding community, the nearest houses of which are about a quarter-mile away.

Your Items: The shop itself has an assortment of sheet metal, wire, angle iron, and other assorted metal parts available. You also have the following: a half-dozen cellphones of different makes and models, a laptop computer with a wireless card, a walkie-talkie that one of the janitors left, and a cordless telephone — though, as we've said, the line it's connected to in the school is down.

Using some or all of this stuff, you need to devise as many ways as you can to try and create a communications link with the outside world. And yes, if you pull it off, it will be counted as extra credit toward your final grade. Good luck!

Send a detailed description of your MakeShift solution with sketches and/or photos to makeshift@makezine.com by May 11, 2007. If duplicate solutions are submitted, the winner will be determined by the quality of the explanation and presentation. The most plausible and most creative solutions will each win a MAKE sweatshirt. Think positive and include your shirt size and contact information with your solution. Good luck! For readers' solutions to previous MakeShift challenges, visit makezine.com/makeshift.

Lee D. Zlotoff is a writer/producer/director whose numerous credits include creator of *MacGyver*. He is also president of Custom Image Concepts (customimageconcepts.com).

Photograph by Doug Adesko

PRIMER

Working with Carbon Fiber

Form, lay up, and cure your own high-performance composites.

By John Wanberg

It seems as though nearly everything "high performance" these days boasts some amount of carbon fiber in its construction. Originally used in aerospace, carbon fiber has moved into the mainstream and can be found in luxury automobiles, mountain bikes, and sports equipment.

Some laptops and cellphones even use printed decals to simulate this lightweight material's cutting-edge look. The good news is that you don't need a state-of-the-art manufacturing facility to work with carbon fiber composites. In fact, you can do it at home.

This article discusses some of the basics of carbon fiber construction and explains how to create a carbon fiber iPod case. All you need are some basic woodworking tools and skills, and the right materials. And because the same process also applies to fiberglass and Kevlar composites, these skills give you multiple ways of boosting your future projects to a new level!

Photograph by Meiko Arquillos

UNDERSTANDING CARBON FIBER

Composites are created from two or more dissimilar materials that act together as one. While concrete and plywood are technically composites, the term *composite* in industry has come to refer to reinforcement fibers held together in a resin matrix and formed in a mold. Carbon fiber is one of several textiles used in this class of materials. When joined together through a procedure called a *layup*, the fiber and the resin form a material with properties that exceed those of either constituent material.

As a rule of thumb, composites offer their greatest strength in the direction the fibers run — similar to how wood is strongest along the grain. Because of this, you can "tune" a composite's strength characteristics by controlling and combining the directions of the fibers. If you want strength over the length of a part, simply align the fibers lengthwise. Likewise, if you're making a tube that needs torsional (twisting) strength, it's best to arrange the fibers helically, like springs, weaving them together with opposite rotations.

The table below describes the three most common forms of manufactured fibers: woven, unidirectional, and filament. Each general type can be produced from carbon, fiberglass, aramid (Kevlar), boron, basalt, and several other materials, which are chosen according to their particular physical properties.

FIBER TYPE	CHARACTERISTICS
Woven (aka "cloth")	Comes in rolls and resembles the thick nylon fabric used in trampolines. Weave styles vary according to the fibers' directional alignment, drape (how well the cloth conforms to mold surfaces), and wet-out (how easily the fibers can be infiltrated by resin).
Unidirectional	Comes in rolls and resembles a very fine, wide paintbrush. Composed of parallel fibers that are intermittently joined over their width by thin resin-coated fibers that keep the strands aligned into an easily usable form.
Filament (aka "roving" or "tow")	Comes on reels as continuous strands of fibers, loosely gathered into a thread. It can be easily unwound and placed wherever necessary in a layup.

Resin is available in hundreds of different types, each with its own chemical and physical characteristics. In general, polyester, epoxy, and vinylester liquid resins are the most widely used in composite constructions.

You can also buy *pre-preg* fiber, which already has the resin mixed in. Pre-preg is easier and less messy to use, and it doesn't waste resin, but it's also more expensive, is more difficult to obtain, has a limited shelf life, and comes in limited resin types. Most pre-preg comes shipped cold and must be used immediately, or can be frozen (to retard curing) but must be used soon. Another type, *elevated-temperature cure pre-preg*, ships normally and is cured in an oven or autoclave. Molding a composite using pre-preg is known as a form of *dry layup*, while starting with separate fiber and resin is called *wet layup*.

When constructing high-performance composites, designers attempt to do several things at once: orient the fibers for highest strength, inhibit delamination (the peeling apart of fibers), and ensure dimensional accuracy. In addition, they try to minimize voids and bubbles in the composite, which weaken the structure, and keep the resin-to-fiber ratio down to somewhere between 50/50 and 40/60, which is the optimal range for strength and lightness. Many automotive enthusiasts who purchase carbon fiber hoods for their cars complain that they often weigh more than the original steel. This arises from too much resin, either in a thick, glossy top coating called a *flood coat*, or in a generally poor layup.

MAKING THE CASE

We're going to create our own composite iPod cover using wet layup, and *compression-mold* it by using C-clamps. This relatively low-tech method proves successful for making flat shapes out of sheets of material. For shallow rounded forms, you can compress the shapes under plastic bags filled with sand, clamped between wooden boards. Since the iPod has a simple rectangular shape, we can create a nicely consolidated, smooth-surfaced shell by using an internal mold (or *mandrel*) with blocks clamped on its large sides. Although this demonstration explains how to make a case for an iPod mini, you can use this same method to make a hard, lightweight case for practically any device that has a uniform cross-section over its length.

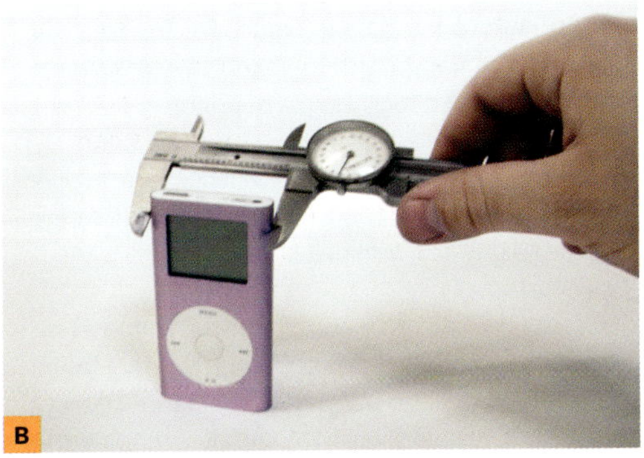

Fig. A: iPod minis and their carbon-fitted cases.
Fig. B: Measure the iPod.

These instructions include a metal belt clip for the case. If you don't want the clip, just skip the steps that involve the screws, the T-nuts, and the metal strip. For diagrams with complete measurements of all the parts, see makezine.com/09/primer.

PREPARE THE MANDREL

1. Use calipers to take close measurements of the iPod and determine the mandrel size (see Figure B). To guarantee a snug (but not too tight) fit for the iPod, the mandrel's width and thickness should measure .02" to .04" more than the iPod's actual dimensions. The iPod mini is only slightly over ½" thick, so you can form a mandrel easily using a ½" thick piece of medium-density fiberboard (MDF) that has been bulked up with tape to the correct dimensions. If you anticipate taking your iPod in and out of the case frequently, you should line the interior walls of the case with felt or velvet to keep the iPod from getting scratched. In this case, it's advisable to oversize the mandrel even more to account for the thickness of the lining.

For the mandrel's height, add 1" to the height of the iPod. This translates to a total height of 4.75", which will allow enough extra carbon fiber material for clean trimming of the final part.

Here's What You'll Need

MATERIALS

1.5" diameter carbon fiber "sleeve" material, at least 2' length but it's good to have more on hand, just in case sollercomposites.com

Two-part epoxy, slow cure, high-strength from a hobby or hardware store

Scrap piece of ½" thick medium-density fiberboard (MDF) about 6" square

Masking tape

Clear packing tape

Polyethylene sheeting 3-5 mil (75-125 micrometers) thick

Clear lacquer spray paint

Rubber band

Button-sized amount of modeling clay

Petroleum jelly

Latex or nitrile gloves and face mask

For the belt clip:
Matching #6-32 threaded T-nuts (2), ¼" hex button-head socket cap screws (2), and ½" slotted set screws (2) from hardware store, or try mcmaster.com or smallparts.com

12"×1"×.028" strip of stainless steel from hobby or hardware store, or mcmaster.com part #8457K49. You only need 3", but 12" is a typical length.

TOOLS

Table saw or bandsaw

Drill press or drill and drill bits

Router with ¼" round-over bit

Dremel with rotary file or cut-off bit

Calipers

Ruler

C-clamps (2) and vise

Flat-blade screwdriver

Hex/Allen wrench to fit button-head socket cap screws for belt clip

Hammer

Pliers

Metal file

Sandpaper

Scissors

Sheet metal snips

Marker

Photography by John Wanberg

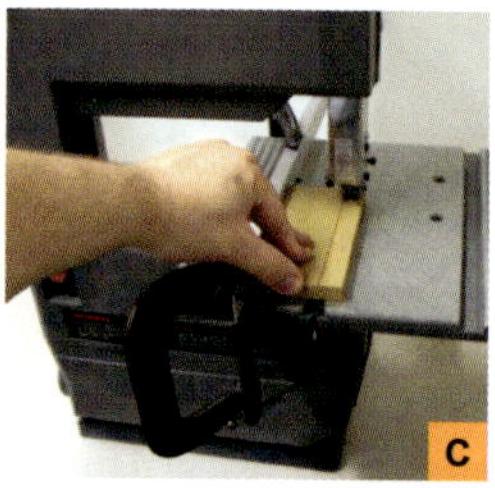

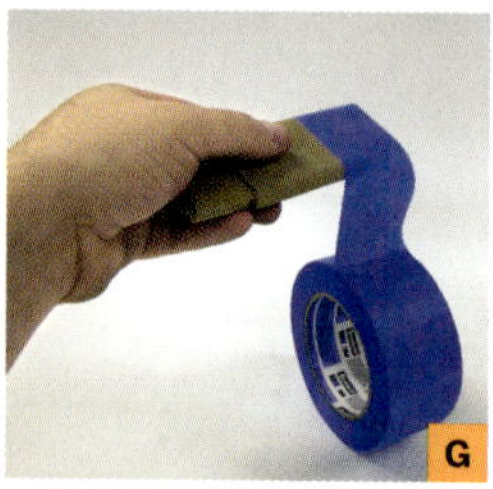

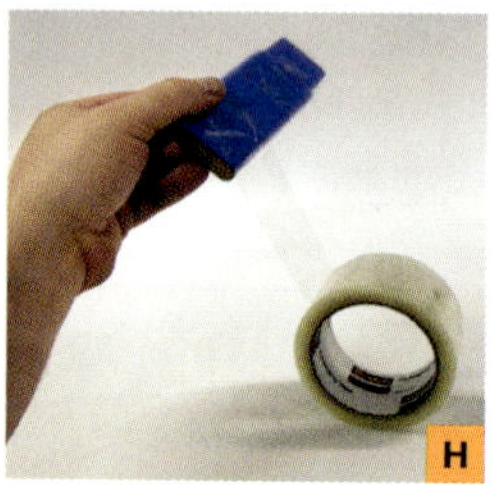

Fig. C: Cut the mandrel.
Fig. D: Replicate the fillets with a router.
Fig. E: Flange cuts.
Fig. F: Vertical flange cuts.
Fig. G: Blue tape wrap.
Fig. H: Clear tape wrap.
Fig. I: The finished plug.
Fig. J: Clamp blocks.
Fig. K: Insert marking.

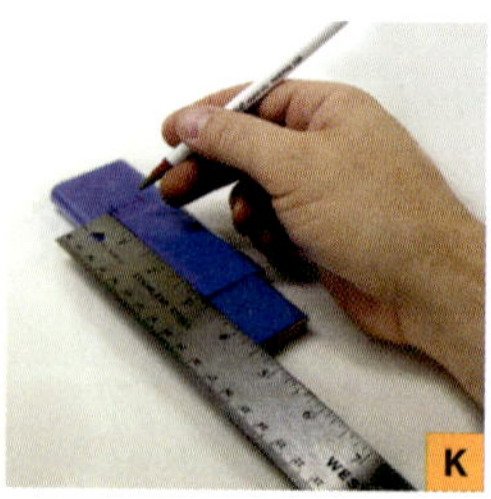

2. Cut the mandrel to size, making sure that it's very straight along its length, so it will slide out of the hardened composite after curing. You can ensure a straight edge by cutting the MDF using a fence on either a table saw or band saw (Figure C). Then use a router with a ¼" round-over bit to replicate the fillets (rounded edges) along the sides of the iPod (Figure D).

3. In order to form the bottom of the case cleanly, the mandrel needs to have a slight flange for the carbon fibers to wrap around. Create this flange by marking its location 1" from the bottom of the mandrel, then cutting a ⅛" deep groove around the mandrel with a band saw (Figure E). Trim off the excess on the flange with the band saw by using a block to hold it squarely in place (Figure F).

4. Wrap tape over the mandrel if necessary, to build it up to the desired dimensions (Figure G). Be careful to keep out any wrinkles or bubbles.

5. Create a removable sheath for the mandrel by wrapping it with either polyethylene sheeting taped to itself, or with clear packing tape (Figure H). I created a tight, removable sheathing by wrapping the packing tape around the mandrel sticky side up, and then wrapping it again sticky side down (Figure I).

6. Cut 2 clamping blocks out of MDF scrap, making them the same dimensions as the mandrel, without the extra 1" of length. Cover the blocks with packing tape to keep them from adhering to the composite during cure (Figure J).

MAKE THE BELT CLIP

1. If you're adding the belt clip to your case, mark and drill two ⅛" diameter holes on one of the clamping blocks, in the location and spacing that you want for the clip's screw mounts.

2. Mark and drill holes in the mandrel at locations that match the ones drilled in the clamping block (Figure K). These are pilot holes for the set screws that hold the T-nuts during layup and cure, so make them slightly smaller than the screws' diameter.

3. Tighten the set screws into the mandrel so they are exposed at least ¼" above the surface (Figure L).

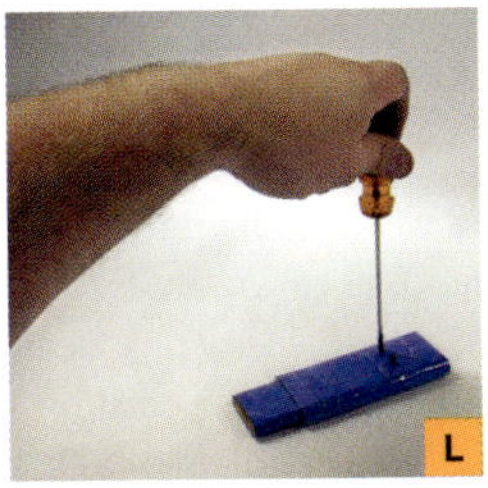

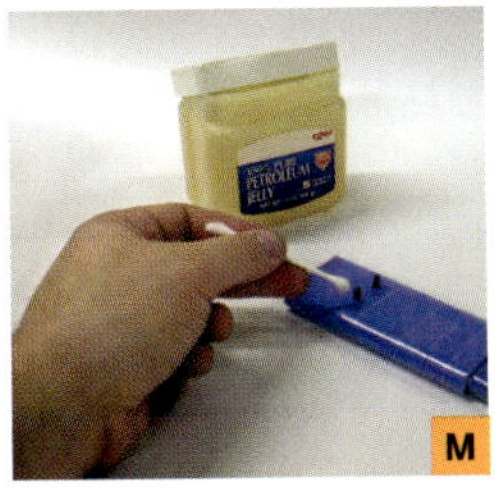

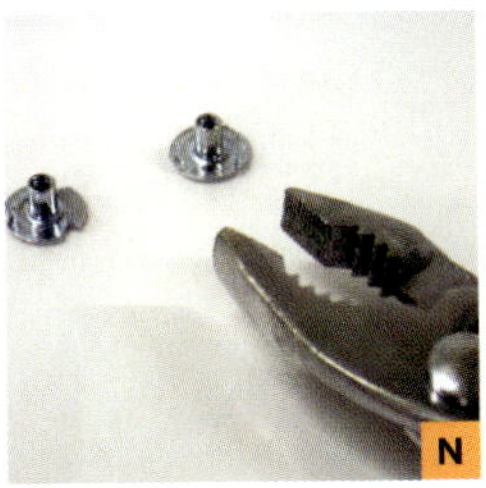

Fig. L: Set screw insertion.
Fig. M: Lubricate inserts.
Fig. N: Nut forming.
Fig. O: Raw carbon trim.
Fig. P: Clay insertion.
Fig. Q: Epoxy spreading.
Fig. R: Clamping.

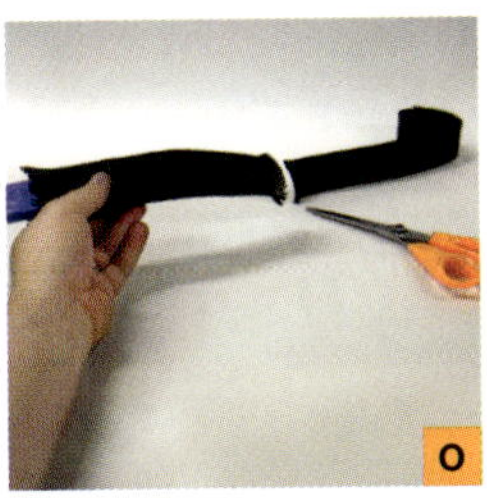

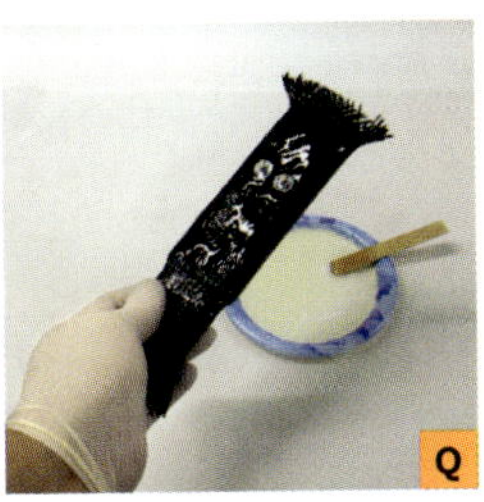

4. Apply a very thin coating of petroleum jelly to the set screws to keep the resin from locking them in place during curing (Figure M).

5. Use pliers to flatten the spikes found on the T-nuts (Figure N). The slightly jagged nubs that are left will grab into the composite later and help keep the T-nuts from rotating when you insert the screws.

LAY UP THE COMPOSITE

1. Cut 2 equal lengths of the carbon sleeve, long enough to allow a little extra material to hang over each end of the mandrel (Figure O).

2. Carefully slide one length of sleeving over the mandrel, and wiggle the fibers around the set screws so that they lay flat against the mandrel with the screws poking through. Screw the T-nuts onto the set screws and pack a small amount of modeling clay into the top of each to prevent resin from filling in the hole (Figure P).

3. Mix up the epoxy according to the manufacturer's directions. Wearing latex or nitrile gloves, apply it generously to the sleeve, working it between the fibers with your fingers to ensure total wet-out.

4. Pull the second layer of sleeve over the mandrel, and, as with the first sleeve, work the fibers around the T-nuts so that the nuts poke through and the fabric lays flat. Work more epoxy into this second layer until it is also completely saturated (Figure Q). Tug carefully on either end of the sleeves to straighten any undulations in the fiber weave, and smooth it down as much as possible to ensure clean aesthetic results.

5. Place the clamping blocks on either side of the layup and mandrel with the holes in the drilled clamping block matching the locations of the T-nuts. Apply the C-clamps carefully and tighten them down while trying to minimize shifting in the fiber (Figure R). Wrap a rubber band around the lip of the flange to pull the fiber around the edges. Smooth away any excess resin from the layup with your gloved fingers; this will save surface finishing time later. Allow the layup to cure overnight.

RESOURCES

Article about current carbon shortage due to increased demand from aerospace and military: fibreglast.com/carbonfibershortage

Soller Composites: sollercomposites.com

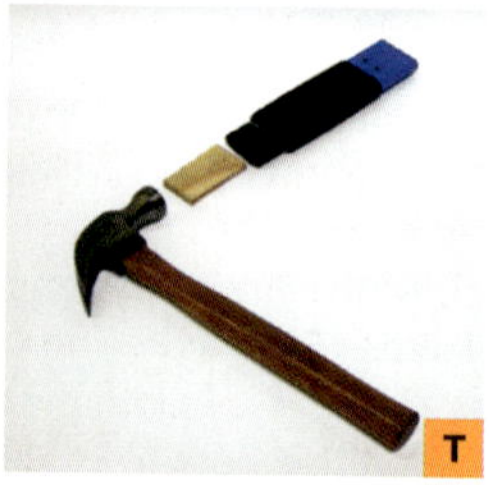

Fig. S: Trimming excess ends with Dremel.
Fig. T: Knocking out the mandrel.
Fig. U: File smoothing.
Fig. V: Final trimming.
Fig. W: Lacquer coating.
Fig. X: Clip forming.
Fig. Y: Finishing the screw tightening.
Fig. Z: Finished clip.

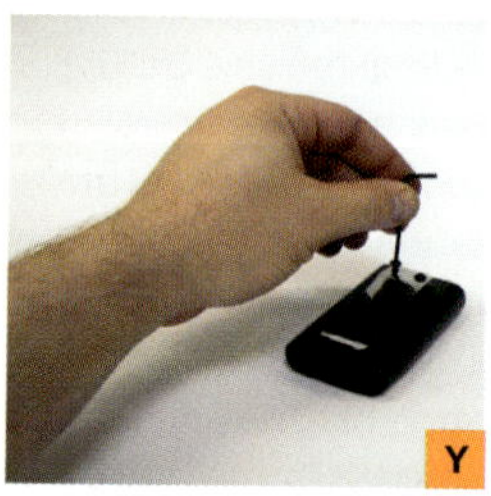

FINISH THE CASE

1. Remove the clamps and pry off the clamping blocks with a flat-blade screwdriver. Use a Dremel with a rotary file bit or cut-off wheel to trim the excess ends from the composite (Figure S). Make sure to use a dust mask; carbon fiber and epoxy produce some nasty dust. Also wear gloves and use caution, because cured carbon fiber slivers are even more heinous than those produced by fiberglass.

2. Unscrew the set screws from the T-nuts. Use a hammer and a small scrap of wood to knock out the mandrel from the flange (Figure T).

3. Pull the plastic sheathing out from inside the case and then trim the case down to size with a Dremel. File and sand the case as needed to create a smooth surface (Figure U). Wet sandpaper works best when smoothing carbon fiber surfaces, and also keeps dust to a minimum. Take care to keep from sanding through one layer into the other, as this will look unsightly and weaken the case.

4. Precisely measure and mark the locations of the screen and buttons on the case (a 1.5" diameter circle template will mark the thumbwheel nearly perfectly). Use the Dremel to cut the holes, and finish the edges with a file and sandpaper (Figure V).

5. Paint the case with several spray coats of clear lacquer to give it a professional finish (Figure W).

FINISH THE BELT CLIP

1. Use sheet metal snips to cut the stainless steel strip down to 3" long, then smooth and round the edges and corners with a file. Then drill holes at one end of the clip with spacing to match the T-nuts.

2. Use a vise to put 2 slight widthwise bends in the strip, one ½" from the end with the screw holes, and the other in the opposite direction, ¼" from the other end (Figure X).

3. Mount the clip to the case with the button-head socket cap screws (Figure Y). Your case is now finished and ready for use (Figure Z)!

John Wanberg is a professor of Industrial Design at Metropolitan State College of Denver who enjoys researching innovative technologies and designing alternative automobiles.

RETROCOMPUTING

An Apple II in a Wristwatch

By Tom Owad

Emulating a classic Apple computer in the Palm OS.

Zoom in on Zork: When you install the Appalm][emulator on a Palm wristwatch, you might need a microscope to read it.

I've occasionally heard claims, usually as a tribute to our technological advancement, that a modern digital wristwatch has more power than some old mainframe or the space shuttle, or some other system involving early computer technology. The Apollo Lunar Module seems to be the most popular comparison. While it's difficult to compare a watch to a navigation system, I think you'd be hard pressed to find a wristwatch with 4KB of RAM and a real-time multitasking operating system.

On the other hand, it's easy to understand the awe of wearing a computer on your wrist that's more powerful than that used to land on the Moon. One wristwatch that can reasonably make such a claim is the Fossil Abacus Wrist PDA. With a 66MHz Motorola DragonBall based on the 68000 core and 8MB RAM, it's not dissimilar to a personal computer from the early 90s.

The operating system is Palm OS 4.1, which is very well suited for a PDA. What I'm looking for, however, is not a PDA but a general-purpose computer — the technology of yesteryear shrunken to miniature using modern technology.

What I really wanted was a Macintosh on my wrist, but the technology isn't quite there yet (at least not for the $50 an Abacus Wrist PDA costs). I did, however, find an emulator for the Apple II that runs on Palm OS. Appalm][(palmapple.sourceforge.net) is intended for full-sized Palms, but the Wrist PDA does claim it can run most Palm applications, so I decided to give it a try.

I copied the files to the watch and launched Appalm. The screen immediately went black and a line of miniature white text appeared at the top: "Apple //e". I clicked on the Palm menu and loaded the test disk image. The disk successfully booted to the menu screen, full of text. The Apple IIe typically had a 12" to 14" monitor. The Wrist PDA has an LCD that's barely 1" in diameter. The tiny white text on the black background was almost impossible to read. Angling the screen so I could make out the text, I successfully loaded a benchmarking program and ran the Memory Read test. It came in at 168 seconds on the wristwatch, compared to 112 on the real Apple IIe.

Satisfied that this would not be an appealing platform for arcade games, I decided to try something simpler: Zork! Within minutes, I was "standing in an open field west of a white house, with a boarded front door." At least that's what I think it said. By this time I was getting a headache from squinting at the small text, so I removed the wristband with a spring bar remover and placed it under a microscope at 20x magnification.

At this magnification, each pixel was clearly visible, but I had to shift the watch to read the entire screen. My eyes no longer hurt, though, so I decided to stick with the microscope for my game of Zork.

Between the microscope and the pen-based data entry (you write over the top of the Apple II screen using Jot or Graffiti to enter text), the game is tedious but nonetheless fully playable and serves as a curious testament to the advancement of technology. Launch the classic 1970s game Lunar Lander and you can use the 66 MHz DragonBall on your wrist to pretend it's 1969 and you're piloting the Apollo Lunar Module.

Photograph by Tom Owad

Tom Owad (owad@applefritter.com) is a Macintosh consultant in York, Pa., and the editor of applefritter.com. He is the author of *Apple I Replica Creation* (Syngress, 2005).

Read about crop circles, make your own temporary tattoos, and take crystal-clear photos with a circular polarizing filter.

TOOLBOX

Tokyo Time Hackers

$100 and up tokyoflash.com

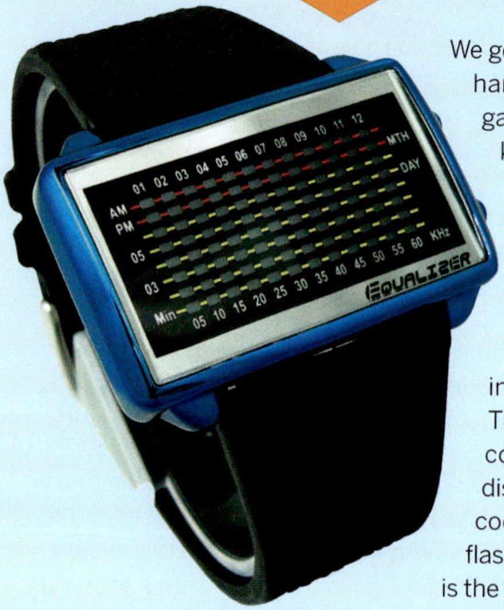

We geeks love to hack the wetware (aka our brains) as much as the hardware and software. Puzzles, brainteasers, math problems, games with complex rules and devilish stratagem, anything to keep those neurons sparky. So when a propeller-head chooses a watch, if given the choice between one that tells time via mechanical hands or digital readouts versus one that rewrites the very rules of time-telling, displaying it in some unspeakably nerdy way, you know what the choice will likely be.

For many a high-dome and wired hipster, the choice is TokyoFlash, the Japanese chronograph merchant specializing in unique, and some might say, insanely odd, timepieces. Their Morse Code Watch tells time by sounding it out in Morse code, "refracting the sound through your wrist," as well as displaying the code on the watch's face. The Scope uses X and Y coordinates and a flashing LED scope-like screen to point and flash out the time. One of TokyoFlash's more well-known watches is the B Version by Twelve 5-9. It uses a radar-like display to indicate the hours, a row of five LEDs for the 10s and nine lights for the intervening minutes. The High Frequency PU uses LCD technology to display a dance of brilliant blue (or green) lights that spike and die like a graphic equalizer. Eventually, all the lights fade away except for the ones that indicate the current time (or month/day).

If you really want to play Play-Doh with your thought forms, get several of these watches, ones that use very different methods of time-keeping, and switch off wearing them. Just be prepared to get the snot kicked out of you if somebody asks you for the time on the subway and you shove one of these weirdo watches in his face.

—Gareth Branwyn

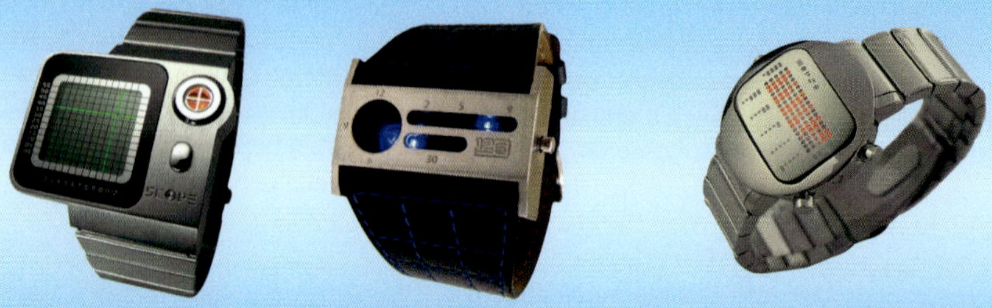

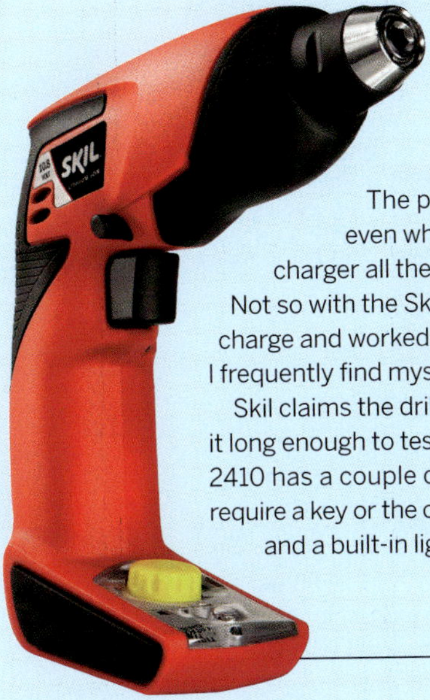

Staying Power

Skil 2410 10.8V Li Ion Drill, $80 skil.com

The problem with most cordless drills is the way the batteries drain even when you aren't using them. Unless you keep the battery pack in the charger all the time, you never know if the drill is going to work when you need it. Not so with the Skil 2410 10.8V Li Ion Drill. It sat in my closet for a month without a charge and worked like a champ when I pulled it out, unlike the NiCad-powered tools I frequently find myself cursing.

Skil claims the drill can hold a charge for 18 months of non-use, but I haven't had it long enough to test. (If you don't trust it, you can store the drill on the charger.) The 2410 has a couple of other handy features, like a quick-change chuck that doesn't require a key or the coconut-crushing grip of an orangutan to make sure bits don't slip, and a built-in light to help you see what you're doing.

—Mark Frauenfelder

Fun with a Fridge

$14 frigits.com

My mother has collected magnets for her fridge as long as I can remember. Whenever we went on trips, all she wanted was a souvenir magnet. So, you can imagine how surprised we were to see the memory slate wiped clean for a game.

Frigits is a great magnet set that lets you build a marble run right on your fridge. Each piece has swiveling magnets on the back, so you can turn them as you'd like. They send you 12 marbles, and you can buy the "Frigits Extension Launcher" so short folks can send the marbles high enough.

I believe that the only real requirement to use it is to understand gravity; I think most Earth residents are familiar with that. Kids love the clicking, clacking, and spinning as the marble makes its way down. We ended up moving the Frigits to the side of the fridge so we would be ready to play at all times. (See my video of Frigits in action at makezine.com/go/frigits.) ***—Brian Stucki***

The Daily Grail

dailygrail.com

I've savored The Daily Grail for years, and like Boing Boing, its mix is hard to describe except to say that it combines many things I'm interested in — archaeology, anthropology, paranormal phenomena, alternative history, natural science, the occult — and approaches them with the same fascination and skepticism that I have. The links are great, and the comments frequently hilarious. As well as keeping you up-to-date on outlier thought, TDG also follows a pantheon of fringe notables like Graham Hancock and Loren Coleman, and performs due diligence on self-assured authorities such as Egypt's celebrity minister of antiquities Zahi Hawass and career skeptic James Randi.

At the end of 2003, the long-running TDG became the watered-down "TDG News Briefs" behind the home page of the ill-fated *Phenomena* magazine. Thankfully the site came back two years later with even more to share. Now it also hosts The Red Pill, a reference wiki that "grew out of Wikipedia's tendency to avoid 'fringe' issues," and the PDF magazine *Sub Rosa*, which takes a deeper look at fringe topics with historical features, profiles, interviews, and artist spotlights.

It all just makes me want to kidnap the bloggers so that we can drink wine and talk until dawn, while taking turns choosing the music to put on the stereo. I'm sad that I have only one lifetime to explore all the fun things that TDG covers. Or do I?

—Paul Spinrad

Editors Phillip Torrone and Arwen O'Reilly talk about kits from makezine.com/blog.

Snow Bike Kit

$300 ktrakcycle.com

Add a ski to the front of your mountain bike and a tracked rear drive, and you're set to be James Bond. Not a bad hack either, if you have the time to build it from scratch.

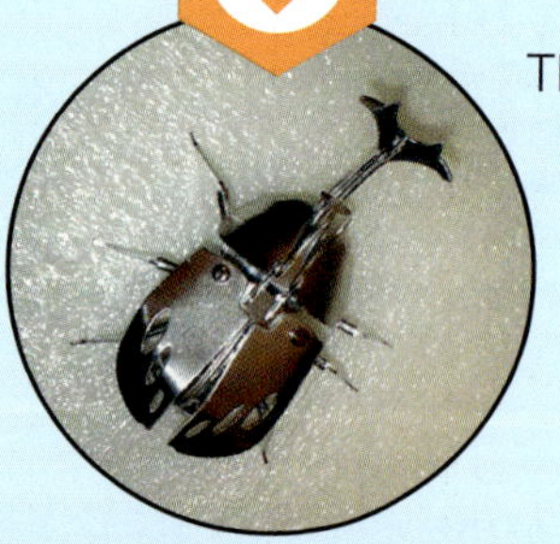

Meet the Beetles!

$50 e-clec-tech.com/gamkitse.html

The flat metal pieces are cool enough, but once assembled you end up with a piece of art: a rhino beetle, scorpion, praying mantis, butterfly, or stag beetle.

Disco Dance Floor Kit

dropoutdesign.com

Hey, those craft engineers who did the Disco Dance Floor have a new company and sell kits now. Check 'em out — yay, more 'lectronic kit makers!

LED Cube Kit

$120 and up

hypnocube.com

Make your own disco with these LED cube kits. Go minimalist with two colors, or go wild with all three!

Photograph of beetle by Izu Watanabe

Hovercraft Kit

$4,696 (or $112 for the plans) hovercraft.com

Wow, forget those paper plate hovercrafts. This is what you want to make. You know ... for your family.

Power Trips

$50 and up newenergyshop.com

Exergia has a ton of great kits: Stirling engines, steam engines (including a tiny steam-powered trike), fuel cells, photovoltaics — everything "interested individuals" would want to build. A great way to start understanding more about alternative energy technologies.

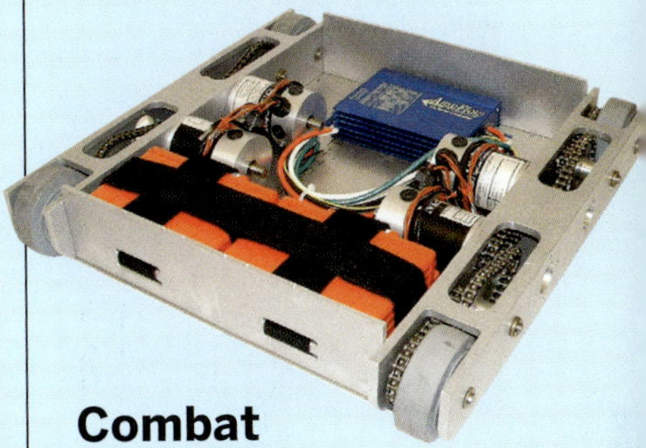

Combat Robot Kits

$500 battlekits.com

Whoa, BattleKits has combat robot kits ranging from 30 pounds to a massive 340-pounder — all based on the extremely successful "BioHazard" bot.

DIY Temporary Tattoo Kit

$15 makezine.com/go/tattoo

Temporary tattoos are always a big hit with kids, but the designs usually aren't that great. Here's a temporary tattoo kit for kids so they can make their own (or just use the stencils). It looks like it's a pen with a motor, so you could make your own, too.

Altair 8800

$1,500 altairkit.com

What a great kit — a fully functional Altair! It's not cheap, but the parts are all high-quality, you get free tech support and repair, and you end up with a reproduction of one of the early computers.

Minds on the Fringe

***Worlds of Their Own* by Robert Schadewald**

$23, SangFroid Press

Bob Schadewald had a lifelong interest in fringe science. This posthumous collection of his published and unpublished materials (skillfully edited by his sister, Lois) is a highly readable account of several varieties of pseudoscience. The unifying theme is "fringe thinkers" who create their own versions of reality, contemptuous of established mainstream science. Bob treats his subjects with respect and even sympathy (he knew many of them personally), but he clearly reveals why their ideas are flawed and misguided.

His chapter on "The Philosophy of Pseudoscience" reveals the common characteristics of these independent thinkers. Here you'll find the stories of historian Immanuel Velikovsky, who rewrote the book on solar system astronomy; John Keely, who claimed he could power a freight train coast-to-coast on a gallon of water; and assorted creationists who freely engaged in "lying for God." This is an informative and entertaining book of continuing relevance, for ideas of this sort never die, perpetually reborn in new clothing.

—Donald Simanek

Roadside Science

***Pomegranate Roads: A Soviet Botanist's Exile from Eden* by Dr. Gregory M. Levin**

$18, Floreant Press

One tidbit to take away from this book is the word *punicology*, the study of pomegranates. If you were a punicologist, you would call its seeds by the name "arils." There are far more varieties of pomegranate than you can imagine: the author collected 1,117 varieties from 27 countries.

Levin calls this book a "montage" and it truly is: part science, part history, part gardener's guide, part autobiography. It's the life story of a Soviet botanist who left his native Leningrad to work and live at the agricultural station of Garrigala: "I witnessed the prime time and the sunset of this little center of culture at the very edge of Soviet influence." Using Garrigala as his base, he explored the entire region, finding wild varieties of pomegranate growing in subtropical gorges. In 1991, Turkmenistan became an independent country and Levin was cut off from any funding. He ended up in Israel, where there's ongoing research. Levin's life story is about science, not as conducted in a laboratory, but out in the field. "The world should be studied by roaming across it," he writes.

There is also a nice backstory to this book, whose publisher heard about Levin from a BBC broadcast and decided to go find him. She got him to write this book, which you can pick apart to find sadness among its many arils of sweetness.

—Dale Dougherty

Liberation Psychology

***Fate* magazine**

$32/year (12 issues) fatemag.com

Fate magazine covers UFOs, ghosts, psychic phenomena, and other topics that you'll also see in the *Weekly World News*, but *Fate* takes a more principled approach. Where the *WWN* complicates its voice by mixing real reports (largely from the third world) with funny exaggerations and *Onion*-like fabrications, *Fate* explores with undistanced curiosity.

Fate is run like a fanzine, and the lively community it has created is evident. A long Letters section serves as a miscellaneous forum, and other sections collect readers' first-person accounts of mystical experiences, proof of survival beyond the grave, and other topics.

Features range from local ghost legends, UFO contacts, and speculative history to celebrity stories, like an account of the 1973 séance where Church of Satan founder Anton LaVey summoned the ghost of Marilyn Monroe. Through it all, intriguing ads promote books, psychics, and even messiahs.

If you're more perceiving than judging, and you relish your role as a contributing neuron to the wise hive mind, then you and your pals will have a great time processing *Fate*'s fascinating supply of edge cases and shades of gray.

—Paul Spinrad

Ground Zero

***Overlook* by the Center for Land Use Interpretation**

$35, Metropolis Books

The Center for Land Use Interpretation occupies a nondescript building along Venice Boulevard in L.A., but its location takes on meaning when you realize that it is next door to, and shares internal walls with, the Museum of Jurassic Technology (read *Mr. Wilson's Cabinet of Wonder*, about the museum's creative take on history and technology, if you haven't already). CLUI, as it is popularly known, is basically a vast database of the ways that humans alter landscape — an "inventory of examples," if you will. They discover the abandoned plutonium dump sites, the covert military installations, the whistle-stop tourist attractions, and the towns flooded by reservoirs (one stunning chapter is devoted to "intentionally drowned" towns). A research institution with a whiff of the art collective about it, CLUI holds exhibitions and runs guided tours, but the book opens our eyes to the spaces we ignore and transforms the way we think about them. A love letter to the outwardly drab outposts of civilization, *Overlook* gives power to the abandoned, the hidden, the unintentionally beautiful, and the just plain weird.

—Arwen O'Reilly

TOOLBOX FAVORITE THINGS

As chosen by John Krewson, an editor at *The Onion*.

Mechanix Gloves Eye protection is great, but *Onion* writers are conscientious about preserving their typing fingers while planning the next heist. We've found comfortable Mechanix gloves deflect impacts, sharp splinters, and hot weld spatter; they also guard against spreading fingerprints around bank vaults. We especially like the Glove Light, with its small but powerful LED flashlight on the back, good for illuminating cramped air ducts or locating loose alarm-panel bolts before they trip the motion sensors.

Dental Tools and Picks You don't have to be a lock-picking enthusiast to love these incredibly sharp, durable, and, we should stress, legal-to-own items — although we're not sure it's ethical to sell them to dentists. We employ them daily in the imprecise fields of "getting gunk off of things" or "cleaning crud out of crannies." People with bad teeth and bad memories will hate the distinctive sound they make, but there's no other downside; they're even cheap.

High-Powered Angle Grinder Anybody can carve their initials in wet cement, but angle grinders let you put your mark on almost anything. Yes, they're perhaps the most dangerous handheld power tool there is: they make showers of sparks, they throw chunks of your material everywhere, and the discs tend to ... well, "explode" is really the only word for worn discs flying apart at 11,000 rpm. Still, nothing's better to have around when you need to get through a barred window or demount a security door, or when some idiot locks their bike to yours. We love the Milwaukee Electric Tools 6155, but anything with enough amps works just as terrifyingly well.

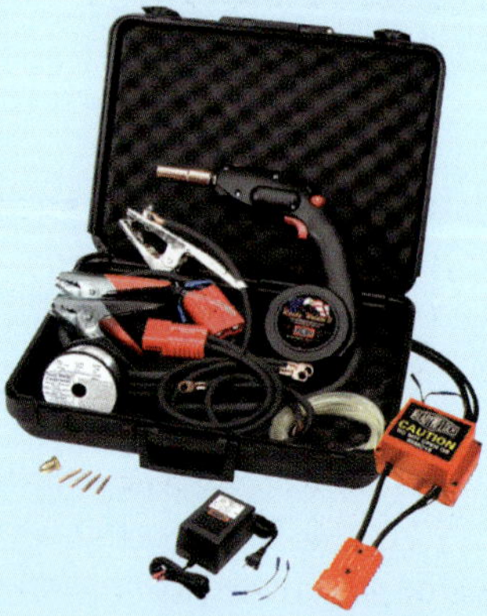

Ready Welder II As MAKE has noted before, one must know welding to be truly capable. And if one wishes to be suave and debonair as well, one must arrive at a moment's notice with one's welder in a ruggedly handsome briefcase, ready to transform an airport shuttle into a fake armored car that's guaranteed to fool sleepy night watchmen. Therefore, one needs the Ready Welder II, a fairly forgiving MIG setup that can run off batteries, welds thin steel even without gas, and packs neatly away in its own ABS case.

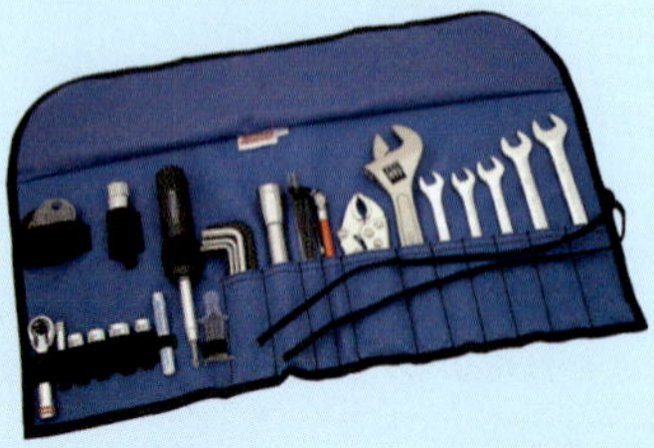

CruzTOOLS Tool Kits Most hobbyists neglect to get the good basic pieces you need to work on everyday machinery like counterfeit printing presses, getaway cars, and the trap doors on elevator ceilings. Instead, they get the $20 all-in-one checkout-line kit featuring tools made from an alloy of sawdust, soda cans, and rat droppings. CruzTOOLS RoadTech tools absolutely don't break; every tool in the roll is solid and dependable, from the combination wrenches to the flashlight to the lovely, jewel-like ¼-inch palm-sized ratchet.

The Onion (theonion.com) is a weekly humor magazine and front organization for an intrepid team of international thieves.

I'll Be Your Mirror

SS Mirror $3.45 for 3
greenbulb.com

The SS (Self Shot) Mirror is something I can get behind, er, in front of. It's a little silver button you can stick on the front of your camera to aid in taking self-portraits, and it's a steal at $3.45 for three — and you'll want one for every device.

It's a very easy installation since it's just double-sided sticky, and if it gets scratched or you're bored of it, it's easy to remove. I compulsively take self-portraits, and though I've gotten good at holding the camera at arm's length with some idea of what's in the background, this works like a charm to make sure the camera is at a good angle and everyone's in the photo. *—Andrea Dunlap*

Power Shot

Holga Film Camera $20
makezine.com/go/holga

Holga sounds vaguely Eastern European, but it was created as a budget snapshot camera in China. Then a funny thing happened: a dud at home, it became a hit in the West, for its artfully terrible optics and low-tech approachability.

Yes, it's the Holga, cult favorite of art students and hipsters everywhere. Search Flickr for "Holga" and you'll quickly see how its signature vignetting and erratic focus bring an atmospheric mystery to ordinary photo subjects. The Holga shoots on big 120 roll film, with most fans choosing the 6×6 square-format setting.

It's also a great kid's camera. No pesky shutter speeds to worry about, and if it tumbles down the concrete steps ... hey, it's $20! (It may even work afterwards.) It's also a subtle way to encourage your future hacker. The Holga's technical deficiencies invite (er, require) a certain amount of remedial engineering, like fixing the back latches, plugging light leaks, and finding a use for the non-functional "sunny" setting. But no worries: the net is overflowing with tips, mods, and other users to give advice. See holgamods.com/mods/mods.html for inspiration. *—Ross Orr*

Photograph by Michael Simmons

Circular Logic

Circular Polarizing Filter
$30 and up tiffen.com

A circular polarizing filter is a wonderful tool that gives me an amazing amount of control over my photographs, yet most amateur photographers don't take advantage of one.

It allows me to eliminate or emphasize reflections on a variety of surfaces including glass, water, many plastics, and even skin, just by rotating the filter on the end of my lens. Amazingly, there's even a band of the sky that can be darkened using the filter without affecting the rest of the scene.

Don't think: "I guess I can't get a shot of the circuit board without a distracting reflection." Get a circular polarizer.

They come in particularly handy when you don't have much control over the lighting. They're sold in sizes that will fit on virtually any SLR lens and can be attached to many non-SLR cameras with an adapter.

A circular polarizer will cost two or three times more than a linear polarizer, but linear polarizers have a nasty habit of interfering with the exposure meter and autofocus system in most cameras, so the extra money is well spent.
—Tanner Beck

Tanner Beck is a necessity-loving nerd photographer in Ann Arbor, Mich.

Gareth Branwyn is the cyborg-in-chief of streettech.com.

Andrea Dunlap is a photographer and filmmaker for seedlingproject.org.

Ross Orr keeps the analog alive in Ann Arbor, Mich.

Donald Simanek's "Museum of Unworkable Devices" can be found at lhup.edu/~dsimanek.

Brian Stucki lives in Las Vegas with his family and writes for freemacware.com.

Have you used something worth keeping in your toolbox? Let us know at toolbox@makezine.com.

BLAST FROM THE PAST

Old-School Hand Tool Hacks

By Mister Jalopy

What I learned from the 1963 Bureau of Naval Personnel Training Course.

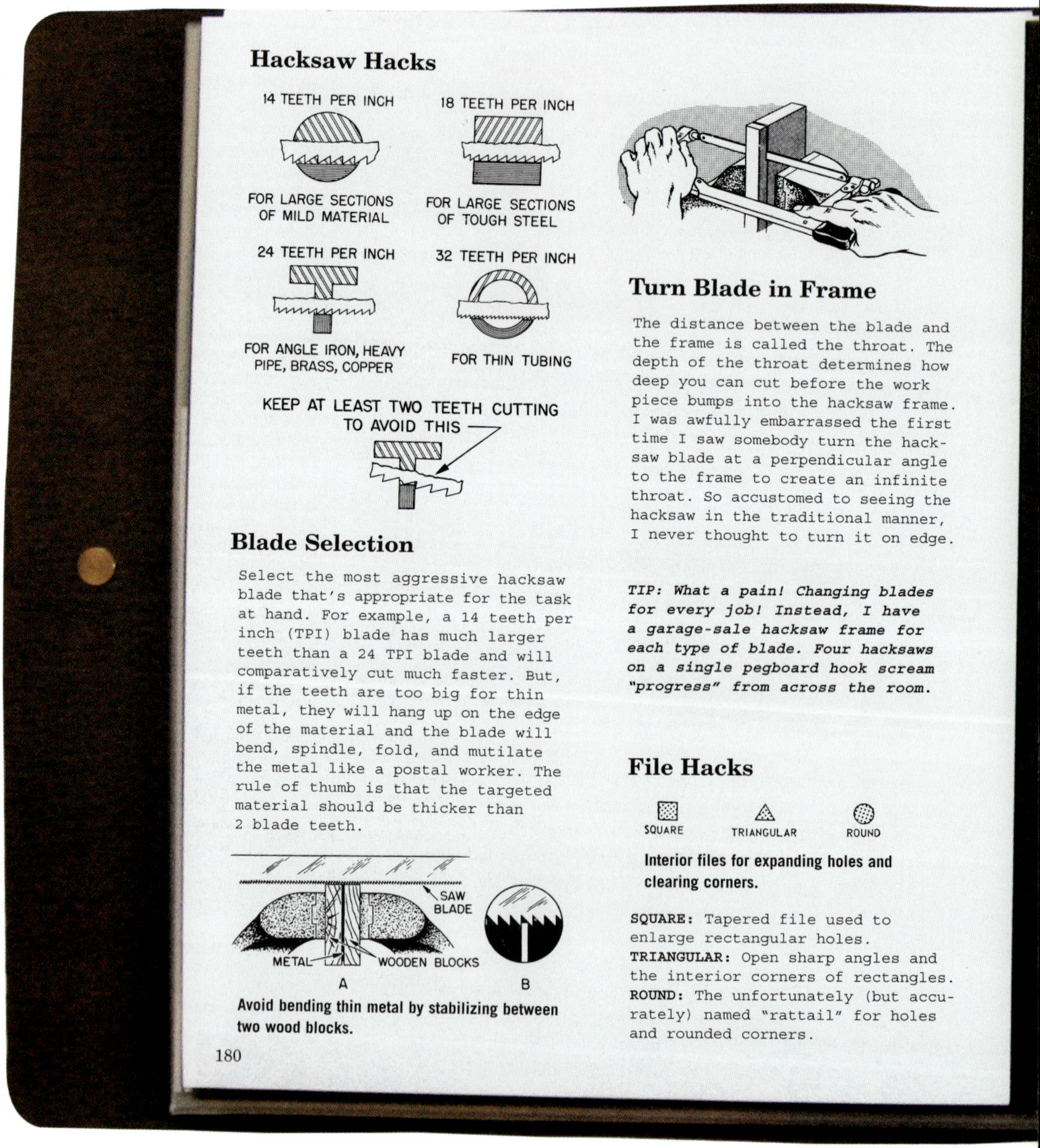

Hacksaw Hacks

14 TEETH PER INCH — FOR LARGE SECTIONS OF MILD MATERIAL

18 TEETH PER INCH — FOR LARGE SECTIONS OF TOUGH STEEL

24 TEETH PER INCH — FOR ANGLE IRON, HEAVY PIPE, BRASS, COPPER

32 TEETH PER INCH — FOR THIN TUBING

KEEP AT LEAST TWO TEETH CUTTING TO AVOID THIS

Blade Selection

Select the most aggressive hacksaw blade that's appropriate for the task at hand. For example, a 14 teeth per inch (TPI) blade has much larger teeth than a 24 TPI blade and will comparatively cut much faster. But, if the teeth are too big for thin metal, they will hang up on the edge of the material and the blade will bend, spindle, fold, and mutilate the metal like a postal worker. The rule of thumb is that the targeted material should be thicker than 2 blade teeth.

SAW BLADE
METAL
WOODEN BLOCKS
A
B

Avoid bending thin metal by stabilizing between two wood blocks.

Turn Blade in Frame

The distance between the blade and the frame is called the throat. The depth of the throat determines how deep you can cut before the work piece bumps into the hacksaw frame. I was awfully embarrassed the first time I saw somebody turn the hacksaw blade at a perpendicular angle to the frame to create an infinite throat. So accustomed to seeing the hacksaw in the traditional manner, I never thought to turn it on edge.

TIP: What a pain! Changing blades for every job! Instead, I have a garage-sale hacksaw frame for each type of blade. Four hacksaws on a single pegboard hook scream "progress" from across the room.

File Hacks

SQUARE TRIANGULAR ROUND

Interior files for expanding holes and clearing corners.

SQUARE: Tapered file used to enlarge rectangular holes.
TRIANGULAR: Open sharp angles and the interior corners of rectangles.
ROUND: The unfortunately (but accurately) named "rattail" for holes and rounded corners.

180

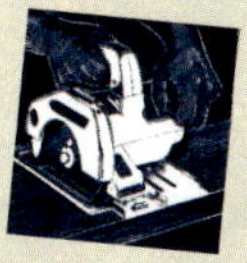

Despite the lackluster title, the U.S. Navy publication *Basic Hand Tools NAVPERS 10085-A* would create a revolution if it were required high school reading. A nation of graduates would be armed with the basic knowledge to change the world around them. Rather than settling for extended warranties and temporal furnishings, individuals could take control of their lives by being handy. It is the rare machinist, farmyard engineer, or Old World craftsman who claims to know how to do everything, as they recognize it is a lifetime of learning. The Navy's *Basic Hand Tools* is an extremely dense, informative, and generous book, as it can save a year in the handy continuum — one hack at a time.

Exterior files for larger, exterior work.

FLAT: General purpose, most often used.
MILL: Thinner, finer, single-cut files for more delicate work.
HALF ROUND: Match the curve radius as closely as possible to the curve of the work.

Cleaning Files

Whether you are using sandpaper or grinding discs, most fine cutting surfaces lose efficacy when they "load up" and fine particles clog the cutting surface.

Pick out the bigger chunks, use the card to brush with the grain of the teeth, and then finish with the brush. Resist oiling files as the added lubrication will negatively impact the cutting ability.

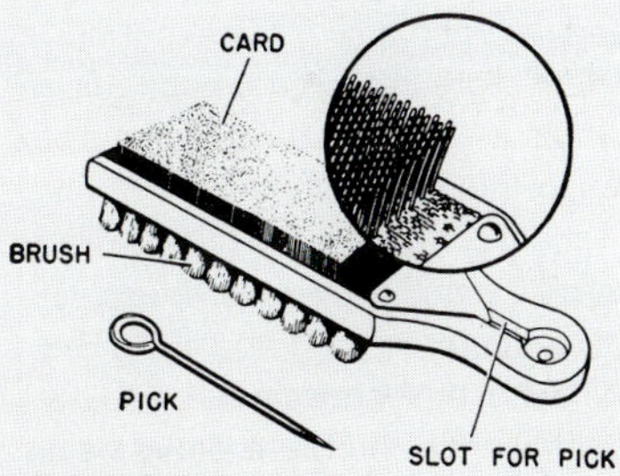

File cleaner.

Cutting Glass Tube

With leather gloves on, use the triangular file to score the glass as you roll the tube. When you have scored the circumference, place a wood matchstick under the glass tube about 1/8" from the scored line. Hold one end securely and press on the other side until it snaps. Clean up the cut with emery cloth wrapped around a file or block of wood. Be careful! Ask to see my gory scar caused by a moment of daydreaming while glass cutting!

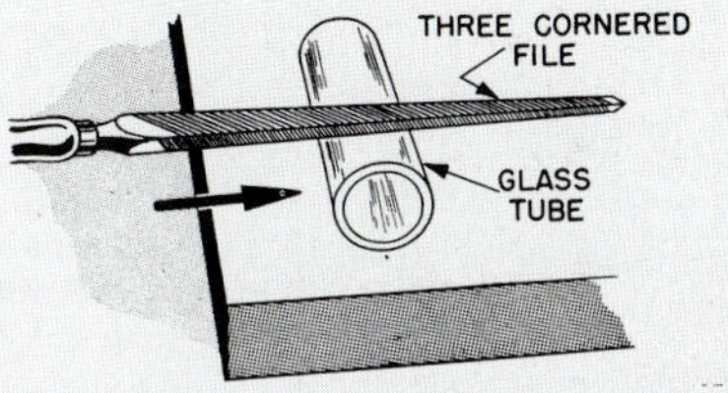

Cutting glass tubing with a three cornered file.

Clean Up Bolt and Pipe Threads

Though technically not a file, I keep my thread restorer with my other files if for no other reason than it looks right in the file drawer! Each face of the thread restorer has teeth for a different thread "pitch." The thread restorer will not perform miracles on heavily damaged bolts, but it does clean up rusty or dirty threads with a little patience. Work slowly and continually check with a nut or female pipe to gauge progress. I will use a smidge of automatic transmission fluid to help matters along when thread-cutting oil is not available.

READER INPUT

Where makers tell their tales and offer praise, brickbats, and swell ideas.

It was great to see the ornithopter project in MAKE 08! For anyone who missed it, an ornithopter is a device that flies like a bird, by flapping wings. I hope MAKE readers find the same fascination and enjoyment I've had from designing and building my own flying bird models over the past 20 years.

Gurstelle's build instructions were great as usual, but the intro focused on the negative. Instead of giving a sense of all the fantastic accomplishments in this field, it was the same rhetoric we've heard so many times before: people can't fly by flapping wings, but you can build a small rubber-band-powered one. The "short history" tells little about what's actually been done. As a result, makers didn't get to hear about all the cool stuff people around the world have been building.

Often the ornithopter concept is judged a failure or impossibility because supposedly no manned ornithopter has flown. This is flawed for two reasons.

First, why should the concept be judged at the manned scale? If we build a device the size of a bird that flies like a bird, then we succeed in imitating bird flight. MAKE readers are smart enough to see where technology is headed. In the future, there will be a lot of unmanned vehicles doing useful work. Moving people around will be far less important than it seems now.

Also, there have been some manned flights. They were dismissed in MAKE out of pure skepticism. The idea that we can't fly is such a deeply internalized myth that people can't accept news of a successful flight after it happens. Of course you can't fly like Daedalus by gluing feathers on your arms, but with the right technology and a powerful-enough engine, we *can* fly by flapping.

Ultimately we can make anything that exists in nature. Even Leonardo understood this, 500 years ago and well before the means were in sight.

—Nathan Chronister

Bill Gurstelle replies: **Mr. Chronister's points are well taken. I, too, look forward to the eventuality of the manned ornithopter.**

Thank you for publishing your magazine! Like many engineers, I got started in engineering by tinkering with ham radio (á là Dilbert), storing static electricity in styrofoam cup capacitors, blowing up power supplies — taking things apart to see how they worked. In short, I got started by engaging in the kind of play you show in each issue of MAKE.

When engineering became a job, somehow the fun got lost. I believe that too much of engineering is treated as an assembly line function — compartmentalized, categorized, commoditized. It's as if creativity can be planned and released in exact, controlled amounts at the exact, appropriate times.

Thanks for bringing back the "messy" human part of engineering. Your magazine should be required reading in every engineering school in the country.

—Jesse Alexander

Editor's Note: **Many readers (and the entire magazine staff) were touched by Colin Berry's memoir (*MAKE, Volume 07, page 36*) about his older brother's soap box derby successes and failures in the 1970s. We also persuaded Colin to record the piece in a quite memorable audio file; you can listen to it at makezine.com/blog/archive/2006/10/the_maker_file_2.html. After listening to the piece, one reader had his own nostalgic tale to add to the soapbox saga.**

Colin,
I just listened to your touching podcast story about your brother and his soap box derby experience. What a compelling story. It took me back to my innocence and youth as I raced in the 1976 soap box derby in Valparaiso, Indiana.

As I listened, I could not believe the race similarities. I too crashed at the end of the race; the brake malfunctioned (pushed too hard and lifted the weight off the wheels), threw my car left, and my rear axle hit the guardrail, and there would be no more racing for me that 4th of July as the axle was bent beyond repair.

My family was there cheering me on amongst the crowd of about 4,000 spectators, and I felt embarrassed and somewhat of a failure at the time to have my dream taken away so quickly and in such a highly public way. But the good news was that everyone remembered me and wanted to look at the damage to the car, and I got interviewed on the radio.

I remember feeling crushed as I thought of all of the hours that I lovingly labored on my passion. The dream was over in a flash, and the only celebration

would be a cake at the house and a party, as I tried to forget the pain that I felt as my dream of making it to Akron was swept away.

As a child growing up, I was obsessed with the race and clearly remember the "Magnet Scandal." How small a world it is to think that this scandal impacted your brother and family's life. Thanks for the great podcast, as it gave me goose bumps to listen to it, and I can't wait to share it with my brother, nieces, nephews, and family. Thanks again.

—Rodney Rumford

I can't believe that nobody has flagged this error in MAKE 07, DIY Circuits, "Hello, World." On page 150, there is a phrase: "the `display 3` routine shows a '3' by turning on all the segments, except for the top-right and bottom-right ones." That will result in an uppercase "E". The phrase should read: "the `display 3` routine shows a '3' by turning on all the segments, except for the top-left and bottom-left ones." Thanks!

—Bruce Cowan

My son and I made the workbench based on the directions I found on makezine.com. First of all, I love the bench, and we had a great time working on it. I came across a neat design enhancement (in my humble opinion) completely by accident. When I was deciding on cutting the legs, I went with 36 inches. This meant that I used an 8-foot 2×4 to make two 36-inch legs, which left a 24-inch piece. I did this twice. When I finished the assembly, I used the leftover 24-inch pieces to further strengthen the top by installing them at 32 inches and 64 inches as braces. Thought you might like that. Now there's no waste (even though I know I could have used the pieces on something else).

Thanks for sharing the design. It made me clean up the garage after.

—Art Trombley

MAKE AMENDS

In the special gift guide section of MAKE, Volume 08, XGameStation programmer Andre Lamothe's name was misspelled.

On page 25 of MAKE, Volume 08, the kinetic sculpture artist Nemo Gould's name was transposed in both the story and the photo credit.

ANDRÉ LAMOTHE'S
XGameStation
WWW.XGAMESTATION.COM
THE WORLD'S FIRST VIDEO GAME CONSOLE DEVELOPMENT KIT.
© 2006 Nurve Networks LLC • 1112 Nancy Lane, San Ramon, CA 94582 • Ph: 925.829.4508 • support@nurve.net

Enhance Your Connectivity
The RCM4000 — Ethernet And Much More
• Ethernet with royalty-free TCP/IP stack
• Up to 25 configurable GPIO
• On-board 12-bit A/D
• Up to 5 serial ports
• 512K flash (program), 32 MB NAND flash (data)
• Rabbit® 4000 at 58.98 MHz
New! Next Generation Rabbit 4000
RCM4000
Complete Development Kit
From $149 Limited Time Offer
RABBIT Semiconductor
Order Online At rabbitkit.com
07064

We give you the tools.
You make the difference.
firstgiving
online fundraising for everybody
www.firstgiving.com

We've got your (iPod's) back.
Parts & Accessories for your Mac + iPod
do-it-yourself Mac & iPod Parts
free online Fixit Guides
thousands of parts and upgrades
use coupon 'MAKE9' for $5 off
iFixit
Mac + iPod Parts & Upgrades
1-866-61-FIXIT iFixit.com

RETROSPECT

String Theory

By George Dyson

High-tensile fastenings stand the test of time.

Lashing and sewing, among the oldest of technologies, remain competitive with the best. String was not invented — it was discovered: as vines hanging in the tropical jungle, sinews exposed when carcasses were scavenged in the savannah, tendons and baleen fibers appropriated by arctic hunters who took sea mammals apart. Net making is even older: witness the highly evolved spider's web.

The puzzle is not why did human beings start lashing things together, but why did they stop? We still sew our clothing, sails, and tents, and our surgeons still sew us, but we no longer sew or lash much else. The few exceptions — like the carbon-fiber skeleton of the *Gossamer Albatross*, Paul MacCready's human-powered aircraft that crossed the English Channel in 1979 — are rare. Ironworkers still assemble the steel backbone of our reinforced-concrete-based civilization with tie-wire, but these temporary lashings are quickly covered by cement. Anything that is lashed together is dismissed as jury-rigged or haywire, makeshift at best. We only tie things together after they break. Lashing survives on the front line of emergency repairs, not because it is primitive, but because it handles loads — especially the sudden loads that break things — so well.

"In pure strength, apart from their flexibility, the lashings, sewings, and bindings used by primitive peoples, and by seamen down to recent times, are more efficient than metal fastenings," argued James E. Gordon in *The New Science of Strong Materials*, the best study ever written on why some things hold together and why other things fall apart.

When Thor Heyerdahl built the *Kon-Tiki* for a voyage across the Pacific, its structure was lashed together, not only to prove a historical point, but because this was the only method of fastening that held any hope of keeping the soft Peruvian balsa logs from falling apart. When Fridjtof Nansen and Hjalmar Johansen left their ship, the *Fram*, frozen in the polar ice at 84°N in March of 1895, to see if they could reach the pole on their own, they took lashed-together sledges and bamboo kayaks, and made it back to Franz Josef Land — 15 months later — alive. Their survival hung by a thread. "When, for instance, a rib had to be re-lashed," explained Nansen, "we could not rip up the old lashing, but had to unwind it carefully in order not to destroy the line."

When I built a treehouse 95 feet up in a Douglas fir on the coast of British Columbia (*see MAKE, Volume 05, page 190*), I used #15 tarred nylon seine twine to tie the framework to 14 different branches, not because I had any objections to pounding nails or driving lag bolts into living wood (working around logging camps had cured me of that), but because I wanted to be sure that house and tree remained attached. The entire structure had to be able to flex wildly in 60-knot winds, and no mechanical fasteners were flexible enough.

The puzzle is not why did human beings start lashing things together, but why did they stop?

Photograph by Beverly Dobbs, ca. 1906, courtesy Baidarka Historical Society

I adopted lashings when I began constructing aluminum-skeleton kayaks, not for historical authenticity — or I would not have used aluminum — but because tensile fasteners made practical sense. When two pieces of high-tempered, thin-wall aluminum tubing are attached with eight turns of 60-pound test twine, you can be sure that it will take something close to 8×4×60, or 1,800 lbs., to pull them apart. There is no risk of metal fatigue, and the joint will flex elastically (and repeatedly) long before it breaks. Shock absorption is built into every joint.

If you sew two pieces of tempered aluminum plate together, there is no loss of temper as in a weld, and none of the corrosion that affects metal fasteners

Left: Siberian trading party in Nome, Alaska, after crossing Bering Strait in their skin boat.

Right: Joe Ziner and Lou Kelly in the author's kayak-building workshop at Belcarra Park, British Columbia, March 1977. In four months, using miles of braided nylon twine, we lashed together six 28-foot, 3-hatch baidarkas (kayaks), and launched them on a voyage north. The frameworks are 6061-T6 aluminum tubing, 6061-T6 aluminum sheet (from salvaged road signs), and Sitka spruce.

Below: The author lashing the bow assembly for a one-hatch baidarka, 1981, using 60-lb. test nylon twine. The holes in the aluminum are carefully countersunk. If the twine breaks frequently you are applying too much tension; if it never breaks you are not applying enough.

Photograph by George Dyson (above); photograph by Ann E. Yow (below)

in the presence of salt. You can safely predict the ultimate strength of the resulting assembly, since the lashings will bear the load equally, and there are no hard stress points to form a weakest link or to propagate cracks. When you sew a seam on a thick nylon kayak skin with four 40 lb. test stitches every centimeter, you can count on the seam taking a load of 200 lbs. per inch without tearing apart.

Not only are lashed fastenings unusually robust and structurally sound, they are pleasant to work with, are forgiving of sloppy craftsmanship, and celebrate the abilities of human hands. If you complete a seam or a lashing and decide it could have been done better, you just undo it and try again. You can stop work at a moment's notice, and resume at any time. You can strive for perfection, but a haphazard-looking lashing will be almost as strong.

Traveling through heavy seas in a loaded and lashed-together kayak, there is a reassuring creak in the vessel's joints, similar to the old days of loosely fastened wooden sailing ships, designed to work in a seaway for months and years on end. In a modern welded or composite vessel (or aircraft), any sounds emanating from the structure are cause for alarm.

With a load-bearing structure that is sewn or lashed together, what you see is what you get. Finite element analysis (FEA) is now used, in association with computer-assisted design, to predict stress, requisite materials, and required fasteners in a given structure under a given load. The empirical design of lashed structures is the ancestor of FEA. Lashings (and the elements they join together) are finite elements, and can be used to handle stress (and predict strength) in a directly analogous way.

If we examine how nature joins high-strength, high-stress materials — mending a broken bone, joining the plates in a baby's skull, healing a cut in an animal's skin, or attaching a mussel to a rock in the surf — we see lashings at work, on a microscopic scale. In designing and building our own structures, we can do no better than to emulate nature. We have, and we will again.

Approximately 35,000 years ago, a small band of human beings launched into the open Pacific Ocean from Malaysia in lashed-together vehicles. The experiment was a success. Why stop?

George Dyson, a kayak designer and historian of technology, is the author of Baidarka*,* Project Orion*, and* Darwin Among the Machines*.*

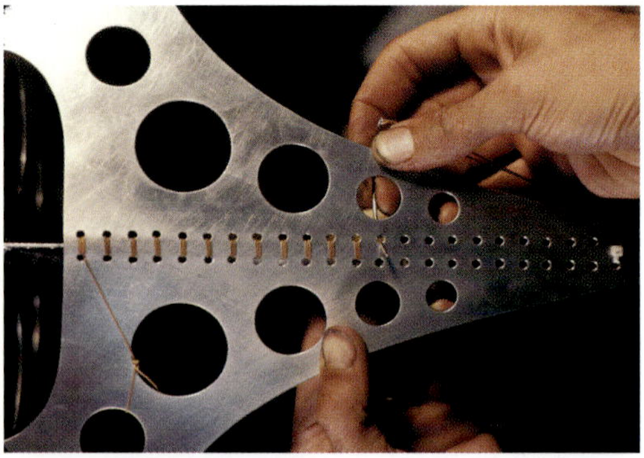

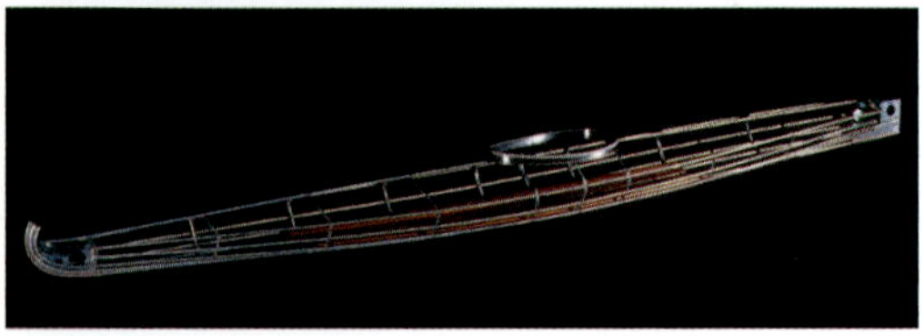

Baidarka (kayak) construction, circa 1986. The completed 5-meter (16.4-foot) skeleton, made of 1⁄16" aluminum sheet and ½" o.d. 6061-T6 aluminum tubing (.049" wall) weighs 18 lbs. The skin material shown here is a 26 oz., (per square yard) double-weave heat-shrinking nylon. The twine is a phenolic-coated nylon obtained at very low cost as remaindered tire cord.

Photography by Ann E. Yow

AHA!

Puzzle This

By Michael H. Pryor

MAKE's favorite puzzles. **(When you're ready to check your answers, visit makezine.com/09/aha.)**

A Matter of Degree

What is the angle between the minute hand and the hour hand at 3:15 on an analog clock? (Hint: It's not 0.) How often does the minute hand pass the hour hand on an analog clock?

Reptilian Repetition

At one point, a tropical island's population of chameleons was divided as follows:

13 red chameleons
15 green chameleons
17 blue chameleons

Each time two different-colored chameleons would meet, they would change their color to the third one. For example, if green meets red, they both change their color to blue. Is it ever possible for all chameleons to become the same color? Why or why not?

100 Lockers

You have 100 lockers in a high school hallway that are all initially closed. You walk down the hallway 100 times starting at the first locker each time. The first time down the hallway, you go to every locker, and if it is open you close it; if it is closed, you open it. The second time you walk down the hallway, you only visit every second locker (lockers #2, #4, #6, etc), reversing the state of each locker you visit. The third time through, you visit every third locker (lockers #3, #6, #9, etc). You continue this until your 100th pass down the hallway when you only visit locker #100. When you are finished, which lockers are open and which are closed?

Illustrations by Roy Doty

Michael Pryor is the co-founder and president of Fog Creek Software. He runs a technical interview site at techinterview.org.

MAKER'S CALENDAR Compiled by William Gurstelle

Our favorite events from around the world March · April · May

MARCH

» University of Illinois Engineering Open House March 9-10, Champaign, Ill.
The University of Illinois Engineering Open House is one of the best showcases of engineering creativity in the country. This year's theme is "Inspiring Innovation." From crushing cars with huge anvils to crunching numbers with tiny microprocessors, EOH is where you can see what the future engineers of the world have in mind.
eoh.ec.uiuc.edu

» Nevada Test Site Public Access Day
March 20, April 24, May 30, 65 miles NE of Las Vegas
A few times a year, the usually off-limits Nevada Test Site allows a busload of visitors to look around. A 250-mile bus tour begins in Las Vegas and takes visitors to the various nuke-created craters, the town of Mercury, and more. www.nv.doe.gov/nts/tours.htm

» ShmooCon
March 23-25, Washington, D.C.
ShmooCon is a popular East Coast hacker convention. Those attending bear witness to some of the most devious minds, code, and gadgets focused on breaking and remaking the technology the rest of the world simply buys and uses.
shmoocon.org

» The Texas Mile
March 24-25, Goliad, Texas
Land speed racers from around the country meet on the long, long, runway of the Goliad Industrial Air Park to test their fastest machines (motorcycles and automobiles). Some attain speeds of more than 250 mph.
texasmile.com

Have a blast at the Nevada Test Site Public Access Day.

» O'Reilly Emerging Technology Conference March 26-29, San Diego, Calif.
ETech frames the ideas, projects, and technologies lurking just below the mainstream radar. Every year, 1,200 technologists, hackers, researchers, thinkers, strategists, and entrepreneurs meet in San Diego to learn how to make technological magic.
conferences.oreillynet.com/etech

» Rube Goldberg Contest Finals
March 31, West Lafayette, Ind.
A Rube Goldberg device is any exceedingly complex apparatus that performs a very simple task in a very indirect and convoluted way. Inspired by machines drawn by the eponymous cartoonist, the contest inspires college students nationwide to design whimsical machines that require skill and imagination. A series of regional events culminates at the national championship held at Purdue University.
purdue.edu/UNS/rube/rube.index.html

APRIL

» Edinburgh Scotland Science Festival
April 6-15, Edinburgh, Scotland
This festival features hundreds of talks, tours, and exhibitions for children, families, and adults in locations around the city. This is one of the world's largest public celebrations of science and technology.
sciencefestival.co.uk

» NOTACON
April 27-29, Cleveland, Ohio
A favorite of Midwestern hackers, this annual computer enthusiast conference held in Cleveland explores and showcases the technologies, philosophy, and creativity associated with hacker life.
notacon.org

» Toledo R/C Model Show April 13-15, Toledo, Ohio
This is one of the nation's premiere events for radio-controlled vehicle enthusiasts. Modelers and suppliers from across the nation come

Mrs. Ohio's big R/C plane.

to this show to see the latest and greatest in R/C models, electronics, gadgets, and parts. toledoshow.com

MAY

»Interactive Telecommunications Program Show May, New York, N.Y.
Over 100 interesting projects at New York University's annual Interactive Telecommunications Program Show draw thousands of visitors. It's a 2-day exhibition of interactive and unusual sights, sounds, and physical objects from the student artists of ITP. This event is free and open to the public. itp.nyu.edu/show

»The Intel International Science and Engineering Fair May 13-19, Albuquerque, N.M.
The world's largest pre-college science competition, with over $3 million in awards and scholarships, will attract 1,500 top science and technology high school students representing the 50 U.S. states and 40 global countries. sciserv.org/isef

EDITOR'S CHOICE

»Maker Faire May 19-20, San Mateo, Calif.
It's the big kahuna of maker-style events, sponsored by MAKE. Over 20,000 makers come together for a weekend of projects, eye-opening demonstrations, entertainment, and fun at the San Mateo Fairgrounds, near San Francisco. From robotics to electronics to pyrotechnics, it is a celebration of the maker ethos encompassing exhibits, seminars, and many hands-on opportunities. Please introduce yourself to the staff of MAKE, who will be on hand. makerfaire.com

»Andrews Air Force Base Joint Services Open House May 19-20, Washington, D.C.
The JSOH pulls in crowds of nearly 400,000 people. The spectacular aerial demonstrations are the biggest attractions. Typical displays include precision aerial maneuvers by the Air Force's Thunderbirds, aircraft fly-bys, parachute jumping demonstrations, and ground displays. andrews.af.mil

»Jet Propulsion Lab Open House May 19-20, Pasadena, Calif.
This event showcases JPL's accomplishments with exhibits and demonstrations about the laboratory's ongoing research and space exploration. Many JPL scientists and engineers will be on hand to answer questions about the lab's work. www.jpl.nasa.gov/pso/oh.cfm

EDITOR'S CHOICE

»World Championship 39th Annual Kinetic Sculpture Race May 26-28, Arcata, Calif. to Ferndale, Calif.
Art and engineering combine to produce imaginative human-powered racing machines. Thousands of spectators watch as dozens of highly beautiful and cleverly engineered vehicles resembling animals, vegetables, and giant eyeballs pedal 30-plus miles over land and water for the simple glory of winning. kineticsculpturerace.org *(See page 28 for a story about last year's race.)*

»Atlas V Rocket Launch May 31, Cape Canaveral, Fla.
An Atlas rocket is scheduled to launch ICO North America's geostationary satellite into orbit in late May. The rocket will fly in the powerful 421 vehicle configuration with a 4-meter fairing, two solid rocket boosters, and a single-engine Centaur upper stage. kennedyspacecenter.com/launches/scheduleStatus.asp

Important: All times, dates, locations, and events are subject to change. Verify all information before making plans to attend.

Know an event that should be included? Send it to events@makezine.com. Sorry, it is not possible to list all submitted events in the magazine, but they will be listed online.

If you attend one of these events, please tell us about it at forums.makezine.com.

HOMEBREW

Our 6,000-volt Wedding Cake

By Heidi Kumao and Michael Flynn

Our rotating, sparking, 6,000-volt wedding cake was the highlight of our wedding ceremony, because our collaboration on this project represented the way we approach our life together.

Our design began with gear-shaped cakes, symbolizing the way our lives have meshed. We agreed that the cakes should actually rotate, so we mounted plywood gears on large lazy Susan bearings, with one of the gears driven by a 10 rpm electric motor.

We decided the traditional cake-topping figures should be recognizable caricatures of ourselves, and that a spark should jump between the hands of the bride and groom to represent passion. The spark was provided by a 12,000-volt neon-sign transformer, which we hid in the huge "Wed-O-Matic" gift box underneath. The electric current traveled from the transformer to the figures through electric motor brushes that scraped together under the rotating lazy Susan bearings. It was important to avoid arcing the current through the ball bearings themselves, so the sliding electrical contact was made on the uppermost disc of each bearing. Jumper wires ran the current from the top plate of the bearing, through our three cakes, up the figures' legs, and out through their arms.

The completed machine was finally ready for its first test run five days before the wedding. We left the drive motor's frame ungrounded for fear of attracting sparks to the motor, forgetting that the neutral wire inside the motor windings is itself a path directly to ground. When we energized the transformer, sparks sizzled and leaped, destroying the wiring inside our specially ordered, slow-rpm motor.

Luckily, Heidi always orders a spare motor for her art projects. We grounded the new frame so that any stray sparks could be grounded without entering the motor's windings. We ran the transformer at half power, and we found that it still produced sparks adequate to jump between bride and groom. This cake was too dangerous to leave plugged in, so we demonstrated its full function only during the cake-cutting ceremony. We played a fun, cartoon-style tune by Raymond Scott, "Powerhouse," and wore oversized safety goggles for dramatic effect.

You can see the video of Heidi and Michael's 6,000-volt wedding cake performance at heidikumao.net/wed.html.

Photograph courtesy of Heidi Kumao and Michael Flynn

Volume 08

Make:

technology on your time™

Special Toys & Games Section

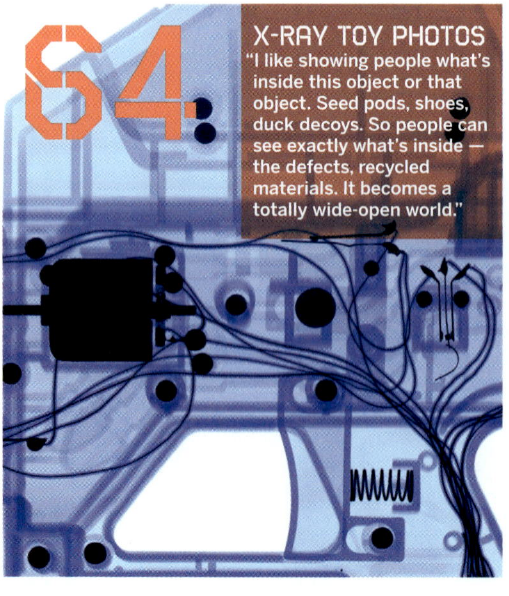

X-RAY TOY PHOTOS
"I like showing people what's inside this object or that object. Seed pods, shoes, duck decoys. So people can see exactly what's inside — the defects, recycled materials. It becomes a totally wide-open world."

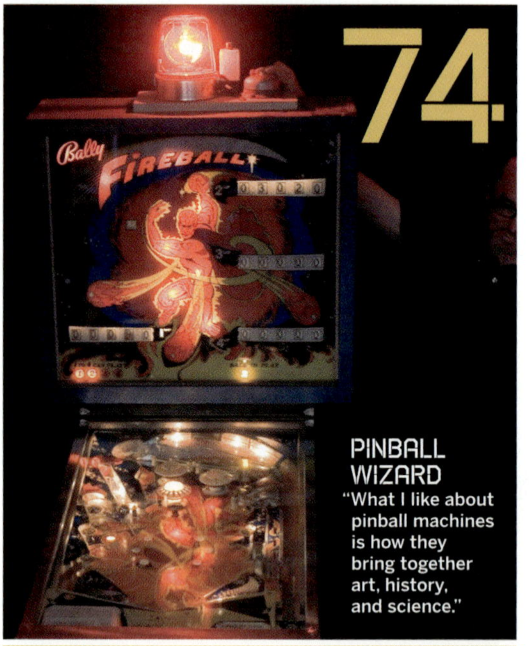

PINBALL WIZARD
"What I like about pinball machines is how they bring together art, history, and science."

ON THE COVER
When MAKE Creative Director Daniel Carter (shown at right) went to the Lucky Ju Ju pinball arcade photo shoot, he was given the task of muffling the bell on the *Fireball* pinball game so it wouldn't bother the beauty salon next door. *Photography by Steve Double*

Columns

Vol. 08, November 2006. MAKE (ISSN 1556-2336) is published quarterly by O'Reilly Media, Inc. in the months of February, May, August, and November. O'Reilly Media is located at 1005 Gravenstein Hwy. North, Sebastopol, CA 95472, (707) 827-7000. SUBSCRIPTIONS: Send all subscription requests to MAKE, P.O. Box 17046, North Hollywood, CA 91615-9588 or subscribe online at makezine.com/offer or via phone at (866) 289-8847 (U.S. and Canada), all other countries call (818) 487-2037. Subscriptions are available for $34.95 for 1 year (4 quarterly issues) in the United States; in Canada: $39.95 USD; all other countries: $49.95 USD. Application to Mail at Periodicals Postage Rates is Pending at Sebastopol, CA, and at additional mailing offices. POSTMASTER: Send address changes to MAKE, P.O. Box 17046, North Hollywood, CA 91615-9588.

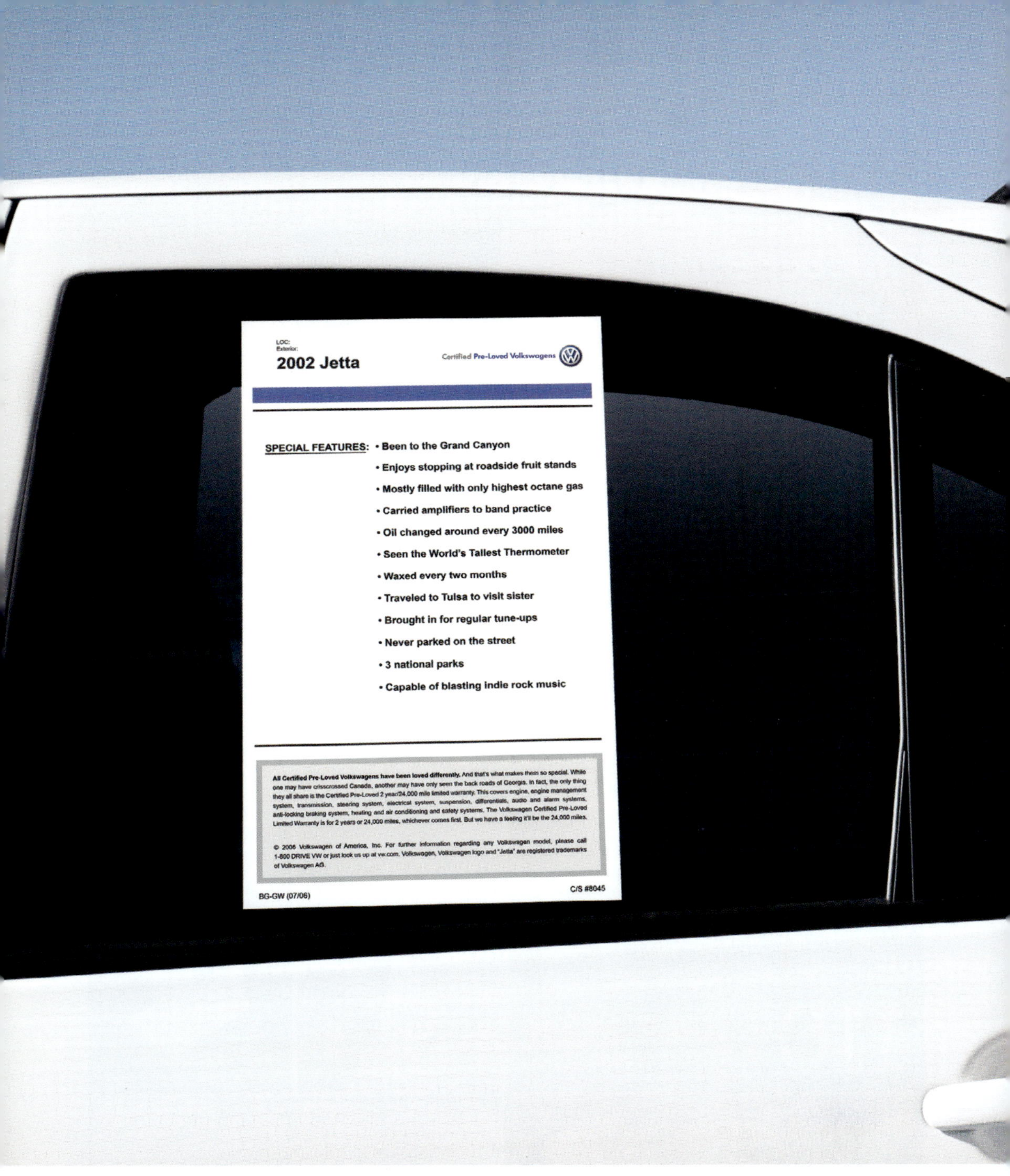

Volkswagen owners love their cars more than other drivers. Whether it's changing the oil or sharing a trip to see a fun part of the world. That's why we call them Pre-Loved Volkswagens. Of course, every one of them has to pass our 112-point inspection to be certified. And we back it with a 2-year/24,000 mi. limited warranty. So head down to your local VW dealer, we think you'll fall in love with one yourself.

Certified Pre-Loved Volkswagens

Make: Projects

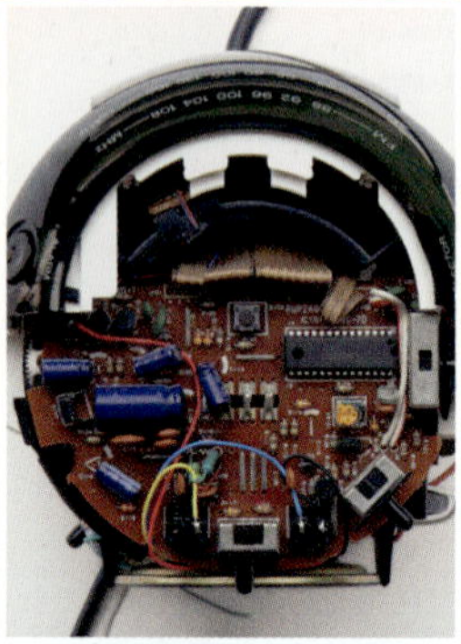

Volume 08

READ ME: Always check the URL associated with each project before you get started. There may be important updates or corrections.

Columns

Maker

123

MAKER PROFILE
Can a machine have a soul? Can it laugh and cry, bug you for a snack, tease you, and dream robotic dreams? Can you sell it for $250?

Make:
technology on your time™

EDITOR AND PUBLISHER
Dale Dougherty
dale@oreilly.com

EDITOR-IN-CHIEF
Mark Frauenfelder
markf@oreilly.com

MANAGING EDITOR
Shawn Connally
shawn@oreilly.com

ASSOCIATE MANAGING EDITOR
Goli Mohammadi

SENIOR EDITOR
Phillip Torrone
pt@makezine.com

PROJECTS EDITOR
Paul Spinrad
pspinrad@makezine.com

STAFF EDITOR
Arwen O'Reilly

COPY CHIEFS
Keith Hammond
Marlowe Schaeffer

COPY EDITORS/RESEARCH
Mary Brady, Phil Dangler, Colleen Gorman, Sanders Kleinfeld, Rachel Monaghan, Lydia Onofrei, Laurel Ruma, Adam Witwer

CREATIVE DIRECTOR
Daniel Carter
dcarter@oreilly.com

DESIGNER
Katie Wilson

PRODUCTION DESIGNER
Gerry Arrington

ASSOCIATE PUBLISHER
Dan Woods
dan@oreilly.com

CIRCULATION DIRECTOR
Heather Harmon

ADVERTISING COORDINATOR
Jessica Boyd

SALES & MARKETING ASSOCIATE
Katie Dougherty

MARKETING & EVENTS COORDINATOR
Rob Bullington

EDITOR AT LARGE
David Pescovitz

ONLINE MANAGER
Terrie Miller

MAKE TECHNICAL ADVISORY BOARD
Gareth Branwyn, Limor Fried, Joe Grand, Saul Griffith, William Gurstelle, Bunnie Huang, Tom Igoe, Mister Jalopy, Steve Lodefink, Erica Sadun

PUBLISHED BY O'REILLY MEDIA, INC.
Tim O'Reilly, CEO
Laura Baldwin, COO

Visit us online at makezine.com
Comments may be sent to editor@makezine.com

For advertising and sponsorship inquiries, contact:
Dan Woods, 707-827-7068, dan@oreilly.com

Customer Service cs@readerservices.makezine.com
Manage your account online, including change of address at:
makezine.com/account
866-289-8847 toll-free in U.S. and Canada
818-487-2037, 5a.m.–5p.m., PST

MAKE is printed on recycled paper with 10% post-consumer waste and is acid-free. Subscriber copies of MAKE, Volume 08 were shipped in recyclable plastic bags.

Please Note: Technology, the laws, and limitations imposed by manufacturers and content owners are constantly changing. Thus, some of the projects described may not work, may be inconsistent with current laws or user agreements, or may damage or adversely affect some equipment.

Your safety is your own responsibility, including proper use of equipment and safety gear, and determining whether you have adequate skill and experience. Power tools, electricity, and other resources used for these projects are dangerous, unless used properly and with adequate precautions, including safety gear. Some illustrative photos do not depict safety precautions or equipment, in order to show the project steps more clearly. These projects are not intended for use by children.

Use of the instructions and suggestions in MAKE is at your own risk. O'Reilly Media, Inc., disclaims all responsibility for any resulting damage, injury, or expense. It is your responsibility to make sure that your activities comply with applicable laws, including copyright.

Contributing Editors: William Gurstelle, Mister Jalopy

Contributing Artists: Melinda Beck, Brian Biggs, Roy Doty, Steve Double, Dustin Amery Hostetler, Jemma Hostetler, Timmy Kucynda, Tim Lillis, Christopher Lucas, Nik Schulz, Damien Scogin, Jen Siska

Contributing Writers: Oona Albertson, Kirsten Anderson, Tim Anderson, Tom Anderson, Wendell Anderson, Thomas Arey, Prof. Greg Arius, David Battino, Barton Blum, Joost Bonsen, Gareth Branwyn, Bill Bumgarner, Christy Canida, Dr. Shawn Carlson, Kiera Collins, Kindy Connally Stewart, Hiram Cook, Travis J. I. Corcoran, Larry Cotton, Cory Doctorow, Andrea Dunlap, George Dyson, James Ernest, Joe Grand, Saul Griffith, Blake Hannaford, Michelle Hlubinka, Roger Ibars, Dean Johnson, Peter Kirn, Mike Kuniavsky, Robert Luhn, Dave Mabe, Merlin Mann, Megan Mansell Williams, Alan Mellovitz, Harry Miller, Robyn Miller, Cassidy Mountjoy, Emmett Mountjoy, Danny O'Brien, Ethan O'Toole, Bob Parks, Charles Platt, Michael H. Pryor, Kristina Reed, John Riney, Matthew Russell, Adam Savage, Mikey Sklar, Eric Smillie, Ewan Spence, Bruce Sterling, Bruce Stewart, Cy Tymony, Wild Planet designers, Tom Zimmerman, Lee D. Zlotoff

Interns: Matthew Dalton (engr.), Adrienne Foreman (web), Jake McKenzie (engr.), Ty Nowotny (engr.)

Founding Design Team: David Albertson, Kirk von Rohr, and Sarah Hart at Albertson Design

Printed in the USA by Schumann Printers, Inc.

Contributors

Steve Double (*Pinball Restoration* photos) has been working as a professional photographer for 21 years. Some highlights are having 90 seconds to photograph Bill Gates, waiting six hours for the Spice Girls, being chided for lateness by Public Enemy, drinking caipirinhas with Nick Cave, blagging an upgrade off Richard Branson, spending five days in a van with Nirvana, and never having to wear a tie. He recently acquired a digital back for his Hasselblad and now embraces techniques from Polaroid image transfer and emulsion lift to completely filmless shooting. He is old enough to know better.

Mikey Sklar (*Making It*) likes to blow things up. It started innocently enough just placing Black Cat firecrackers in dog poop and running. This technique basically covers you in crap. Twenty years later he has created an open source fire-shooting trampoline. When he is not thinking about fire, his time is well-spent programming microcontrollers, cooking up vegan meals, and watching how-to documentaries. He is presently building a self-balancing electric vehicle and a bed-and-breakfast out of garbage.

Roger Ibars (*Toy Gun Alarm Clock*) remembers: "Once when I was a kid, my dad could not open a door. The key somehow got jammed into the lock and he tried everything ... twisted it, banged it, pushed it. Nothing. I was 3 years old and I remember I had a strange feeling. I felt I could open that door! ... So I waited until my dad left, I put my hands on the key and ... magic! I opened it! This picture shows the outfit I was wearing that day."

Larry Cotton (*Small-Batch Coffee Roaster*) is a "musical, energetic, creative, happily married, semi-retired power tools engineer-turned-math teacher" (try saying that in one breath), lives in New Bern, N.C., with his wife, Sylvia, and "one irrationally exuberant dog." He is a compulsive tinkerer who is currently playing with "weird clocks based on an el-cheapo time-base device" and who loves lasagna, bright colors, and his 1950s vintage Shopsmith. His dream project is the "Pixbox," a hardware USB digital photo editor loaded with switches and knobs, a randomize button for creative types, unlimited undo, and in Google's honor, an "I Feel Lucky" button. It doesn't exist yet, but Larry is open to collaboration.

Brian Biggs (*Flipbook* illustrations) was born in Little Rock, Ark., in 1968, the son of a Portuguese train bandit and a mystical one-armed sailor with secret powers. He grew up mostly in Pasadena, Texas, where he learned to talk to cows. He went to Parsons School of Design with the mistaken notion that he would be a magazine art director. Later, he realized that he was not cut out for regular jobs. Brian now lives in Philadelphia, where he rides bikes, draws robots, plays the accordion, teaches illustration, and hangs out with his two kids (who are magical raccoons and know how to fly). Visit him at MrBiggs.com.

Adrienne Foreman (MAKE web intern) is a Jill of some trades, and master of none. She assists with web production, helps keep the MAKE website up and running smoothly, and adds the entire Table of Contents online for each issue. A college student without a major, she generally has a hard time making up her mind. She is currently in the middle of a love affair with the color green and chocolate-covered PayDay bars.

Simply put, **Adam Savage** (*Moldmaking*) makes stuff. He's constructed everything from spaceships to Buddhas, puppets to rifles, sculptures to toys. His fascination with creating things started when he began building his own toys at age 5. In addition to co-hosting Discovery Channel's *MythBusters*, Adam also teaches advanced model making in Industrial Design departments and somehow finds time to devote to his own art. His sculptures have been showcased in more than 40 shows in San Francisco, New York and Charleston, W.V.

vol. 05
Make:
technology on your time
Tree-to-Tree Protocol: Build Your Own Backyard Zip Line
» Ultimate Treehouse
» Electric Wind Turbine
» Aquatic Robot Race
» Free HDTV
» Jam Jar Jet
Launch a High-Powered WATER ROCKET
100 FEET OR HIGHER!
Street Legal Electric Cars
60
78
90
102
O'REILLY
makezine.com
makezine.com
Give a gift, get a gift
Make:
Get your own MAKE T-shirt when you give a gift subscription to MAKE magazine. Just $34.95 for a full year (4 issues) devoted entirely to DIY technology projects, plus a MAKE T-shirt for you.
Subscribe online at makezine.com/gift
Enter code: 46GIFT
"Watch out, corporate America. The freedom to tinker has a posse." — Wired
Make:
technology on your time™

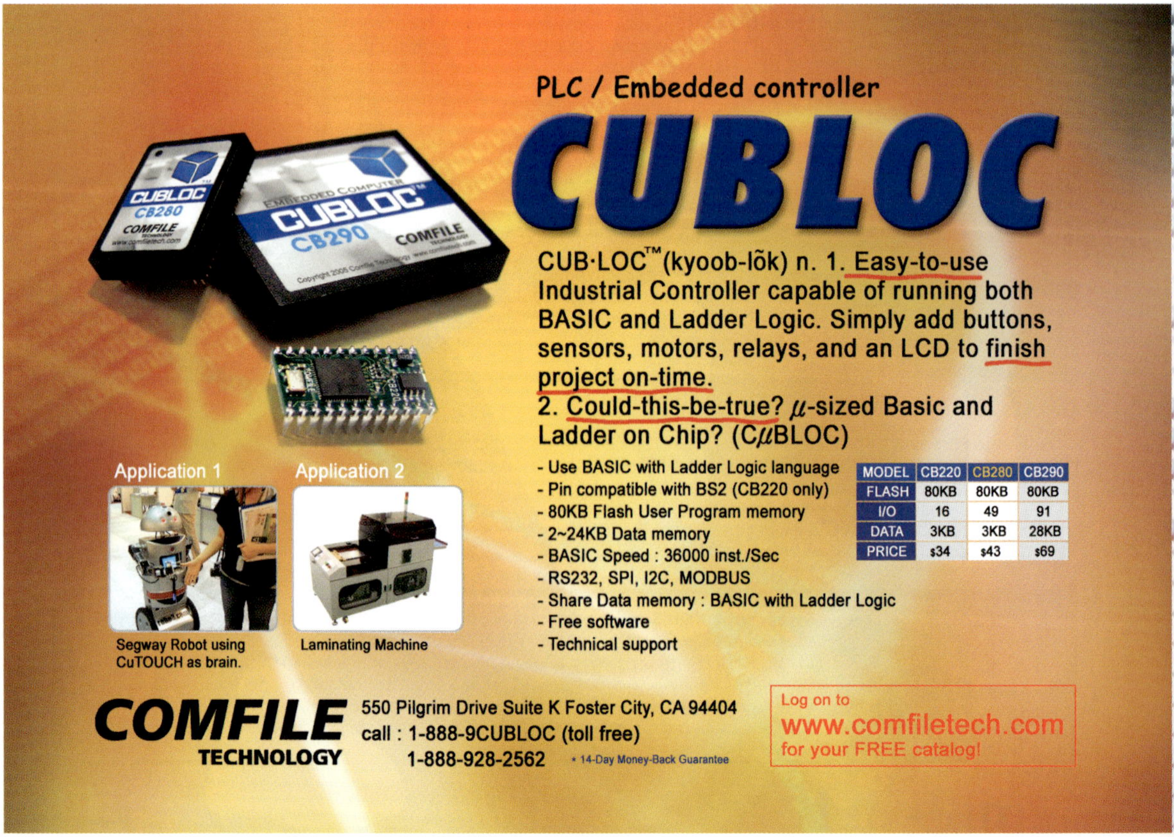
PLC / Embedded controller
CUBLOC
CUB·LOC™(kyoob-lŏk) n. 1. Easy-to-use Industrial Controller capable of running both BASIC and Ladder Logic. Simply add buttons, sensors, motors, relays, and an LCD to finish project on-time.
2. Could-this-be-true? μ-sized Basic and Ladder on Chip? (CμBLOC)
- Use BASIC with Ladder Logic language
- Pin compatible with BS2 (CB220 only)
- 80KB Flash User Program memory
- 2~24KB Data memory
- BASIC Speed : 36000 inst./Sec
- RS232, SPI, I2C, MODBUS
- Share Data memory : BASIC with Ladder Logic
- Free software
- Technical support
MODEL CB220 CB280 CB290
FLASH 80KB 80KB 80KB
I/O 16 49 91
DATA 3KB 3KB 28KB
PRICE $34 $43 $69
Application 1
Application 2
Segway Robot using CuTOUCH as brain.
Laminating Machine
COMFILE TECHNOLOGY
550 Pilgrim Drive Suite K Foster City, CA 94404
call : 1-888-9CUBLOC (toll free)
1-888-928-2562
* 14-Day Money-Back Guarantee
Log on to www.comfiletech.com for your FREE catalog!

FOLLOW THE BOUNCING BALL

By Dale Dougherty

PULL BACK THE LAUNCHER AND LET GO. The silver ball rides up the side, follows the curve at the top, and then begins to descend. This is the much-anticipated moment that makes playing pinball so great. You just don't know what the ball is going to do.

Michael Schiess of Lucky Ju Ju Pinball Gallery (ujuju.com), profiled in our pinball feature (*see page 74*), gave me a brief history of the game. Pinball started as a French game known as bagatelle, something the aristocracy enjoyed playing.

The playfield was a raked wooden board with fixed pins. The ball was launched with a cue; it would bounce off the pins and eventually drop into holes, which had number values for scoring. During the American Revolution, the French brought bagatelle to the United States, where it became popular. Then, during the Depression, this pin game became a machine, fully mechanized; then electrified; and finally digitized. Manufacturers competed by offering compelling art themes and new features; for instance, flippers were added in 1947.

The first pinball machines were placed in the corner grocery store but were soon clustered in arcades, where teenagers could hang out and act depraved like French nobles. Schiess and others emphasize this social side of pinball, which was lost when kids stayed home to play video games.

In his book *The Pinball Effect*, techno-raconteur James Burke uses the unpredictable path of the pinball as a metaphor for technological innovation. His book is not about how things work, but "the crazy way the pinball of change works its magic, bouncing here and there across time and space." This book, like his well-known TV show *Connections*, attempts to follow the nonlinear path of discoveries and new ideas and how they jump across borders, industries, and disciplines.

From Burke's book, let's take the real-life example of Gideon Sundback, a Swedish-born Canadian engineer who works for the Universal Fastener Company in the early 1900s. Sundback works for an engineer named Judson, who's been trying to solve what I'll call a "boot sequence" problem. Burke sets up the linkages: Goodyear's synthetic rubber is used to make better bicycle tires, and bicycles become so popular that even women want to ride them, which causes women's fashion to change and hemlines to rise. Women's boots follow suit, rising to cover more of the leg. The number of hook-and-eye fasteners on a boot was at 20 and growing, a time-consuming number to fasten.

Judson had patented the "C-curity Fastner" but it didn't take off. Sundback improved it. His version had 10-11 teeth per inch to Judson's four. He called it the "Separable Fastener" in his first patent, which was awarded in 1913. He also created machines to make these fasteners. But his invention was also slow to take off, perhaps because his fasteners weren't yet fashionable.

In 1923, the BF Goodrich Company created rubber boots for the military that used Sundback's fastener, which they named the Zipper. Only later, in the 1930s, did zippers begin to be used widely in clothing as a convenience.

Now I'll take Burke's example one step further, jumping beyond technology. In 1971, Andy Warhol designed an album cover for the Rolling Stones that used a real zipper. The zipper had to be left open partway so that the back of it would line up with the record label, and not the record itself, which it might scratch. *Sticky Fingers* transformed zippers from a functional item into a pop-art icon of desire.

But wait, there's more. The same year, a new pinball game called *Fireball* introduced a new feature: "Zipper Flippers." The flippers actually slide together to catch the ball and prevent it from dropping down the hole.

Even pinball machines are subject to the Pinball Effect. That's just crazy.

Dale Dougherty (dale@oreilly.com) is editor and publisher of MAKE and CRAFT.

News From the Future

GAMES WITH A PURPOSE

By Tim O'Reilly

IN JULY OF 2006, CARNEGIE MELLON professor Luis von Ahn gave a tech talk at Google entitled, "Games with a Purpose." In it, he described his ESP Game, which matches up two people selected at random over the internet and asks them to guess the same label for photos that are shown simultaneously to each of them.

Participants have a good time — players report a great feeling of satisfaction as they get "in sync" and make similar guesses. People have played for 15 hours at a stretch, and many regularly play as much as 20 hours a week.

The outcome: 75,000 players have provided 15 million labels for images. And labels generated by agreement between two independent players are remarkably accurate and useful. Von Ahn estimates that if the game were played by 5,000 simultaneous players — a number common on popular gaming sites — it would take about two months to label all the images in Google Images. In September, Google launched the Google Image Labeler, putting customers to work in much the way von Ahn has proposed.

This story is interesting, not only as an illustration of one of the themes I've written about previously — that our software is becoming bionic, fusing human and machine intelligence rather than achieving artificial intelligence — but also because it highlights the importance of play as an outlet for human energy.

In his talk, von Ahn pointed out that in 2003, about 9 billion human-hours were spent playing solitaire. By contrast, only 7 million human-hours (6.8 hours of solitaire) were spent building the Empire State Building, and only 20 million human-hours on the Panama Canal. That's a staggering amount of play, and that's on only one small game. Von Ahn waggishly pointed out that harnessing humans to play games, especially games that solve computer problems that AI cannot yet solve, would have been a far more plausible pretext for the AI of the Matrix to keep humans around. In fact, he's committed to just that goal, saying: "We're going to consider all humanity as an extremely advanced and large-scale distributed processing unit that can solve large-scale problems that computers cannot yet solve."

I liked von Ahn's phrase, "games with a purpose," but of course, all games have a purpose, not merely those that put us to work helping out our computers. Play is so central to human experience that historian Johan Huizinga suggested that our species be called *Homo ludens*, man the player. (Notably, in the same paragraph, he also mentioned *Homo faber*, man the maker, as another alternative to the familiar *Homo sapiens*, man the thinker.)

What makers understand is that play is as important as work. It's not just how we pass the time, it's how we learn and explore.

As I mentioned in my last column, back in August, in Wired News John Seely Brown told the story of how status as a *World of Warcraft* guild master helped Stephen Gillet win a senior engineering management position at Yahoo.

"The process of becoming an effective *World of Warcraft* guild master amounts to a total-immersion course in leadership. ... To run a large one, a guild master must be adept at many skills: attracting, evaluating, and recruiting new members; creating apprenticeship programs; orchestrating group strategy; and adjudicating disputes. ... Never mind the virtual surroundings; these conditions provide real-world training a manager can apply directly in the workplace."

Perhaps even more importantly, he writes: "Unlike education acquired through textbooks, lectures, and classroom instruction, what takes place in massively multiplayer online games is what we call accidental learning. ... Accidental learning relies on failure. Virtual environments are safe platforms for trial and error. The chance of failure is high, but the cost is low and the lessons learned are immediate."

We play to learn. What we make when we play is ourselves.

Check makezine.com/08/nff for related stories.

Tim O'Reilly (tim.oreilly.com) is founder and CEO of O'Reilly Media, Inc. See what's on the O'Reilly Radar at radar.oreilly.com.

Life Hacks

WHY THOSE KIDS FROM PODUNK ARE KEEPING YOU DOWN

By Merlin Mann and Danny O'Brien

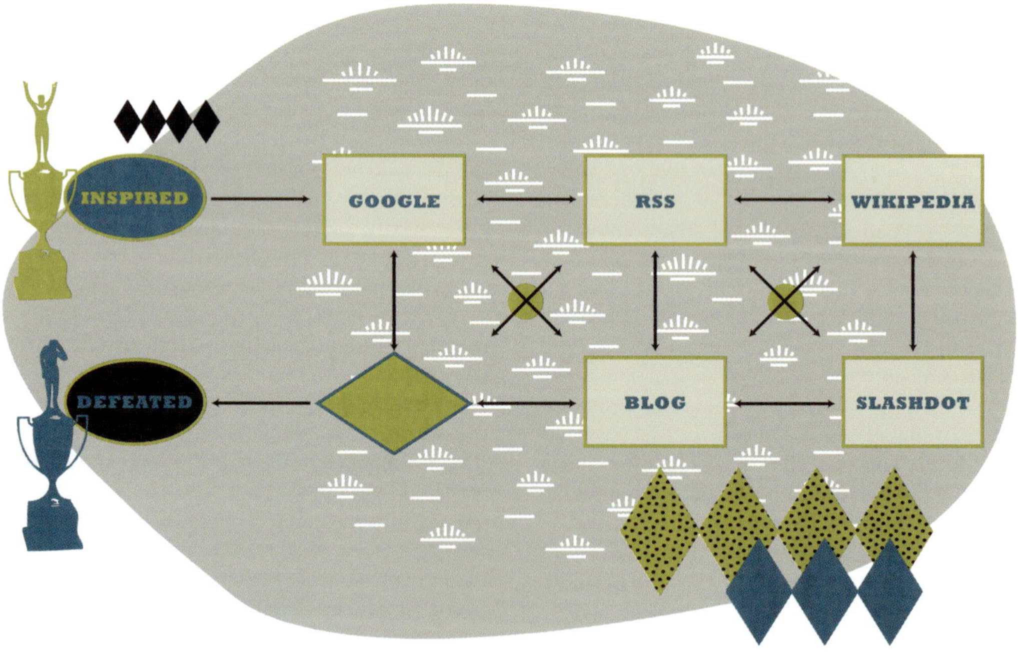

Illustration by Jemma Hostetler

GLENN DAVIS, THE GUY WHO FOUNDED Cool Site of the Day, once told Danny what he thought made the net great. He said, "Suppose there's this young kid in Podunk, Connecticut. He runs a site about a niche, weaving wicker baskets, or naked piano-tuning, or what have you. In another age, that kid would probably have toiled in obscurity, but now he can discover that he is the best wicker-weaver or nude piano-tuner in the world. And he learns, as the rest of the world learns, through his website."

Now, to Glenn, that was a good thing. Such was the youthful optimism of the net.

Well, we've been monitoring what can slow down a maker, and we're not so sure of this whole "meritocracy equals good" thing. We think it's those kids from Podunk who are keeping you down.

Neat organizing tricks contribute about 30% to efficiency and productivity. The other 70% is about motivation. And, frankly, constant exposure to all the world's most wonderful hackers can act as a great demotivator to even the most iron-hearted beginner. How many times have you had a cool idea, googled it, realized that about 50 other people have implemented something similar, so didn't bother to take it any further?

The problem is that over time, the web has become a Massively Parallel Distributed Network of Kids in Podunk. Everywhere you look, people are doing things of such great coolness that you hardly feel you can keep up. This magazine is full of these sorts of things. Your RSS aggregator is dripping with such excitement.

Depending on your mood at the time, the whole idea of a world of excellence might fill you with sickening dread. At those dark moments, the web becomes that horrid machine that Douglas Adams once conjectured, the "total perspective vortex" where, at any one time, you know exactly how many people were ahead of you in the queue when God gave out the ability to play poker, learn the accordion, hack Xboxes, or lip-sync to Romanian pop songs.

Well, speaking as the 43rd and 72nd world experts in cheering up dispirited geeks, respectively, allow us to speak to your horror, and thereby lift your spirits.

First, let's add some perspective of our own. As part of the Life Hacks research, we met and talked to some of these paragons of productivity. Let us reassure you, they're precisely as screwed up as you are. Possibly more so, actually.

Behind every successful geek is a trail of barely mentioned (yet spectacular) failures. We never see this, because most of us, even in our most confessional bloggings, put a little bit of shine on the work that's being done.

Don't try to compete with the entire internet. Your RSS aggregator is called that for a reason — it's glomming together the best of millions. You don't have to be as good as all of them.

If you're ever frozen into inaction by the kids from Podunk, here are a couple of practical tips to switch your "spectator/participant" toggle switch.

First: Hack the net's tendency to catch both the good and bad. Don't try and learn from your geek heroes as they are now. Almost certainly, they'll be less intimidating if you look at their work from five or ten years ago, when they were blundering around as much as you and had their own heroes that they could never live up to.

Back in Glenn's time, the kids from Podunk didn't have a history on the net. But now we all do, and the Internet Archive's Wayback Machine is your friend. So, if you're awed by Tim Berners-Lee, Glenn Davis, Dave Winer, Guido van Rossum, or anyone in the contents of this magazine, do a little exploring. Find out how they started and emulate that — if you can.

Second: Hack your awe of someone else for your own use. A friend of ours really wanted to get into Linux Kernel development but had no idea how. Rather than just file a bug he found in the kernel, he included a request asking where he should start if he wanted to fix it. Alan Cox, who plays deputy to Linus Torvalds, emailed back with a short, sweet suggestion. It wasn't much, but our friend was so in awe of Cox that it made his week and kickstarted his coding.

As long as you're prepared to take on a share of the work (rather than just fawning or drive-by flaming), the heroes and heroines of the net are happy to point you in the right direction. And the narrower your hero's focus might be, the more impressed he'll be that anyone is paying any attention to him. You never know, you might become his hero for being so interested in his hobby. We've seen it happen. Kids from Podunk are just looking for other kids to play with.

Learn how to reel in your mind at Danny O'Brien's lifehacks.com and Merlin Mann's 43folders.com.

HIGH-DEFINITION EQUALS HIGHLY DEADLY

By Cory Doctorow

WHO KNEW THAT HIGH-DEFINITION television would prove so deadly? The way I see it, HD could precipitate great harm to open source (and the tinkerers who love it), competition, free speech, and the right to remix, not to mention the bottom lines for the studios and today's TV stars.

Standard-def TV might be a little fugly (NTSC: Never Twice Same Color!), but it has this virtue: anyone can implement it. There are no patents lurking in SD, no copyrights, no license agreements. In 1979, I plugged an Apple II+ into a standard TV set. Today, I can connect a Slingbox to it. The world is full of these cheap and dirty displays, and innumerable revolutions owe their success to this fact, from Pong to the PC.

High-def is different. Between DRM and patents, it's nearly impossible to make an HD set that receives content from the widest pool of HD devices — like DVD-HD and Blu-Ray players, and many HD consoles — without a lot of butt-kissing. If you want to go near this stuff, prepare to build neutered devices that won't upset anyone's apple cart. You have to please media executives who think that skipping commercials should be a crime.

It gets worse. The studios are bent on bringing something called the Broadcast Flag (in the U.S. and Canada) and DVB-CPCM (in the rest of the world) to HD. This would require anyone building an HD receiver to ensure that it couldn't be used in connection with open source ("user modifiable") software, and that any implementers must license an "approved" anti-copying system that will come loaded with all sorts of conditions about how you can and can't build a device. Forget plugging in a PC, Pong console, or even a Slingbox.

Because PCs are potential HD tuners, this rule will encourage manufacturers of commodity components — like hard drives and video cards — to follow the same rules so that their products can be used in HDTV. Goodbye open platforms, hello monopoly lockware.

But will the public buy HD sets in large numbers? Maybe not. After all, contemporary programming looks pretty weird in high-def. Check out an HD episode of *Friends* — Courteney Cox does not improve at higher resolutions. Every pore gapes, the hair-sprayed coiffure looks as stiff and unnatural as a corset, and the standard-def cute moues and pouts become grotesque comedy masks at high-res.

If HDTV has a future, it's in the invention of a new art form that exploits the inherent characteristics of the medium. Rather than trying to bridge the "uncanny valley" in HD, a savvy director might instead stick several low-res pictures on the screen

Forget plugging in a PC, Pong Console, or Slingbox. Goodbye open platforms, hello monopoly lockware.

at once, overlapping them and bringing parts of them forward and back, like the hyperactive display on a customized *World of Warcraft* environment or your own PC desktop.

And if that's too weird, there's plenty of room at the bottom: tiny HD screens forgive many of the sins that are starkly revealed by wall-hogging monster TVs.

Cory Doctorow (craphound.com) is a science fiction novelist, blogger, and technology activist. He is co-editor of the popular weblog Boing Boing (boingboing.net), and a contributor to *Wired*, *Popular Science*, and *The New York Times*.

NO POSTAGE
NECESSARY
IF MAILED
IN THE
UNITED STATES

BUSINESS REPLY MAIL

FIRST-CLASS MAIL PERMIT NO 865 NORTH HOLLYWOOD CA

POSTAGE WILL BE PAID BY ADDRESSEE

O'REILLY

Make:

PO BOX 17046
NORTH HOLLYWOOD CA 91615-9588

1+2+3

Custom Travel-Game Mod

By Harry Miller

Make a travel edition of your favorite, and otherwise housebound, board game.

You will need: Magnetic metal boards (available at office supply stores), clear Contact paper, duct tape, adhesive magnetic paper or magnets, plastic clips (to hold cards or paper money together)

Can you take your favorite board game with you on a trip? If not, follow my instructions to make your own travel edition. I made a travel version of Rivers, Roads and Rails (by Ravensburger). The original version has to be played on the floor or on a table. You can take mine in a plane, a car, or a boat. When choosing your game, keep in mind that the fewer the pieces the better.

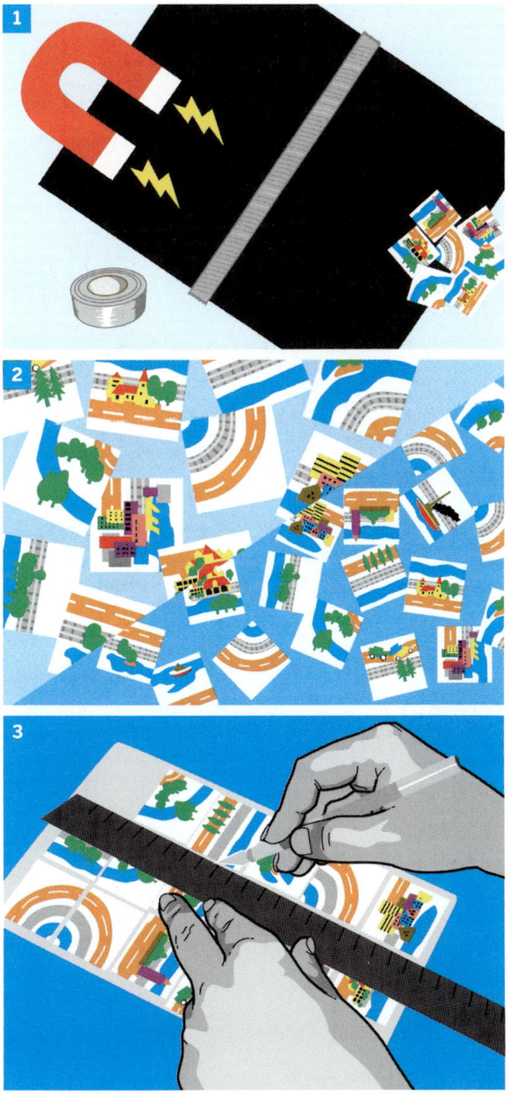

1. Make a magnetic board.

Rivers, Roads and Rails is composed of 140 2"×2" cardboard tiles, on which are drawn a river, a road, a railroad, or a combination of any of the three. The pieces for a travel edition need to be able to move without disrupting the game. Also, the playing space needs to be more confined. I found that 2 magnetic metal boards (about 9"×14") held together by a piece of duct tape were a small-enough space to play on if I shrunk the tiles down by half.

2. Shrink game pieces and playing field.

I used my printer/copier to shrink the tiles down by half to 1"×1" squares. If your game has an illustrated playing field, you'll have to do the same thing for the game board, and then stick it to the magnetic boards.

3. Make the game pieces magnetic.

Take a sheet of flexible adhesive magnet paper and stick the tiles onto them. Then cover them with clear Contact paper and cut them apart using a cutting board. (My mom helped a bit.) I sent a letter to Ravensburger with my design plans, hoping they will produce it. (Maybe I can even get some credit.)

For extra credit

+ A zip-lock plastic bag might work just fine as the travel bag, and some plastic potato-chip clips can keep your game cards together. On the other hand, a hand-sewn bag of vintage fabric and an old silver cigarette case may be a more elegant card holder.

You're done! Play the game to find any flaws. If there are problems, go back and make necessary adjustments until it's perfect!

Illustrations by Dustin Hostetler

Harry Miller is a 12-year-old home-schooler who loves to invent, build, and destroy things. On sunny days he can be found terrorizing plastic soldiers with a magnifying glass.

MADEONEARTH

Report from the world of backyard technology

Mr. Cranky

After you've made what George Lucas called his favorite *Star Wars* parody, how do you keep turning culture on its ear? Retired filmmaker **Ernie Fosselius** (*Hardware Wars*) has been cranking out a wonderful menagerie of mechanical woodcarvings.

Like his hit 1977 movie spoof, which turned eggbeaters into spaceships and crusty rolls into Princess Leia's braids, Fosselius' sculptures poke fun with quick, broad strokes. The face of *Bi-Polar Bear* changes from happy to sad with the turn of a crank; when you turn the crank on *The Politician*, the smirking statuette does ... nothing.

Part of the joy is watching the clever mechanisms Fosselius puts in plain sight, like the spinning drill bit that simulates a stream of coffee. "I like to be the magician who reveals all his secrets, so you want to go home and try it yourself," he explains.

He takes that same exploratory approach as a maker. "I just keep myself open to what [a piece] wants to do," he says. "I never have a blueprint, only a rough cartoon. It's really that *process* of building and tinkering and refining that's the most interesting, rewarding, and fun part. Finishing is anticlimactic."

Fosselius crafts his characters out of basswood, doing the initial shaping on a band saw and then whittling them down with a pocketknife, a carving knife, and palm gouges. "If you look closely at the boxes and bases, you'll see that I'm not much of a woodworker," he claims. "I don't like rulers. And I hate to sand. If you can put the joke over in the simplest way, why bother? Use that time to make another piece.

"I feel that I'm still making movies. I'm creating little characters, situations, and stories that hopefully make people laugh. But it's *direct* entertainment. Unlike films, I get to be there when people experience it."

—David Battino

Photography by Mark Frauenfelder

Read an interview with Ernie Fosselius at makezine.com/08/made_fossel

Ernie Fosselius: contraptions@msn.com
en.wikipedia.org/wiki/Ernie_Fosselius

Belly of the Beast

It's not often that studying physics leads to the creation of huge, self-propelled sand walkers that seem more like the living skeletons in *Jason and the Argonauts* than sculptures made of conduits and wire. But for **Theo Jansen**, science has a long, bony arm. He never finished that physics degree, becoming a painter instead.

"In the 80s the physics came back again," he says, and he started tossing around more sculptural ideas. He launched *UFO*, a 14-foot black disc, over the town of Delft, Netherlands. He dabbled in performance and photography, creating light sculptures of the track of an ultralight flight and of a drill suspended high over the street, swinging back and forth.

Then, in 1990, he came across some cheap conduits in a store, started playing around with them, and has been playing ever since. Called the *Strandbeesten* ("beach creatures"), his creations are built from yellow plastic tubing, recycled wood, string, plastic sheeting, and vast amounts of imagination. Each creature can take up to a year to build. Propelled by wind and inertia, they stumble, creak, and race across the ground.

Jansen has displayed his sculptures in galleries (and, more recently, in London's Trafalgar Square), but their native home is clearly the beach. While some look more like giant armored spiders or mechanical, many-legged rhinos, others are elegant spiral skeletons that sweep along the sand with surprising agility. Sometimes huge, fan-like sails catch the wind and start the motion, lending the entire scene the look of a futuristic pirate fleet. Or soda bottles pumped full of air will jump-start the creatures, like some sort of elaborate, earthbound bottle rocket.

Extremely low-tech, futuristic, and visionary (Jansen's CAD drawings look like the cockroaches that would survive Armageddon), the creatures have an aesthetic that is humble and handmade, and at the same time sophisticated and highly technical. Clearly a lot of time and effort goes into each *Strandbeest*, but when asked about the plans for his next sculpture, Jansen says: "Theo must survive. So I am thinking very functional."

—Arwen O'Reilly

>> **Theo Jansen's website:** strandbeest.com

Photograph courtesy of Theo Jansen

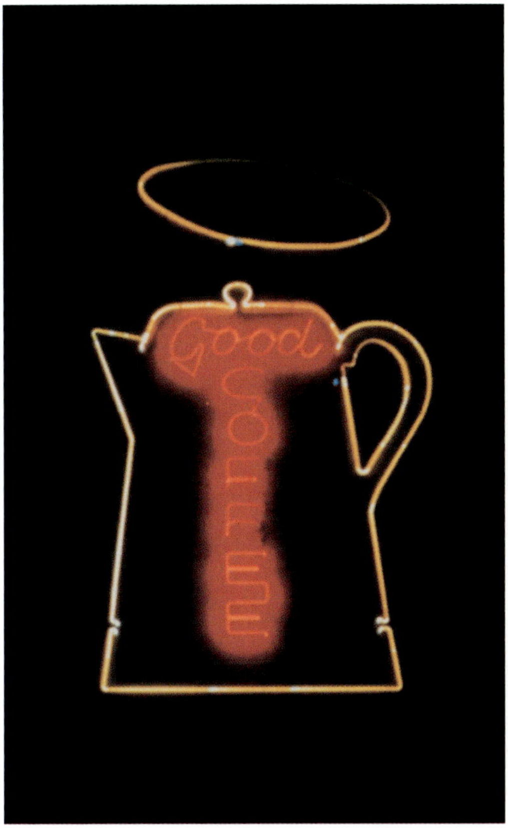

Camera-on-Wheels

Famed landscape photographer Ansel Adams once said, "You don't take a photograph, you make it." According to San Francisco photographer-sculptor **Jo Babcock**, that's not good enough. Babcock isn't satisfied unless he's also made the camera itself.

During the last 25 years, he's crafted more than 200 of them, turning everything from coffee pots to suitcases into the DIY darlings commonly known as pinhole cameras. But the most ambitious of Babcock's creations are the contraptions he's fashioned from motor vehicles, including a VW van and an Airstream motor home. What's next, a 747?

The van-cam works on the same general principle as Babcock's other pinholes, except that it's on wheels. A needle-sized hole in the side wall acts as a mini aperture. Light pours in, projecting an image onto photo paper (you can't find film that large) taped to the opposite wall. No lenses required. But the van also doubles as a proper camera obscura ("dark room" in Latin), thanks to the addition of a lens and mirror on the roof. Because of the tiny aperture, the first van-cam took four hours to expose a pinhole pic. Using the lens, Babcock can send an image to the vehicle's floor and develop a paper negative in a fraction of the time.

Babcock, who taught photography at the San Francisco Art Institute and the University of California, Santa Cruz, among other institutions, is a firm believer in the merits of creating his tools from found objects. "You don't necessarily need newer or high-tech equipment to make better art," he says.

Fittingly, Babcock's new book, *The Invented Camera*, which chronicles his "symbiotic" photos over the years (the coffee-pot pinhole shoots a dreamy coffee shop sign, the suitcase a fuzzy bus station, etc.), mirrors his autonomous approach. To get the book out there, Babcock learned Quark and Adobe Photoshop, worked through the night for weeks assembling the digital page files, then raised the money to print the books by working as an electrician. Up next? A clicker converted from a 2005 Scion xB.

—Megan Mansell Williams

>> Jo Babcock's website: jobabcock.com

Photography by Jo Babcock

Open Source Open Water

Those with doctorates in artificial intelligence are never the best stewards of houseplants. Programmer **Bryan Horling** says he's killed whole swaths of greenery inside and outside his rural western Massachusetts home. But at least one plant will survive, thanks to a computer-controlled plant watering system — a simple network of plastic tubing and an aquarium pump to keep the Wandering Jew plant in his living room alive.

"You can do this project even if you're too lazy to water your own plants," Horling says. His computer activates an aquarium air pump, which displaces water in a soda bottle, sending it to the plant. He started with an X10 appliance controller from RadioShack, an aquarium pump, length of tubing, 2-liter plastic soda bottle, and aquarium check valve. He drilled two holes in the soda bottle cap, inserted the tubing, and sealed them with a glue gun. A metal binder clip holds the tube in place, and the check valve keeps the water from backing up the pipe after the plant is watered.

Horling, 30, uses hobby projects as a release from his difficult studies. His research on multi-agent systems is funded by the U.S. Army, which wants to use software agents to help filter and analyze information received by commanders in the field. To take a break from his thesis, he created a recipe for baby wipes that involves paper towels, no-tears shampoo, and a table saw.

In this case, the houseplant waterer is controlled by a Linux PC in his den. A freeware application for Linux allows the computer to control the X10 device. Then Cron, a built-in application in Linux, lets him run scheduled tasks. The script tells the air pump to turn on for five seconds every day.

A 2-liter bottle holds enough water to keep the plants happy for two weeks. Originally, he got it because his wife Maura and he traveled constantly. Now it comes in handy because they're distracted by a new baby. "It created a whole other problem," says Horling. "It's just enlarged the scope of our forgetfulness. Now we forget to fill the water bottle."

—Bob Parks

» **Computer-controlled watering:** makezine.com/go/plant

Photograph courtesy of Bryan Horling

Photograph courtesy of Matt Jones

The Secret Life of Death Clouds

Matt Jones contemplates life by building moving sculptures that fail to replicate it. A graduate student in art at Stanford University, his investigations have led him, among other things, to use an air compressor to animate a respiratory system fashioned from old bicycle inner tubes, and to motorize a carpet of zip ties laced with LEDs to approximate a pulsing, gently respiring, furry hide. His goal: to tease out the vital essence that makes the living live.

It took a giant garbage bag full of hot air to teach him to appreciate the life coursing through his creations. To create the grandiose piece *Black Cloud* for a death-themed Land Art show in the cactus-studded desert of central New Mexico, Jones needed little more than a pair of scissors and a lot of tape.

He cut out black garbage bags, sealed their edges to each other, and then rigged a squirrel cage fan with ducting to fill the vessel with sun-heated air, floating it several feet above the ground. Once aloft, the Suburban-sized balloon seemed to find a mind of its own in even the gentlest breeze. Trying to steer the cloud with fishing line before a crowd of spectators, says Jones, was "like trying to drag in an orca — an orca that insists on jumping into cacti." Long patching sessions followed each brief and otherworldly flight.

Despite the difficulties and constraints inherent in making kinetic sculpture (it has to work, after all), Jones says it pleases him more than traditional media. "Besides color, line, and solid shapes," he explains, "there are entire regions [of the brain] devoted to detection of movement, areas untouched by static art."

Certainly, Jones' kinetic works breathe life into many regions of the mind — especially when they're cooperating. "After the showdown in the desert," he says, "I came to cherish those moments when my work wasn't broken."

—Eric Smillie

>> **Matt Jones' website:** ojdingo.com

Blue Spark

"A custom-made life is better than a Wal-Mart world," says **Sean Barrow**. Tall, dark, and tattooed, Barrow looks more like a rock star than the avid eco-design aficionado he is.

His post-apocalyptic appearance at first glance seems at odds with the elegant, minimal, Japanese-inspired aesthetic he studies and employs in his sustainable furniture making. But both display his practical approach to 21st-century salvage: to reveal rare beauty and utility from former chaos.

Case in point: the sleek *Electron Monument*, a bewitching handmade side table that hides an array of outlets for electronic devices and their chargers. "'Charging station' sounds so unsexy — hence the name," laughs Barrow, who installed six outlets, capable of handling power blisters as well as standard plugs, in the table's inside base.

Sitting high on salvaged metal legs, the box is made from spalted pin oak that Barrow snagged from a dying tree (which creates the zigzagged black segments in the wood grain), and held together by wooden finger joints. The removable top is reclaimed and sanded zebrawood and rosewood, with sides angled in at 13 degrees (a favorite angle he uses in much of his furniture), finished with nontoxic Osmo oil.

The most compelling feature of the *Monument* is the hypnotic, softly glowing cobalt light in the front of the box, with machined metal pieces added to create an abstract power-outlet motif. Using almost no energy, it is illuminated with an LED plucked from an old night light.

The combination of traditional Asian design, sustainable resources, and the sci-fi hieroglyph glowing from within the *Electron Monument* make it a perfect example of Barrow's work, and one of the coolest ways to hide your electronic clutter.

—Kirsten Anderson

>> **Sean Barrow:** skrewgun.com

Photograph by Sean Barrow

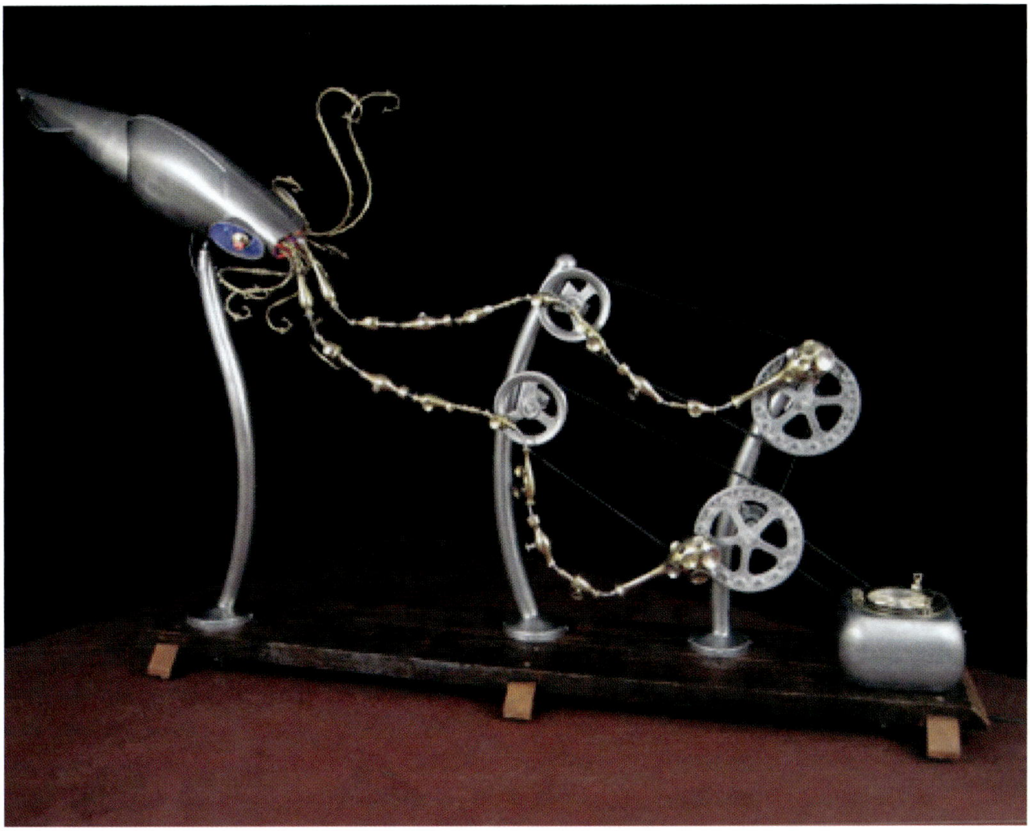

Kinetic Fauxbot Sculptures

Raised by two artists, **Gould Nemo** started drawing as a young child and received a traditional art school education. But once he started creating art with the heaps of scraps and junk he collects, drawing became unsatisfying and there was no turning back. Now Nemo creates a wide range of robot-like sculptures out of recycled materials that move and gyrate in interesting ways.

Nemo calls them "fauxbots" because they don't really count as robots, at least not to the more technical robotics crowd. And he's fine with that; he readily acknowledges that his creations are sculptures first. While many of them are inspired by robots, Nemo admits they are not necessarily helpful or useful in any practical sense.

Nemo's sculptures are made almost entirely from recycled materials. He only purchases a few technical components, like LEDs and bearings. To find parts, Nemo haunts a favorite local salvage yard and relies on garage sales, curbside finds, and the occasional dumpster dive.

One of Nemo's more notorious creations, a 10-foot-tall robot sculpture named *Goliath*, resides famously in San Francisco's ultra-ritzy Pacific Heights neighborhood, where it has attracted neighborhood scorn, the attention of vandals, and delighted tourists. *Goliath* is now a popular fixture on the local tour bus route. The gasoline-pump nozzle that serves as Goliath's male appendage is one of his more attention-getting features.

Not all of Nemo's creations are based on your typical robot. His latest and most challenging sculpture is a truly giant mechanical squid (more than 6 feet tall and 10 feet long) that writhes and undulates in a variety of ways.

"I am really fond of curved lines and surfaces. Aligning belt wheels, bearings, motors, and non-essential squid-looking stuff in an organic form that does not simply fall apart was a real thrill."

—Bruce Stewart

Fauxbots: nemomatic.com

Photograph by Gould Nemo

EGGHEAD

INVENTOR NIKOLA TESLA SAW REALITY IN HIS OWN WAY.

By Bruce Sterling

IT'S A TOY IN A MUSEUM: AN EGG IN A bowl. The egg shifts restlessly. It rocks and it rolls. The wandering egg explores the bowl's limits. It dances arcs within the bowl's tall rim.

Then, just when it seems about to free itself and fly out of the bowl toward its viewers, the egg hesitates. Somehow, mystically, the egg wriggles upright onto one of its ends.

The egg is spinning like a top. It has achieved some kind of post-chaotic feat of stability, and now drifts smoothly back to the spot where it started: the center of the bowl. There, dead center, it spins like crazy, self-contained and untouchable, defying gravity, friction, and common sense.

On its aging shelf inside Belgrade's Nikola Tesla Museum, the *Egg of Columbus* spins like a turbine. The egg *is* a turbine — the conceptual ancestor of pretty much every electrical motor in the world.

Technologists tinker. They engage with the grain of the material, testing empirical realities, pushing for higher performance. Scientists seek out unifying laws and principles that make sense of the natural world. Nikola Tesla, an inventor, did neither of these things. He saw reality in his own way.

Tesla's testament, *My Inventions*, written in later years when he was drifting through fleabag New York hotels and feeding pigeons, is a painfully sincere account of what life felt like for him. The work is intensely psychedelic: visual hallucinations were always the most exciting and significant things that happened to Tesla. A true visionary, Tesla experienced mind-bending flashes of insight, sometimes accompanied by tongues of fire.

Tall, good-looking, and sturdy, Tesla was capable of handling various and sundry tools: he could wire a ship, fix a fire engine, and dig a ditch. But Tesla did not hack; he didn't tinker. He invented through his immensely detailed and vivid mental visions.

Although Tesla blew through college in mere months, absently memorizing entire books and deftly speaking multiple languages, he didn't consider himself a genius. Tesla had an older brother who was much brighter, more capable, and more ambitious than himself, but who (maybe luckily for the human race) died young.

There may not be adequate words to describe the extreme events that took place within Tesla's skull. The shattering flashes of light that paralyzed him were likely migraines. The childish bedtime trips into imaginary Oz-like cities, where he befriended strange people and flew through the air, were lucid dreams. His existential loathing for pearls was probably a phobia. Tesla forced himself to read all 100 volumes of Voltaire's complete works, in French, which indicates that he likely suffered from obsessive-compulsive disorder. He had hallucinations on tap as an everyday mental utility. Tesla could visualize imaginary machines down to the last detail, even the rust and wear-marks.

Tesla could visualize imaginary machines down to the last detail, even the rust and wear-marks.

Except for his invention of the alternating-current induction motor, a huge advance that transformed civilization and later contributed to what is known today as the greenhouse effect, Tesla's genius was mostly a personal matter. Wizards prize spiritual advancement over mere cash and power. Poverty and privation were an adventure for Tesla. Even when his pockets were brimming with money (rather often), he took a positive pleasure in systematically depriving himself of food, booze, gambling, cigarettes, and women. In the long decades of his adult life, Tesla never changed weight: he liked to volumetrically

Photograph by Bruce Sterling

The Serbian 100-dinar bill features Nikola Tesla, his favorite motor, and the equation that defines the scientific unit of measurement, the "tesla."

measure his food and stop when he had eaten enough cubic centimeters. Tesla never married, probably because he could never find a girl like dear old Mom — an illiterate genius who could hand-tie three knots in a human eyelash.

This year — 150 years after his birth — Tesla still enjoys the reputation of a man ahead of his time. Tesla's sesquicentennial finds him a 21st-century household name. Tesla coils, transformer coils that beautifully spit giant writhing sparks, are his signature mad-science toy. The tesla (T) is an international scientific unit of measurement, good for the ages. A sports car and a rock band carry Tesla's name, while about a million goofy New Age cranks are still enthralled by Tesla's musings.

Tesla was certainly a genius, but not a "neglected, forgotten genius" — that is a myth. It seems that almost everybody who met Tesla knew at once, on some deep primal level, that he was freakishly smart. It took people a while to figure out that Tesla was beyond the aid and comfort they could offer.

Although he was a paranoid hypochondriac, Tesla lived until his mid-80s. And he wasn't the picture of misery. Tesla always had an excellent opinion of himself and his own doings. He had little money because he was basically unemployable. He ignored the profit motive and never did boring grunt-work under duress. Medals, honors, doctorates, banquets, publications, and patents arrived in heaps for Tesla.

Romance had no relevance for him. Besides his mother, brother, and maybe his father, no human being ever impressed him much. Though he may well have invented the remote-control drone, Tesla was no robot: he was a living being of wild passions and eerie psychic extremities. Tesla had a sense of humor — enough of one to recognize that Mark Twain was a genius and a man he could befriend.

Tesla was not recklessly improvident. He was probably as good at finance as most engineers, but World War I wrecked his prospects for attaining great wealth. It seems the war proved to Tesla that the human race was crazy. In the darker years after the Great War, Tesla scarcely pretended to be normal. He still dressed in his Broadway-dandy uniform — his tailored black Prince Albert coat, top hat, and spotless white gloves — but his tics, phobias, and delusions took free rein. In the real world, he simply went through the motions. Tesla gave some pronouncements in the press, he opened his fan mail, but he could no longer be bothered to construct huge wireless power-towers on Long Island, create death rays for the U.S. Navy, or build magnetically levitating ultra-high-speed global airships.

Tesla could still see these wonders in his mind's eye, probably down to the paint flecks and serial numbers. But spinning deep within his own mental bowl, he no longer bothered to climb over the rim.

Tesla's spinning egg toy is a truly superb proof of concept. It came from a time in Tesla's existence when he had a pressing need to prove his concepts to the outside world. With the passing of years, that need passed. There is one other smooth bronze egg in the Tesla Museum in Belgrade. It is entirely spherical, polished and unblemished, stagily lit, and poised on a pedestal. Some siftings are sealed inside of it — black ashes. That's Nikola Tesla, egghead inventor. He's still in there.

Bruce Sterling (bruce@well.com) is a science fiction writer and part-time design professor.

I, PLEO

How Caleb Chung went from street mime to toy-robot maven.

INTERVIEW & PHOTOGRAPHY BY ROBERT LUHN

Can a machine have a soul? Can it think? Can it laugh and cry, bug you for a snack, tease you, or curl up on your couch and dream robotic dreams? Can you build such a machine? And can you sell it for $250? »

It's alive! Puppet master and puppet — or rather, Caleb Chung and his autonomous robot companion, Pleo.

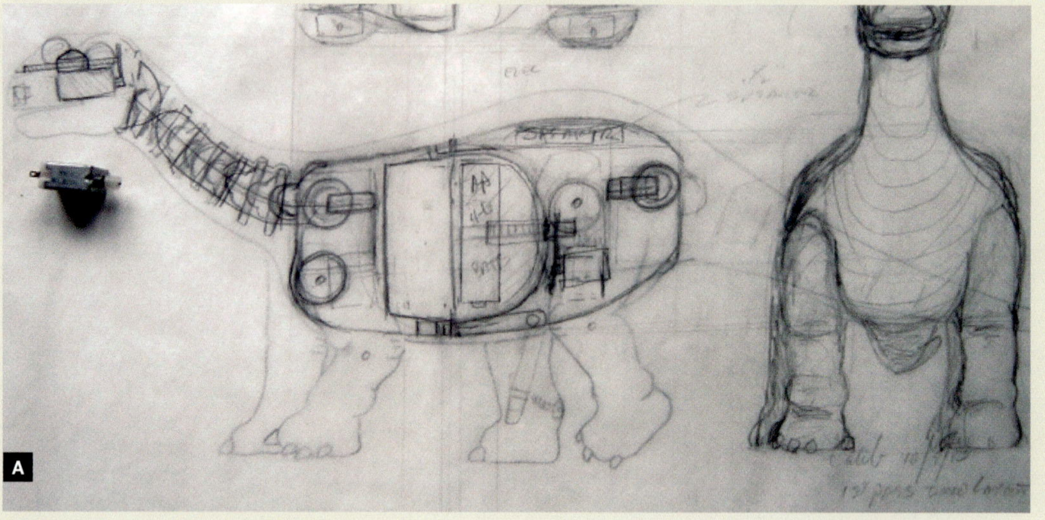

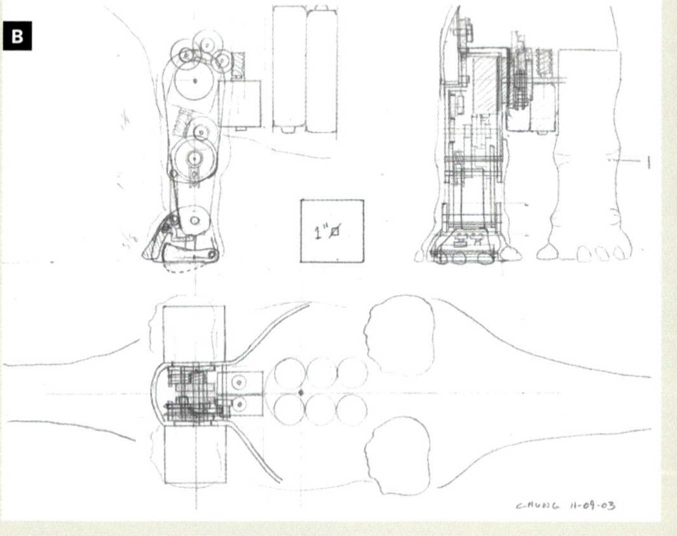

A. Pleo is the size of a week-old baby camarasaurus, a dinosaur from the late Jurassic period. **B.** A camarasaurus' naturally stocky legs are perfect for containing motors and gears. **C.** Chung and his colleagues studied camarasaurus fossils to create an accurate model for Pleo. **D.** The head contains an IR communication device, an object detector, a light sensor, a mouth object sensor, and a motor to control jaw, nose, and mouth movement.

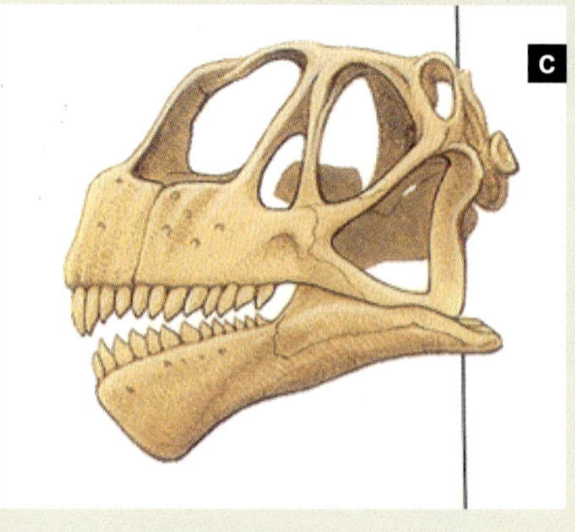

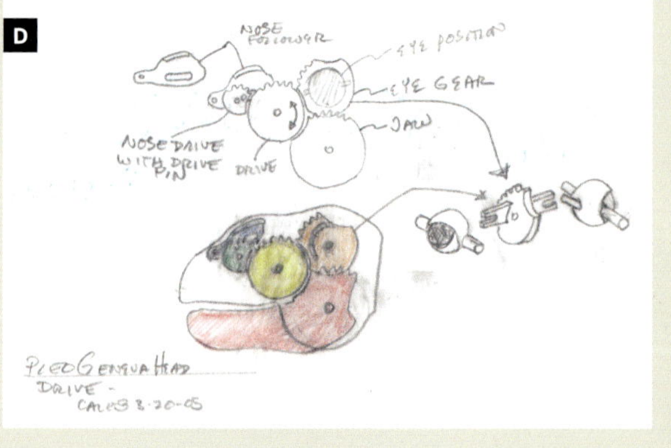

Those are just some of the challenges facing Caleb Chung and his cohorts at Ugobe, Inc. ("*You! Go and be!*" Get it?) Chung doesn't need the tsuris or the money. As one of the developers of the Furby, which has sold an estimated 40 million units since 1998, Chung never has to work again. But there's something gnawing at Chung, something beyond conquering the toy world again or making enough money to buy his own country.

"Humans have a fundamental need to nurture — it makes you a better person," says Chung. "And one of the roles of digital life forms is to bring out the best in people."

The embodiment of Chung's passion? An unassuming baby dinosaur — the size of a cocker spaniel — called Pleo. Inside this camouflage-green camarasaurus is a ton of technology: 14 servo joints; 38 touch, sound, and light sensors; a half-dozen digital signal processors; a gaggle of electric motors; a camera installed in its nose; an infrared transmitter for communicating with other Pleos; a flash-card slot; and a real-time operating system that oversees Pleo's lifelike movements and its personality.

Outside, Pleo is sheer experience, performance art combined with emotional sleight of hand — just what you'd expect from Chung, a former mime. Pleo ambles along in amazingly lifelike fashion, smiles and shrugs, frowns when he's unhappy, and yowls for "food." According to Chung, Pleo is not only autonomous, but evolves over time based on your interactions with him. Unlike his robotic brethren, Pleo isn't a battle bot; he's a love bot, designed to be your pal and to summon your latent (or overactive) nurturing instincts.

But something more primal drives Chung: the desire not only to manipulate life, but to create it from scratch. To create a robotic companion that thinks and, maybe, even feels. Call it the Geppetto Syndrome.

It's Alive! (Kind Of)

To meet the puppet and the puppet master, I fly up to Chung's lab in Boise, Idaho. As I stride out of the spacious new terminal and behold the stunning mountains in the distance, a worn, white Lexus pulls up to the curb. Out jump Chung and John Sosoka (chief technology officer and key artificial intelligence guru), both clad in Hawaiian shirts and flip-flops — standard high-tech attire in Boise. After a jovial exchange of handshakes and jokes, we're off — and so are Chung and Sosoka, chatting animatedly about the joys of Boise, the nature of robotic emotions, synthetic versus organic life, and more. A fancy lunch follows, and then we're off to the lab.

Alas, there are no bubbling beakers, giant sparking electric generators, or even a tiny tar pit for Pleo to play in. The "lab" is a nondescript box in a faceless office park, manned by a handful of engineers and designers. I meet the staff and poke around, examining Pleos in various stages of undress. One Pleo has returned from a demo in China in tatters, apparently chewed by a jealous panda. Another is stripped down to its motor/gear/wire skivvies, as three engineers fiddle with a locomotion problem. I snap pictures of Caleb in silly poses with Pleo and talk with one of the engineers about the joys of QA testing Pleo's tail.

Later, at a local cafe, I chew the robotic fat with Chung and Sosoka and finally meet a more-or-less working Pleo face-to-face. To say I'm skeptical is an understatement. The company's website and literature make some mighty big claims: "[Pleo is the] genesis of a whole new era in robotics. If you can imagine it, Ugobe can create it," "Pleo is the first truly autonomous Life Form capable of emotions that allow personal engagement," and "Pleo ... can feel joy and sorrow, anger and annoyance ... even dream." The Pleo I meet runs a six-minute script that showcases his range of motions and emotions, but it's hardly autonomous.

Still, for a construction of rubber, gear trains, CPUs, and wire, Pleo is pretty engaging, with a sly smile on his face that's pure Chung. Chung puts Pleo through his paces, the baby dino showing fear (complete with trembling), happiness (wagging tail, dancing, little yippy sounds), groaning for food, playing dead (like a kid, dramatically falling over with a clunk), sneezing, and more. Pleo's motions aren't perfect, but that's the idea.

Illustrations courtesy of Ugobe

"Pleo's movement has 'noise'," says Chung. "That's what makes it organic. And because he moves naturally, his movement becomes invisible — and that's when you become interested in what Pleo can do, what he's thinking."

Thinking Pleo

The question is, what can Pleo do or think about? At this point, it's a little theoretical because Pleo is still a prototype. But according to Chung and Sosoka, Pleo will be autonomous first and foremost — which means no remote control.

"When you leave for work in the morning, your dog doesn't sit by the door all day. He explores; he does stuff. Pleo will be doing something when you're gone because he's alive", says Sosoka, without a trace of irony.

Pleo will also remember and adjust his behavior accordingly; in other words, he will learn. If you twist Pleo's leg, he'll know you're abusing him. He'll cry out and limp, and you won't be able to play with him for a while. He'll store that grudge, cataloging the event and assigning it a value. The next time you try to play with him, he may snap at you. Conversely, the more you reassure and nurture Pleo, the quicker his naturally sunny nature will reappear.

As you'd expect, Pleo is programmed with strategies for solving certain problems, such as finding food. But ultimately, says Chung, you can't program every possibility. "You have to let him choose. Pleo keeps a record of what works and what doesn't — filtering experience the way all living things do."

Like his Furby forebears, Pleo will be able to communicate with his kin, although Pleos will do more than swap colds. "When two or more Pleos get together, there'll be some kind of meeting ritual and they'll pick the 'alpha' — and that Pleo will transmit behaviors and tricks and moods to the others," says Chung. But that's just a start.

You can plug Pleo into your PC via USB, run a little program, and adjust Pleo's personality by moving sliders or writing scripts akin to JavaScript or Logo. Want to really hack Pleo? In the future, there may be a C/C+-style developer's kit that lets you mod the dino to the max. Whatever personality you create, you'll be able to back it up and share it with others.

"Pleo is not a closed product. At every step we ask, 'How can other people play with Pleo, modify it, personalize it?'" says Sosoka. Eventually, users will upload new personalities and tricks to Ugobe's site that others can use. "Pleo is ultimately a platform, and creating new Pleo personalities is like making a movie." Pleo, in short, is as much art form as life form.

Making Pleo

Building that art form, however, involved some real-world tradeoffs. For example, why build a robotic baby dinosaur? Why not a popular computer-game character?

It was a sound business decision, says Sosoka. "Dinosaurs are great, they're already a brand, and you don't have to pay royalties on them." More important, says Chung, is that you'd never be afraid of a baby dinosaur like Pleo — he's small enough to sit on your desk. And a baby anything is easier to animate. Adult creatures have sublime, complex movements; babies falter and stumble, and they have a smaller emotional palette. One plus even a baby camarasaurus offers: big, blocky legs, perfect for holding motors and gears.

And these motors and gears have been notable headaches. "Exotic magnets that go all the way around a motor are really powerful, really efficient, and really expensive," says Sosoka. The solution? Write a sophisticated control program that makes cheaper, less accurate motors work better. Slap a 12-cent 6502 processor on the motor, says Sosoka, and you've got a smart servo that knows force feedback, self-clutching, and self-calibration. "We do a lot of localized intelligence, just like your body," he adds. Off-the-shelf gearboxes were noisy, so the company turned to an "unrelated" industry for custom units.

Ugobe won't disclose much about its real-time "Life OS" that oversees Pleo's movements and cognition. Basically, Pleo is built around a simple, one-layer neural net; something happens and the event is weighted and remembered. Over time, the connection strength between a stimulus

A. **Pleo is a proportionately correct baby camarasaurus robot. B. An exoskeleton provides support. C. Motors, gears, and linkages provide realistic movement. D. Sight, sound, and touch sensors are hidden under the hide.**

Illustrations courtesy of Ugobe (A, B, C). Illustration courtesy of Mack Fraga Studios (D)

(like turning Pleo upside down) and a response (Pleo laughing or crying) changes. "It's almost like muscle memory," says Sosoka. It doesn't dictate how Pleo will react every time an event occurs, but Pleo's past experience will influence his future choices.

The other key factor in Pleo's reactions? Emotions. (Well, what passes for emotions in a $250 robot.) The factors that really control Pleo are his drives — hunger, bonding, fatigue — and the goals associated with them. If Pleo's hungry, the goal is food, and the preprogrammed strategy Pleo picks to meet this goal is based on which one worked best in the past. It's kind of like fuzzy logic: rules combined with sophisticated probability with a little sloppiness allowed. In an unusual situation, emotions — colored by Pleo's success or failure, someone hitting or petting him, etc. — act as a secondary control system. It's why Pleo should have many different reactions to a given stimulus over time.

From Mime to Machine

Pleo's ultimate stimulus, of course, is Caleb Chung. Chung has pushed, prodded, and pulled Pleo into existence with a scary singlemindedness. "He's a character — a mad scientist," says Ivy Ross, a former Mattel executive who worked with him on the Miracle Moves Baby doll. "He has a vision and relentless drive to bring something alive."

Gary Schwartz, longtime friend and Chung's mime partner, says simply, "He's the irrepressible kid, the incarnation of Tom Hanks in *Big*, pure inspiration, pure yin, but with the emotional savvy of a 50-year-old."

How did Chung get from street mime to toy-robot maven? At first glance, Chung's resume looks like a hodgepodge of careers. He went from performing mime (as part of the Schwartz & Chung comedy team in the 70s and early 80s), to voicing live-action TV cartoon characters (he was QT the Orangutan in *Dumbo's Circus*).

By chance, he applied for and got a job devising cutting-edge toys for Mattel's R&D group. Later, his knack for creating props led to work designing mechanical effects for such films as *Total Recall*. Since the late 90s, he's consulted with various toy companies, helped develop Furby, and ultimately, decided to tackle Pleo.

“Pleo’s movement has ‘noise.’ That’s what makes it organic.”

In his years as a performer, Chung became a master of props and mechanics. “We got a gig on the *Queen Mary* in Long Beach to entertain the tourists,” says Schwartz. “They couldn’t afford animatronics, so they hired mimes.” It was here that the duo created their first notable bit, “Caleb 9000.” Chung did a spot-on robot routine while Schwartz, offstage, supplied his voice. At the end of the bit, Caleb 9000 says, “Absolutely nothing can go wrong ... wrong ... wrong” and then blows up, courtesy of a remotely triggered explosive pack Chung designed.

So what’s the connection between mime and robotics? “Caleb is a superb actor,” says Schwartz. “It’s a key skill that game and robotics animators need to learn. As a mime, you’re also super-aware of your body, of motivated movement, the physical cues to emotion.” Chung’s ability to tell a story via movement, to play a cartoon character, to suspend disbelief, and to think like a kid is uniquely suited to dreaming up innovative toys or convincing you that an animatronic dinosaur is emotionally aware.

Not surprisingly, the inspiration for all of Chung’s work is his fervent belief in, yes, magic. “The first books my mother read to us were Tolkien’s,” he says. “Fantasy and magic were an integral part of my upbringing. They’ve become an integral part of my product design. I want to transmit that wonder.”

Not just via robots, mind you, but via random acts of surrealism. Take Chung’s interest in, um, fairies. On an island near his house, Chung collects sticks and makes little fairy furniture that he leaves all over the island — a little chair, a tiny broom, maybe a bassinet with a walnut “baby” inside. With a diamond-bit drill, he even carves pseudo-hieroglyphics onto stones. “If you actually found one, you’d never think someone put it there, because it’s too frickin’ obsessive. But I want to change how people view the world.”

A Tale of Two Furbys

After visiting Chung and meeting the lovable Pleo, it’s easy to take a sip of Ugobe’s Kool-Aid. Chung is intelligent, charismatic, and clearly dedicated to making Pleo a breakthrough. The question is, can he deliver? Ugobe expects to release Pleo in March 2007, but the units I saw — only seven months before release — didn’t seem very close to prime time. Still, robotic innovation hardly runs like clockwork, so ship dates are always fluid.

But the dispute over Furby — the robotic toy that made Chung’s reputation and fortune — casts a shadow over Chung’s current efforts. Chung, who was responsible for the design of Furby’s mechanics, tells one story about how Furby came to be.

David Hampton, the co-inventor responsible for the programming and electronics, tells a rather different one, which is mostly corroborated by several former executives from Tiger Electronics, the company that licensed Furby. (Tiger was bought by Hasbro the day after Furby’s debut at the 1998 International Toy Fair in New York.) The engineer brought in to “save” the project contributes still another angle.

What most everyone (including Chung) agrees on is that 1) the mechanics for the Furby prototype that was going to be demo’d at the Toy Fair didn’t work; 2) mechanical engineer Richard Maddocks’ last-minute assistance helped get Furby working in time; and 3) after the Toy Fair, Chung was bumped off the project.

The rest of the story has multiple versions as to who did what, when, and why. On one hand, it’s clear that Chung bit off more than he could chew. While he was instrumental in developing the Furby concept and designing the mechanics, execution was another thing. Chung couldn’t use off-the-shelf parts; the tiny pulleys, gears,

Photography by Robert Luhn; Furby drawing by Kelly Heaton

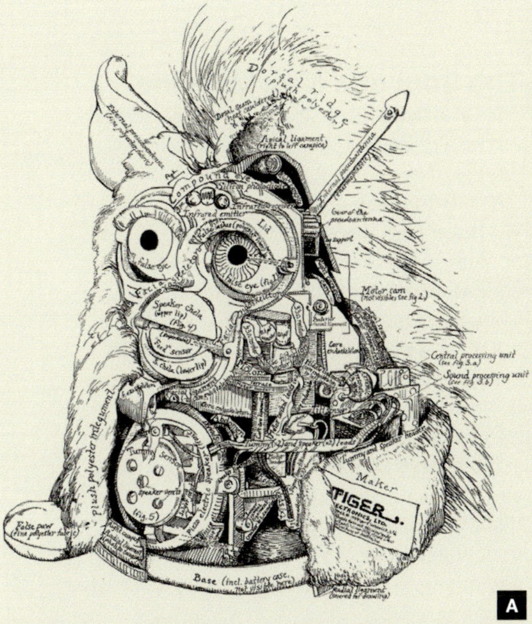

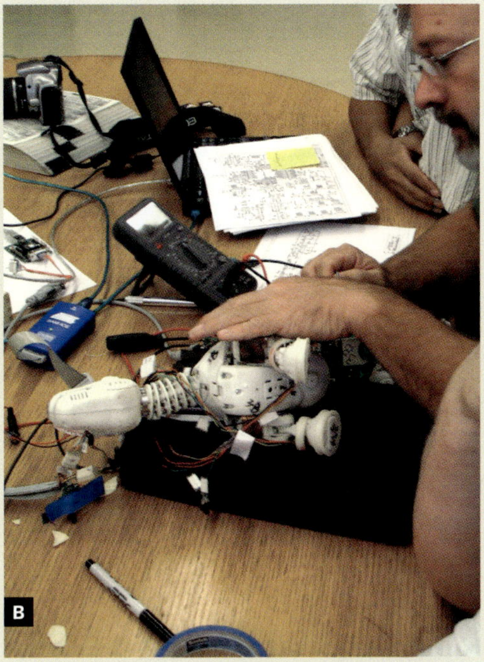

A. ***The Anatomy of the Furby*** **by artist Kelly Heaton, who once made an interactive Furby wall. B. A Ugobe engineer troubleshoots a locomotion problem with Pleo. C. Caleb Chung's lab is an unassuming box in a small office park outside of Boise. D. A Pleo tail-testing unit. E. Pleo in a classic pose.**

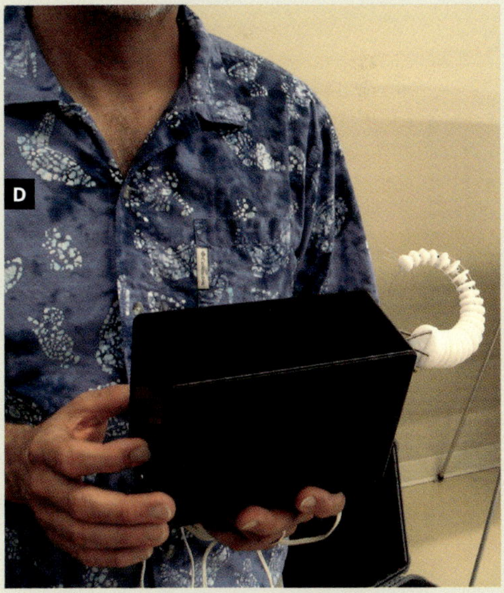

cams, and other parts had to be custom made to very fine tolerances. If a part didn't work, Chung modified it until it did — but that introduced friction that could stall Furby's motor. "[Chung's] design was brilliant and sound in principle," says Richard Maddocks, now a senior principal designer at Hasbro. "But there was a huge amount of stuff packed into a very small form factor. No one had ever done this before. I didn't redesign Furby — I just helped it run reliably."

First, Maddocks put in a bigger motor to overcome the friction. Then, he created new parts to replace Chung's oft-modified (and worn) parts. After some final tweaking, he got Furby working reliably enough to debut at the Toy Fair.

Chung also readily admits he was lousy at overseeing his time and that of the engineers and machinists he hired. (At the same time, he sings David Hampton's praises: "He's a virtuoso in digital design and programming — a great talent who, until Furby, never got the credit he deserved.")

Tiger, too, can be faulted for slashing Chung's prototyping budget (from $60,000 to $40,000) and not keeping a closer eye on the progress of Chung's work, which, after all, involved making something that no one had ever made before. "We were undermanaged," says Chung. "God knows, I needed a manager."

What did Chung learn from the Furby?

"I thought I could run the process and manage a group, but I made some bone-headed decisions," says Chung. "I never made that mistake again. At Ugobe, I invent, I come up with ideas — but I let someone else manage and do the tooling and engineering."

Apparently the lesson stuck. Chung's next project, creating the Miracle Moves Baby for Mattel, won high fives all the way around. Ivy Ross, the senior VP in charge at the time, lauds Chung's dedication and ability to execute.

"Caleb really understood the emotional connection [the doll] needed to have with kids. We didn't have this kind of expertise in house. Caleb showed up and collected a whole group of experts who made it happen," he says.

The Soul of a New Machine?

The Great Furby Debate may never be settled. As the old saying goes, "Success has many fathers." The acid test for Chung will come when Pleo hits the market. Will it work as advertised? Will buyers embrace Pleo as their robotic pal, or dismiss it as a novelty? Chung's reputation and his worldview are riding on the results. But Chung is simultaneously bullish and realistic.

"I think we have something magical," says Chung. "When I showed off Pleo at the Demo 2006 show [see makezine.com/go/pleo], I realized this moment in time would be remembered in the history of man-made life-forms. They'll say, 'These people didn't have all the tools, and there were all these restraints, but they tried to do this for real.' Still, I don't have to prove something to the world. Pleo is going to be what it's going to be."

But if Pleo lives up to half its promise (and a fourth of its hype), it will raise a host of intriguing and troubling issues. Is Pleo an autonomous life-form? Is he sentient? Can he feel?

"All that matters is what the user perceives," says Chung. "Once you create organic movement and add just enough emotive cues, people will suspend disbelief — they'll fill in the rest and fall in love with Pleo." Besides, he adds, he's not even sure it is suspension. "I think Pleo really does these things."

But if Pleo or his successors develop even a tincture of self-awareness, don't Chung and company have a huge obligation to these new creatures and their human owners?

"Of course we do — especially for a creature that plays to the human heart," says Chung. "An invention like Pleo can fundamentally change society. That's why we want to be first, to set a responsible tone. These machines are extensions of us, how we think and feel. As much as we're going to get monsters and creatures we don't want around, we'll also get beauty and magic from artists who really know how to create," he says. "Great art tackles the great issues of our time. That's why we're digging here — we're digging in the right place."

Grandpop's Shop

By Robyn Miller

I JUST RETURNED FROM PHILADELPHIA, the city where most of my family was born and raised. Many of them still live there, including Grandmom, who's still sharp as a tack.

My grandfather repaired elevators and dumbwaiters. He was well-loved, and since his death, his machine shop under the house has been kept almost as a shrine to his memory. My uncle uses the shop for small projects, but after ten years, not much has substantially changed: my grandfather's toolbox rests open, his scrawled notes and plans are scattered all around, and even his smell — machine oil — is everywhere.

The shop was always a magical place for us kids. Grandpop would invite us in, show us a strange mechanism (usually his lathe), and begin to create a trinket right before our eyes. A miniature baseball bat, a Plexiglas lightsaber. I'd watch him shape metal as if it were putty. Turn square blocks into curved cylinders. Nothing seemed impossible.

Then he'd invite us to try our hand at it. I was most apprehensive with the lathe, and would stand there thinking: this hunk of spinning wood is definitely going to pop out and smack me right in the forehead. Grandpop always stood directly to my right, encouraging me while I tried to cut my marvelous colonial chair leg. Overambitious plans for a 9-year-old, but I'd just watched him make the same thing with freehand ease!

Now, whenever I visit, I always go down and poke around, but not too much. It almost feels like he's still there.

Robyn Miller is most popularly known as the co-creator of *Myst* and its sequel, *Riven*. He continues to work on a variety of projects and has most recently finished an album titled *1000 Years and 1 Day*. Visit him at tinselman.com.

Photograph by Robyn Miller

Heavy Lifting

Placing huge towers up a mountain is just the start to reaching Troy Caldwell's ski-resort-on-a-budget dream.

By Dale Dougherty

SIXTEEN YEARS AGO, TROY CALDWELL went to buy 10 acres from Southern Pacific and got a sweet deal on 400 acres of land, most of it mountainous, near Lake Tahoe. A ski bum who left college to learn the sport, Caldwell eventually became part of the U.S. Ski Team in the 1970s, and he says he's been part of the ski industry (which he pronounces "SKI-in DUS-TREE") ever since.

Caldwell's dream is to build his own private ski facility. Lacking the big bucks, he decided to build it himself, and he's been working the past six years on designing and constructing his own chairlift. Some of the work has been bartered as "Tahoe trade-outs." He made office cabinets in exchange for the structural engineering of the chairlift.

Last fall, he placed 17 towers up the mountain, with the help of a helicopter and 30 volunteers who hope to ski on the property one day. The towers, which weigh up to 3,000 pounds, were built in his garage using a series of pulleys and hoists. This allowed him to move these heavy objects himself, as he built stairs and platforms and welded them to the large pipe. "You have to do the towers right," Caldwell says, "because you don't ever want to have to do them again."

Caldwell has been challenged by lawsuits that drain his bank account but not his enthusiasm. "You want your mind to be focused on the positive things

Photography by Dale Dougherty

and not the real-world problems that can make you bitter," he says.

Weather has also been an issue. Two early-season snowstorms of 20 inches seemed to wipe out his chance to place the towers last autumn, as the footings were full of snow. However, his volunteer crew encouraged him to continue by offering to dig out the 7-foot-deep holes and make them ready.

Caldwell says that when the helicopter arrived on site to move the towers, he could tell he had a nervous pilot who had worked on forest fires but not this kind of heavy lifting. He had to direct him not to try to drop the tower into the hole but just to hold the weight of the tower and let the ground crew move the tower into place.

One of the volunteers, Ken Gracey, vice president of Parallax (makers of the BASIC Stamp microcontroller), says: "You've seen how large this tower is

> "You have to do the towers right because you don't ever want to have to do them again."

on the ground and then you see it coming at you, as the helicopter is lowering it, and you're aware that all you've got to protect you is a hard hat."

Caldwell has tried to leave the mountain close to how he originally found it, not removing trees and boulders. He hopes to have a ski mountain that looks natural without several big scars down the side of it. Of course, doing it this way has its challenges. "Here's a mile-long job site that I can't get any heavy equipment to," he says. "I've had to figure out how to use hand tools to move massive weights, such as lifting a 5,000-pound rock out of its hole. You get to find out what real leverage is." He adds, "The other problem is that if I don't have the right tool with me, it's three hours to go back and get it."

Caldwell's goal is to prove he can create his own ski resort on a low budget in an environmentally sound way. He hopes to be up and running in the winter of 2008. With the eager participation of volunteers, his dream seems about to be realized in the manner of an old-fashioned barn raising.

Dale Dougherty is editor and publisher of MAKE.

Caldwell says you design the towers for the wind load on the cables that run between them, not for the people on the chairlift. The towers are also pitched on the mountain at various degrees of lean. Caldwell welded these towers in his garage, which also houses his snowcat (shown at left).

CHARLIE ASQUITH'S JET DORY

By Tim Anderson

Charlie is 78 years old and he's been fishing for mullet in Hawks Nest, Australia, for over 60 years.

Today, like most days, he's sitting on the beach, watching the water through polarized sunglasses. He hopes to see a dark shadow moving across the bay, a school of mullet coming in from the ocean. Each year the mullet come to spawn in the estuaries that feed the bay. Then he and his son Les will launch their boat to try and catch some in their net.

He calls his boat a "dory." It's a greatly modified runabout with a jet ski power plant. He added a cowling at the front, along with the controls from the jet ski. The engine and water jet impeller drive are in the stern, covered with a smooth housing.

Their truck has a trailer hitch on the front as well as the back. To launch the boat into surf, the boat is put on the trailer as shown (top) and pointed at the water with the skipper on board. At the proper moment, the driver of the truck drives toward the water. He hits the brakes just before getting there. The boat slides off the trailer into the water, and the skipper drives it out through the waves. To get back in through the waves, the skipper rides in just behind a big wave, guns the engine, and runs the boat up the beach to dry sand. A perfect system.

The fishing nets are piled on top and nearly fill the boat. Behind the driver there can be no projections that might snag a net. When the boat reaches the school of fish, he throws out an anchor attached to one end of the net. The anchor starts pulling the long net out over the stern of the boat. As the net is pulled over the stern, he drives the boat around the school of fish, attempting to encircle them.

It's a six-week fishing season here, and then the fish move north. The local fishermen don't follow the fish; they go back to trapping lobster. Charlie and Les built their own lobster trawler as well as fiberglass rowing skiffs, based on traditional shapes, for when the jet dory is overkill.

Tim Anderson, founder of Z Corp., has a home at mit.edu/robot.

1. Charlie Asquith and his jet dory. 2. The jet nozzle and exhaust at the stern of the boat. The boat goes faster than the jet ski did and gets better gas mileage. This jet ski engine is about 100 horsepower. The white truck (left in photo) has a snorkel so the engine can still run while the truck is nearly underwater. 3. The skipper stands in the front with the controls from the jet ski.

Photography by Tim Anderson

Making Trouble

UNHINDERED CREATIVITY

ENABLE KIDS TO INVENT THEIR OWN GAMES.

By Saul Griffith

LOOKING BACK AT THE TOYS I LOVED most as a child, I find that the memories that lasted, and the adventures that stuck, were with homemade toys and homemade experiences. I of course remember playing with Legos and a fabulous wooden train set that I was given when I was 5, but few other toys actually stick in my mind.

My parents gave me the luxury of inventing my own toys and games from a very young age; sometimes my games started with a ready-made toy or the odd materials my parents had sitting around. Most were fueled by imagination and sheer curiosity — either that or the particularly delightful boredom that is characteristic of childhood.

Shooting my sister with a water pistol was fun, but the most distinctly proud moment I recall was a game I invented for myself when I was 8 that entertained me for hours. With water pistol in hand I'd shoot a squirt at the very top of a window. The stream would dribble down the window, typically dividing into two or three paths. I'd then shoot the bead of water at the head of each path, which would in turn create two or three more paths.

Quickly, exponentially, the number of paths would increase and I'd be frantically shooting at a fractal of water droplets, trying to score the most direct hits before the first drop dribbled to the bottom of the window. This was analog *Asteroids*, that fabulous Atari video game, only higher resolution and more compelling than its computer counterpart. I can still feel the sense of achievement of having invented a cool game that soon was played by my other friends.

At 9, I became fascinated with grappling hooks. I do not remember which comic book introduced me to the idea, but as soon as I had it in my head that a grappling hook could take me wherever I wanted to go, I simply had to make one. I never made it really work, but I can remember dozens of trees and walls I tried to ascend with the various prototypes. Like an X Prize project, I didn't necessarily reach the end goal, but I learnt lots of other applications on the way.

I had to learn to tie knots. I learnt the material properties of all the hooks I tried (wood, plastic, aluminum, steel). I learnt that falling pieces of steel on the end of string are subject to the laws of gravity, and that sharp points may effectively grip branches, but they also pierce your skin. I learnt that you need a certain diameter of rope to support your weight and not cut into your hands as you try and haul yourself up with it, and I learnt that certain tree branches can't support your weight.

By 10, I was on to inventing jet-powered helicopters. Do you like fire? I like fire. Fireworks were illegal in Australia when I grew up, except for the Queen's Birthday weekend celebrations, which conveniently fell near my sister's birthday, so we always had lots of fireworks (my sister was the princess in my family on her way to being queen). I loved fireworks, not for the colors and noise so much as for the raw materials.

After a few years of making jet-powered Matchbox cars, I realized I could put these precious black powders and cardboard canisters of energy to a better use: a helicopter. I must have spent three days hand-carving and gluing a 3-foot-diameter rotor blade out of balsa and bamboo from the garden. The idea was to redirect all the thrust of the Roman candle out through an aluminum-can-and-electrical-

tape nozzle at the tips of the rotor. Thus spinning, the helicopter would lift off. I'm not going to tell you it worked great, but it did in fact work — getting the two fuses to ignite the two candles simultaneously took hours of hugely fun experimentation. The chopper flew about 8 feet up and about 30 feet sideways until it hit the clothesline. It only singed one bed sheet before I managed to turn the garden hose on it. At least that time I remembered to keep the hose handy.

The thing I find fascinating about these escapades is how strongly etched each of these experiences is in my memory. Not in my parents' memory, of course, as I often went to great lengths to conceal the actual goings-on. It's amazing how many injuries via failed inventions could be passed off as bicycle falls, climbing trees, or another water fight with that nasty kid up the street. Poor kid; more than once my mother castigated his mother for a purported wrongdoing.

This rambling is my plea for encouraging the "other" toys and games, not the ones you buy, not even the ones you learn to build through MAKE. It's by providing the environment (and turning the blind eye) that you enable kids to use their unhindered creativity to invent their own games. Keep an extra box of cardboard around, a jar of screws, a coil of rope. It may not look like invention to you, but for the kid at that moment it is their entire world; it is their idea and inspiration; and their intensity and devotion to it should be celebrated. I took no photos of these things I built, yet they stick as firmly in my memory as anything else I did in my childhood. That sort of recollection can only be attributed to the intensity of pure joy.

If this column helps you recall the triumphant toy inventions of your youth, I'd love to hear about what they were (and are)!

Getting some huge, gnarly air as an 8-year-old using a bike ramp I built myself. Note: I didn't even injure myself ... this time.

Photograph by Dr. Ross Griffith

Saul Griffith thinks about open source hardware while working with the power-nerds at Squid Labs (squid-labs.com).

Citizen Scientist

HOMEBREW MAGNETOMETER

BUILD A TORSION BALANCE TO MEASURE TINY CHANGES IN THE EARTH'S MAGNETIC FIELD.

By Dr. Shawn

MOST OF US ONLY THINK ABOUT THE Earth's magnetism when we need to find our way in the wilderness. When you look at a compass, the Earth's magnetic field appears to be a steady guide. But in reality, the magnetic field fluctuates in response to all sorts of aetheric influences, some of which come from beyond our world.

For example, the sun boils off a steady stream of electrons and protons into outer space, and whenever a large solar prominence erupts from the surface that's facing us, a hot torrent of charged particles rushes outward and eventually slams into our cold and relatively quiescent ionosphere. The collision between these two wispy plasmas sets into motion turbulent oceans of electric currents thousands of miles across, which, in turn, create undulating magnetic fields that stretch down through the atmosphere to reach us here on the ground.

Although huge in scale, these magnetic fields are rarely more than a percent of the strength of the Earth's home magnetic field. Therefore, although these effects are quite common and last from minutes to hours, the magnetic disturbances are notoriously difficult to detect.

Until now, the high cost of tracking these signals meant that professionals had exclusive access to the field. However, anyone can now easily study these magnetic micro-pulsations by building the remarkable magnetometer described here. The simple device requires less than $50 worth of parts and can be built in an afternoon. Yet it can easily capture truly tiny distortions in the Earth's magnetic field, as well as the dramatic effects of a magnetic storm high in the ionosphere.

The magnetometer uses one of the most sensitive instruments in science — a torsion balance. It measures a force by seeing how far it can be made to twist a fine filament. The thread gently resists rotation with a torque (a twisting force) that grows as the fiber twists, until it just balances the torque created by the applied force. The resulting angle of deflection, which is found by bouncing a light beam off a tiny mirror attached to the filament, is proportional to the force under study. With the beam from a laser pointer and a match-head-size mirror, you can, in principle, resolve deflections as small as one ten-millionth of a degree.

Most torsion balances that you'll find in professional laboratories use fine quartz fibers, which are incredibly strong and insensitive to changes in humidity and temperature. Sadly, quartz fibers are difficult for ordinary mortals to come by. But nylon fibers work almost as well as quartz and can be easily extracted from any soft nylon cord.

BUILDING THE SENSOR

If you want to use carbon fiber, check your local hobby stores; I prefer nylon because it's much easier to get. To use nylon, start with silky, multi-filamented nylon cord — which you can purchase at any hardware, craft, or boating store — and cut one cord

YOU WILL NEED:

- Multi-filamented soft nylon cord
- 5-minute epoxy
- Small piece of aluminized Mylar At least 5mm thick
- Spring pin (clothespin)
- Clear plastic DVD cases (2)
- Penny, minted prior to 1982
- Sheet of corrugated cardboard or particleboard
- Rare-earth magnets (2 packages)
  - RadioShack part #64-1895
- Doughnut-shaped magnets (4)
  - RadioShack part #64-1888
- Laser pointer

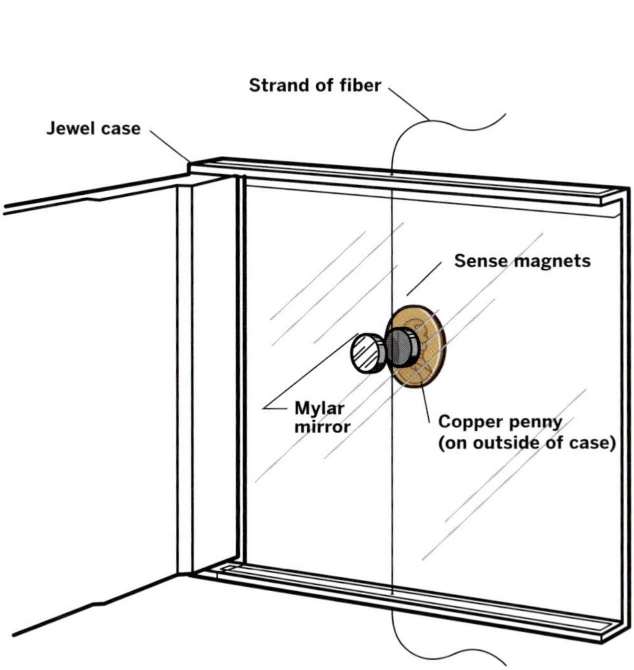

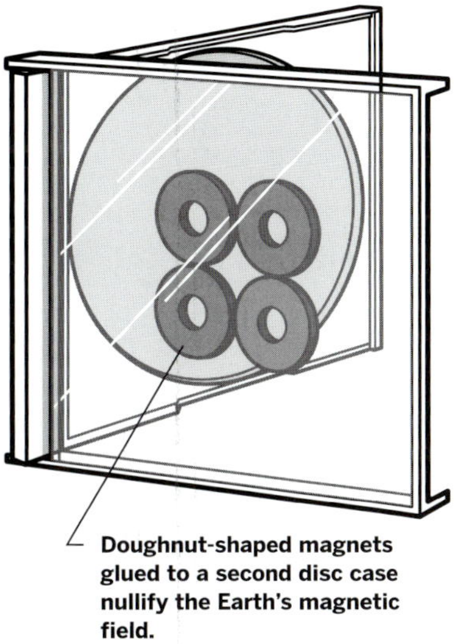

The nylon fiber (left) should be between the 2 round magnets. A copper penny glued to the outside of the disc case cancels unwanted currents generated by the magnets.

30cm (1 ft) in length. Next, gently unravel it and use tweezers to select the finest strands, which should be about 25 microns (0.001 in) thick. A slow and steady pull will drag one fiber away from the others. But be careful: these individual minute miracles of modern chemistry are almost impossible to see, and, when left in the open, they often disappear as if by magic. So, if you need to set it down, make sure to secure your harvest in a plastic sandwich bag so you can find the fiber when you need it.

Your next task is to install the nylon filament into a case that is impervious to changes in humidity and isn't too sensitive to shifts in temperature. A clear plastic DVD jewel case works great. DVD cases come in several thicknesses, and you want to get the deepest case you can find. Look for a plastic insert that holds the DVD, or a jewel case made for two DVDs at a time. Pop out the inner plastic DVD holder and keep just the outer case.

With the case standing on its side, install the nylon fiber so that it runs vertically through the exact center. To do so, you'll need to create 2 tiny holes, with the finest drill bit you own, on exact opposite edges to thread the fiber through.

Working with the nearly invisible fiber takes a great deal of care and persistence. First, delicately thread the fiber through the eye of a fine sewing needle using tweezers and a steady hand. Then, pass the needle up through the bottom of the case and out the top. Tape the top end of the fiber to the outside of the case and use a toothpick to completely fill the hole with a dollop of 5-minute epoxy. Smear some on the fiber that lies against the case for good measure. To provide the necessary tension, tape 2 quarters to the other end, which is threaded through the bottom of the case, and lay the case flat on the edge of a table with the coins hanging free. Now, completely fill the lower hole with epoxy and smear some on the fiber on the outside of the case. Let the epoxy cure thoroughly. Then, use a pair of tweezers to break off the ends of the fiber at the epoxy.

To coax the nylon to twist in response to minute changes in the ambient field, you need to affix a powerful magnet to the fiber. Because a massive magnet would respond sluggishly, the ideal attractor would possess a powerful field and yet be extremely lightweight. Such magnetic miracles exist; they are called rare-earth magnets. These marvels are tiny and yet harbor magnetic fields at their surface that are 10,000 times stronger than the Earth's. Best of all, you can pick up a pair of them at many RadioShack stores for less than $2. In fact, buy 2 pairs because you'll need to use one pair to position the other pair on the fiber.

But first you'll need to install a tiny mirror on one of your tiny magnets. To do so, dollop a smidgen of 5-minute epoxy on the face of one of the magnets.

Illustrations by Damien Scogin

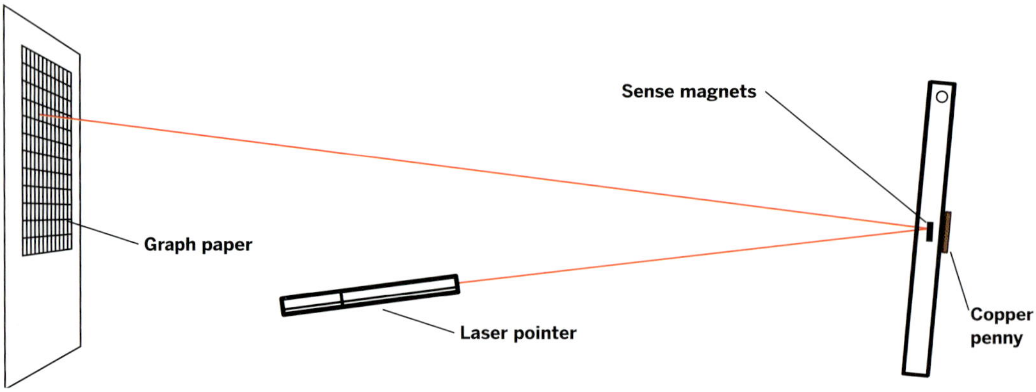

Next, place the doped side against a small square of aluminized Mylar, at least 5mm thick, and then clamp it in place by snapping a second rare-earth magnet to it on the other side of the Mylar sheet. After the epoxy has thoroughly cured, use a pair of nail clippers to trim the Mylar neatly against the magnets. You can then separate the magnets using your fingernail. A quick swab with a Q-tip dipped in rubbing alcohol should provide a clean surface that will adequately reflect a laser beam for this project. If the spot size is too diffuse for your liking, it can be made smaller by delicately darkening the perimeter of the mirror with a fine-point Sharpie pen.

The magnetometer uses one of the most sensitive instruments in science — a torsion balance.

Now it's time to connect your magnets to the fiber (see diagram, previous page). First, to get them both correctly oriented, snap a magnet onto the back of your mirrored magnet. These will be your 2 sense magnets. Then, separate them and, making sure not to turn the back magnet over, position it so that it is directly centered on the fiber. Next, clamp this magnet against the plastic case by positioning your other 2 rare-earth magnets directly behind it. By tweaking the location of these 2 exterior magnets, you can make whatever fine adjustments you need to the position of the sense magnet. When it is perfectly centered on the fiber, carefully lower the mirrored magnet — mirror side up, of course — directly down on top of the back magnet until they snap together. Just remove the exterior magnets and you'll have yourself a working magnetometer.

Now test it. Pick up the apparatus and slowly rotate it 180 degrees and back again. The magnets should remain fixed on the Earth's magnetic field as the case moves around them. Note: It is vital that the magnets be free to rotate completely without ever touching the case. If your magnets wedge themselves against the walls, you need a wider jewel case.

Next, glue a solid copper penny — minted before 1982, when the purity of the copper was still high — on the outside of the case, just behind the magnets. Whenever the magnets move, they induce electrical currents (called eddy currents) in the metal to produce their own magnetic fields, which oppose the motion of the magnets. This clever trick quickly damps unwanted oscillations, making the magnetometer much easier to read.

Now, mount the entire assembly vertically to a smooth flat base. For example, you can epoxy the device to a piece of corrugated cardboard or particleboard.

Your sensor is now an accurate compass. As you walk around, the magnets should align to magnetic north and display little oscillation.

NULL IT GOOD!

Because the Earth's relatively large magnetic field absolutely overwhelms our coveted micro-pulsations, you must first null the instrument, or cancel out the Earth's field, before it will register the faint signals we are looking for. For that, you'll need 4 doughnut-shaped magnets.

Epoxy the 4 doughnut-shaped magnets side by side to the center of the second DVD case so that their magnetic fields are all facing the same way. That's easy to do. Just place 2 magnets side by side. If they repel each other, their magnetic poles are facing the same way. Make sure each magnet repels the other magnets as you glue it down and hold it in place while the epoxy sets. Just make

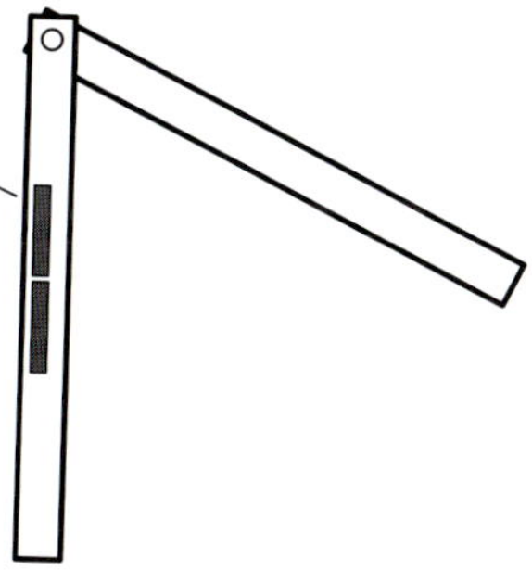

The final magnetometer setup, as viewed from above. Place the disc cases on the floor, pinch the laser pointer with a clothespin, and set it on a book. Position the laser at an angle so you can measure the reflection on a wall.

certain that the center of the assembly is at exactly the same height as the sense magnets in your magnetometer. If you open the jewel case a little and set it on its edge as shown, it will stand up and stay put wherever you want it.

Nulling out the Earth's field is easily done. First, put the nulling magnets far away (like in the next room) and note the orientation of the magnetometer magnets. They will be aligned with the Earth's magnetic field; that is, one will face magnetic north and the other will face magnetic south. Then, position the nulling magnets about 3 feet away, directly behind the sense magnets along the north-south line. Slowly move the nulling magnets in toward the sense magnets. Watch carefully for the moment that the sense magnets start to rotate toward the nulling magnets. (If it doesn't happen, flip the nulling magnets around and do it again.) This is the instant that the pull of the nulling magnets cancels the pull of the Earth's magnetic field.

With the nulling magnets at this exact location, your magnetometer will be exquisitely sensitive. For example, you will find that any small twists of the nulling magnets will be perfectly tracked by the sense magnets. A rare-earth magnet can be detected from many feet away.

WATCHING THE MAGNETIC WORLD GO BY

The bright beam of a laser pointer completes the instrument. Position the laser so the beam shines through the plastic case and bounces off the mirror and onto a distant wall. A clothespin will not only hold the pointer in place, it will pinch the button into the on position for totally hands-free operation.

Ripples in the Earth's magnetic field will show up as deflections of the beam. I think of my own magnetometer as a work of art as much as I do a scientific instrument, like a brass barometer set in teak. The bright laser spot makes an elegant display of a fundamental phenomenon of nature. However, if you should feel the urge to quantify your observations, tape a piece of graph paper to the wall, centered wherever the beam happens to fall, and mark the beam's location.

Watch the beam as it slowly shifts about the paper. As it starts to drift, mark its location with a number, 1, 2, 3, etc., at regular intervals ranging from seconds to minutes, depending on how fast things are evolving. By using a ruler to measure the distance from each of these numbers to the starting point, you'll be able to graph the displacement versus time. Over time, you can identify the relative strength of these micro-pulsations in the Earth's magnetic field.

Although it's exciting to watch the spot of light dance about as it responds to what are often extraterrestrial influences, this sort of hand operation quickly becomes tedious. Fortunately, serious makers — who want to use their home computers to continuously monitor the dynamic nature of the Earth's magnetic field — can learn how to do so from a CD-ROM I publish: *The Amateur Scientist*.

It contains more than 1,000 cutting-edge science projects taken from the pages of *Scientific American*'s now defunct "Amateur Science" column (I wrote that column from 1995 to 2001). My explanation of how to connect this style of magnetometer to a PC to create a professional-grade magnetometer observatory at home can be found in the March 2000 article. (The CD is available from brightscience.com.)

Dr. Shawn (Shawn Carlson, Ph.D.) is the founder and executive director of the Society for Amateur Scientists (sas.org). He won a MacArthur Fellowship for his work helping ordinary people do extraordinary science.

Shaker Flashlight Modification

1+2+3 By Cy Tymony

Powering small electronics with your muscles.

You will need: **Shaker flashlight, wire, small Phillips screwdriver, soldering iron, solder**

How a Shaker Flashlight Works

A shaker flashlight has a bright LED, a NiCad battery, and an electrical generator made from a coil of wire and a cylindrical magnet. When you shake the flashlight, the magnet slides within the coil to charge the battery. Shaker flashlights provide about 2 to 4.5 volts but produce a very low output current, so there will be limits on their ability to supply power to devices. Recently Walgreens stores offered one for just $5.

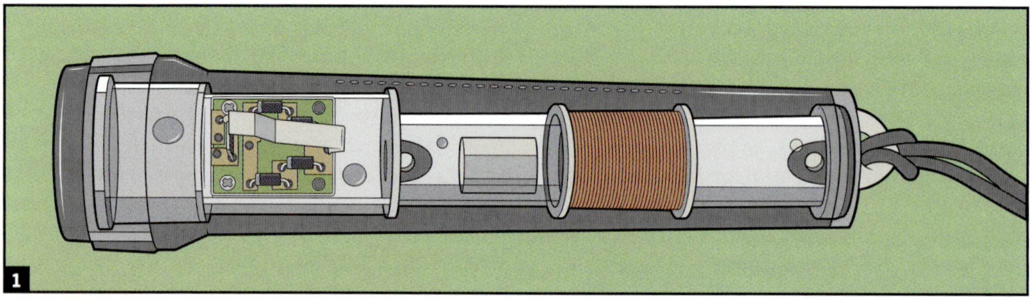

1

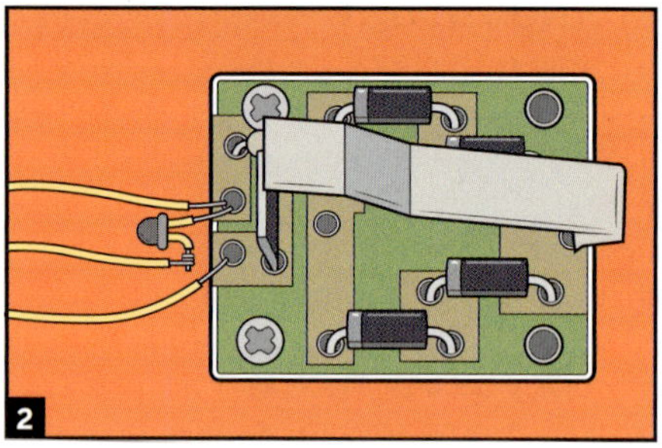

2

3

1. Modify the flashlight.

Figure 1 shows the basic parts of the flashlight — the case, coil, magnet, battery, and printed circuit board. Unscrew the lens assembly and remove the screws that secure the PC board to the case. Remove an LED lead and solder 3 wires in place (Figure 2), so you can still use the flashlight normally when you're through powering other devices.

To power a device, connect the 2 wires attached to the flashlight's PC board, with the correct polarity, to the device's battery clips. When you need to use the flashlight normally, connect the LED wire and one of the PC board wires together (use color-coded wire or labels for easy identification).

2. Adapt a device to connect to flashlight.

Select a device that requires 1½ to 3 volts at low current, like a small electronic LCD video game. (I'm an avid collector of the free handheld games that fast food outlets supply with kids' meals. They require very little power if you turn off their sound option.) Wrap the bare ends of 2 wires tightly around the clips. Figure 3 shows the electronic game connected to your shaker power supply.

3. Go further.

A larger NiCad battery can be substituted for more capacity. You should be able to power other low-current devices such as a small Walkman-type AM radio, travel clock, or micro radio-control car. You might also (possibly) use the modified flashlight as an emergency cellphone power supply.

Illustrations by Timmy Kucynda

Cy Tymony (sneakyuses.com) is a Los Angeles-based writer and is the author of *Sneakier Uses for Everyday Things*.

THE FIRST ANNUAL

Make: HOLIDAY GIFT GUIDE

Gift Ideas for Incorrigible Makers and Inquisitive Tinkerers

All Ages

MAKE KITS

The MAKE Controller

$149.99
makezine.com/controller

Born in the fire and smoke issuing from Survival Research Labs' ramshackle compound near Potrero Hill in San Francisco, the MAKE Microcontroller traces its lineage to the digital controllers used to control SRL's robots and artistic weaponry.

Engineers Michael Shiloh and David Williams designed the hardware and software that controls SRL's robots and gives them savage intelligence. While brainstorming ideas for something more powerful and easier to use than available microcontrollers, they came up with what evolved into the MAKE Controller.

The MAKE Controller is a microprocessor unit, a computer-on-a-chip designed to handle a small set of related tasks. But the MAKE Controller has a more powerful processor than typically available on most microprocessors, plus built-in networking capabilities, an onboard USB port, and large memory. With the MAKE Controller, the maker's world opens up, enabling unlimited opportunities for automation, motion control, and artistic expression.

Straight out of the box, the controller is ready to connect to sensors, motors, and your own killer robot. It comes with common functions preprogrammed for immediate experimentation.

Daisy MP3 Player

$114.95
makezine.com/daisy

In 2001, artist and designer Raphael Abrams went looking for a new challenge. After some long and careful consideration, he came upon the idea of designing and building his own open source MP3 player kit.

His criteria? First, it had to be easy to build. Second, it had to be open source. Finally, and most importantly, it had to be more than just a handheld device — it had to connect easily to many interfaces, everything from simple button-pushing to parallel ports to very powerful serial modes.

It took several iterations, but eventually he came up with the Daisy, an easy-to-build, pocket-sized MP3 player. Daisy sounds as good as an iPod, can access 65,000 tracks, and plays 48kHz WAV files as well as MP3s. And, unlike an iPod, you can change its battery.

But the big thing about Daisy is the ease with which it interfaces with so many devices, including the MAKE Controller. It is the perfect MP3 kit for makers, for it is easily integrated into kiosks, displays, art installations, or just about anything else.

MiniPOV2

$17.99

makezine.com/minipov2

The MiniPOV2 is a persistence-of-vision device that rapidly blinks eight LEDs on and off so that when waved through the air, an image or message appears to float in front of the viewer.

MIT engineer Limor Fried originally cooked up the MiniPOV2 as an easy-to-build demonstration showing how microcontrollers work. She wanted something that was easy to make, inexpensive, and simple. This little kit, says Fried, teaches several not-so-little lessons:

- **How to solder**
- **How to assemble simple kits**
- **How to program microcontrollers**

The kit includes a microcontroller, sockets, resistors, LEDs, connectors, a battery case, and a printed circuit board. Add some basic tools, a PC with a parallel port, and a little programming, and the MiniPOV is ready to blink out your deepest thoughts.

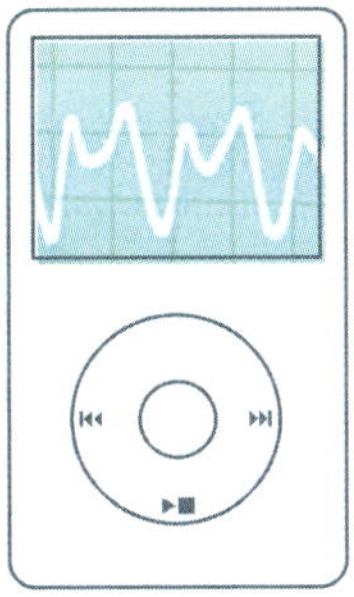

The iBump

$99.95

makezine.com/ibump

"Freshman volume" is the sound level metric introduced to college students in their first year — it's the highest sound level possible at which there is nobody screaming at you to turn the music down. When Ben Anderson made his first attempt at freshman volume, he used his iPod, several guitar amps, and a regular stereo. Although loud, the sound fidelity was surprisingly low. What was wrong?

Big speakers work best with low notes; little speakers work best with high notes. Ben needed a way to separate lows from highs before amplifying his music.

Designed by Ben's dad, Wendell, the iBump is an audiophile-quality active crossover (which separates the highs from the lows). It is inserted between the source (iPod) and the amplifiers. Routing the iBump subwoofer output to a bass amp enables clear, earthmoving low notes. The iBump left and right channels provide high, undistorted volume.

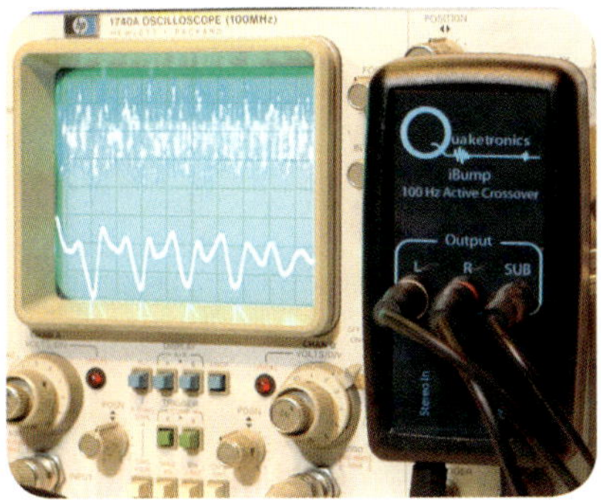

Now, Ben uses an iBump to make sure his neighbors understand the true concept of freshman volume. So can you.

Minty Boost

$19.99

makezine.com/mintyboost

"People put a lot of interesting stuff in Altoids tins," says one maker. "Usually, it's either drugs or condoms." But besides such pedestrian uses, Altoids tins have been used in electronics projects for decades.

At Maker Faire last April, Limor Fried gave her latest spin on the Altoid box project concept. The Minty Boost is a small, simple, but powerful USB charger for MP3 players, cameras, cellphones, and other gadgets that plug into a USB port to charge.

The AA-cell-powered Minty Boost is the Cadillac of battery-powered USB chargers, lasting twice as long as most other Altoid tin chargers. Plus, it's a simple project requiring only easy soldering. All materials (except the Altoids box and batteries) are included.

Make: TOYS & GAMES

A challenging and exciting collection of articles and projects that's fun for the entire family! (Batteries not included.)

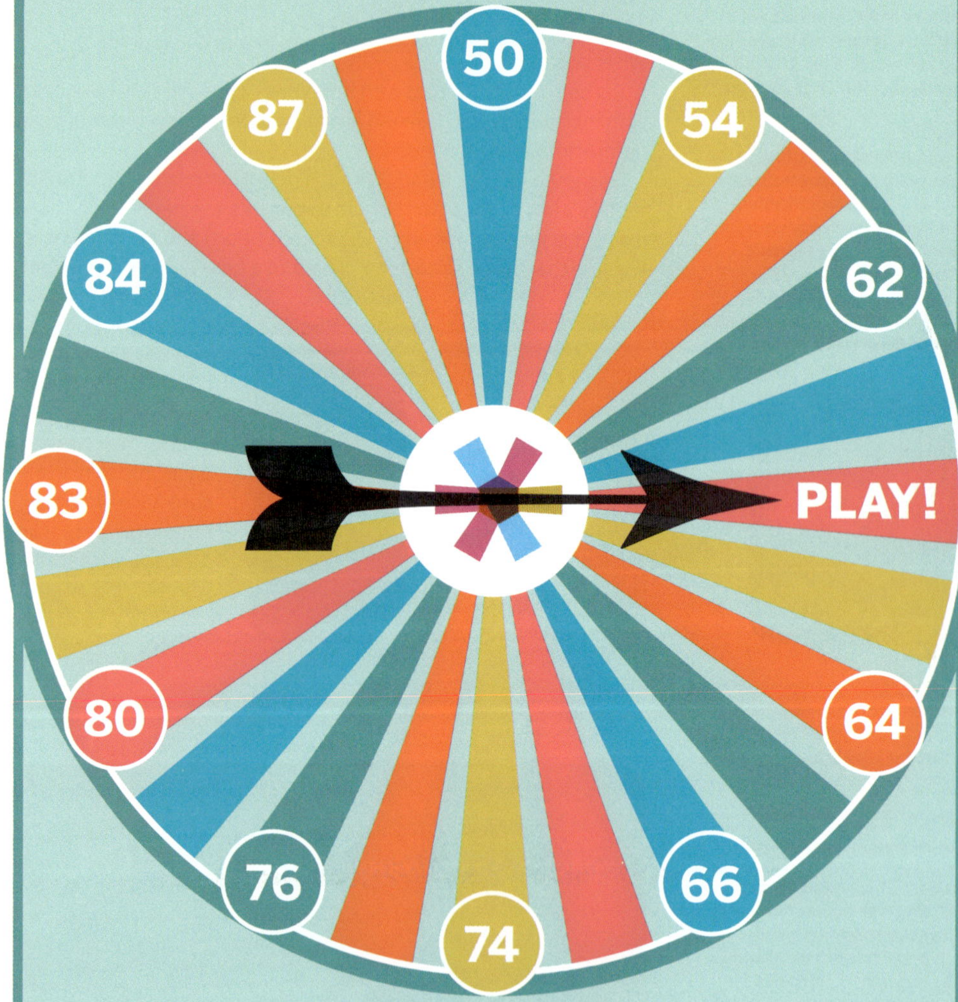

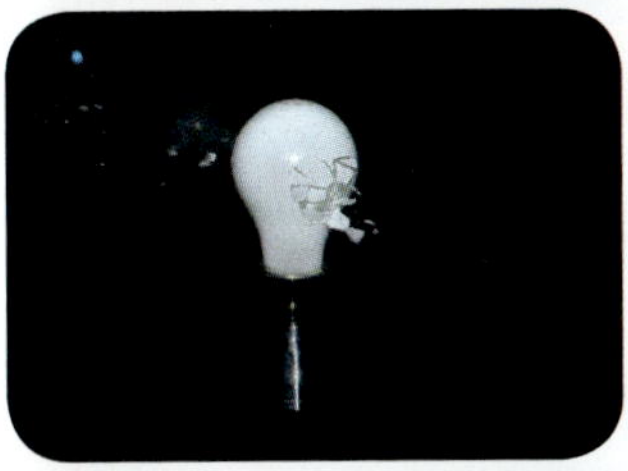

High-Speed Photography Kit

$99
makezine.com/flashkit

Back in 1972, when light shows were all the rage, Tom Anderson visited the Edmund Scientific Company. At the time, Edmund had a terrific light-show display, with mirrored balls, black lights, fiber optics, color wheels, and strobe lights.

Inspired, Anderson's dad, a prototypical maker, designed and built a strobe light for him, adding an external trigger that fired the strobe via a microphone. Using his Polaroid, Anderson and his dad took lots of high-speed photos in their basement: popping balloons, splashing water drops, and breaking glass.

Thirty years later, Anderson shared his secrets in a MAKE Volume 04 article describing the technique and materials for high-speed flash photography. The key is the Flash Controller, a circuit board with the requisite sensors, amplifiers, and timers. The unit includes a high-speed flash, and a flash trigger that synchronizes the flash with the high-speed event. The Flash Controller comes fully assembled and includes a 6-foot connector cable and disposable camera.

Build Your Own Electronic Game Kit

$39.95 U.S., $49.95 International
makezine.com/gamekit

Originally designed for the Maker Faire by Grand Idea Studio, the Build Your Own Electronic Game Kit is a custom-created kit intended to introduce you to the world of electronics and soldering. When successfully assembled, the kit becomes a version of the popular memory game, Simon, with a few optional twists.

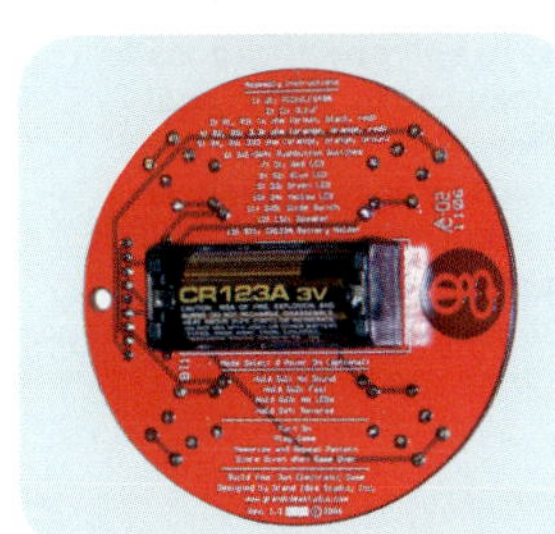

MAKE BOOKS

MAKE Magazine Collector's Edition Boxed Sets

$59.95 Each
MAKE Year 01 (Volumes 01-04)
MAKE Year 02 (Volumes 05-08)
makezine.com/store

With the publication of Volume 08, MAKE magazine has two full years under its belt. MAKE's famed projects run from practical to whimsical, to maybe even a little scary.

In each of the eight issues, we've showcased the great things people are making in their basements, garages, and backyards with the technology at hand. But most importantly, each issue contains hands-on instructions on how to build stuff yourself — practical or whimsical or scary, but always cool.

MAKE Pocket Reference

$12.95
makezine.com/pocketreference

The perfect stocking stuffer for the workshop warrior, the *MAKE Pocket Reference* is both fun and useful.

Page through and you'll find data on everything from physical constants to bolt specifications to resistor color codes. Add in cool stuff such as first aid, crane hand signals, and the encyclopedia of knots, and you can see why it might take up permanent residence in your shirt pocket.

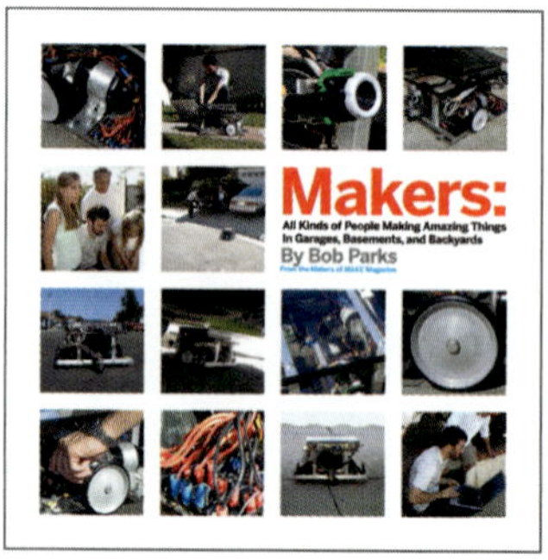

Makers: The Book

$24.95 *MAKERS: All Kinds of People Making Amazing Things in Backyards, Garages, and Basements*
makezine.com/makers

For more than ten years, Bob Parks has explored the world of the technology addicted, the gadget fetishists, the personalities behind the latest technologies.

In his book *Makers*, Parks profiles a brigade of citizen engineers and inventors, men and women who make their own HERF (High Energy Radio Frequency) guns, fusion reactors, night-vision scopes, and tree houses.

Makers celebrates digital tinkering, hardware hacks, and DIY of all stripes, profiling 100 people and their homebrew projects, in a handsome hardcover perfect for coffee breaks or coffee tables.

Backyard Ballistics

$16.95 *Backyard Ballistics: Build Potato Cannons, Paper Match Rockets, Cincinnati Fire Kites, Tennis Ball Mortars, and More Dynamite Devices*
makezine.com/backyardballistics

Sure, electronics projects are cool, but sometimes you just want to make something go ka-boom.

Well, *Backyard Ballistics* was written for just those times. William Gurstelle has been collecting information on build-it-yourself projects that go whoosh, boom, or splat for 20 years. From his trove comes this selection of 13 ballistic devices that people can build in their garage or basement workshops.

Instructions, diagrams, and photographs show how to build projects simple (a match-powered rocket), complex (a scale-model tabletop catapult), and wonderful (a combustion potato cannon). With an emphasis on safety, the book explains the physics behind the projects, and profiles experimenters such as Alfred Nobel, Robert Goddard, and Isaac Newton.

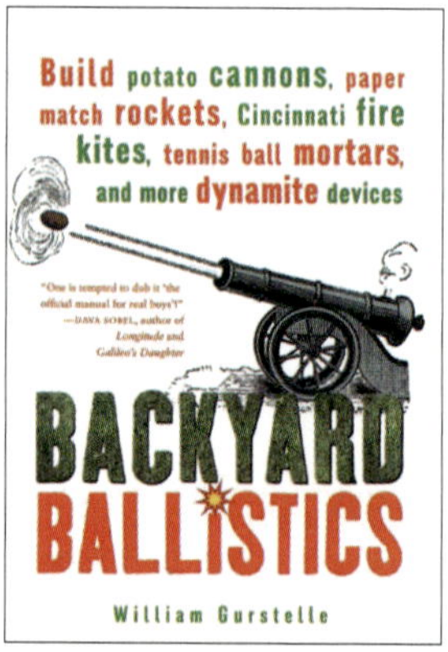

Game of Life Board

$19.99
makezine.com/gol

In 2005 a group of brainy MIT students, the Dropout Design Team, decided to try something really novel and clever. They wanted to invent an electronics kit that would look cool, be simple to build, and be accessible to everyone.

At the time, some of them were toying around with the idea of "cellular automata" — a collection of cells on a grid whose colors change according to what's going on in adjacent cells. (The prime example is British math whiz John Conway's Game of Life, in which a collection of cells lives, dies, or multiplies based on a few mathematical rules.)

"Hey, let's build a huge wall of LEDs that would enact the Conway Game of Life rules," they said one night. "The wall could be split into identical modules, and each freshman could assemble their own module, then add the module to the collective wall."

DDT designed an easy-to-solder kit that is cheap and scalable. Each Game of Life board contains 16 LEDs in a 4×4 grid, a microcontroller, and a communications and power distribution network. Boards can act alone, or can be plugged together, border to border, to create a larger display.

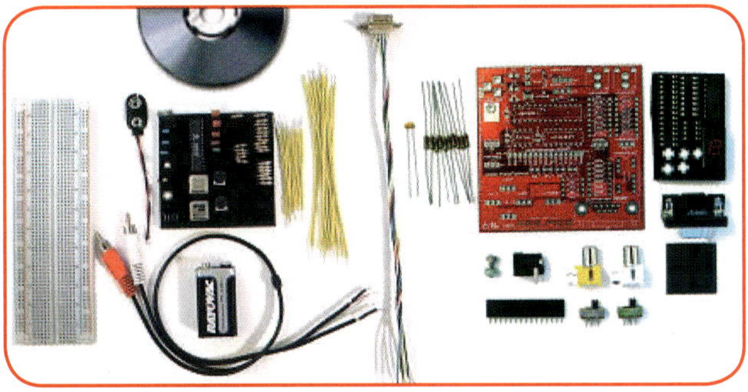

XGameStation Pico Edition 2.0

$79.95
makezine.com/xgamestation

Twenty years ago, says Andre LaMonthe, game programming was a real art. LaMonthe, a programming maven, remembers when 32 kilobytes on machines with clock speeds of 1 MHz was state-of-the-art. Yet those programmers of the 1980s created classic video games with perfectly tuned mechanics and unbelievable details.

LaMonthe's XGameStation is a hardware platform developed for educating a new generation of hardware and software hackers in the nitty-gritty, low-level world of hardcore game development.

Combining modern technology with the bedrock-solid design philosophies of the past, the XGameStation Pico Edition 2.0 is a build-it-yourself game development kit based on the technologies of its bigger brother, the XGameStation Micro Edition.

Solder your unit together and you'll have a completely portable and reprogrammable embedded game system. Complete instructions covering assembly, architecture, and programming of the Pico Edition are included on CD.

The Pico Edition includes an 80MHz Ubicom SX28 microcontroller, a solderless breadboard, a 7-segment readout, 15-pin interface, A/V jacks, and a built-in directional game pad for a completely portable mini game console.

> If you have a spare evening, weekend, or even an hour, there are projects here for you. No doubt about it, a year's worth of MAKE is a lot of information, discovery, and fun.

Homebrew Game Design

Turning wacky ideas into fun board games.

By James Ernest

SO, YOU'D LIKE TO INVENT YOUR OWN board game? Great idea! Here's how to do it — or how I do it, anyway.

1. Determine Who Your Audience Is

Know your audience. If you're making a game to play with your family and friends, that's great! That's the best way to start. But if you aim to sell to a game publisher, it's never too early to think about how to position your game in the marketplace.

For example, when I started Cheapass Games in 1996, I knew I wanted to target "core hobby gamers" like my friends and me — college-age males who frequent game stores. Because gamers already have their own dice and game pieces, I knew we could have fun just with new sets of rules and cards or a board. This simplicity meant that games could be sold cheap, in plain envelopes, and displayed near the cash register of the game store to serve as impulse purchases. That's how Cheapass was born.

2. Outline the Story

My games sell on story, so the first thing I do is try to come up with a good one: zombie fast-food workers fight over a single brain (*Give Me the Brain*) or infected mad cows graze fields to discover unexploded landmines (*Unexploded Cow*). A game's theme is the first impression a new player will have of the game, so it's important to nail it.

Once you choose the theme, think about what game mechanics will fit it — things like turns, dice rolls, and auctions. Some designers base their games around the mechanics, but for me, "Do you want to play a game about pirates?" is more compelling than "Do you want to play a game with a unique movement point system?"

3. Imagine the Ideal Experience

What is the playing experience? What are players "doing," and how do they win? For example, in a space-mining game, do you want to spend more time flying spaceships or managing inventory?

How much luck, skill, or complexity do you want? "Skill factor" games have no luck at all, just pure intellectual competition, as with chess. At the opposite end, other games let players passively follow where the cards or dice take them, don't make you work too hard, and with a little luck, let anyone win.

4. Decide on the Format

The next choice is the overall format of the game. Do you want to design a card game? A board game? Something else? Here is a quick rundown of the most common formats. There are other formats, too, and you can always invent your own.

Board games: Games like Monopoly, Scrabble, or chess, that use a central board for static, common information, such as a map, or for tracking game progress, such as score, army size, and so on.

Card games: Cards handle randomizable, quantized, or secret information. Randomizable means that the cards can be shuffled; quantized means that the game content is broken down into pieces rather than being presented all at once; and secret means that players can hold information that isn't known by everyone. Card games tend to be simpler

Illustration by Melinda Beck; Flipbook illustrations by Brian Biggs

Start
AUDIENCE...
Story
Experience
testing
the End
format
Rules

and quicker than board games, though not always.

Collectible card games like Magic: The Gathering and Pokemon can be complex because the card sets are huge. As randomizers, cards work well if you want hidden results, complex information, or control over how often a particular result shows up.

Dice games: Dice are another randomizer and are better than cards for outcomes that are openly visible and statistically independent from one another. Some games rely purely on dice (Yahtzee), while others use them within a larger game (Monopoly). Spinners and wheels can function in the same way.

Role-playing games (RPG): A paper-based RPG is an adventure-style game in which one player (the gamemaster) leads the others through a campaign setting that he controls. It's not a competition between the gamemaster and the other players; he just wants to give the players a fun experience.

Party games: Party or "parlor" games are simple games such as charades that can be played by a large group. Party games usually don't need much equipment, but mass-market party games (Pictionary, Balderdash, Time's Up) include cards and other components so that there's something in the box. With so few rules and minimal equipment, these may seem easy to design, but simple games are often the hardest to get right.

5. Sketch Out Some Rules

Write an outline of the game rules. Even if they all change, you need a place to start. Explain the number of players, the game components, setting up, and the order of play. Describe what you are doing, game length, and how you win.

Also, remember to make the game fair. This is one of the reasons we like games — they're a sanctuary, perhaps, from an unfair world.

> A good game is a road made of doors, with a clear goal and multiple ways to achieve it.

I like to say that there are two kinds of novice game designers: those who make houses out of doors, and those who make roads out of walls. The "house made of doors" game has no clear goal but offers many ways to get there. The "road made of walls" has a clear objective, the end of the road, with obstacles along the way. But a good game combines these two — a road made of doors, with a clear goal and multiple ways to achieve it.

First-timers also get "first game syndrome," which is like making a salad with every ingredient in the kitchen. They include all the rules they can think of, and the game bogs down. So, after writing down all the rules you want to try, pare them down until there are just enough to make a playable game.

If you're stuck for game mechanics, consider your story and what it suggests. To help you along, here are three basic types of mechanics, based on the "road made of doors" metaphor. Robust games mix these together to produce complex choices.

Two doors (random): A player is offered two identical doors, side by side. One leads forward, the other leads back, and that player has no way of

BUILDING A PROTOTYPE

At some point, you will want to build a prototype. When you should start building it depends on the game. For example, real-time games with a speed component need art designed early and made very legible because readability is important. Here are a few pointers:

Artwork: There's a huge assortment of clip art online. For homemade games, or even prototypes for game companies, you don't need to buy your art. And for a game you want to sell, licensing clip art costs next to nothing.

Cards: These can be hand-cut from paper or made from index cards, but if they will be shuffled a lot, use sturdy, flexible paper and cut them to exact dimensions with a decent cutter. Rounded corners also help.

You can also make them out of existing playing cards, which are printed on laminated card stock — a plastic-coated paper that you unfortunately can't buy in an art store. Your local card room probably gives away old poker decks for free, and any game store probably also has a huge stack of common, low-value cards from the latest collectible card game on their way to the trash.

For cards with black and white artwork, I laser print on 110 lb. index, a smooth card stock that comes in lots of colors. For color cards, I print label sheets and stick them on old poker decks after cutting the labels down to fit. This gives you custom color artwork on a very sturdy deck. You can also print on normal paper, and insert it into plastic card-collector sleeves, available at game stores, along with a playing card as a backing sheet.

knowing or influencing which is which. Such luck-based mechanics can paradoxically make a game both less challenging and more entertaining.

Hurdle and mud pit (statistical): You have a choice in how to move ahead: you can try to jump the hurdle or walk through the mud pit. The hurdle requires a die roll — if you roll a six, you clear it and move on; otherwise, you make no progress. The mud pit, on the other hand, takes exactly four turns to cross.

This type of choice is interesting because the "right" answer varies with the situation. Statistically, both choices are about the same (you have about a 52% chance of rolling a six within four turns), so the optimal strategy changes based on the game state. For example, if everyone else is already in the mud and you absolutely have to be the first player across, you need to take your chances with the hurdle.

Taxicabs (political): Several taxicabs are available, but sharing a ride with another player is better than traveling alone. A game mechanic like this gets players to negotiate with each other to their mutual benefit, while still trying to win the game individually.

6. Play-Testing (Repeat)

Game design is an iterative process that needs testing with real people, who won't always act the way you expect! In the first tests, be prepared to make sweeping changes. If a game isn't working, abandon it mid-game, fix the problem, and start again.

In a game with secret information, it helps to start play-testing "open handed," where players show their secrets and talk about what they are doing, explaining their reasoning along the way.

When something doesn't work, resist the urge to add a new rule to fix it. Rules for special cases are hard to remember and tend to create new problems. It's better to change the underlying cause. For inspiration, look back at your story and try to identify other aspects that you can bring out.

Here's an example: Suppose you're writing a dice-based horse-racing game, and you find that halfway through, the player in the front almost always wins. (In a random race, odds strongly favor the early leader.) You might be tempted to introduce a "headwind," the code name for a kludgy rule that arbitrarily slows down the front-runner.

But horse races aren't about the weather. The real problem is that random races are too predictable and have no strategy. What if instead of moving the horses, the dice gave them "energy" that players could spend early to sprint ahead or conserve for later? Using the dice in this way makes the game state more complex, and lets players come from behind, thanks to wise resource management instead of a patchy and artificial rule.

For play test groups, it's good to mix casual gamers with experienced game mechanics people. Naive users are useful for judging overall fun, and the hardcore gamers are good at analyzing the rules and coming up with fixes. Sometimes an astute gamer will suggest debugging a rule in a way that no one else understands. In this case, try playing the suggested way and see if it works.

The last step in the play-testing process is blind-testing, in which you give the rule book and components to a group that's never seen the game before, and watch them play. If the group understands the rules and has fun, you've got a winner!

James Ernest is the president and lead designer for Cheapass Games (cheapass.com) in Seattle. His portfolio includes Kill Doctor Lucky, Give Me the Brain, and Unexploded Cow.

Cards should be easy to read. Don't handwrite them unless your handwriting is very good. If players will be holding a hand of these cards, "index" them by putting key information in the upper-left corners (held both ways) so that it's visible when the cards are fanned.

Boards: Mass-produced game boards are made of a printed label stuck to a sheet of cardboard. You can get the same effect by printing on full-sheet labels or using spray glue on matte board or other heavy cardboard. For larger boards, print your graphics as several sheets and tile them together, or find a large-format printer.

To waterproof your board (inkjet prints run if they get wet!), cover it with a sheet of clear contact paper, and then trim the contact paper, print paper, and backing all at once for a clean edge. Make the board foldable by cutting it in half and reattaching the back seam with packing tape or cloth tape. For a four-fold board, pre-cut the board along one dimension before you add the contact paper, which will form the forward-folding seam. Then finish the board, cut it in half the other way, and use backing tape for the backward-folding seams.

Other components: Game components can be salvaged from other games you own or from games at the local thrift shop. Things like dice, play money, counters, and chips are widely available, and game stores also sell other components. You may have to make some pieces by hand. I frequently use Lego to build prototypes, but that's just because I have a lot of it.

The Secret History of *MYST*

Co-creator Robyn Miller reveals why it became the best-selling adventure game of all time.

"**HOW DID YOU MAKE *MYST*?"** This has to be the question we still hear the most. So we begin to talk about the production: how we created mountains out of grayscale grids, or how we attempted to create a sense of detail using textures instead of geometry. But typically people are not overly interested in these things. What people usually want to know is, "How did you come up with the ideas for a game like *Myst*?"

I'm always left fumbling for a sound-bite answer to this question. Something quick and easy like, "Designing a world like this is largely an intuitive process ... we made decisions because they felt right." And although this is true, practically speaking, it explains nothing.

In fact, there was an evolution to many of the concepts behind *Myst*. Some seemingly odd sources gave birth to core *Myst* ideas (some of which we almost took for granted). And all these years later, it's relatively easy for me to look back and see the enormous impact these sources had on *Myst*.

First, we have to go back number of years. Maybe to 1981-82, when I was doing time at Henderson High in rural east Texas, and my brother Rod routinely got together with friends to slip into the skin of a Wizard — or a Paladin. They'd sit around a big table and pretend to have adventures in trap-ridden dungeons where some horrifyingly grotesque monster or demon hides behind each corner. This role-playing game, called Dungeons & Dragons, was growing in popularity around that time (and still has quite a following).

Every once in a while, I'd sit in on one of these games. They were curious. And fun. These times I played D&D were something like mini-vacations, or in some ways, even better. I could explore ancient castles. Or dig through the ruins of some futuristic city, long dead. Hey, this wasn't half bad! Especially when Jeff Zandi (who was later immortalized in the *Myst*-related *Uru*) acted as Dungeon Master. He told a lot of jokes, ignored dice rolls that weren't in the players' favor, and generally kept things moving along expeditiously. This was all vital because usually the game moved like molasses, and the rules were so numerous that they filled up three heavy books (faithfully brought to every game). A bad Dungeon Master spent half the game looking through his books; a good Dungeon Master (like Jeff) would just confidently pretend to have it all memorized.

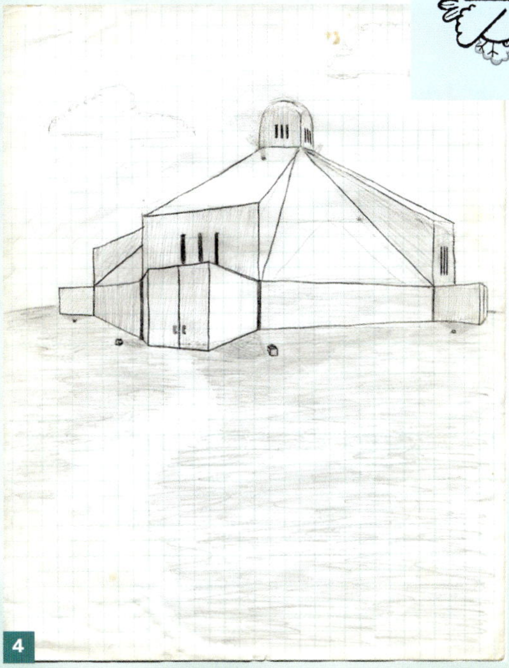

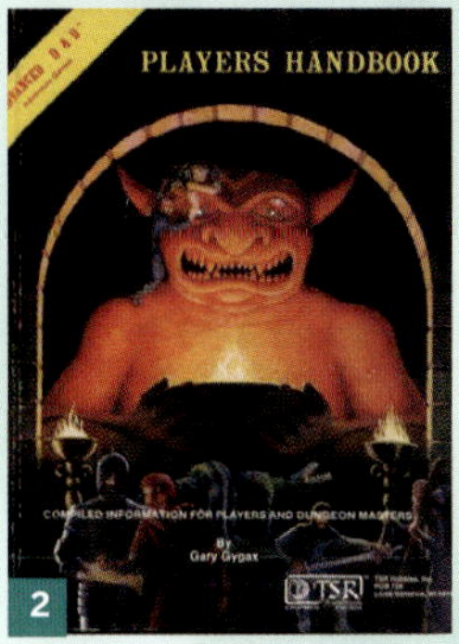

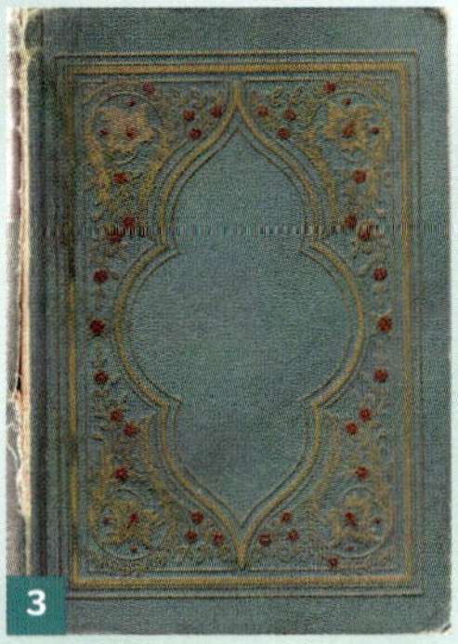

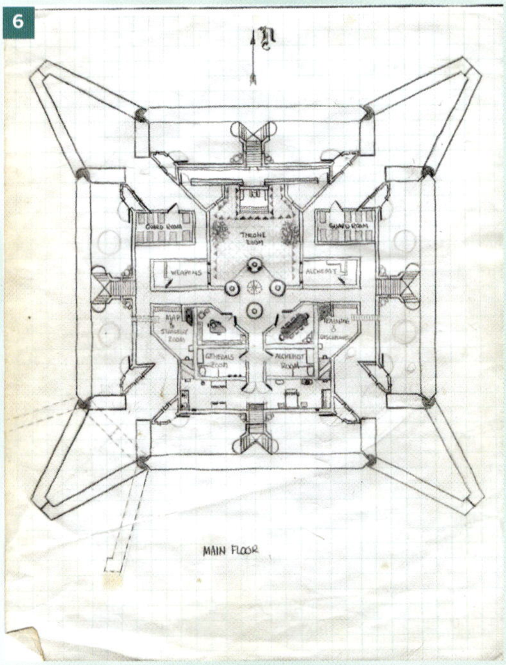

1. Rand delights in torturing Robyn, in Dallas. **2.** This was one of the oft-referenced rule books of D&D: the *Players Handbook*. These rule books presented another hurdle to losing oneself completely in the fantastic D&D environments. **3.** Many of the books used in *Myst* were scanned directly out of my personal library, like this copy of Robert Louis Stevenson stories (which most notably contained *Treasure Island*). **4.** Rand's D&D fortress worked shockingly well. He'd even show the players the occasional drawing, like the one above, as a way of placing them more powerfully in the world. **5.** Rand teaches Robyn how to mow the lawn, in Philadelphia. **6.** Rand's fortress, which he tested as a D&D-like game back when I was in high school, ended up having a direct impact on the game play of the Mechanical Age in *Myst*.

Images courtesy Robyn Miller

> **"From the moment that closed manhole appeared on the monitor, I was dying to know what would happen when I lifted the cover. What kind of world was down below?"**

No Dice, No Avatars

My oldest brother Rand also sat in on a few of these sessions, and later, he and I enjoyed discussing what we saw as D&D's various successes and failures. The good stuff was obvious: the adventure and the sense of escapism. But we'd also talk about how we disliked the slow speed and endless rules. We felt it was a poor interface that only the most devoted fans would ever bother to commit to memory. One of the qualities that especially seemed to hinder the experience was the assignment of a false persona, or avatar, to every player. The player's avatar would even be given characteristics that were wholly not his own: dexterity, intelligence, strength, and so on. To us, this detracted from the sense of escapism the game offered. These puppet-like characters were nothing like us, so we were left with a sense that they were the ones having the adventure in that environment.

Lastly, one had to ignore the poor "visual design" of the typical D&D environment, which rarely rose above the level of a second-rate fantasy novel. It almost never attempted to be provoking or original. Each dim hallway looked the same. Each ghoul or gargoyle died exactly the same way as the last ghoul or gargoyle.

It was right around this time that Rand, wanting to test an improvement of D&D, created his own intriguing and complex D&D castle. He then hosted Rod's group of friends through it. Rand played by none of the tired D&D rules: no dice rolls, no predetermined characters, no hit points or rolling for damages. Instead, all the players had to do was collectively decipher their way through what was essentially a *Myst*-like world. I sat in for the first session and just watched. I'd never seen them have so much fun. His test was completely successful! A D&D without the rules and avatars worked better than the D&D with them.

This successful test of Rand's D&D game was the first step in the evolution toward *Myst*. The second step came a few years later, in my parent's basement.

The First-Person Interface

I was married now. My stunning wife and I lived in Seattle, where I attended the University of Washington and freelanced the occasional illustration and design job. Rand called from Texas to tell me he had discovered a development environment that would allow a programming novice (like myself) to create relatively complex applications (or "stacks," as they were called). The name of this environment was HyperCard.[1] As Rand began to explain its simple structure of "links" between "cards," I had to stop him in the midst of his excitement to remind him that I didn't own a Mac.

And this is where my parents' basement comes in. They lived close by, so a couple of weeks later, I found myself sitting in their basement, taking HyperCard out of its shiny white packaging (it's beautiful!) and slipping it into their recently upgraded Mac SE30. A few moments later, I discovered the HyperCard drawing tools and began to sketch a manhole cover.

Rand had proposed a multipage children's book with interactive "hot spots" on every virtual page, but as soon as I began to sketch, it became obvious that there was a more appropriate model. From the moment that closed manhole appeared on the monitor, I was dying to know what would happen when I lifted the cover. What kind of world was

1. HyperCard was the brainchild of Bill Atkinson, one of the early designers of the original Macintosh team. The tool he and his team created was phenomenal because it was simple enough for a child to use, and yet it could also create relatively complex applications (such as *Myst*). But HyperCard didn't seem to fit into any known software "category," and this is probably why Apple left this revolutionary hypermedia construction set to flounder and disappear (instead of evolving it into a simple and highly intuitive web-media application).

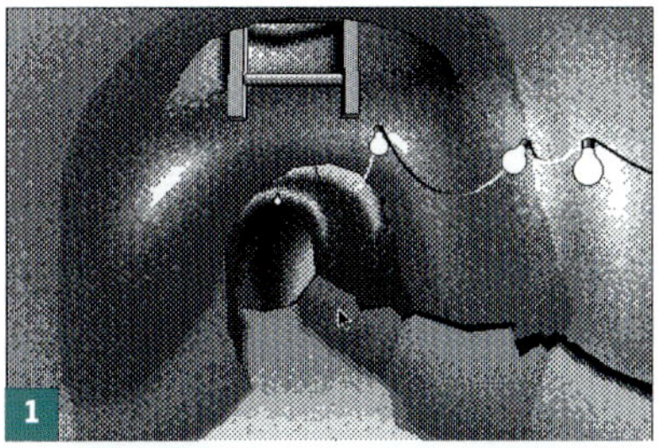

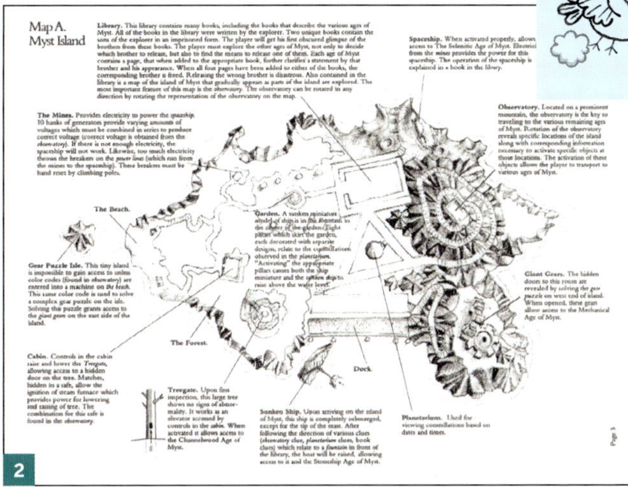

1. A drawing from a long-discontinued children's product called Cosmic Osmo and the Worlds Beyond the Mackerel.
2. A page from our *Myst* proposal to Broderbund, the company that published *Myst* in the United States. Our successful meeting with them was one of the first hints we had that *Myst* might be something different.

down below? And to how many different places was this going to lead me? All at once it was a world and not a linear book.

This later became the beginning of our first children's product, called simply *The Manhole*. Within a matter of a few hours, I created a navigable segment of that world, which operated using the same basic first-person interface. The interface wasn't a matter of conscious thought; it simply sprung to mind as if it were a necessity of the medium, and I'm now convinced that our history with D&D inspired it. In other words, our vision of an ideal interface had already been floating around in our heads for some time. We'd even tested it with Rand's D&D castle. So now, presented with a tool like HyperCard, that non-avatar interface could not help but resurface (for our style of interactive world).

After that initial project, we applied the same first-person interface to all our children's products, and with each one of them, attempted incremental improvements. We tried to keep those products as free as possible of keystrokes or graphical embellishments — windows, menus, maps, packs, avatars — anything that might create a barrier between our users and the virtual environments we were creating. Our goal was for players to lose themselves in these self-contained "worlds" in the same way we'd so often been lost in the self-contained worlds of films or books. And there was no obvious interface interrupting a film or a book (though both have beautifully evolved interfaces), so, as much as possible, we pushed to strip away all computer-like interface elements, leaving nothing but the navigable world of sight and sound. We felt that this was not only the route to achieving a more cinematic experience, but also the player's most uninterrupted and direct link to our visual/aural environments (especially for new users or children).

Around this time I began picking up old Jules Verne novels. I enjoyed his writing, not just because of his predictions (though they were often astonishing) but because the environments he created were weird — fantastic yet believable. And better yet, all of this was couched in the unique aesthetic sensibilities of the Victorian age! Jules Verne had never visited many of the places about which he wrote, so he tended to let his imagination run wild when describing places like the ocean floor or the surface of the moon. And not surprisingly, the ocean he described is the one I found myself wanting to explore. Verne had never journeyed to the center of the Earth, but his endless caverns would be more enticing to almost any explorer than a mere Mammoth Cave or Carlsbad Caverns.

My interest in Verne didn't seem to fit with anything we were doing. Certainly not yet, and certainly not with the kids' stuff. But when I read his *Journey to the Center of the Earth*, I was inspired, and I began to write a novel in my spare time about a boy named Matt Dunny.

The gist of it was this: Matt lives in a gigantic shack in the middle of nowhere with his eccentric family and his insane grandfather. Through a strange series of events, he discovers that the shack sits atop a trapdoor. He's attracted to the trapdoor for whatever reason, so down he goes. He eventually discovers an entire kingdom far underground called Dunny. The enlightened Matt now

"We pushed to strip away all computer-like interface elements, leaving nothing but the navigable world of sight and sound."

realizes he's always been a citizen of this kingdom (and he has amazing adventures in Dunny). I had no idea that this story would end up saving us enormous amounts of time during our *Myst* development, and, along with Jules Verne's influence, would become an integral piece of the *Myst* game and the larger *Myst* "universe."

A Call from Japan, a Blank Slate, and a Side-Room in a Mobile Home

Around 1991, a Japanese company we hadn't heard of, Sunsoft[2], called us out of the blue. They wanted to know if we had ever considered putting together something a bit more sophisticated and larger-scaled. Something along the lines of what we'd already been doing, but goal-oriented and aimed at an older audience. We could barely contain ourselves! A couple of years earlier, we'd proposed an almost identical concept to the young Activision, but had been turned down. Now we were getting a chance to attempt that very thing: a world with a goal and a story.

So here it is: the blank slate. Of course this always entails a certain amount of trepidation. However, at this point, much of that blank slate was already filled. We'd been slowly growing large chunks of that blank slate, and we saw *Myst* as just another step in our slow evolution.

We began designing *Myst* at Rand's office, a side-room of the mobile home (talk about luxury!) where he and his wife lived. It was not so much a discussion as a vomiting of ideas. Being brothers helped during these sessions because we not only had the freedom to tell one another when an idea worked, but we could also lambaste one another (in our special brotherly way) when an idea was hopelessly bad. So we talked and talked and sketched and sketched, and slowly, *Myst* began to take shape.

It was a very rare and exciting experience — working together like that and watching those ideas form. We never could have predicted that *Myst* would be such a "big deal," so we had the freedom to make substantial decisions in a seemingly off-the-cuff way, not necessarily focusing on "the consumer" or "the fans" (because there were none), but only on whether we reacted in a positive way (one that we often couldn't even articulate to each other).

It didn't take long for us to affirm the assumptions we'd made about our first-person interface. Even though we were aiming for a more adult audience, we still felt that the first-person interface we used for *The Manhole* — an interface with few computer graphic elements — would also work for *Myst*. We wanted *Myst* to be as cinematic as possible, so we naturally kept anything at bay that would take players out of the world and place them in front of their computer screens. The *Myst* world had to be believable; a player had to know that he or she was really walking around Myst Island — listening to the waves in the distance, peering through the lonely mist — and any "computer-like" elements might easily destroy that believability.

In addition, Rand and I were not "gamers," so we were naturally distracted by the same style of adventure-game graphic elements: avatars, packs, maps, etc. We believed that non-gamers would face a small hurdle learning and adjusting to graphics like this, and many would find them unappealing or downright distracting. An audience of gamers did not seem to mind deciphering the rules for a new game, but we never imagined that *Myst* was specifically for gamers. We never dwelled on who the *Myst* audience might be, but we ourselves were non-gamers and usually didn't bother learning the latest interface for the latest computer game. There had to be others like us. An adventure game should feel like a world, immediately obvious from a sensory point of view with no barriers to learning it — even for the most inexperienced computer user.

2. A division of Sun Electronics from Nagoya, Japan, Sunsoft was a complete delight to work with throughout the *Myst* project, and we remain friends to this day.

1. Our concept and our hope was that even a non-game-playing public could be dropped into the *Myst* world without any rules, directions, or explanations, yet would be immediately able to intuit their way forward. 2. An example of one of *Myst's* "puzzles": this generator powered an antiquated battery. 3. This luxurious hidden bedroom in *Myst* displays a Verne-like visual influence. 4. During *Riven's* development, Richard Vander Wende and I were inspired by a slew of visual sources, but Verne's presence filtered through from the original *Myst*.

The First Game for Non-Gamers

So the decision to go with the first-person interface was almost automatic. And it proved to be the right one. The transparency of the interface was a feature that (unexpectedly) helped to transform *Myst* into a best-selling piece of software by opening the doors to an entirely new computer-game-playing population. Non-gamers, many of whom had very little patience or time to learn how to play a game, turned out to be the vast majority of our audience and enjoyed the almost instant learning curve that *Myst* offered. We were even surprised and very pleased to learn that our growing *Myst* audience contained a number of people who'd never even touched a computer; *Myst* was their introduction!

When it was time to design the actual game-play, we had no experience under our belts. Even so, after letting our D&D background percolate for all those years, we had a good idea of what we were after. Between us, we had one real desire for game-play: we didn't want anything to feel like a clever mind puzzle or an arbitrary trick. We simply wanted to build impediments into the environment. Stumbling blocks. We hoped that puzzles would feel nothing like puzzles, but would instead seem to belong in the *Myst* world. If we did our job right, simple observation would suffice. The only thing players would have to do is pretend they were really there.

We invented most of these logical "impediments" as we sat around in Rand's office (probably gorging ourselves on chips). But we began scratching our heads after awhile; it wasn't so easy and we were quickly running out of ideas. So at a certain point, we went ahead and got out Rand's old D&D dungeon. We began looking through his notes and maps, thinking, "Where can we use some of this stuff? These great puzzles? Hmm ...." We eventually took the whole layout, shifted it around a bit, made it spin, added and took away some things, and called it Mechanical Age. This was D&D's most direct influence on *Myst*; it only takes a quick glance at Rand's castle map to spy a clear correspondence to the rotating Mechanical Age.

As it turned out, we were relatively successful with the game-play in *Myst*, although a number of arbitrary-feeling puzzles slipped through the cracks — even when we attempted to hold them at bay. Later on, during *Riven*'s development, we did a

> "An adventure game should feel like a world, immediately obvious from a sensory point of view: no barriers to learning that world — even for the most inexperienced computer user."

better job in this respect, though the puzzles were often so difficult that the vast majority of players lost patience with the game and gave up far before they could reach any of the endings (it's a difficult balance).

It never occurred to us to allow our players to die. This would have destroyed their experience. One does not die in reality and then get a second chance. That was not believable, so it was never considered. Incidentally, along the same lines, we initially were not going to allow the player to "save game." This also felt like a break from the believability we were trying to establish. Instead, we'd originally reasoned, the game would simply be saved every time the player moved forward. If he quit (or if the computer bombed), he could easily return to the exact spot in which he'd last been playing. By taking away any convenient option to save, we thought that players would feel a much greater story tension and would therefore not make any decision lightly. Eventually, we were talked into adding a "save game" feature, but I'm not sure I'd do the same thing again.

Most of our goals didn't seem impossible. We were almost positive we could carry off an environment with little or no learning curve. We felt it would be a bit more difficult to successfully place non-arbitrary "puzzles" into that environment. We had a clear approach, yet there was one lingering obstacle that was still a complete *Myst*ery to us: story. Rand and I thought it might be possible to create a user-driven story within the realm of the "computer game." Not just a story in which the player moved from A to B, but one that also moved its players, provoking them intellectually or emotionally in some small way.

Could we do this? Could we create an interactive work that evoked similar emotions to the more linear games our audience was already familiar with? Attempting this was largely an experiment, and we went to great lengths to set up scenarios in which the player would be forced to make crucial choices at set story junctures, many of which grew more climactic as the player's adventure progressed. We kept it simple. And we hoped our audience might feel something more than the typical computer game.

Myst did provoke. But not in the way we expected! The story we created seemed to have a fleeting impact compared to the islands themselves. The *Myst* audience surprised us by expressing a bond with the islands. The Island Ages became, in a manner of speaking, the "characters," and players felt the impact of these characters in ways Rand and I could never have imagined! I came to recognize the more linear story of the brothers and their father as almost ancillary, as another vehicle simply to distract players from what was much more vital to the *Myst* experience — the landscapes.

Jules Verne's Contribution

As Rand and I designed, the look of those islands evolved. Our inspirations probably came from a wide gamut of outside sources ranging from *Star Wars* (of course) to Disneyland, but on *Myst* I was finally free to let my current favorite, Jules Verne, run a bit amok. Verne's fantastic Victorian aesthetic had been piling up in my brain for a while and, before I knew it, his peculiar imagination had seeped its way into our design. This was most obvious in the way that Victorian-era mechanical technology was contrasted against small pockets of finery. Odd metal spaceships against pipe organs. Underground rail cars against lavish bedrooms. Of course,

3. The name "*Myst*" was indirectly taken from the Verne book *Mysterious Island*, about a group of men who are essentially dropped out of the sky onto a strange, unexplored island. Rand and I hit on the name over a phone call while discussing the book.

we could never begin to equal Verne's imagination, but the *Myst* world did take on that flavor. Like another Nemo, our main character had constructed his own small kingdom of weird wonders, replete with touches of luxury (and even a Nemo-esque library of books).[3]

Finally, we had all our sketches and designs in hand and were almost ready to get to work when we realized we were missing a key part of the *Myst* universe: backstory. We hadn't built a *Myst* history into our world; it felt incomplete. We needed a backstory and we needed it fast. Fortunately, we had a story sitting around, waiting to be used — my unfinished novel about Matt Dunny.

We took that story and, in a matter of a few minutes, our "Dunny" history was complete. "Fire-marbles" stayed; they were an excellent source of power. The idea for a separate Dunny language continued, though we annihilated the thriving Dunny kingdom and ecosystem, leaving it empty of both people and plant life.[4] At the time, we felt that the Dunny kingdom would remain nothing more than an insignificant backstory. We were very wrong.

So, as far as I can tell, that's how we came up with the ideas for *Myst*. It didn't come to us in some great moment of inspiration. It didn't form itself out of the ether. It grew out of our shared experience over the course of many years. The ideas for *Myst* evolved. And now that we had all our designs, sketched and scattered in a few yellow legal pads, we were ready to move into production (which would end up being a completely different story).

The Spreading *Myst*

But none of these ideas seemed like a big deal. None of it hit like a bolt of lightning. Perhaps that's why we were shocked by the overwhelming success of *Myst*. We hoped to make a small profit. Instead, *Myst* became the best-selling computer game of all time (though this record has since been broken by *The Sims*). We had expected *Myst* to enjoy a level of success similar to our previous products — in other words, just enough to fund our next thing. But instead, *Myst* fired a large enough spark to start something of a domino effect (which, I assure you, is a surreal thing to observe from the inside looking out).

It's still exciting for me to see how the *Myst* world continues to evolve, now in even more unexpected ways. During the development of *Riven*, Matt's hut became the desert "cleft," and his grandfather became a grandmother. The Dunny name was changed to D'ni, and after that, the convoluted history of the thing began to evolve, expand, and seemingly take on a life of its own.

This has to be one of the oddest yet best parts of my *Myst* experience: to witness a broad community become so interested in an alternate reality that many of them know it more intimately than Rand and I do. And yet saying they "know" it is not nearly enough; D'ni has become too large a concept to be conveniently controlled.

What started out in the form of one game and one novel is now scattered throughout thousands of websites. And this broad web community, in pondering a fictional reality, often begins to fill in gaps about which its creators have not yet begun to think. They conjecture. They join D'ni "guilds." They write D'ni stories. They create their own words using the D'ni language. They draw D'ni art. In so doing, they too have become not only participants but creators of the *Myst* universe. *Myst* is no longer any one computer game — it's a community. In this respect, D'ni and *Myst* now belong to their users.

Which is the best idea of all!

Resources:

You can see the entire proposal for *Myst* on Robyn Miller's blog, which includes renditions of every island in readable resolution:
makezine.com/go/mystproposal

Here's a *Myst* spaceship built in Second Life:
makezine.com/go/SLrocketship

4. My Matt Dunny story was inspired both by Verne's *Journey to the Center of the Earth* and by *The Lord of the Rings*, and I had ambitions for it to be "epic." After we used it for *Myst*, I trashed my only copy, my outlines, and my map of the world. No big loss: as far as I can remember, it was mediocre and it found a much better place in *Myst*.

Robyn Miller is best known as the co-creator of *Myst*. After his work on *Riven* (the sequel to *Myst*), he left Cyan, the company he founded with his brother Rand, to pursue independent projects.

1966: A Big Year for Video Games

An afternoon with Ralph Baer, the "Father of Video Games." By Joe Grand

I MET RALPH BAER A FEW YEARS AGO. I had just finished writing my book *Game Console Hacking*, and, given that Ralph is regarded as the "Father of Video Games," he was an obvious choice to seek out to write the foreword. To my surprise, Ralph accepted the challenge and we've been friends, co-inventors, and business partners ever since.

At 84 years of age, Ralph's thirst for inventing shows no sign of being quenched. He has invented hundreds of video games and toys, including the Magnavox Odyssey, Simon, and Computer Perfection.

What follows are selected questions and answers from an afternoon I spent with Ralph in his Manchester, N.H., home in June 2006. For the story of how video games were invented and developed in the early age of computing, along with pictures and schematics, be sure to check out Ralph's recent book *Videogames: In the Beginning* (Rolenta Press), and his website ralphbaer.com.

Can you give us some background on where you grew up and how you got involved in electronics?
The short story of a long story: born 1922 in Germany, left Germany in 1938 because of the Nazis. I came to New York and worked in a factory. Started studying correspondence course for radio and television services and graduated as a radio service technician from the National Radio Institute. Got drafted in 1943 and wound up in the same old place that I came from [Germany].

After I came out of the service in 1946, I worked for Emerson Radio in Long Island fixing broken radio sets as they came off of the production line (which was almost every one). I eventually ended up at Sanders Associates in 1956, a defense electronics company. We developed systems such as radar and electronic countermeasure equipment. In 1966, I came up with something that had nothing to do with my job, namely the concept of using a television set for playing games.

That was the year you designed *Chase Game*, your first television video game prototype. Can you describe it?
A simple device in which two spots on the screen could be controlled by two players, the object being for one spot to chase the other, catch up with it, and wipe it out. We also had light-gun, target-shooting games right from the start.

This led to the development of a game console in 1968, called the Brown Box (after the fake wood-grain contact paper you applied to the metal case), which was the basis for the Magnavox Odyssey. What are some challenges you faced in licensing this novel invention?
The biggest challenge was, "What the hell do we do with it?" What did I know about licensing at the time? I was running an engineering department and building radar systems. Then, it dawned on me: the components inside our Brown Box were identical to

Ralph demonstrating the first-ever light-gun game. He was an expert on military small arms in the U.S. Army during World War II.

Photograph by Joe Grand

those inside of television sets, so obviously television manufacturers could make this thing.

We started ringing bells at all the TV manufacturers we could find in the U.S. (and there were a lot at the time). RCA was interested but came back with an inch-thick licensing contract, so we walked away. At Magnavox, we faced a bunch of engineers, marketing people, and management sitting around a long, wooden boardroom table looking very glum, like they weren't too happy about getting involved in this. But when Gerry Martin, the VP of marketing, saw the demonstration he instantly said it was a "Go!" and that started a whole year of futzing around written agreements between us.

Ultimately, the Magnavox Odyssey wasn't released until 1972, four years after the Brown Box had been built, but it was a great success.

Editor's Note: 100,000 units were sold in the first year and more than 350,000 Odysseys were sold by early 1975.

You also invented the smash hit Simon, in the late 1970s, and many other successful games for Marvin Glass Associates (the famous game design studio). What sorts of projects are you working on now?

The recreation of early cable games [using the cable television system to display a graphical overlay on the screen in a time before computer-generated graphics were available] and building some replica Brown Boxes for use in museums.

Do you have any final words for aspiring inventors, makers, hackers, builders, artists, or other do-it-yourself enthusiasts?

Get born to rich parents so you can afford to become a hacker and play around and have a real good time without having to worry about making money. Otherwise, reality sets in real soon because you find out that creativity and talented engineering capabilities and all that are only a small part of the big picture. There's always all this nasty stuff [like sales and marketing] that can get in the way.

When not inventing new consumer electronics, toys, and video-game accessories for his company, Grand Idea Studio, Inc. (grandideastudio.com), Joe Grand collects classic video games and plays ping-pong on his Magnavox Odyssey.

Photos from the Inside

Toys sent through the X-ray machine give new insights. By Shawn Connally

TED KINSMAN MADE THE DECISION MORE than a dozen years ago to take photos of stuff that's "sciencey" and sell it to the general population. He's accomplished this by using a variety of techniques — close-up macros of snowflakes, infrared and time-lapse photography, even X-rays — and he says, "I always break even ... or lose money."

His time-lapse images are seen in many nature documentaries, and he's one of a handful of people doing serious infrared photography. If you're watching a flower bloom, there's a good chance Kinsman took the images. With two or three cameras working for him at home, he goes off to his *other* full-time job as a high school physics teacher. "I have a lot of fun with it," he says. "I want to teach young kids that science is out there and it's cool. And I think I'm doing pretty well."

Kinsman's X-ray photos recently traveled from his home in Rochester, N.Y., to Alameda, Calif., for a special exhibit. Of particular interest to Kinsman are the plastic toy guns. "When you look at the X-rays, whoa, it looks exactly like a gun. Some of the toys are quite graphic and look like something else," he says.

"I like showing people what's inside this object or that object. Seed pods, shoes, duck decoys. So people can see exactly what's inside — the defects, recycled materials. It becomes a totally wide-open world."

1

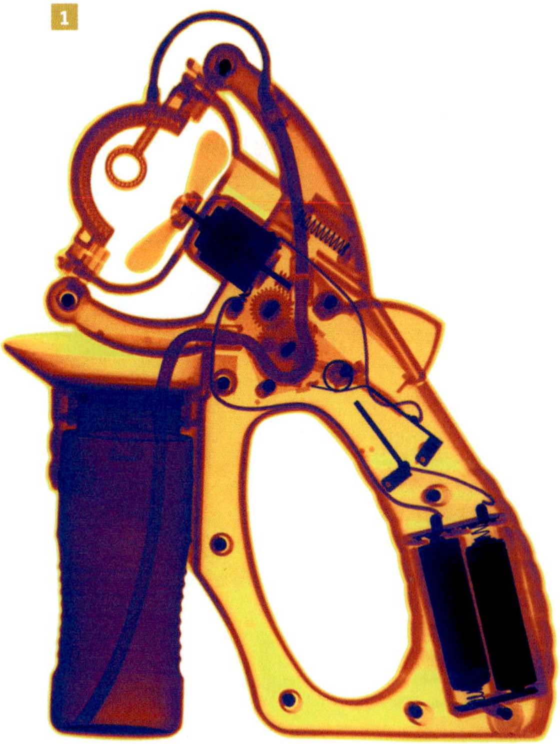

2

Photography by Ted Kinsman

3

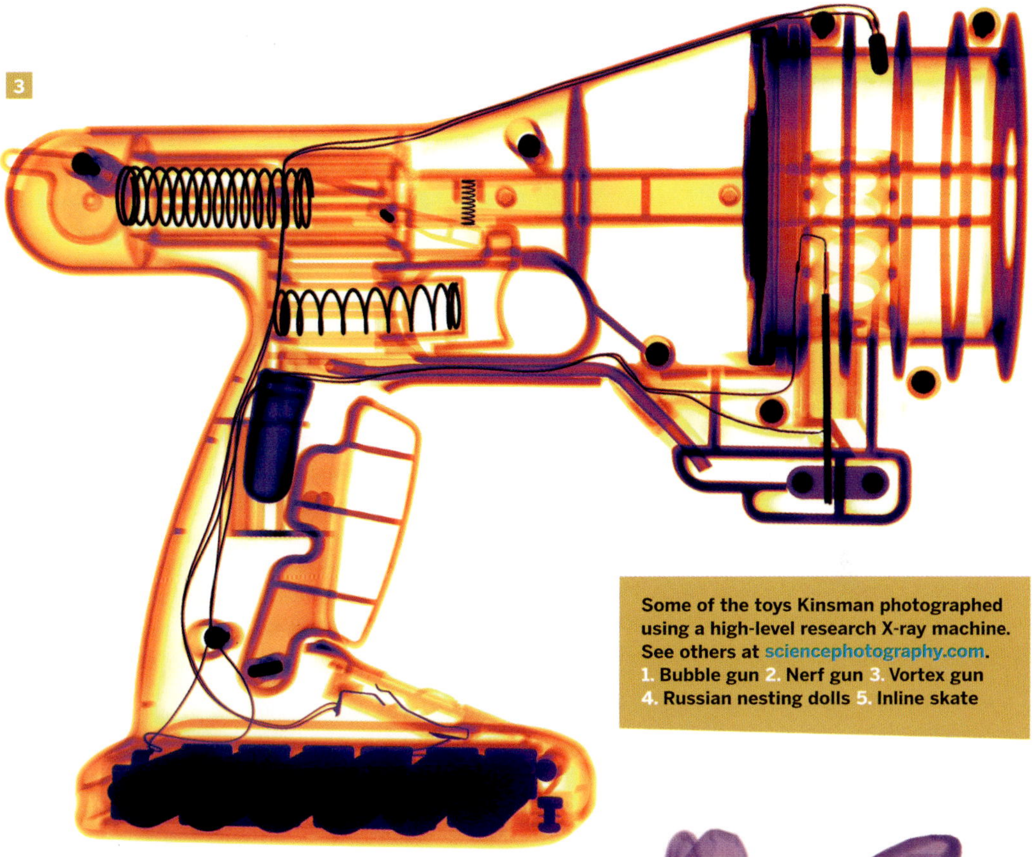

Some of the toys Kinsman photographed using a high-level research X-ray machine. See others at sciencephotography.com. 1. Bubble gun 2. Nerf gun 3. Vortex gun 4. Russian nesting dolls 5. Inline skate

4

5

Pinball, Resurrected

Restoring a crusty, beat up *Cyclone*.

By Bill Bumgarner

THE FLIPPER'S SWEET SPOT STRIKES THE steel sphere, sending it through a gantlet of colorful obstacles. Electrical relays fire, bells ring, lights blink, sounds play, and the world vibrates with a dizzying barrage that only sharpens your focus. You parse the steel ball's trajectory, then adjust its course by nudging the 300-pound machine underneath; there's nothing "virtual" here. You are playing pinball, and it envelops you in a sense-filling physicality that video games can't begin to approach.

Even though the once-ubiquitous pinball machine is a rare sight in arcades and bars today, you can still fulfill your wildest pinball dreams by buying and restoring one yourself. The machines and the parts are out there, and many games in good or reparable condition are collecting dust, waiting for a loving home. Old pinball machines are tough, designed to survive with little maintenance in rowdy, smoky roadhouses. They're also built for repair; different models share many of the same mechanical and electrical subsystems, which you can easily swap around. I've enjoyed restoring several pinball machines, and my most recent adoption was a Williams *Cyclone*, a great solid-state pinball machine from 1988. Here's how I did it, along with more general advice on pinball restoration.

Find a Machine

Pinball machines can cost anywhere from $100 for an as-is electromechanical to nearly $10,000 for a particularly collectable game in good condition. Prices vary wildly, so unless you absolutely must have a particular, hard-to-find game, you should be prepared to look at many before finding a good one to buy. For first-timers, even those with backgrounds in electronics or mechanics, I recommend spending the extra dollars to buy a machine that's at least in decent shape. There will still be plenty of things to tune.

You can buy pinball machines in better condition from restorers, who range from hobbyists to full-time professionals. The most pristine are "home use only" machines, which were never operated in public places. You can also pick up machines from arcade operators and "route operators" who place them in bars and restaurants. In my experience, you should be more careful with these. I have seen evidence of numerous dubious ad hoc "repairs" done to on-route machines in order to avoid the effort and expense of taking them off-location for proper maintenance. The worst offense is the use of spray lubricants such as WD-40. Pinball machines were designed as sealed boxes with mechanical systems that don't require oiling, so they could work maintenance-free in smoke-filled bars for months. Extraneous lubricant sucks up dust and smoke, and turns into a thick, damaging sludge.

To find machines for sale, I scanned craigslist.org, the rec.games.pinball forum, and *Mr. Pinball* classifieds (see makezine.com/08/pinball for more resources). I prefer late-80s and 90s Williams/Bally pinball machines, and my search led to Thursday Night Pinball (pinballnight.org), a group in San Jose that meets in a battery shop. The group's organizer had a Williams *Cyclone* for sale for $1,000. *Cyclone* is one of my favorites; it's a brilliant machine from

Photography by Bill Bumgarner

CYCLONE
JACK POT
MYSTERY WHEEL
Williams
PINBALL AMUSEMENT PARK
EXTRAVAGANZA

1988 with a fairgrounds roller-coaster theme, a kitschy midway soundtrack and voice-overs, ramps, a Ferris wheel, and a spinning wheel of chance on the backglass.

I played the machine and found no catastrophic problems. All parts were present, though extremely dirty, and the playfield had minimal paint cracking and no bare spots. But the machine would still need a lot of work. The Ferris wheel didn't spin, rubber bumper bits were broken or missing, one string of lights flickered intermittently, the flippers were weak, and the Player #3 scoring display didn't work. Pretty typical for a machine that had spent time on route.

I handed over the cash, and readied the machine for transport by folding the backbox down and unscrewing the legs. Like this, it measured 53" by 29" square, and just fit into my Subaru Forester.

Initial Play Test

After transferring the machine safely to my garage, the first step was to play it for a few days and note any problem areas — which is also a good way to meet the neighbors and their kids. When play-testing a neglected machine, you should watch for any relay coils that are energized all the time.

Pinball machines have dozens of these coils, which are what make playfield objects move. Any feature that's stuck in the wrong position — like a bumper always popped out, a flipper stuck up, or a target that won't drop — indicates an always-energized coil underneath, which will also typically buzz and be warm to the touch.

If you find one, it probably means either a short circuit or a blown transistor is leading to the coil. You should turn the machine off immediately and try to locate the problem using the schematic and a multimeter (*see "Special Tools," page 73*). Pinball machine coils are not designed to be on all the time; when they are, they overheat, strain the power supply, and can damage other components or scorch the playfield.

As you play-test, learn the rhythm of your machine. If you feel something change — like a flipper or bumper losing strength, or a ramp starting to wiggle — this is often the initial symptom of something that could grow to become a major problem later.

Get Documentation

Before going further, you need the manual and schematic for the game you're restoring. Fortunately, my *Cyclone* came with both. If it hadn't, I would have looked for a printable copy online, at Marvin's Marvelous Mechanical Museum or the Internet Pinball Database, or else ordered a reproduction manual from Pinball Resource or Bay Area Amusements (see makezine.com/08/pinball). Marvin's site is a fantastic resource in general; it has saved me countless hours with its repair guides and other helpful info.

You will refer to the schematic often during restoration, but, unfortunately, it isn't always the gospel truth. A game's circuitry may be tweaked during its manufacture run, making it differ in small ways from the published schematic.

Run Diagnostics and Check Components

Check the manual or repair guide for how to run your machine's diagnostic routines. Every solid-state machine has these. On *Cyclone* and other solid-state Bally/Williams machines, a single dot appearing after the number of game credits indicates that the diagnostics software has found a problem. Switches inside the coin door let you enter diagnostic mode, and from there you can navigate menus to identify what's wrong.

Diagnostic mode also lets you test individual subsystems of the game, all the way down to individual lights and coils. With multimeter and schematic in hand, you can use the diagnostic mode to check everything out. Whether or not you are taking guidance from the diagnostic software, you'll want to check for proper voltages from the power supplies (there are several), check for bad bulbs, bad diodes on coils, or shorted/broken coils.

For the rectifiers, probe the output voltage of that rectifier's power supply section. With the multimeter in continuity testing mode, put one lead on ground (metal strap in the backbox is easiest) and touch the metal tab on each transistor. Any transistor that reads 0 ohms or causes the meter to buzz is bad, dead shorted, and will cause the corresponding coil to be energized all the time. Coils will typically have a diode across their two contacts and this diode will need to be unsoldered for full testing of the coil and diode, something I only do for coils that look particularly beat.

NOTE: Like any device that plugs into a wall, pinball machines can bite. Be careful when working inside the machine. Some electrical components may remain live while the machine is not in play, or even while it's turned off! Also, ground faults are common, so things that should not shock you

INSIDE THE CYCLONE

[A] Knocker that fires when you win a free game

[B] Sound board

[C] Fuse holder

[D] Mystery Wheel (spinning wheel of chance in the backglass)

[E] MPU/driver board. Contains CPU, software, switch matrix input, and driver output.

[F] Evil batteries that leak all over the main board. Moved to external battery holder in a Ziploc bag.

[G] Driver transistors that control the coils and flashlamps on the playfield. Flippers are not controlled by these in this model of machine.

[H] Auxiliary power board

[I] Wiring harness that connects the playfield to the MPU

[J] Bridge rectifier. One of many, but this one runs particularly hot.

[K] Big fat power supply capacitor

[L] Left speaker (midrange). Bass is in bottom cabinet.

[M] Display cable. Don't plug this one in backwards.

[N] Display board

[O] Right speaker (tweeter). Stereo effects are achieved by pitch, not true stereo.

sometimes will. Mechanically, pinball machines move a very heavy steel ball at high speed, so you can imagine what they will do to your finger.

Replace the Batteries

Like other solid-state pinball machines, *Cyclone* has batteries in the backglass that save settings and high scores when the machine is off. These batteries can leak acid and damage the CPU board. With the machine unplugged, our first task is to unscrew the backglass, open it up, and replace the batteries.

In my *Cyclone*, the battery holder was damaged from an acid leak, but the CPU board was still OK. I ripped the holder off the board and soldered in the wires of a new, off-board battery holder. I enclosed the new holder in a Ziploc bag and zip-tied it to a screw I drove into one interior side of the backbox. This arrangement will prevent any future acid-related damage, and I recommend it for any arcade machines that have batteries on their CPU boards.

Raise the Playfield

Remove the playfield glass by opening the coin door on the front of the machine and sliding the little lever inside that is just above and to the left. This unlatches the lock-down bar that spans the glass at the front of the machine. The glass will then slide out of the machine easily. Lean it out of the way, with the bottom edge resting on something soft like a towel.

It's important to remove the balls before flipping the playfield up all the way, lest they drop and break lamps and other components. To do this, lift the playfield a little, reach underneath, and flip the metal bit that kicks the balls into the launcher. With all the balls removed, you can swing the playfield up until it leans vertically against the backglass.

Initial Breakdown and Cleaning

I start cleaning at the bottom of the playfield, the flipper end, and work my way to the top. The bottom is simpler, and addressing it first gives me a feel for what kind of grime and other recurrent issues I'll be facing. With *Cyclone*, I quickly encountered a number of stuck screws, and it was better to practice handling these in a simple, open area.

With all the dirt, it soon became clear that I would need to remove every component from *Cyclone*'s playfield. I grabbed my digital camera and took over 100 pictures of the playfield and various assemblies in close-up, to help me reassemble everything later.

After documenting, I started by stripping the lower-left quarter of the playfield, the simplest quadrant. Using warm water and cotton rags, I gently took up a thick layer of dust and grime, only to reveal a more difficult cleaning problem underneath: the playfield was covered in thousands of tiny paint blobs, probable evidence of an old spray-painting mishap. No major discoloration, but definitely noticeable.

For the components themselves, I used warm water. I treated especially grungy plastic parts with Novus #1 or #2. If you don't completely replace the rubber rings, treat the grimy ones with rubber cleaner.

Playfield Detail

There are three kinds of pinball playfields: painted wood, Mylar, and Diamondplate. Each requires a slightly different cleaning technique. With a painted wood playfield, artwork is screened onto the wood with paint that's quite tough, but that steadily wears down under the abrasive powers of a rusty pinball or dragging flippers. To prevent this, some manufacturers began protecting the playfield with clear, adhesive-backed Mylar. And as of about 1990, Bally/Williams began covering the entire painted wood playfield with a thick coating of Varathane, branding it Diamondplate.

My *Cyclone* had plain painted wood, and happily the paint was still firmly bonded to the playfield. This let me remove most of the grime and spray paint with plain water. I used Wildcat Playfield Cleaner for my initial cleaning, followed by Mill Wax and Playfield Cleaner, a less harsh wax-based cleaner.

Note that Wildcat destroys plastics, so it should not be used to clean ramps or used near plastic parts. Mylar playfields become cloudy if you clean them with harsh cleaners like Wildcat. Diamondplate requires a good plastic polisher like Novus #1 or #2. All cleaners mentioned here are available from pinball parts suppliers.

New Lamps and Rings

While the playfield was stripped of parts, I replaced all the burned-out bulbs and changed all of the "general illumination" lamps (the ones that are on all the time) from #44 to #47. The latter use less current, which eases stress on the power supply. Cleaning the machine makes up for the slight decrease in light. Using brighter #44 lamps for the computer-controlled illumination makes special lighting effects stand out more.

After cleaning each section of the playfield surface and its resident components, I reassembled them, substituting new rubber rings from the replacement kit for the old ones.

1

2

4

The art, playfield, and mechanics of *Cyclone*, mid-restoration. Every pinball machine is an antique model never to be made again. 1. Cameo appearance by Nancy and Ronald Reagan on backglass. 2. Left flipper exposed for cleaning. 3. Flipper mechanism with flipper shaft removed. 4. Newly cleaned playfield under removed left flipper.

3

Upper Playfield

Cleaning the upper playfield was just like with the lower playfield, but with more features to remove. *Cyclone* has a network of clear plastic ramps in its top half, many of which have integrated switches. Some of the switch leads disconnect under the playfield, which lets you remove sections of ramp with their switches attached. But they don't all work like that. Along with three separate ramp assemblies, I removed the ball launching assembly and several plastic-on-posts assemblies. Underneath these toys I found a mess of rubber bits, lots of dirt, and even a few seed pods (no idea).

After cleaning and reassembling, I readjusted the ramps. With the playfield down, I took a ball, rolled it around the ramps, and fine-tuned their positions so that it wouldn't get stuck between the upper and lower ramp.

Next came the Ferris wheel. I found that it didn't spin because its drive belt was broken. Simple enough, but at $6, this special rubber belt turned out to be considerably more expensive than any other rubber pieces. Installation wasn't easy, either; I had to disassemble the wheel completely to put the belt around its hub, then pass the belt around a piece of metal that prevents the ball from falling out under the playfield, and finally attach it to the motor underneath.

Clean Underneath

The underside of the playfield is pretty straightforward to clean. On *Cyclone*, a ball guide connects the various under-the-playfield features together. It is made from the same plastic as the topside ramps, and it should be clear, though mine was close to opaque when I first opened the machine. A bit of Novus #2 cleared that up.

If my *Cyclone*'s paint were in better shape, I would have stripped the playfield bare on both sides and coated it with an automotive clear coat. As it was, this wasn't worth doing, but I am on the lookout for an NOS (New Old Stock) bare playfield that I can clear-coat and swap in for that gleaming-perfection look.

> "A pinball machine is like a mid-70s British roadster: it may play just fine, but there is always something you can fix, clean, or tweak to make it work better."

Flipper Rebuild

Every time a flipper flips, a metal pawl is pulled through a relay coil and slams against a stop. After thousands of repetitions, the end of the pawl and the stop both change shape; and if the pawl isn't replaced, it eventually mushrooms and fragments around the edges. This can rip apart the coil, causing a short and possibly a power-supply failure. Also, *Cyclone* uses "Fliptronics I" flippers, in which the buttons directly switch on the 50V DC coil current. This results in considerable arcing and wear on the switches, which need to be replaced as frequently as the mechanical components. It is possible to upgrade Flip I's to software-controlled Fliptronics II, which use low-voltage switches, but that's another project.

Rebuilding a flipper involves taking it apart and replacing all of the components that wear out with new ones, usually from a flipper rebuild kit. The tricky part is putting the flippers back together so that they're both at the same correct angle when in the "down" position and never grind against the playfield surface. My *Cyclone* machine showed some wear around the clown's mouth from improperly adjusted flippers, but it didn't go through the paint entirely, and I touched up the damage with some paint pens.

Finally, it was time to put the playfield back together, install the ball, and fire up the game. It's good to do this before replacing the glass because there's invariably something else that you need to tweak before you seal it all back up.

I have yet to figure out why *Cyclone*'s Player #3 display is dead, but the machine is otherwise in great condition: bright, responsive, and fully playable. Frankly, given that the machine works so well, fixing the display was bumped down the priority list.

Ongoing Maintenance

A pinball machine is like a mid-70s British roadster: it may play just fine, but there is always something you can fix, clean, or tweak to make it work better.

Maintaining a restored pinball machine is easy. Clean the playfield regularly with the proper cleaning compounds. Grime builds up in a pinball machine, and if a metal flake ever gets stuck in the grime, it can easily be driven into the playfield, creating a seed for more significant damage later. On a clean playfield, they just vibrate out of harm's way.

Conventions, Modifications, and Upgrades

There are pinball swap meets, conventions, and competitions held around the world. See the *Pinball News* site for upcoming events. There is also a huge pinball community online, including many of the original game designers. Pinball hackers have even produced an open source pinball emulator and control board, PinMame and PinMame-HW, that let a PC control a long list of pinballs — an easier way to restore if an original board is missing or trashed.

A lot of machines also have relatively easy hacks and upgrades. In some cases, the original machine design included extra switches and coils that offered more advanced play or special effects. These were sometimes removed during manufacturing to reduce costs, but the software was typically left untouched.

Custom versions of software also add features to some popular hobbyist machines. For example, a popular home version of *Twilight Zone*'s software adds a "pause" feature: if you catch the ball on a flipper and hit a certain button, the machine will go silent and hold the ball until you are ready to resume the game! With a bit of research, you can find what additional features your game might support.

Because pinball machines are so modular, many of their control circuits and mechanical systems can be broken out to do real-world tasks. You can also buy these at reasonable prices, and full schematics and documentation are online. Pinball machine subsystems can perform some powerful tasks in the real world, as I hope to describe in another MAKE article.

You can spice up a machine quite a bit even without making any functional or permanent modifications; for example, many an *Addams Family* owner puts a funny character in the game's electric chair.

Bill Bumgarner plays with high voltage, cooks with fire, hacks code, corrals bugs with his son, and tries to make stuff do things that were never intended. See friday.com/bbum.

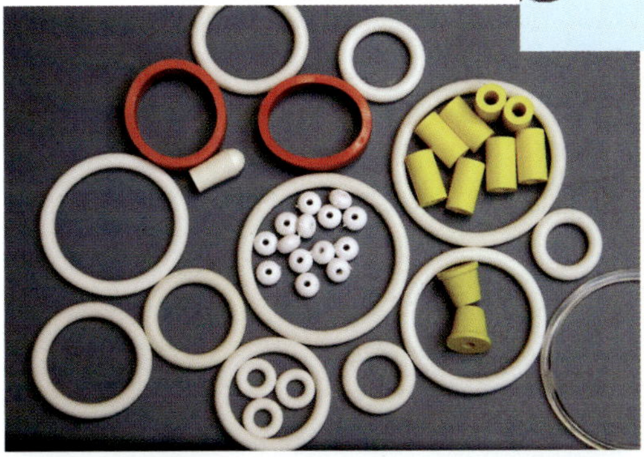

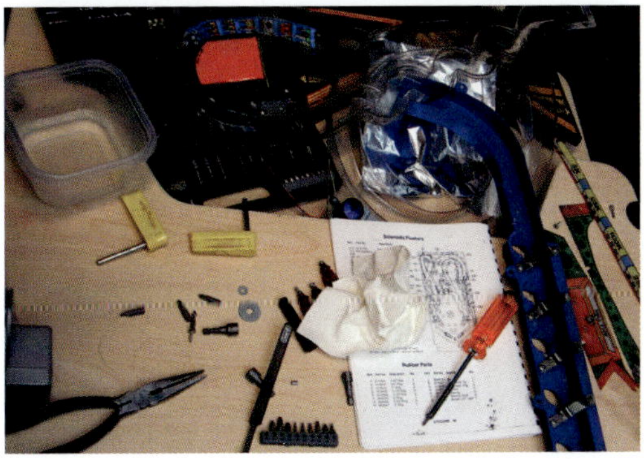

SPECIAL TOOLS

A multi-bit magnetic screwdriver, preferably with a telescoping pick-up tool. This is your most important tool by far. Pinball machines have countless screws in hard-to-reach locations. Worse, if you drop one while the playfield is rotated up, it will invariably "pachinko" its way through the maze of wires and parts, only to fall in some tiny crevice.

A multimeter that can handle both low and high voltages. Modern pinball machines use 5V DC to handle the logic and sensors, 18-50V DC to drive lamps and coils, and upwards of 400 volts to drive the gas-plasma scoring displays.

A big box of slow- and quick-blow 5×20mm glass fuses, of many different amperages. Pinball machines run within tight power consumption parameters, so you should never substitute fuse types or deviate from the proper amperages.

A full rubber ring kit for the machine you are restoring (pictured above). Many rubber rings are buried under hardware, making them difficult to access. Replacing them during restoration will save frustration later.

Flipper rebuild kit for the machine you are restoring. As with the rubber ring kit, these are available from pinball parts retailers. See makezine.com/08/pinball.

Pinheads in Oddball Places

Inside the electromechanical underground.

By Dale Dougherty

MICHAEL SCHIESS / LUCKY JU JU

Seventy-five years ago, in the city of Alameda on the San Francisco Bay, an amusement park called Neptune Beach was a rival to Coney Island. Now all that's left of it are some postcards hanging in the Neptune Beach Amusement Museum, also known as Lucky Ju Ju Pinball.

In this oddball headquarters, 20-some pinball machines crowd a two-room space that shares an off-street entrance with a beauty salon. Proprietor Michael Schiess calls Lucky Ju Ju an "Arcadium," short for Arcade Museum, but it's also an art gallery. It's only open on weekends, and Schiess charges $10 at the door, with all the machines on freeplay.

In his early 50s with a gap-toothed grin, Schiess is an affable caretaker and willing custodian. Before establishing Ju Ju, he composed electronic music and worked at the San Francisco Exploratorium as an exhibit builder and maintainer. But his long-time love of pinball art led him to start collecting machines. While taking me on an informal museum tour, Schiess explains that he prefers the art of Christian Marche, whose prolific output included many science-fiction-themed machines; he also admires Jerry Kelly's work, such as in the Beatles-themed *Beat Time*. Schiess calls attention to Dave Christiansen's *Captain Fantastic and the Brown Dirt Cowboy*, the artwork of which was censored in mid-production by the manufacturer. This led to an X-rated follow-on story that I won't repeat.

Schiess is particularly proud of *Fireball*, a 1971 machine that he says is "everyone's favorite" because it integrates superhero art by Christiansen with deft skill shots and a whirlwind spinner that

Photography by Steve Double

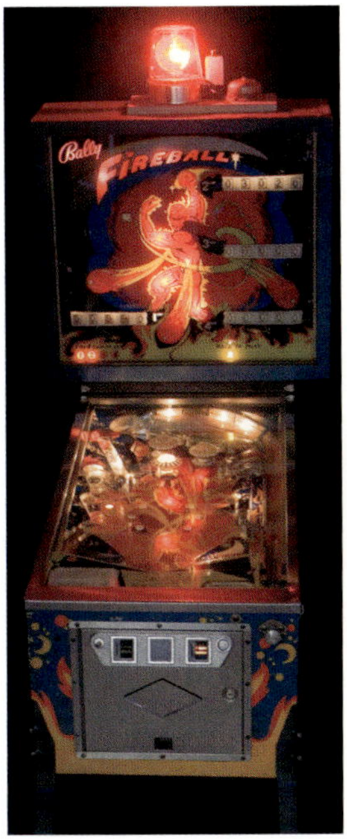

Lucky Ju Ju's ***Fireball*, *Teacher's Pet*, *Beat Time*, and the guts of *Faces*, a 1976 machine from Spanish manufacturer Sonic.**

changes the ball's direction as it approaches the flippers. "This is such a well-themed machine," says Schiess. "On the backglass, it's got this guy throwing balls at you, and that's just what it does on the playfield. The function follows the art, which is rare."

Schiess hopes his collection of classic pinball machines will be the centerpiece of a larger vision that restores the old Ferris wheel from Neptune Beach, and invites kids to have fun while learning how these wonderful machines work. With his combined pinball and Exploratorium experience, Schiess may understand the educational possibilities of amusement machines better than anyone. "What I like about pinball machines," he sums up, "is how they bring together art, history, and science."

TIM ARNOLD / PINBALL HALL OF FAME

A mile or two east of the bright lights of Las Vegas, there's a dimly lit strip mall with an aging movie theater, some ethnic restaurants, and the Pinball Hall of Fame. Inside this darkened space filled with pinball machines, I found owner Tim Arnold making his nightly rounds, caring for the machines he loves. The glass is off the one he is working on, the fully electromechanical *Gold Rush*. Arnold replaces a bulb, then wipes down the playfield before replacing the glass and testing the machine. He is as quiet as a monk, but soon the machine's chimes are ringing — a good sign.

With thick glasses and ageless blond hair pulled back in a pony tail, Arnold mirrors the hybrid personality of pinball itself: part idealistic hippy, part road-hardened truck driver, and part nerd. Arnold sees every pinball machine as a hooker with a heart of gold — a storyline from the West that could be a pinball game itself. "These machines don't belong in the home; they belong where people come to hang out and have fun," he explains.

Years ago, Arnold worked as a route operator, selling and servicing amusement machines throughout the Midwest. He sold his business, moved to Vegas, and set up the Pinball Hall of Fame. Arnold runs it as a nonprofit, with extra revenue going to the Salvation Army. One day, Arnold hopes his Pinball Hall of Fame will become a popular oddball Las Vegas attraction like the Liberace Museum down the street.

Michael Schiess and Tim Arnold know each other. "Beware of Mr. Schiess," Arnold tells me jokingly. "He's charging you a flat price for all you can play but when you are done, how do you know it is ALL YOU CAN PLAY?"

Schiess stays with Arnold when he visits Las Vegas. "Tim's been around pinball machines a lot longer than me," Schiess explains. "He keeps telling me that I'm still in the romantic phase, and I'll get over it."

Hear an audio tour of Lucky Ju Ju Pinball Arcadium at makezine.com/08/pinball.

Chris Ware's ACME Papercraft

Comics you can build! By Gareth Branwyn

WHEN YOU PICK UP A CHRIS WARE "comic book," such as his critically acclaimed *ACME Novelty Library* series, you know you're about to encounter something extraordinary, even before you crack it open. The covers themselves immediately work to tickle our too frequently jaded fancies. Ware's books are usually oversized or off-sized and sport evocative art and snazzy embellishments (such as foil stamping, paper banding, and pockets hiding mini-comics) designed to communicate difference: this is not your ordinary comic book.

Peering *inside* a Ware comic is when the psychoactivity begins to peg your novelty meter. As one reviewer put it, "You don't read a Chris Ware book so much as open it up and fall into it." The unique landscape onto which you tumble is populated by a weird and wonderful cast of characters, whose often breathtakingly mundane or tortured lives are meticulously rendered in strips such as *Rusty Brown*, *Jimmy Corrigan: the Smartest Kid on Earth*, *Quimby the Mouse*, *Rocket Sam*, and *Tales of Tomorrow*. The look and feel of the strips are very much a throwback to *Krazy Kat*, *Little Nemo*, and other early comic strips. The breathtaking neglect, alienation, existential cravings, and the sense of sensory overload are decidedly modern themes.

There is a profound tension between the pristine virtuosity of Ware's renderings — which feel like they exist in a kind of Zen garden quiet — and the almost dizzying density of all the content he packs into each of these books. There's stuff going on everywhere: multiple strips per page, mock comic book advertising, corporate reports from the ACME Novelty Company, bizarre essays (e.g., a field guide to "The Collector"), actual glow-in-the-dark star maps, animated flip book frames. It all seems to beg for interactivity. To consume a Ware book, you end up turning it in every possible direction to not miss a single graphic or buried treasure. On the cover of the *ACME Novelty Library* anthology, published last year by Pantheon, there was even a comic strip printed on the cardboard-thin leading edges of the cover (billed as the "World's Smallest Comic Strip").

One of the most delightful features of Chris Ware's *ACME Novelty Library* books has always been the inclusion of do-it-yourself activities, namely the cut-and-fold toy models. Looking at these projects, reading the thoroughly detailed teeny-font instructions, one can't help the temptation to actually try putting them together. You too can build a model of Rocket Sam's retro ship, a Lilliputian library of Ware comic books, complete with an ornate book cabinet to store them in, and even a working, hand-cranked 3D movie viewer.

Like everything Ware does, the project plans look so well designed and drawn out that they appear as though they'd really work. Are we supposed to actually put them together? Is this an image of a project — another painstaking put-on, like the fake ads, the fake ACME company histories, the fake coupons and contests — or a real DIY project? While few readers have likely gone beyond questioning the integrity of these models, a handful of hardy souls have actually put scissors to the hallowed pages of *ACME Novelty Library* in a devoted effort to bring

Niem Tran, one of the few brave souls who actually makes the paper models in cartoonist Chris Ware's *ACME Novelty Library*.

Photography by Jen Siska

Ware's work into the third dimension.

"I don't know of too many people who've built the actual models," says Niem Tran, one of the few fans who's actually done it. He's built them all, in fact, and shows off his efforts on his website (niemworks.com). So why did he take on the challenge? "I'd been collecting Chris Ware's comics for a while and really loved them, including the paper toy designs. The models looked like they'd be fun to build, but I never really gave too much thought to actually constructing them myself.

Over the years, as my appreciation for Ware's work grew, so did my curiosity about assembling the models. After fruitless net searches for photos of others' attempts, I decided to try making them myself. One project led to the next, and now I've made just about all of them from the *ACME* series."

Another builder of Ware's often tricky papercraft is Dan Raeburn, author of *Chris Ware*, the 2004 Yale Monographics edition about Ware and his work.

"I built the models in order to get more intimate with his work," says Raeburn. "By building them, I hoped to feel some of what he felt making them. I'm not sure if it worked, but it provided a great way to procrastinate from working on the book itself. It was 'research.'" Like Tran, Raeburn knows of few other readers who've actually built the models. "Of all the people I know, I think only Niem and I are geeky enough to try," he jokes.

> "The covers themselves immediately work to tickle our too frequently jaded fancies."

Believe it or not, Chris Ware himself doesn't even build the models he so scrupulously designs, draws, and then instructs readers on how to properly construct and display. "I've only ever built a prototype for one of the models," says Ware. "For the rest of them, I simply draw them out on the page and hope

Niem Tran in his home workspace. When working papercraft, the blade and a good straightedge are everything — along with a pot of fresh paper cement.

that they function if anyone actually ever tries to cut them out and put them together."

When asked if he created the models with the expectation that anyone would actually construct them, he replies: "I suppose I'd hoped that someone might, but one never has any idea what kind of reaction one's work is going to get, let alone when it requires the use of scissors, tape, and glue."

For devoted fans of Ware's books, the cringe-worthy question to ask builders like Tran and Raeburn is: Do you actually cut up the books or make color photocopies? They both say they've cut up actual copies of the books, but then this was years ago, before decent scanner and printer technology, making it somewhat of a necessity. "If I were to make the models today," offers Tran, "I'd try printing them directly onto Epson Matte Paper Heavyweight."

So, what other words of advice would these pioneering papercrafters offer those who'd like to attempt building Ware's *ACME* models? "Be patient and have plenty of time to spare," cautions Tran. The 3D Movie Machine (the most challenging and impressive model, found in *ACME* #15) took him about 24 hours of build time. Most of the others took between 2 and 12 hours of concentrated work. If you just want to get your feet wet, Rocket Sam's ship from *ACME* #7 is a good model to start with. "I like the rocket ship the most," says Raeburn, "mainly because it was relatively easy to build and reminded me of the toys I made as a child."

+ CHRIS WARE'S *ACME* MODELS AND OTHER PAPERCRAFT RESOURCES:

Niemworks niemworks.com/else/acmetoys.html

The *ACME* Novelty Archive kempa.com/blog/archives/000120.html

The *ACME* Novelty Warehouse quimby.gnus.org/warehouse

Chris Ware's Flipbook Images as GIF Animations davidcihla.com/ware.htm

Iceberg Paper Model Links www.peterjvisser.demon.nl/indexe.html

In the Studio: Visits with Contemporary Cartoonists, by Todd Hignite (Yale), chapter on Ware and his work studio

1

2

1. The Robot from *ACME* #1. 2. The elaborate 3D Movie Machine from *ACME* #15. 3. The Lilliputian library of Ware books, complete with ornate book cabinet, from *ACME* #7. 4. The house from *ACME* #13, also reprinted in it. All from Niem Tran's collection.

4

3

SPECIAL TOOLS

Tran and Raeburn offer these additional papercrafter tips for *ACME* builders:

- Use a self-healing cutting mat and have fresh hobby knife (X-Acto) blades handy.
- Use a metal ruler when cutting. Don't press too hard.
- Score all lines before folding. Use a metal ruler and a blunt sewing pin, applying slight pressure as you trace the lines.
- Make all your folds before applying any glue.
- Don't squeeze the white glue out directly from the container onto the tabs. Instead, place a puddle of it onto a dish or scrap sheet of cardstock and dip a sewing pin into it. Use that pin as your applicator, and you'll have more control over the application.
- On larger models, such as the Rocket Ship, use 3M Spray Mount or Spra-Ment and mount the parts onto card stock.

Dan Raeburn adds: "Glue your models to bases or at least keep them away from drafts, especially in the summertime when the windows are open."

Is it just me, or does the image of a collection of pretty paper toys, lovingly constructed by some doting fanboy, flying away on a indifferent current of air, have a decidedly Chris Ware beauty to it?

Tabletop Terrains

That's no pile of trash; it's my asteroid mining colony! By Gareth Branwyn

SAY THE WORD "GAMING" AND MOST people these days will think of the first-person shooting and role playing that takes place on a computer screen. Or maybe a family board game. But to a relatively small and dedicated cadre of players, "gaming" brings to mind a rich and diverse skill set and activities including carpentry, painting, sculpting, mold-making and casting, scale modeling, and environmental simulation. This is the realm of the tabletop miniature wargamer.

There are tabletop wargames based in many genres, but the most common are historical, fantasy, or science fiction. The process for making terrain boards and pieces is basically the same across the genres, but in this article, I'll talk about sci-fi terrain — the type with which I've had a lot of personal experience.

What Is Tabletop Miniature Wargaming?

Before we talk about the terrain making, we need to understand what the terrain is for. The mechanics of each tabletop wargame are different, but play basically boils down to "fire and movement": moving your toy soldiers ... er ... troops into position (how far you can move is determined by dice rolls and measured with a tape) and firing your weapons at your opponent's models (determined by dice rolls and the rules for the weapons depicted on the model). Nearly all tabletop wargames are WYSIWYG (What You See Is What You Get), so the models must actually show the weapons and defenses they are using in play (e.g., you can't have a model holding a projectile weapon and claim that it's a plasma pistol). Terrain (woods, swamps, mountains) becomes very important because it affects movement rates and "line of sight" rules. So, the buildings that gamers spend dozens of hours constructing — sometimes to nearly museum display quality — are not just for ambiance and the cool factor. They also provide much-welcomed cover when your ground forces need a place to duck behind, or your sniper team needs high points in which to nest.

Terrain for Beginners

Getting started in miniature wargaming can be an expensive and time-consuming proposition. Like paper-and-pencil role-playing games and map-based wargames, the rules can be very intimidating to newcomers. There are lots of hit, wound, and save tables to memorize, terrain and other dice modifiers to understand, special rules to remember. On top of that, miniatures mean toy soldiers: dozens, even hundreds, of little plastic and metal soldiers to buy, assemble, paint, and base.

Once all of that's done, the last thing you want to do before playing is to spend months more building a terrain board and scratch-building dozens of buildings, barricades, rubble piles, and the like.

Luckily, you can get away with some pretty groovy-looking terrain made from little more than kitchen trash. The trick is developing a terrain builder's eye. Look carefully at what goes into your trash can. Any metal can, turned over, can become a storage tank of some sort. A bunch of them together can serve as a fuel depot, a worthy objective for a game. A cut-up egg carton becomes a field of alien gestation pods. The plastic dividers in boxes of cookies, crackers, and candy, flipped over, can serve as futuristic army barracks, power stations, command bunkers, and anything else your imagination can dream up.

Of course, none of this stuff is going to look very

Sean Patten's scratch-built Thunderhawk Gunship, a veritable coral reef of plastic and metal bits surrounding a styrene plastic form.

Photograph by Sean Patten

believable in its native packaging colors. This is quickly and easily dealt with by spraying everything with matte black paint and then drybrushing it to bring out details. Drybrushing is when you put paint on a brush, remove most of it on a rag or paper towel, and then work the "dry" brush over the target area. Any raised details on the piece pick up some of the drybrushed color, adding texture and dimension. For concrete and rock effects, drybrush successive layers of darker to lighter gray and then white. For a metal look, use metallic paints from darker to lighter shades. You'll be amazed at how effective this technique is once you get the hang of it.

Meet the *Gomi No Sensei* of Terrain Makers

In William Gibson's collection *Burning Chrome*, he describes the character Rubin as "a master, a teacher, what the Japanese call a sensei ... he's the master of ... garbage, kipple, refuse, the sea of cast-off goods our century floats on. *Gomi no sensei*. Master of Junk."

The masters of terrain building have an uncanny ability to look at the mundane objects around us and visualize what they'd look like turned upside down, twisted sideways, glued to some other piece of detritus, cut, improvisationally embellished with bits of plastic at hand, painted, and lit by twinkle lights and some AA batteries.

One such venerated master is Sean Patten. A computer game designer by day, he got into tabletop wargaming in college. At first, the terrain was an afterthought, with cardboard boxes sheathed in building façades photocopied from books serving the utilitarian need. When Games Workshop came out with *Warhammer 40,000* (aka WH40K), their tabletop miniatures sci-fi game, in 1987, Patten says he was "immediately attracted to its gritty art style and dark gothic-industrial setting."

It is this dense, baroque junkyard world — where a far-future dark age sits on top of the strata of all periods of history — that serves so many of the *gomi no sensei* of terrain, like Patten. He's built terrain for other game systems, such as *Mage Knight* (where he was commissioned to build a couple of boards for its maker, Wiz Kids), *Mechwarrior*, *Mordheim*, and others, but he's most in his element plying his trade in the 41st century. He became something of a legend in the

WH40K community years ago when he built a Thunderhawk Gunship, an iconic vehicle in that game universe, which, at that point, had rarely been seen in three dimensions. His version of the ship took three months to build and used hundreds of parts, including *Star Wars* toys and models, other sci-fi and army toys, sheet plastic and lead, clothes pins, belt buckles, and spent printer cartridges.

Mark Zimmer is another master of the terrain maker's art. Like Patten, he got on board with WH40K in 1987, but wasn't that interested in terrain until he got back into the hobby years later and discovered amazing online resources for terrain crafting, such as the venerable TerraGenesis. "Discovering Gary James's excellent do-it-yourself site got me really excited about the idea of building my own terrain."

> Look carefully at what goes into your trash can. A metal can, turned over, becomes a storage tank. A cut-up egg carton becomes a field of alien gestation pods.

Zimmer now runs Parasitic Studios, a site where he shows off his work, offers pointers, and sells his terrain. Like most accomplished builders, he sells pieces via eBay and does commission work. "My original goal was to get this hobby to pay for itself. Over the last few years, I've done considerably better than that," says Zimmer. Like Patten's Thunderhawk, Zimmer also has at least one project that got away from him. "I built this massive, modular, 35-square-foot *Space Hulk* table (a WH40K game that takes place on a derelict spaceship). It was comprised of 88 individual pieces. The rooms and passageways could be arranged in different layouts (like a big puzzle). It was insane."

Terrain Making Tips

Here are some words of wisdom on getting into making tabletop terrain, gleaned from my own terrain-making experience and that of Patten and Zimmer.

Start a collection of plastic and metal containers and packaging rescued from the trash. Save stuff that looks even mildly interesting.

Keep a "bitz box," a collection of smaller parts, pieces from old scale models, jewelry, junked toys, wargame miniatures, etc. Buy any and all models at yard sales. It may be a car or a plane or goofy *Battlefield Earth* toy *now*, but repurposed and painted properly, it can become almost *anything*.

Learn the way of the blueboard. Blueboard (aka insulation foam, or pink insulation foam) is used in construction. You can get it at any home/building supply place. It is, literally, the ground upon which most terrain projects sit. You can cut it with a hot wire, sculpt it, sand it, paint it, and turn it into nearly any shape you want. Take a few pieces of this board, tape them together, cover with glue and dyed-green sawdust, and you've got yourself a gaming board.

Get some specialty tools. Besides the usual building and hobby tools, you'll want a hot-wire foam cutter (you can get these cheaply at a craft store), a hot glue gun, and a Dremel rotary tool.

One of Patten's secret ingredients for great terrain is Pre-Mixed Concrete Patch (found at building supply stores). "You can use it for adding texture, filling gaps in terrain pieces, and building up ground surfaces."

Zimmer is partial to kitty litter. "By mixing kitty litter with white glue, you get a chunky, spreadable cement mixture that makes incredible-looking rubble/debris when it dries. Be sure you buy the cheap clay litter."

Zimmer also recommends what's known as "granny grating" in the trade: "These lightweight plastic grids are actually designed for use in cross-stitching, but are excellent for making fencing and floor grating."

+ More resources at makezine.com/08/terrain.

Gareth Branwyn writes about the intersection of technology and culture and is a member of MAKE's Advisory Board.

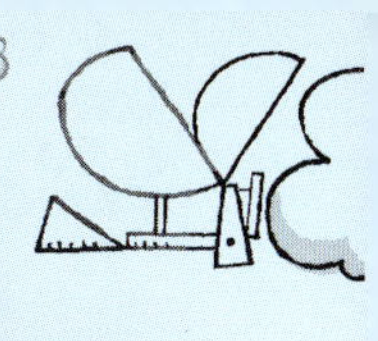

Image courtesy of Dean Johnson

Making Your Own Video Game

Microsoft's XNA Game Studio Express opens up game development.

By Dean Johnson

ALMOST ANYONE WHO'S EVER PICKED up a joystick or controller to play a video game has thought about creating a game. When I was very young, I drew out levels for Super Mario Brothers with crayons and construction paper. As I progressed through school, I taught myself a lot about programming. I knew I wanted to create games. The problem was that I was a one-man show. When video games first emerged around 1961 (the year Steve Russell created *Spacewar* while attending MIT), it was possible for a single programmer to create a simple yet engaging game. But now expectations are much higher. A professional game development company must hire dozens of people in order to create a single console game.

Where does this leave students and hobbyist developers like me? Behind a mountain of barriers. But there's good news: in late August, Microsoft released the beta version of XNA Game Studio Express, a framework for creating games. With very little knowledge of programming, you can create games for Windows (and, in later versions, for the Xbox 360).

Included in Game Studio is an updated 3D *Spacewar* game, along with its source code. Right away, you can pull apart the code to see how it works. Then, you can start modding the game. One of the first changes I made was to add triple shot to the ship; the ship can now send a wave of bullets toward the enemy. Having a full game right out of the gate allows you to see the results of your programming quickly.

After you play around with *Spacewar*, you might want to start creating your own games. I suggest starting with some of the classics. One of the first games I ever wrote was a clone of *Pong*. Creating a full game such as *Pong* will allow you to learn all of the aspects of programming necessary to create games, including graphics, input, sound, collision, and AI. After *Pong*, try creating other classics, such as *Arkanoid/Breakout*, *Asteroids*, *Space Invaders*, and *Tetris*. Once you've finished creating a few of the classics, begin working on your dream game.

To get started with XNA Game Studio, visit msdn.microsoft.com/xna and download your free copy. I also suggest signing up for the forums and posting any questions you might have. You can view some starter tutorials and samples at a community site I helped create: xnaspot.com.

Dean Johnson created his first game using BASIC on his parents' PC. He now attends Full Sail in Orlando, Fla., studying game development.

TOYS & GAMES

Pummer, Dude!

Part robotic plant life, part techno-sculpture, these desktop toys are easy and fun to make.

By Gareth Branwyn

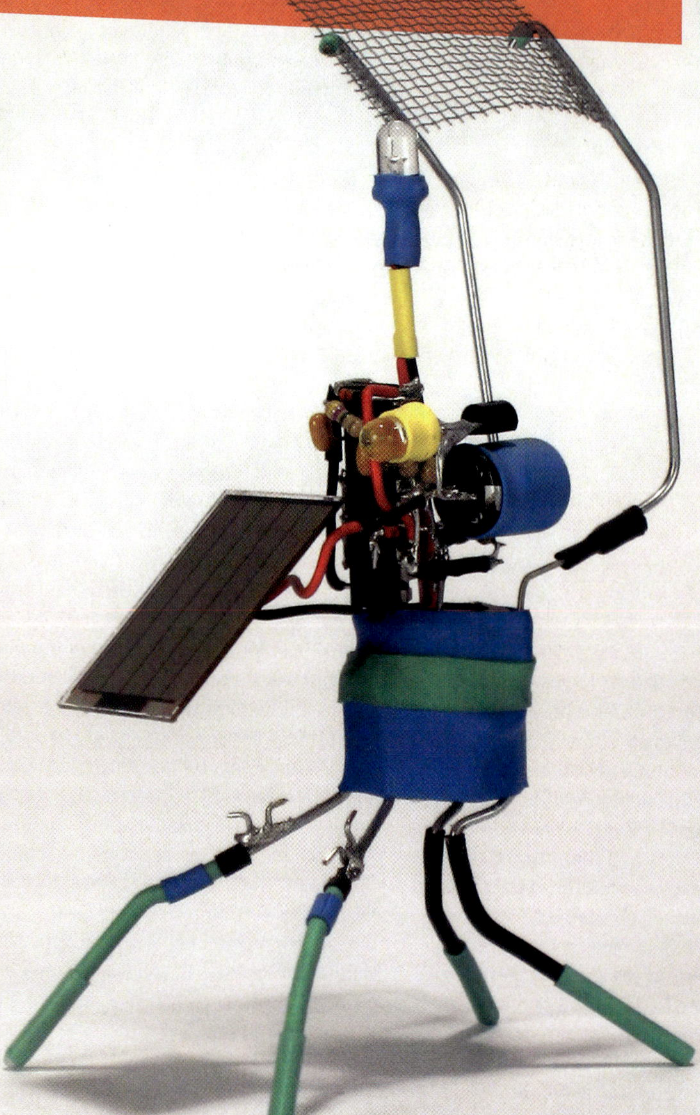

Photography by Zach DeBord/M27.com

IN MAKE VOLUME 06, I WROTE ABOUT BEAM, a branch of robotics built on low-end, mainly analog electronics that is inspired by biology. I described how to build two types of bots in the BEAM taxonomy: Solarrollers and Symets. One of the more obscure members of the BEAM family tree is the Sitter, an immobile robot with few or no moving parts.

One of my favorite types of Sitters is the Pummer, a nocturnal, robotic plant that soaks up the sun during the day; stores that energy in batteries or capacitors; and then, when it senses darkness, feeds power to a light which pulses, or "pumms," away in the dark. Since the electronics are simple and minimal, you can have fun with the design of your Pummer, creating a swanky piece of high-tech art that will intrigue everyone who sees it adorning your geekosphere.

MATERIALS

To build this Pummer circuit, you'll need:

Solar cell that can deliver 3V at 20mA (I recommend the SCC2433a from Solarbotics)
74AC240 Octal Inverting Buffer IC
AAA NiCad batteries (2) or you can use 10F "gold" capacitors (2)
0.22μF capacitors (2) often marked with "224" on the cap
1000μF capacitor or 3300μF for a longer fade-away
1kΩ resistor
4.7MΩ resistor
LED any color, high-intensity LED recommended
Diode A low-voltage type, such as the 1N5818 Schottky or a germanium diode, is best, but a silicon one works too.

How a Pummer Works

In "A Beginner's Guide to BEAM" (*MAKE, Volume 06, page 54*), we talked about different types of Solarengines (SE), which are simple power circuits for actuating miniature robots. We mentioned the nocturnal type of Solarengine. This is the variety of SE used in many Pummers. All SE circuits work in much the same way: the solar cell captures light energy, converts it to electrical energy, and sends it to storage, either in capacitors or rechargeable batteries. When a trigger value is reached, the stored energy gets sent off to do some sort of work. In a voltage-triggered SE, the trigger is a set voltage ceiling. In a nocturnal SE, the trigger is a threshold value of light.

Looking at the circuit diagram on the following page, you might be asking yourself: where is the sensor that tells the Pummer that it's lights out and time to get with the pummin'? Ingeniously, the solar cell and the circuit itself serve this purpose.

During the day, when light hits the cell and the cell is sending juice to storage, the diode in the circuit keeps the enable line set to high. When the level of light/current reaching the cell/circuit falls below a certain value (as set by the value of the parallel resistor), the enable goes low, triggering the discharge cycle and the pumming of the LED(s). The diode, being a sort of one-way valve in a circuit, prevents the current from flowing back into the charging part of the circuit; it has no place else to go but along the discharge path.

Pummer Circuits

There are a number of different Pummer circuits you can use, from simple ones that power a single LED, to more sophisticated ones designed to maximize power collecting and discharging, and ones that can power multiple LEDs. The one shown here, used in the Solarbotics Bicore Experimenters BCP Applications Project (see makezine.com/08/pummer), balances simplicity with circuit efficiency and bang-for-buck; i.e., it makes a pretty damn cool Pummer without too many building headaches.

This nocturnal SE circuit makes use of another hallmark BEAM circuit, the bicore, which is the basic "neuron" of BEAM "intelligence" (*see MAKE, Volume 06, page 54 and page 58*). Here, the two-state oscillator is used to create the flashing/pumming behavior. The C1 and C2 caps are used to set the blink/pause rates, and C3 handles the "decay" rate of the pumms. You can play around with these rates by trying different capacitor values on a breadboard.

Other Pummer circuits, including those that can handle multiple LEDs, can be found on Solarbotics.net, in */library/circuits*. Costa Rica BEAM (costaricabeam.solarbotics.net) has a fairly thorough library of schematics for Pummers, including a circuit for making a Type 1 Solarengine (which uses a 1381 voltage trigger) into a darkness-activated power circuit.

Pummer Designs

One of the cooler aspects of a Pummer is that, because it's a Sitter and has no moving parts and no concerns over weight, etc., the design and aesthetics of the robot can take center stage. You can build Pummers to look any way you want. A lot of builders, inspired by the idea of Pummers being a sort of robotic plant life, put the LED(s) on a long stalk or on multiple stalks. But Pummers have also been built in the shape of modern sculptures, hexagons, triangles, cubes, even a dragon with solar

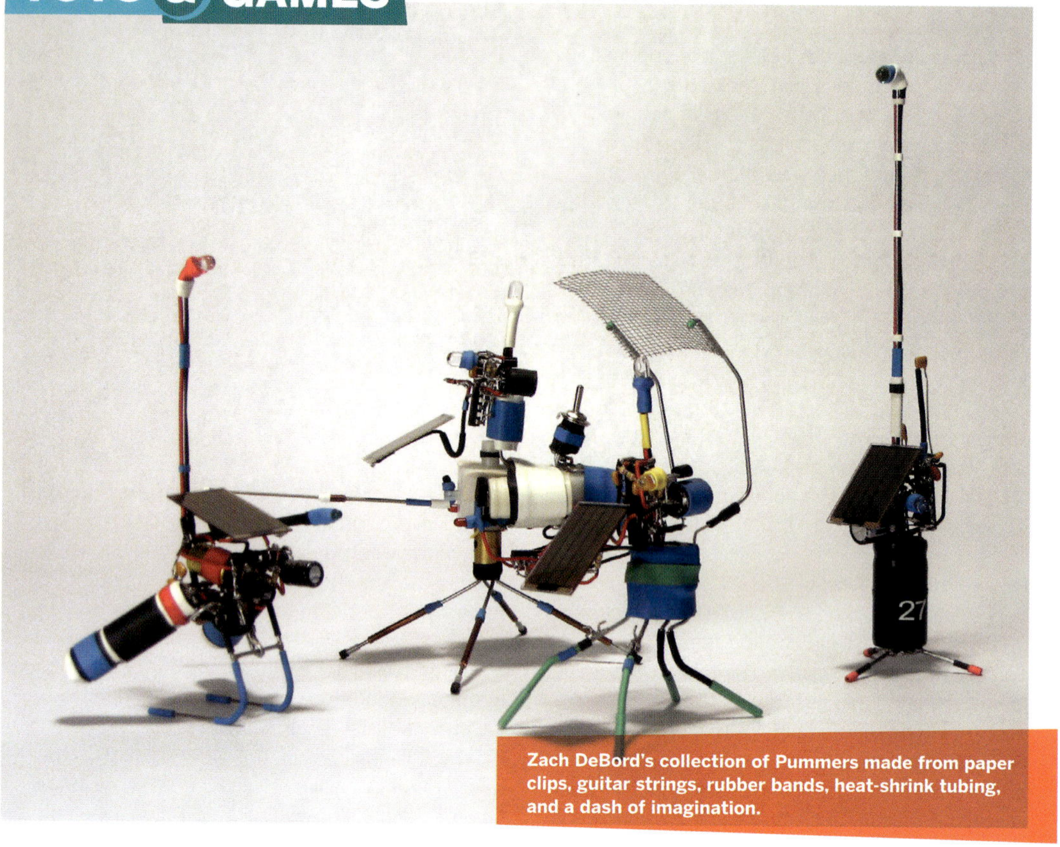

Zach DeBord's collection of Pummers made from paper clips, guitar strings, rubber bands, heat-shrink tubing, and a dash of imagination.

Single-LED High-Efficiency Pummer

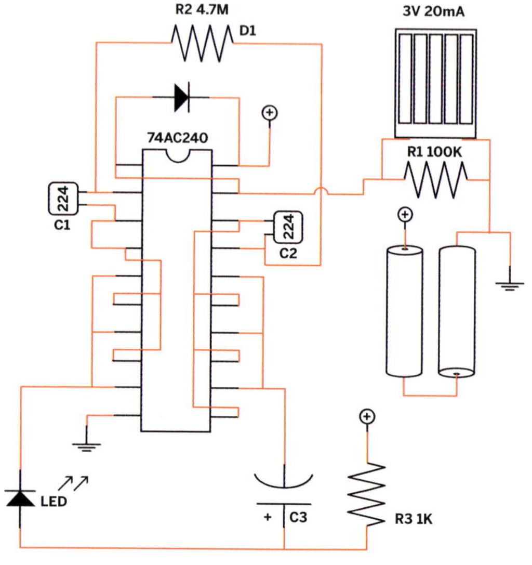

cells on the wings and glowing LED eyes. Really, your imagination and building skills are the only limitations.

A large majority of Pummers are built using paperclips as the main building component. Zach DeBord, a BEAM builder from Chicago (whose Pummers are pictured here) writes: "Buy a pack of jumbo and regular sized paper clips. For the $2 you spend, you'll be able to build a whole fleet of robots. I almost exclusively use paper clips and guitar strings for my creations."

Other common structural components are rubber bands and heat-shrink tubing. "An assortment pack of heat shrink (available at RadioShack and other places) goes a long way," says DeBord. "Not only are your bots more interesting looking, but you can use tubing in key places to reinforce weak joints."

For more Pummer resources visit makezine.com/08/pummer.

Gareth Branwyn writes about the intersection of technology and culture for *Wired* and other publications, and is a member of MAKE's Advisory Board. He is also "Cyborg-in-Chief" of Streettech.com.

Roachball Goes Open Source

In this fast-paced lunch-hour sport, changing the rules is part of the game. By Mike Kuniavsky

Illustration by Dustin Hostetler

JASON WILSON DIDN'T START OUT TRYING to invent an open source sport. While living in Brooklyn, he and some friends wanted an incentive to go outside during the cold months, so they formed a winter kickball league. One game day, only five people showed up (because it's winter in New York), which is not enough to make two kickball teams. Not wanting to give up on their original idea of getting out of the house, the group started wandering around Brooklyn. Walking through the snowy streets, they found a bocce ball court. Putting the court and the kickball together, they improvised a new game, which they called Roachball (because, well, it's Brooklyn).

They started playing standard kickball on the bocce court but quickly had to modify the traditional rules. The bocce court was too small for three bases, so they appropriated cricket's two-base system (but kept kickball's baseball-inspired system of strikes, outs, and innings). From soccer, they borrowed the idea of throwing the ball back in when it goes out of bounds, which led to another important change. As Wilson says, "you can [really] bean somebody" when you throw a ball in the close quarters of the bocce court, so they introduced rules from dodgeball.

Borrowing ideas and adapting the rules became part of playing the game. With spring, bocce season started and the courts were in use on the weekends, but the bank drive-through wasn't, so it became a legitimate court, and the rules changed again to accommodate it. As Wilson puts it, "It was never declared that it was going to have a collaborative rule-making process, but that's what it was."

Fast-forward a couple of years. Wilson moved to Portland, Ore., was working long hours and blowing off steam playing Roachball with his co-workers. They noticed the resemblance of the collaborative, improvisational nature of open source software to Roachball, and decided to rename it OSBall (pronounced "ossball" or "ozball"). They created an OSS-like structure for modifying the game.

Wilson is OSBall's project leader and serves as the editor of the rules wiki (osball.org). Although the attitude is lighthearted, he sees OSBall as a serious alternative sport, with advantages over other sports.

"The challenge," he says, "is motivating enough people to play existing ball games." OSBall can be played with a smaller number of people than other sports, so it's easier to get a group together. It's fast-paced, so you can play a couple of innings over lunch. It's a little dangerous because of the close quarters ("But not too bad," Wilson says). And, like skateboarding or parkour (a freestyle urban-navigation activity), the rules encompass available terrain, so you don't need formal playfields, and it can be played in crowded cities.

OSBall players are encouraged to wear funny costumes during a game.

In addition, the OSBall team hopes to take the ideas beyond a ball game. They're planning a library of ideas for collaborative, improvisational activities they've had, and creating "a packet of stuff that you can use for something totally different."

Mike Kuniavsky is a San Francisco- and Portland-based ubiquitous computing and user-experience consultant and writer who blogs at orangecone.com.

OSBall: Game Play and Rules

OSBall borrows from existing sports. From cricket it borrows the two-base system of running and scoring. From soccer it borrows the idea that hands can only be used while throwing into the court from the sidelines. From kickball it borrows the notion of a pitcher who rolls the ball to a kicker. From dodgeball it borrows the ability to get someone "out" by hitting them with the ball. From bocce ball it borrows the court.

- Two teams of 3 players.
- Play a set number of innings (3 innings is a good introductory length, 6 is a fairly intense game, 9 is tournament length).
- Flip for offense or defense choice.
- All players must be in the court at all times.
  - ➡ Defense team can position anywhere, but it's advised to have one pitcher, one catcher, and one outfielder (1 team member at each end and 1 in the center).
  - ➡ Offense team congregates around home base.
- The ball is pitched, hands OK.
- Kicked ball must remain inside court; if it goes out this is a strike.
- If the kicked ball goes out of bounds and hasn't touched the ground yet (airball) a defense player can catch it (outside the court) for a catch-out.
  - ➡ 3 strikes you are out.
  - ➡ 3 outs = ½ inning (switch offense to defense).
- Kicker runs to other base, avoiding any contact with the ball.
- Physical blocking of the runner is prohibited.
- Runner may not step outside the court, and may not step on top of the court's retaining wall.
- Defense players can exit the court during play only if the ball has already left the court.
  - ➡ Defense players can use their hands while outside the court to throw the ball at runners.
- If any body part is touching the base "end" wall (home base or other base) runner is safe.
- The runner may decide to return to home base in the same turn (home run), but this is purely optional.
- When the second player kicks, the first runner has the option to stay put (2 runners can share a base, but 3 cannot, as this would cause a lack of a kicker).
- A point is scored for every player returning home before the moment of the third out.

Photograph by Jason Wilson

Make: Projects

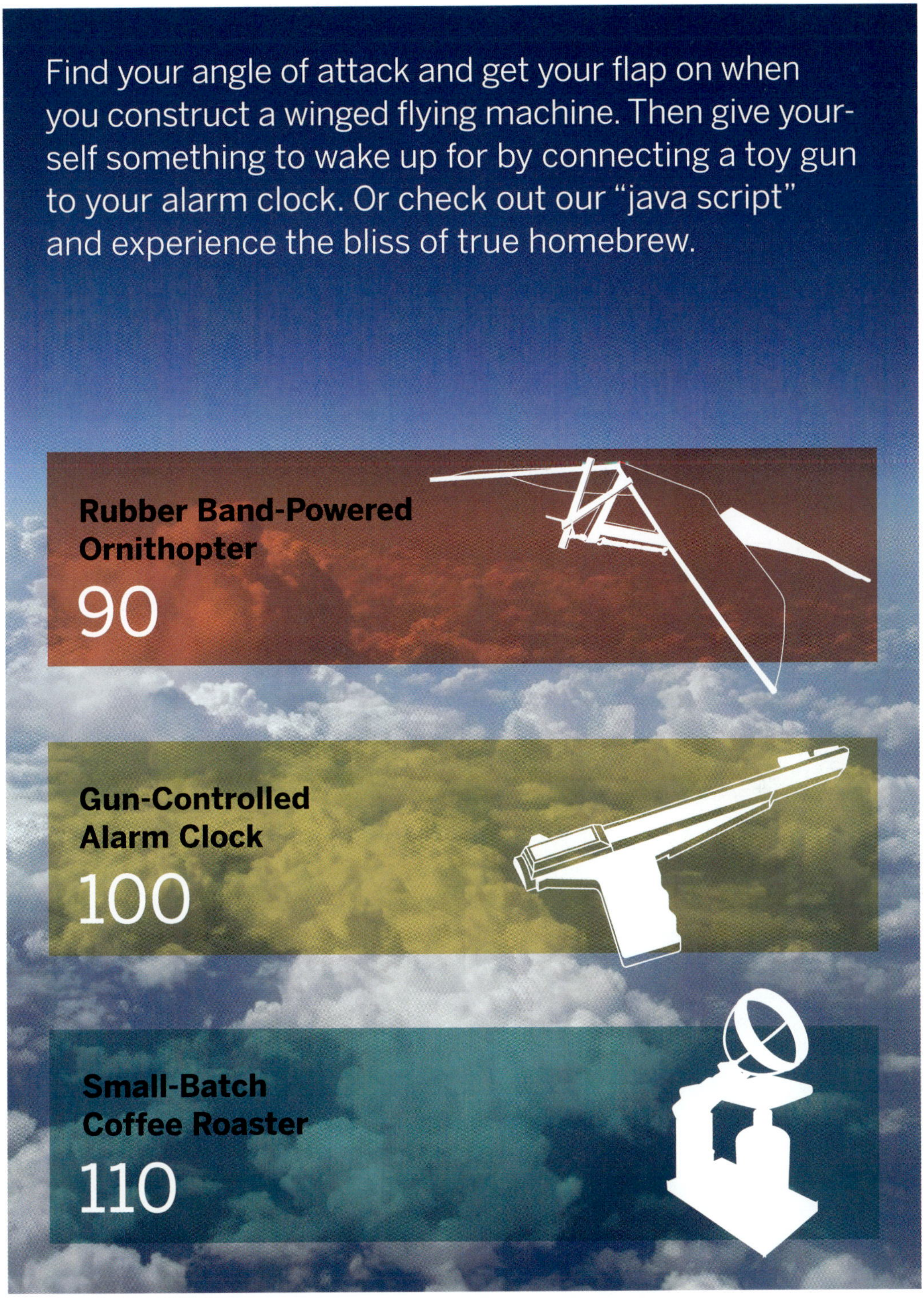

Find your angle of attack and get your flap on when you construct a winged flying machine. Then give yourself something to wake up for by connecting a toy gun to your alarm clock. Or check out our "java script" and experience the bliss of true homebrew.

Rubber Band-Powered Ornithopter
90

Gun-Controlled Alarm Clock
100

Small-Batch Coffee Roaster
110

BUILDING AN ORNITHOPTER

By William Gurstelle

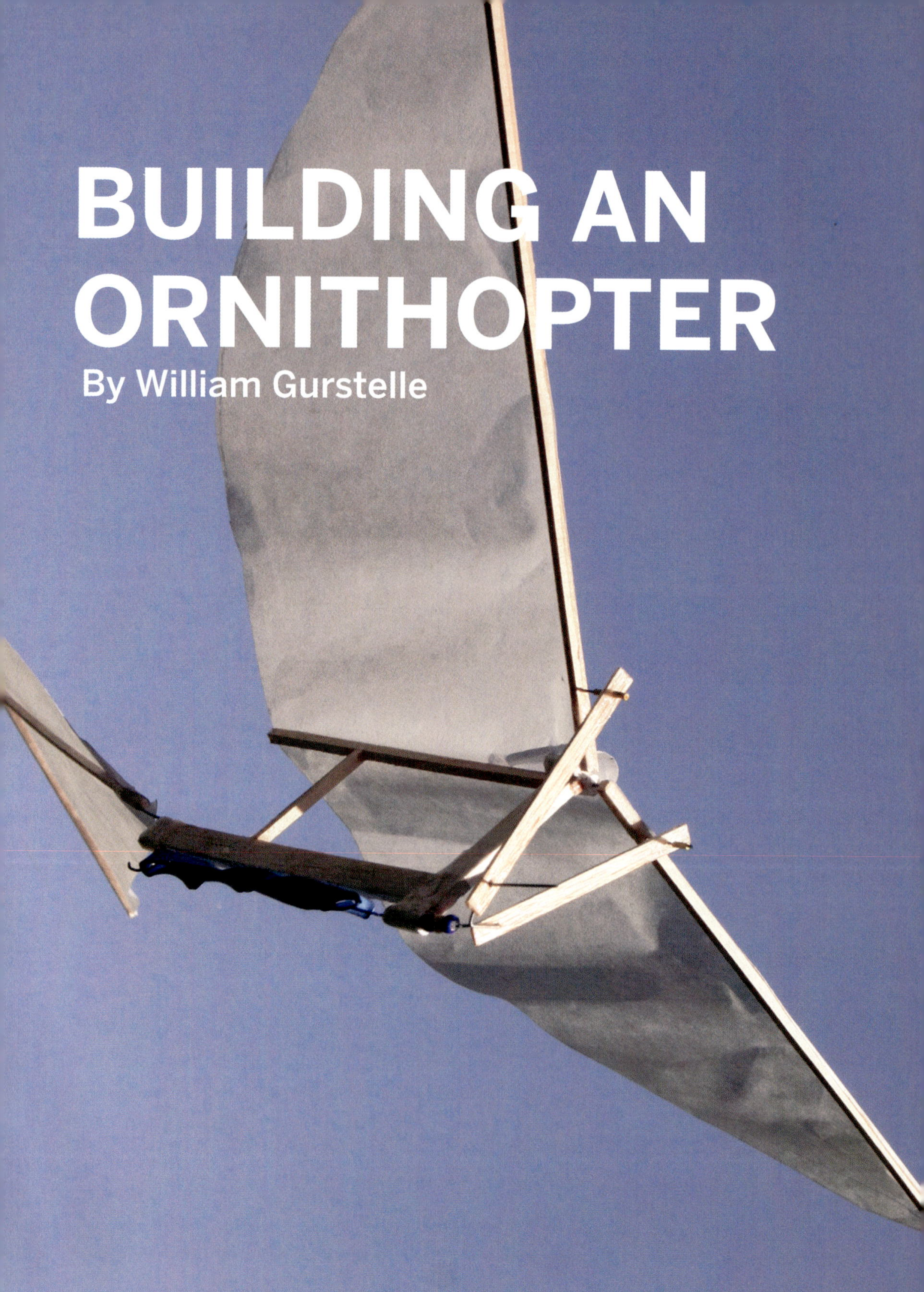

FLIP, FLAP, FLY

For millennia, men and women have studied birds, bats, and beetles, observing and experimenting, attempting to determine what humans must do to fly by flapping.

But people can't fly by flapping: not with wings covering their arms; not with pedaled, chain-driven wings; and, so far, not with internal-combustion engines, either. Nonetheless, the concept of manned ornithopters continues to hover on the periphery of aeronautical engineering. This project shows you how to build a small, rubber band-powered ornithopter we call Orly.

There are many types of ornithopter designs. Orly is a simple monoplane, meaning there is a single wing mounted above the motor-stick, and its motion is similar to a bird in flight.

Photograph by Gerry Arrington

Set up: p.94 Make it: p.95 Use it: p.99

William Gurstelle is a contributing editor of MAKE and is the author of *Backyard Balistics* and *Adventures from the Technology Underground*.

HOW DO ORNITHOPTERS FLY?

Get up on the downstroke.

How do ornithopters fly? According to Nathan Chronister of the online Ornithopter Zone, "The ornithopter wing is attached to the body at a slight angle, which is called the angle of attack. The downward stroke of the wing deflects air downward and backward, generating lift and thrust.

"Also, the wing surface is flexible. This causes the wing to flex to the correct angle of attack we need in order to produce the forces that we want to achieve flight."

The mechanics of flapping flight are far more complicated than that of fixed-wing flight. For an aircraft with fixed wings, only forward motion is necessary to induce aerodynamic lift. But for flapping flight, the wing not only has to have a forward motion, but also must travel up and down. This additional dimension means the wing constantly changes shape during flight.

Illustration by Timmy Kucynda

FROM ICARUS TO ORLY: A SHORT HISTORY OF ORNITHOPTERS

Without doubt, even the earliest humans watched birds fly past and felt, well, rather envious. Thus when Thag, a Pleistocene caveman, looked up and saw flocks of ducks and geese soaring above, he might have gathered together a few palm fronds, lashed them around his arms with a vine, and leapt off a tree. Poor Thag never got airborne, or at least he didn't live to record the episode in petroglyphs on his cave wall.

Later, from ancient Greece, comes the legend of Daedalus and Icarus. Daedalus was a skilled engineer who angered King Minos. Minos ordered him imprisoned in a tower.

According to *Bulfinch's Mythology*, "Daedalus contrived to make his escape from his prison, but could not leave the island by sea, as the king kept strict watch on all the vessels.

"So he set to work to fabricate wings for himself and his young son Icarus. He wrought feathers together, beginning with the smallest and adding larger, so as to form an increasing surface. The larger ones he secured with thread and the smaller with wax, and gave the whole a gentle curvature like the wings of a bird."

Unfortunately for Daedalus, the attempt at flight didn't entirely work. His son, Icarus, flew too near the sun, melting the wax that held the wings together. Icarus fell out of the sky and drowned in the ocean.

To a large extent, that's been the typical outcome of human flapping flight experiments, right up to modern times.

The New York Times has run many stories over the years:

» "Ornithopter Somersaults — Captain White Hurt in Crash" (June 1928)
» "Inventor Tries to Soar Like a Bird; Narrowly Escapes Drowning" (March 1932)
» "100,000 See French Birdman Die in 9,000 Foot Fall" (May 1956)
» "[University researchers] named their ornithopter 'Mr. Bill' after the perpetually maimed character on the television show *Saturday Night Live*." (May 1992)

Around 1490, Leonardo da Vinci was carefully studying the mechanics of avian flight. From his bird-watching came perhaps the first blueprint for a human-carrying ornithopter. More of a theoretician than a true maker, Leonardo never got his flying machine off the paper in his notebooks. Had it been built, it likely would not have flown. Still, experts say his design is clever, and embodies modern aerodynamic principles developed hundreds of years later.

Interest in flapping flight took off again in the 1870s. Building model ornithopters became fashionable in Europe, and a number of enthusiasts — among them Alphonse Penaud, Hureau de Villeneuve, and Gustave Trouvé — built internally powered birds that soared over the fields of France and Flanders. Soon, ornithopters powered by rubber bands, gasoline, electricity, and even gunpowder were flapping away, but as scale models, not people-carrying aircraft.

Since then, many people have tried to build a manned ornithopter, but none have yet succeeded. There are unconfirmed reports that the Germans made one during World War II and that the Soviets flew one during the Cold War, but solid evidence is lacking. Today the University of Toronto is making a game attempt.

Why bother with ornithopters at all? Because flappers can do things other aircraft cannot. They probably have the best maneuverability of any aircraft. Unlike fixed-wing drones, ornithopters, at least in theory, can stop and hover like a hummingbird, which makes them extremely versatile, and they need less space to maneuver than a helicopter. Couple all that with their ability to fly at very slow speeds, and ornithopters may be the perfect surveillance vehicles. The military applications for unmanned ornithopters are numerous.

Ornithopters have practical applications in civilian life, as well. For instance, the Colorado Division of Wildlife uses an ornithopter to research a hard-to-capture endangered species called the Gunnison sage grouse. This skittish bird flies away at the first sign of danger but will stay on the ground if it sees a hawk flying above. So state biologists use a motorized, radio-controlled ornithopter painted like a hawk to keep the flighty grouse on the ground long enough for them to capture it.

SET UP.

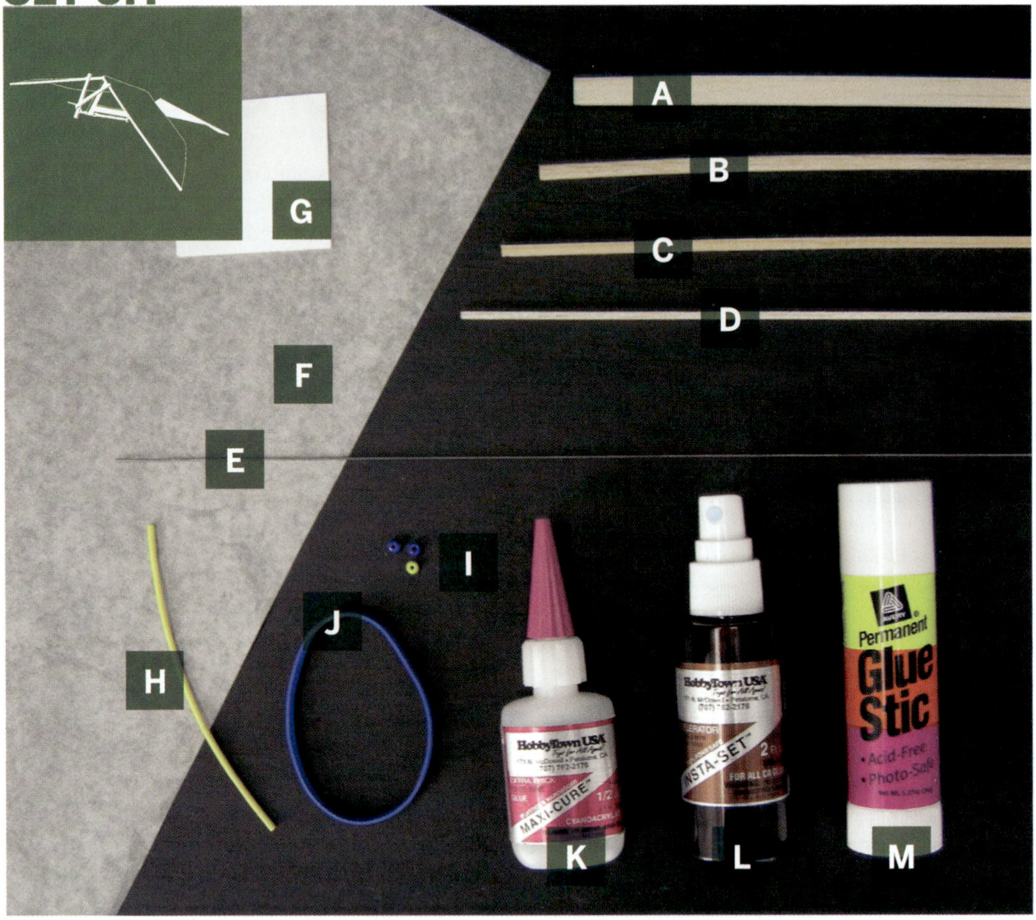

MATERIALS

[A] Balsa rod 5⁄16"×1⁄8", 7 inches Cut into lengths of:
- 5" motor stick
- 1½" front vertical connector

[B] Balsa rod 3⁄16"×1⁄16", 6 inches Cut into lengths of:
- 2¾" connecting rods (2)

[C] Balsa rod 1⁄8" square, 24 inches
Cut into lengths of:
- 8" wing spars (2)
- 5" top wing attachment member
- 1½" back vertical connector
- ½" crank standoff

[D] Balsa rod 3⁄32" square, 14 inches
Cut into lengths of:
- 7" tail members (2)

[E] Music wire .032" diameter, 10 inches
Thicker wire is too heavy and can adversely affect Orly's performance.
Cut into lengths of:
- 3" tail/rear motor attachment wire
- 2½" crank/front motor attachment wire
- 2" wing spar wires (2)

[F] Sheet of tissue paper, about 18"×18" Made specifically for modeling, about .04 ounces per 100 square inches. Regular tissue paper is comparatively heavy.

[G] Square of 16 lb. paper, 2"×2"

[H] Heat-shrinkable tubing, 1⁄16" diameter Cut into 3 pieces each 1⁄8" long

[I] Small beads (2) With inside diameter just large enough to accommodate the .032" diameter music wire. From bead or craft stores.

[J] Model airplane rubber 12" long Tied in a loop to make a big rubber band. From hobby shops.

[K] Cyanoacrylate (CA) glue

[L] Cyanoacrylate drying accelerator

[M] Glue stick

Vegetable oil (not shown)

TOOLS
(not shown)

Needle-nose pliers (2)

Utility knife To cut the balsa wood to size

Ruler

Scissors

Photography by Matthew Dalton and Ty Nowotny

MAKE IT.

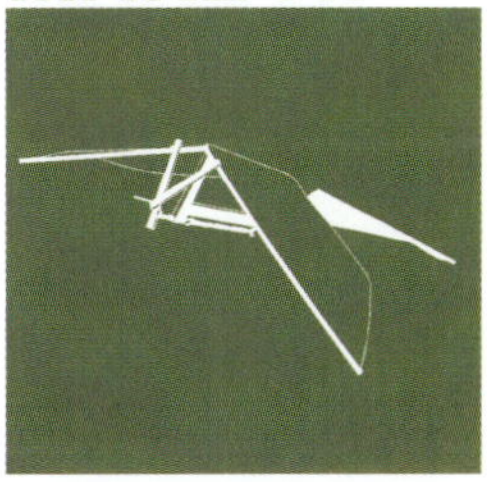

BUILDING ORLY THE ORNITHOPTER

START Time: A Day Complexity: Medium

1. MAKE THE FUSELAGE

1a. Form a hook in the tail/rear motor attachment wire as shown. Carefully push the wire through the center of the motor stick at a point ⅜" from the tall end. Then make two 90-degree bends in the wire as shown, and glue into place using CA adhesive. Reinforce the wire-to-balsa joint by placing a tissue paper cover over it. Dab the joint with a thin layer of CA. Spraying CA drying accelerator on the joint makes the process faster and less messy.

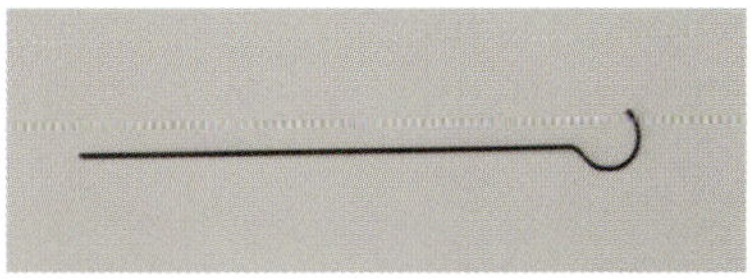

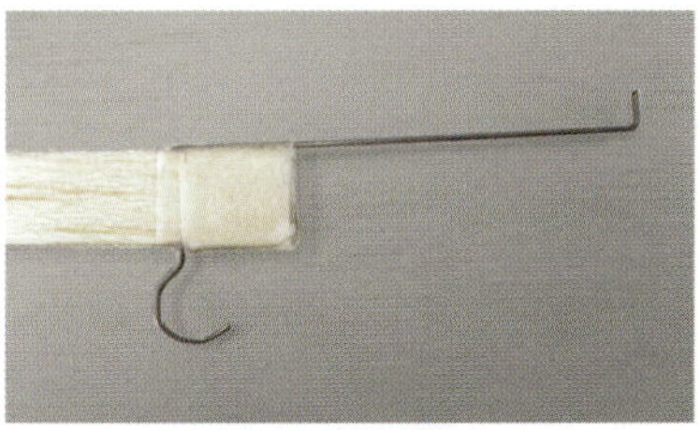

1b. Glue the fuselage together as shown.

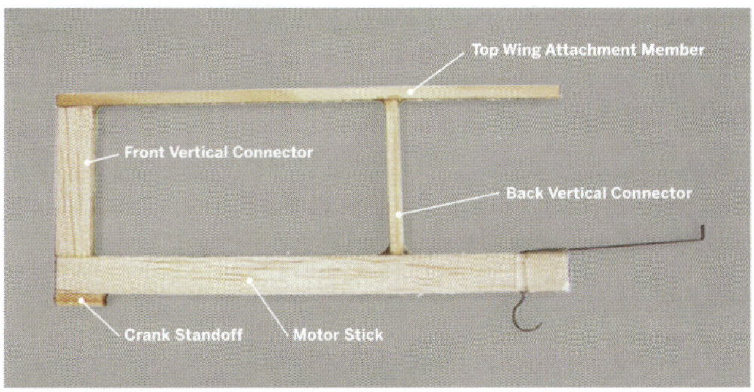

1c. Roll the 2"×2" paper into a narrow tube using the music wire for a mandrel. Remove the music wire and carefully daub the tube with CA, taking care to maintain the tube's openings. Spray with CA accelerator. Cut into three ½" long tubes and discard the remainder.

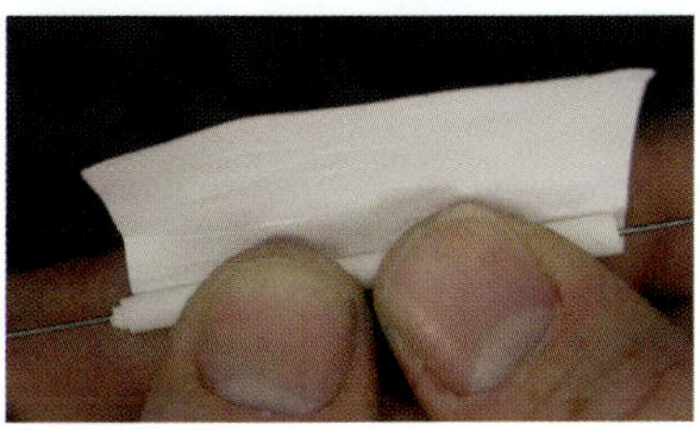

1d. Attach the 3 tubes to the fuselage as shown, with CA and accelerator. Make certain the tubes are aligned with the long axis of the fuselage.

1e. Use the needlenose pliers to bend the wire so the crank appears as shown. Insert the crank wire through the paper tube glued to the crank standoff. Place the 2 beads on the wire. Create a bend in the back end of the wire to serve as the motor hook.

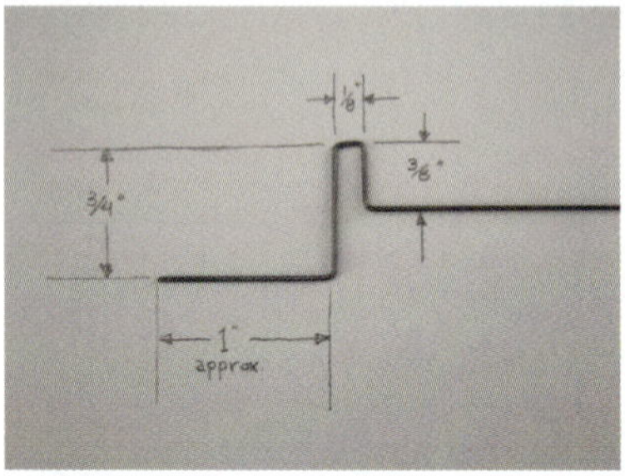

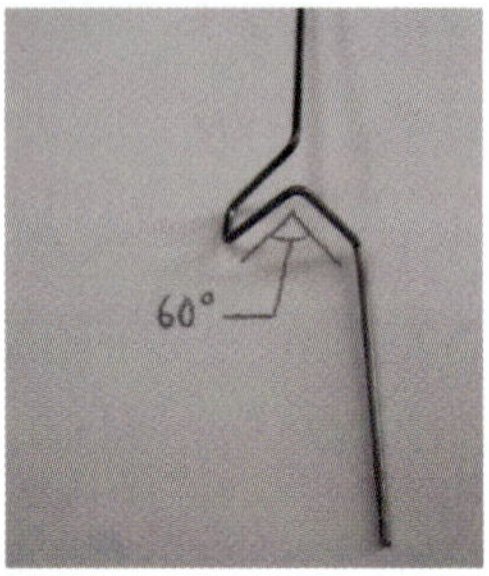

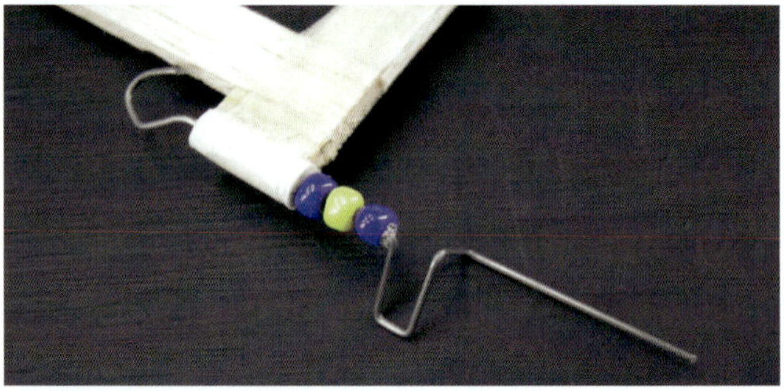

2. MAKE THE WING SPARS

2a. Bend the music wire as shown. Carefully push the wire through the wing spar at a point ¾" from one end. Glue into place using CA. Reinforce the joint by wrapping a layer of tissue paper around the joint and coating with CA.

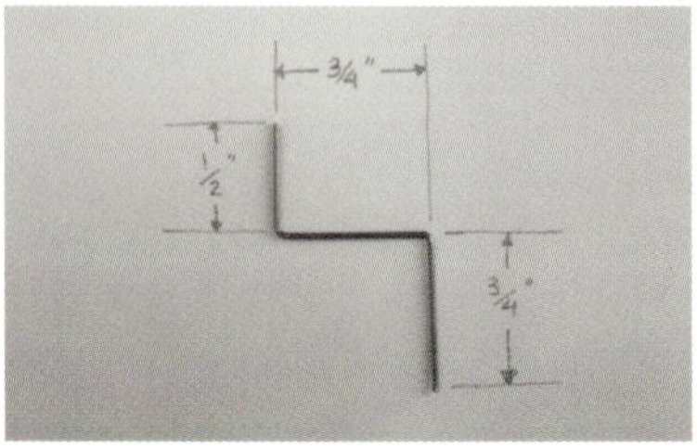

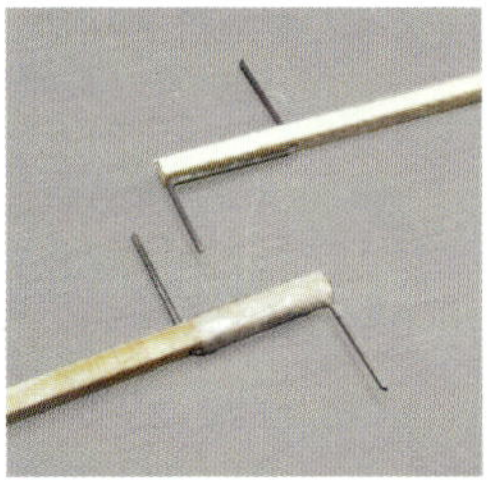

3. MAKE THE TAIL

3a. Use CA adhesive to glue the balsa rods into a T shape. Reinforce the joint by covering it with tissue paper soaked with a thin layer of CA.

3b. Poke the end of the fuselage tail attachment wire into the balsa tail member; then glue the assembly with CA. Reinforce by wrapping with tissue and CA.

4. MAKE THE CONNECTING RODS (CONRODS)

4a. The conrods undergo considerable stress. Harden the rods by coating the last ½" of each end with CA.

4b. Make 2 holes for the music wire in each conrod, ¼" from each end.

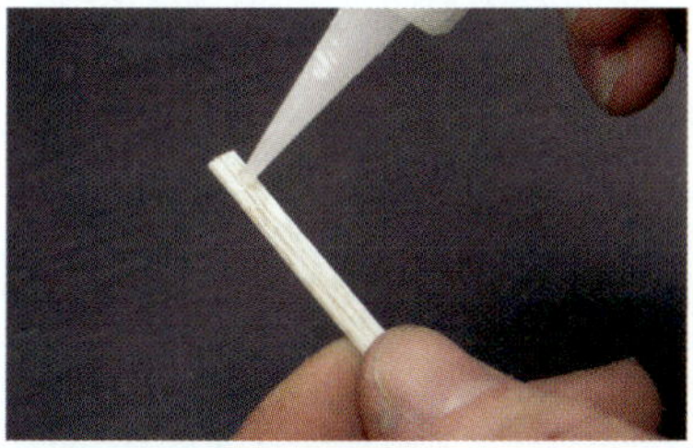

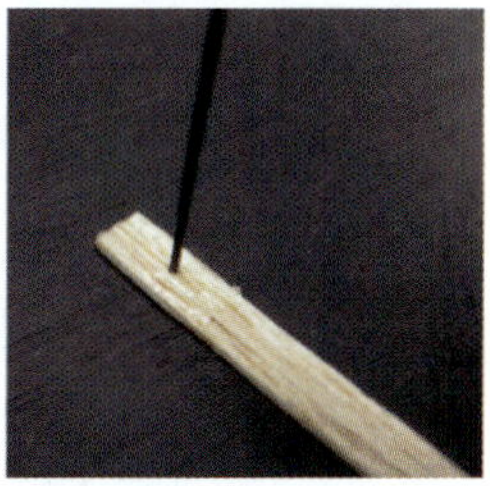

5. FINAL ASSEMBLY

5a. Cut out the tissue paper wings and tail using the templates.

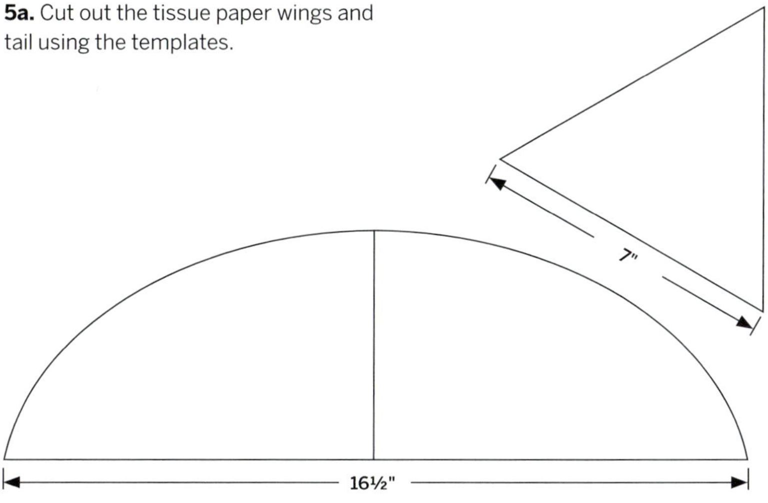

NOTE: Full-scale templates are posted at makezine.com/08/orly.

5b. Glue the tissue paper wing to the wing spars and the top wing attachment member. Glue the tissue paper tail to the balsa T frame with a glue stick.

5c. Connect the conrods to the wing-spar attachment wires and the crank. Adjust the spacing of the conrods so the crank turns smoothly. Place heat-shrinkable tubing over the crank and wing spar wires to maintain alignment, and carefully heat with a match to shrink the tubing.

WARNING: The tissue paper, the balsa wood, and the CA catch fire easily! Use great care in this step.

5d. Bend the tail up so it is at about 15 degrees from the plane of the motor stick.

6. SENDING ORLY AIRBORNE

6a. Double the rubber band and place it over the front and rear motor attachment hooks.

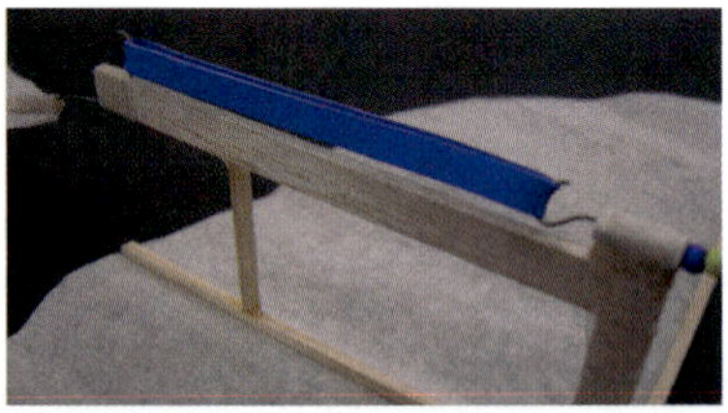

NOTE: To accommodate a longer rubber band, double it into 2 loops and place it over the front and rear attachment hooks.

6b. Rub the band with a little vegetable oil for lubrication.

6c. Wind up the rubber band motor by turning the crank at least 35 turns.

6d. Angle Orly's nose up slightly and release gently.

FINISH X

NOW GO USE IT »

USE IT.

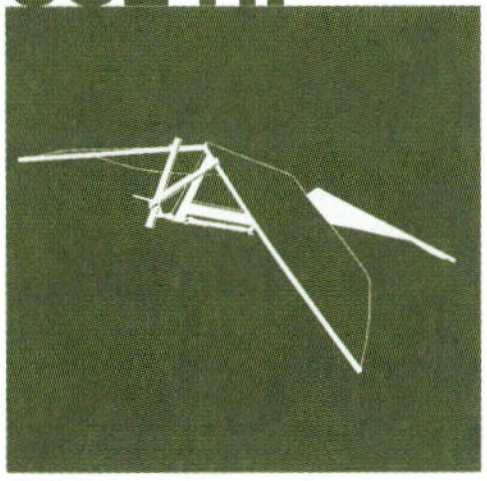

FLY, BE FREE!

TROUBLESHOOTING

Ornithopters can be difficult to fly. Common problems include stalling, nosediving, and veering, in various combinations. If your ornithopter doesn't fly well, try the following:

1. If the ornithopter dives and veers, winding the motor in the opposite direction may help.

2. Balance is important. Make certain the action of each wing is the same. Make the conrods and crank carefully to ensure balanced wing operation.

3. If Orly has a tendency to flip or roll in flight, you need to improve your craft's stability. Try lengthening the distance between the motor stick and the tail, or adding a rudder (a vertical stabilizing surface on the tail).

4. If the ornithopter veers consistently in one direction and then nose-dives, add a small wire weight to the end of the wing tip on the side opposite the direction of the veer.

5. The angle of the tail is important. Bend it slightly up if the flapper nose-dives, and bend it down if it stalls.

6. If Orly goes through a series of stalls before ultimately diving into the ground, your tail may be mismatched to the rest of the aircraft. Fix this by decreasing the size of the tail. If that doesn't help, extend the length of the tail boom — that is, increase the distance between the wing flappers and the tail.

7. A direct head first plunge to the ground may be a signal to increase the size of the tail stabilizer.

8. If your ornithopter flaps vigorously but won't gain altitude but lowers into the ground tail first, try to move the center of gravity forward. It is best to make the rear lighter instead of making the front heavier.

9. Bank and spiraling problems are common in ornithopters, and can be tough to correct. If your ornithopter starts out with few good looking flaps, but suddenly banks or rotates around its longitudinal axis and then spirals down, try the following:

• Reapply the tissue paper to the wings, making sure the paper is not applied too loosely or too tightly stretched. Both wings should have the same amount of tension.

• Bend a small rotation in the tail plane relative to the longitudinal axis of the ornithopter.

• Add a small weight to the outside of the wingspar opposite the direction of the bank.

Fixing the bank and spiral problem can be difficult. You may need to try a number of fixes in combination before the problem clears.

EXPERIMENT!

You may be able to extend the duration of flight by making a few changes to Orly's design. For instance, you can experiment with rubber bands of differing lengths and thickness. You can also vary the shape, sweep angle, and size of the wing.

Sometimes, adding wing gussets made from a thin piece of transparent tape improves performance. Finally, experiment with the length and width of the tail.

RESOURCES

There is an active community of ornithopter enthusiasts, with a lively online forum at ornithopter.org. This excellent site also provides plans, kits, motorized models, and advice.

For the latest news on the University of Toronto manned ornithopter project, visit ornithopter.net.

GUN-OPERATED ALARM CLOCK

By Roger Ibars

Photography by Roger Ibars

KILLING TIME

Hack a retro gaming light gun with some tilt switches to control a vintage digital clock radio. After the alarm wakes you up, you can grab the gun and kill it off. Isn't that what you've always dreamed of doing?

Clock radios are everyday hated devices that designers seem to ignore, judging by how little their features and user interfaces have changed. This project adds a new capability, letting would-be sleepyheads enjoy a human-machine interaction of a different sort. Don't worry, we're not going to connect your clock to an MP3 player and play mellow New Age sounds. We're going to shoot the alarm off. With a gun. Wake-up time is now payback time.

We'll base the project around a digital clock radio and a light gun for gaming; huge selections of both of these are available inexpensively second-hand, with many beautiful and well-designed examples. To enable our FPSI (First Person Shooter Interface), we'll outfit the gun with five tilt sensors, arranged at different angles on a small circuit board. A cable tethers the gun to the clock and carries your tilt and trigger signals to the clock's time and alarm control button contacts.

Set up: p.103 Make it: p.104 Use it: p.109

Roger Ibars lives and works as an Interaction Designer in London. He is interested in how people understand technology and how technology understands people. See more of his work at **selfmadeobjects.net**.

IMPLEMENTING GUN CONTROL

Our control circuit uses tilt switches to detect the gun's position.

Tilt switches contain a metal ball or a conductive liquid (such as mercury) that rolls inside a small capsule. When the switch's contacts point downward, the conductor bridges the contacts, closing the switch. Turn the switch upside down, and the conductor falls away from the contacts, opening the connection.

The tilt switches are arranged in a plane perpendicular to the gun's barrel, to detect the approximate rotation of the handle when the gun is aimed forward (its "roll" in aeronautical terms). The gun doesn't need to point at the clock to work (but it's more satisfying that way).

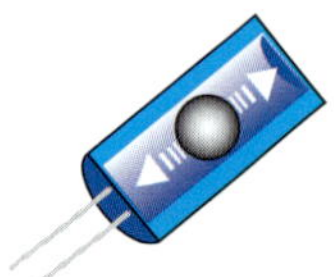

Tilt switch: A rolling ball (or some mercury) makes or breaks the connection.

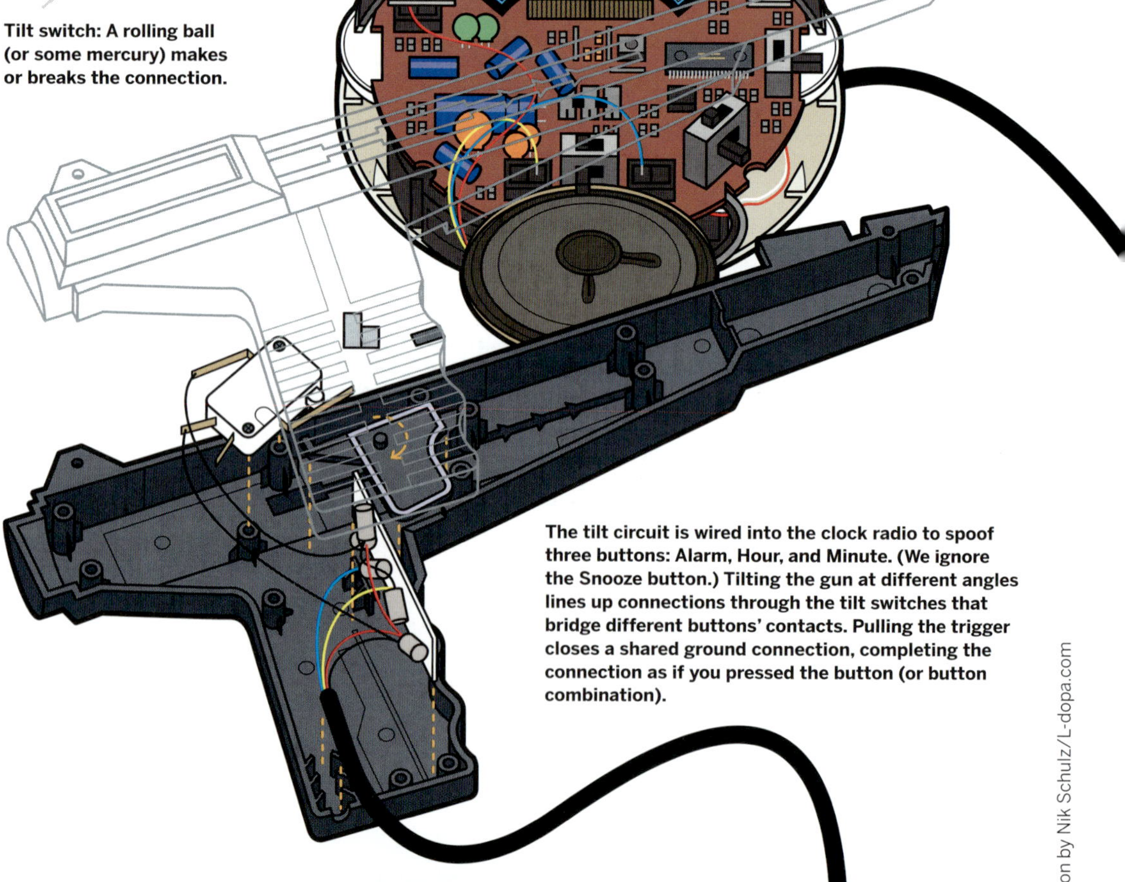

The tilt circuit is wired into the clock radio to spoof three buttons: Alarm, Hour, and Minute. (We ignore the Snooze button.) Tilting the gun at different angles lines up connections through the tilt switches that bridge different buttons' contacts. Pulling the trigger closes a shared ground connection, completing the connection as if you pressed the button (or button combination).

Illustration by Nik Schulz/L-dopa.com

WARNING: Do not work on the alarm clock while it is plugged in, or you risk serious and possibly fatal electric shock.

SET UP.

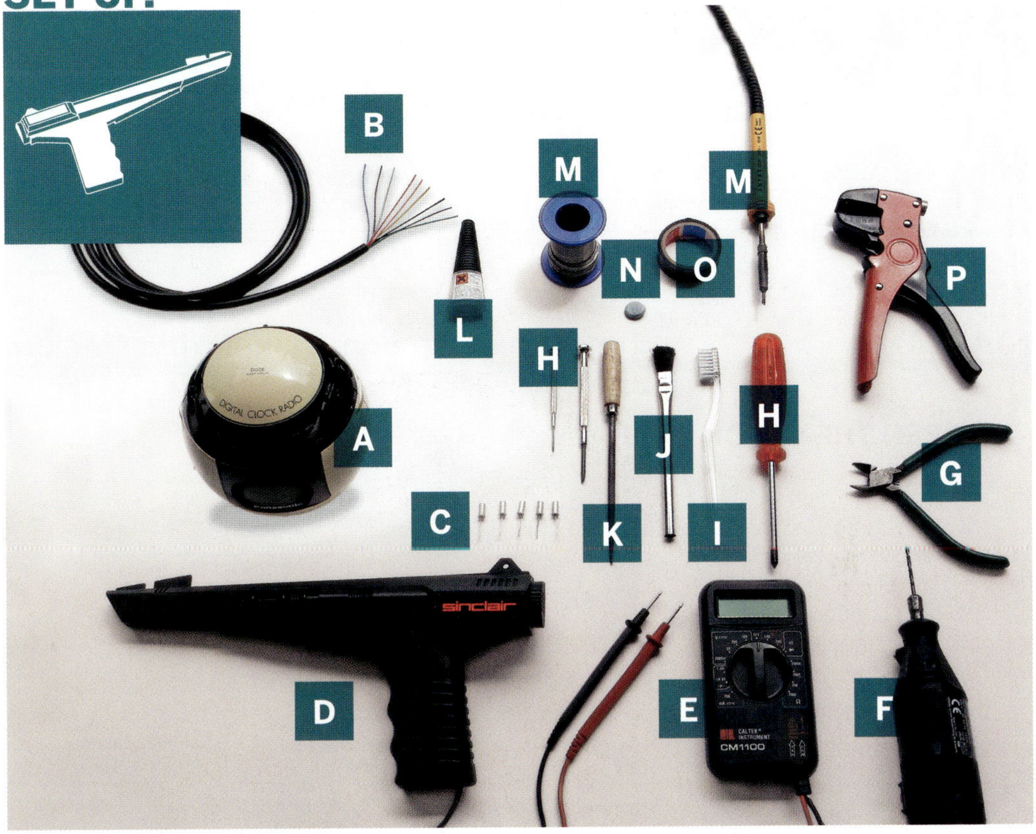

MATERIALS

[A] Digital clock radio Almost all of these operate the same way and will work. If you go retro, don't go too far back; 15 years old is a good limit. Make sure all the buttons, the alarm, and the screen still function. Check for cracks in the case, and inspect the underside to see how much the color of the plastic has yellowed or faded. Finally, play with the radio's volume to make sure it doesn't sound like frying eggs. I chose a vintage, spherical Panasonic RC-70.

[B] Multi-way cable with at least four conductors, about 5' long I used a fancy SCART cable consisting of 9 color-coded, stranded wires wrapped in a grounded, metallic Mylar screen, all sheathed in black PVC to an overall diameter of 6mm. This top-quality choice can be found in hi-fi stores and will give you a nice curvature of the cable.

[C] Small tilt switches (5) I recommend non-mercury switches for environmental reasons. These cost about $1 each, and are available from electronics suppliers such as Farnell (farnell.com), Newark InOne (newark.com), Rapid (rapidonline.com), and RS (rswww.com). For models with just one lead, the case works as the other contact.

[D] Light gun Many are available secondhand. I particularly like the Nintendo Zapper, the SEGA Light Phaser, the Atari G1, the Konami Justifier, and the Sinclair Magnum (my choice). Make sure there's enough space inside the grip to fit the circuit that we are going to build.

Small perfboard Available from electronics suppliers, including RadioShack (not shown).

TOOLS

[E] Multimeter

[F] Rotary tool and bits

[G] Side cutter

[H] Screwdrivers

[I] Toothbrush

[J] Artist paintbrush

[K] Round metal file, no more than 5mm thick

[L] Strong glue (or glue gun and hot glue)

[M] Soldering equipment

[N] Poster putty

[O] Electrical tape

[P] Wire stripper

[NOT SHOWN]

Spray plastic polish The kind used to shine your car dashboard works perfectly.

Kitchen soap

Cotton cloth

MAKE IT.

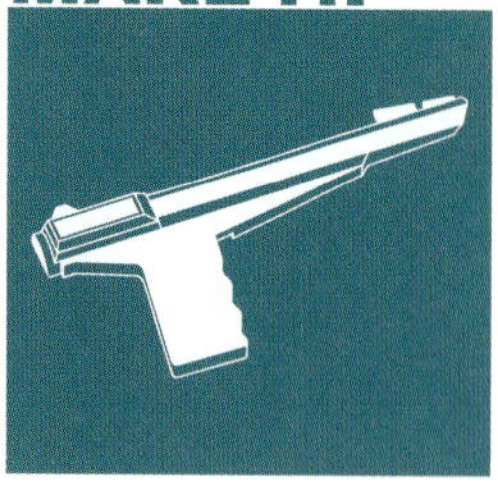

BUILD YOUR GUN ALARM CLOCK

START **Time: An Afternoon Complexity: Medium**

1. OPEN THE CLOCK RADIO

1a. Unplug the clock radio.

1b. Disassemble the case. Carefully unscrew the four screws concealed by wells on the back, and separate the electronics from the plastic parts. To avoid damaging the plastic, choose a screwdriver that fits well, press down firmly, and turn slowly.

Marvel at the design details and quality! The shell is extremely well-crafted, with neat cavities for the screws and an elegant, raised icon for the power cord. The controls even resemble a face. This clock is full of design generosity, which is quite rare nowadays.

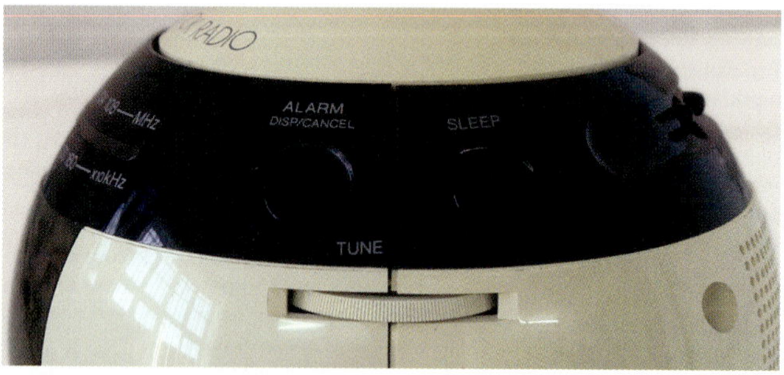

1c. Untie the electric cable, and separate the electronic parts from the plastic shell.

Remember exactly how you took apart the electronics block and untied the electric cable, so you can put it all back together later.

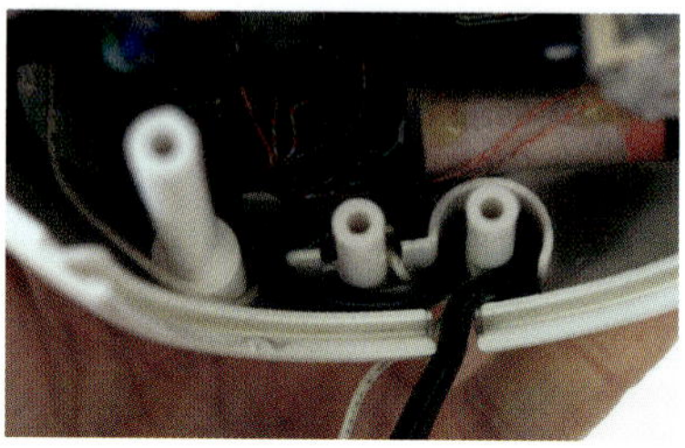

2. CLEAN AND SHINE IT

2a. Use the paintbrush to clean the dust accumulated on the electronics.

2b. Clean the plastic case with the toothbrush and kitchen soap. The bristles won't damage or scratch the surface. Take your time to enjoy cleaning every corner! Let all pieces dry thoroughly.

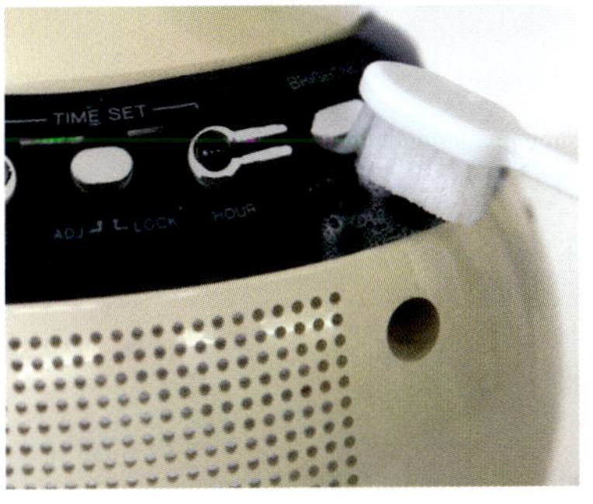

NOTE: After the pieces dry, spray them with plastic polish, following the instructions on the can.

Wait a few minutes after spraying, then polish the surfaces with a clean cotton cloth.

You'll be amazed how new they'll look! Now your alarm clock is ready for more serious work.

3. DRILL THE HOLES

3a. On the front half of the plastic shell, measure and mark a point to drill near the base and centered below the display. Protect the surface around the hole area with electrical tape, in case your drill skips away from its proper destination.

3b. Find a drill bit that's a bit thinner than your multi-conductor cable and drill the hole.

NOTE: This is the riskiest part of the work: drilling holes for the cable through the clock's case. (I see some vintage collectors out there raising their hands in objection.)

3c. Widen the hole with the round file. Keep widening slowly until the cable can just go through the hole with a bit of pressure. We don't want a hole bigger than the cable.

3d. Drill and widen a second hole in the plastic shell about 1" to the left and slightly back from the first hole, in the interior wall that holds the battery case (see picture below right). The cable will enter the first hole and make a sharp left turn into this hidden hole as it routes its wires around to the back of the clock.

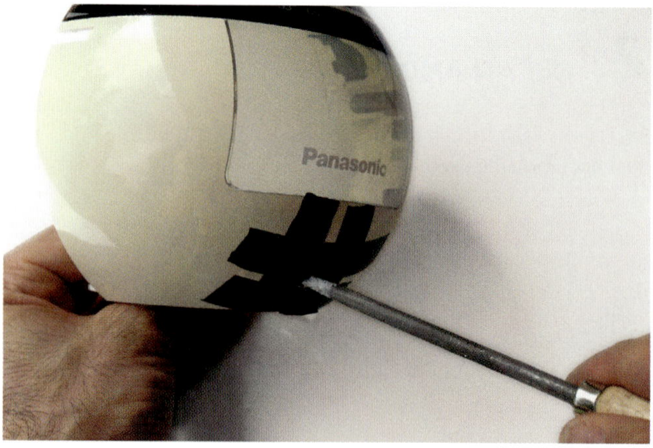

4. ATTACH THE CABLE

4a. Strip about 16" of sheathing off of each end of the cable, to reveal the color-coded wires inside. Pass one end of the cable through the holes you just drilled.

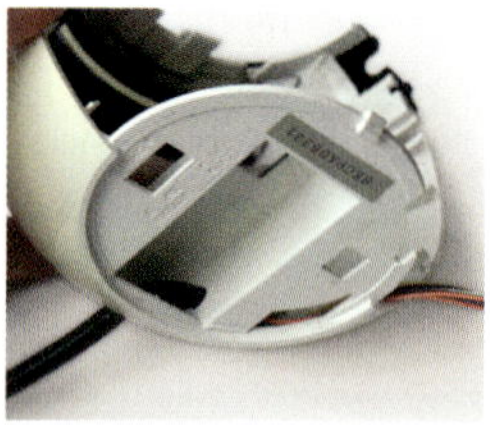

4b. Use hot glue or another strong adhesive to fix the cable firmly in place, keeping the individual wires free inside the main compartment. Leave it neat and strong since this cable is going to be used.

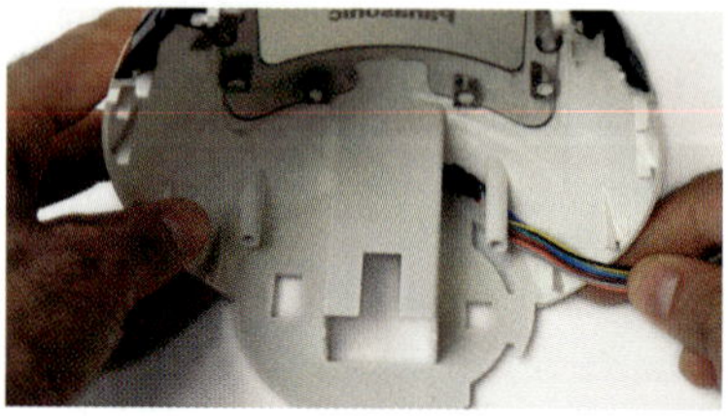

4c. Put the electronics block back into the front half of the case. Now we are ready to hard-wire the color cables to the clock switches.

5. HARD-WIRE THE CLOCK

If you aren't modding a Panasonic RC-70, don't worry, because almost 99% of digital clocks work the same way. Just remember that each button has 2 connections, which come into contact with each other when you press the button. You want to extend these connections so that, instead of closing the circuit with the button, you'll be using the light gun.

5a. Find the buttons that control the essential functions: set time and alarm (hours and minutes), and alarm off. The RC-70 uses just 3 buttons for these: Alarm, Hour, and Minute. The Alarm button does double duty, shutting off the alarm and switching the Hour and Minute buttons from "time-set mode" (the default) to "alarm-set." Some clock radios use a switch instead of a button to change between time-set and alarm-set modes.

5b. Locate the essential buttons' contacts, 2 per button, and use a multimeter to follow each of them out to a solderable connection point. With the RC-70, all 3 buttons shared a common ground, so I needed to find a total of 4 connection points.

5c. Solder wires from your cable to the contact points, using the color coding to track what goes where. Following convention, I connected the cable's black wire to the shared ground, and designated colors for the 3 button-specific connections.

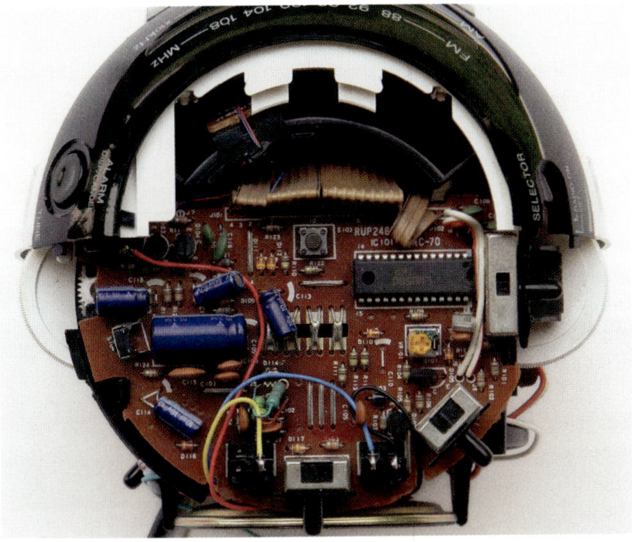

5d. Organize and fix all the color cables between the free spaces of the clock. Trim extraneous cables to get them out of the way. Avoid placing cables near parts that get warm, such as the power converter.

5e. Reassemble the clock and screw it back together.

5f. Test the clock by plugging it in and touching together the other ends of the cable wires you just soldered. Confirm that bridging the wires mimics the functions of the buttons you connected them to.

If you didn't make any big mistakes, everything should work! Don't worry about touching the wires going to the buttons with your fingers, because the current going through them is very low. Once it's working, unplug it again.

6. HARD-WIRE THE LIGHT GUN

6a. Take the light gun apart, and clean and shine it up the way you did with the clock radio in Step 2.

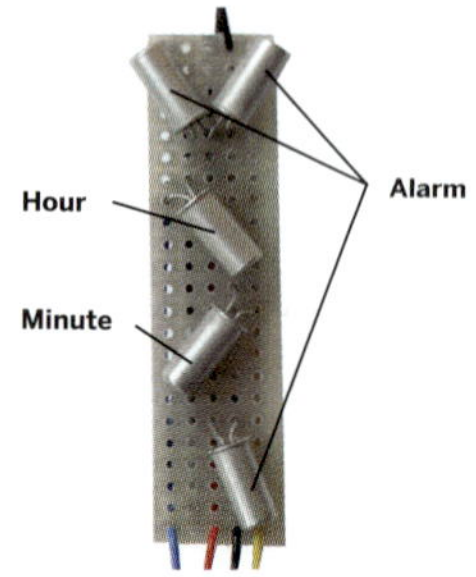

6b. Cut a rectangle of perfboard that's big enough to carry the 5 tilt switches flat, but narrow enough to fit facing backward inside the handle. For my Sinclair Magnum, my board was about ⅝"×2⅜".

6c. If your clock radio is an RC-70 or work-alike, arrange the tilt switches on the perfboard as shown at right. Leave some extra room around each switch to let you bend and fine-tune their positions later.

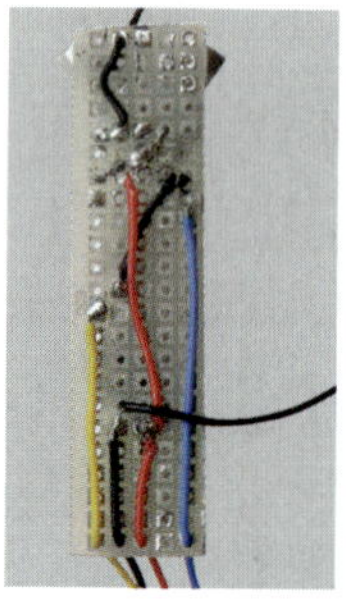

6d. Following the wiring diagram online (makezine.com/08/alarmgun), solder the tilt switches into place and connect them to the cable wires. Don't worry about which sides you connect; the switches are functionally symmetrical.

6e. Connect the ground to each sensor section and route it through the trigger switch contacts in the gun. To make my circuit more readable, I routed all wiring from the cable and trigger through the front of the board, at the edges. I also split 1 ground coming from the trigger and connected it in 2 places.

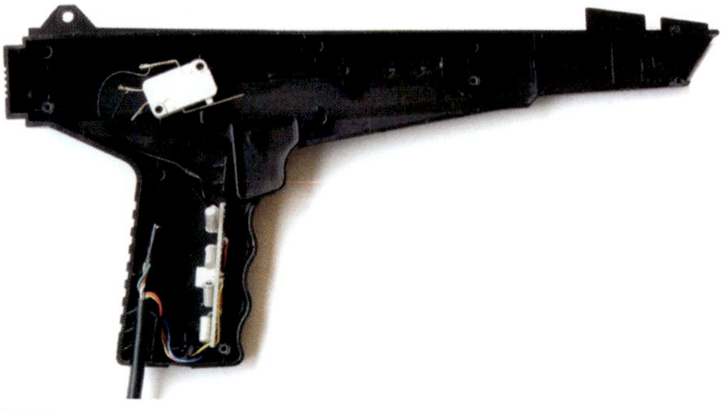

6f. Use poster putty to hold the tilt circuit in position inside the gun-half with the trigger. Plug in the clock radio, and test the interface by tilting and firing. Experiment with different firing angles, and bend the tilt switches around to refine their operation.

6g. When the tilt switches work together properly, glue them in place, and then glue the circuit board into the gun handle. Re-assemble the gun.

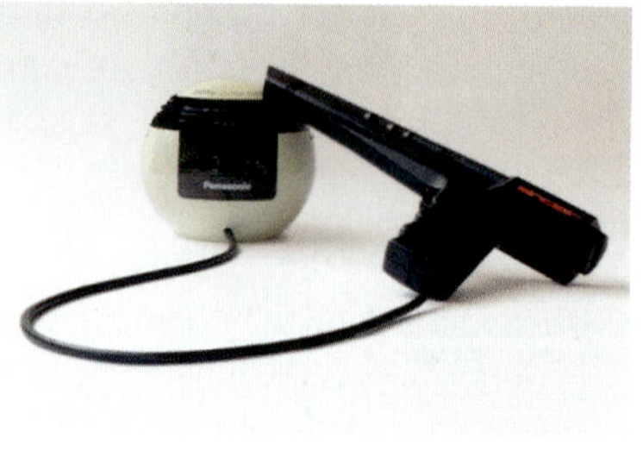

6h. Set the time, set the alarm to +1 minute, wait a minute, and FIRE!

Now send a photo of your design to roger.ibars@gmail.com, and I'll send you back an exclusive preview of my latest projects.

FINISH X

NOW GO USE IT »

USE IT.

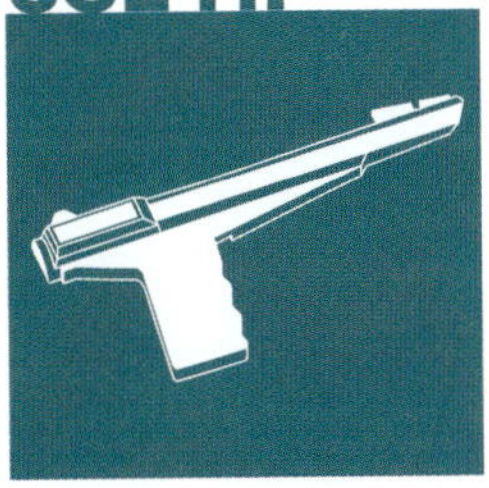

DO YOU FEEL LUCKY? WELL, DO YA CLOCK?

OPERATION

The Panasonic RC-70 has a switch that locks the time set and disables the Hour and Minute buttons from changing the time. For our gun to work, this switch needs to be in the unlocked position.

To use this new hard-wired device, imagine that the clock is a circle of degrees (not difficult with a spherical clock). A straight up-and-down shot kills the alarm. Tilting left sets hours, and tilting right sets minutes. Small tilts set the time, and big tilts, below the horizon, set the alarm.

+ To see how tilt sensors work together, see makezine.com/08/alarmgun.

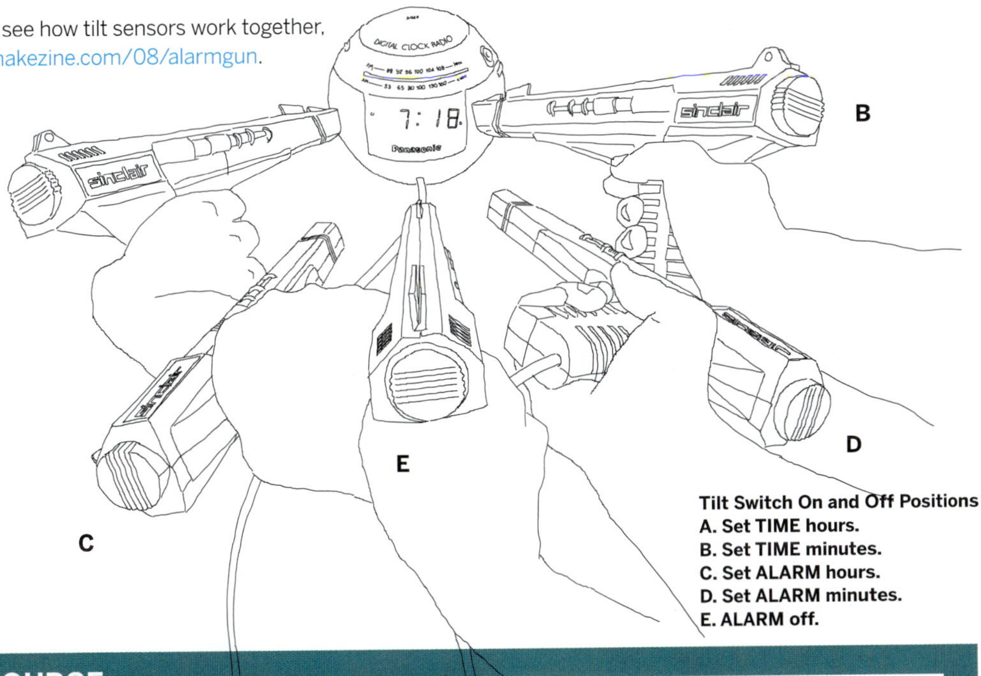

Tilt Switch On and Off Positions
A. Set TIME hours.
B. Set TIME minutes.
C. Set ALARM hours.
D. Set ALARM minutes.
E. ALARM off.

SOURCE

HARD-WIRED DEVICES BY ROGER IBARS

This project comes from my Hard-wired Devices collection, which pays tribute to great consumer electronics designs from the 70s and 80s. They are remanufactured vintage devices that blend two cultures of interface: computer games and household appliance design.

In these pieces joysticks, light guns, and controller pads from the classic age of computer games connect actively via cable to digital alarm clocks and other iconic devices.

You can see the full collection of hard-wired devices at selfmadeobjects.net.

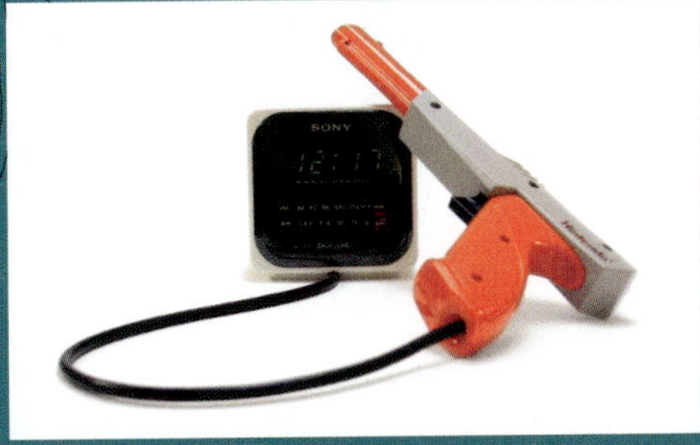

Illustration by Roger Ibars

COFFEE ROASTER

By Larry Cotton

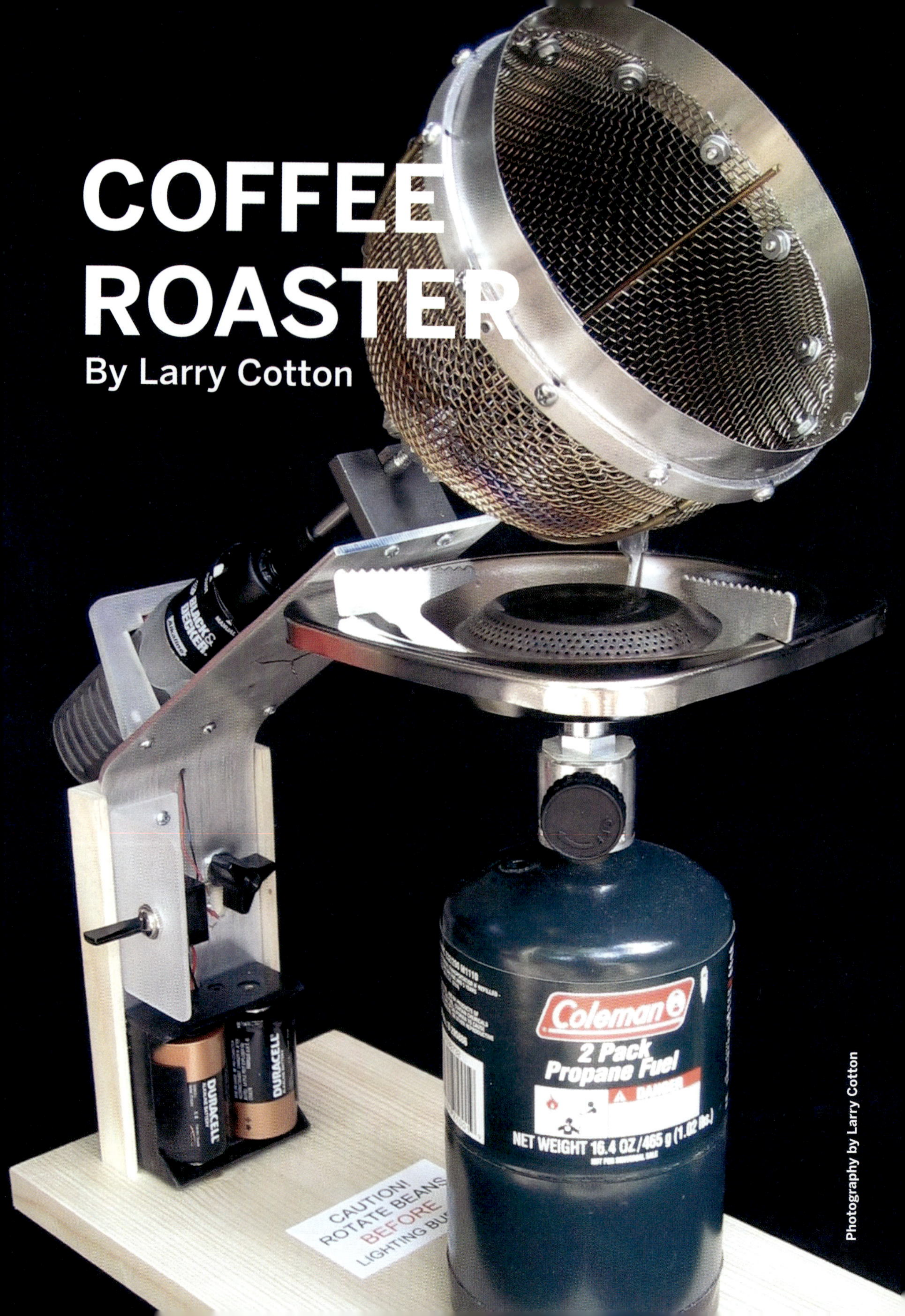

Photography by Larry Cotton

THE NIRVANA MACHINE

Lots of folks think that quaffing a cup of coffee from boutique beans comes close to nirvana, but roasting your own beans will bring you even closer. That's why I call this roaster the Nirvana Machine.

I didn't drink coffee for most of my life, and I even survived without it in the Navy. But when my son introduced me to a cup of legendary West Coast java (OK, Peet's), I began to understand what all the fuss was about. Soon, I too became fussy about excellent coffee.

Beans lose flavor after they're roasted because all of those delicious but volatile aromatic compounds dissipate and break down. Roasting your own beans guarantees ultimate freshness, putting all the flavor into the cup; you will not drink a fresher, more satisfying cup of coffee!

Home roasting is easy and inexpensive. Top-quality green beans cost less than roasted beans and have a much longer shelf life. You can have fun fine-tuning your roast's darkness and developing your own blends. Indeed, a growing selection of countertop roasting appliances are now sold, but they're pricey, non-portable, and very noisy.

Set up: p.113 **Make it:** p.114 **Use it:** p.121

Larry Cotton is a retired power-tool engineer, musician, part-time math teacher, and full-time coffee devotee who lives in eastern North Carolina.

JAVA IMPLEMENTATION

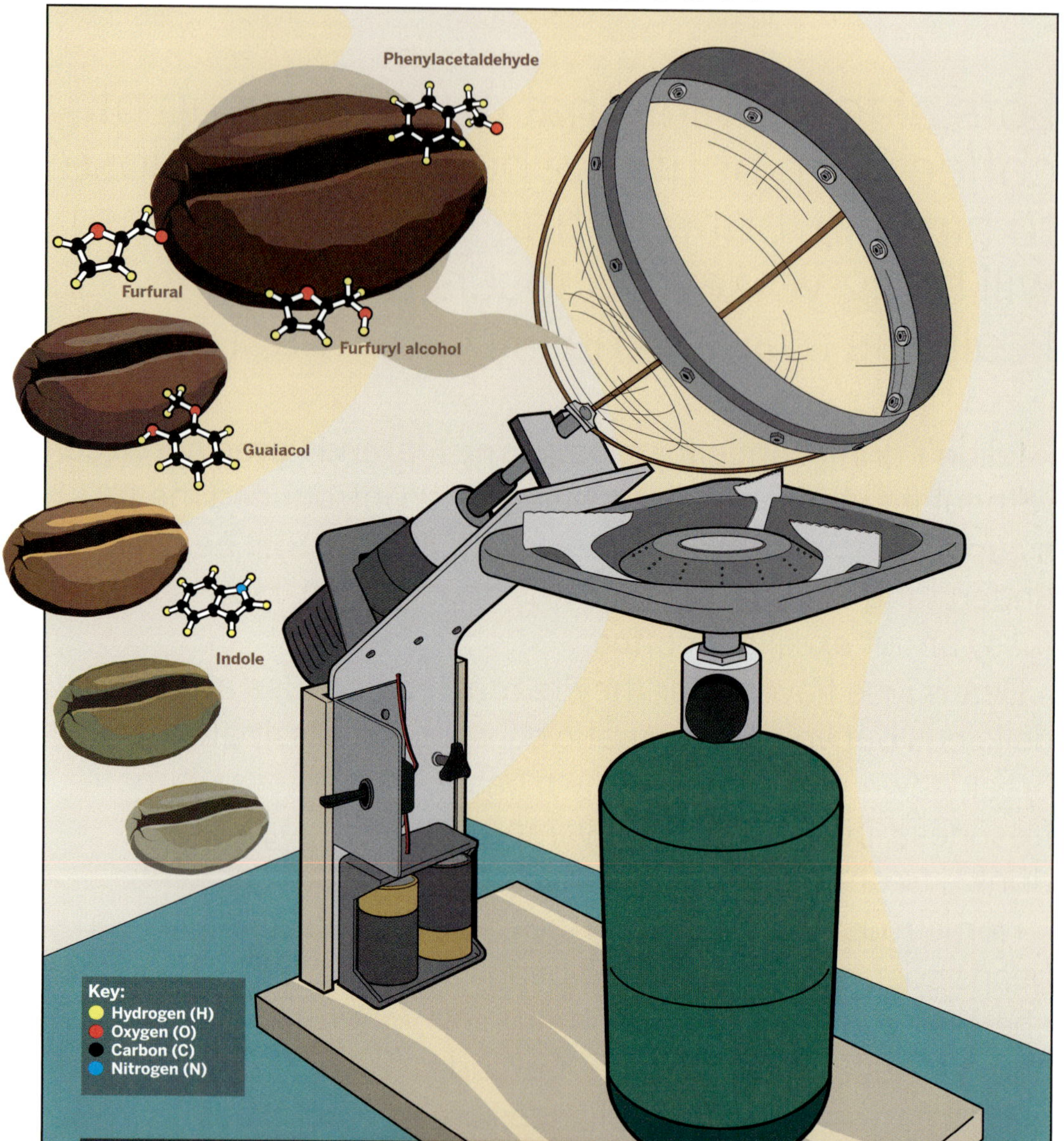

The Nirvana Machine uses a battery screwdriver to tumble beans over a camp stove. Unlike commercial machines, the basket is tilted and open-ended, which makes it easier to load the beans and to see and smell their progress during roasting.

When you roast coffee, sugars caramelize, water vaporizes and makes the beans expand, and aromatic compounds are produced, including the ones shown above. The compounds develop at different rates, so varying the roasting time and temperature lets you balance which you want more or less of, to determine the resulting flavor.

The Version 1 basket was a cheap strainer, but the beans broke through that and caused a fire!

Illustrations by Tim Lillis

SET UP.

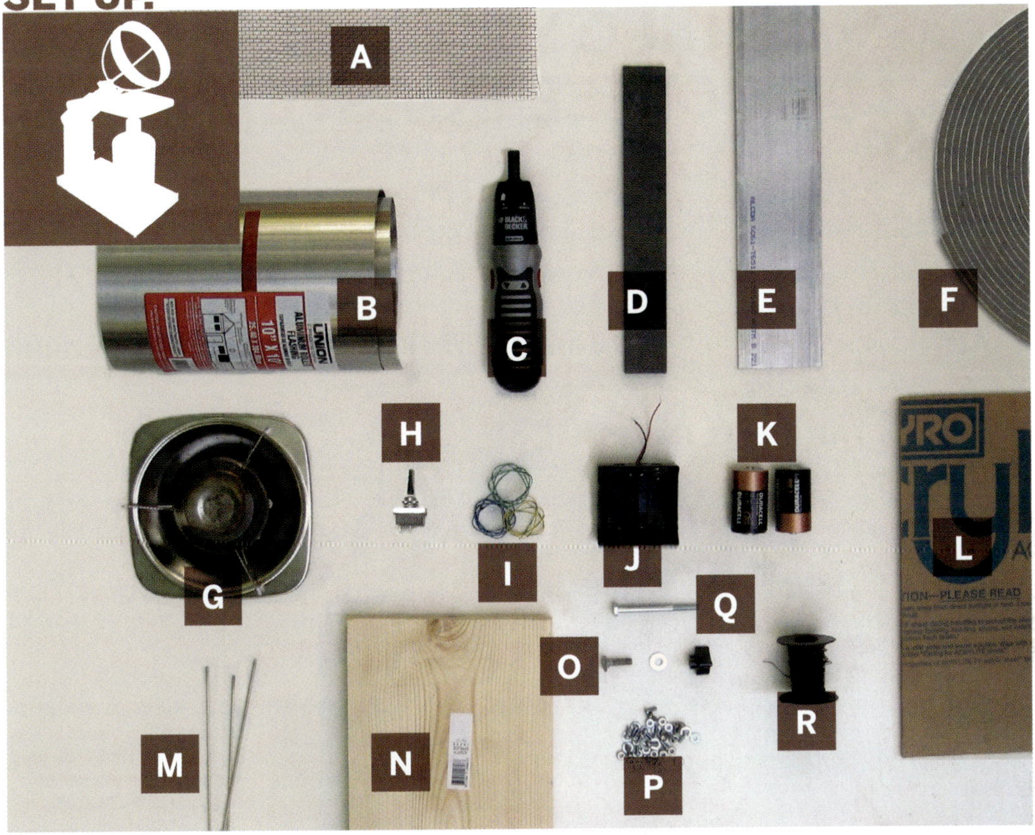

MATERIALS

[A] 1' square stainless steel wire cloth, .025" wire diameter Do not substitute. McMaster-Carr part #9238T523 mcmaster.com

[B] Unpainted aluminum flashing 10'×10"×.010"

[C] Battery-powered screwdriver I used a Black & Decker AS600.

[D] 1½"×¼" piece of aluminum, at least 4" long Cut from McMaster-Carr part #6023K231

[E] 3"×1/8" aluminum bar, 3' long Cut from McMaster-Carr part #8975K833

[F] A few inches of adhesive foam weatherstripping

[G] Single-burner propane camp stove The Century 2058 Trail Scout is best.

[H] Simple on-off (SPST) switch RadioShack part #275-701

[I] 2' insulated wire, 22-gauge minimum Telephone cable wire works well.

[J] Twin D-cell battery holder RadioShack #270-386

[K] D-cell alkaline batteries (2)

[L] Two pieces of acrylic or other hard plastic, at least 3" square each, one 1/8" thick and the other ¼" thick Cut from McMaster-Carr part #8589K41 and #8589K81

[M] 3'×3/32" brass brazing rod From a welding supply house

[N] Scrap plywood 8"×8"

[O] Matching 1" long carriage bolt, washer, and knob, ¼" threaded

[P] Matching 3/8" long machine screws, washers, and nuts, 6-32 threaded (16)

[Q] Matching 3½" long hex-head bolt and nut, ¼" threaded

[R] Uninsulated wire, 24-gauge solid McMaster-Carr part #8870K16

[NOT SHOWN]
Standard 16.4 oz disposable propane cylinder

¼" long machine screws, 6-32 threaded (4)

Wood screws, ½" long flathead #10 (2) and ½" long #6 (8)

Scrap 1×8 lumber or plywood At least 18" long

Scrap wood block About 6" cube

8" wood or plastic dowel, ¼" diameter

Wood glue or 5-minute epoxy

Paper For templates and caution label

See a list of tools for this project at makezine.com/08/roaster.

MAKE IT.

BUILD YOUR COFFEE ROASTER

START >> **Time: A Weekend** **Complexity: Medium**

1. MAKE THE METAL PARTS

This project has blueprints with complete measurements of all the parts at makezine.com/08/coffeeroaster. Start by downloading and printing them all. Then you can easily follow along and make the metal parts, the wood and plastic parts, and the bean basket.

1a. For the support plate, saw a 14⅛" piece off of the 3' aluminum bar and drill and tap 12 holes as shown in the support plate blueprint. File the cut surface smooth after this and all other cuts.

1b. Clamp the plate in the vise between two 2×4s and use the C-clamp to bend it widthwise, 6⅞" from the bottom, to an angle of about 127°.

NOTE: You could also start by making the basket (Step 3, page 117). That's the hard part, and then you can go back and make the other parts to fit afterward.

1c. To cut the height adjustment slot, drill four or five ¼" or 5⁄16" holes as close together as possible, then saw or file the slot until it's even.

1d. Make the drive shaft bearing. Drill a ¼" hole in one end of 1½"×¼" aluminum bar, then saw or drill and file the slot in from the end. Saw the piece off the bar to measure 1¼" — it's easier to machine the slot before trimming the bearing down to final size. File the cut end smooth, and drill and tap two 6-32 holes.

NOTE: Refer to bearing blueprint.

1e. The trickiest bit of metalwork is the triangle piece, which connects the basket to the driveshaft. To start, I drew a paper template with the dimensions shown on the blueprint, and taped it to the leftover ¼" thick aluminum bar.

NOTE: Refer to triangle blueprint.

1f. Drill 4 holes in the aluminum, tap the large one, and saw the 3 slots to the small holes before cutting the triangle to shape. This piece is small and fairly precise, so go slowly and be safe. Then saw the triangle shape, file smooth, and put it in a vise to drill and tap the 3 clamping-screw holes.

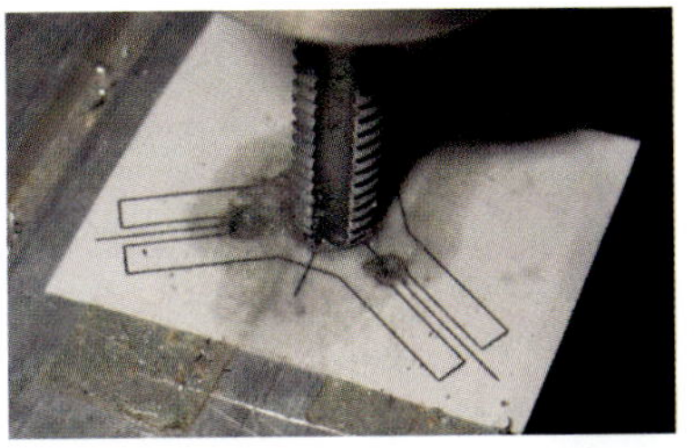

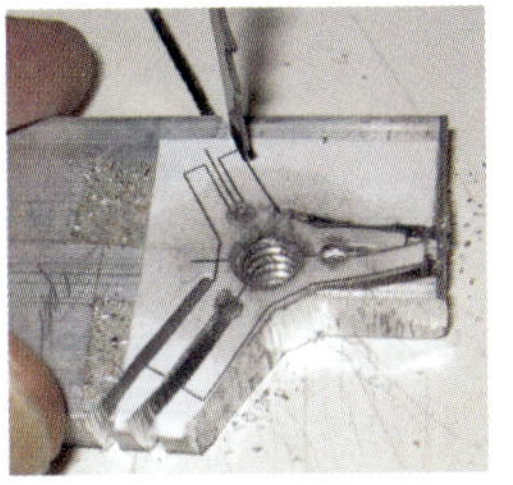

1g. Thread 3 screws into the part and check its clamping action on small pieces of the 3⁄32" brazing rod. It should hold the rods parallel and tight.

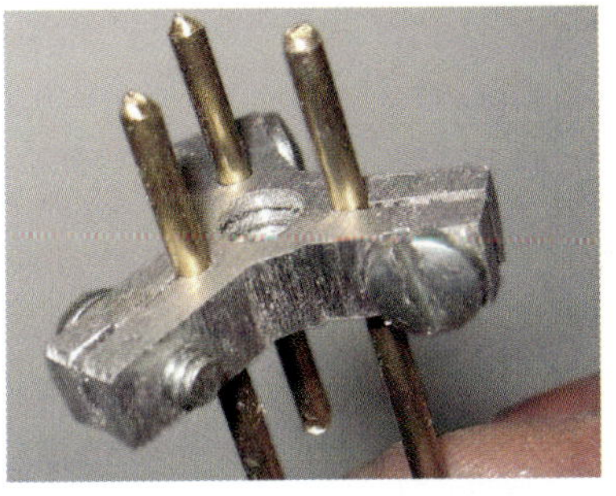

NOTE: Hold the triangle in a vise for drilling and tapping.

1h. For the reinforcing ring, draw a 5¼" circle as a template. Cut a strip of the ⅛" aluminum bar, about ⅜" wide and 19" long, and slowly hand-bend it into a circle, comparing against the template as you go.

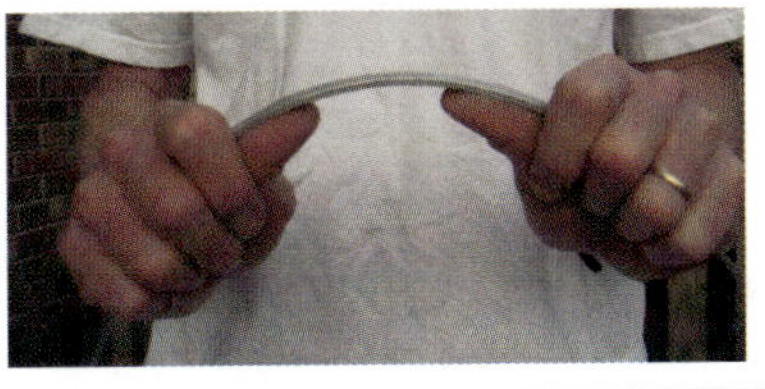

NOTE: Use a vise to get a good grip for bending the very ends.

1i. Saw or file an overlapping joint at the ends, then drill a ⅛" hole and use a pop rivet tool and a short rivet to join the ends together.

1j. Using a compass, mark at least 6 evenly spaced hole locations in the reinforcing ring. Drill the holes to 9⁄64".

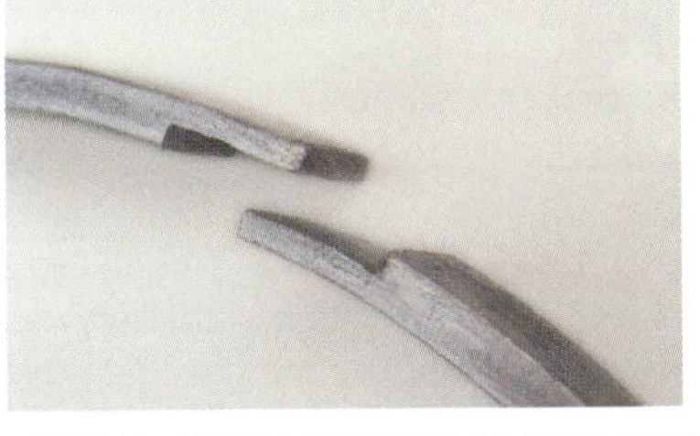

1k. For the appearance ring, use metal snips to cut a strip of the aluminum flashing about 20" long and 1" wide, or wider to make the basket hold more beans. Mark the strip to match the locations of the holes on the reinforcing ring, and snip pairs of ¼" notches to flank each hole. Bend the metal back and forth in the notches until the tabs break out.

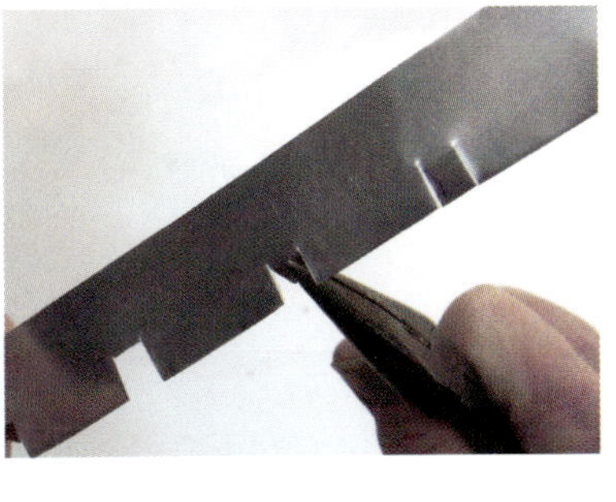

2. MAKE THE WOOD AND PLASTIC PIECES

2a. For the base, cut a 13" length of the 1×8 and drill and countersink two 3⁄16" holes 2½" apart and 1⅛" from one end, with one hole 1¹³⁄16" from one side.

NOTE: Refer to base blueprint.

2b. For the height adjustment block, cut down another piece of the 1×8 to 6⅝"×3½". Use a table saw or router to cut a 3" channel ¼" deep down the block, to just fit around the support plate.

NOTE: Refer to height adj block blueprint

2c. Nest the support plate in the height adjustment block, stand them up together on their bottom ends, and mark the very top of the slot on the block by tracing through plate. Drill a ¼" hole through the block at this mark, for the height adjustment bolt.

2d. On a piece of ¼" thick plastic, trace the screwdriver hold-down shape from the blueprint online, and saw it down to shape. The interior dimensions are fairly critical, to securely hold the screwdriver. Clamp it in a vise, and drill and tap two 6-32 holes matching the pair of holes at the top of the support plate.

NOTE: Refer to hold-down blueprint.

2e. For the switch bracket, cut a 3"×3" square of ⅛" thick plastic. Saw or file 2 edges round, and drill two 5⁄32" holes matching the holes near the slot on the support plate, and a larger hole on the other side to fit your switch.

NOTE: Refer to switch bracket blueprint.

2f. For the bracket, wrap foil around the plastic to shield all but a ⅛"–3⁄16" wide bend-line. Holding the plastic with a gloved hand, heat it gradually over a stove. Press on it periodically, and when it's soft enough, remove it from the heat, bend and hold it at a right angle, then run cold water over the bend to harden the plastic again.

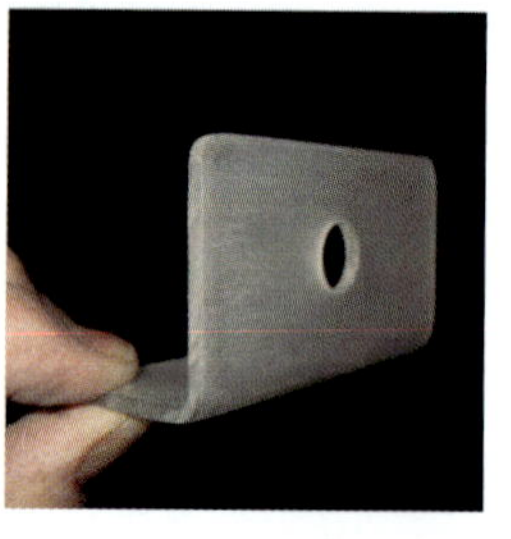

2g. Remove the battery chamber of the powered screwdriver and drill a clearance hole in the end, angled to miss all internal parts. Thread two 1' lengths of insulated wire through the hole and solder one to each battery contact. Note the polarity of the wires and replace the battery chamber. Test the connections by touching the wires to one or two 1.5V batteries in series and switching the screwdriver on.

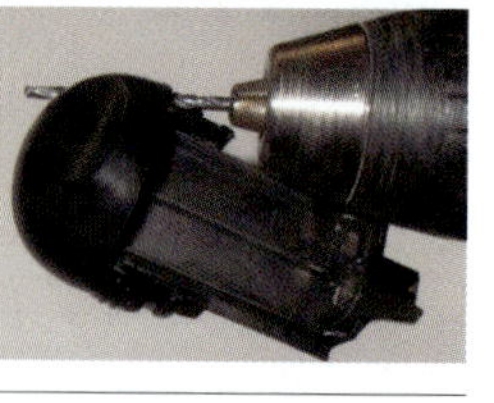

2h. For the drive shaft, carefully mark the edges of the bolt's head with a marker, then use a grinding or cutting wheel to cut it down so that it fits the chuck of the screwdriver, about ¼" across. The new hex-head must be concentric with the shank of the bolt.

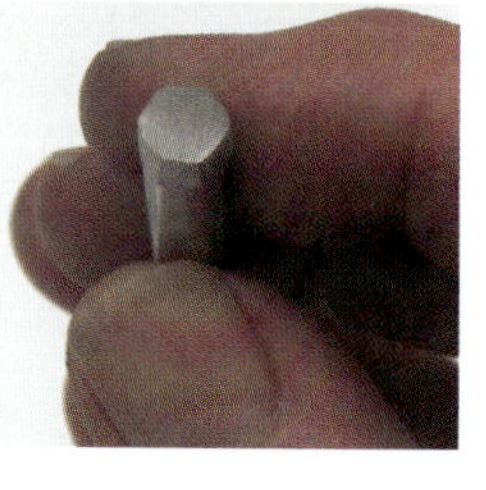

3. CONSTRUCT THE BEAN BASKET

3a. Make the pressing ring tool using the circular template from Step 1h. Cut a 5¼" diameter circle out of plywood by drilling in the center and then sawing out. File the inside edge of the ring on one side to make it round and smooth.

Then make a round form out of a wood block, 5" diameter and approximately spherical or hemispherical. You can turn the form on a lathe, but the shape doesn't have to be perfect, so you could also just cut down and shape it with a saw and a coarse rasp.

NOTE: The form also needs something in the bottom that you can clamp in a vise. I epoxied a ¼" dowel into the hole in its flat face.

3b. Draw an X in the exact center of the 1' square wire cloth. Mark the front tip of the form, then drill a hole and screw the center of the cloth to the tip of the form with a ½" #6 wood screw.

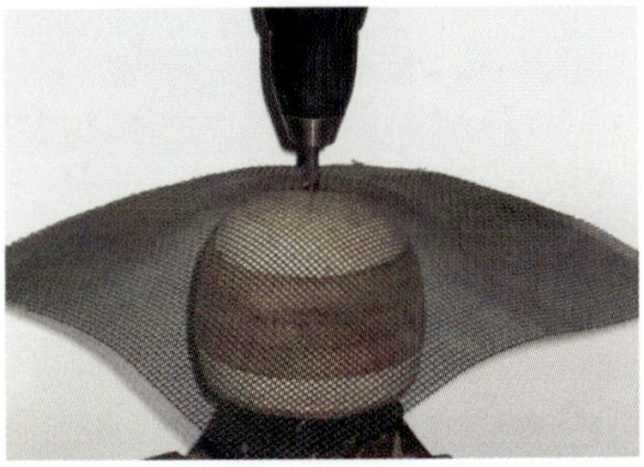

3c. Center the pressing ring, rounded edge down, above the form and wire cloth square. Apply downward pressure until the cloth looks like a bean basket. This is easier said than done. Gradually push a little, guide a little, and compress the wire cloth a little below the ring to help it go down.

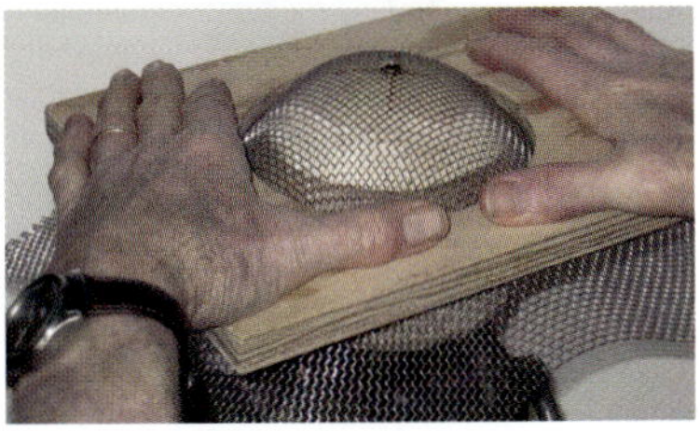

NOTE: The square holes in the wire cloth will deform into diamonds of various size and shape; this is as it should be.

3d. Towards the end of pressing the wire cloth, it will help to have a friend drive more small #6 screws in along the way, to hold the screen against the form. It also helps to cut off some of the excess material at the bottom with metal snips, but leave about 1" all around, past the flat end of the form. Now take a break!

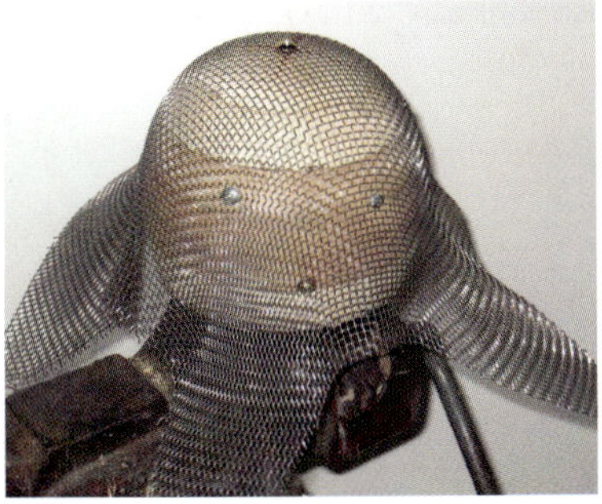

3e. Remove all the shape-holding screws, leaving just the one at the tip. It's OK if your basket is distorted at this point; the reinforcing ring will circularize it. Lay the pressing ring and reinforcing ring on top of wide-open vise jaws. Flip the form and wire cloth assembly upside down and re-press it through the rings. Attach the reinforcing ring to the bean basket with 6-32×⅜" machine screws, washers, and nuts.

For each attachment point, wedge a piece of scrap wood between the form and the basket and drill a 9⁄64" hole through both the reinforcing ring and the basket before installing the screws. The basket should be a minimum of 4" deep, tip to ring.

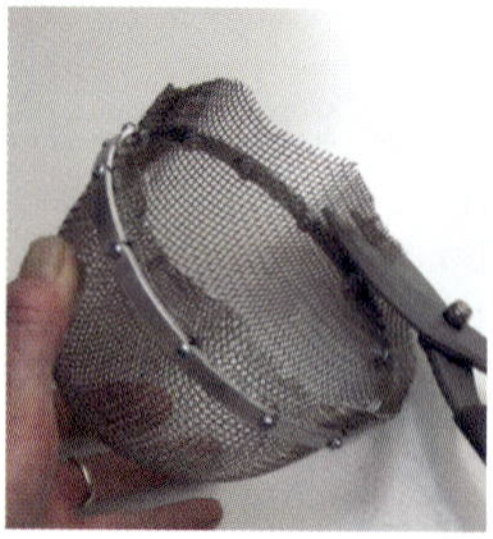

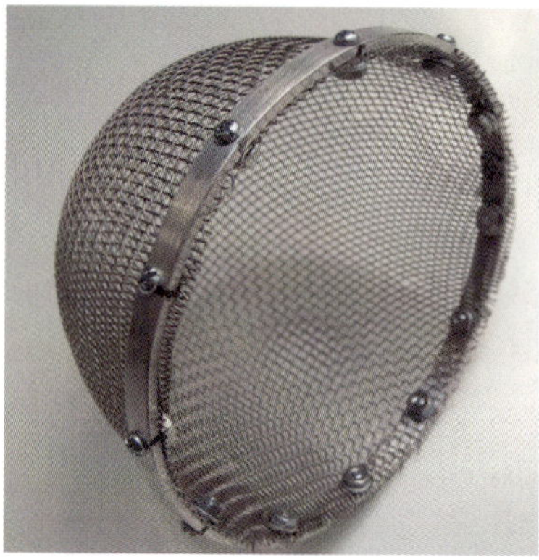

NOTE: With all screws tight, trim the excess wire cloth as close as possible to the ring.

3f. Cut and bend the 3⁄32" brazing rod into 3 pairs of ribs that reinforce the basket, extending out from its center point inside and out. On the 3 outer pieces, leave an extra bend as shown, to fit into the aluminum triangle piece.

Attach the rod pairs to the basket and each other by twist-tying them together with short lengths of uninsulated wire. Use at least 3 wire loops per pair. The pairs should converge close to the center inside and out, forming three 120° angles.

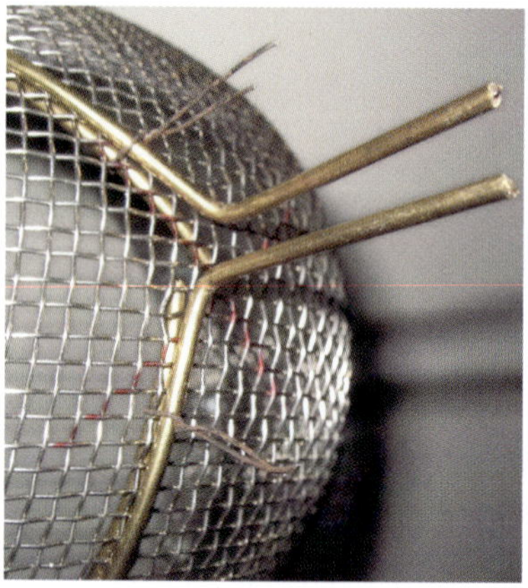

NOTE: Refer to the brazing rod piece blueprint.

3g. Push the 3 outside pieces into the triangle, and tighten the 3 clamping screws securely. Make sure the bean basket assembly cannot be pulled off! Cut and file the ends of the rods flush with the face of the triangle.

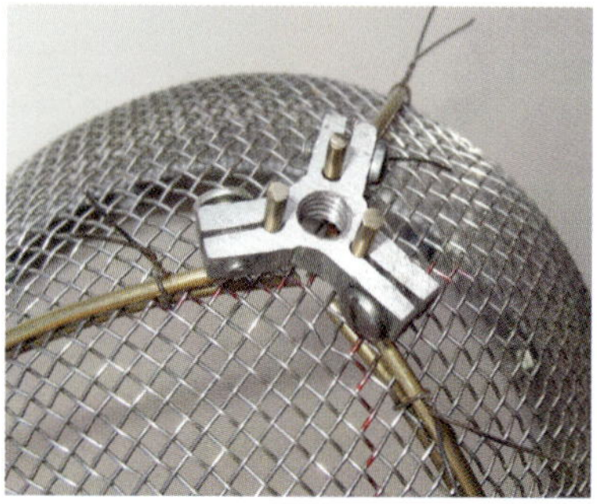

3h. Thread the ¼" nut onto the driveshaft bolt until at least ¼" of thread protrudes. Using pliers to keep the nut from turning, thread the triangle onto the driveshaft until it jams tightly against the nut.

3i. Loosen (but do not remove) all the reinforcing ring screws, and insert the appearance ring between the reinforcing ring and the raw edge of the wirecloth. You may need to shorten it first. Re-tighten the screws.

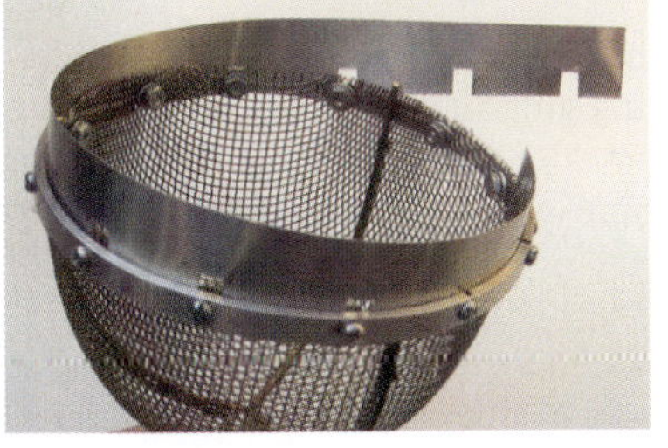

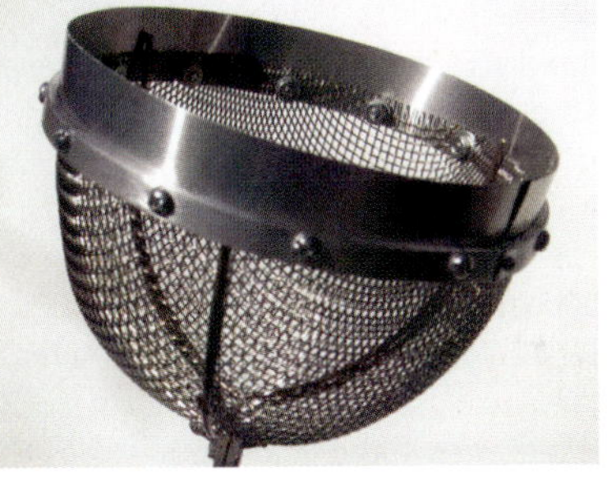

4. FINAL ASSEMBLY

4a. Stain, paint, or varnish the base as you wish, then attach 4 small feet cut from adhesive foam weatherstripping under the corners.

4b. Make a caution label (or print out the one online), and glue it to the base.

4c. Screw the height adjustment block to the base with two #10 ½" long flathead wood screws.

4d. Attach the D-cell battery holder, wires pointing upward, to the bottom end of the support plate with two ¼" machine screws. You may have to enlarge the holder's mounting screws.

4e. Mount the switch in the switch bracket, and mount the bracket to the support plate with two ¼" machine screws.

4f. File off any screw threads that protrude from the back of the support plate.

4g. Mount the bearing to the support plate with two ⅜" machine screws.

4h. Attach the support plate assembly to the height adjustment block with the carriage bolt, washer, and knob.

4i. Thread the 2 screwdriver wires through the hole in the support plate's bend. Use the hold-down and two ⅜" screws to mount the screwdriver to the plate, with its chuck pointing up. Make sure the sleeve adjacent to the screwdriver chuck is set to "Power" position.

4j. Thread a ⅜" screw into the bottom of the support plate, under the screwdriver's central switch button. Tighten it enough to fully depress the button. This screw keeps the screwdriver switched on internally, and helps hold the tool. Briefly touch the 2 screwdriver's power wires to a battery to make sure that it runs.

4k. Slide the bean basket's shaft through the slot in the bearing and into the screwdriver chuck until it stops. You may have to adjust the bearing's position slightly so the shaft doesn't bind.

4l. Thread a piece of uninsulated wire through the support plate and wrap it around the screwdriver near the chuck. There's even a convenient groove for this. Finger-twist and trim the wire ends. Avoid over-tightening, which would put an extra load on the motor.

4m. Lean the assembly on its back and, noting the wires' polarity, solder the switch in series with the battery holder and the screwdriver wires.

4n. Put 2 fresh D-cell alkaline batteries in the holder.

4o. Make sure the camp stove is off, then screw it onto a standard 16.4-oz. disposable propane cylinder and center it onto the base with one of the burner's flat sides toward the support plate. Position the propane flow-control knob so you can reach it easily.

4p. Flip the switch on, and check that the bean basket rotates freely, doesn't wobble, and seems sturdy and safe. Use the height adjustment knob to position the basket at 1" above the burner. Your Nirvana Machine is ready to roast!

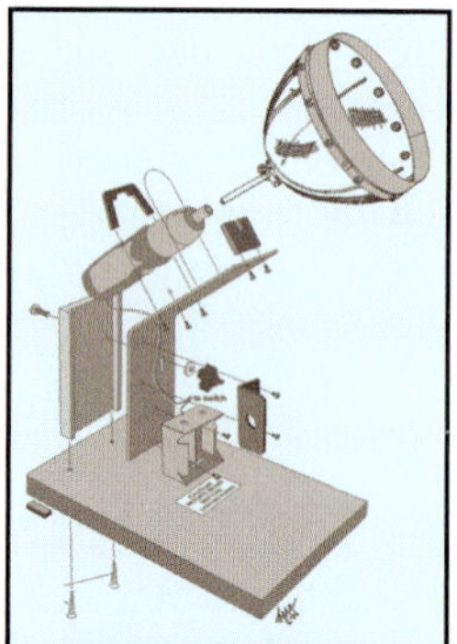

Support plate holds assembly together and slides along the height adjustment block attached to base in back.

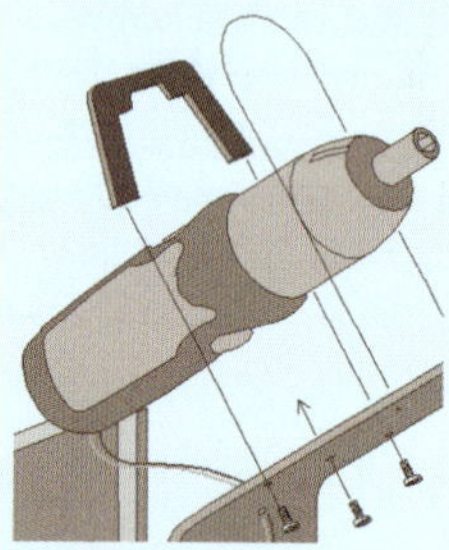

Plastic hold-down and wire loop secure the screwdriver while another screw keeps its On button pushed in.

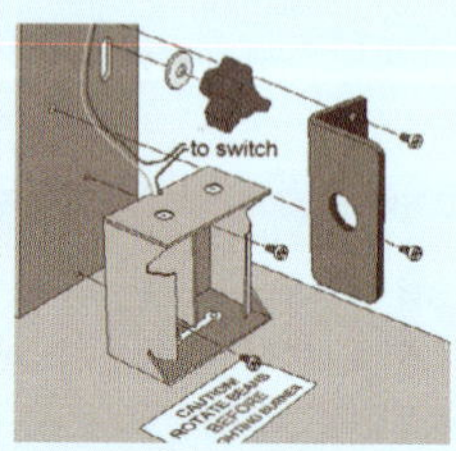

Knob adjusts basket height while switch (not shown) turns the screwdriver on/off to rotate the basket.

Illustrations by Larry Cotton

FINISH X

NOW GO USE IT »

USE IT.

ROAST YOUR COFFEE BEANS

PROPER NIRVANA PROCEDURE

1. Take the Nirvana Machine to a well-ventilated place, preferably outdoors.

2. Make sure the batteries are fresh, and put a drop of oil on the shaft bearing.

3. Load up to ½ cup of green coffee beans into the bean basket.

4. With the burner off, switch the screwdriver on and observe the beans' rotation pattern. The 3 ribs inside should cause the beans to tumble nicely. The bulk of the tumbling beans should be centered above the burner. Green beans expand as they roast, so they should stay well away from the opening, to leave room for expansion.

5. If the beans tumble properly, light the propane burner and set the gas-flow knob on low. The beans should begin to yellow in a couple of minutes, and should roast in 10 to 15 minutes.

NOTE: Never roast coffee unattended, keep kids at a safe distance, and use plain old common sense.

As the beans darken, they will make cracking sounds and shed thin, parchment-like chaff, which drifts out of the opening, making somewhat of a mess. To me, this just adds charm to the experience, but your significant other's opinion may vary.

If you roast outdoors in cool or breezy weather, you may have to crank up the heat, reduce the clearance between the flame and the basket or shield the basket from the elements. I've used aluminum flashing, but any windbreak should do.

As the beans turn even darker, you'll hear lighter crackling. This is about when I think the beans are perfectly roasted. If you keep going, the beans smoke more, caused by the oil that seeps from the beans (this oil also makes the beans shinier). The smoke may smell great to you, but the odor can linger, which is another good reason for ventilation. Be careful not to scorch the beans!

When the beans are done to your liking, turn the screwdriver off and remove the burner. Beans like to be cooled fairly rapidly, so let them continue to

rotate for a few minutes in the cooler air.

Dump the beans into a container and seal it. Purists say to wait anywhere from 4 to 24 hours before grinding the coffee. You will be amazed at how good it is. Good luck!

RESOURCES

Green bean sources: sweetmarias.com, burmancoffee.com

Coffee and home roasting information: coffeegeek.com, thecoffeefaq.com, homeroaster.com, ineedcoffee.com/roasting

Roasting color guides: sweetmarias.com/roasting-visualguidev2.html, breworganic.com/coffee/vowtoroast.htm

1+2+3 Streamerator 2000

By Professor Greg Arius of The Phenomenauts

This favorite stage prop launches streams of toilet paper into mid-air, or can completely mummify someone in the front row.

You will need:

Leaf blower We use the Toro Power Sweep ($30) in red.

31" length aluminum rod, ⅜" diameter

24" length galvanized steel bar, 1⅜" wide, 1⁄16" thick, perforated with ⅜" holes

Sheet metal or machine screws (5+) and washers (2)

Rolls of toilet paper

Brackets (2) We used conduit holders, with water heater strap to go around, but any thin bendable metal will do.

Phenomenauts stickers (2) Available from interpunk.com

1. Attach the roll holder.

Bend the aluminum rod at a 90° angle, 23" from one end. This will be the roll holder.

Use 2 brackets and screws to attach the rod to one side of the leaf blower nozzle, screwing directly into the plastic. Attach one bracket about halfway down the nozzle, and the other about 1" from the base. The bend that holds the TP roll should be about 3½" in front of the tip of the nozzle, and aligned with its lower edge so that the air stream hits the top of the roll, spinning it forward and carrying the paper away.

Carefully drill a pilot hole into the aluminum rod behind the bottom bracket, and use another machine screw to secure the rod to the nozzle. This will keep the roll holder from detaching during use.

2. Attach the stabilizer.

On the galvanized steel, cut a channel from the side of the bar into one of the holes 3–4 holes down from one end. It should be wide enough to accommodate the roll holder rod.

Use a couple of screws and washers to attach the steel bar to the other side of the leaf blower nozzle, with the cut-through hole in front. The end of the roll holder should pull back and nestle into the stabilizer's channel and hole, under some tension.

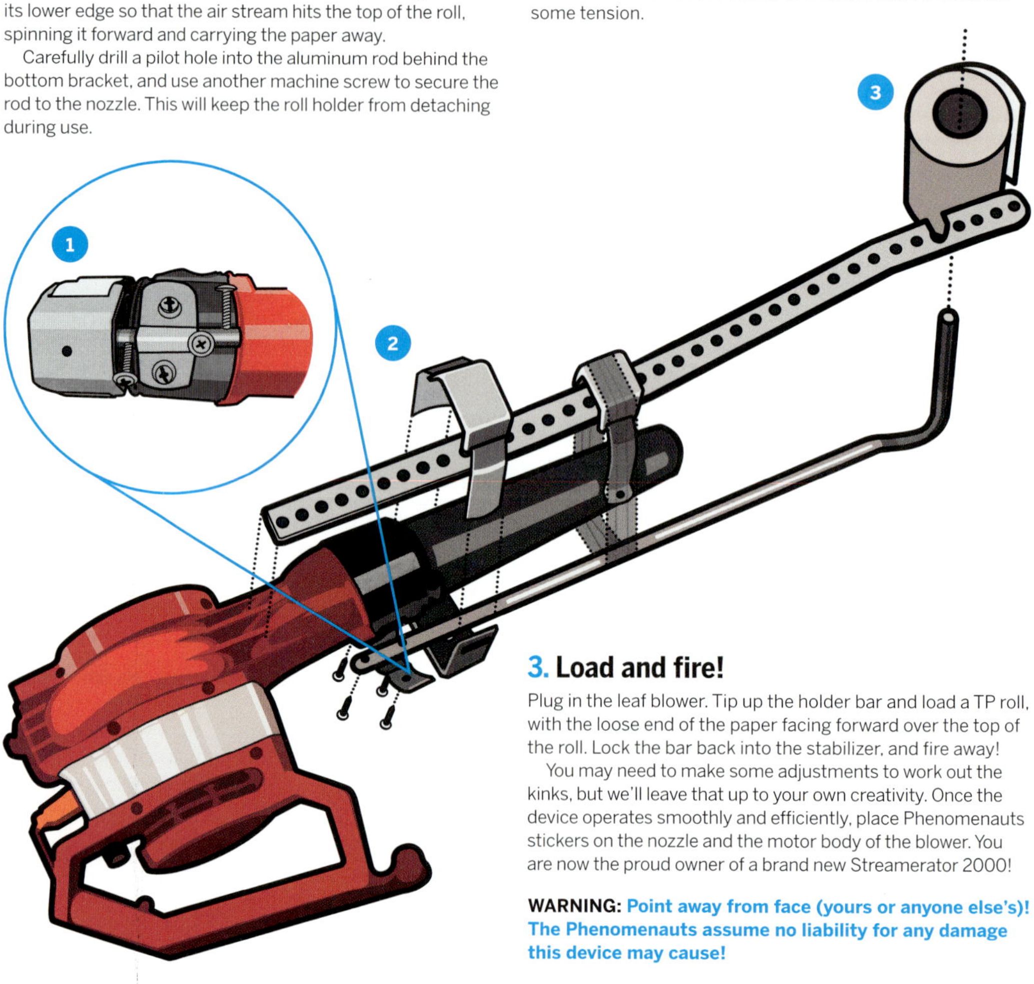

3. Load and fire!

Plug in the leaf blower. Tip up the holder bar and load a TP roll, with the loose end of the paper facing forward over the top of the roll. Lock the bar back into the stabilizer, and fire away!

You may need to make some adjustments to work out the kinks, but we'll leave that up to your own creativity. Once the device operates smoothly and efficiently, place Phenomenauts stickers on the nozzle and the motor body of the blower. You are now the proud owner of a brand new Streamerator 2000!

WARNING: Point away from face (yours or anyone else's)! The Phenomenauts assume no liability for any damage this device may cause!

Illustrations by Damien Scogin

In the amazing year of 2002, Professor Greg Arius was asked to join the Phenomenauts for a special Rocket Roll mission. Upon completion, he became the band's arsenal manager, sound effects engineer, and sometime Science Officer.

TV-TO-SYNTH INTERFACE

Triggering sound from video images.

By Tom Zimmerman

In 1922, a 16-year-old farm boy from Idaho sketched the idea for TV on his high school teacher's blackboard. An avid reader of science fiction (he would have loved *MAKE*), Philo Farnsworth got the idea of scanning an electron beam across a phosphor screen while mowing hay fields.

In this project, we'll stick phototransistors to a TV screen to trigger musical notes on a synthesizer. A peak detector circuit removes the 60Hz pulses caused by Farnsworth's scanning technique. When a bright object passes under the phototransistor, a CMOS switch closes, triggering a musical note. Any video source displayed on a CRT or LCD can be used, including movies, TV, live cameras, and computers. I originally built this circuit to generate sound from a recording of dancers that was filmed with a Fisher-Price camcorder. The circuit emulates a mechanical switch, so it can be wired to any low-voltage circuit (like a remote control), allowing movement in a video image to start a CD player or a VCR, for example.

Illustration by Damien Scogin

Install a D25 Connector to a Keyboard Circuit Board

Get an inexpensive musical keyboard. Remove all the screws from the back of the keyboard and remove the case, being careful not to pull any wires loose.

Locate the cable that connects the main circuit board to the keyboard. The keys are multiplexes, so the cable should contain about a dozen wires. Prepare a D25 solder-type socket with enough wires to connect to every wire in the cable, plus two for power and ground. Using a Dremel or

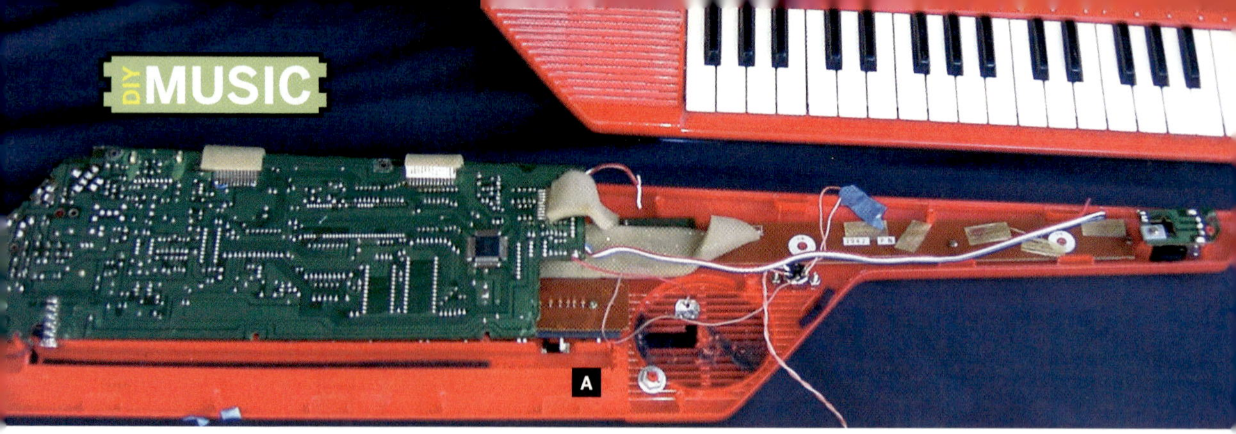

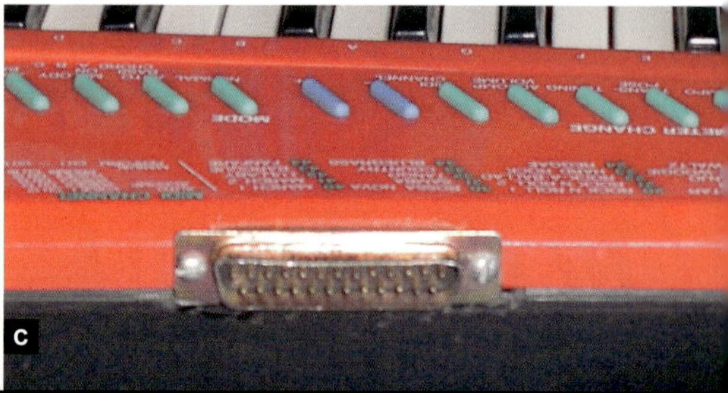

Adding a connector to the keyboard. Fig. A: Carefully open the case of any inexpensive keyboard synthesizer. Fig. B: Prepare a D25 solder-type socket with enough wires to connect to every wire in the cable, plus 2 for power and ground. Solder each wire from the D25 connector to the circuit-board pins on top of the keyboard cable. Fig. C: Check the connections, close the case, and attach the D25 connector to the back of the keyboard.

MATERIALS

Label	Value	Comment
R1	100kΩ	Converts phototransistor current into a voltage
R2 pot	10kΩ	Trigger sensitivity — lower voltage means less light is required to turn on CMOS switch
R3	10kΩ-1MΩ	Pulls C1 to supply voltage. Sets hold time: 10k = instant, 1M = 10 sec
R4	220Ω-1kΩ	Sets LED brightness. Use 220Ω for 5V supply, 1k for 12V supply
R5, R6	10kΩ	Sets middle voltage for output comparator
R7	15kΩ	Positive feedback creates hysteresis to remove noise from comparator output
C1	22 µF	Retains voltage for peak tracker
C2	0.1 µF	Filters power supply
C3	10 µF	Filters power supply
D1	1N914	Part of peak tracker
D2 LN85RP	LED	Trigger indicator. Side-view LED (Digi-key #P444-ND)
D3, D4, D5, D6	1N914	Prevents cross-talk among key multiplex lines
Q1	LTR-301	Side-view phototransistor (Digi-key #160-1065)

file, create a hole to mount the D25 solder-type socket to the synth case. Solder each wire from the D25 connector to the circuit-board pins on top of the keyboard cable, starting with pin 1. Be careful not to short pins together with solder or disturb the connection between the keyboard cable and main circuit board. Find a place on the circuit board that has a ground and a switched power (hint: locate a big capacitor), and check it with a meter. Hook the ground to pin 25 and the switched power to pin 24 on the D25 connector.

Check your soldering job to make sure you didn't cause any shorts. Turn on the power to confirm that the keyboard still works. Close it up and admire your work.

Decode the Keyboard Matrix

Now comes the fun part. Make a matrix tester with an LED and 1kΩ resistor. See circuit (Figure H).

Attach the resistor to the ground (pin 25) and touch the LED anode (positive end) to every pin (Figure J). Write down which pins light up the LED — these are the scanner Source pins. The other pins are the Sense pins. If the LED never lights, hook the bottom of the 1kΩ resistor to the positive end and search again for Source pins. (If this

Photography by Tom Zimmerman

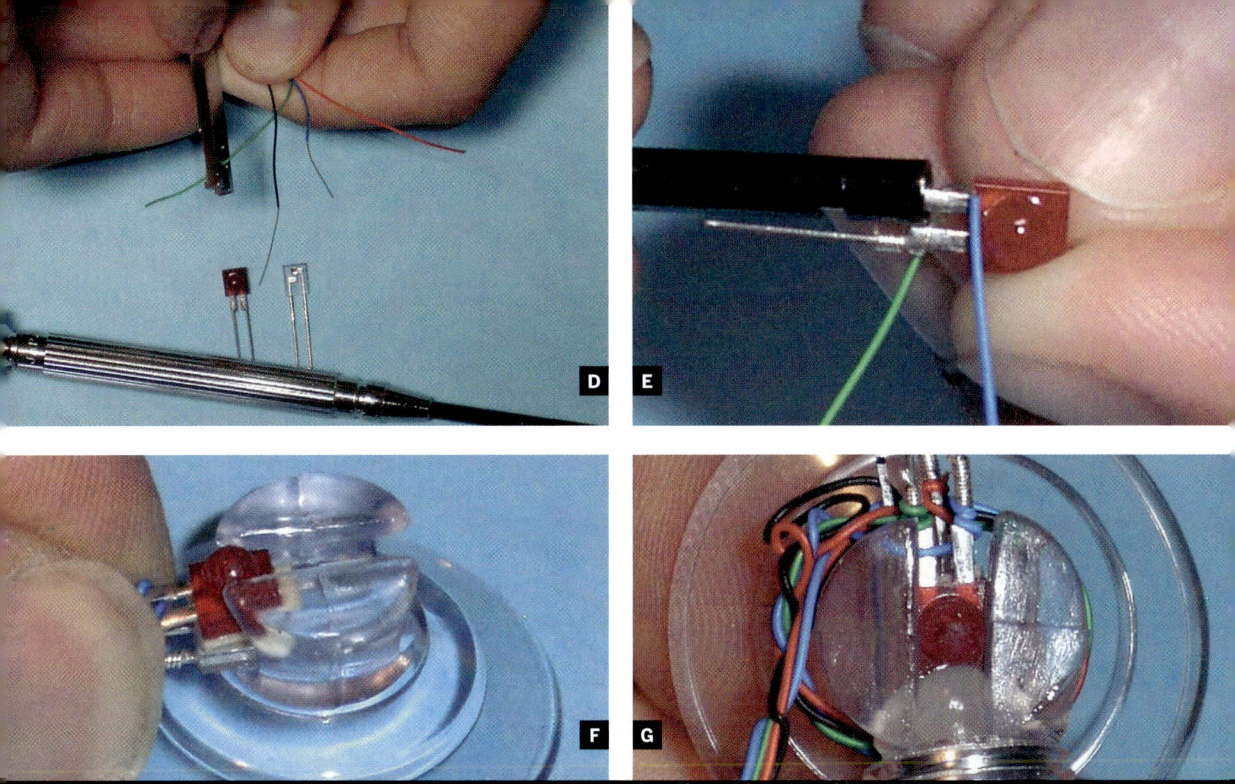

Assembling a suction cup sensor. Fig. D: Each suction cup sensor consists of an LED and a phototransistor. Fig. E: Use a wire wrap tool to attach connector wires to component leads. Fig. F: Stick LED and phototransistor together with tape and insert into suction cup. Fig. G: Apply silicone glue to the assembly.

works, you'll need to reverse the directions for diodes D3, D4, D5, and D6.)

Touch one side of the 1kΩ resistor to any Source pin and the other end to any Sense pin. You should hear a note play. If you're an organized person, write down what note is played for every combination. Otherwise, just marvel at your ability to play notes by touching a 1kΩ resistor to pins on the connector.

Prepare Side-View LEDs and Phototransistors

A clear suction cup holds the phototransistor to the TV screen. The best suction cups are "Light Holders" from Adams Manufacturing Corporation in Portersville, Pa., available at Home Depot. The suction cup has a molded, slotted socket that is perfect for holding a phototransistor and LED. Phototransistors come in many packages, but a flat side-view package phototransistor sandwiched with a flat side-view LED is ideal. The LED indicates when the phototransistor detects a bright object.

Determine the polarity of the phototransistor with circuit (Figure I). The voltage should be high in the light and decrease to ground in the dark.

Cut 4 different-colored 3-foot lengths of #30 wire. Bundle the ends of the wires together, twist a loop at one end, and put that end into a drill chuck. While a friend holds the other end, pull tight and spin the wires carefully into a tight, twisted cable. Keep the cable taut while running your hand over it a few times to keep it wrapped and not tangled. You've just made the thinnest four-conductor cable available!

Strip 1 inch of insulation off the 4 wires and use a wire-wrap tool (RadioShack #276-1570) to wrap them around the phototransistor and LED leads. Document the color that goes to each lead, and be consistent.

Cut some double-sided foam tape into tiny squares, slightly larger than the phototransistor. Stick one side to the back of a phototransistor. Peel off the backing tape and stick the other side to the back of an LED, making a phototransistor/LED sandwich. Cut the leads of the phototransistor and LED slightly below the end of the wire and bend them away from each other so they don't touch.

Insert the phototransistor/LED sandwich into a suction cup. Wrap the cable around the suction cup and loop it into a knot to relieve strain. Inject

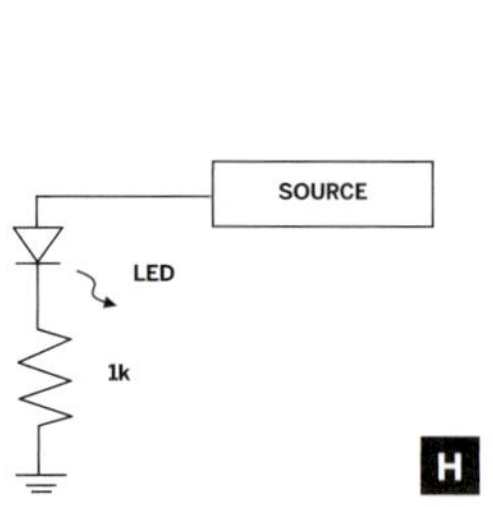

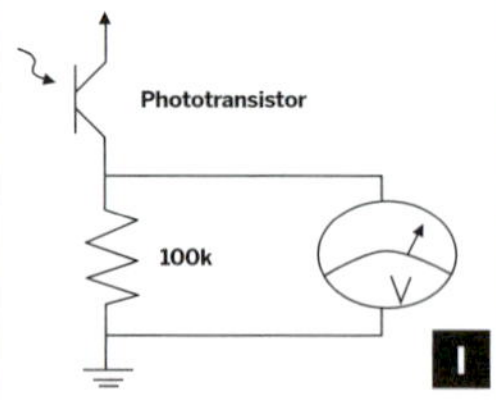

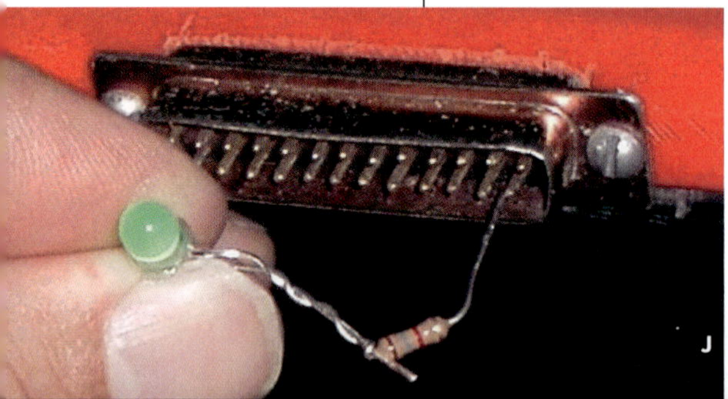

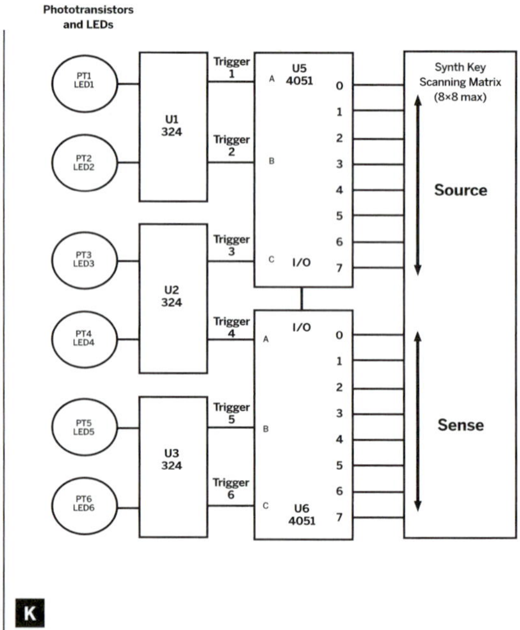

Fig. H: The matrix tester is made from an LED and a 1k resistor. Fig. I: Use this circuit to determine the polarity of the phototransistor. Fig. J: Attach the resistor to the ground pin and touch the LED anode to each pin, writing down the pin numbers that light up the LED. Fig. K: Triggering up to 64 monophonic notes using two 4051 CMOS switches.

clear-silicone sealer glue into both sides of the slotted socket by pushing the tube nozzle against the slot opening. It's OK to get glue on the wires and LED — the light will shine through the clear silicone. Now repeat the process to make a total of 6 suction cup sensors using the same wire-color code, and let them dry overnight. The next day, strip and solder the other end of the 6 cables (24 wires) to a D25 solder-cup plug.

Build the Detector Circuit

Using Figure N, we see that each detector uses half of an LM324 quad op-amp. The op-amps are used as comparators. When the TV screen under the phototransistor (Q1) gets bright enough to exceed the reference voltage set by R2, the output of comparator U1a pulses at 60Hz, thanks to Farnsworth and his scanning electron beam.

This pulls C1 low, which can only rise through R3 due to diode D1. R3 sets the hold time; the larger the resistance, the longer it takes for the trigger output to shut off after the bright object leaves the phototransistor. Using positive-feedback resistor R7, comparator U1b cleans up the output, making sure the trigger is a sharp binary signal (no noise during transition). The LED D2 is mounted on the suction cup next to the phototransistor to indicate when a bright object has been detected.

The circuit operates from 5 to 15 volts and is powered by the synth to avoid batteries and power cubes. If you're going to use batteries, remember to hook the ground of the circuit you're controlling to the batteries, and never hook to anything that plugs into the wall! Wire a battery-powered remote control instead.

Trigger 4 Polyphonic Notes

Each detector independently triggers a 4066 CMOS switch (see Figure M), providing a polyphonic sound (several notes at once). This is useful if you want a sensor to trigger a specific function, like the Record button on a VCR.

The CMOS switch is electrically equivalent to a mechanical switch. Since we are tapping the keyboard scanner before the keyboard switches, we must include diodes D3 to D6 to prevent switch cross-talking.

Trigger 64 Monophonic Notes

The previous circuit used 4 channels to control 4 notes simultaneously. Now, using two 4051

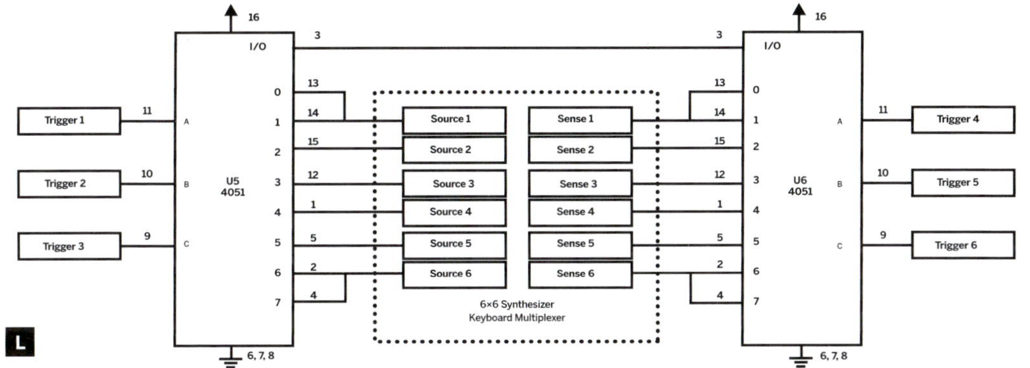

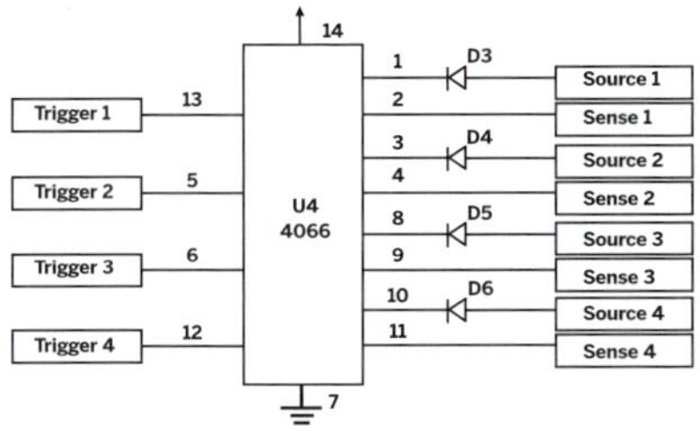

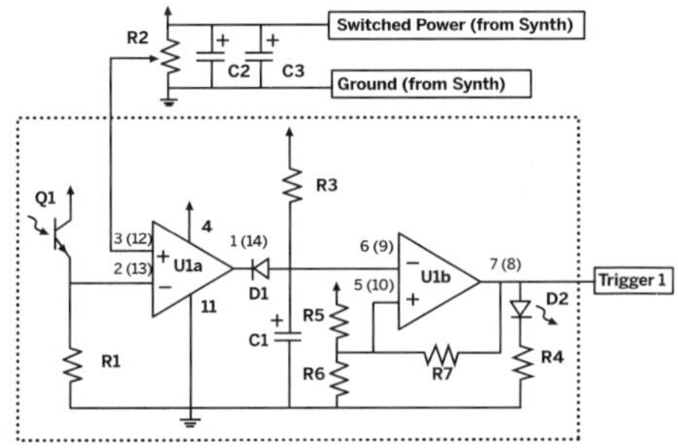

Fig. L: Doubling up switches on each 4051 CMOS switch. It's recommended that you wire-wrap this circuit, or make a printed circuit board (See MAKE Vol 2, page 164 for a printed circuit board primer) Fig. M: Triggering 4 notes at once to create a polyphonic sound. Fig. N: The detector circuit. Each detector uses half of an LM324 quad op-amp, which are used as comparators.

CMOS switches (see Figure K), we'll make a circuit that can select up to 64 notes (8×8), one at a time (monophonic).

The synth I use has only 6 Source and 6 Sense lines for a total of 36 notes, so I double up 2 switches on each 4051 (see Figure L). I recommend you wire-wrap the circuit, or, if you are more adventurous, lay out a printed circuit board.

Conclusion

With a video camera and this circuit, you can perform electronic magic, controlling audio/visual equipment by moving your hands into hot spots that correspond to suction cups on the monitor. The circuit can trigger sound effects, perfectly synchronized to video action.

If you're an installation artist, this circuit is an easy way to create an interactive performance space. By applying light sensors directly to the video screen, we avoid all the complications of video-signal processing.

We have Farnsworth to thank for the scanning TV, and now we have suction cups to thank for holding our simple sensors to his ingenious invention.

QUICK WIRE TWISTING

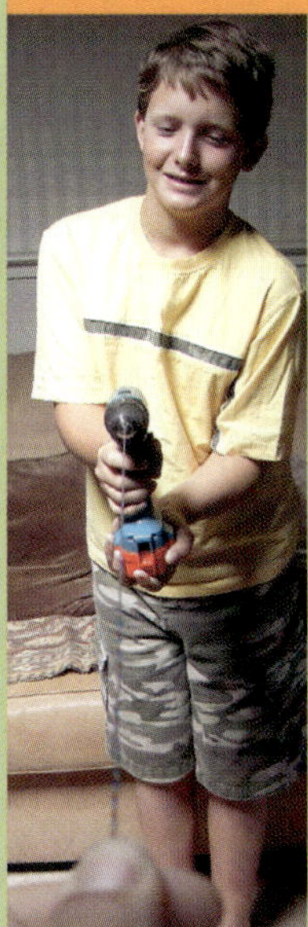

You can use a power drill to quickly and neatly twist wires into lengths of conductor cable.

1. Gather the wires together and make a loop at one end.

2. Put the loop into the drill chuck.

3. Another person can hold the other end tightly while you spin the wires carefully into a tight twisted cable.

4. Keep the cable taut while running your hand up and down the twisted cable a few times to train it to stay wrapped and not twist upon itself.

Tom Zimmerman is an inventor, educator, and researcher at the IBM Almaden Research Center who loves gadgets, LEDs, synthesizers, and hooking people up to computers.

Photograph by Wendell Anderson

WORLD'S LOUDEST IPOD

iBump crossover lets you crank it up without distortion. By Tom Anderson and Wendell Anderson

Getting a big sound from portable music players should be easier. Fancy new plastic mini-speakers are everywhere, but what if you want to plug into a big old stereo and truly rock the house?

An inexpensive Y adapter cable can split a player's headphone mini-jack into two RCA plugs for left and right channels, but you'll get better sound if you put the low frequencies into a subwoofer and send just the mids and highs to the stereo speakers. That's because the large movements that a speaker cone must make to produce low frequencies will distort the mids and highs — especially at high volume. Our iBump improves this situation with an active crossover circuit, which separates audio into high- and low-frequency channels. Connect the low end to a powered subwoofer, and you'll have proper thump!

The iBump's crossover circuit is based on the classic Linkwitz-Riley audiophile design, which you can read about at Siegfried Linkwitz's site (linkwitzlab.com/filters.htm).

Some newer audio amplifiers include separate outputs for powered subwoofers and filters to keep the lows from the main speakers. We tested these and found that our iBump provides much better sound. Try it and see for yourself. This crossover sounds natural without being boomy or leaving a gap between the highs and lows.

If you don't have a powered subwoofer, you can make one using our article "Resurrecting This Old Amp," (*MAKE, Volume 02, page 110*). Start with an eBay bass amp, add large speakers, and presto: you'll have yourself a 'quake. Or buy powered subwoofers from Klipsch, JBL, and others.

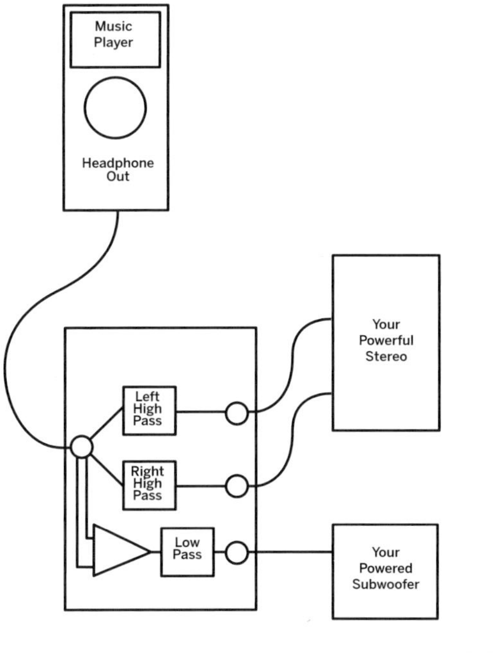

A

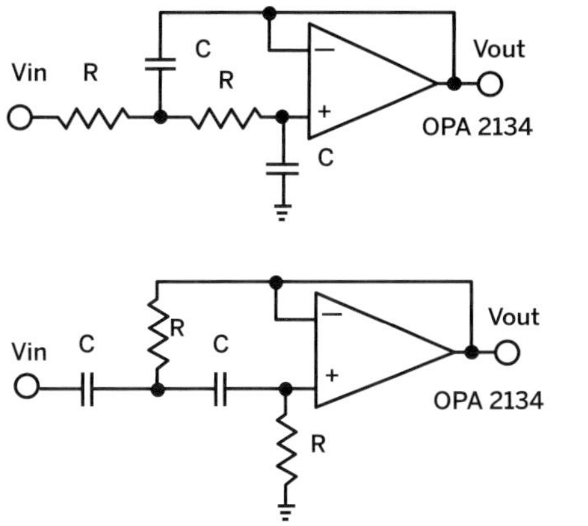

B

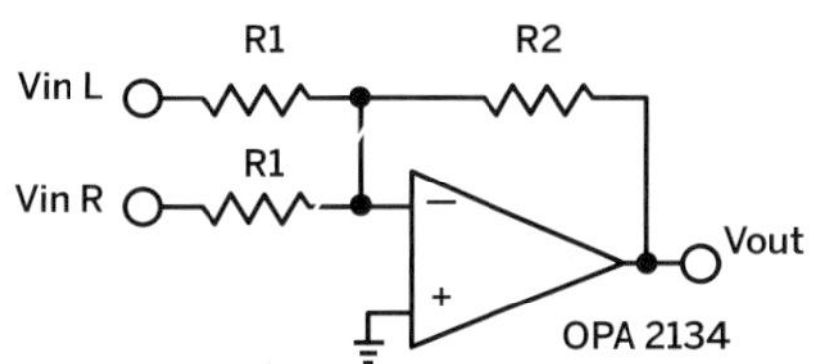

C

Fig. A: iBump splits each input channel in two. One half of each is fed through its own high-pass filter and sent to the stereo speakers. The other halves are summed, routed through a low-pass, and sent to the subwoofer.

Fig. B: The high- and low-pass filters share the same formula for transition frequency and identical layouts, but with resistors and capacitors swapped. Fig. C: The summer circuit combines left and right channels.

Build It

The iBump is an assemble-and-solder project on a single printed circuit board (PCB), with no moving parts. You can build the circuit from our schematic at makezine.com/08/ibump. You can also order the iBump bare PCB, or buy the board pre-assembled and packaged at the same URL. The iBump kit is designed for learning audio electronics, with clear labels and test points that make it easy to probe the circuit with a multimeter or oscilloscope. Once you've assembled the circuitry, enclose the board in a small case.

How It Works

The iBump's all-analog circuitry uses matching high- and low-pass filters to block low and high frequencies. Given resistor value *R* and capacitor value *C*, the formula for where they cross over is: $1/(2\pi RC)$. Our iBump uses 23.7kΩ resistors and 0.068μF caps, which gives a crossover frequency of 99Hz, which is close to a bass guitar's G string.

The summer circuit combines left and right inputs before sending signal through the low-pass and to the subwoofer. Human ears are no good at locating low frequencies, so mono is fine here.

Our circuits are all based on the OPA2134 op-amp from Texas Instruments. This is a great amplifier chip with low noise and 0.00008% distortion. The 5532 op-amp from Philips is a less expensive choice, and the high-end OPA2134 may be overkill, but that's the whole point, right?

High Volume Fun

CAUTION: Do not deafen yourself or anyone else with the magnificent power of your iBumped stereo. Wear ear protection.

To test and adjust the iBump, download and play the test tones from makezine.com/08/ibump and use a sound-pressure level (SPL) meter. You can be an audiophile and tweak the levels to achieve a flat frequency response. Or you can just crank up the subwoofer until things start falling off your shelves.

Challenge your neighbors to see whose stereo is loudest when you listen from the street. When the police come, prove that you're not exceeding local ordinances by showing measured sound-pressure levels from your SPL meter!

Tom Anderson and Wendell Anderson are engineers who like to develop audio hardware and software projects.

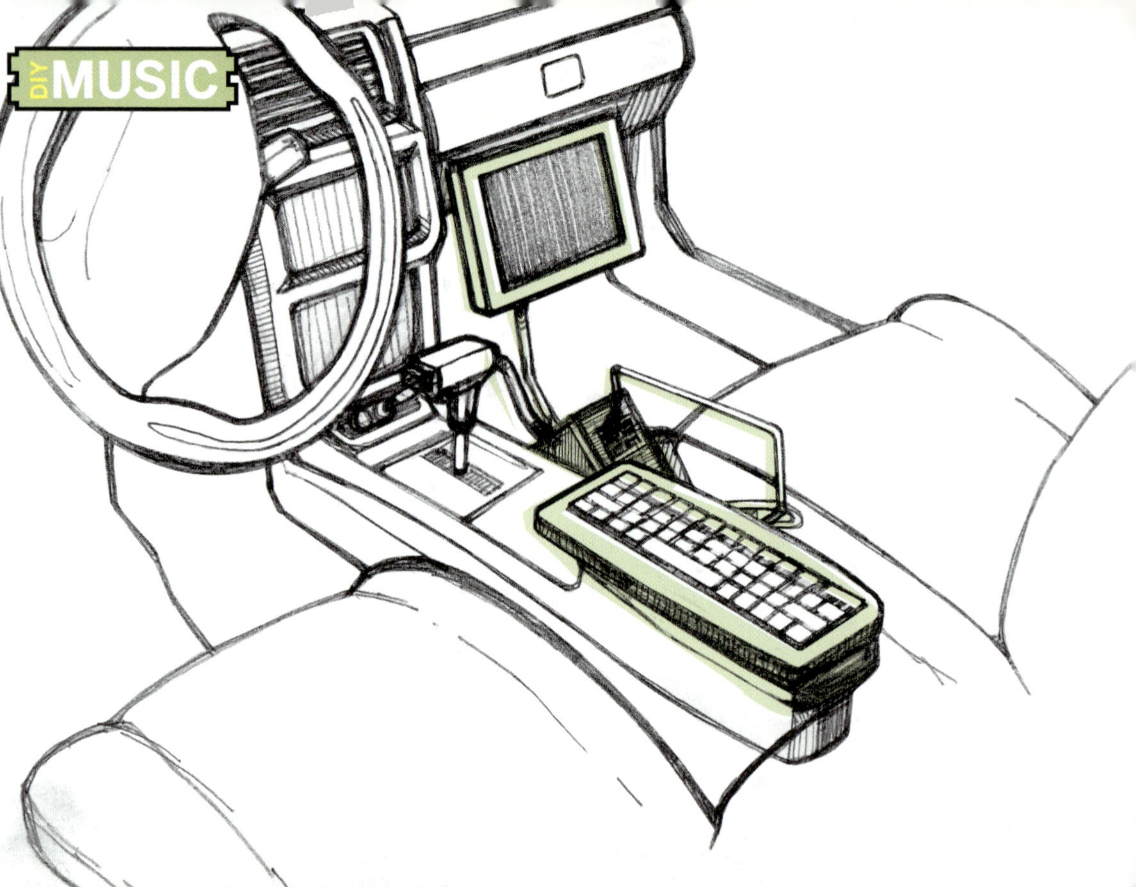

PROJECT REDSHARK

Turn your Xbox into a mobile media monstrosity.

By John Riney

I wanted a system for my car that could play my entire music collection. A sensible human being would have purchased an iPod and plugged it into the tape deck. But I had a dusty Xbox lying around, and I figured I could turn it into the foundation of a powerful car computing environment — something that I could play music with now and add more features to later.

The Xbox is just a PC wrapped up in fancy plastic, and I knew that hackers had gotten Linux working on it. Thus began Project Redshark, named after my old car — which, in turn, was named after the red convertible in *Fear and Loathing in Las Vegas*.

After some beer-assisted research and de-velop-ment, I got it to work. Here's a summary of how I did it, and you can see more details on my website (rineysoft.com/xbox).

1. Xbox, meet Linux. Linux, meet Xbox. First, I discovered that Microsoft doesn't want people installing other operating systems on their game console! Who knew? Fortunately, the Xbox-Linux project (xbox-linux.org) describes several workarounds, including one hack that boots Linux on an Xbox without ever cracking its case. Sold!

The technique comes from the Xbox game *MechAssault*, the first version of which has an exploitable bug in its load/save code. If you load a properly crafted savegame into *MechAssault* from an Xbox memory cartridge, the box will crash, allowing you to run foreign code. (You can also load the fatal savegame from a USB stick, using an Xbox-to-USB adapter cable.)

The payload of this exploit then runs a Linux kernel from the Xbox's hard drive or from a disc

Photography by John Riney / Illustration by Damien Scogin

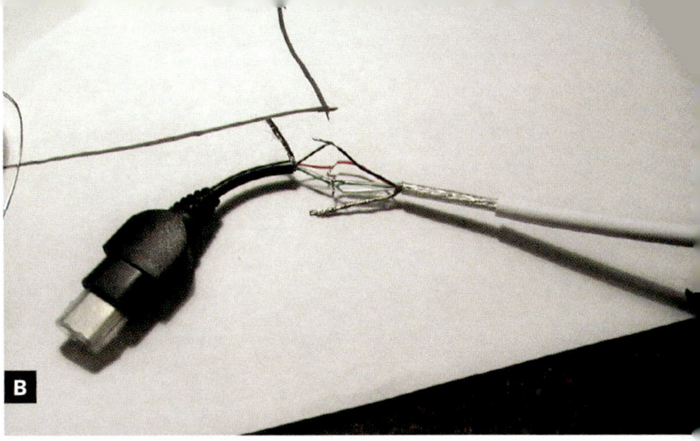

Fig A: Remote switch for Xbox grounds a wire running to the second pin from the bottom of its front panel PCB.
Fig B: Spliced USB-to-Xbox controller adapter cable.
Fig C: Soldering across tiny bridge on main PCB makes Xbox's BIOS chip writeable.
Fig D: Mini board steps 5V down to 3.3V for Xbox power.

in the DVD drive, on top of the Xbox's own BIOS. A minimal Linux running on the Xbox BIOS isn't a very robust combination, but it's a beachhead from which you can continue the Linux takeover.

Note that using Xbox Live closes the *MechAssault* loophole. In this case, you must follow a procedure that includes installing a modified chip.

2. Add a keyboard. Xbox doesn't come with a keyboard, which isn't surprising since there's no English composition involved in playing *Burnout 3*. Fortunately, those plastic Xbox controller connectors are just USB ports in disguise — same pins, different shape. So I made my own Xbox-to-USB adapter cable by splicing together an el-cheapo Xbox controller cable and a USB-B cable, matching up the 4 wire colors. (Microsoft must not require Xbox controller manufacturers to scramble cable wire colors.) Attach a USB keyboard with this cable, and you can type!

3. Flash the BIOS. The Xbox's BIOS resides on something called a TSOP chip (I really don't know what that stands for), and you need to overwrite it with a Linux BIOS. To make the chip writeable, you have to open the box, remove the motherboard, and solder across 2 pairs of absurdly tiny points, one on each side. My cheapo soldering iron couldn't cut it, so I purchased a Weller WP35 iron, which has a 1⁄32" conical tip, and soldered while looking through a magnifying glass.

With the TSOP chip writeable, you use a flasher utility named Raincoat (rimshot, please) to install a replacement BIOS named Cromwell (thanks — I'll be here all week). Cromwell gives Linux full control of the system (with no Restoration).

4. Upgrade the hard drive. Next, you can slam a new hard drive into the mix. I picked a 200GB Maxtor with a parallel IDE interface that I got cheap. Unscrew the stock drive from the plastic carrier you removed earlier, and screw in the new one. This is the easiest part of the project.

5. Install Linux in full. Now you have to decide which Linux distribution you want for the full, final OS. One minor wrinkle is that depending on what flavor of Xbox you have, its DVD/CD drive may be picky about what recordable media it can read. Mine likes Maxell DVD-Rs; your mileage may vary.

I'm a Gentoo fan, so I went with that, but there are Xbox-specific versions of Fedora Core (called XFedora), Debian (called Xebian), and dyne:bolic, which may streamline your install process. The standard Gentoo install CD knows nothing about booting onto an Xbox, so I burned a Xebian install

Junction box powers and switches Xbox, LCD display, and USB hub from 12VDC cigarette-lighter power. At left, the USB, phone jack (for remote switch), video, and power all connect to a massive "cable snake" that runs to the Xbox in the rear of the vehicle. The 4-port USB hub connects to the Xbox USB, letting you plug a keypad, USB thumb drives, and other fun things into the Xbox.

CD, booted from that, and then ran the Gentoo CD to complete the install.

6. Replace the power supply. An Xbox power supply may look similar to a standard PC ATX power supply, but the Xbox's standby voltage is +3.3 volts rather than +5V. So I got a car ATX power supply and ran its output through a mini board that I built to step it down to 3.3V. I also built a cable to carry +12V in, ground, and an ignition sense signal (not currently used), and ran it out the hole where the Xbox's AC cord used to go.

7. Build the front end. For the user interface, I picked up a 7" color LCD panel from a pawnshop and bought a little numeric keypad. Then I wrote my own little player front-end in Python, using the Pygame library. My app, Redshark Player, is basically a file browser with large type. You can download it from my project website.

8. Install the Xbox. I set the system up so I could hide the Xbox out of the way and keep the LCD and keypad up front. This took a lot of cabling: USB, video, and audio cables, Ethernet cable, and two-conductor wire for power and a remote starter switch. I wired a junction box to handle all the interfaces up front, and then used several bags of zip ties to tie all the cables together into one long snake, to feed the Xbox in back.

I originally planned to mount the Xbox in the trunk, but it gets toasty in operation and needs ventilation. So I ended up just sitting the box on the back seat. I'm glad I don't do this for a living.

Conclusion

An Xbox is perhaps not the easiest computer to put in a car, but it's rewarding to drive around town, listening to music coming out of a system you put together! That big, black slab of plastic is good for something other than *Halo* after all!

Thanks to Nafees bin Zafar for editorial assistance with this article.

Resources

Project Redshark web page: rineysoft.com/xbox
The Xbox-Linux Project: xbox-linux.org

John Riney is an engineer, writer, and generally nice guy living in Charleston, S.C.

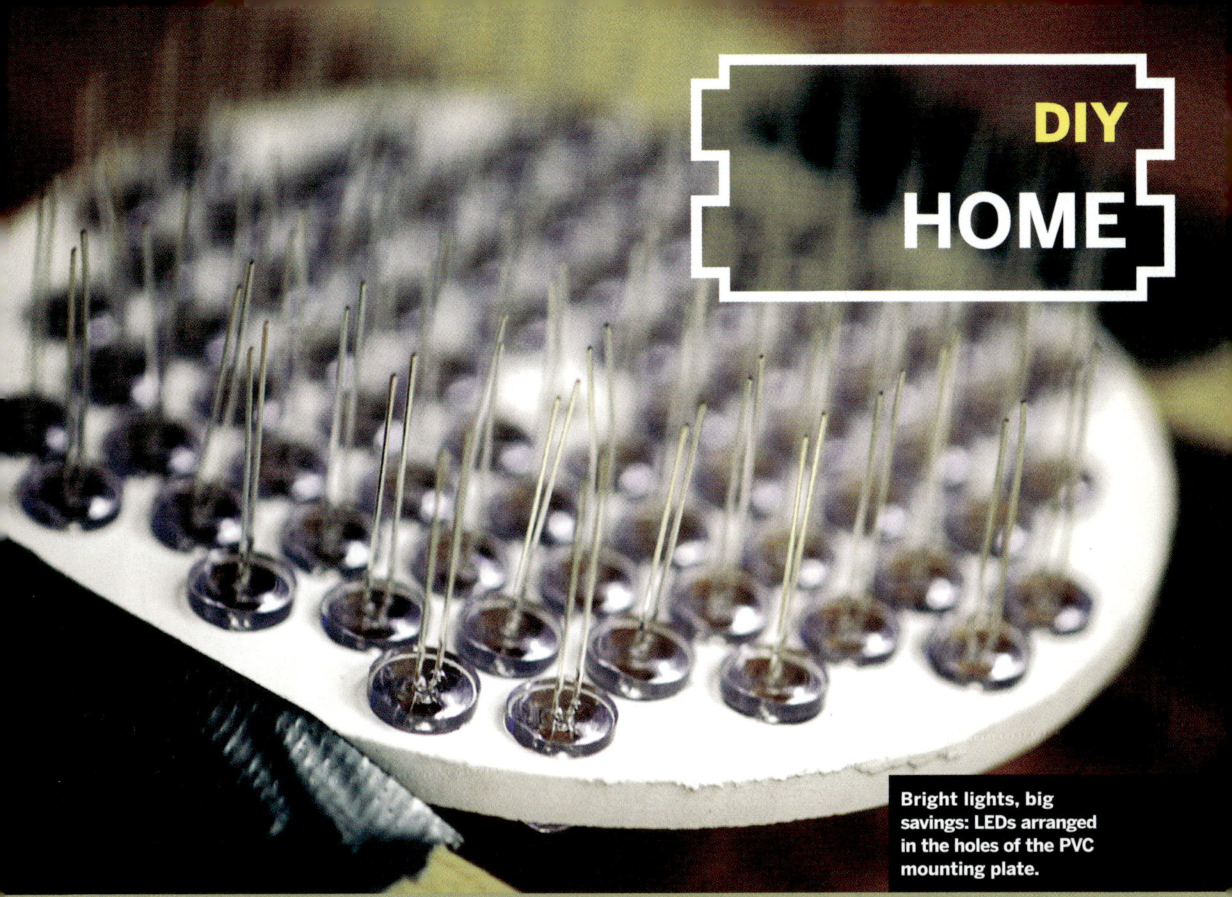

Bright lights, big savings: LEDs arranged in the holes of the PVC mounting plate.

MY LOVE AFFAIR WITH LEDS

Build a bright, low-powered desk lamp.

By Charles Platt

Photography by Charles Platt

Quartz halogen lamps are more efficient than old-fashioned incandescent bulbs; fluorescents are more efficient than quartz halogens; and white light-emitting diodes (LEDs) are the most efficient of all (if you consider the ballast that is needed by fluorescents). So why haven't white LEDs become as fashionable in homes as a Prius in the driveway?

One reason may be that a white LED isn't quite white. It has an eerie purplish tint. This doesn't seem to bother people when the product is a flashlight, but it has been a barrier to the application of LEDs in domestic settings. Personally, I like freakish technology that looks as if it came straight out of *The Jetsons*, so a purplish tint is fine by me. With this in mind, I set out to be the first on my block to have an LED desk lamp.

After an unproductive attempt to find ultra-high-power LEDs using Froogle, I tried eBay and hit pay dirt. Hong Kong manufacturers are now accepting PayPal for state-of-the-art LEDs ordered by the hundred, and are charging less than 25¢ per diode. Diodes are 1cm in diameter and have a claimed light output of at least 100,000mcd.

Now, you may be wondering what this "mcd" measurement is. It stands for "millicandles," which is confusing, since old-fashioned light bulbs are rated in lumens. A lumen measures the total amount of light emitted in all directions, whereas a millicandle is a measurement of intensity per unit of area — more appropriate for an LED since it has a lens that emits a tightly focused beam, often as narrow as 5 or 10 degrees.

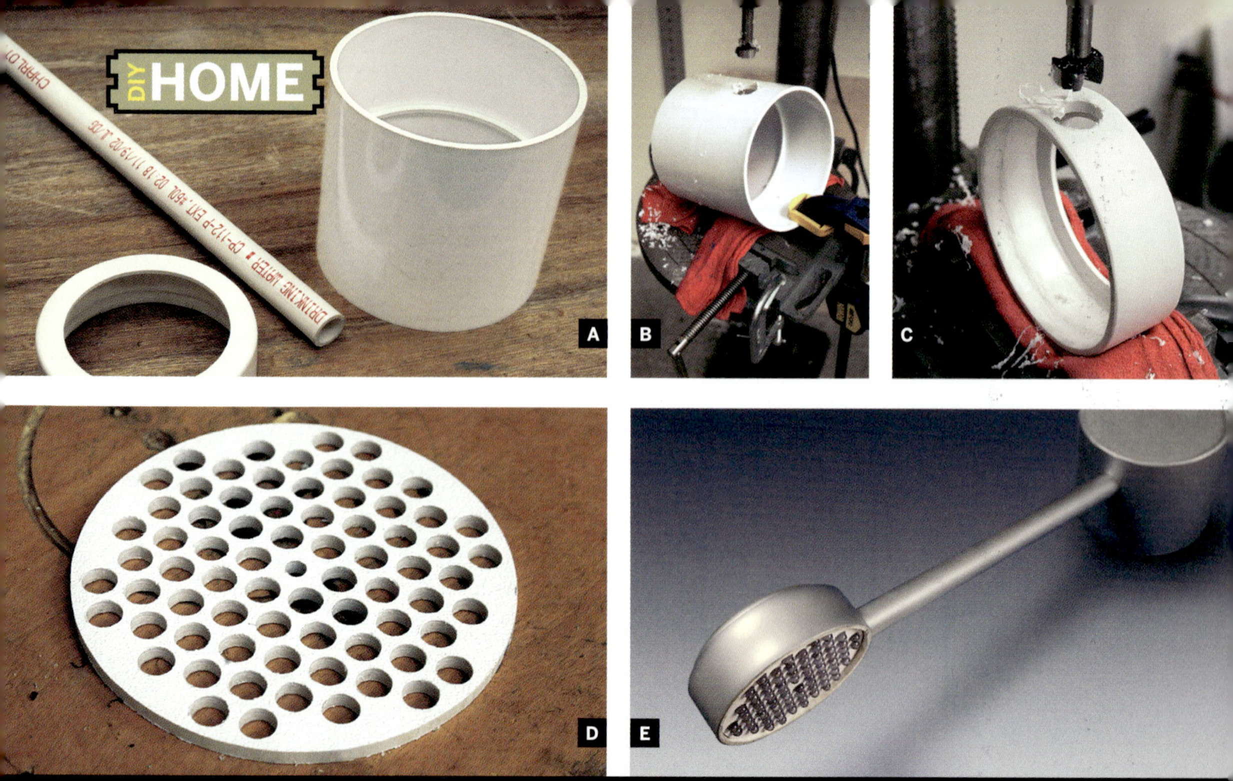

Fig. A: The body consists of PVC pipe pieces glued together, sanded, and spray-painted silver. Fig. B: The base is drilled at an angle to accommodate the stem and filled with sand. Fig. C: The stem's other end is glued into a 1" hole in the head. Fig. D: The mounting plate is ABS plastic drilled with 72 holes. Fig. E: The finished lamp, with the head tilted to show the LEDs. Total cost: about $40; total power consumption: about 5W.

It can best be compared with a reflector bulb, but since those aren't rated in millicandles, the only way to evaluate LED brightness conclusively is by buying some and trying them for yourself.

When my Hong Kong package arrived, I opened it and found not only LEDs but also little brown blobs on wires — resistors to be put in series with the lights. Evidently, a common application of raw LEDs is to trick out cars with underbody lights and other effects. So, manufacturers helpfully deliver LEDs with resistors that reduce the 12V from a car battery to the 3.5V preferred by a typical diode.

This was not so helpful from my point of view, since a resistor dissipates excess power as heat. Fortunately, there are a couple of ways to avoid using this wasteful device in your ultra-efficient lighting system.

If you want to use an LED light source at home, you can simply buy an AC power adapter that converts house current directly to the 3.5V DC that you need. However, I wanted my lamp to work equally well in an RV, so I wired my LEDs in groups of three. Because each LED has its own internal resistance, you no longer need a wasteful series resistor. When you put three together in series and attach a 12V supply to the ends of the chain, each LED requires 4V. You can add more chains like rungs of a ladder, as shown in Figure B on page 135.

Because LEDs are fussy about getting precisely the correct amount of power, you do still need a very small amount of resistance in order to optimize the voltage. For this purpose, I placed a wire-wound potentiometer (variable resistor) at the bottom of the ladder, to adjust the supply for all of the LEDs simultaneously.

I used a multimeter (available for about $20 from RadioShack) to check the voltage across a few sample LEDs while adjusting the potentiometer to the correct setting. By turning the potentiometer back from this point, I could make it do dual duty as a lamp dimmer.

The natural next question is "How do I wire everything together?" You will need a soldering iron — the low-wattage type designed for electronic components (15W is typical) — which you can find at RadioShack. To avoid roasting the diode, you must apply this to your LED wires very briefly. I place an alligator clip on the wire just

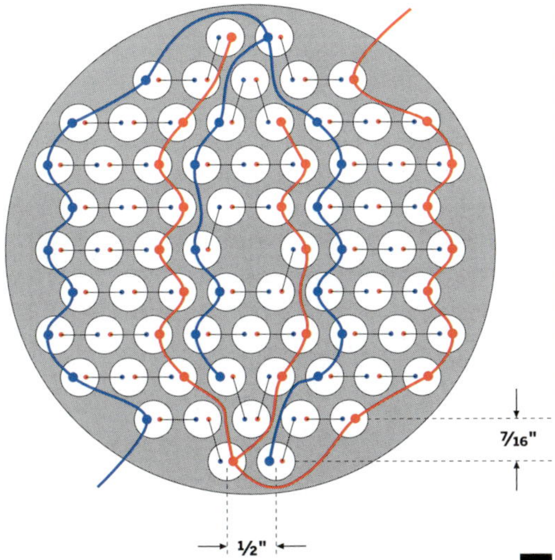

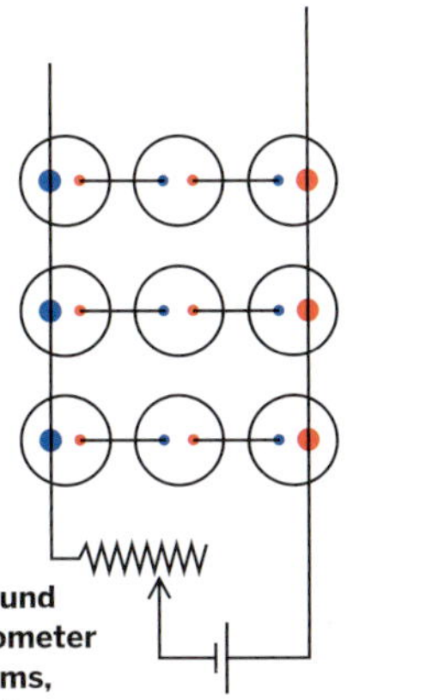

Fig. A: One possible pattern for a lamp using LEDs wired in threes. The result is 24 chains, with a total of 72 LEDs powered by 12V.

Fig. B: Basic schematic for wiring LEDs in chains of three, driven by a 12-volt DC power source and with a potentiometer to fine-tune the voltage (usually 3.5 volts across each LED).

below the soldering iron as a heat sink.

(Soldering guides will tell you that "real solderers don't use heat sinks," but since the difference between a diode that works and a diode that doesn't work can be 6 seconds of soldering time instead of 5, why not take an elementary precaution?)

Most importantly, you will need to remember that the long wire of the LED goes to the positive side of your DC supply, whereas the short wire goes to the negative side. If you wire the LED the other way around, you may burn it out, depending on how much voltage you apply.

With all of these details, you may be wondering whether LEDs are just too much trouble. They're really not. I composed a kind of sunflower pattern for my 72-LED lamp using Adobe Illustrator, then printed it and used it as a template for drilling holes into a piece of ¼"-thick, white ABS plastic.

If you don't happen to have any metric drill bits in the house (I certainly don't), you can use a 25⁄64" bit, which is almost exactly equivalent to the 1cm diameter of the LED. After you drill the holes, poke the LEDs into them, glue them in place (I smothered them with silicone caulking), and solder them together. The whole project took me an afternoon.

If you have a recreational vehicle, you may like the idea of ultra-efficient lighting that won't add to the power drain on your electrical system. In fact, I'm expecting RV owners to go for white LEDs long before homeowners.

Once your system is working, you will enjoy its peculiar economic benefits: a chain of three 100,000mcd LEDs lasts about 100,000 hours and uses a mere 20mA (0.02A) of electricity. If you follow my lamp design, using 24 chains (a total of 72 LEDs) powered by 12V, the whole assembly will consume approximately 5W, while emitting about as much light as a 60W incandescent bulb.

Of course, you will have to get used to that freaky purplish glow, but early adopters should be willing to embrace that kind of thing.

Charles Platt is a freelance journalist who wrote the profile on cold fusion researcher Ed Storms *(MAKE, Volume 03)*.

Direct TV: Motorized rotating stand lets you point the screen to where it's needed most.

TV SPINNER

Motorized lazy Susan aims the screen where it's needed. By Alan Mellovitz

Space can be a precious commodity when you live in a shared dorm room. However, if you allocate space efficiently, you can transform a 10'×10' room from a confining cell to a desirable and pleasurable living area. It can be a challenge on move-in day to create a solution that satisfies the needs of two strangers. More likely than not, the room unintentionally becomes split into two longitudinal sections, and each roommate must try to respect the other's sector, accommodating TVs, computers, video games, tools, wires, large boxes of Dunkaroos, live panda bears, and/or any other strange possessions and hobbies that have the power to annoy.

I know firsthand that the typical hardworking college student vitally needs entertainment. Yet under cramped dormitory circumstances, amusement sacrifices must sometimes be made. To reduce the need for such sacrifices, I recalled that my 10'×10' room had a third dimension: an 8' ceiling. I thought, why not suspend a lightweight flat-screen TV/computer monitor in this unused airspace? And while I was at it, why not make the screen rotate to any viewing angle anywhere in the room, with more than 180 degrees of rotation? That way the screen could generously face the roommate who needed it, wherever and whenever circumstances demanded. Here's how I built my motorized rotary LCD TV mount, after dusting off an old box of Legos.

Selection and Placement

First, consider what piece of equipment you would like to suspend and rotate (computer screen or TV), and where you would want to view it. I bought

Photography by A.J. Kane

MATERIALS

LCD TV or monitor with wall mount

Lazy Susan swivel, steel ball-bearing, 6" diameter
From home and/or hardware store

2×4 lumber: 10' length or (2) 8' lengths
From home and/or hardware store

2×2 lumber: 2' is plenty Find a scrap piece if you can

Adhesive EPDM rubber door and window seal weatherstripping

Small door hinge Strap & T style

L-shaped metal brackets: 6" (2) and 5" (2)

½" particleboard, 2'×2'

¾" plywood, cut to 6¼"×13" or so

Small box of #8 2½" screws
I used some brass ones I had

Small box of #8 ½" screws

4" all-purpose spring With low spring constant

Cable ties

Lego pieces: Medium gears (3); Technic axles (8); Technic gears: 8-tooth, 40-tooth; Technic bush ½ Type I; Technic pins with friction (12); Technic bricks, perforated: 1×4 (4), 1×8 (5), and 1×12 (4); 4×12 plate; 9V motor with gear reduction; electric brick pair 2×2, ⅔" thick with wire; generic bricks to complete assembly

Double-pole double-throw (DPDT) slide switch
From RadioShack

24 gauge Cat5 wire From RadioShack

Electrical tape

TOOLS

Hammer

Screwdriver

Drill and wood drill bits (⅛", ¼", and ½")

Tape measure

Circular saw with adjustable miter angle

Jigsaw

Hacksaw

Carpenter's square/straightedge

Level

a TV especially for this project: a Dell W1900 19" LCD TV, which is wall-hangable and has a built-in TV tuner and both DVI and VGA inputs. Whatever flat-screen you use, this project requires it to be under about 25" wide, including the speakers. As for placement, I cantilevered the TV over the foot of my bed, so I could watch TV and use the computer from the comfort of my cushions and pillows. The stand is attached to the bunk bed itself.

Assemble the Stand

Measure the base height for the stand — the distance between the floor and the bottom of the monitor. Mine was 6' 3". Cut one of the 2×4s to this length, and use the long 2½" screws to attach the 6" L-brackets to the top of the 2×4, leaving ¾" between the top of the L-bracket and the top edge of the 2×4. Drill pilot holes before sinking the screws, and use a hacksaw to trim the points sticking out the other side.

Cut a piece of ¾" plywood to 6¼"×13" or so, a bit more than half of the TV's total width. Use the long screws to attach the plywood to the free side of the L-bracket running perpendicular to the 2×4 the long way. For the angled brace that helps support the plywood shelf, take the leftover piece of 2×4 and measure out the plywood length plus about 3". Then cut the 2×4 to this length (16" in my case), with 2 opposing 45-degree miter cuts. Screw this brace onto the long 2×4 and the plywood shelf. As with all wood-to-wood joins in this project, drill pilot, clearance, and countersink holes before attaching the screws.

Decide which side of the 2×4 upright you want to mount the motor on (probably the side nearest the wall). Cut a 2×2 to about 1' and screw it onto the 2×4, running parallel up along your motor side and ending flush with the top. Cut a 2½" piece of the 2×2, and toe screw 2 screws partway into this block, angled in to attach one end of the block horizontally to the top of the stand later. This piece will hold the motor, but don't attach it yet.

Build the Rotating Base

The TV base consists of a big circle of particle-board with a 2×4 that points upward, offset from the center, and supports the screen as if it's mounted onto a wall. A larger circle gives the little Lego motor more torque with which to rotate the TV, so its diameter should be at least as big as the total width of the TV. For my Dell LCD TV, that's 24".

Use a screw, string, and pencil to trace a circle on the ½" particle board. Then cut the circle; I used a jigsaw. Once it's cut, draw 2 perpendicular lines through the screw hole. The Lazy Susan swivel should have 4 mounting holes on each half; center the lazy Susan's top half on the circle by positioning all 4 holes so that the perpendicular lines show through, then drill and screw-mount it to the particleboard.

Drill a ⅛" hole through the center mark on the particle board, then measure and mark a point on the top side of the board that's the same distance out from the center as the lazy Susan mounting screws on the underside, but offset from them radially so that there's no metal on the other side.

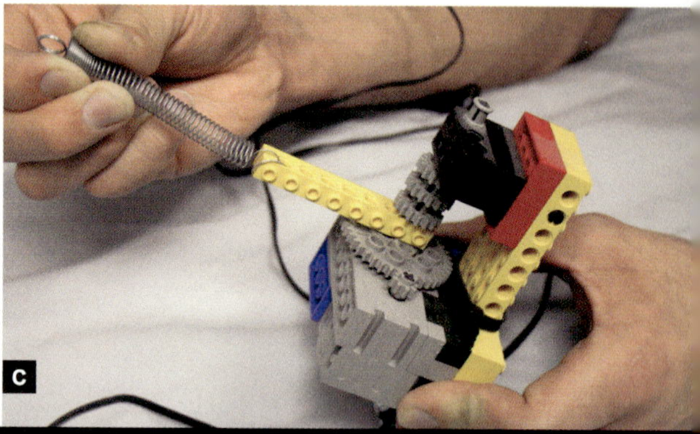

Fig. A: Gear reduction on Lego motor gives sufficient torque while 3 medium gears on the axle engage the rubber weather stripping on the spinner's edge as a friction drive.

Fig. B: Underside of TV spinner, showing lazy Susan, brackets, brace, and motor mount. Fig. C: Tension spring pulls the axle against the edge of the stand.

Drill a ½" hole through this point; this is an access hole that will let you screw the rotating base assembly down onto the stand later.

Mount the TV

Using the top center hole as a reference, figure out where the TV should sit on the base. To minimize stress on the motor, the TV's center of gravity should fall on or near the center of the base. Mark the base directly below the mounting holes on the back of the TV.

Examine and measure the mounting plate and hardware supplied with the TV's wall mount and decide how high you want your TV to be from the particleboard base. Add this height to the height of the mounting plate plus the distance between the bottom mounting holes on the back of the TV and the bottom edge of the screen, and cut another piece of 2×4 to this total height or a little more (mine was 10"). Use the two 5" L-brackets and 2½" screws to attach this piece of 2×4 perpendicular on the base. Then use the heavy-duty hex screws supplied with the TV mount to mount the plate at the proper height onto the 2×4. After the mounting plate is secure, trim any screws sticking out the bottom of the particleboard base.

Clamp the stand to a workbench; this will make things easier. Using ½" screws and working through the ½" access hole, mount the other side of the lazy Susan to the ¾" plywood at the top of the stand so that the circle hangs about ¼" over the back of the main upright 2×4. Wrap around the edge of the particleboard with a strip of door and window seal, and reinforce it with staples for extra durability.

Build and Install the Motor

Use the pictures above as a reference to build the motor out of Lego. It does not have to be exactly like mine, but you do need an 8-tooth gear on the axle driving a 40-tooth gear to step it down, and 3 medium-sized gears on the same rod as the 40 to friction-drive the edge of the TV base. There should also be a 1×8 Technic brick through the axle to attach the tension spring, which will pull the motor against the base.

Once the motor assembly is complete, hold it on top of the 2½" piece of 2×2 and mark where the 2×2 should attach to the stand. Screw the 2×2 block on, and attach the small hinge to the

The author Alan Mellovitz's dorm room at the University of Illinois. It's no wonder Alan thought to use the upper airspace in his room; he spent years living in a three-story tree house before he headed off to college.

back side of the block, opposite the base. Use pliers to make a 90-degree bend in the free side of the hinge to support the motor. Mount the motor to the 2×2 by drilling three ¼" holes in the 4×12 Lego plate and screwing it on with ½" screws. Use a cable tie to anchor the motor to the hinge. Attach the spring to the free end of the 1×8 Technic brick, and sink a ½" screw into the ¾" plywood to anchor the other side of the spring so that it pulls the motor's gears up against the base. Crafting the motor does not have to be exact, just so long as it works.

You can use a slide DPDT switch to control the motor. Install the switch in a convenient location, and wire its 2 middle terminals to the wire pair from the Lego electric brick. Then wire any 9V power supply or battery in both directions on the switch's 4 remaining terminals; solder the (+) and (-) one way on one side of the switch, and in the reverse direction on the other side.

For a fancier solution, you can even build a remote control to run the motor off a relay board. The kit I used for this was the 12 Channel IR Remote w/Relay Board (item #CK1615) from Carl's Electronics (www.electrokits.com).

Install It

Now you can fasten the TV spinner to your bunk bed. Use a level and have someone hold it in place while you make ⅛" pilot holes. Use the 2½" screws to fasten the stand to one of the bed's uprights. Then mount your TV to the mounting plate and connect all the cables. Bind the cables together with cable ties, leaving a large loop of slack to allow the TV to rotate around at least 180 degrees. Then staple the cable ties to the 2×4.

The rotary TV mount can also be useful for attaching other necessities to it, like fans, hooks, or more wires! You have the option to trim off the front edge of the circle in case it blocks your view of the TV, but if it doesn't bother you, it does make a nice nightstand.

+ A schematic for the motor and switch can be found online at makezine.com/08/diyhome_spinner.

Alan Mellovitz is a University of Illinois sophomore studying mechanical engineering. He has his own handyman business, likes to create gadgets, and plays the drums.

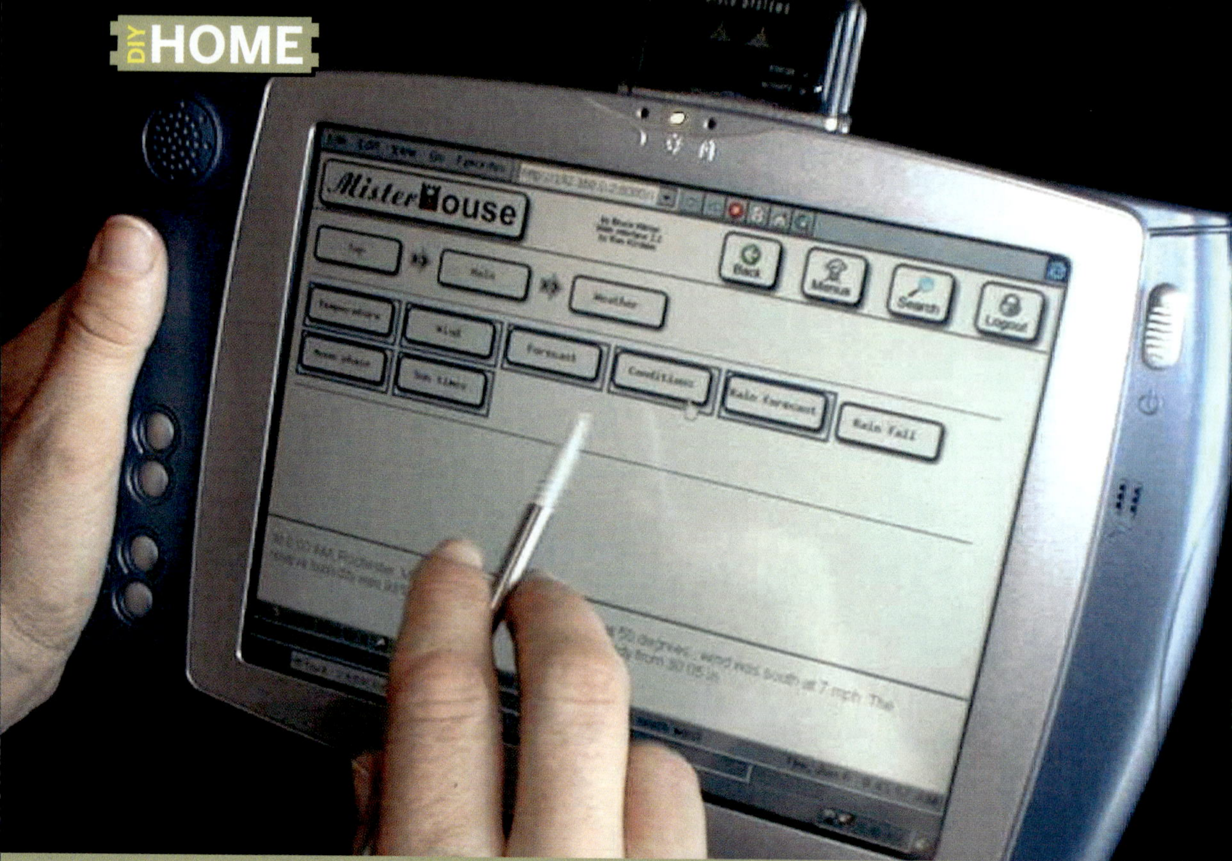

SMART HVAC

Energy-efficient A/C knows when you're in the room. By Dave Mabe

When I began working from home, I converted a shed in the backyard into an office. The shed wasn't heated or cooled, and summer was bearing down, so I needed to find an HVAC solution quickly. I didn't want to extend my home's HVAC system because this would have been lot of trouble, quite costly, and wasteful; I knew the office would remain unoccupied much of the time. Being a home automation enthusiast, I saw this as a perfect opportunity, and I came up with a cheap, efficient solution that sets the temperature to different levels depending on whether I'm in the office.

My office is about 20 feet square, so all I needed for cooling was one energy-efficient window-unit air conditioner. As with the main house, I decided to use X10 home automation, which transmits through home electrical wiring. X10 hardware is easy to install: you just plug the controller modules into the wall and then plug your appliances' power cords into the modules.

The peripheral components in my system were the X10 appliance module on the A/C unit, an X10-readable thermometer, and an X10 motion sensor. The thermometer I used was part of a multi-function weather station that takes indoor temperature as one of its readings, but you just need any cheap digital thermometer that can communicate over X10. To control the system, I used the excellent home automation program MisterHouse and a serial adapter that translates between MisterHouse messages on the computer and X10 signals on the power lines. (My weather station connects to a different computer in my network from the one that runs MisterHouse, but you can also plug the thermometer directly into the MisterHouse computer if it has a spare port.)

Photograph by Dave Mabe

MATERIALS

Air conditioning unit Or fan, heater, or other HVAC appliance
X10 appliance module (model A466) One for each appliance, from x10wirelesshome.com, or just click on the company's notoriously ubiquitous pop-up window
Digital thermometer that's readable via X10 I took readings from my Oregon Scientific WM918 weather station via HomeSeer software, but the MisterHouse project site reports that you can also use an iButton DS1920-F5 sensor in a DS1402D reader ibutton.com with an iButtonLink Link12 serial adapter ibuttonlink.com.
X10 motion sensor (model MS14A) x10wirelesshome.com
Wireless transceiver (model TM751) x10wirelesshome.com
X10 serial computer interface Such as an X10 Activehome kit, smarthome.com/1140.html
MisterHouse software Open source project misterhouse.com

I plugged the A/C unit into power through the X10 appliance module and set it to a lowest temperature of 60°F, since my MisterHouse virtual thermostat would be turning it on and off anyway. Once this was set up, the rest of the "thermostat" was some simple Perl code executed by MisterHouse:

```
$office_ac = new X10_Appliance('C1');
# C1 is the X10 code I set on the appliance module
$desired_temperature = 75;
$wiggle_room = 1.5;

if ($Weather{TempIndoor} > ($desired_temperature +
$wiggle_room)) {
 set $office_ac ON;
}

if ($Weather{TempIndoor} < ($desired_temperature
- $wiggle_room)) {
 set $office_ac OFF;
}
```

I placed this code in a file and saved it to the MisterHouse code directory, where it would be executed continuously, several times a second. In the code, `$Weather` is a special variable that MisterHouse sets automatically when it checks for a weather station, and `$wiggle_room` defines some leeway in maintaining the desired temperature. Increasing `$wiggle_room` saves energy by causing the A/C to switch on and off less frequently.

To add an occupancy check, I used a battery-powered X10 motion detector near my desk chair, which talks to an X10 wireless transceiver plugged into the wall. When the office is unoccupied in the summer, the desired temperature gets set higher (for winter, with a heater, it would be the opposite). To accomplish this I added the following code to the same file:

```
$desired_temp_unoccupied = 85;
$desired_temp_occupied = 75;
$office_occupancy_timer = new Timer;
$office_occupancy_timeout = 10;
$office_movement = new X10_Item('N9'); # this is the
motion detector

if (state_now $office_movement eq ON) {
 # motion detected in office!
 $desired_temperature = $desired_temp_occupied;
 set $office_occupancy_timer ($office_occupancy_timeout
* 60);
}

if (expired $office_occupancy_timer) {
 # no motion for 10 minutes
 $desired_temperature = $desired_temp_unoccupied;
}
```

MisterHouse's `Timer` variable type is perfect for this. When the motion detector senses movement, the code sets a timer for 10 minutes and sets the virtual thermostat to 75°F. If no motion is detected after 10 minutes, the timer expires and sets the thermostat to 85°, because I'm not in the office anymore (or maybe I'm just asleep!).

Because all this control is in software, there are limitless opportunities to customize your setup. For example, you could change the desired temperature remotely via MisterHouse's built-in web interface. You could track and determine your A/C's cooling rate at different outdoor temperatures. Even the fanciest "programmable thermostat" hardware would be put to shame by the flexibility you have with this system.

Dave Mabe is the author of *BlackBerry Hacks* from O'Reilly Media, Inc., and lives in Chapel Hill, N.C.

DIY

CIRCUITS

Shock the monkey: Sensor-equipped sock monkey changes its tune as it twists and contorts.

MIDI CONTROLLER MONKEY

A/V monkeyshines with flex sensors and a MIDIsense board. By Peter Kirn

Who said input devices have to be hard and mechanical? Here's one that's as soft, bendable, and easy to play with as a plush toy. In my MIDI primer (*MAKE, Volume 07, page 158, "MIDI Control"*), we saw how to use the MIDI data specification, originally designed for music, to interconnect both musical and non-musical hardware and software. Now, we'll use this approach to construct a sock monkey instrument you can use to control visuals and sound.

The monkey has flex sensors sewn into its limbs, and it wears a sensor interface board as a backpack. The sensors detect the monkey's movements, and the board converts the readings into MIDI data. This lets the monkey conduct audio-visual symphonies and perform other MIDI magic.

Circuit Bending

Designers have long used off-the-shelf sensors to translate real-world movements into digital realms. For this project, we'll use one of the most commonly used types: a flex (or bend) sensor. This type of sensor was designed by Abrams/Gentile Entertainment and used in the infamous Mattel PowerGlove (a game controller for the Nintendo Entertainment System), as well as in the more recent Essential Reality P5 gaming glove.

The concept behind a bend sensor is simple: its electrical resistance changes as it bends. This lets it work in a circuit just like other resistive sensors, such as light sensors or potentiometers. Variations in voltage that result from bending can be converted to a digital value and trans-

Photography by Peter Kirn

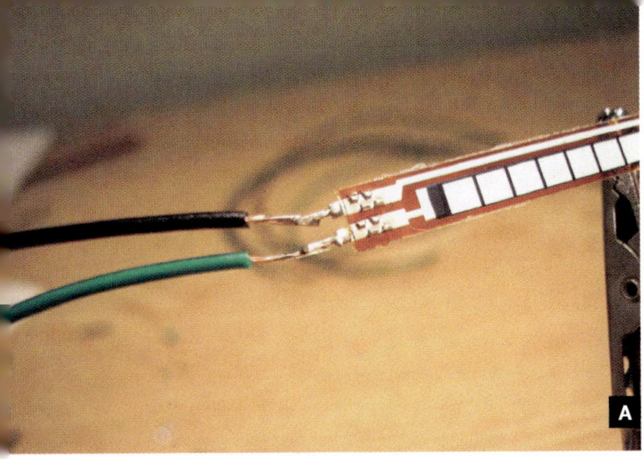

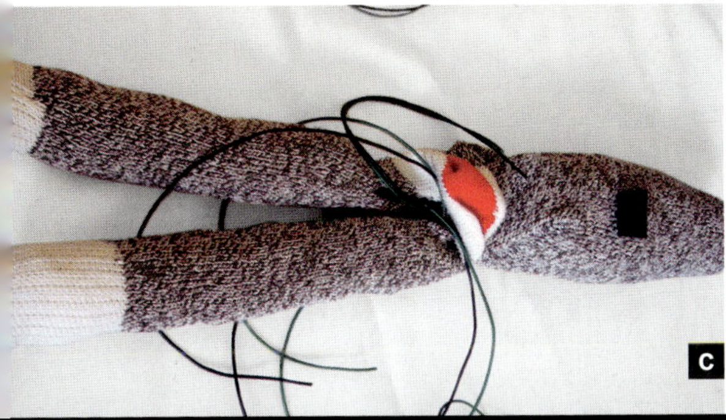

Fig. A: Two leads soldered to the end of the bend sensor. Fig. B: Bend sensors slip into muslin sleeves sewn onto the insides of the sock monkey's limbs.

Fig. C: Partially assembled sock monkey, with sensor wires and reinforcement sewn at base of monkey's back. Fig. D: Sensor wires connected to MIDIsense board.

mitted as data to a computer or other device. The typical bend sensor application is a data glove, which uses one bend sensor for each finger. Here, we'll try something different.

Monkey See, Monkey Do MIDI

Constructing sock monkeys is well-documented and easy to do. All we need to add are the bend sensors, some wires, and a way of translating the movements into MIDI. There are several off-the-shelf microcontroller boards that could perform this translation, but Limor Fried's open source MIDIsense board, which we also used in the MIDI Control primer, is ideal for our monkey. It's small, inexpensive, and designed to read sensors and nothing more. The MIDIsense takes input from six resistive sensors and has a MIDI jack output rather than serial or USB. This makes it easy to use the monkey with standard standalone music and video hardware, rather than only with computers.

For this design, I collaborated with my sister Anne Kirn, who has a knack for designing and making durable plush toys. Instructions for making sock monkeys are included in the packaging for Rockford Red Heel socks and are also widely available online, so I'll focus on the elements that are unique to this project.

Make the Monkey

1. Cut up the socks. Follow standard sock monkey instructions for cutting up the 2 socks. Basically, 1 sock forms the legs and body, and pieces of the other are used to form the tail, arms, and mouth.

2. Prep the sensors. For each sensor, cut an 8" length of a two-contact ribbon cable or a pair of hookup wires. Strip the ends and solder 1 end to the sensor contact prongs (Fig. A).

3. Reinforce the connection. Place a small piece of heat-shrink tubing over each connection between the wire and sensor. Make sure you have a size that fits loosely. Use a heat gun or hair dryer to shrink the tubing around each joint, and then reinforce the end of the sensor itself with electrical tape. (We found that all the heat caused some of the backing to peel away from the sensor; the tape made sure it continued to bend properly.)

4. Prepare the appendages and sensor sleeves. Sew the edges of the arms and legs per sock monkey instructions, so you have empty, inside-

MATERIALS
- **Rockford Red Heel socks** 1 pair, men's extra large
- **Polyester fiberfill** Or other stuffing
- **Muslin fabric** For the "sensor sleeves"
- **Felt** To reinforce attachment points
- **Thread** To match sock color
- **Buttons (2)** For eyes
- **Fleece or other non-raveling fabric, narrow double-fold bias tape, mini anorak/jacket snaps, and snap setter tool** For backpack
- **Bend/flex sensors (6)** Jameco part #150551 jameco.com
- **MIDIsense board** Available at ladyada.net/make/midisense
- **Heat-shrink tubing** RadioShack sells an assortment
- **Electrical tape**
- **Solder** Rosin-core/lead-free, any finer size
- **Insulated copper hook-up wire** Or two-contact ribbon

TOOLS
- **Wire stripper** Jameco #159290 is a good size
- **Wire cutter**
- **"Helping hands"** Or vise (optional)
- **Heat gun** Or hair dryer
- **Soldering iron with a fine tip** Weller is a good brand
- **Soldering stand and de-soldering tool or braid**
- **"Micro" flat-head screwdriver**
- **Computer or other MIDI-to-USB interface** Such as the Edirol UM-1EX or M-Audio Uno
- **Sewing machine** (optional)
- **Needle and scissors**

out appendages. Cut 5 sensor-length sleeves out of muslin, and sew them to the outsides of the inside-out legs, arms, and torso. For the torso, turn the uncut sock inside out and attach the sleeve midway between what will become the head (the sock's toe) and waist (just above the heel). You won't need a sleeve for the tail sensor, since the tail is smaller than the other appendages.

5. Insert the sensors. Turn the various parts of the monkey parts outside out as per the sock monkey instructions. Then gently push the sensors and wires into their sleeves (Fig. B). To make it easier later, you can label by body part the wire ends of each sensor.

6. Add the stuffing. Stuff the monkey with the fiberfill. Make sure you add enough so that the monkey feels firm and not floppy, but not so much that he refuses to bend easily.

7. Assemble the parts. At this point, you'll have some stuffed limbs and a lot of loose wires. Follow the sock monkey instructions to finish the assembly. Then cut a hole just large enough for all the wires at the base of the monkey's back, just above the tail. Reinforce the hole by sewing on a small piece of felt (Fig. C).

8. Feed the wires. Label the sensor wires, if you haven't already. Feed the wires out of the base of the monkey's back so they can connect to the sensor board. Pull the wires through the body so they all meet on the way out of the hole, and then pull them out together.

9. Make the sensor-board backpack. Assemble the MIDIsense board according to the enclosed instructions. Then sew together a monkey-sized backpack with a board-sized pouch and a hole in the bottom to allow the wires to feed into the board from the monkey. Small straps let the monkey wear the board snugly. We made a pack with a snap closure out of fleece trimmed with double-fold bias tape.

10. Attach the wires to the sensor board. Strip the wires from the sensors and attach them to the 6 screw terminals on the MIDIsense board. Each terminal has 2 holes, one for each wire. Insert the wires, and screw the terminals down tightly using the micro screwdriver (Fig. D). We found that the wires would sometimes pop out when torquing the screws, so we added a bit of solder to make the connections more permanent. Carefully put the board and wires into the backpack.

11. Add some finishing touches. Eyes give the monkey personality, and using 2 buttons of slightly different size give it even more character. From there, accessorize at will. Now the monkey is ready for action!

Configure the Board

1. Set up the sensors. Connect the monkey to your computer via the MIDI-to-USB interface. Use the MIDIsense software to enable all 6 sensors, and select Control Change (CC) type messages for each of them to transmit. There are many MIDI CC messages, but they all transmit continuous values rather than on/off, which is what we want. For best results, use one of the message types with a range of 0 to 127.

If you're planning to use your monkey with external MIDI hardware rather than computer software, make sure you choose messages that your hardware can interpret; most hardware documentation includes a MIDI Implementation Chart that you can check for this.

2. Calibrate the sensors. Prep each sensor so it will transmit the full range of data it's sensing. To do this, use the MIDIsense application to calibrate the low, middle, and high end of the sensor's values. Bend each sensor to its maximum, then partway, and then fully straighten while tracking the corresponding slider position onscreen. Then set the green, yellow, and red arrows to those locations. You may need to invert the high and low values so that the sensor sends 127 when bent and 0 when unbent.

Sample Application

Now our monkey MIDI is all set up. But what can you do with it? The answer, as we saw in the MIDI Primer, is anything you like; you can control music and visuals software as well as any hardware with MIDI ports. Once MIDIsense has configured the ports, you don't even need a computer; you could hook your monkey up to a hardware beatbox and use the monkey in place of the box's own knobs and sliders.

Using the free Quartz Composer tool included in the developer tools that ship with Mac OS X 10.4 and later, I rigged up a simple MIDI-controlled, animated monkey for testing and demonstration purposes.

The Quartz Composer patch I assembled takes an image of a banana and assigns it to a Particle System, which is a way of animating lots of objects onscreen with similar but independent movements. (In this case, the bananas are the "particles," spurting out in clouds of floating fruit.) Quartz Composer translates MIDI input from each of the monkey's limbs into the size of the bananas in the particle system; if the sensors are straightened, the bananas are hidden, and as you flex the sensor, the clusters of floating bananas appear and grow. You could add sound for an audiovisual interactive sock-monkey installation. For more on this patch and the calibration and programming procedures for the monkey, with videos, see makezine.com/08/diycircuits_monkey.

Of course, you can also ignore this monkey business and do your own thing, putting the sensors into another object, or using them to control something else. That's the whole fun of this. But the next time you see a stuffed monkey in a DJ booth, you'll know why he's there.

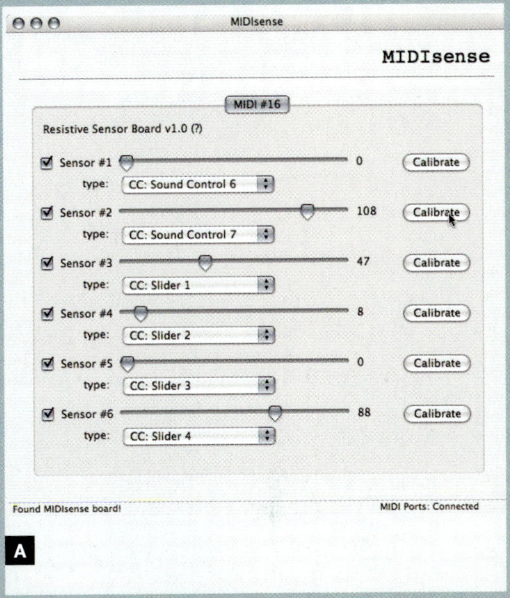

Fig. A: The MIDIsense board supports 6 sensors, which you configure and calibrate through a simple window. Set sensors as MIDI continuous controllers so that they generate a constant, reactive stream of message data.

Fig. B: A patch built in Apple's free Quartz Composer lets the MIDI monkey trigger on-screen bursts of bananas, which are animated via a particle system — delicious!

Composer and media artist Peter Kirn is the author of *Real World Digital Audio* (Peachpit, 2005), and editor of createdigitalmusic.com.

One person's trashed dishwasher is another person's source of treasured components.

CLEAN OUT A DISHWASHER

Salvaging components from unwanted appliances. By Thomas Arey

Nothing looks more forlorn to me than a piece of electronic or mechanical equipment set out to the trash.

I know that many folks who live by the maker philosophy share this view. So when I saw this early 1990s GE dishwasher sitting out for disposal, I thought I might be able to give some of the parts inside a second chance to perform some service. As with most disassembly projects, I not only discovered some neat parts, I also learned something about how things work.

As with any attempt at trash picking, make sure you are not trespassing, and take steps to ensure your personal safety. It doesn't take very long for indigenous flora and fauna to turn a curbside find into a permanent residence. Gloves and safety glasses at a minimum, please!

I began disassembly by turning the unit upside down and having a look-see. I quickly discovered that damaged impeller blades had sent this dishwasher to the curb. Probably repairable, but such parts can be expensive, so the owner sought replacement. His loss, my gain.

That said, a working motor and impeller pump have many uses. My first notion to turn those parts of the unit into a pond fountain was put aside as I began to strip the dishwasher for other future projects.

My first discovery was how simple a device a basic dishwasher is. It performs all of its basic functions by way of a single motor and a pair of solenoid-controlled valves. The single motor/ impeller pump serves to wash the dishes, as well as fill and drain the unit throughout its operating

Photography by Thomas Arey

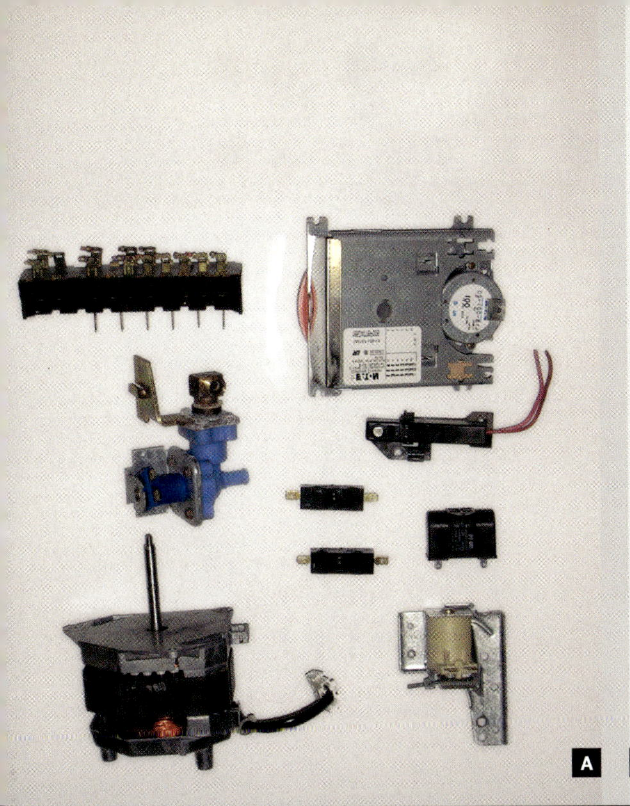

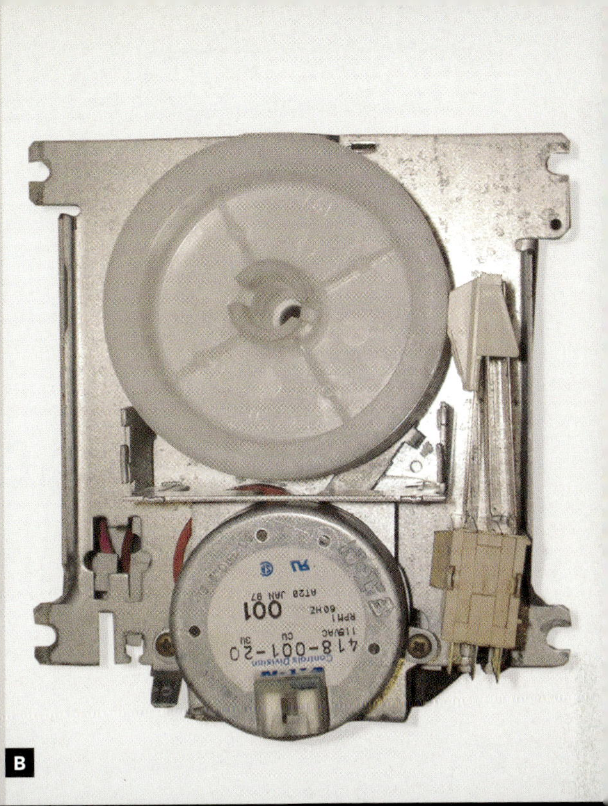

Fig. A: A few minutes' work with common hand tools and a generous application of maker philosophy yields motors, switches, valves, and solenoids, all taken from my curbside find. Fig. B: The "brains" of the dishwasher, a cam stack and 1 rpm synchronous motor. Sturdy, fail-safe electromechanical-timed switching that can be applied to hundreds of possible uses in projects where more complex electronic switching is unnecessary.

cycle. Also, all functions are performed with standard AC house current, so the salvaged parts might be useful in any project meant to be plugged into a wall socket.

The first pass at stripping out the big parts gave me a Siemens thermally protected motor that routinely shows a "Buy Me Now" price of $15 on eBay. Such fractional horsepower AC motors have dozens of practical uses for an experimenter.

The two previously mentioned solenoid-controlled valves will still likely end up in a pond fountain project next summer. But anyone who wants to control fluid flow by remote means will find such devices useful.

The unit also provided three heavy-duty momentary switches and a terminal block containing half a dozen NE-2 style neon lamps and their accompanying current-limiting resistors. Again, there are dozens of potential uses for such ubiquitous components.

The dishwasher contained a 70°C (158°F) thermostatic switch that could serve the same protective function in another project. The unit's intact heating/drying element is an item that someone more adventurous than I might find an application for. I chose to let it go.

The final item in the big parts picture is the switch unit. Control functions on this dishwasher were all electromechanical. Cycling through the unit's functions was performed by a 1 rpm synchronous motor geared to a plastic cam stack.

As the cams turned, they activated a switch assembly consisting of 6 SPDT switches. Of course, the simple 1 rpm motor can be used in many basic timing applications by itself, but since I was in the mood to take things apart, I decided not to stop there with this subassembly.

Further examination of the motor, gearing, switches, and cam stack led me to think of more advanced timing applications. Modification of the existing cam stack or construction of a new stack out of layers of plastic would allow the recovered parts to perform a maximum of 12 switching functions — all conducted without the use of ICs or PIC controllers.

Lots of food for thought there for a dedicated maker. There are many applications in which electronic circuits are not practical and a mechanical switching function is desirable.

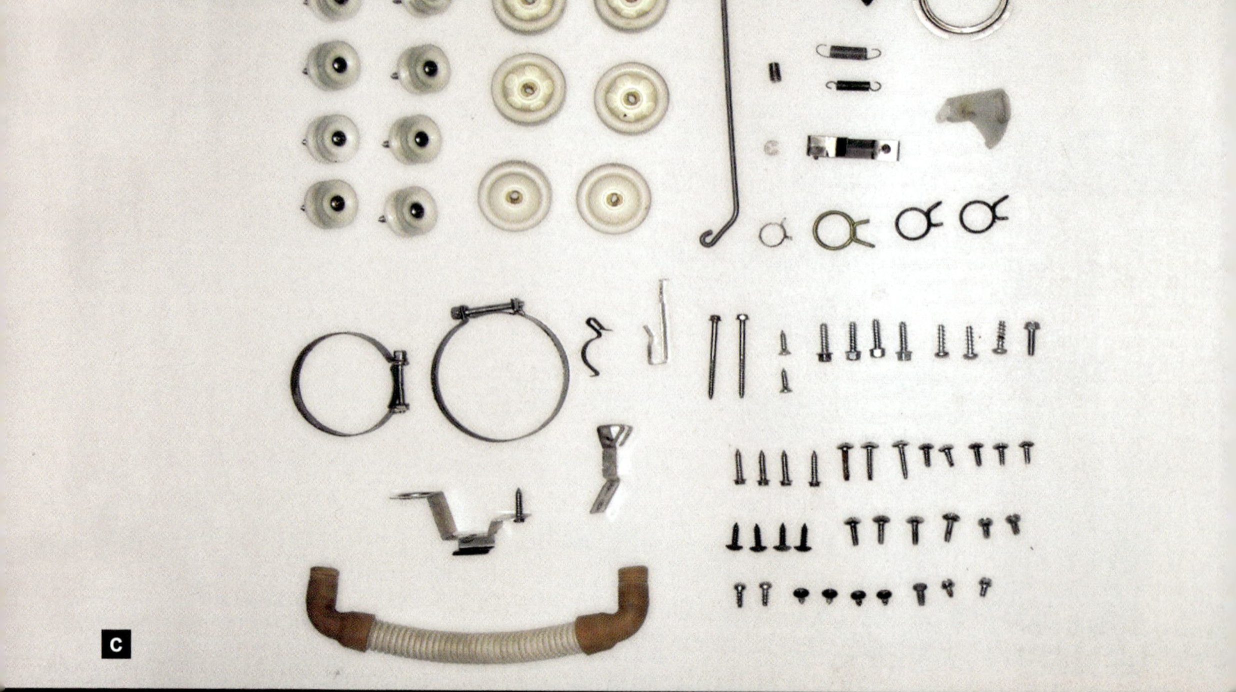

Fig. C: "Big Box" hardware stores will charge you 10 cents each or more for most common fasteners. Stripping down any appliance will turn up several dollars worth of screws, nuts, and bolts. You will also find specialty hardware including springs and clamps. The dishwasher racks provided 6 small wheels and 8 grooved wheels that can be used as pulleys. All these liberated parts will find their way into future projects.

Of course, more disassembly will always turn up a host of smaller hardware that can be used for future projects.

This dishwasher was no exception. Most of the rubber hoses were too deteriorated for reuse; however, I was able to find one plastic hose that has potential. I also added a number of hose clamps, springs, self-tapping screws, nuts, and bolts to the hardware cans under my workbench.

The main hard plastic tub that lined the inside of the unit, measuring out at over 20 gallons, gave me several ideas. The pond project came to mind once again. Another possibility would be a parts cleaning tub for projects around the garage. My wife added yet another idea: drill the tub liberally with 1" holes and turn it into a composter.

The dishwasher's dish racks also contributed to my future projects in the form of a number of small plastic wheels. When scrounging, almost everything has a potential use and salvaging appliances for future projects is always time well spent. You will not only save many useful parts from your local landfill, but also learn much along the way.

BEGINNER'S SALVAGE TIPS

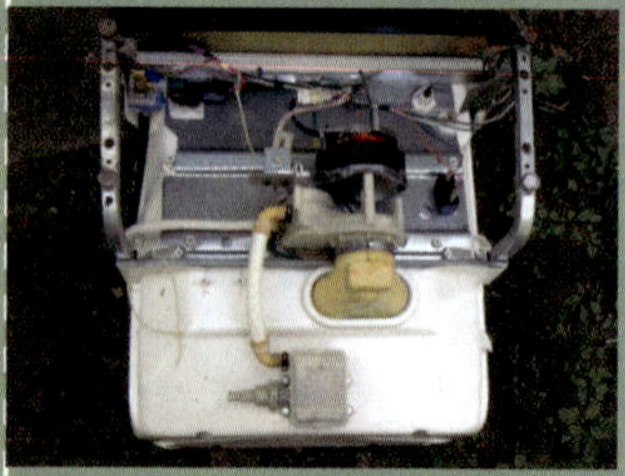

Wear gloves and goggles when working with salvaged equipment.

1. Specialty fasteners can be drilled out if you don't own the proper tool.

2. Fasteners are often hidden under moldings. A small pry bar can be your most useful tool.

3. A web search on the model of the unit may turn up information that can help with disassembly and parts identification.

4. If the part ain't broke ... save it or swap it.

T.J. "Skip" Arey N2EI has been a freelance writer to the radio/electronics hobby world for over 25 years and is the author of *Radio Monitoring: A How To Guide*.

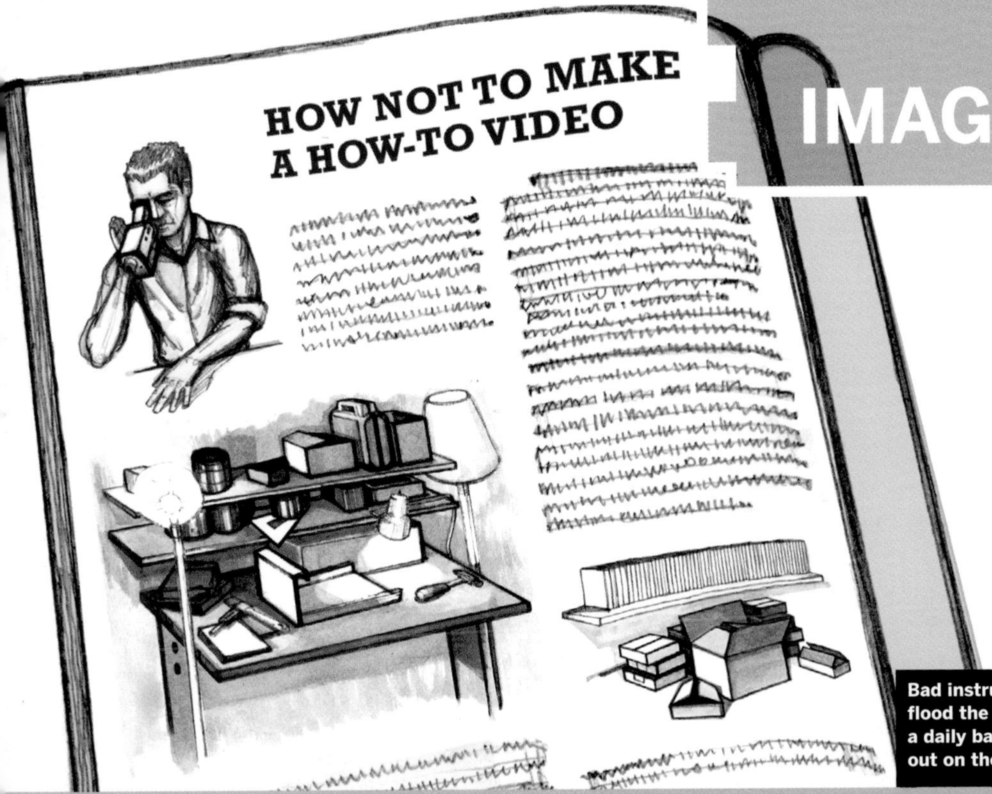

Bad instructional videos flood the internet on a daily basis. Don't miss out on the fun!

HOW NOT TO MAKE A HOW-TO VIDEO

Ignore these handy rules and your instructional video will turn out great! By Travis J. I. Corcoran

Many people have noted the collision of the open source/hacker ethic with the growing interest in hands-on projects. Add the declining cost of digital video cameras to the mix, and the result is a huge surge in folks making their own instructional videos.

If you can imagine a project or a skill, there's probably someone who's made a how-to video about it — everything from smelting metal to brewing beer to creating goth cemetery scenes to lifting large stones.

Perhaps you're interested in making your own how-to video. It's a great idea, and there's a huge market! In my job of running a how-to video rental website I've seen lots of good videos, but also lots of bad ones. After watching all these videos, I've developed a list of rules you should definitely break if you want to make a great video.

Let's take a look at some tips from the low-fi playbook, but remember: Use these tips only in reverse!

1. Hold the camera yourself.

Decent quality, consumer-grade tripods can cost as much as $45. Save that money, and have a friend hold the camera for you ... or better yet, hold the camera yourself, as you operate machinery.

2. Film with dim lighting.

Halogen shop lamps can cost as much as $15 each at a big-box retailer, and diffusers and reflectors made out of poster board could cost

Illustration by Damien Scogin

another $6. Skimp on these. Either use only a few small lights to help you cast shadows on your work, or use fluorescents to give everything a greenish, underwater tint.

3. Mumble.

Having good sound means using a decent quality microphone, placing it correctly, and practicing your delivery a few times. It might even mean shooting a scene more than once, to get the delivery and sound right. You want to skip these steps — do one take only, and speak softly.

4. Be poorly organized.

Sometimes when folks tackle projects they're poorly organized — the parts aren't ready, the tools are scattered, and the workbench is cluttered. For an air of verisimilitude, reproduce this in your video. Use a dark-colored bench. Do not lay out your tools and materials cleanly. Do steps out of order. Do not name or explain your tools or techniques.

5. Do not edit.

The process of building something often has long boring stretches, such as waiting for metal to melt, doing dozens of passes on the drill press or table saw, etc. Make sure to document each and every repetitive step in your finished video. If it's boring to do, it should be boring to learn!

6. Reproduce the video yourself.

Bulk duplication services will reproduce your video on DVD for just a few dollars per copy. Save this money by buying blank media, DVD-shaped stickers, and empty DVD cases, and duplicate your videos one at a time on your PC. Your defective DVD rate will be higher this way, and you'll have to process lots of returns from customers, but it gives your product that handmade look.

7. Do your own distribution.

You could sell your video via customflix.com and Amazon, but why should some faceless company have all the fun of putting things in boxes? Instead, sell the video yourself. Buy a roll of stamps and shipping boxes, and do shipping on your kitchen table on the first Tuesday of every month.

8. Keep your video a secret.

Do not use Google Adwords. Do not mention your video on discussion boards or your blog. Do not post excerpts of the video to Google Video, YouTube, or other free video websites — doing so might result in snippets being forwarded around the internet, thus exposing your product to thousands of interested hobbyists.

9. Have a bad website.

Ideally, you would sell your video only by snail-mail. If you must have a website, make it hard to navigate. Offer your video only as part of a package with tools that customers might already own. Use HTML features that do not work in certain browsers. Consider having an archive of "stupid customer" emails, where you ridicule people who wrote to you with questions.

10. Price your DVD too high.

Back in the 1970s, Hollywood sold movies to consumers for $100 each and made millions. Today, DVDs sell for $12 each at Target, and the studios make billions. Your task: Put on a pair of bell-bottoms, turn on the 8-track, and price your video appropriately.

There are lots of other advanced bad-video techniques I didn't cover, but these are the basics. If you reverse every one of these hints, you've got most of what you need to demonstrate your special skill or project to thousands of folks in a well-done video, and have fun doing it.

Travis J.I. Corcoran (known to friends as "TJIC") hacks software, wood, and metal. He is the founder and president of smartflix.com.

Why bother paying for advertising when you can drive around and broadcast your message?

VAN TV

Big sights and sounds hit the streets.

By Ethan O'Toole

Some time ago, I had an idea for a project that would lend itself well to marketing applications. I had previously created a few local websites and needed to promote them. After contacting the traditional media (newspaper, radio, TV, etc.) for advertising rates, I came to the conclusion that I could probably get more bang for my buck with something original. It was time to build "The Truck."

The Vehicle

First, I needed a suitable vehicle. After looking at a few different models, I stumbled across ex-U.S. Postal Service trucks on the surplus market. These trucks are based on the 1987 Chevy P-30 chassis, which is used by a number of work/utility vehicles and RVs. Of all the step vans, it is definitely my favorite in terms of the body style. It's not too large, and it's fun to drive.

The Sight

I tested various materials for rear projection, settling on translucent plexiglass with a matte texture on one side. The size of plexiglass sheets from the local plastics vendor set the limits on the screen size, so I ended up with a piece measuring 6' wide by 4' high, or 54" diagonal. That doesn't sound very large for an outdoor display, but it fits the truck well and is very impressive in person.

The first step to mounting the screen was to cut a hole out of the side of the van, which I accomplished (after many sweaty hours) with an angle grinder (the Dremel just couldn't hack this

Photography by Joe Legeckis

Here's a view of the inside of the van with the back doors closed. I set up both the projector and the mirror installation to be removable so the truck can still be used to move large equipment.

one). I had to be careful of the pocket doors while cutting through the wall of the truck; these are the reason the display is not widescreen. I cut the hole a little smaller than the plexiglass sheet, to allow for easy mounting and sealing. After installing the plexiglass, it turned out that too much light was going through it, which bleached out the picture. So, I used a white sheet stretched behind the plexiglass to act as an added projection surface.

The projector connects to the computer via a VGA cable. My original projector was an older model with a halogen lamp. But after testing a friend's projector that used a metal halide lamp, I knew I needed to upgrade. On eBay I found a 1500-lumen Sony projector with zero hours on the bulb, similar to my friend's, for $300. Its resolution is 800×600, which has fallen out of demand, but it's more than enough for this project.

With the stock lens, the projector required a long throw distance. To facilitate this, I took a mirror from none other than an old rear-projection TV. It's a standard mirror, with coating on the back of the glass, and not a front-surface mirror like those used in laser projection, but it works just fine. I set up both the projector and the mirror installations to be removable, so the truck can still be used to move large equipment.

With truck, screen, computer, and projector, I was off and running. Since then, I have put the van through numerous upgrade cycles, including moving its control to a LAN accessible via touchscreens. Here's a tour of its current, enhanced configuration.

The Sound

The audio is handled by recycled components, including 6"×9" speakers, a subwoofer, a 40-watt, 4-channel amplifier, and an active crossover that ensures that low, potentially damaging frequencies don't reach the 6×9 speakers. The system is quite loud and the sound is clean, although upgrades may take place in the future.

The Brains

The main computer in the rear of the van is a 1GHz white-box PC running the FreeBSD operating system, and it handles many functions. It runs video playback on the projector, using the MPlayer and Video LAN Client video applications. It also stores all the content, along with some scripts that help manage it. For control, it serves a private web page that lets you run everything, hosting it from an Apache web server, and it also runs a browser and XWindows to display the page to touchscreen terminals in the front of the van.

Meanwhile, the PC also hosts other applications including XMMS, which is used for MP3 playback with visualizations, and XScreensaver phosphor module, which displays text on the screen in a retro style.

The Controls

The main control interface to the system is via three touchscreen LCDs in the front. The touchscreens are recycled 486-133MHz point-of-sale terminals that I converted from Windows 98 to the NetBSD operating system. They communicate with the main computer over an Ethernet LAN, and since they aren't powerful enough to run standard browsers themselves at usable speed, they render the control interface by running XWindows to remotely display the browser window executing on the main machine.

In addition to these "cockpit control" touchscreens, a second Ethernet jack on the main computer connects to a wi-fi access point, in order to provide wireless remote control. Apache is configured to serve the private control page over both the Ethernet and wireless networks. With this arrangement, you can control the van's projection and audio via a wireless-capable laptop or PDA; using the two physically separate interfaces for the wired and wireless networks ensures that people can't hack the terminals via wireless.

The Live Feed

For an outdoor concert experience anywhere, you can also wirelessly stream live A/V content to the truck. For this, I connect the camera to an MPEG2 encoder box, which connects to a wireless-enabled, Linux-based laptop via USB. If there's a live music show, the truck can be parked on a nearby street and show video that's happening in the venue. I've tested this in a parking lot but haven't done it at a show yet.

The Power

The truck still maintains its original alternator. Power is fed from the alternator into a dual-battery isolator, which splits it between two separate sets of batteries. The truck itself uses two standard car batteries, and I have three class-24 marine deep-cycle batteries mounted in the rear for the computer and A/V equipment. The deep-cycle batteries in the rear are connected to a low-quality 2200-watt power inverter/charger. I originally tried to power the system with smaller UPS units, but they didn't work well with the projector (and the larger UPSs use 24V, which is incompatible with standard 12-volt vehicle power).

The Ground Effects

The finishing touch was LED ground-effect bars. I originally set out to build my own, but after realizing how much time and effort it would take, I contacted a manufacturer in China named CodeLED and purchased LED tubes and a controller. I am currently working on reverse engineering the communications protocol that the controller uses to signal the tubes. My plan is to integrate it with the computers, so the touchscreens can also manage the ground effects. In the meantime, the supplied controller is fairly flexible in programming and capable of producing many colors by using pulse-width modulation. I hacked them up a bit and am using four long tubes on the truck (which aren't normally available with the factory-shipped kits).

The Future

Future additions planned include a HDTV receiver and a GPRS/SMS card so people may text-message the truck from their mobile phones. I plan to feed the text-message data into the phosphor program.

The truck can often be spotted around southeastern Virginia. I generally take it and park in downtown Norfolk, downtown Virginia Beach, and the Virginia Beach oceanfront.

When parked in a well-lit store parking lot, the display is clearly visible. The truck rides high enough that the screen clears many cars. It gains a good amount of attention in any situation, so I would call the project a success.

Ethan O'Toole is a creative Unix and network admin who lives in Norfolk, Virg., and likes to start lots of projects. (Completing them is often the challenge.)

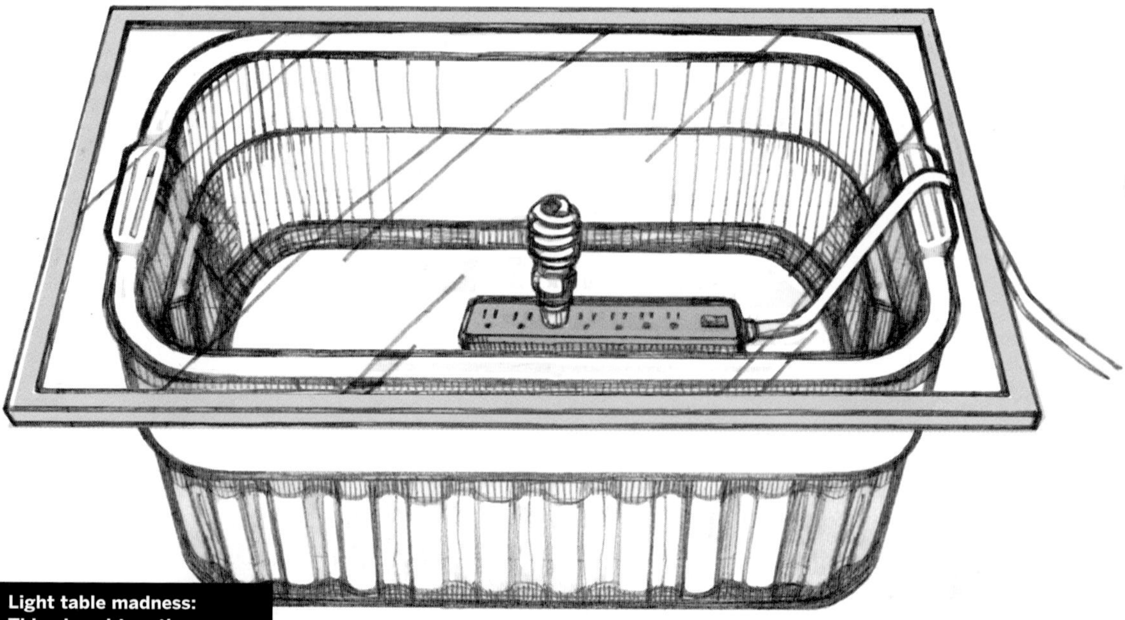

Light table madness: This pieced-together light table works like a charm.

QUICK AND DIRTY LIGHT TABLE

A storage bin, a pane of glass, and fluorescent light saves hundreds of dollars. By Hiram Cook

A couple of weeks ago my wife asked me to make her a 2-foot by 3-foot poster to display at a retirement party for a friend from work. I created the image on the computer and printed it using a graphics program that sections large, banner- and poster-sized images into pieces that a home printer can handle. But after trying to line up the sections to neatly tape them together, I realized that I needed a light table to get the alignment right.

Necessity is the mother of invention. In a hurry and not wanting to spend big bucks on a light table I might never need again, I looked around for light table makings. For the tabletop, I used the storm window from our back door. It measured 25 by 36 inches — big enough to hold the finished poster. I laid the glass flat on a storage bin without its top, and placed the bin on top of a bench to bring the tabletop up to workable waist height. For the light underneath, I used a 23-watt compact fluorescent bulb screwed into a light adapter and plugged it into a power strip, all inside the bin.

This improvised light table worked like a charm. I taped the poster together in no time, it came out looking great, and my wife was happy — which is what really counts, isn't it?

Hiram Cook is a CNC programmer from Allentown, Penn., who enjoys solving problems with low-tech solutions.

Illustration by Damien Scogin

BY SAUL GRIFFITH, NICK DRAGOTTA & JOOST BONSEN '06

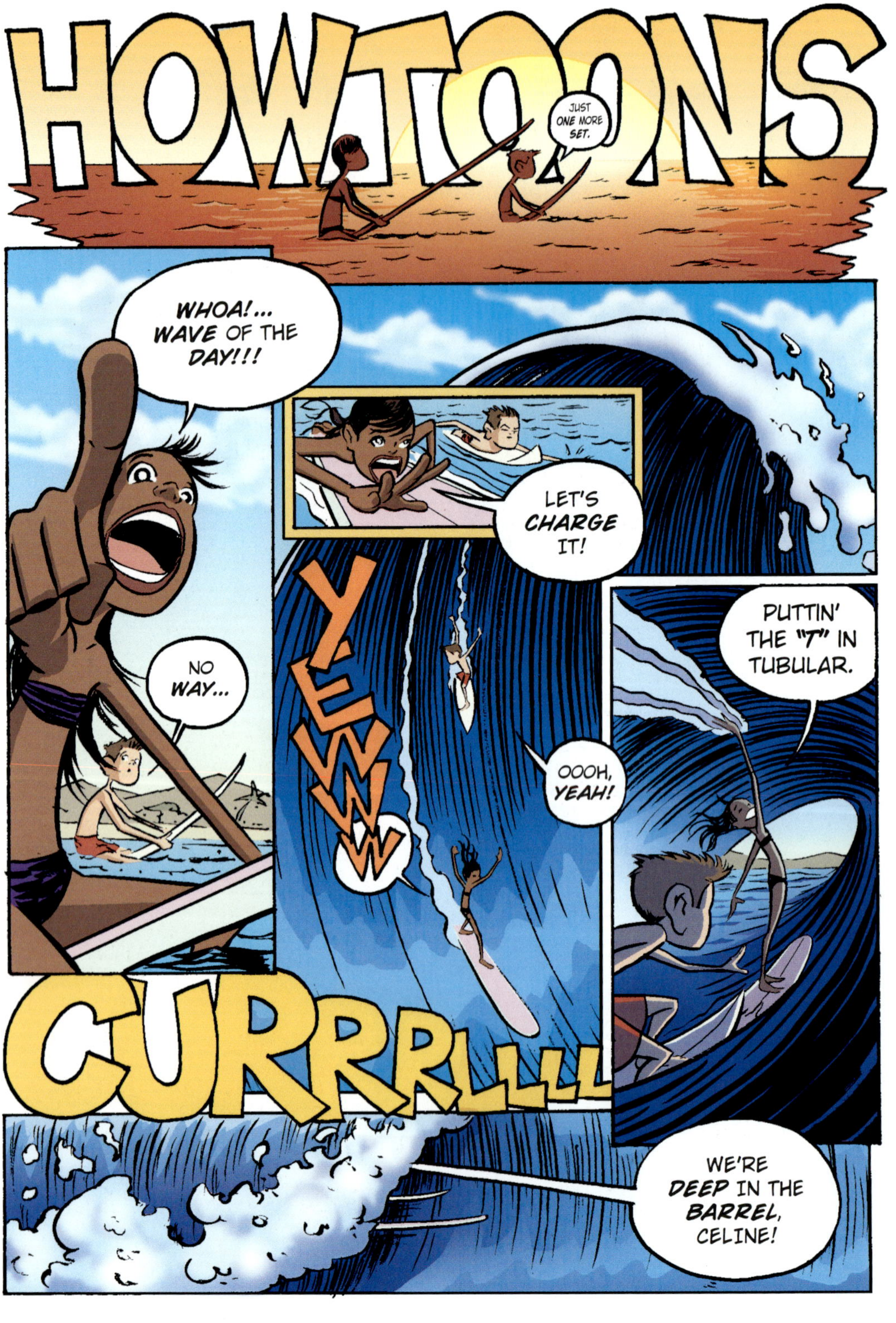

NOW LET'S CARVE IT UP.
WOOOOOOH
STICK THAT STALEFISH TUCK!
HANG LOOSE CELINE, ROLA BOLA!
KILLER SESSION, CELINE.
YEAH DUDE.
Build Your Own... ROLA BOLA!
PVC PIPE SHOULD BE A COMFORTABLE HEIGHT AND WIDER THAN THE DECK!
MAKE SURE YOUR DECK IS 30-36 INCHES LONG...
...AND WIDER THAN YOUR FOOTSIES.
ATTACH A STOPPER FOR SAFETY!
USE MANY SCREWS FOR STRENGTH!
END FOR NOW!

MakeShift

By Lee D. Zlotoff

There are few fears more primal than being cast away on a deserted island in the middle of who-knows-where. Maybe it's a throwback to our eviction from the Garden of Eden, or the nagging realization that Earth itself is just such an island in the endless ocean of space. Whatever the reason, it has been and remains a mainstay of the collective imagination.

The Scenario: You are on a small sailboat in the South Pacific when a freak wave of biblical proportions swallows your craft. You awake to find yourself on the rocky, sandy beach of — what else? — a tiny and deserted tropical island. As the fierce equatorial sun beats down on you, you realize that the boat is gone, but a large section of the white, waterproof nylon sail has washed up on the beach.

Relieved that you are alive and have sustained no major injuries, you quickly scout the island. There is a cave for shelter, an abundance of vines and vegetation, but no trees to speak of. You see enough sea birds and marine life to provide a subsistence diet, but there is no source of fresh water! And the rainy season is still months away.

The Challenge: Come up with a reliable way to produce potable water until you are rescued — or the meaning of life is revealed to you, and being rescued no longer matters.

Your items: You have only the nylon sailcloth and what you were wearing when you washed ashore: a dark, waterproof windbreaker; a T-shirt; and shorts, in which you find your Swiss Army knife (or Leatherman tool) and a pack of waterproof matches. If it provides additional motivation, feel free to be cast away with the fantasy celebrity of your choice — but this person is still counting on you to provide drinkable water. And if you're looking for extra points here, forget the pack of matches.

Good luck, and rest assured that we're all out there looking for you.

Send a detailed description of your MakeShift solution with sketches and/or photos to makeshift@makezine.com by Feb. 23, 2007. If duplicate solutions are submitted, the winner will be determined by the quality of the explanation and presentation. The most plausible and most creative solutions will each win a MAKE sweatshirt. Think positive and include your shirt size and contact information with your description. Good luck! For readers' solutions to previous MakeShift challenges, visit makezine.com/makeshift.

Lee D. Zlotoff is a writer/producer/director among whose numerous credits is creator of *MacGyver*. He is also president of Custom Image Concepts.

Photograph by Christopher Lucas

PRIMER

Molds can be made of rubber, steel, clay, brass, wood, plastic, lead, and even sand, and people use them to cast objects ranging from 40-foot bronze Buddhas to microscopic cell structures. This article focuses on casting small objects in rubber, which is a reliable technique that's employed throughout the film and special effects industries, as well as the jewelry industry.

Photograph courtesy of Adam Savage

Moldmaking

How the pros replicate objects.

By Adam Savage

In principle, moldmaking is a simple process, but with every object you want to replicate comes a new series of pitfalls, innovations, and solutions.

This article explains how to make a two-part, underpoured block mold, which is a versatile and beginner-friendly type that's great for small, detailed objects such as jewelry, game pieces, masks, picture frames, and figurines. I learned this technique by apprenticing under some of the great moldmaking masters in the special effects industry, and this article reveals their unpublished tricks. I hope they don't get mad.

We'll make our mold out of silicone rubber, an excellent casting material, but it costs about $100 per gallon. This process uses as little of it as necessary, and it's important to follow all of these instructions, because a mistake can be costly. Then we'll cast our duplicates in opaque urethane resin (clear resin requires a more difficult process).

Why a Two-part, Underpoured Mold?

Two-part molds can handle more shapes than one-piece molds, which work only for simple, completely convex objects. And underpoured molds minimize problems with bubbles in the resin.

Underpouring means that you pour the resin into a main intake vent (or *sprue*) that curves around to fill the mold up from below, rather than simply pouring into the top (Figure A, opposite). Meanwhile, smaller vents on top allow the displaced air to escape. As you pour, resin splashing down forms bubbles which can stay in the main cavity and ruin the surface of the casting. These bubbles also tend to collect in fine-detail areas, where they are the most difficult to deal with.

The advantage of under-pouring is that it generates fewer bubbles, and lets them rise up into the vents where they won't cause trouble. Top-pour molds are sometimes acceptable, but pouring from underneath is generally worth the extra silicone required.

Design the Mold

The first step in designing a mold is to visualize your object upside-down and inside-out. Imagine it as a cavity being filled with resin, imagine the path that resin will take, and what happens to it along the way. Where will bubbles form? Where should the main vent go? How should the object be positioned? (Probably upside down, so most cleanup takes place on the underside.) What will be the hardest part to clean up? Where do I need to put vents? Should the piece be cast as one part, or broken into several parts?

Thinking through these questions makes everything easier later, when you clean up the casting. This means cutting off and sanding any excess resin from the vents and seam, and filling in any bubbles. A good casting has few bubbles and is easy to clean up, and this starts with good mold design. It's always easier to sand away material than to fill voids.

As an example, we'll cast an angel figurine (Figure B). I decided to remove her arms and cast them separately because they looked like trouble; they extend far from her body, and the hands have delicate detail. Broad, thin parts are also difficult to cast, so I removed the wings and replaced them with new ones I made separately on a vacuuform machine, which is excellent at producing flat forms.

1. Set up your object on a piece of foamcore and build up the *pouring gate* and *vents*. After removing the angel's arms and wings with a Dremel tool, I set up the body upside down; the bottoms of her feet are flat, so they made an ideal point to pour into. I sprayed the body with primer in order to see the surface better where I had made the cuts (Figure C).

Form the pouring gate using a plastic hemisphere, and make the sprue and main outlet vent using armature wire. Use cyanoacrylate (CA) glue with an accelerator to piece everything together. To make clean-up easier, the contact between main chamber and vent should be as small as possible; I sanded the wire to a point where it touched the angel's head. I then used .06" square styrene strips to add smaller vents from the folds of her skirt and sleeves, which seemed bubble-prone.

2. Use a Sharpie to mark along the *parting line*, the line along which you'll cut the mold open (Figure D). It should extend across the base itself and all the way around the model, following any corners that exist, or along high points and open spaces to define a seam. *Flash* (excess material) from mold separation can form along this seam, so you should place it where the flash will be easy to sand and fill.

3. Build the mold box around the object using one large piece of foamcore (Figure E). Score and fold the foamcore without cutting through it, roughly following the object's shape. Keep at least ¾" between the model and the outside of the mold. Thinner, and the silicone can deform under the poured resin later; thicker, and you're wasting silicone and the part will be harder to cut out.

✱ Taking Care

***Many mold materials are time-sensitive, so after each step, you lay out all the things you'll need next.**

***Molds are too often destroyed by impatience; not spending an extra 2 minutes mixing the silicone (for example) can throw away several days' work.**

***Some materials, such as the thickening fillers, are also toxic, so wear eye protection, gloves, and a smock or apron, and work in a well-ventilated area.**

***Finally, no matter how careful you are, be prepared to mess up until you have more experience.**

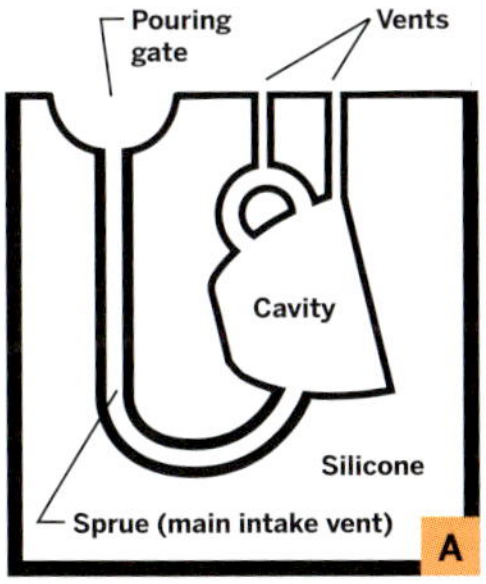

Fig. A: Cross section of underpoured mold. Fig. B: The object we're going to cast, an angel figurine. Fig. C: The angel being set up for molding. The sprue runs to her feet, which are a conveniently flat surface. Fig. D: Marking the parting line, along which the mold will be cut later. Fig. E: Building the mold box using a single piece of foamcore. Fig. F: The mold box, ready for sealing with hot glue.

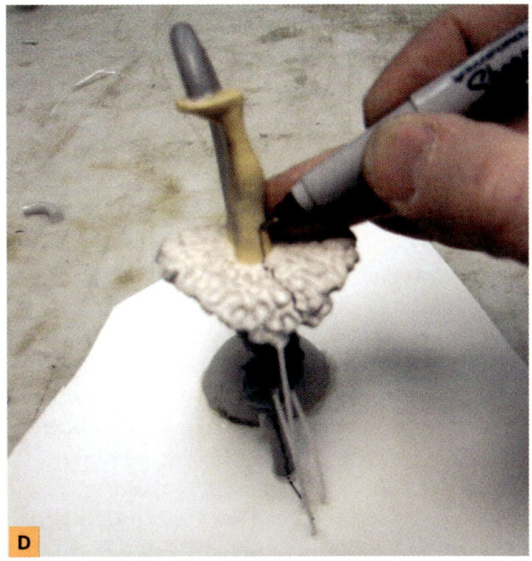

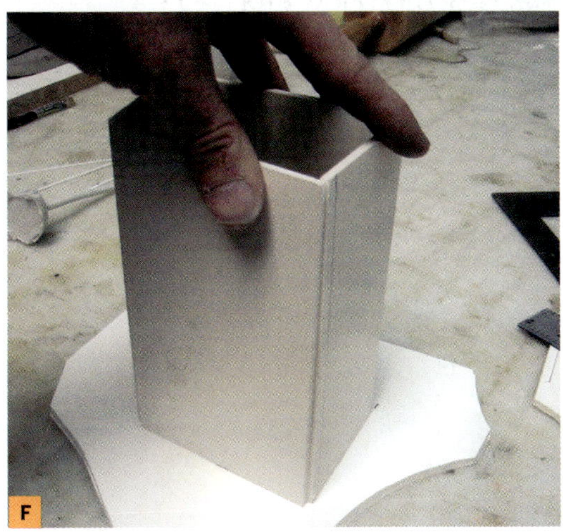

Photography by Adam Savage

4. Use a hot glue gun to secure the box to the base, and to glue the (single) seam of the box (Figure F). Carefully examine your entire glue seam to make sure it will hold the liquid silicone, which will leak out of the smallest of holes.

Pour the Silicone

1. Figure out how much silicone you need by calculating the volume of your box. I do all my moldmaking in grams instead of ounces, since it's so much easier to multiply and divide by 10. Remember, *volume = length × width × height*. Measure in centimeters, and you'll get a volume like 755 cubic centimeters. Silicone is close to the specific gravity of water, so thanks to the convenience of metric, this is the same number as the weight in grams of silicone you'll need. If you measure in inches, multiply your cubic inches volume by 17 to get the rough equivalent in cubic centimeters or grams.

2. Measure the silicone and activator. Almost all silicones use a ratio of 10:1 silicone to activator. You can also add an accelerator, in the amount of 2-10% of the total weight, but the resulting silicone won't be as durable and will tear more easily. Curing times will be 24 hours with no accelerator, 4 hours with 2% accelerator, and 1 hour with 10% accelerator.

3. Mix the silicone, activator, and accelerator, if used (Figure G, opposite). The activator is blue to give you a visual cue when you're done mixing. If you see any streaks, you're not nearly done. Make sure you're using a mixing stick that's strong enough, and scrape the sides and bottom, where it's hardest to get a good mix. I usually mix by pushing the mixing stick down the sides of the bucket, and dragging it across the bottom, turning the bucket as I go, for as many as 6-10 full revolutions. Silicone is thick, so it helps to have someone else hold the bucket.

Beginners often make the mistake of not mixing enough; a good rule is to mix until you think you're done, then mix for another 3 minutes (Figure H). There's nothing worse than wasting a mold because of inadequate mixing.

Here's What You'll Need

This is a lot of stuff, but it's all essential; the price of a missed step is often catastrophe. Also, you can use these tools and supplies to make many more molds.

Silicone RTV (room temperature vulcanizing) rubber and activator Silicone is soft (low durometer), durable, and accurate for detail. Many varieties and colors are available, which will all work for our purposes, and I've found moldmaking suppliers to be helpful about which products are best suited to different projects.

Urethane resin or other casting material

Talcum powder or cornstarch

Wax and shellac (optional) It is necessary to seal the pores and keep the silicone from sticking if your object is unglazed terra cotta, wood, or other highly porous material.

Styrene strip or equivalent Most hobby stores carry Evergreen or Plastruct brands.

Hobby knife with a curved blade and regular #11 blades

Matte knife Or snap-blade knife

12" or 18" straightedge

Armature wire 3⁄16" or 1⁄4" At any art store

Plastic hemisphere(s), at least 3⁄4" diameter From plastics supplier or art store

Cyanoacrylate (CA) glue Such as Zap a Gap or Krazy Glue

CA glue accelerator (or baking soda) This makes the CA glue "kick" almost immediately; regular baking soda works well for this, and with no smell!

Paint mixing sticks

Tongue depressors or popsicle sticks

Mixing cups

1-gallon paper paint bucket

Hot glue gun and glue sticks

3⁄16" or 1⁄4" foamcore

Rubber bands

Small cloth sack

Permanent marker Such as a Sharpie

Scale Digital is best, but a triple beam balance will do

Dremel tool with fine cutting wheel

Orbital palm sander Or large massage vibrator

Needlenose pliers

Rubber or nitrile gloves From drugstore or medical supply

Safety glasses or goggles

Respirator Silicone is relatively benign, but a respirator is a good idea

Wet paper towel Or water-based clay

And for those more serious:

Pressure pot

Evacuator/vacuum chamber

Air compressor

4. Now the silicone is ready to pour into the box. Don't just pour it willy-nilly; this is a surefire way to trap bubbles in parts of your mold. Instead, pour in one location, away from your part, and let the silicone slowly fill up the mold around it (Figure I). Pour as thin a stream as you can, about as thin as a pencil lead. This starts to remove some of the tiny, invisible bubbles formed during mixing. Use the stir stick to scrape every last bit of silicone out of the bucket.

Should you discover silicone leaking from a small hole in the box, stop it with a wet paper towel or, better, a small piece of water-based clay. The water helps the silicone coagulate. Trying to patch the hole with hot glue might just make it wider.

5. In a professional shop, bubbles are removed by putting the bucket into an evacuator (vacuum chamber), a piece of equipment which can cost $1,200. On a budget, you can *degas* your mold by agitating it with an orbital palm sander. Just hold the sander against the table, right next to the mold, for about 5 minutes. A layer of bubbles should rise to the surface.

6. Wait until the mold "kicks," 24 hours for most RTV silicones without an accelerator.

Cut the Mold

1. Remove the box and silicone from the mold base by inserting your X-Acto knife blade at an angle from just above the hot-glue seam around the bottom (Figure J, next page). This frees the mold, but keeps almost all of the box intact. We like this box. It will keep our mold aligned and accurate.

2. Give the box a yank to separate it from the base. There will be some stickage of silicone and you will tear loose the glued-down pouring gate, but nothing major. After you free the base, cut an *alignment key* into the edge of the box and the newly exposed top

Fig. G: The silicone before mixing. Fig H: The silicone after mixing. If you see any white streaks, you're not nearly done. Fig I: Pouring the silicone. You should pour thinner than this, unless you have an evacuator to put the mold into immediately afterwards (like I did).

G

H

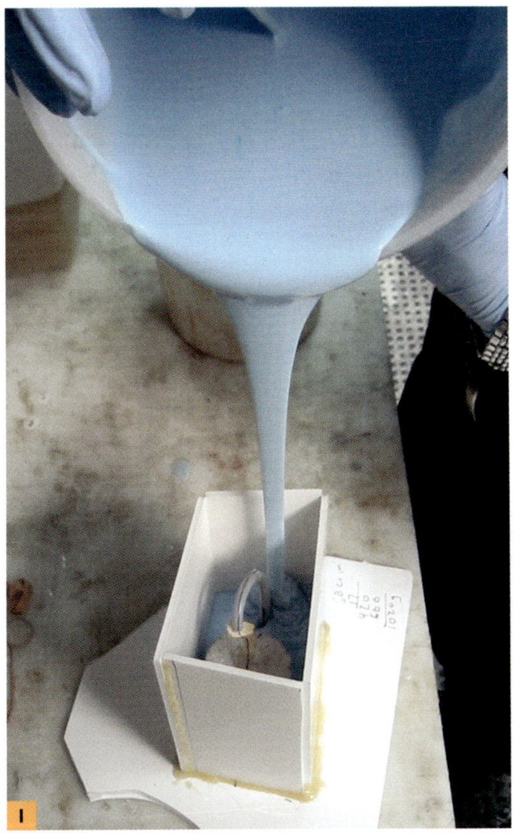

I

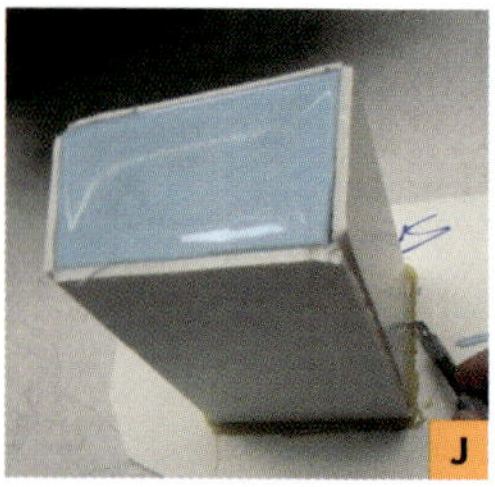

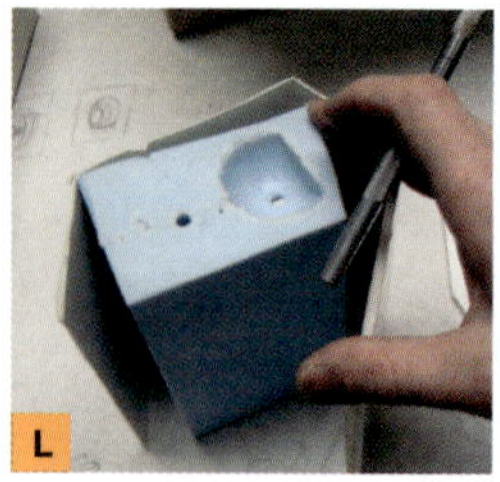
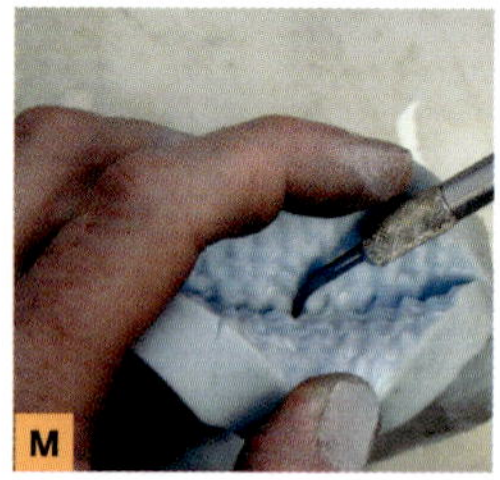

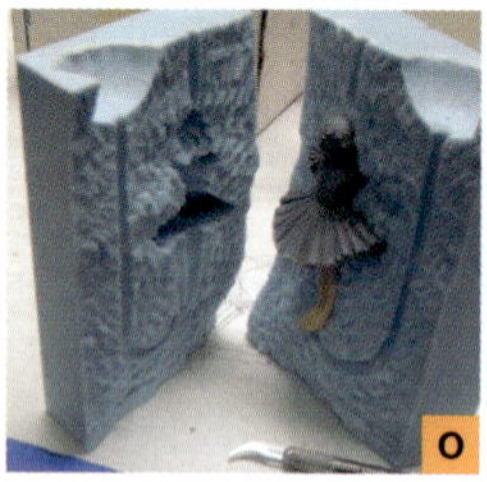
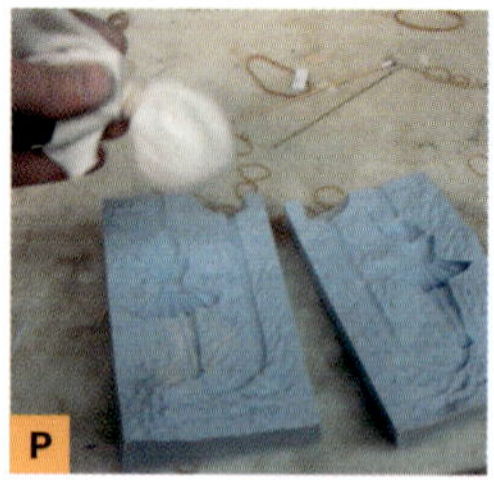

Fig. J: The mold, ready to be cut out. Fig. K: Cutting the box seam. Fig. L: The mold, released. Fig. M: The jeweler's cut. Fig. N: Prying the halves apart. Fig. O: The angel revealed (with vents, etc. removed). Fig. P: Dusting the mold with release agent (cornstarch or talcum powder).

of the mold. This is a marker notch that lets you exactly replicate how the mold fits into the box. Lining up this key prevents deformed castings when the box isn't precisely symmetrical (which it isn't).

3. Cut open the seam by inserting the blade perpendicular to the butt joint of the foamcore and slicing downward (Figure K). Try not to damage the box, because as I said, it will support the silicone later, during casting. Release the mold (Figure L).

4. Now that our mold is released, it's high time we released the object inside. This involves what may be the toughest skill in the moldmaker's arsenal: the dreaded *jeweler's cut*. You're going to cut the part out by cutting the silicone block in half, but you also want the 2 halves to line back up just the way they started, or "key together" in moldmaker's terms. The jeweler's cut creates bumps and ridges that force the halves to key together (Figure M).

On the silicone block, find the line of ink that transferred from the section of parting line that you drew on the base. Using a curved X-Acto blade, start cutting at this line from the inside out, never deeper than 3/16" (3-4mm), following the line drawn on the object out to the edge of the block.

As your hand drags the knife through the silicone, twist the curved blade side to side with the tips of your fingers. This twisting action creates the hills and valleys that define the jeweler's cut. After you cut your first twisty line, pry that slice as wide as you can with your fingers (or use mold spreaders, a tool that's like reverse pliers), and make another twisty cut at the bottom of the gully you just created. This is where the curved blade comes in handy: a straight blade, encountering a bump from a previous cut, will cut right through it, whereas the curved blade moves around it!

Some pointers: Make your hills and valleys deep, but no wider than 3/16". Areas where the model comes closest to the edge of the block have the most danger of leakage, so you want at least 2 complete passes of hills and valleys between the model and the block's edge.

Stagger your cuts. Don't cut your hills all in a line, because this will let the mold halves slide along one axis relative to each other. If your mold can move, I guarantee you it will. Also, if you're slicing through the hills from a previous pass, you're not spreading your mold apart enough. Always cut from the inside out, and go slowly!

This takes practice; I won't lie to you. It takes a time to get a feel for it, and spreading the silicone is hard work. But careful attention here can make the difference between a casting that comes out ready to paint, and one that needs a day of sanding and filling.

5. Keep cutting and cutting until you can separate the halves and remove the object (Figures M, N, and O).

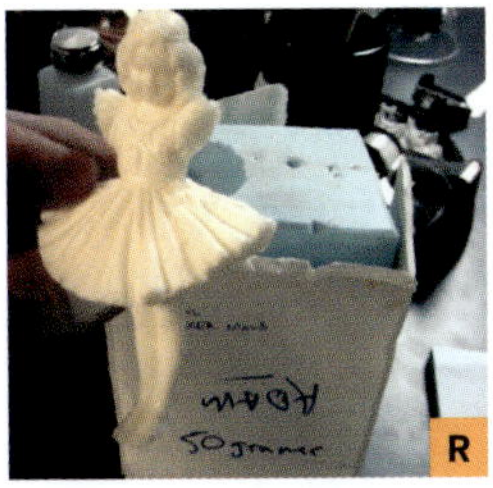

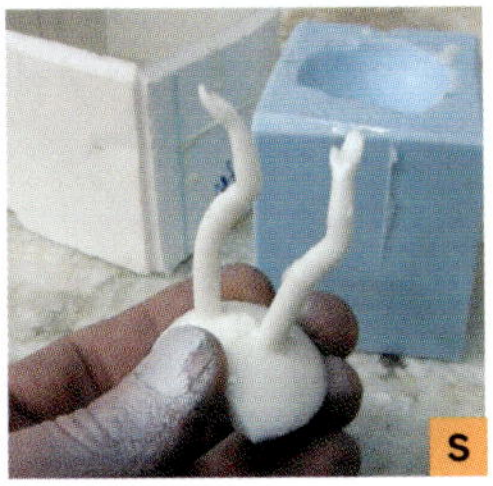

Fig. Q: Pouring the resin. Fig. R: The main casting unpacked from the mold. Fig. S: The arms removed from their mold. Fig. T: The finished casting, cleaned up, glued together, and painted with primer.

Cast It!

1. Remove the original part, sprue, vents, and pouring gate, and dust the inside of each mold half lightly with cornstarch or talcum powder. I put my powder into a small cloth sack, secured with a rubber band, to spread it evenly without clumping (Figure P). Silicone rubber gets a slippery tack to it, as the silicone gradually sweats out. This means that silicone molds generally need no *mold release*, because the silicone itself prevents sticking. But a light dusting still helps the two halves align together perfectly, and also acts as a sponge to the resin, drawing it into the fine details of your mold and inhibiting small bubbles which can gather at the high points.

2. Put the mold halves back together, box them in their original box, and secure it all with rubber bands, pouring side up.

3. Mix the 2 components of the resin, following package instructions. I used a common, fast-setting urethane resin called Insta-Cast, from Douglas and Sturgess, which has a pot life of only 60-90 seconds. This means that you have to mix it fast! In general, I mix by counting 30 strokes with the mixing stick, taking care to scrape the sides and bottom as I go.

4. Pour the resin into the mold (Figure Q). The hemispherical pouring gate provides a spill-proof cup for the resin (imagine trying to pour accurately into a tiny hole). When you see the resin come up the vents, you're done. I usually keep an old simple mold at my pouring bench, so that when I have extra resin in a cup from a pour, I can dump it somewhere useful. If resin doesn't come up the vent, the vent may be blocked, or the main sprue needs to be thicker. If that's what you suspect, let the resin set, and use a knife to widen the vents or the sprue on the open mold afterwards. I often add vents where I find unexpected bubbles in my molds. Moldmaking is all about improvisation, and as long as you prepare well, there are many problems you can solve even after casting.

5. Wait for the resin to set, and unpack the casting. To remove the extra resin from inside the vents, break them off with needlenose pliers, then sand the surfaces smooth. On my angel, I was right about the vents in the dress and the shoulders, as she needed no clean-up at all (Figure R). For her arms I just made an overpoured mold (Figure S) and shoved a Q-tip down the hole after pouring the resin to free up bubbles. Then I glued the arms on, installed new wings, and painted it all with primer (Figure T).

Another tip: After pulling the parts out of the molds, I weigh them and write the number of grams on the foamcore mother mold. This lets you know exactly how much resin it takes, so you can mix only what you'll need.

Adam Savage is, above all else, a collector of skills. He makes a lot of stuff you might know about, and a bunch of stuff you don't. He lives in San Francisco with his wife and twin sons.

Create your own computer games, discover some magnetic attractions, and become a paper airplane champion.

TOOLBOX

Stop Buying Hideous Board Games

Prices vary ebay.com

Compared to 30 years ago, today's board games are pieces of junk. Last year I bought three games I remember loving as a kid: Life, Kerplunk, and Trouble. I wanted my 8-year-old daughter to experience the fun I had playing them. But when I opened the boxes, I was disgusted with the shoddiness of the materials. The straws in Kerplunk were made of astonishingly flimsy plastic, making it difficult to push them though the holes. The pieces on the Life gameboard constantly fell off, forcing us to stop playing every so often to reattach components. And the cylindrical game pieces in Trouble bounced out of the holes every time you pressed the Pop-O-Matic die tumbler.

I couldn't believe how lousy these formerly terrific games had become. I went to eBay to look at pictures of the old versions and discovered that you could buy the original games there, often at lower prices than the new, junky versions. A 1967 model of Kerplunk, with stiff sticks and a cylinder that doesn't fall apart every time you touch it, can be yours for under $10. A vintage 1965 edition of Trouble, with pieces that stay put, is just $5. And a 1960 edition of The Game of Life, featuring Art Linkletter's hearty endorsement and board pieces that don't fall out, is $15.

The other bonus you get from buying old board games is the cheerful, well-designed package and gameboard graphics. The modern versions are so ugly that they depress me. From now on, eBay is the only place I'll ever buy a popular board game.

—Mark Frauenfelder

Photograph by Tracey Jipson

Battling Tops

$20–$30 ebay.com

Battling Tops was my favorite game in the 1960s, and its gladiator-like action is still popular today, inspiring other wind-up top games like Beyblades. It can be appreciated by little kids (as long as someone helps wind their tops) and adults alike. The secrets of winning are well guarded (they might include variations in winding or in wrist action, but I'll never tell). The original 1968 Ideal version remains my favorite among the several versions produced since then; its rink is larger and the graphics on the labels are classic. It can be found used on eBay for $20-30 (or more for rare-colored rinks that are collectors' items). A new version called Original Battling Tops by Mattel is available in stores, but it has a smaller rink that pales in comparison to the original. Whatever version you find, the name of the game is still last top standing!

—Barton Blum

Box-a-holic

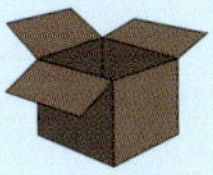

Cardboard boxes, Free

The first thing that comes to my mind when I think of my favorite toys is a big cardboard box, like a refrigerator or washing machine box. Let's face it. Boxes are the coolest toy on the planet. You can do anything with a box. Need a ride to Mars? All aboard the cardboard-box rocket ship. Looking for a good robot to play with? Stuff one of your lil' friends in a cardboard-box robot suit. Playing house? Box. Love to paint murals? Box.

As attention span stretchers, boxes are the only disposable fun I've ever seen, and they easily outlive a roly-poly trapped in a jar. Bugs could have topped the list of favorites, had torturing them not been deemed cruel and unusual by the insect world.

But I can't write about a box now, can I? That would be simply ridiculous because, in the real world, toys are not really considered toys until they are purchased at a retail store. Boxes, as we all know, are usually free. But that's where makers come in. Cherish your boxes. Give them respect. Think inside the box for once.

—Kristina Reed

Easy as Pie

PyGame, Free pygame.org

If you've ever pondered the idea of creating a computer game but thought that the learning curve was just a little too steep, consider giving PyGame a shot. While you may not make the most cutting-edge game the world has ever seen, PyGame does have several important advantages: it's written in Python (a very easy language to pick up); doesn't cost anything; gives you an easy entry point into the paradigm of game development; and makes tasks such as rendering graphics, playing sounds, and interfacing to the mouse and keyboard as easy as ... py.

At the end of the day, these advantages make a tremendous difference in your productivity, and they allow you to spend more time having fun. Think about it: not sinking hours into figuring out where to start, and then getting bogged down with an unimaginative task, means you get to spend more time on interesting activities that will make your game more enjoyable to play. You can download PyGame and try out some of its excellent tutorials. It's an easy install and it only takes a couple of hours to create one of the example games. Why not get started?

—Matthew Russell

Editor Phillip Torrone talks about kits that caught his eye on makezine.com/blog.

Build a Better Mousetrap (Car)

$8–$20 docfizzix.com

Remember that mousetrap car you had to build back in grade school? Well, Doc Fizzix has taken it to the next level, with a website full of mousetrap vehicle kits (even balloon-powered ones) and a book of tips and advice for the fledgling maker.

Embedded Linux Starter Kit

$399 emacinc.com/trainers/linux_starter_kit.htm

Adding to the list of interesting kits, EMAC has an embedded Linux starter kit. A lot of equipment uses embedded computers these days, from soda machines to incubators. This kit will get you familiar with embedded Linux systems.

Change Your Point of View

$37.50 makezine.com/go/pov

Check out these incredible LED bicycle wheel animations using the SpokePOV kit. Mario scoring points, a running mouse, rock 'em sock 'em bots, Pac-Man eating a ghost, and Duck Hunt!

Too Hot to Be Cool

$50 makezine.com/go/logger

If you need to log temperatures, this looks like a pretty interesting kit that uses iButtons. The logger automatically wakes up and measures temperature at user-programmable intervals from 1 to 255 minutes; programmable high and low temperature alarm-trip points are available. Find out which part of your backyard is the warmest, or make sure your fridge is keeping an even temperature.

Top Chopper

$22 makezine.com/go/heli

Build a working helicopter out of wire, balsa, and rubber bands. The kit also "teaches basic helicopter theory." Clearly, a lesson for life.

Homemade Electronics Lab for Kids

Free makezine.com/go/kids

OK, here's a great how-to on making a fun experimenter's board for your little one. It's never too early to learn about electricity!

I Want My MP3

$72–$115 teuthis.com/html/kits.html

Here are a couple open source MP3 player kits; PCB/schematics and source code are all available. You get a pre-programmed PIC, but it's reprogrammable. There's also free email support if you need it. A MAKE-branded kit is on its way! (Check the MAKE store: makezine.com.)

Tube Time

Crystal Radio Kit for the Modern Era

XS404 Big Brother Crystal Radio Kit

$35 midnightscience.com

In this modern world of radio communications and surface-mount component design, it's easy to dismiss radio kits as a remnant of the past. However, nothing could be further from the truth. This Xtal Set Society kit is designed to bridge the gap between the basic crystal receivers of yesteryear and the solid-state component receivers typically used today.

Offering molded coils — as well as the option to wind your own — the kit features an antenna tuner, a variable capacitance diode for station tuning, and a transistor amplifier for the included earpiece. Unlike the simple crystal receivers (and their accompanying frustrations) we all grew up with, the XS404's tuner aids in suppressing nearby signals while the transistor serves to boost weaker ones.

The kit, based on a 1N34 crystal diode, also includes the obligatory printed circuit board, Fahnestock clips for attaching the earpiece and antenna, two varactor diodes, two panel-mount 10K variable resistors, other miscellaneous resistors and capacitors, two black pointer knobs, and a front panel. A 9-volt battery is required.

Thanks to the efforts of the family-owned and -operated Xtal Set Society, kit enthusiasts can once again put solder to the pad and enjoy a great (not to mention functional and educational) radio build.

—Joseph Pasquini

The Viper Is Coming

$140 makezine.com/go/viper

William Cox wrote a great review about the Viper kit from Microbric. Looks pretty good; you get a lot of components for the price (he even goes so far as to call it a "steal") and his only complaints are minor. Plus, it's called the Viper.

Chosen by the designers at Wild Planet

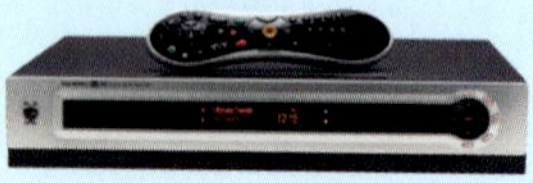

RadioShack 10-Watt Handheld Powerhorn
Wild Planet's product design department uses the Handheld Powerhorn to summon a meeting's latecomers. While this tactic may sound a little harsh to outsiders, it's all done in the name of good, clean fun!

Hi-Tec-C Pilot Pen with 0.4 Tip These Japanese pens are highly sought after in many design departments, and guarded intensely at Wild Planet headquarters. Some of our finest designers claim "it writes and draws so well that it is therapeutic!" The highly pigmented, insoluble gel ink dries fast, doesn't smudge and has a super-fine point for drawing and writing precision. The pens are normally available only in stationery stores in Asian countries, but they've been known to pop up occasionally on websites like eBay.

Wild Planet's Spy Video Car We hate to toot our own horn, but the spy video car is a must for keeping office time interesting. A video camera attached to a remote-controlled car allows you to drive into top-secret territory without being detected. You can also see in the dark: infrared light illuminates real-time video transmitted to your headset's private eyepiece. Where was this technology when we were kids?

TiVo HD Series Three At $800, this technology doesn't come cheap, but it's truly awesome. The new TiVo records in HD, so not only can you watch your favorite shows at your leisure, but you'll see them at the same quality as if you were watching the program live, with movie-quality sound. TiVo keeps our designers in touch with pop culture, even when they're working late nights at the office.

Sharper Image Sound Soother 20 The design department is all about keeping a soothing, Zen frame of mind, even when hectic deadlines are looming. The high-quality speaker masks the office din and makes cube life a little less stressful as you listen to all-digital recordings of peaceful sounds like the rainforest, the ocean, a heartbeat, or chirping crickets.

Dimension BST 3D Printer This printer builds functional 3D models from the bottom up — one layer at a time. The use of ABS plastic ensures that the prototype models are durable, which, as you might guess, is pretty important. Affectionately named Hal 9000 by Wild Planet's design and engineering teams, the Dimension BST produces high-quality 3D forms in a fraction of the time it would take using conventional processes.

Founded in San Francisco in 1993, Wild Planet creates kid-cool products that garner parent approval. Wild Planet playthings focus on fun and imagination, and are available worldwide. Through Toy Opinion Panels and Kid Inventor Challenge contests, Wild Planet encourages children to invent, and then turns their ideas into real toys. wildplanet.com

Flex Your Plexus

Superplexus toy

$15 ebay.com
www.superplexus.com
en.wikipedia.org/wiki/superplexus

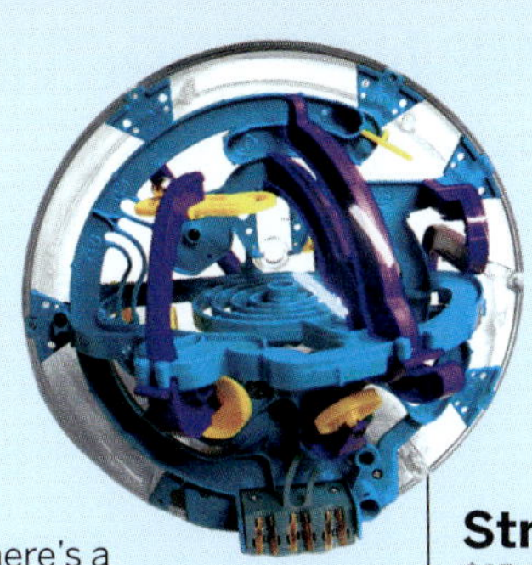

The Superplexus is a clear plastic handheld sphere the size of a head of cabbage. Inside, there's a crazy path — a series of ramps, tubes, chutes, and grooves — through which you must deliver a small steel ball by tilting the sphere. The goal is to get the ball to the end without having it fall off the path.

Along the route, you must pass 100 numbered points, each of which represents an obstacle to navigate. Every time I pick this up, I get immersed inside Superplexus' twisty 3D maze. I become the little steel ball. Despite having had this puzzle for over a year, I've never made it all the way to the end, but I'm not discouraged, because I've never gotten the feeling it's impossible to solve.

The Wikipedia article on Superplexus has a couple of cool concept sketches and prototype photographs, as well as inventor Mike McGinnis' notes on its design and history (he made the first prototype while he was in junior high school). You can learn even more about McGinnis' terrific puzzle at the Superplexus website. I'm astonished that it's no longer in production.

—Mark Frauenfelder

If You Can't Open It ...

OpenX clamshell box opener

$5 myopenx.com

You can hurt yourself opening clams and oysters, even if you have the right tool. I once jabbed the point of an opener in my palm trying to make oysters in the half-shell. You can also get hurt trying to rip open clamshell packages, the sealed plastic packaging that encases electronics such as batteries, flash memory, and game accessories.

Also known as blister-packs, these clamshells are frustratingly difficult to open, whether you use scissors or your teeth, and you wonder if the manufacturer ever intended you to open it. Another special feature of this packaging is that you can't reclose it. Not knowing this is part of the difficulty in opening them; I try to open them in a way that allows me to put the two sides back together when I should just shred the plastic.

OpenX is a special tool with two different blades designed to slice into and separate the two pieces of plastic so you can get to what you want inside. There's no disputing that OpenX works better than scissors or a kitchen knife, but the fact that I need a tool bothers me, even if it costs under $5. Then again, I wouldn't try to pry open an oyster without something other than my teeth.

P.S. David Pescovitz pointed out on Boing Boing the irony that OpenX arrives in a clamshell package.

—Dale Dougherty

Straws and Connectors

$15 constructiontoys.com/store/straws.php

Cheap toys that thoroughly entertain kids for longer than 30 seconds are rare. I find it impossible to keep children interested in pretty much anything for more than a couple of minutes, and that includes playtime. But Straws and Connectors are the exception to the rule.

Cheap, easy, and fun (no, not me), Straws and Connectors are awesome when you're stuck with twenty 5-year-olds on a rainy San Francisco day, praying for a break in the weather so you can boot their little hyper-hypo behinds out the door for five minutes of peace and quiet. They love this game.

It's also a team builder. There are lots of games the kids love to fight over, like Legos and puzzles. Kids have a tendency to get territorial over things, and it's completely normal. I have a great deal of respect and appreciation for any toy that encourages kids to work toward a collective goal, like building the tallest building in the world or the longest train on the tracks. It's heartwarming to see them work so hard to help each other find the right pieces. That's too cool for school. So cool that grown-ups love them, too. They're actually perfect for helping to think in 3D.

And if you want to go even cheaper, you can make your own set with a $1.99 package of drinking straws and some of those lil' mini potatoes ... or big, fat grapes. Your pick.

—Kristina Reed

By Michelle Hlubinka

MAGNETS

Playful, mysterious, surprising, they provide a creative force for many makers. A few magnetic playthings stand out in their field.

Shogun

boardgamegeek.com/game/2043

This game operates on a similar principle to Touché — except this time it's a variation on checkers rather than tic-tac-toe. As you advance, the number of spaces the piece can move on your next turn changes dynamically as a number between 1 and 4 swings into alignment over a hidden magnet.

Uber Orbs

uberorbs.com

These highly stylized super-duper magnets look like a pair of pets from another planet. An accompanying training manual includes special tricks you can have your orbs perform, such as leaps and wiggles and curious chirps.

One important caveat for households full of geeky gadgetry: orbs must not be allowed to sneak near any digital device. We store ours on a low bookshelf among our Clarke and Dahl paperbacks.

Touché

boardgamegeek.com/game/2763

In this suspenseful variation of tic-tac-toe, magnets flip some markers' loyalties from white to black, or black to white. I found this at a yard sale in Jamaica Plain, Mass., en route to a housewarming party. Soon after arrival, I had a roomful of twenty-somethings oohing and aahing as I and my opponent took turns gingerly placing our pieces on the grid, never knowing which polarity would lay beneath our chosen square.

Polarity

boardgamegeek.com/game/380

In this unique game, players float magnetic discs off other nearby discs' fields. It intrigues. It frustrates. Half an hour into the game the sudden jumps — as nearby pieces snapped together — still startled us.

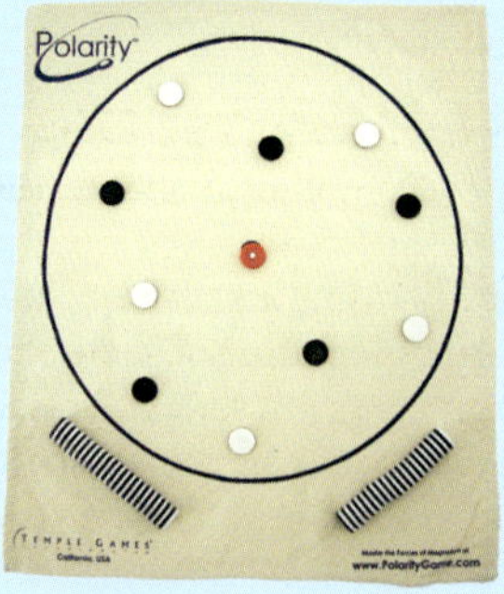

Busy Bugs Music Box

schylling.com

Wind it up, and two cheerful, wobbly ladybugs spin around the top, guided by magnets just beneath the surface, creating a delightfully erratic dance step. And you can even pluck one off, and its partner continues its joyful reverie, and nobody gets hurt. (This is great for the longevity of the toy, as a mechanically attached bug would be easier to destroy!)

Magnetic Play Theatre

makezine.com/go/theatre

This Czech mini-stage is the commercial, prince-and-princess version of a clever DIY project I first saw in an elementary school in St. Paul, Minn. In the homespun original, Mrs. T waved unseen magnet-wands below her stage to guide the kid-made magnet-mice around a small wooden set (made of a fruit crate).

Michelle Hlubinka is education and public programs manager at Zeum, an arts and tech museum for children, and owns more than 200 board games.

As Good As It Gets

The Great International Paper Airplane Book **by Jerry Mander**

$3.29 and up, Simon and Schuster

When I was a kid, I'd bother my father all the time to steal office supplies. I knew that, in the momentous bureaucracy of his university, there were sheets of colored paper perfect for paper airplanes.

I must have folded thousands of them. If only I had known then of the existence of *The Great International Paper Airplane Book*, I could have stood on the shoulders of giants, jumped off, and flown farther.

Everyone from schoolboys to professional aerospace engineers participated in *Scientific American*'s "First International Paper Airplane Competition" in 1966, and the variety and beauty of the final products is astounding. I'd bet that this book did more for innovation in aviation than any before or after it. This is the book for any maker's coffee table, a table hopefully covered in a pile of tax, insurance, and modern bureaucracy forms that are looking for a more important purpose. Take to the wing.

—Saul Griffith

How-to Tales

Toys and Tales with Everyday Materials **by Sudarshan Khanna, Gita Wolf, and Anushka Ravishankar**

$20, Hushion House, Tara Publishing

Written by professors at India's Toy Centre for Research, this book is inspired by folk and vernacular toys in India. Each project has a simple how-to for younger kids, a technical explanation for teens, and a short essay on the politics and sociology of play for adults.

Covered in brown paper with pasted-in project photos, it has a fun, homemade feel. Whether it's a section on practical joking or a snazzy spin toy used to show how wings work, it's a great book for getting kids interested in playing with materials and building things, and a wonderful resource for anyone interested in learning how to connect the things kids do with the workings of the world.

—Arwen O'Reilly

Not Disposable

Making Heirloom Toys **by Jim Makowicki**

$14, Taunton

Are there kids in your life who deserve better than those expensive, battery-powered, disposable plastic toys you find at the mall? *Making Heirloom Toys* is a serious woodworker's guide that seems like it was written by grandfathers for grandfathers (or grandmothers). This book presumes you already own and regularly use a lathe, router, and drill press. (If you don't, try *Riding Machines for Kids*, by Edward Baldwin, instead.) These are beautiful, small, indoor-rug toys with a focus on high-quality wood, construction, and durability, and tips on making age-appropriate design modifications. There's a refreshing emphasis on longevity and reuse. So go on: build any of these toys for a child you know, and prevent cheap toys from becoming landfill.

—Christy Canida

I have liked Legos ever since I was 4; before that, I was into regular wooden building blocks. It was like the next step for me. One reason I like Legos so much is that they let your imagination run wild. I cannot tell you how many countless hours my friends and I have spent building our own super-cool Lego armies and then fighting with them. Or maybe we build our own high-tech racing vehicles and then race them. We might even build our own spaceships and race, or fight, in space. We build sport fields and play Lego baseball, or have downhill snowboard competitions. Sometimes we play poker and bet with our Lego vehicles. I've liked Legos for a long time, and I hope they keep making them in cool sets like *Star Wars* and Harry Potter.

—Kindy Connally Stewart, age 11

I like Risk because, in the game, you can own different countries. It's fun to own places like Africa or China. It's been around for a long time so lots of people know how to play, so you can play with your grandfather. It gets you all worked up because it's about world domination.

—Emmett Mountjoy, age 11

I had teddy bears since I was zero. I really like them because they are cute and they are very soft. Me and my friends dress them up, and we play with them. We make up stories, and I imagine that I am in a forest and there are a lot of animals. I can use my other stuffed animals to pretend they are forest animals, too! I love them when I sleep with them, and I love them in the morning when I play. They help me with my feelings. I hug them when I am sad. I play with them when I am happy, and I squeeze them when I am asleep. Teddy bears are my best friend for always.

—Kiera Collins, age 6

I love playing chess with our *Star Wars* chess set. It's a very exciting game with all my favorite characters. It is from *A New Hope* so it has R2D2, C3PO, Princess Leia, and more. Chess is one of my favorite games.

—Cassidy Mountjoy, age 8

Pac-Man is cool. I like it because you can eat up points and you have to get away from a light pink ball that will eat you up. I like to play it on the TV. There are little ghosts, and if you eat them, they give you points, but if they hit you from behind, you just lost a life. The Pac-Man figure looks like the circle from the book *The Missing Piece*. And that's why it's my favorite game.

—Oona Albertson, age 9

Lines of Action

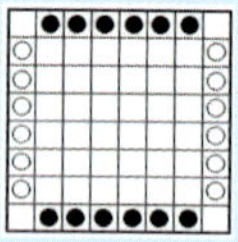

Free wikipedia.org/wiki/lines_of_action

Try something new the next time someone questions your board-game skills by introducing Lines of Action (LOA), an abstract strategy game that two people can play with an ordinary checkers set. Apart from using the same board and pieces and being really easy to learn, LOA is nothing like checkers.

A game of LOA starts by setting up the board and picking your color. The object is to connect all of your pieces together sooner than your opponent, and there are only a few rules to remember: 1) you can move your piece in any straight line on the board, but only as many spots as there are pieces in that particular "line of action," 2) you can jump your own pieces but not those of your opponent, 3) if you land on an opponent's piece after a move, your opponent loses that piece and it gets removed from the board, and 4) you can't land on your own pieces.

You can read more about LOA on Wikipedia, but the best way to learn is by playing. Dig that old checkers set out of the closet and go challenge someone to a game!

—Matthew Russell

Barton Blum is VP of American Soil and Stone, and lives with his family and his seven sets of Battling Tops.

Christy Canida is a distracted biologist who builds things in Oakland, Calif.

Andrea Dunlap is a photographer and filmmaker for seedlingproject.org.

Ross Orr keeps the analog alive in Ann Arbor, Mich.

Joseph Pasquini is an avid amateur radio operator as well as a shortwave and scanner listener.

Kristina Reed is a kindergarten teacher in San Francisco by day, and a crazy Yelp reviewer by night.

Have you used something worth keeping in your toolbox? Let us know at toolbox@makezine.com.

RENEW, REPURPOSE,

FROM MAKER TO MAKER

Who doesn't appreciate a really good tip now and then? Especially the kind, as one reader put it, "that changes your life." Whether it's something as easy as getting permanent marker off your whiteboard or as clever as repurposing bathroom hardware to help ripen bananas, we all rely on our friends and neighbors to tip us off to what's new and good. *—Arwen O'Reilly*

Banana Hacking

When Sarah and Daniel Drucker moved into their new apartment, chaos reigned. "After our first grocery trip, the kitchen counter was a mess and there was nowhere to put the produce. But there was a box of shower curtain rings on the bathroom floor. We used a hinged, circular ring that snaps closed. Slide the open ring between the stems at the top of a bunch of bananas, and close it around any accessible bar or hook where it can swing freely, so the bananas will ripen evenly. If the ring is closed all the way you should be able to yank a banana off the bunch without undoing anything."

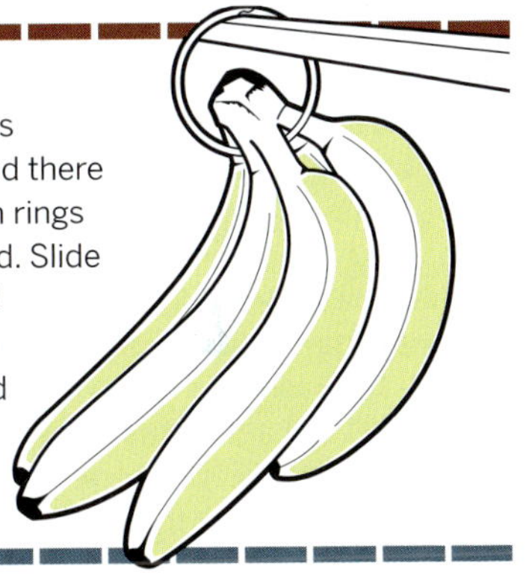

Tools in a Jumble?

Corwin Hardham of Squid Labs suggests: "Try using a big carabiner as a cool means to store your box-end wrenches. The beener keeps all the sizes in order (or more precisely, you can keep them in order on the beener), and the whole mess clips to your belt for easy transport ... and for making more jingling noises than your high school janitor."

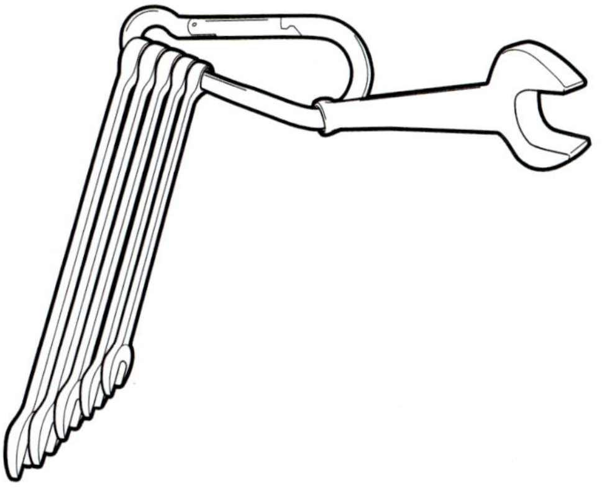

Have a tip for MAKE readers? Send to tips@makezine.com.

Lock It Up

Wendy Boswell, over at the Lifehacker blog, has a great tip for remembering your lock combinations: "Just add your birthday or other important date (the month, day, and last two digits of the year) to the lock combination numbers and write it in permanent marker on the back of the lock. No one else will have a clue what the numbers mean, and if you forget, all you have to do is subtract."
makezine.com/go/lockit

The Sharpie Incident

Now in my opinion, Sanford, the makers of Sharpie and Expo dry erase markers, haven't done enough to differentiate their products. It's all too easy when you're caught in the fire of a new idea to pick up that permanent marker and scrawl all over your whiteboard, only to find out afterward that no amount of scrubbing with the "eraser" will get it off. But wait, there is a solution! And it's easier than you think! Just write over the permanent marker with the dry erase marker, and your scrawl will disappear like magic. I first discovered this by accident, but there's a great post on wikiHow.com about it, too.
makezine.com/go/sanford

Illustrations by Tim Lillis

Gaming on the EDSAC and PDP-1

By Tom Owad

From the earliest computers came the first computer games.

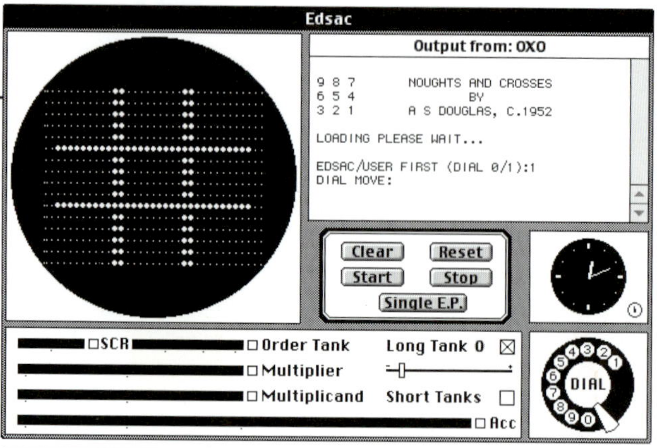

The Edsac Simulator lets you recreate what may be the first computer game: *OXO*, a tic-tac-toe game written by a Cambridge University grad student.

Electronic Delay Storage Automatic Calculator (EDSAC).

DEC Programmed Data Processor-1. There's something in the names of these pioneering computers that suggests they may not be the greatest gaming platforms. Nonetheless, in between searching for prime numbers and analyzing elliptic curves, A.S. Douglas found time to write what may have been the first computer game. *OXO* was a tic-tac-toe game Douglas wrote as a demonstration for his doctoral thesis in 1952. The board was displayed on a 35×16 pixel CRT and allowed players to match wits with one of the earliest digital computers. Today, if you don't have the EDSAC in your garage, you can download Edsac Simulator (www.dcs.warwick.ac.uk/~edsac), which includes *OXO*.

The next well-known digital computer game came in 1962, when MIT's Stephen "Slug" Russell wrote *Spacewar* for the PDP-1. *Spacewar* was a two-player game in which the objective was to shoot the opposing spaceship and avoid being captured by the gravitational pull of a nearby star. User input was via switches on the front panel. Missiles and fuel were limited. A hyperspace mechanism allowed players to jump out of the paths of incoming missiles, but on the third hyperspace jump, your ship exploded.

Spacewar quickly became very popular and was eventually ported to countless mini and microcomputers. While the only remaining operational PDP-1 is at the Computer History Museum, it's still possible to capture the feel of the original by running on a similar 1970s minicomputer. Fred White has done just that with his Digital Computer Controls (DCC) D-116 and an old oscilloscope, which he exhibited at the Vintage Computer Festival East in 2004 in Burlington, Mass.

Oscilloscopes were commonly used as minicomputer displays, using a digital-to-analog converter card. White didn't have one, so he built his own using a Burr-Brown DAC7624. The board connects to the oscilloscope using three wires: X, Y, and intensity. *Spacewar* loads the X and Y registers with the desired position of the spot and then pulses the intensity line to make it light up. This can be performed a few tens of thousand times per second — enough to display several hundred pixels without flicker. If you'd like to read more about White's D-116 projects, visit d116.com.

In 1977, Cinematronics released an arcade version of *Spacewar*, faithful to the original. The hardware for Cinematronics' arcade system is emulated in the Multiple Arcade Machine Emulator (MAME), and with the proper ROM it's possible to play this early version of *Spacewar*. MAME is capable of creating a very authentic game-playing experience, and many enthusiasts have built custom controllers and arcade cabinets to house their MAME PCs. Original *Spacewar* arcade machines may be impossible to find, but with some craftsmanship and MAME, it's possible to create a reproduction with all the tenor of the original.

+ More information on Cinematronics' version of *Spacewar* can be found at mameworld.net/maws/romset/spacewar.

Source code to the original 1963 version of *Spacewar* can be found at applefritter.com/spacewar.

See photos of Fred White's D-116 at d116.com.

Tom Owad (owad@applefritter.com) is a Macintosh consultant in York, Pa., and the editor of applefritter.com. He is the author of *Apple I Replica Creation* (Syngress, 2005).

AHA! Puzzle This

By Michael H. Pryor

MAKE's favorite puzzles. (When you're ready to check your answers, visit makezine.com/08/aha.)

Gold Chain

A man has a gold chain with seven links. He needs the service of a laborer for seven days at a fee of one gold link per day. However, each day of work needs to be paid for separately. In other words, the worker must be paid each day after working, and if the laborer is ever overpaid he will quit with the extra money. Also, he will never allow himself to be owed a link. The laborer will keep the links he is paid in his pocket until the end of the week.

What is the fewest number of cuts the man can make to the chain to allow the laborer to be paid?

Monty Hall Problem

Monty Hall, host of the game show *Let's Make a Deal*, presents you with three doors. One door has a million dollars behind it. The other two have goats behind them. You do not know ahead of time what is behind any of the doors, but Monty Hall does.

Monty asks you to choose a door. You pick one of the doors and announce it to him. Monty then counters by showing you one of the doors with a goat behind it and asks you if you would like to keep the door you chose, or switch to the other unknown door.

Should you switch your choice, and if so, why? What is the probability of finding the million dollars if you don't switch? What is the probability if you do?

Chicken Nuggets

You can go to a fast food restaurant to buy chicken nuggets in 6-packs, 9-packs, or 20-packs. Is there such a number N, so that for all numbers bigger than or equal to N, you can buy that exact number of chicken nuggets using 6, 9, and 20 packs?

Michael Pryor is the co-founder and president of Fog Creek Software. He runs a technical interview site at techinterview.org.

Illustrations by Roy Doty

BLAST FROM THE PAST

Selections from the 1948 Union Hardware and Supply Catalog By Mister Jalopy

Old catalog provides glimpse of how modern civilization was built without laser levels or pneumatic nail guns.

THE POST-WORLD WAR II ECONOMIC BOOM in Los Angeles required tools and materials to meet the latent demand for everything from housing to airplanes. Union Hardware sold it all. The 1948 Union Hardware and Metal Supply Catalog is a wonderfully illustrated peek into how things were done yesterday. Though some of the tools seem quaint, like a butter churn, there are some great ideas that are still applicable today.

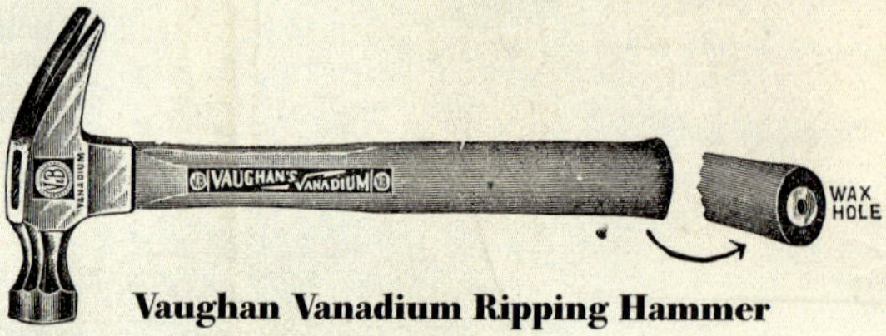

Vaughan Vanadium Ripping Hammer

Certainly an elegant-looking hammer with the octagonal neck and white hickory handle, but what is vanadium? Thanks to Wikipedia, we learn the element vanadium (V, atomic number 23) is added to steel to create a rust-resistant alloy. And what about the wax hole? When hammering a nail into hardwood, the wood will sometimes split instead of yielding to the nail. Not surprisingly, there is an elegant old-world trick to avoid this problem: lubricate the nail with beeswax. Clever! As soon as I saw this picture, I immediately drilled a hole in a hammer handle to create a handy wax reservoir.

Hammer with beeswax reservoir, 1948..........$3.00
Today's dollars..........$23.51

VANADIUM, HICKORY HANDLE HAMMERS ARE AVAILABLE TODAY.

Barnes Junior Mini-Hacksaw

Instead of trying to unscrew a ridiculously corroded galvanized pipe elbow, professional plumbers will often gently cut the offending pipe off with a close quarter saw. With patience and attention, you can remove the old component without damaging the threads on the underlying pipe. I can't tell you how many pipes I have broken off in the wall before my hometown's best plumber gave me the killer tip. (Thanks, John!) I should have asked years ago. The trick is the very thin, very hard, and very sharp blade that makes a regular hacksaw look quite crude in comparison.

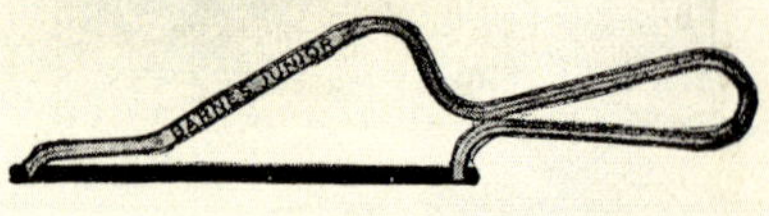

Mini-hacksaw, 1948..........$0.41
Today's dollars..........$3.21

AVAILABLE: $10.15 LENOX CLOSE QUARTER SAW FROM *amazon.com*

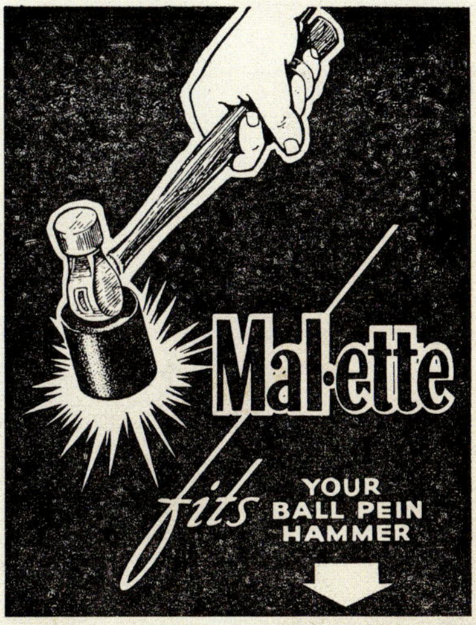

Mal-ette

Besides the knockout graphics, this is a great idea! Rubber mallets take a heck of a beating as they are usually hammering something harder than the rubber head. Long before the handle wears out, the rubber face on a typical mallet is gnarled like a deformed walnut. It doesn't appear that the Mal-ette is currently manufactured, but I would buy one the day it becomes available!

Rubber head for hammer
1948..............................$.80
Today's dollars:.................... $6.27
Not currently being manufactured.

Yankee Handyman

As I own the Yankee Handyman, I can report that the ratcheting screwdriver is a concept that makes sense in theory alone. In my experience, applying adequate turning pressure is always enough to push the driver right out of the screw, resulting in gouging your work. However, when you fit the Handyman with the included drill bits, it is a remarkably elegant solution to drill perfect holes with the slow, deliberate precision that a hand-powered tool allows.

Push drill/ratcheting screwdriver
1948..$3.85
Today's dollars..............................$30.17
Not currently being manufactured; last manufactured by Stanley.

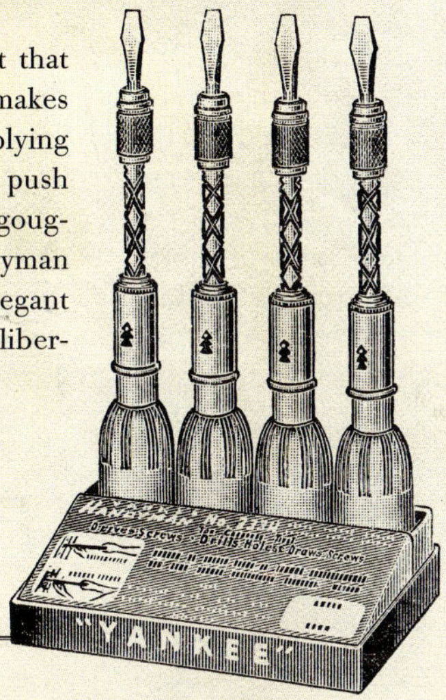

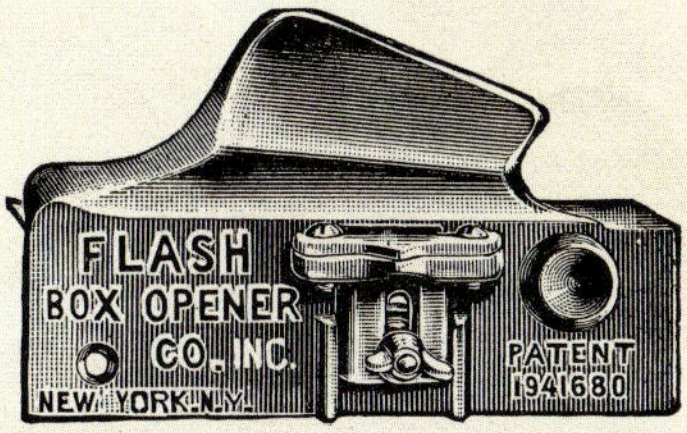

Flash Box Opener

Sure, there are box cutters available for pennies, nickels, and dimes, but they don't have an adjustable depth guide to keep from damaging the box contents. How often have you seen boxes of Cocoa Pebbles that have been cut with a blade? Smart!

Adjustable cardboard carton opener, 1948......................................**$2.01**
Today's dollars...**$15.75**

AVAILABLE: $18.75 FROM *postagetape.com*

Note: "Today's dollars" prices (2004) were calculated based on CPI, using the handy and free NASA cost-estimating website at www1.jsc.nasa.gov/bu2/inflateCPI.html.

1+2+3

Paper Water Bomber

By Ewan Spence

Winged origami missile with front-load tank delivers wet payload.

You will need:

Piece of A4 or letter-size paper, knife or scissors, water, target

Key:

Mountain fold **Valley fold** **Crease line**

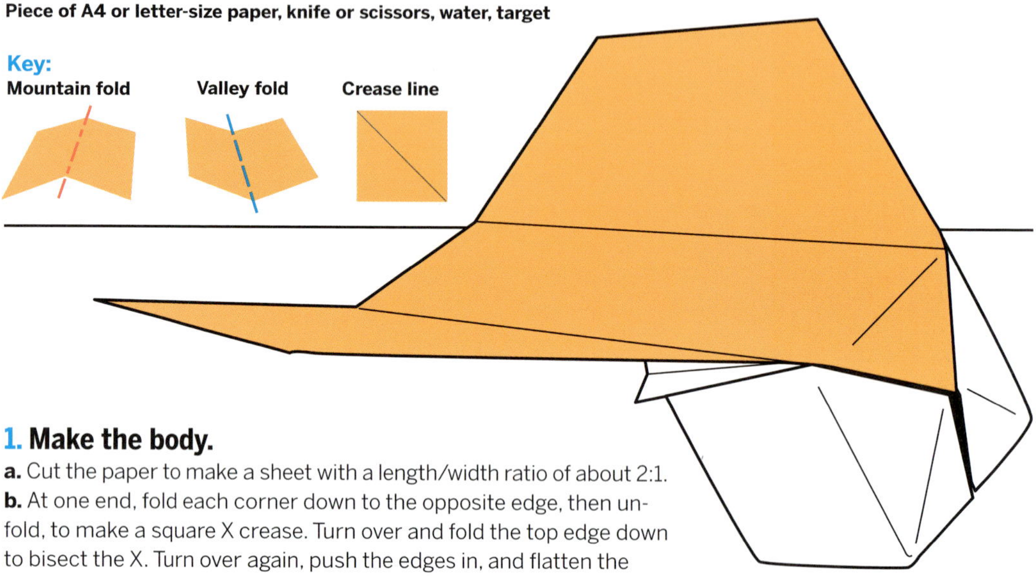

1. Make the body.

a. Cut the paper to make a sheet with a length/width ratio of about 2:1.
b. At one end, fold each corner down to the opposite edge, then unfold, to make a square X crease. Turn over and fold the top edge down to bisect the X. Turn over again, push the edges in, and flatten the paper so there's a point at the top.
c. Turn over, and fold the triangle down along its bottom edge. Fold each entire top corner down along the centerline and unfold just the back corners. Leave the front corners folded down.
d. Invert the crease you just made on the back corners, so they lie underneath and inside, making the top pointy again.

1a

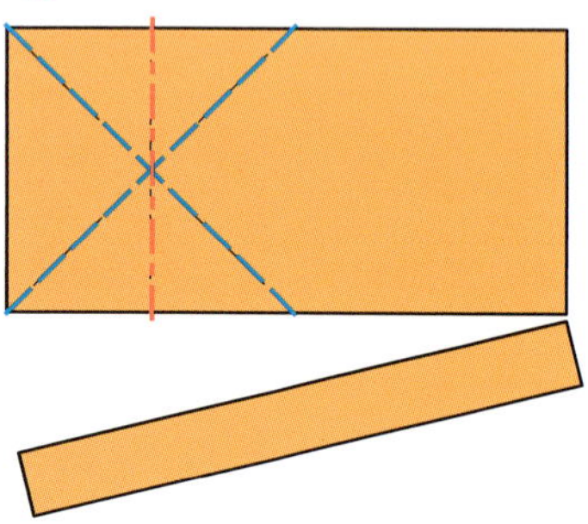

1c

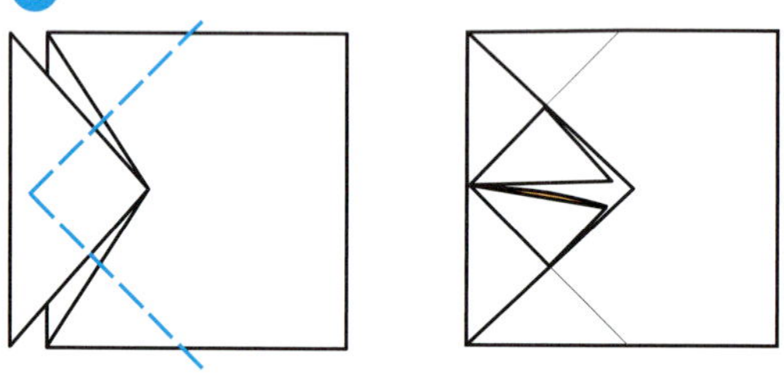

1b

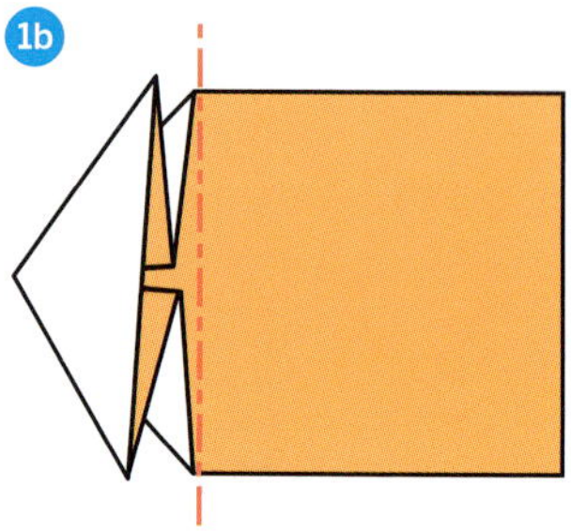

1d

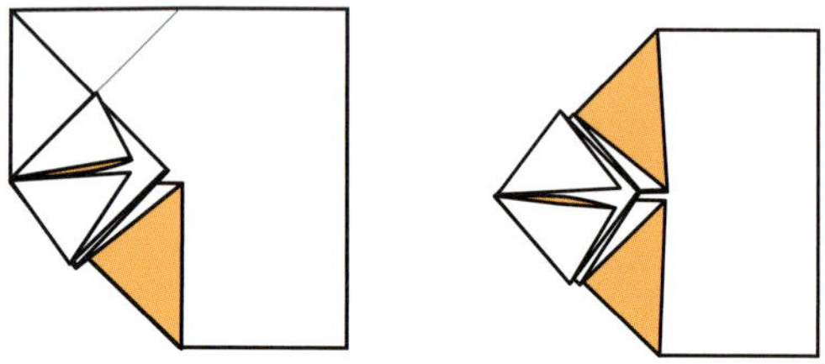

Illustrations by Gerry Arrington

2. Fold the cargo bay.

a. Fold the side of each flap in front to touch the centerline.
b. Fold the bottom point of each flap up to touch the same point.
c. The tricky part: Fold the bottom points up again diagonally, and tuck them inside the pockets you made in the previous step (see 2d).
d. Fold each side corner of the diamond back to touch the centerline behind. Fold the top and bottom corners in front to touch the center-line. Unfold all four folds.
e. Crease the wings to line up with the edges of the cargo section.
f. Blow into the nose of your plane while gently pulling up to inflate the cargo section into a cube.

2a

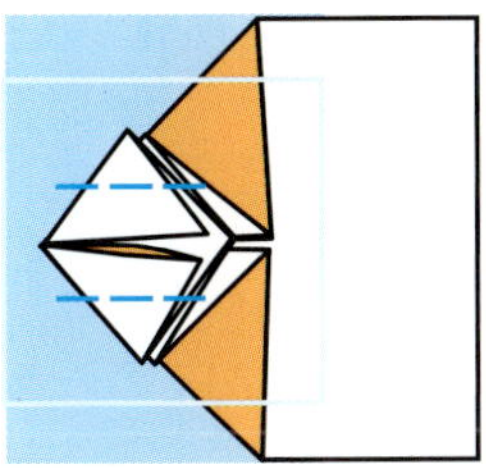

2b

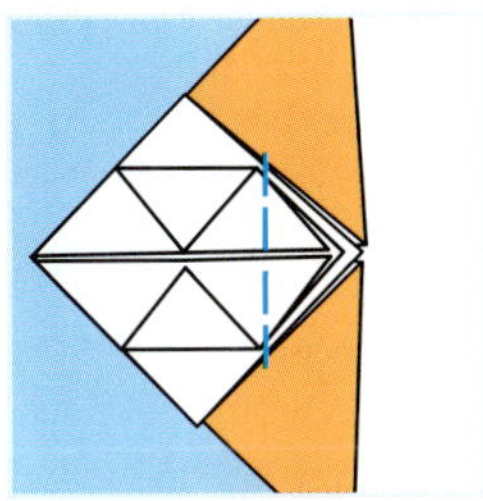

2c

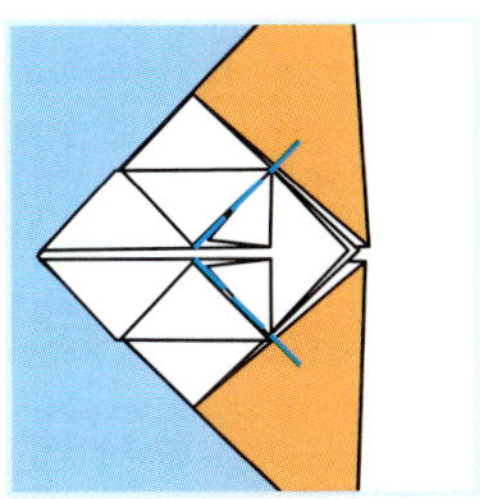

2d

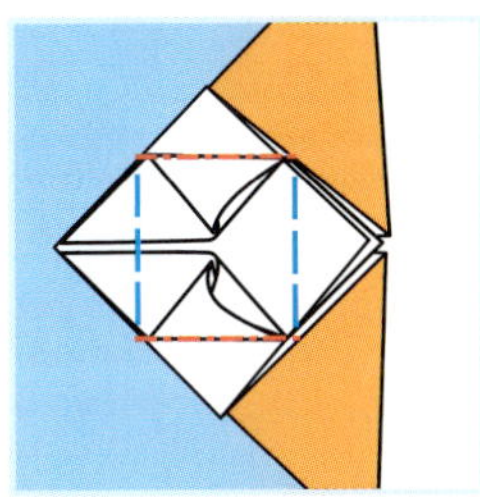

2e

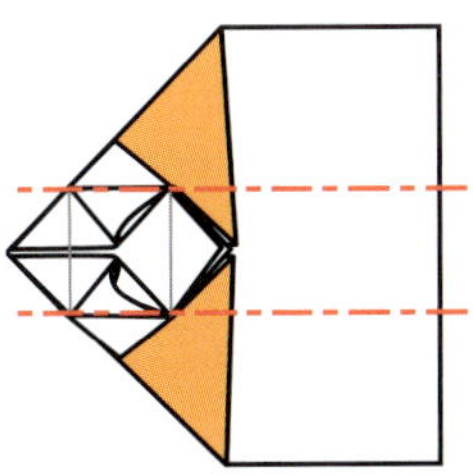

2f

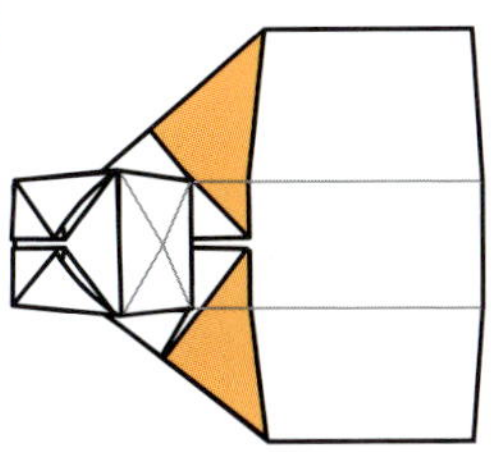

3. Bombs away!

a. Fill the cargo section with water, pouring carefully into the hole in front.
b. (Optional) Wait until no one is looking.
c. Chuck the bomb at your target or victim of choice.

3a

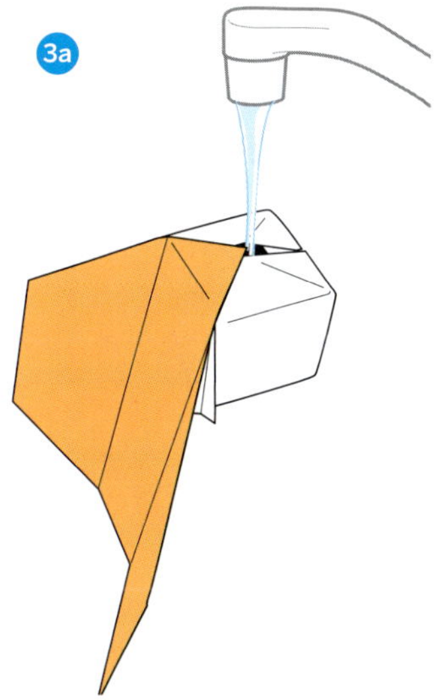

3c

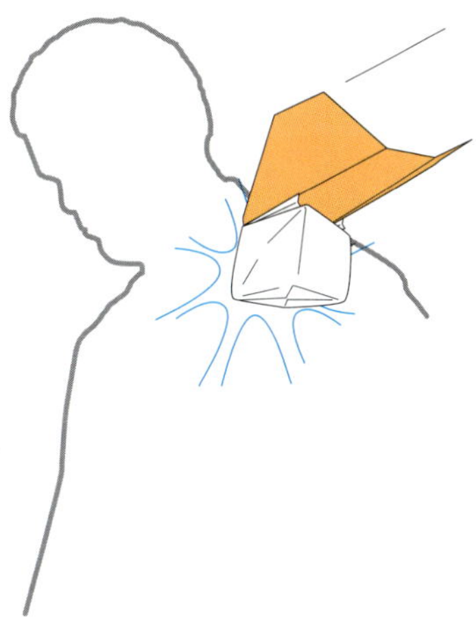

Ewan Spence (ewanspence.com) is a podcaster, blogger, reporter, new media junkie ... generally a hoopy frood who's fun to be with.

READER INPUT

Where makers tell their tales and offer praise, brickbats, and swell ideas.

I'm sure a thousand people have already mentioned this, but in MAKE 07, on page 17, the penny-powered LED, there's a comment about older pennies working better, and it says you don't know why. I don't know the year offhand, but at some point the rising price of copper caused the mint to switch pennies to a copper-coated zinc core, instead of pure copper. Cut one in half, or more fun, hit one with a propane torch. By the way, unbelievably great magazine. Keep it up!

—Dave Schaum

I've been playing fangirl for MAKE. [My friend] and I met a 15-year-old at the dog park recently who is really into making potato guns (and his mom is very supportive, bless her, but he's going to try out marshmallows next because he's afraid of hurting someone). I suggested he check out the MAKE website for more fun projects. The next time I ran into him at the park, he told me that the MAKE site is "the coolest website I've ever seen." I had an extra copy of a MAKE quarterly in my car, so I gave it to him and he immediately started devouring it. It made me feel really good and I hope it does you, too!

—Becky Carella

MAKE has reawakened my creativity. After seeing a blurb on a blog somewhere, I picked up my first copy and remembered just how much fun I had as a kid tearing things apart to see how they work. Since then, I've built several projects — including the spider gun and the cigar box guitar — and I've launched into several other things as well. MAKE gets the thanks for that, too.

—Andrew Setters

Just read the "Life Hacks" in Volume 07. Don't forget, XP [extreme programming] espouses pair programming. In my case I am lucky to have a great wife who sits with me and makes sure the unit tests pass....

—James Warfield

I just subscribed to your magazine and I wanted you to know it was because of your podcast that I did so. Although I'm relatively new to podcasting

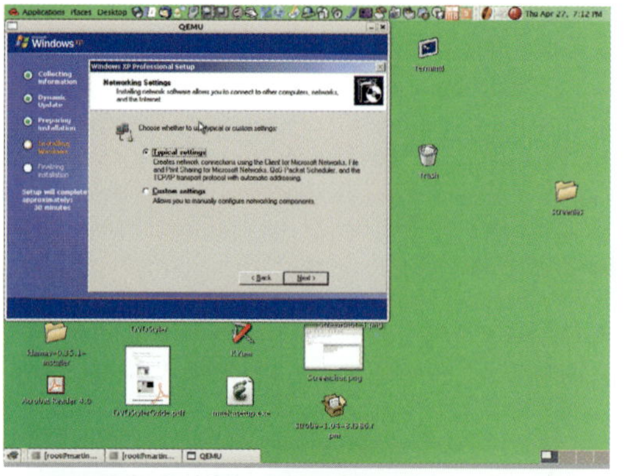

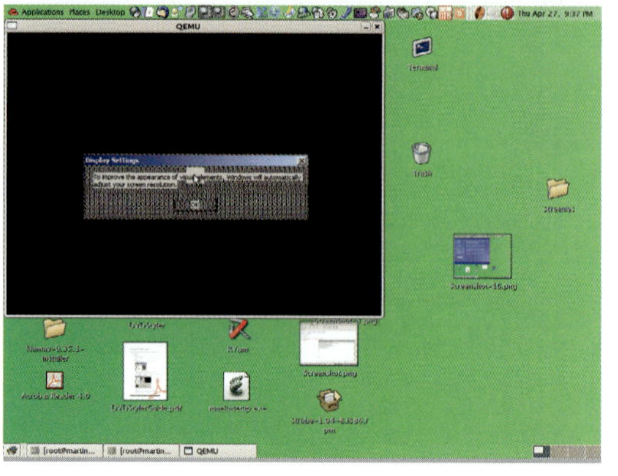

This is in regard to pages 124–125 of Volume 06. DUAL BOOTING? (sigh) ... There's a better way.

Emulators will run a guest OS without the need for repartitioning and re-installing anything. A good commercial emulator is VMware (vmware.com).

Us geeks, however, like a more free and open source solution, which is why I use QEMU (fabrice.bellard.free.fr/qemu).

It is my opinion that QEMU is superior to VMware, which could just be my open source bias, but I have used both and QEMU was better. I have included some screen shots of XP running on my Linux box thanks to QEMU.

—Rev. Martin S. Murphy

I really love what your team is doing. Thanks for the mind candy.

—Howard Horkey

So having seen MAKE before and finally deciding to pick up an issue, I was pleasantly surprised to see that it's published by O'Reilly. I have been a fan of your computer books for a while now, and thus, my expectations were high. In almost every way, as I read through the issue, those expectations have been met. I have, however, one complaint.

There are one or more websites listed for each project. I plan to visit them all at a later time. But nowhere can I find a card where all of the sites are listed for me. I will now have to go back to each article and write down the sites that I want to remember for later. Is it critical? No. Will it prevent me from reading the magazine again, and probably subscribing? Not in the least. Will it prevent me from trying anything? Absolutely not. Will it bug me a little bit? Yeah. Providing the directory (in a tear-out card would be even better) would make it easier to go online after reading a few articles instead of having to jump to the computer after each article to note the website.

This magazine is everything that I expect from an O'Reilly product. It's extremely well-written (even if I do find myself a bit over my head sometimes, but as a perpetual beginner, I have gotten used to that with your materials, as I find them written to appeal to users of all levels) and enjoyable. It is a perfect way to sit down and spend a few quiet hours, planning ways to spend some not-so-quiet ones later in the week. I look forward to future issues.

—Justin Meyer

Editor's Note: **We thought Justin's idea was a good one, and would be even better if it was online and the links were already live. So now, every URL mentioned in each issue of MAKE is available on one page. For this volume, go to makezine.com/08 and click "View a list of all links referenced in this volume" near the top of the page. *Thanks, Justin!***

I really enjoy MAKE but I do have a gripe.

You publish electronic projects without including schematics. Why? Do you think your readers are "dumb" and wouldn't understand? I don't think that is the case. In the latest issue [Volume 06], on page 155 you almost get it right. Almost. Why the pictorial of the LED?

On page 81, "Build Your BeamBots," you have some great macro photography pictures. If they had been annotated with part names and had an included schematic, it would have been *much* better. In my mind I can see little markings on the pictures like "U1," "U2," "R1," and "C4" in nice contrasting colors. On page 85, you could have used a real technical drawing of a TO-92. Take a look at: fairchildsemi.com/ds/2N/2N3904.pdf. It is right on the first page. I hope (expect?) that any transistor manufacturer would let you use their drawings if you asked to.

Please keep up the good work. I will continue to subscribe (even if you don't take my suggestions <sigh>). Even go monthly (insert an <evil grin> here, because I think I have a grasp of how much work *that* would be).

—Louis Taber

Editor's Note: **Louis is right. Schematics are important and we're starting to include more of them in the issues, and online if we're tight on space. Be sure to check the corresponding webpage to all projects and DIY articles for extra information, possible corrections, and more illustrations. *Thanks, Louis!***

MAKE AMENDS

On page 17 of Volume 07, the "Penny-Powered LED" article implied there was zinc on the outside of the dimes, when in fact there isn't any zinc on the exterior surface of any U.S. coins — it's all nickel. Zinc is too soft and reactive, but there is still enough potential between the nickel and the copper to make a battery. It would, however, work better with zinc.

In Volume 07, the "Home Molecular Genetics" article contained a few errors:

• The batteries shown in the "Gel Box" illustration on page 66 should be shown connected in series, not in parallel. The text does explain setting the batteries up in series and should be followed.

• On page 67, in the "Prepare the DNA" section, the amount of running buffer should read "75 µl (3 droplets)."

• In the "Stain the Gel" section (page 67), the dilution ratio for 2.3% Methylene Blue should be 115:1.

• In the diagram on page 69, mineral oil only surrounds the reaction tube; the PCR reaction mix inside the tube is a water-based solution. See scq.ubc.ca/MAKE for a full discussion of the project.

In the "Tips, Tricks, and Traps" section on page 74 of the "Hack Your Plants" article in Volume 07, the fifth bulleted item should express degrees in Fahrenheit, not Celsius. The sentence should read: "Keep your scions cool. 'Dormant scion wood — stems from pear, apple, peach, cherry, and other trees — is best stored between 30 and 38 degrees [Fahrenheit].'"

In Volume 07, page 92, the direction of rotation of the pulleys is incorrect in the "Two-Can Stirling Engine" article. A corrected PDF of the page is available for download at makezine.com/images/07/p92new.pdf.

ANDRÉ LAMOTHE'S
XGameStation™
WWW.XGAMESTATION.COM
THE WORLD'S FIRST VIDEO GAME CONSOLE DEVELOPMENT KIT.
© 2006 Nurve Networks LLC • 1112 Nancy Lane, San Ramon, CA 94582 • Ph: 925.829.4508 • support@nurve.net

We give you the tools.
You make the difference.
firstgiving
online fundraising for everybody
www.firstgiving.com

We've got your (iPod's) back.
Parts & Accessories for your Mac + iPod
› do-it-yourself Mac & iPod Parts
› free online Fixit Guides
› thousands of parts and upgrades
› use coupon 'MAKE8' for $5 off
ifixit
Mac + iPod Parts & Upgrades
1-866-61-FIXIT iFixit.com

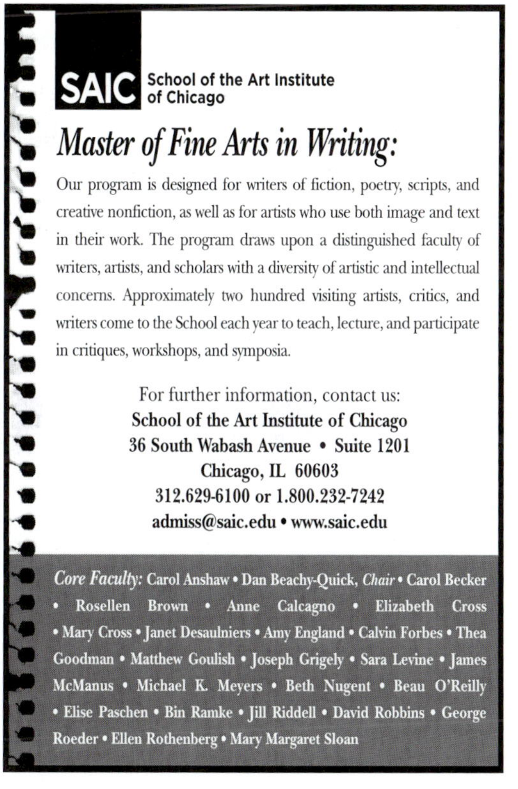
SAIC School of the Art Institute of Chicago
Master of Fine Arts in Writing:
Our program is designed for writers of fiction, poetry, scripts, and creative nonfiction, as well as for artists who use both image and text in their work. The program draws upon a distinguished faculty of writers, artists, and scholars with a diversity of artistic and intellectual concerns. Approximately two hundred visiting artists, critics, and writers come to the School each year to teach, lecture, and participate in critiques, workshops, and symposia.
For further information, contact us:
School of the Art Institute of Chicago
36 South Wabash Avenue • Suite 1201
Chicago, IL 60603
312.629-6100 or 1.800.232-7242
admiss@saic.edu • www.saic.edu
Core Faculty: Carol Anshaw • Dan Beachy-Quick, Chair • Carol Becker • Rosellen Brown • Anne Calcagno • Elizabeth Cross • Mary Cross • Janet Desaulniers • Amy England • Calvin Forbes • Thea Goodman • Matthew Goulish • Joseph Grigely • Sara Levine • James McManus • Michael K. Meyers • Beth Nugent • Beau O'Reilly • Elise Paschen • Bin Ramke • Jill Riddell • David Robbins • George Roeder • Ellen Rothenberg • Mary Margaret Sloan

Making It

How I went from cubicle slave to full-time maker.

By Mikey Sklar

THIS ARTICLE SHOULD PROBABLY BE called "Making It Out of IT." I spent the last ten years as a Unix admin in IT, and now I'm out. I climbed the ridiculous corporate ladder to the highest technical position that I could reach. I rode my bike to Wall Street each morning and back home to Williamsburg, Brooklyn, each evening. This commute provided a fleeting feeling of freedom that would end once I arrived at my cubicle. I did collect a large paycheck, but it came with strings attached, of course.

I crammed my evenings with the things that I wished I had the time to do during the day. As my interest in IT wavered, a love of electronics emerged and occupied my nights. I wanted to make circuit boards, program microcontrollers, and design gadgets of questionable usefulness all the time. I was no longer interested in writing high-level-language Perl scripts that monitored things I cared little about. I wanted to work in the physical world, and the IT gig was clearly not the way.

Unsure where the road away from IT led, I quit and made my first unemployed decision. My new goal: connect microcontrollers with some force from nature. I chose fire. Looking around my apartment, I realized that it was time to expand. Sure, I could make wearables and mess with RFID implants in my apartment, but imagine what

Photography by Wendy Tremayne

I could do with real space and low rent!

Just two weeks after leaving my IT job, I left New York City and its high cost of living for a piece of the New Mexico desert. Here I planned to work with metal, fire, and electronics. I became a full-time maker living in a town full of artists in the desert. For the trade, my rent in New York ($3,220 per month) secured me a great house ($78,120 including closing costs), space to work, and freedom to make, build, or invent whatever came to mind, full time (12–14 hours of "work" per day) — with money to spare.

In the six short weeks after the move, my girlfriend and I designed and built *The High-Lighter* — a fire trampoline that demonstrates how to misuse common backyard items: a trampoline and a BBQ propane tank. A custom circuit board, an ultrasonic distance sensor, a solenoid valve, and many long nights of programming brought the trampoline to life. *The High-Lighter* shoots fireballs vertically based on how hard a person jumps on it.

I built in an RFID reader that responds to a tag I had implanted in my hand months before departing New York. The RFID reader adds an additional level of safety to the device (and gave me another excuse to use my implant) — if I'm not present, nothing works.

The High-Lighter's first performance was at the 2006 Maker Faire's fire show. It shared the stage with fire artists Lucy Hosking's *Satan's Calliope* and Therm's *Darwin*. Since then, I've revamped *The High-Lighter*: the flame ball is nearly three times its original size. Our next stop was Burning Man.

My future plans include building a workshop where I can teach people how to work with fire interactively via microcontrollers. But because New Mexico has been on high fire alert since I moved here in March, I've had to look at options that offer work that is more than seasonal.

Building environmentally clean buildings, growing food, and generating power are also calling to me. The desert environment is ideal for working with adobe, straw bales, rammed earth, papercrete, and used shipping containers. My LED work is likely to develop into indoor food-growing environments that offer year-round growing with minimal power draw. And all with no strings attached.

Mikey Sklar is an inventor in Truth or Consequences, N.M., who creates uncommercializable electronics with open source tools.

My fire trampoline, *The High-Lighter*, was part of the Truth or Consequences Fiesta, an annual parade and celebration that commemorates the town adopting the name of the popular 1950s game show.

RETROSPECTIVE

Barricelli's Universe

By George Dyson

And the evening and the morning were the fifth day...

At 10:38 p.m. on March 3, 1953, Italian-Norwegian mathematical biologist Nils Aall Barricelli (1912-1993) inoculated a 5K digital universe with random numbers generated by playing cards drawn from a shuffled deck. A viral geneticist with a background in physics, Barricelli was granted access to the computer at the Institute for Advanced Study (*see Make, Volume 06, page 190*) to investigate the role of symbiosis in the origin of life.

"A series of numerical experiments are being made with the aim of verifying the possibility of an evolution similar to that of living organisms taking place in an artificially created universe," he announced in the Electronic Computer Project's monthly progress report for March 1953.

These experiments were hosted by mathematician John von Neumann, who, after the success of the Manhattan Project at Los Alamos, decided that building a digital computer was next. "I am thinking about something much more important than bombs," he explained in 1946. "I am thinking about computers." He was also thinking about self-reproducing automata and the origins of life.

The Institute's computer was used for meteorological and other scientific problems during the day and ran bomb calculations for "the AEC boys" at night. Barricelli took whatever time was left. He often operated alone, without engineers in attendance. Though his log entries recorded the usual problems ("Something is wrong with the building air conditioner. One of the compressors seems to be stuck and the smell of burning V-belts is in the air."), he had an unusually high success rate, despite the running battle between coding and engineering. ("Code error! Machine not guilty!" and "Dr. Barricelli claims machine is wrong. Code is right.") Often he was still at work when the engineers showed up at dawn. "Machine worked beautifully. Off!" is his last entry in the early morning of May 31, 1953.

Barricelli's universe was designed to appear unbounded to any of its inhabitants. "The universe was cyclic with 512 generations ... and the code was written so that various mutation norms could be employed in selected regions of the universe," he explained. "Only 5 out of each 100 generations were recorded during reconnaissance. Interesting phenomena were then reinvestigated in more detail." These included symbiosis, incorporation of parasitic genes into their hosts, and fruitful crossing of gene sequences, which suggested sex.

"But," he reported in August 1953, "in no case has the evolution ... led to higher and higher organisms. Something is missing if one wants to explain the formation of organs and faculties as complex as those of living organisms. No matter how many mutations we make, the numbers will always remain numbers. They will never become living organisms!"

Two essential characteristics of real biology were missing: environmental diversity and the interplay between genotype and phenotype that allows Darwinian evolution to advance. "In 1954, new experiments were carried out," he reported, "by interchanging the contents of major sectors between three universes. The organisms survived and adapted themselves to different environmental conditions. One of the universes had particularly unfavorable living conditions, and no organism had been able to survive in that universe previously during the experiment." To make the leap from genotype to phenotype, he concluded, "we must give the genes some material they may organize and may eventually use, preferably of a kind which has importance for their existence."

> "The distinction between an evolution experiment performed by numbers in a computer or by nucleotides in a chemical laboratory is a rather subtle one."

Barricelli spent much of the remainder of his life circulating among computer facilities in the United States and Europe, seeking the CPU cycles his numerical "symbio-organisms" needed to grow. "The distinction between an evolution experiment performed by numbers in a computer or by nucleotides in a chemical laboratory is a rather subtle one," he noted in 1962. He began to develop the functional equivalent of a phenotype by interpreting moves in

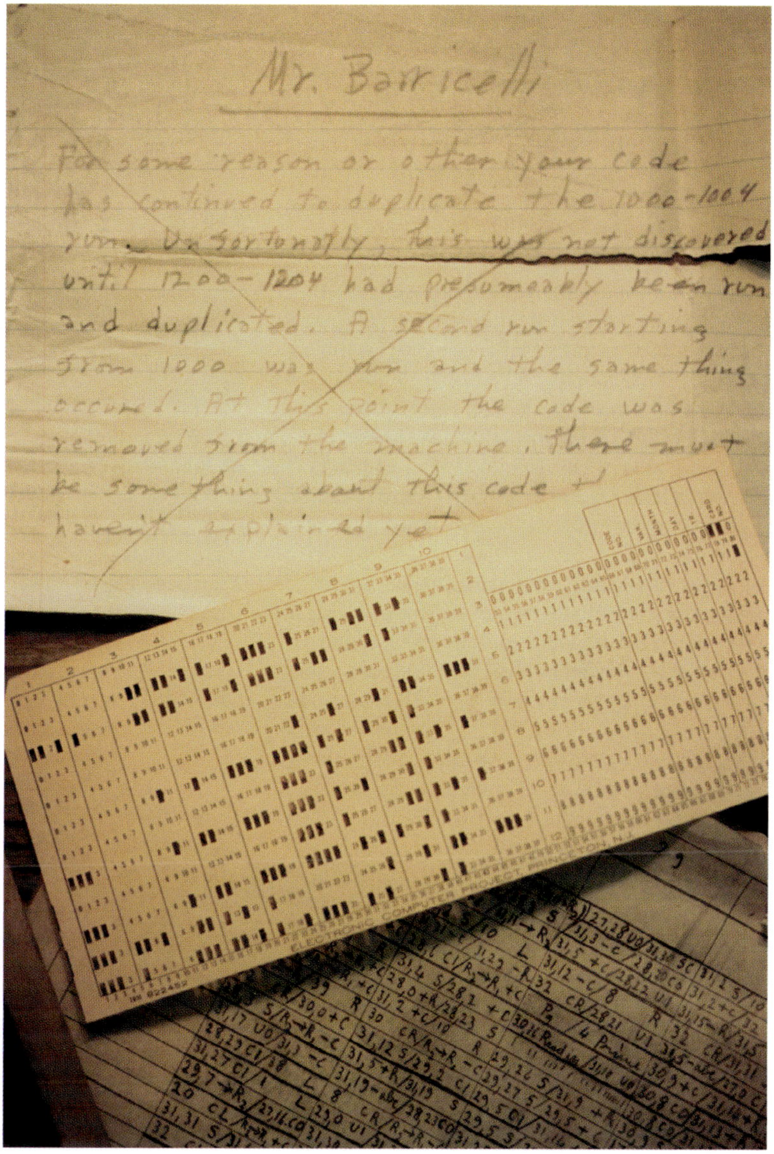

DIGITAL PALEONTOLOGY: Engineer's note, output card, and input code from a numerical evolution experiment run on the Institute for Advanced Study computer, ca. 1954. The note to Barricelli reads: "For some reason or other your code has continued to duplicate the 1000-1004 run. Unfortunately, this was not discovered until 1200-1204 had presumably been run and duplicated. A second run starting from 1000 was run and the same thing occurred. At this point the code was removed from the machine. There must be something about this code that you haven't explained yet."

Photography by George Dyson, courtesy of the Institute for Advanced Study

various board games, including chess, via a limited alphabet of machine instructions to which the gene sequences were mapped. The process mimicked the way that nucleotide sequences code an alphabet of amino acids to translate proteins from DNA.

"Perhaps the closest analogy to the protein molecule in our numeric symbio-organisms would be a subroutine, which is part of the symbio-organism's game strategy program, and whose instructions, stored in the machine memory, are specified by the numbers of which the symbio-organism is composed," he explained.

Barricelli's insights into viral genetics informed his understanding of computers, which in turn informed his understanding of the origins of the genetic code. "The first language and the first technology on Earth was not created by humans," he wrote in 1986. "It was created by primordial RNA molecules — almost 4 billion years ago. Is there any possibility that an evolution process with the potentiality of leading to comparable results could be started in the memory of a computing machine?" He titled his final paper, published in 1987, "Suggestions for the starting of numeric evolution processes intended to evolve symbio-organisms capable of developing a language and technology of their own."

RETROSPECTIVE

"Are they the beginning of, or some sort of, foreign life forms? Are they only models?" he asked. "They are not models, not any more than living organisms are models. Unless some other severe limitation is imposed by the conditions of the experiment or the type of universe in which the organism exists (computer, planet, or test tube), there is no a priori reason for assuming that other classes of symbio-organisms could not reach the same complexity and efficiency characteristic of living organisms on this planet." Barricelli knew this would take some time. "A question that might embarrass the optimists," he warned in 1954, "is the following: If it's that easy to create living organisms, why don't you create a few yourself?"

In November 2000, while excavating in the basement of the Institute for Advanced Study, I turned up a cardboard box abandoned since the shutdown of the computer project in 1958. The smell of burning V-belts still permeated a layer of black, greasy dust that had settled in the box. At the bottom was a carton full of IBM data-processing cards, accompanied by a note written on a half-sheet of disintegrated lined paper that identified them as "Barricelli's Drum Code." (A 2,048-word high-speed magnetic drum was added to the computer in 1954.) The box also contained three sheets of ledger paper filled with the original coding for the universe that was preserved, in a state of suspended animation, on the cards.

"Mr. Barricelli," the note read, "there must be something about this code that you haven't explained yet."

The note writer, an engineer, was right. Barricelli's numerical organisms had roamed an embryonic universe of 1/200 of a megabyte, running at 16 kilocycles for an hour or two at a time. Fifty years later, we have only their fossilized imprint, confined to a box. What if they were brought back to life? What if they had continued to evolve? What if they had escaped?

At the same time as Barricelli's first announcement that "we have created a class of numbers which are able to reproduce and to undergo hereditary changes," a similar class of numbers — the order codes — had already taken root and seized control in the von Neumann universe. Order codes constituted a fundamental replicative alphabet that diversified with the proliferation of different metabolic hosts. In time, successful and error-free sequences of order codes formed into subroutines — the elementary units common to all programs — just as a common repertoire of nucleotides forms into strings of DNA.

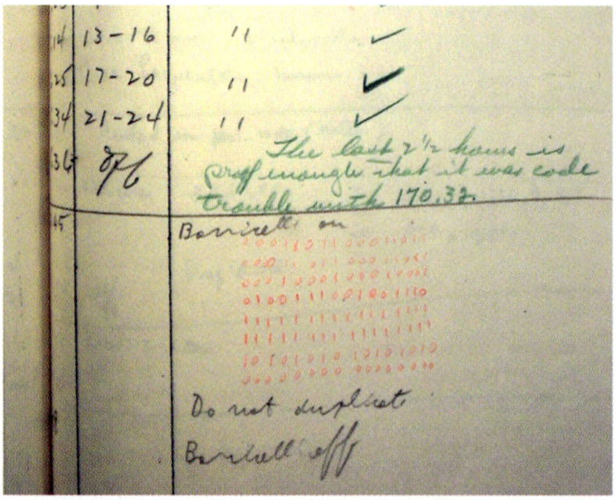

MACHINE LOG, 22 NOVEMBER 1954. Barricelli's annotation "Do[es] not duplicate" indicates that the computer failed to duplicate and suggests hardware trouble. The note in green pencil reads: "The last 2 1/2 hours [without hardware trouble] is proof enough that it was code trouble with 170.32."

These were numbers that did things: they manipulated other numbers, represented text, modeled nuclear weapons, and made fortunes for IBM. They organized themselves into an expanding hierarchy of languages, which influenced the computational atmosphere as pervasively as the oxygen released by early microbes influenced the subsequent course of life. They formed collective structures such as operating systems, which now amount to millions of lines of code. They accounted for money, and so became money; they represented music, and so became music. They learned how to divide into packets, traverse the network, correct any errors suffered along the way, and reassemble themselves at the other end.

These numbers have infiltrated every facet of our existence and have now begun to form complex metazoan structures that are distributed across the internet. The search engines that allow us to collectively scour computers are allowing computers to complete their collective search of us. The barriers between their universe and ours are breaking down completely as digital computers begin to read and write directly to DNA.

Barricelli's universe is no longer confined to a box.

See more images at makezine.com/08/dyson

George Dyson, a kayak designer and historian of technology, is the author of *Baidarka*, *Project Orion*, and *Darwin Among the Machines*.

MAKER'S CALENDAR

Our favorite events from around the world
Compiled by William Gurstelle

January/February 07

2006 was a banner year for makers who love to join like-minded tinkerers to show off their projects and get ideas for new undertakings. Millions of people attend technology-centric events ranging from kinetic sculpture races to air shows to robotics conventions.

The Maker's Calendar editors maintain a database of hundreds of technology events, ranging from small and intimate specialist gatherings to gigantic affairs that exceed a half million people. And the database grows each month.

So how to decide which events to attend? Here MAKE's staff gives you their picks of unique and entertaining Maker Events for 2007. Some are extremely popular, so plan ahead.

Cabin Fever Expo

» Jan. 20–21, York, Penn.

An expo of men, metal, and machines. Exhibits include home shop machine tools, miniature engines, handmade vehicles, and similar creations. A model-maker's dream come true. cabinfeverexpo.com

Maker Faire

» May 19–20, San Mateo, Calif. Other locales and dates TBA

At Maker Faire, 20,000 makers come together for a weekend of tech-centric projects, eye-opening demonstrations, and DIY fun. It's a celebration of the maker ethos encompassing exhibits, seminars, and hands-on opportunities galore. makezine.com

ITP Show

» May and December, New York City

The Interactive Telecommunications Program Show is a 2-day exhibition of interactive and unusual sights, sounds, and objects from the student artists of ITP. itp.nyu.edu/show

Kinetic Sculpture Race

» May 26–28, Arcata to Ferndale, Calif.

25,000 happy people converge on two small Northern California towns to watch builder-racers maneuver their elaborate human-powered kinetic sculptures across 33 miles of land, sand, muds, & suds. kineticsculpturerace.org

DEF-CON

» Summer, Las Vegas

DEF-CON runs for several days and includes many computer- and hacking-related speakers as well as social events and contests. defcon.org

Experimental Aircraft Fly-In

» July 11–15, Arlington, Wash.

The fly-in is a celebration of aviation technology featuring commercial exhibitors, forums, workshops, aircraft demos, and flypasts. nweaa.org

Kinetic Sculpture Race

Speed Week

» Aug. 11–17, Bonneville Salt Flats near Wendover, Utah

There's acceleration and horsepower galore when speed junkies bring everything from blown fuel streamliners to vintage Corvettes to the desert. scta-bni.org

The World Championship Punkin Chunkin'

» November, Millsboro, Del.

Hundreds of punkin chunkers bring forth giant hurling and shooting machines to see who can fire a pumpkin the farthest. punkinchunkin.com

Important: All times, dates, locations, and events are subject to change. Verify all information before making plans to attend.

Know an event that should be included? Send it to: events@makezine.com. *Sorry, it is not possible to list all submitted events in the magazine, but they will be listed online.*

If you attend one of these events, please tell us about it at forums.makezine.com.

January 2007

» Manitou Springs Fruitcake Toss

Jan. 6, Manitou Springs, CO

Hurling enthusiasts take note. A fruitcake toss is held at Manitou Springs Memorial Park near Colorado Springs, Colo., each January. Participants use mechanical devices such as catapults and oversized slingshots to rid the town of unwanted fruitcakes. manitousprings.org

» Light in Winter

Jan. 26–28, Ithaca, N.Y.

Ithaca hosts a midwinter festival of science and art. This year's performers and presentations include NPR's Bruce Adolphe, *The New York Times'* Andrew Revkin, String Theory, Wine and the Mind, Birds That Roar, and more. lightinwinter.com

February 2007

» Western WinterBlast

Feb. 16–19, Lake Havasu City, Ariz.

The Western Pyrotechnic Association's annual event is the premier pyrotechnic event of the winter. westernpyro.org

» Engineers Week

Feb. 18–24, Nationwide

Engineers reach out and touch their communities with numerous events to increase awareness and appreciation of the engineering profession. eweek.org

HOMEBREW

My LED Heirloom Clock

By Blake Hannaford

In the summer of 1975, I had just completed a rigorous course in digital logic design and decided to build an LED clock to last the rest of my life. To start, I partitioned the design into three printed circuit boards: display, clock logic, and power supply. I bought copper-plated boards and ferric chloride at RadioShack, signed each card with my name and the date, and etched out the boards in the backyard of my parents' house. I bought discrete red LEDs from a surplus dealer; two green LEDs made the blinking colon between hours and minutes. The shop manager at Chicago's Institute of Design helped me cut out the plexiglass pieces for the case on a table saw. The clock's design mirrors Mies van der Rohe's famous Crown Hall, the building in which it took shape. I ordered the MM5313 clock chip (the closest analogy today to the power of that chip would be building your own 60GB video iPod) and power supply parts and was off to the races.

Since then, the clock's little green colon has blinked almost a billion times. The fuse I installed for safety has never blown; the capacitors in the power supply still seem to be fine. Nothing inside its plexiglass walls ever changes. Every three to five years, some of the segments begin to dim, so I remove the display board and clean the copper pads with just a pencil eraser.

Sometimes all that is required to keep the clock running smoothly is to pick it up a couple of inches and drop it. We all have our favorite old pieces of retro electronics, but how many have been powered up and working almost continuously for 31 years?

I recently came into possession of my family's heirloom grandfather clock. The owners of the clock have signed their names inside the front door in pencil, starting with John Gates, who built the clock in the early 1800s. I don't know anything about him, but presumably he bought the mechanism and built the clock with hand tools out in his barn — a process not really all that different from my own. I doubt that my clock will serve as long as the grandfather clock has, but the lesson is clear: when doing your next project, put your heart into it. Design it to last. Build it well. You will be glad you did.

Blake Hannaford directs the Biorobotics Lab at the University of Washington, Seattle.

Photograph by Blake Hannaford

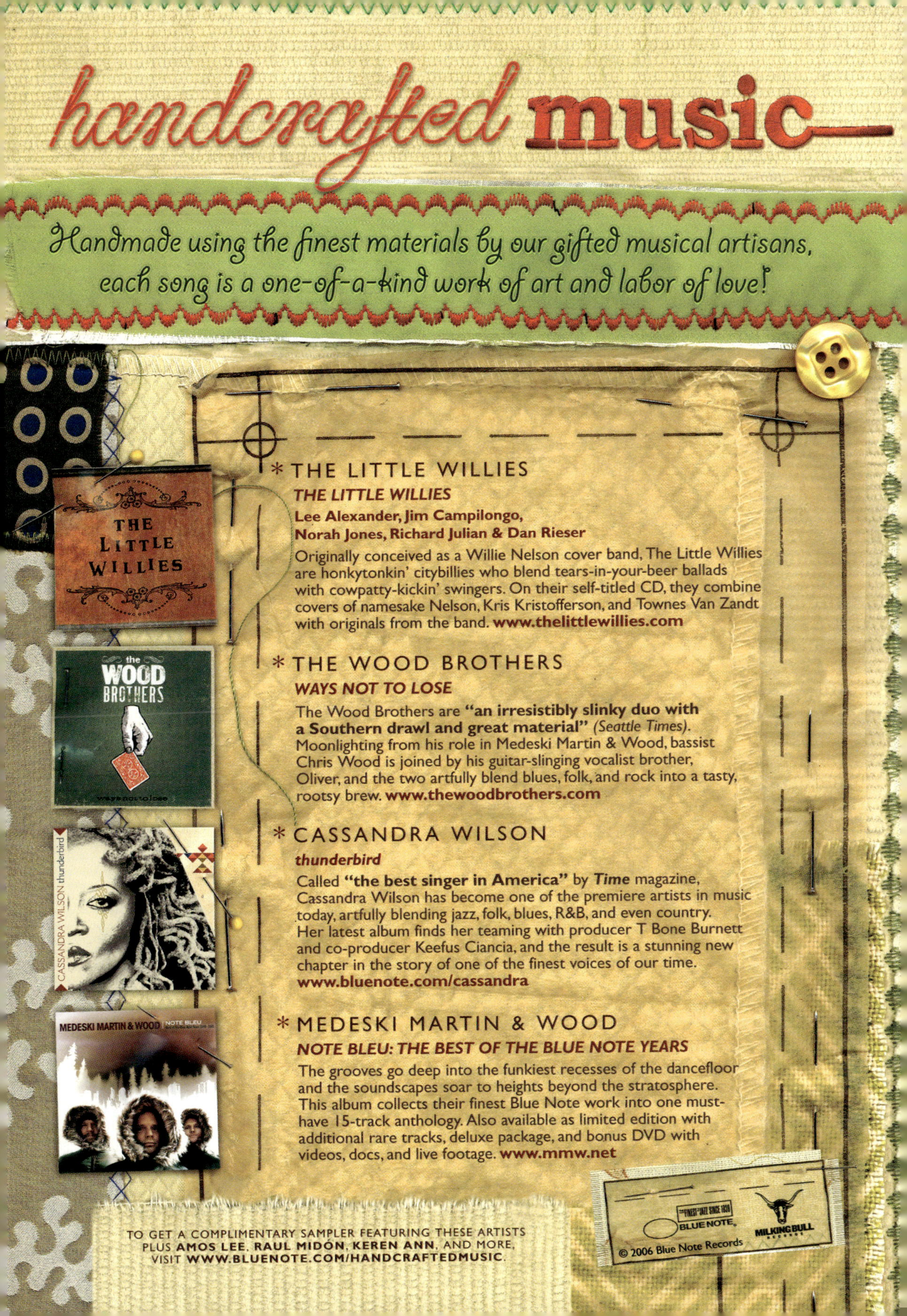
handcrafted music—
Handmade using the finest materials by our gifted musical artisans, each song is a one-of-a-kind work of art and labor of love!
* THE LITTLE WILLIES
THE LITTLE WILLIES
Lee Alexander, Jim Campilongo, Norah Jones, Richard Julian & Dan Rieser
Originally conceived as a Willie Nelson cover band, The Little Willies are honkytonkin' citybillies who blend tears-in-your-beer ballads with cowpatty-kickin' swingers. On their self-titled CD, they combine covers of namesake Nelson, Kris Kristofferson, and Townes Van Zandt with originals from the band. www.thelittlewillies.com
* THE WOOD BROTHERS
WAYS NOT TO LOSE
The Wood Brothers are "an irresistibly slinky duo with a Southern drawl and great material" (Seattle Times). Moonlighting from his role in Medeski Martin & Wood, bassist Chris Wood is joined by his guitar-slinging vocalist brother, Oliver, and the two artfully blend blues, folk, and rock into a tasty, rootsy brew. www.thewoodbrothers.com
* CASSANDRA WILSON
thunderbird
Called "the best singer in America" by Time magazine, Cassandra Wilson has become one of the premiere artists in music today, artfully blending jazz, folk, blues, R&B, and even country. Her latest album finds her teaming with producer T Bone Burnett and co-producer Keefus Ciancia, and the result is a stunning new chapter in the story of one of the finest voices of our time. www.bluenote.com/cassandra
* MEDESKI MARTIN & WOOD
NOTE BLEU: THE BEST OF THE BLUE NOTE YEARS
The grooves go deep into the funkiest recesses of the dancefloor and the soundscapes soar to heights beyond the stratosphere. This album collects their finest Blue Note work into one must-have 15-track anthology. Also available as limited edition with additional rare tracks, deluxe package, and bonus DVD with videos, docs, and live footage. www.mmw.net
THE LITTLE WILLIES
the WOOD BROTHERS
CASSANDRA WILSON thunderbird
MEDESKI MARTIN & WOOD NOTE BLEU
TO GET A COMPLIMENTARY SAMPLER FEATURING THESE ARTISTS PLUS AMOS LEE, RAUL MIDÓN, KEREN ANN, AND MORE, VISIT WWW.BLUENOTE.COM/HANDCRAFTEDMUSIC.
BLUE NOTE
MILKING BULL
© 2006 Blue Note Records

Make:
technology on your time

Volume 06

Features

40

PROTO

"Balloons provide a unique observation platform for doing planetary science that you simply can't do any other way," says Jeffery Hall, who leads JPL's aerobot research. "One of the great attractions of balloons is that it's fairly easy to make them. Then you get to take them outside and fly them around."

78

BEAM ROBOTS

The BEAM approach to robot design creates nimble robots from simple components, with no programming required. Many BEAMbot behaviors are controlled by "nervous nets." These are simple control circuits that emulate low-level peripheral nerves in a spinal column.

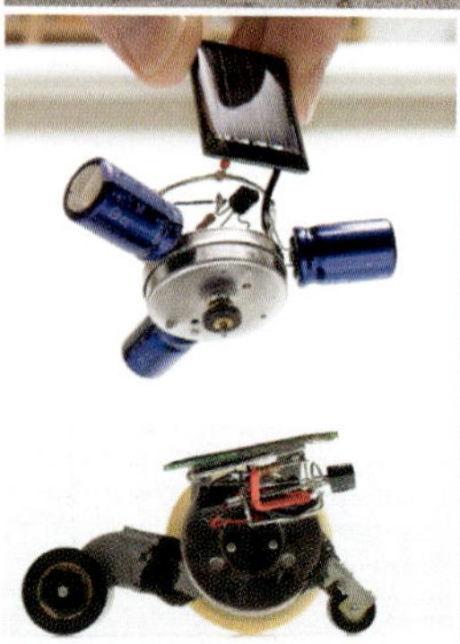

ON THE COVER

When MAKE Art Director Kirk von Rohr took these solar-powered robots (provided by solarbotics.com) outside for a photo shoot, they wouldn't sit still. Like a Hollywood cockroach wrangler, von Rohr was busy keeping the electronic insects from wandering. Cover photographer Douglas Adesko wisely confined the critters to an indoor setting.

Columns

Make: Projects

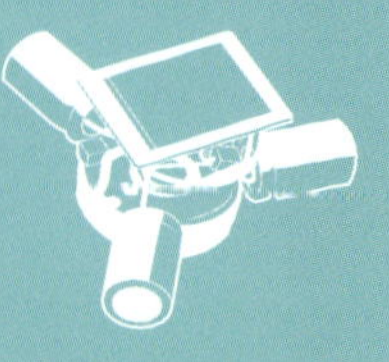

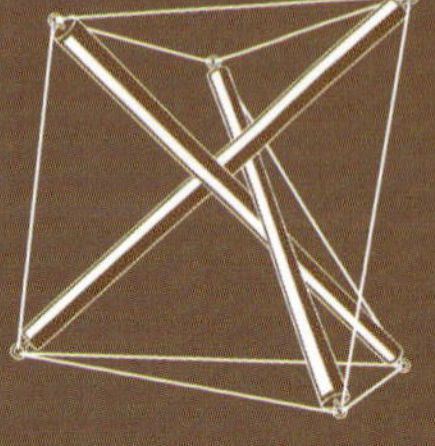

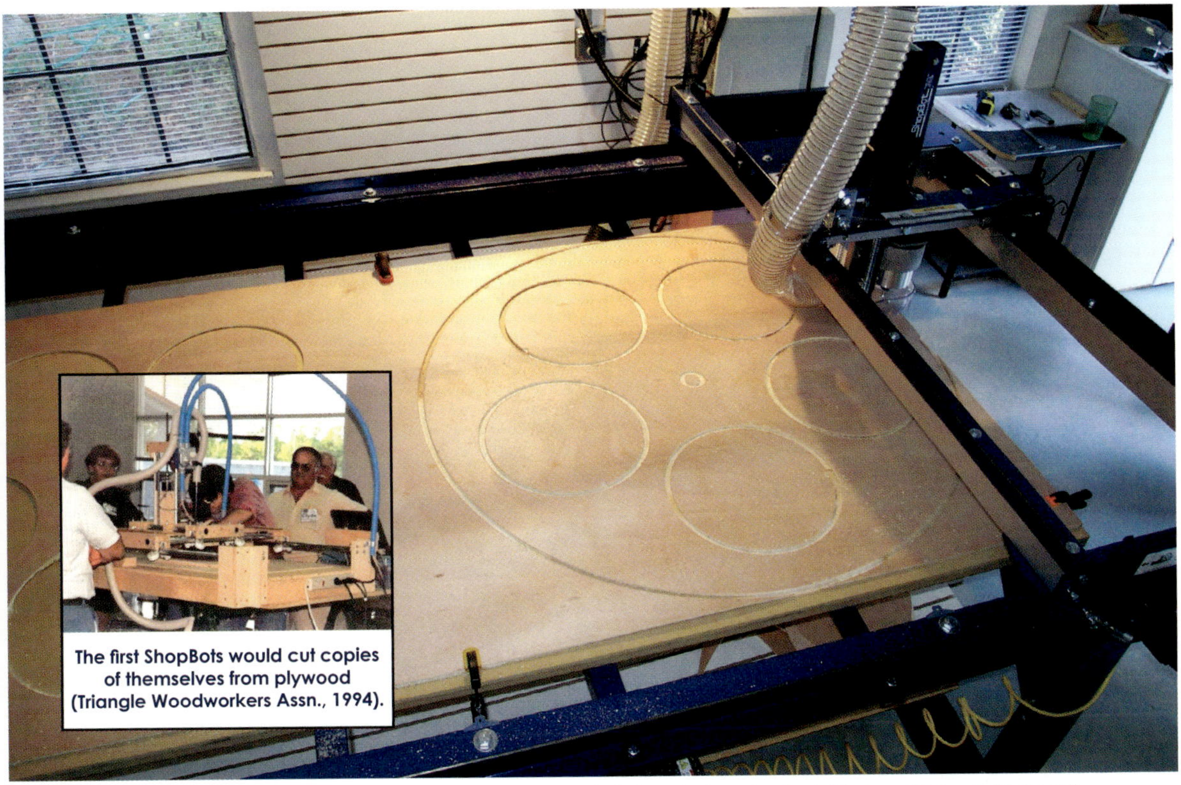

The first ShopBots would cut copies of themselves from plywood (Triangle Woodworkers Assn., 1994).

ShopBot in Santiago Laverde's garage shop showing his first project, a circular table top with insets (Katy, TX).

Volume 06

READ ME: Always check the URL associated with each project before you get started. There may be important updates or corrections.

Maker

117

MAKER PROFIILE
Steven Roberts' super-modified boat, christened *Wordplay*, looks more like a starfighter out of a science fiction TV show than something meant for the water. This analogy isn't too far off, considering the onboard technologies it packs: satellite and cellular phone, ham radio, and marine VHF.

Make: technology on your time™

EDITOR AND PUBLISHER
Dale Dougherty
dale@oreilly.com

EDITOR-IN-CHIEF
Mark Frauenfelder
markf@oreilly.com

MANAGING EDITOR
Shawn Connally
shawn@oreilly.com

SENIOR EDITOR
Phillip Torrone
pt@makezine.com

PROJECTS EDITOR
Paul Spinrad
pspinrad@makezine.com

EDITOR AT LARGE
David Pescovitz

STAFF EDITOR
Arwen O'Reilly

COPY CHIEF
Goli Mohammadi

COPY EDITOR/RESEARCH
Keith Hammond

CREATIVE DIRECTOR
David Albertson
david@albertsondesign.com

ART DIRECTOR
Kirk von Rohr

DESIGNER
Sarah Hart

PRODUCTION
Gerry Arrington

ASSOCIATE PUBLISHER
Dan Woods
dan@oreilly.com

CIRCULATION DIRECTOR
Heather Harmon

ADVERTISING COORDINATOR
Jessica Boyd

MARKETING & EVENTS COORDINATOR
Rob Bullington

ONLINE MANAGER
Terrie Miller

MAKE TECHNICAL ADVISORY BOARD:
Gareth Branwyn, Joe Grand, Saul Griffith, William Gurstelle, Bunnie Huang, Tom Igoe, Mister Jalopy, Steve Lodefink

PUBLISHED BY O'REILLY MEDIA, INC.
Tim O'Reilly, CEO
Laura Baldwin, COO

Visit us online at makezine.com
Comments may be sent to editor@makezine.com

For advertising and sponsorship inquiries, contact:
Dan Woods, 707-827-7068, dan@oreilly.com

Printed in the USA by Schumann Printers, Inc.

MAKE is printed on recycled paper with 10% post-consumer waste and is acid-free. Subscriber copies of MAKE Volume 06 were shipped in recyclable plastic bags.

Please note: Technology, the laws, and limitations imposed by manufacturers and content owners, are constantly changing. Thus, some of the projects described may not work, may be inconsistent with current laws or user agreements, or may damage or adversely affect some equipment.

Your safety is your own responsibility, including proper use of equipment and safety gear, and determining whether you have adequate skill and experience. Power tools, electricity, and other resources used for these projects are dangerous, unless used properly and with adequate precautions, including safety gear. Some illustrative photos do not depict safety precautions or equipment, in order to show the project steps more clearly. These projects are not intended for use by children.

Use of the instructions and suggestions in MAKE is at your own risk. O'Reilly Media, Inc., disclaims all responsibility for any resulting damage, injury, or expense. It is your responsibility to make sure that your activities comply with applicable laws, including copyright.

Vol. 06, May 2006. MAKE (ISSN 1556-2336) is published quarterly by O'Reilly Media, Inc. in the months of February, May, August, and November. O'Reilly Media is located at 1005 Gravenstein Hwy. North, Sebastopol, CA 95472, (707) 827-7000. SUBSCRIPTIONS: Send all subscription requests to MAKE, P.O. Box 17046, North Hollywood, CA 91615-9588 or subscribe online at makezine.com/offer or via phone at 866-289-8847 (U.S. and Canada), all other countries call (818) 487-2037. Subscriptions are available for $34.95 for 1 year (4 quarterly issues) in the U.S. Canada: $39.95 USD; all other countries: $49.95 USD. Application to Mail at Periodicals Postage Rates is Pending at Sebastopol, CA and at additional mailing offices. POSTMASTER: Send address changes to MAKE, P.O. Box 17046, North Hollywood, CA 91615-9588.

Contributors

Lee Zlotoff (*MakeShift*), screenwriter for the original *MacGyver* series, has been making things since he realized both his hands were attached to the same brain. "No question, the inspiration for the MacGyver character came from cellar sojourns with my father, who could pretty much make *anything* work." He lives on a hill in Topanga, and when he isn't surfing or designing furniture and fountains, he makes television and film for Hollywood. Most recently he's launched his own media production outfit, Custom Image Concepts. "The muse said it was time to rock the boat. Who am I to argue?"

Danny O'Brien (*Life Hacks*) lives in San Francisco with his wife and daughter. By day, he fights crime at the Electronic Frontier Foundation as its activism coordinator. By night, he also fights crime, but only by neglecting to get around to it.

Matthew Russell (*Lego Soccer project*) is a Mac zealot, a skydiver, and a Krispy Kreme addict (he once ate 26 in one sitting). He graduated from the Air Force Academy in 2003, is a computer scientist by trade, and has 450 skydiving jumps under his belt. An inhabitant of the DC Metro area, he's a frequent contributor to O'Reilly's Mac DevCenter and likes to dabble with open source software for Mac/Linux. He has two pet ferrets, loves a freshly mixed black and tan from the tap, and always has too many irons in the fire.

Howard Wen (*Maker profile*) is obsessed with the weather — fretting over how hot it is or how cold it is. That's because he's from Dallas, where it's not unusual for the temperatures to change as fast as it takes you to read this paragraph. For his profile on tech-nomad pioneer Steven Roberts, Howard braced himself for the dreary gray skies and chilly rain of the Seattle winter — but the weather was sunny and warm when he visited. So he fretted about how hot it was.

Roy Doty (*Aha! illustration*) has been a cartoonist since the fourth grade. A degree in fine arts and a stint in the U.S. Army during WWII didn't distract him, and he still draws nonstop. Named Illustrator of the Year six times by the National Cartoonists Society, he has illustrated 176 books (29 of which he wrote), and his drawings have been published in so many magazines that the list would fill this entire page. His "Wordless Workshop" for *The Family Handyman* is in its 50th year. His favorite big lie? "That I'm 45 years old." His proudest achievement? "That I have never had a job for a single day in my life."

Contributing Writers:
Tim Anderson, Joost Bonsen, Gareth Branwyn, Mark R. Brown, Shawn Carlson, Bill Coderre, Travis J.I. Corcoran, Larry Cotton, Cory Doctorow, Nick Dragotta, George Dyson, Dan Gonsiorowski, Graffiti Research Lab, Joe Grand, Saul Griffith, William Gurstelle, Tom Igoe, Brian Jepson, Steve Johnson, Haje Jan Kamps, Peter Kirn, Mike Kuniavsky, William Lidwell, Matt Lind, Greg Lipscomb, Merlin Mann, Mister Jalopy, Annalee Newitz, Quinn Norton, Danny O'Brien, Tim O'Reilly, Ross Orr, Tom Owad, Dave Prochnow, Michael H. Pryor, Joel Raedeke, Michael Rosenblatt, Tyler Rourke, Matthew Russell, Bob Scott, Liam Staskawicz, Bruce Sterling, Damien Stolarz, Dan Strunk, Robert Bruce Thompson, Howard Wen, David Williams, Tom Zimmerman, Lee D. Zlotoff

Contributing Artists:
Douglas Adesko, Jake Appelbaum, Nathan Arnold, Kevin Bain, Scott Beale, Ricardo Blanc, Roy Doty, Nick Dragotta, eboy, Mike Fields, John Granen, Alex Handy, Dustin Amory Hostetler, Harold R. Ilano, Timmy Kucynda, Chris Leschinsky, Tim Lillis, Christopher Lucas, Dave McMahon, Jim Mullins, Visa Parviainen, Joe Reinhardt, Seth Schoen, Damien Scogin, Shawn Sinyork, James Sooy, Robin Tafel, Travis Thatcher, Jay Townsend, Susan Williams

Interns:
Adrienne Foreman (web)
Jake McKenzie (engr.)
Ty Nowotny (engr.)

JOIN THE BEAM TEAM*
WITH THE HEP CATS AT
SOLARBOTICS.com
* Afro not required

It is your... Gaming Sidekick
- Chat in game using IM's
- Media player control
- Song search engines
- Email scan
- TeamSpeak support
It is your... Investment Partner
- Stock ticker
- Global weather forecasts
- Explosive features to follow
It is your... Internet agent
- Systems monitoring
- RSS feed aggregator
- Real time sports scores
- Connection monitoring
IT IS YOUR... NEXT GREAT IDEA
- Business Solutions
- Design Pertelian plugins
- Design Pertelian apps
- Linux and Mac ready
PERTELIAN X2040
EXTERNAL USB LCD DISPLAY
www.pertelian.com

Dale Dougherty

TONES DEM TONES, DAMN RINGTONES

AT THE WEB 2.0 CONFERENCE, A PANEL of teens talked about their habits and attitudes for a wide range of gadgets and services. All of them said they almost never pay for digital music, and seemed incredulous that anyone would. Then the moderator asked them about cellphones. One of the teens confessed that he spent $50 a month on ringtones. FIFTY DOLLARS A MONTH! A red light on my maker radar went off.

Shouldn't people make their own ringtones, not buy them? They already own the music. Don't they know how to get it on their cellphones?

The business press is happy to repeat the phrase, "the blah-blah billion-dollar ringtone market worldwide" (the number they say is $4 billion, but when they use big, unsubstantiated numbers so freely, what I hear is "blah-blah"). This means lots of people are making money on ringtones, and they do so mostly by pickpocketing teens and their parents.

Imagine if websites could automatically charge your ISP for activities that your children engage in online. Ringtone vendors, which include carriers, phone manufacturers, and third-party vendors, are taking advantage of the automatic billing relationship behind every cellphone. Increasingly, the game is to get teens to initiate recurring monthly fees hidden from parents in a large, complicated phone bill.

One San Diego parent has brought a class action suit against Jamster, charging fraud and false advertising. This Verisign subsidiary offers a "free" ringtone and then enrolls anyone who asks for it in a service costing $5.99 a month. They advertise on MTV, Nickelodeon, and the Cartoon Network, also promising a free ringtone if you send them a text message. What they don't advertise is that they automatically sign you up for their costly service.

Another infuriated father set up a webpage (www.sleaze-mobile.com) to rant about what happened once he noticed that his 11-year-old daughter had been duped into a $3.99-a-week ringtone service. His screenshots show how a teen could easily think she is saying yes to a free download, since the message saying that she is actually signing up for the service is on a later screen.

Why would anyone pay more for a 30-second ringtone than for a song on iTunes? One answer is that teens see ringtones as personalizing their phones while music downloads are just pure entertainment. Isn't this a terribly shoddy view of personalization?

Do you personalize your phone by signing up for a $5.99 monthly plan to download "I'm N Luv (Wit A Stripper)" by T-Pain, the number one ringtone on Jamster today? Or by downloading Beyoncé's "Check On It" from Cingular's Media Mall for $2.49, where you can't get to the song without seeing promos to "Idolize My Phone" by downloading "Yo Dawg" and other "famous and fun sayings" from *American Idol*?

And why do teens buy so many ringtones? Apparently, teens assign a distinctive ringtone to each friend. The more friends, the more ringtones you buy. Or is it the more ringtones you buy, the more friends you have? It's a sign of your social network.

Of course, DIY ringtones are the best option for true personalization, but don't expect to find good information in your phone's manual or on your carrier's website. They don't want you to know how.

You *can* own your own ringtones, and we'll do our best on makezine.com to cover the tools and techniques for making them on your own.

Dale Dougherty is editor and publisher of MAKE.

Tim O'Reilly

NEWS FROM THE FUTURE

WE HAVE MET THE ARTIFICIAL INTELLIGENCE, AND HE IS US.

REMEMBER THE 1970S TV SHOW, *The Six Million Dollar Man*? It featured Lee Majors as an injured astronaut rebuilt with technology that made him faster, stronger, and more capable. He was a cyborg, a fusion of man and machine.

Anyone with prosthetics — even someone with eyeglasses — is a cyborg of sorts, but what really excites the imagination is today's news of direct connection between brain and machine.

Last fall, University of Florida researcher Thomas Demarse grew a culture of 25,000 rat neurons, then watched as it taught itself to control an F-22 flight simulator. In February, Klaus-Peter Zauner at the University of Southampton, U.K., hooked up a slime mold to a six-legged robot. The biologically controlled robot scrabbles away from bright lights, emulating the behavior of the mold. Last year, a monkey with a brain implant learned to control a robotic arm with its thoughts. In human clinical trials, paraplegic Brian Nagle has succeeded in playing video games with a similar implant.

If it's possible for neurons to control a machine, it should also be possible for machines to control the brain. DARPA researchers recently presented a plan to remotely control hammerhead sharks via a neural implant. The hope: stealth spies able to track enemy ships.

But what has put bionics and the man-machine interface on my radar lately isn't just these modern echoes of old science fiction, but rather the idea that the latest web applications are forging a more subtle, but no less profound, merger of man and machine.

Boxxet founder You Mon Tsang recently introduced a new meme into my vocabulary: *bionic software*. Boxxet is an example of such a system. It's a collaborative news site (a kind of "my.digg.com") in which customers can put Boxxet's spiders to work collecting data on any subject — and then site users rate the results to train the spider. The result is a tool that harnesses both computers and humans to deliver better results than either can do alone.

When You Mon described Boxxet in this way, it struck me that this "bionic" aspect is critical to many of the most successful web applications. Back in 2003, I began using an illustration of von Kempelen's *Mechanical Turk* in my talks, to emphasize the point that one of the things that distinguishes web applications from PC-era applications is the fact that web applications actually have people inside them, working daily as part of the application. Without the programmers running the crawl at Google (and updating the anti-spam algorithms), without the users feeding the spiders by continuously linking to new sites, the application stops working. In a profound way, the users are part of the search engine. This turns out to be true in one form or another for almost every breakthrough web application.

I generalized this idea into one of the key principles of Web 2.0, namely that Web 2.0 applications are systems for harnessing the collective intelligence of their users. But the term *bionic systems* gives a new twist on this concept.

I was talking about the idea of bionic software with venture capitalist Tom Shields a few weeks ago, and explaining how I thought old dreams of artificial intelligence are being replaced by this new model, in which we are creating more intelligent systems by using humans as components of the application. Shields neatly summed up the paradigm shift: "AI becomes IA" (Artificial Intelligence becomes Intelligence Augmentation). We have met the AI, and he is us.

Now that I understand that we're building a next generation of bionic systems, I'm seeing them everywhere. I'd love your thoughts. Where else are you seeing this fusion of human and computer to build capabilities beyond the reach of either alone?

Check makezine.com/06/nff for related stories.

Tim O'Reilly (tim.oreilly.com) is founder and CEO of O'Reilly Media, Inc. See what's on the O'Reilly Radar at radar.oreilly.com.

DEATH BEFORE DULLNESS!

HOW TO DO BORING TASKS, OR WHY ROBOT-RUN DYSTOPIAS CAN BE FUN.

By Merlin Mann and Danny O'Brien

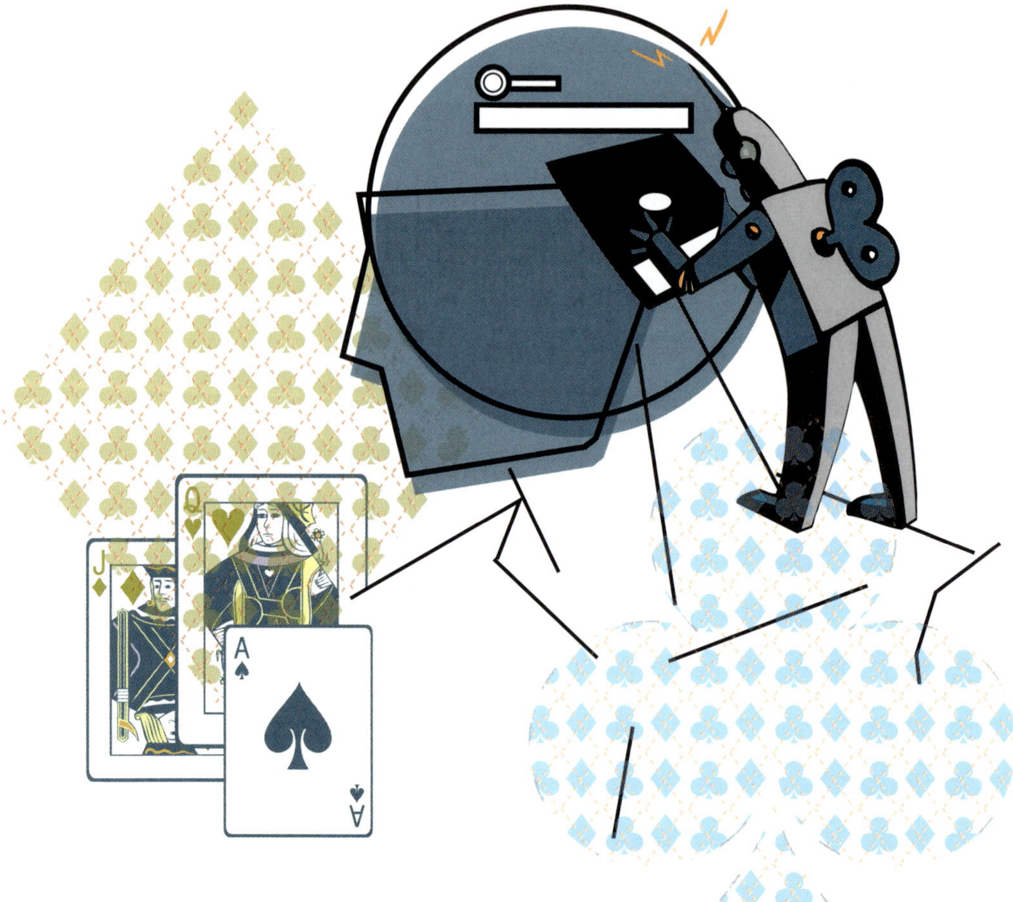

Illustration by Sarah Hart

I THINK MANY LIGHTLY ORGANIZED GEEKS suffer from an overpowering fear of tedium. It's part of being a novelty-seeking kind of person: when you get substantial amounts of pleasure from uncovering new knowledge, any task where you're not learning — where you're just *repeating* — is unbearably painful to pursue.

If you're a programmer, there's an additional level of suffering in drudgery. Doing the same thing over and over while coding, like copying and pasting the same function into different functions? That's never a demonstration that you're doing good work. It's usually a sign that something is wrong with your code. You're paid to remove redundancies. Programmers look at a task like weeding the lawn and think, "I've already done this. Why can't I wrap it in a loop, and just run that?"

As a result, geeks often get themselves into terrible trouble because they cannot bear to do simple, everyday tasks — like pay bills on time, or fill in a tax form, or wash the dishes. It's an aesthetic and moral offense. Better, says your subconscious, to be bankrupt and on the run from the IRS, eating from plastic cups and plates, than to have to do some brain-numbing piece of repetitious nonsense, even for a moment. Death before dullness!

Imagine my surprise, then, as I researched super-organized geeks and I discovered how many apparently mind-numbing routines they indulge in. They have scripts, sure, and little automated hoo-hahs, but they also perform certain repetitious actions, no matter how easy it might be to optimize them away.

The repetition serves to keep certain facts fresh in their minds. Plenty of alpha geeks have a complex system of repeated behaviors for filing a new document or creating a blog entry. Even when they keep their to-do lists on the computer, they still monkishly copy it out onto paper first thing in the morning. The ritual of filing or blogging helps them remember what they're saving. Copying the to-do list fixes the tasks in their minds, and helps to keep the list small.

My favorite story regarding this skill is from the legendary coder John Carmack, creator of *Doom* and *Quake*. Carmack went to Las Vegas for a break and, Vegas being Vegas, decided to have a wager. Carmack being Carmack, he beat the house. The game he played was blackjack, and the technique he used was card counting — which lets you beat the house odds as long as you persistently follow a strictly defined set of rules. As he wrote:

"Playing blackjack properly is a test of personal discipline. It takes a small amount of skill to know the right plays and count the cards, but the hard part is making yourself consistently behave like a robot, rather than succumbing to your gut instincts."

Consistently behaving like a robot is the challenge of many everyday tasks. The hard part for many smart people is to stop thinking, and start acting as though you're running a program *on yourself*.

For many of us, that goes against our guttiest of instincts: we feel as though we're deliberately making ourselves more stupid. But as a strategy, sometimes you really don't want, or need, to be smart.

Many geeks use games to soften the blow. They take boring, monotonous tasks, and by turning them into a challenging game, eke some tiny squib of enjoyment from them.

The games you play depend on your personality. If you're naturally competitive, you'll want to have

Stop thinking and start acting like you're running a program *on yourself*.

some external measure to constantly attempt to beat. If you're someone who enjoys music, match your actions to an internal, complicated rhythm.

I'm about as competitive as anyone always picked last in gym, and music is all bleep bleep bleep to me. But I've gotten through many a terrible job by tapping into the rich bureaucracies of science fiction and thriller shows. Programs like *CSI*, *Battlestar Galactica*, and *Star Trek: The Next Generation* gain some of their appeal from the occasional glimpse of their heroes conducting a tremendously boring task, or minions milling around busily in the background. This leaves opportunities to place you in that world, only temporarily stuck doing the dull Space Taxes Forms.

This can work in almost any otherwise depressing scenario. As science fiction author Rudy Rucker once said: "For me, the best thing about cyberpunk is that it taught me how to enjoy shopping malls, which used to terrify me. Now I just imagine the whole thing is two miles below the moon's surface, and half the people's right brains have been eaten by roboticized steel rats. And suddenly it's interesting again."

Learn how to reel in your mind at Danny O'Brien's lifehacks.com and Merlin Mann's 43folders.com.

Bruce Sterling

HANDS ON: ELEGANT INNOVATION

FROM LAMPS AND CHAINSAWS TO SAVING THE PLANET.

CERTAIN AREAS OF THE WORLD HAVE an innately MAKE-like approach to life. Northern Italy (where I'm writing this installment of my column) is one of those places. Case in point: famed Milanese architect Michele De Lucchi.

For the Italian designer outfit Artemide, De Lucchi created the legendary Tolomeo work lamp. The Tolomeo is bright, sleek, and ductile, with long columnar metal arms and tendon-like wire. It silently bends and swivels at a touch, and stays poised in any position you may place it in. Since its creation in 1983, the Tolomeo has been the number-one work lamp that designers themselves buy for design labor. It's the designer's designer lamp and has been selling merrily for decades.

Oddly enough, De Lucchi is not a designer. He's a "radical architect" from the 1970s, when young Italians rebelled at the constrictions of their discipline and exploded laterally into postmodern home décor, weird laminated bookcases, couture, electronics, graphics — in a word, most anything hackable. This eclectic approach has many practical benefits.

At Olivetti, where De Lucchi worked for 20 years, he involved himself in the production of some 240 products, finishing his career there as the corporation's creative director. He also designed door pulls, tape dispensers, laptops, chairs, vases, interior décor for banks and hotels, and much more.

So far, so good: we're describing a world-famous, multitalented Italian designer at the top of the profession. Now comes the really interesting part: explaining why De Lucchi spent much of 2005 making conceptual art with chainsaws.

Explaining the fondness for chainsaws comes easily enough. To make his point, De Lucchi produces a cherished 12-year-old Italian fountain pen from his immaculate jacket. It occurred to him that although pens and pencils are used with great grace, precision, elegance, and tenderness, no one has extended this approach toward the humble, industrial chainsaw. Why not? Are chainsaws less worthy than pencils? There is no alternative to the presence of industrial objects in modern life.

A large tree fell near De Lucchi's home. That incident required a chainsaw. This was a chance to learn. Once he had his goggles and gloves on, De Lucchi knew that the chainsaw had been radically underexploited as a means of creative expression. The 55-year-old maestro soon made it his business to own and master a variety of chainsaws.

Like many architects, De Lucchi spends much of his professional life making small-scale models of housing. So he decided to refine his chainsaw skills by making model homes straight from the dead tree. No fussy stickler for mere handicrafts,

"Every project is a voyage from idea to realization. There is an ocean of compromise in the middle."

De Lucchi also added telling model details with a laser cutter and a water jet.

These chainsawed model homes look like they were whittled into shape with a giant's jackknife, but the unique models sold at once to eager art collectors. Some of the models were botched. Those, he discarded and wrote a book about: *Twelve Stories About Little Houses*. These chainsaw failures were too ugly to show in public or to display as art, but the effort to make them taught him useful lessons. This resulted in a good set of design war stories.

"Every project is a voyage from idea to realization," he tells me in careful English. "There is an ocean of compromise in the middle."

The architectural lessons from the chainsawed homes are now reflected in De Lucchi's ambitious Japanese eco-village development, outside Osaka. This is a big effort, an entire Japanese suburb, but he has learned, he says, to seek his inspiration for

Photograph by Bruce Sterling

These chainsawed model homes look like they were whittled with a giant's jackknife.

bigness in that which is small, simple, and intuitive. Big, corporate research-and-design teams are all very well in their place, but they are big by nature, and concerned with big resources. So, they are always anxious to avoid big mistakes.

One cannot experiment properly in a state of anxiety. Creativity is closed off by fear. It's even worse to fail to be anxious at a big scale. It's wrong to arrogantly experiment with the lives and fortunes of a company's employees and stockholders — as if those many people didn't matter.

By their nature, big companies and mass production will "commercialize, marketize, banalize, and globalize." But if industry is to improve the world, industry needs something truly good to work on.

Therefore, De Lucchi has divided his own work into sets of physical scales. First, there are the small things he does alone in a home office: "experiments, searches, and fun." By design, these efforts have no deadlines, no clients, no deliverables, no budget, and they are done without commitment to anybody.

At the next level comes a small company called Produzione Privata (Private Production). This atelier features De Lucchi himself, his design assistant, a bookkeeper, and a producer, whose job it is to outsource the manufacturing of De Lucchi's designs. Produzione Privata is deliberately small, but it sells real products and it has a real budget. The next and final step is the De Lucchi architecture firm, aMDL, which does large-scale urban work in Germany, Russia, Japan, Italy, and elsewhere.

These different levels of creative scope do not conflict. Instead, they support and refresh one another. One level is no more or less "serious" than the next. They are a creative ecosystem, where the scale and muscle of the bigger firm can shelter the little greenhouse of the new, and where the small innovative experiments can provide a unique edge and unheard-of innovations for the bigger outfit.

"There is no alternative to industrial organization," says De Lucchi. "But we must also believe that we have the chance to reach a better world through industry. An industry is more than a public investment. If man believes in industry, but industry fails to believe in humanity, the planet is finished."

I don't know about the planet, but having met De Lucchi, I know that Milan and he are the polar opposite of "finished." They have found the means, motives, and opportunities for elegant innovation.

Bruce Sterling (bruce@well.com) is a science fiction writer and part-time design professor.

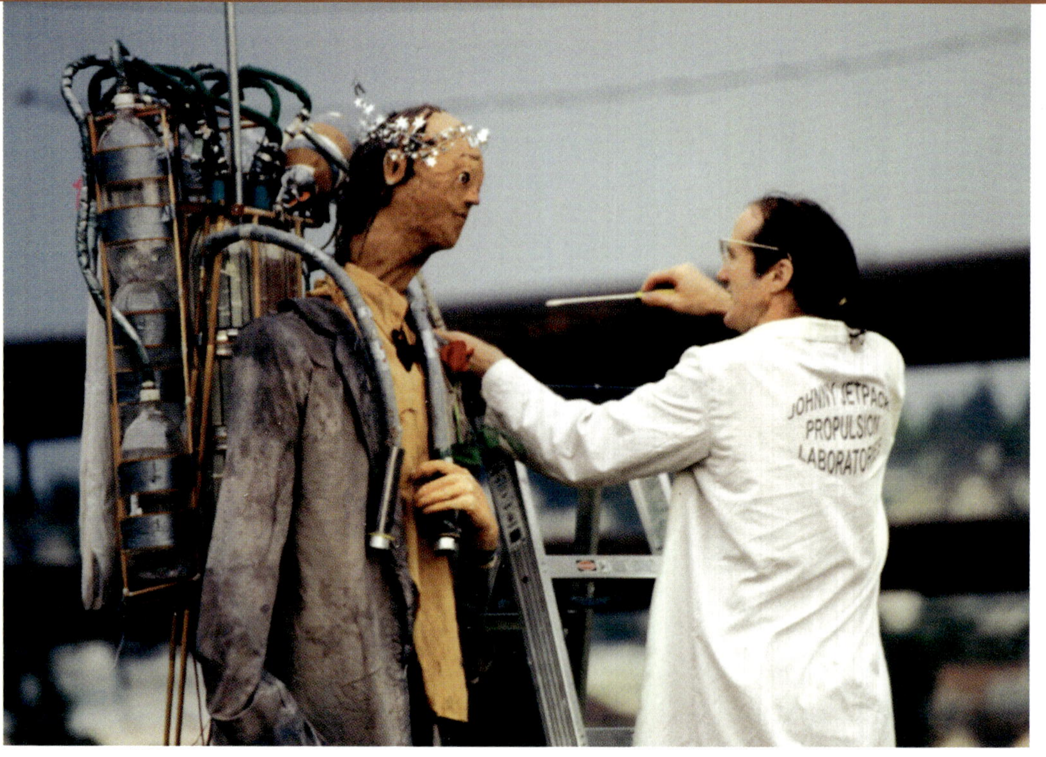

Johnny Jetpack

It's difficult to match the sophisticated and subtle humor of throwing a dummy off something really high and watching it fall to the ground, but to be fair, it's a low-tech laugh.

Luckily, **Nathan Arnold** of Seattle has discovered that strapping a dummy to a compressed-air jetpack, launching it more than 100 feet in the air, and then watching it fall to the ground is three times as funny and provides some geek appeal on the side.

Arnold's performance-art dummy show and his eternal desire to maximize mayhem led to the idea of launching the thing — Johnny, by name — as high in the air as possible in 1983. "It turned out to be more spectacular than I thought it would be," he says, and after more than 100 launches, he hasn't stopped laughing.

Johnny looks surprisingly substantial on the launch pad, so much so that you'd never guess that he weighs in at a svelte 8 pounds. The jetpack — a hodgepodge of carbonated beverage bottles and tubing strapped to a bamboo frame — tips the scales at 30 pounds, but the majority of that weight is liquid accelerant, which will be expelled forcefully within a second or so of launch.

"The hardest part is getting a good seal," says Arnold. "Once you have a seal, you're 90% there."

After Johnny and the jetpack are set up, pointed skyward, and a good seal allows the bottles to be charged with water and compressed air, Arnold has only to pull a pin to initiate the launch.

Johnny accelerates rapidly once the jetpack is activated. His head, filled with cattails, lolls forward from the force, and a Rube Goldbergian sequence of events is set in motion. As the liquid drains from the main chamber, the drop in pressure triggers the parachute ejection device. The removal of the parachute initiates the separation of Johnny from his jetpack, and an "explosion" of cattails blows Johnny's head off. More compressed air chambers in Johnny's chest cavity cause his arms to flail while a parachute deploys to return the jetpack safely to the Earth. Meanwhile, Arnold and whoever is lucky enough to be present laugh at it all.

Thirteen years ago, Johnny Jetpack made his first flight, and over 100 launches later Arnold refuses to stop tinkering on what will forever be a work in progress. You have to protect the technology, of course, so the decision over who gets the parachute — Johnny or the jetpack — is easy. But the jetpack is fragile, and even with a gentle landing, each launch requires roughly 40 hours of preparation. When that kind of time is involved, it's easy to write it off as a back-to-the-drawing-board situation every launch.

What's coming next? Water, compressed air, and soda bottles have served Johnny Jetpack well for a while now, but there comes a time when the laws of physics draw a line. Striding over that line, Arnold picks up a canister of liquid nitrogen. "It's a possibility I've been looking at," he says.

The MythBusters television guys tried to find a way to launch something more mortal than Johnny with 2-liter bottles, but without a good background, they failed despite mounting 15 bottles to a pack. Arnold thinks they just weren't trying hard enough.

"Improper line of force," he says, "They concluded that there was no way to launch a person with soda bottles. I don't agree with that."

—Dan Gonsiorowski

Nathan Arnold's website: johnnyjetpack.com

Photography courtesy of Nathan Arnold

Big Heads

"You maniacs! You blew it up! Ah, damn you! Damn you all to hell!" Oh, my mistake. This isn't a post-apocalyptic world where apes rule, but rather the parking lot behind artist **David Adickes**' studio near downtown Houston.

Adickes is from the bigger-is-better school of making. In addition to a 67-foot-tall statue of Sam Houston in Huntsville, Texas, and 43 busts standing 20 feet tall at Presidents Park in Williamsburg, Va., and the Black Hills of South Dakota, Adickes has big plans for a 36-foot-tall statue of the Beatles, a giant four-person bust of historic figures alongside a busy highway he'll call *Mount Rush Hour*, and a 280-foot-tall cowboy, which would be the largest statue in North America (the Statue of Liberty is about half as tall).

A young 79, Adickes fell in love with gigantism after visiting Mount Rushmore. But he soon found out that creating Rushmore-sized sculptures was no easy feat. He started Sam Houston in 1992 without really knowing how to pull it off.

"It was really a work of engineering without any blueprint," he says. So Adickes learned to apply a simple rule of thumb to making monuments of extreme proportions: "when in doubt, make it stout." And stout they are. His statue of Sam Houston uses five layers of concrete poured over steel mesh attached to a welded steel framework — over 30 tons of material.

Given the extra challenge of scale, why the fixation on gigantic statues? "What I am trying to prove," says Adickes, "is the same thing Frederic-Auguste Bartholdi, the creator of the Statue of Liberty, was trying to prove: think big, do it right, people will come. Today, 108 years later, the Statue of Liberty is an icon, and two million people take a boat to visit it every year. Two years after Bartholdi, Gustave Eiffel built a tower in Paris. Vehemently criticized, it later became the symbol of France."

Whether Adickes' works will endure the test of time like Bartholdi and Eiffel remains to be seen, but there is no question that his titanic figurines make for big attractions and big impressions — just follow the line of tourists. *—William Lidwell*

List of related sites: makezine.com/06/made

Photograph by William Lidwell

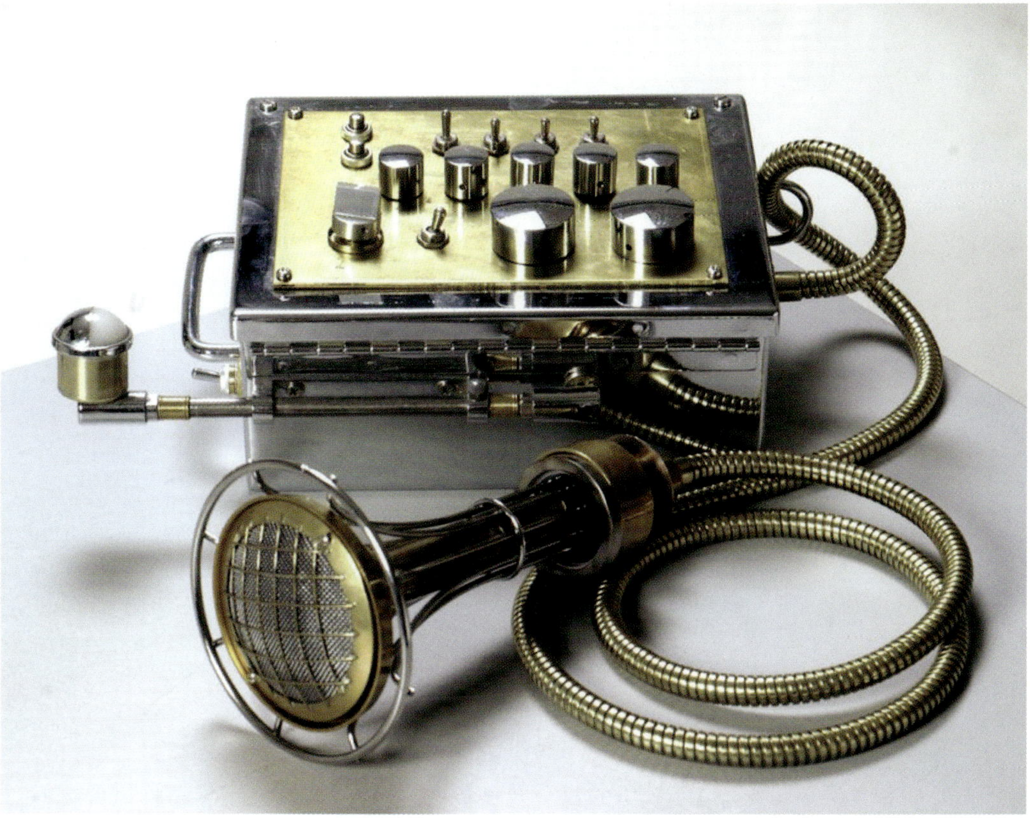

Photograph by Kevin Bain

Sound Made Art

Beeps-a-Lot Box, *Arcane Device*, and *Stylophonic Device* each answer a question I hadn't considered: What would have happened if H.R. Giger had decided to teach band?

These graceful sculptures are working musical instruments produced by Gulf Coast artisan **Mike Ford**. Designed to evoke curiosity as well as admiration, each is heavy with mysterious controls, indicators, and attachments, all beckoning to be explored.

Growing up in a family of Gulf fishermen, Ford learned the value of inventiveness early on. Even as a tyke, he embraced this heritage, secretly designing and building a control panel for an imaginary rocket ship. Sadly, his space exploration dreams were cut short after some critical knobs had to be unglued and returned to his grandmother's TV set.

A skilled stringed instrument builder, Ford was inspired by an article on circuit bending to start modifying discarded electronics boards. Dissatisfied with common project boxes for these creations, he began developing cases as exotic as his circuits.

Finding this new challenge artistically satisfying, he pursued refinements to his design process and metalworking skills while an art student at the University of Mississippi.

Now a full-time artist, Ford melds the electronics of his instruments with cases that, despite their fluidly seamless look, are largely composed of repurposed or found pieces that he works with hand tools. His design is decidedly deco-industrial, a look that has captivated him since he first glimpsed the retro-futuristic gadgets in movies like *Dune* and *Brazil*.

This inspiration is easy to see in the wonderfully strange effectors on his instruments, whose functions seem both obvious and inscrutable, giving his gear the look of something from a not-quite-parallel world (a world where they use a lot of chrome).

Besides building up his stock of sculptures, Ford is now constructing microsynthesizers based on vintage analog integrated circuits. I'm guessing those won't be in plastic boxes. *—Bob Scott*

Mike Ford's website: mikefordsculptures.com

His work is also featured at: getlofi.com

Fly Boy

Look! Up in the sky! Is it a bird? A plane? No, it's **Visa Parviainen**. Last October, Parviainen, sporting a birdman suit and custom jet boots, dove face first out of a hot air balloon high above Lahti, Finland, and took off into the wild blue yonder.

Parviainen is one of a growing number of sky divers who wear wingsuits during their dives. The fabric spanning the legs and arms enables the free-fallers to glide a bit before popping their parachutes. Boosted by his rocket boots, though, Parviainen was able to zip along at 2,000 meters for several minutes.

To build his jet-powered flight suit, Parviainen and his cohorts at BirdMan, Inc. attached a pair of off-the-shelf microturbines to a pair of old hockey skates. Fueled by kerosene, each engine spits out about 16kg of thrust. Tests at a nearby university's wind tunnel convinced them that the aerodynamics should work out. The trick was figuring out a fuel tank system that was lightweight and durable enough for a twisting, turning, windswept human body.

"The solution was to use hot water bottles as fuel tanks," Parviainen says. "Since they're flexible, it's also easier to squeeze every last drop of fuel out of them when you're flying."

Next, the team built a special launch platform to suspend from the side of the balloon canopy. The platform had two purposes: one, it kept exhaust from the boots away from the balloon and passengers as Parviainen revved the engines before takeoff. Two, it was a "nice lounge" for the ride up.

Parviainen knows his way around a machine shop, souping up cars and motorcycles for racing and hacking mounts for helmet cams and other skydiving gear. He's spent the winter working on a new rev of the flight apparatus, substituting a different set of microturbines and tweaking the engines for more reliable operation at chilly temperatures. On his next flight, he also plans to wear a black box recorder of sorts.

"Someday, I want to take off from the ground and land too," Parviainen says. "It's far in the future, but I do think it's possible. Right now though, this is all just good fun." *—David Pescovitz*

BirdMan: bird-man.com

Photograph courtesy of Visa Parviainen

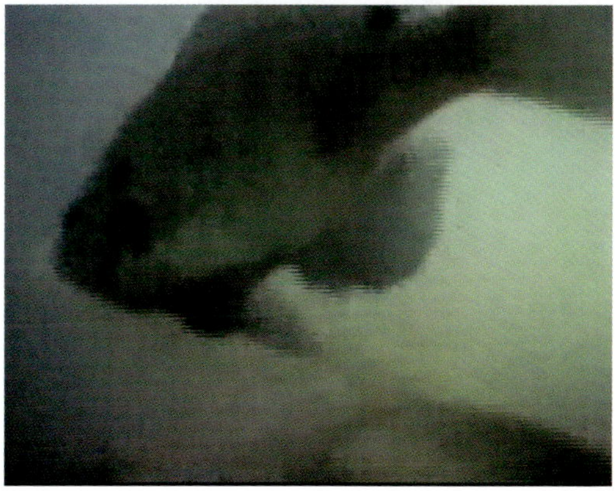

The Eye Aquatic

Cousteau and the undersea world. Cameron and *Titanic*. And now, **Joe Reinhardt** and **Mike Fields** and the depths of Lake Moraine in upstate New York.

Every summer, Reinhardt and brother-in-law Fields tackle a new DIY project at the family's lakeside camp. In 2005, the pair built their own underwater ROV (remote observation vehicle) with two video cameras feeding live images to a shipboard laptop — all for about 100 bucks.

Making things is second nature for Reinhardt, 24, a computer tech in digital imaging. This time, he got to indulge his underwater fascination: "I always wanted to be a marine biologist," says Reinhardt. "I love the water, and ships ... and watching Discovery Channel with the real ROVs exploring the *Titanic*."

Reinhardt and Fields built their homebrew ROV's frame out of PVC pipe, and its transparent camera housing out of scrap quarter-inch-thick acrylic tube from the local plastics supply (milled to watertight tolerances on a friend's lathe). They joined the two with simple but strong carpenter's ratchet clamps.

The B&W video camera was $29 from Harbor Freight, complete with infrared LEDs for night vision, power supply, and 80 feet of RJ11 cable. They scored a Chinese color "Spy-Cam" for $1 on eBay (plus $35 shipping), and ran its video signal up the audio wire in the RJ11 cable.

After an embarrassing misfire with ballast tanks ("We put 'em on top, so it sank upside down every time"), the explorers improvised a solution ("a big hunk of concrete and a bungee cord") and lowered their ROV to the lake bottom to capture video of sunfish, perch, and muskie sporting in the wild. The rig proved watertight to 40 feet.

This summer, they're going deeper: their 2006 model has thrusters for true independent ROV mobility, using watertight 12VDC motors coupled to propellers by super-powerful neodymium magnets. It'll be rated to 200 feet, good enough to dive quarries or wrecks on Lake Erie, Reinhardt says. James Cameron might want to check his rearview mirror.

—Keith Hammond

Homebrew ROV photos and video: makezine.com/go/ROV

Photography by Joe Reinhardt andMike Fields

Good Ship Popsicle Stick

He used to break bones. Now former Hollywood stuntman **Robert Mc-Donald** uses popsicle sticks to break world records.

McDonald has built three Viking-ship replicas out of ice cream sticks. All have been seaworthy, including his latest beast, built from 15 million popsicle sticks over three years. He's now working to break another record by sailing the ship across the Atlantic Ocean in true Viking fashion.

"I have a dream to show children they can do anything," he says. "If they can dream it, they can do it."

In fact, that's what started McDonald down this popsicle path — he wanted to encourage his 8-year-old son to aim high and believe he could succeed, all the while making the world a better place. He is adamant about creative recycling — all the ice cream sticks he used were previously used or imperfect, and were donated by the Ola ice cream company in Europe. McDonald's home port is in the Netherlands.

"[We're] demonstrating how amazing objects can be created from everyday, recycled goods," he enthuses.

"Creative" and "fun" pepper his conversations. And he lives what he speaks. In April 1986, McDonald rocked his way into the world record book by rocking in a chair for 340 hours. Last year, he grabbed another record by sailing a ship made from 370,000 ice cream sticks, the *Baby Ola Bison*.

The bigger replica is 50 feet long and weighs in at a hefty 13 tons, including more than two tons of glue. Named *Mjollnir* (mil-ner) — the Viking god of thunder — she is an open craft with no protection for the sailors whatsoever. The 6-person crew sleeps in true Viking style: hammocks strung across the deck. Her voyage across the Atlantic began in mid-April.

McDonald heads the Sea Heart Ship Foundation, a group spreading fun to kids in hospitals around the world. Captain Rob (as the kids call him) recently returned from a hospital tour of Florida, the Gulf Coast, and New Orleans, where he gave away 28,000 stuffed animals in 14 days (yet another record).

—Shawn Connally

» **Great Atlantic Crossing Adventure:** obvikingship.com/index.php

Photography courtesy of the Sea Heart Ship Foundation

Duet with a Robotic Drummer

Gil Weinberg is having trouble with his drummer: he's trying to get both of Haile's arms to work at the same time. That would be an unusual problem, except for the fact that Haile is a robot.

Haile's microprocessor-controlled, motorized arms are able to play ordinary acoustic drums, with expressive control over timbre and dynamics. But while the movements of its anthropomorphic, wooden body are impressive, Haile's listening ability is as important as its playing. Using custom computer software developed in the Max/MSP multimedia environment, Haile can analyze the performance of a human drummer and respond in real-time.

"We've tried to create a new musical experience — to surprise you," says Weinberg. Haile's responses range from simple imitation to variation and even intelligent accompaniment. The results vary: sometimes, the algorithms simply don't work, or don't work in a way that makes sense to human ears. But Weinberg, who was a jazz pianist for years before he became interested in computers, is most excited by the moments at which Haile feels like an equal musical partner, and plays in ways no human would.

"I've played with people for years," says Weinberg. "But especially when you're in a particular genre, you know what to expect. Here, we are in uncharted territory."

While Haile's human-robot communication skills continue to evolve, it's also getting a chance to facilitate human-human communications. Next, Haile travels to Jerusalem to play on a program featuring collaborations between Jewish and Arabic percussionists. The composition is called *Jam'aa*, or "gathering" in Arabic. Drawing on the communal tradition of Middle Eastern music, Haile will interact with professional darbukah and djumbe players, transforming what they play. At least, once the remaining technical bugs get solved.

"Talk to me Monday," says Weinberg; by then, he hopes Haile will be able to play with both arms.

—Peter Kirn

Photgraph by Travis Thatcher

Robotic Drummer: makezine.com/go/robotdrum

Cory Doctorow

A HOUSE DIVIDED

HOLLYWOOD FEELS THE STING OF ITS OWN COPYRIGHT LAWS.

FOR CASUAL USERS, ITUNES DRM doesn't look so bad, and for many of them, it's a pretty loose set of chains. A lot of info-civilians don't own more than five CPUs (the limit on the number of boxes an iTune can play on), and iTunes is new enough that not many of us have had the opportunity to sink a big investment in the technology and try to take it with us to a new vendor.

For the recording industry, iTunes represents a chance to charge more for less. An iTune can't be sold on as a used good, something that we can do with our CDs. You can't turn an iTune into a ringtone, nor make other uses that the record companies might want to charge us money for. It's a neat and easy way to gouge us for the stuff we used to get gratis.

But for Apple, iTunes is a great way to lock us into using its products. No one is allowed to make a compatible iTunes player without Apple's permission (when RealNetworks tried to make a Real player that ran on the iPod, Apple threatened a lawsuit under the DMCA).

Apple doesn't really care how many CPUs you listen to a track on. Apple doesn't really care how many people you stream a song to. That's just a sop to the record industry. Apple cares about you sinking an investment into an integrated chain of technologies (iPods, CPUs, iTunes metadata) and media (songs) that will raise the switching costs if you decide to give your business to someone else.

From the recording industry's perspective, this is a disaster. Once Apple controls the relationship with music listeners, Apple can name its price and set its terms for selling music. The record companies are trying desperately to raise the cost of some tracks and lower the cost of others because they believe they'll sell more that way, and Apple has basically laughed them off. Why? The record companies will never stop selling songs for use with iTunes. Apple controls the iPod, and uses that control to lock Rhapsody and Napster songs out of the most popular portable player on the market. The music industry can't afford to blow them off.

The entertainment industry, having accidentally created cozy monopolies for the likes of Apple, Microsoft, and Macrovision, is now busily trying to put itself back in charge of its own destiny. Next-generation crippleware DVD standards, like Blu-Ray and DVD-HD, are being defined by studio-crafted, fragile alliances of bitter competitors from the technology industry, companies that will be hard-pressed to present a unified front when it comes time to negotiate terms with Hollywood.

Once Apple controls the relationship with music listeners, Apple can name its price and set its terms for selling music.

While monopolists rearrange the deck chairs on their sinking superliners, we makers are busily building, acquiring, and using open technologies and file formats that let us move our media freely from one device to another, keeping us from being locked into an entertainment company or a DRM vendor.

Let the dinosaurs go for each other's throats — we can see the tar rising around their feet.

Cory Doctorow (craphound.com) is a science fiction novelist, blogger and technology activist. He is co-editor of the popular weblog Boing Boing (boingboing.net), and a contributor to *Wired*, *Popular Science*, and *The New York Times*.

1+2+3 T-Shirt Laptop Pouch

By Ross Orr

Reincarnated tee "iShirt" protects your computer and deters thieves.

You will need: Sewing machine, marker, pins, old T-shirt. A smaller T-shirt works best for this, unless your laptop is enormous. An old favorite with holes and burn marks will just add to the case's theft-deterring stealth. A pocket large enough to store your laptop's power brick is a nice touch.

1. Mark and pin shirt.

a. Lay the shirt face up and then fold the bottom half upwards over the chest. Check by touch that the bottom of the pocket (if the shirt has one) is a bit above the fold. Set your laptop on top of the shirt and then draw vertical marker lines about an inch from both of its sides.

b. Along each marker line, use straight pins every few inches to hold all the layers of fabric together. Pin crossways so that the pins won't get snagged on the sewing machine's foot.

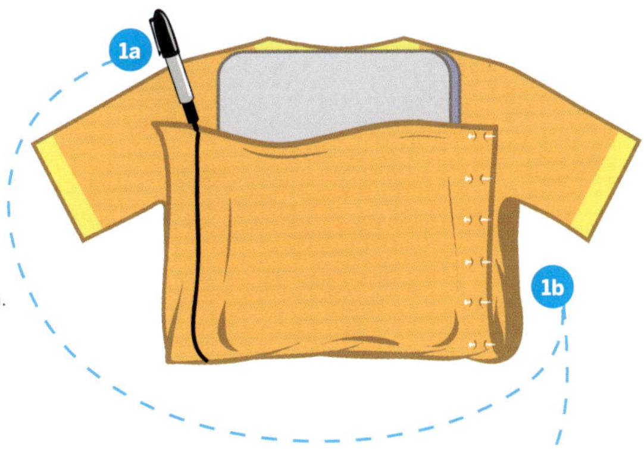

2. Sew and cut excess.

a. Sew a straight seam along both marker lines. For reinforcement, double up the stitching near the open edge of the pouch.

b. Cut off the sleeves and trim the excess cloth around the seam, leaving about 2 inches all around. This extra fabric outside the seam will provide additional padding.

3. Use!

Turn the pouch inside out so that the pocket is on the front again. Slip your laptop into the middle compartment. The neck opening is great for stashing stray cables; the bottom flap holds papers or your ever-present copy of MAKE.

For Extra Credit

+ Take one of the leftover sleeves and sew it into a pouch for your iPod and earbuds!

Ross Orr hacks low-tech gadgets and invasive plants in Ann Arbor, Mich.

Illustrations by Dave McMahon

Maker

Tech-Nomading From Shore to Ship

Twenty-three years ago, he went on a bike trip and never returned.

INTERVIEW BY HOWARD WEN
PHOTOGRAPHY BY JOHN GRANEN

Steven Roberts' workshop is a mess. Nestled in the quiet woods of Camano Island, a small community situated by Puget Sound about 90 minutes north of Seattle, the 3,000-square-foot building is overflowing with miscellaneous electronics, computers, and gadgets. Tools are everywhere — on shelves, on worktables, on the floor.

FORTH
System Documentation
Microship Project
GENOA QUICK COUPLINGS
Clippard
Fiberglass
AMPRO
Vorsicht: Laser in Betrieb
McMASTER-CARR
OLIMEX
ATHER BE
LING!

"There's an old saying: The smaller the boat, the bigger the adventure."

Roberts tells me he's in the process of eBaying most of these things. I notice a stack of Macintosh computers. They're the really old "classic" models, the ones with the tiny built-in black-and-white monitors. The genial Roberts, a tall and bearded fellow, says I can take one or more of them if I'm interested. I politely decline, being one of those people who avoids collecting items for which I have no use.

I'm a little surprised by all of this stuff in his place. When I first contacted him over the summer of 2005, Roberts told me how in 1983, then 31 years old and living in the suburbs of Columbus, Ohio, in a three-bedroom ranch-style house, he worked as a freelance writer (reporting on technology and electronics), and felt trapped. "I was working my ass off to pay for things I didn't want, a lifestyle I didn't want. I was doing things I didn't enjoy," he recalled.

Seeing his workshop, however, I was reminded of an email in which he lamented: "Almost everyone I know is bogged down by 'complexity': Creaky but familiar tools one doesn't dare replace; new toys not yet learned; incompatible power supplies; unlabeled mystery cables clogging drawers; lost documentation; and the vulnerability of it all becoming utterly useless."

Back in 1983, back with all those things he didn't want, Roberts yearned for simplicity and adventure. He had not planned on it, but he was about to become a pioneer of the "tech-nomadic" life: a man who used mobile technology to live on the road, to stay in touch with the world wherever he was, and to free himself from a dreary existence.

The Souped-Up Canoe

Roberts became interested in electronics when he was 8 years old and growing up in Louisville, Ky. His father was a mechanical engineer, and Roberts was inspired by him to build machines.

Roberts' road warrior box, the Shacktopus contains three transceivers, GPS, wi-fi, environmental telemetry, solar power, speech synthesis, audio recording, Bluetooth, a deployable antenna, and more.

As a child, he made numerous electronic projects, which he usually entered into the school science fair: an induction magnet that could pick up aluminum, a Morse-code translator, and a speech synthesizer based on his own vocal tract, taken from an X-ray of his head.

He shows me his latest project in his workshop, a souped-up canoe. Actually, "souped-up" is an understatement. It's been totally overhauled with the addition of two smaller hulls, one connected at each side to the main hull where the pilot sits. Eight blue solar panels, four on each side, are set across from the smaller hulls to the main hull. This water vehicle even has retractable wheels. Why does a boat need wheels? Roberts felt it would be more convenient to move from shore to water if it had wheels, so he spent years designing and building this elaborate mechanism.

This super-modified boat, christened the Wordplay, looks more like a starfighter out of a science fiction TV show than something meant for the water. This analogy isn't too far off, considering the onboard technologies it packs: satellite and cellular phone, ham radio, and marine VHF. And this doesn't even include the video cameras and other gizmos that aren't mounted to it the day I visit. There's a circular, triton-looking antenna set toward the craft's bow that Roberts explains is an "ultrasonic transducer." Once a second, each of the three forks communicates with the other in order to collectively measure the surrounding air mass, wind speed, and wind direction.

Yet Wordplay travels by decidedly low-tech means: by wind with a sail.

The Winnebiko

Roberts shows me around the rest of his cluttered workshop. One item catches my eye; it looks like a control panel ripped out of a jet fighter cockpit. It served as the control panel for his first tech-nomadic vehicle, the Winnebiko.

Back in Columbus, 1983, Roberts went to a party out in the country. That night, he stared into a campfire and then things "just all snapped into place": why not combine the things he was most passionate about — computers, writing, travel, bicycles, and romance — into a new life?

He ordered a custom-built recumbent (a type of sit-down bicycle), then grafted on a RadioShack TRS-80 Model 100 laptop, a Hewlett-Packard HP-110 portable computer, CB radio, and a 5-watt solar panel to power these gadgets. He named the resulting vehicle the Winnebiko.

Starting from Columbus, Roberts biked 10,000 miles, passing through small towns along the southern East Coast, through Florida, the South, Texas, and the Southwest, and ending in Silicon Valley in California about 18 months later. Throughout this trip, he continued to earn a living by writing articles on his laptop. He also wrote about his journey, which eventually caught the notice of people in the media, who wondered about this man bicycling across America on a "computerized bike."

As fulfilling as this really long bike ride had been, he found it frustrating that he could not write while riding at the same time. "I had all this mobility, but I was just watching the words flow away, knowing by night, when I was camping or whatever, I would not capture these thoughts," he says.

Roberts upgraded the Winnebiko for an encore trip in 1986. The Winnebiko II added packet radio for email access, a security system with motion detection and voice synthesis, and a new, more sophisticated control panel. To enable himself to write as he pedaled, he hacked apart the keyboard of the TRS-80 Model 100, and rewired the keys to the bike's handle controls.

His second bike tour ran from Seattle, along the West Coast, and across the country to the East Coast. He was accompanied by his girlfriend at the time, Maggie Victor, who rode her own recumbent. Together, they traveled 6,000 miles.

Behold the BEHEMOTH

The attention Roberts got from the media led to interest from corporate sponsors. From 1988 to 1992, he threw more things onto the Winnebiko II. A lot of things. Much like a succeeding version of a Microsoft application, the bike quickly became bogged with too many features, to the point of absurdity and uselessness. So many components were put on it that a trailer had to be designed to hold them, and for the bike to tow.

Renamed BEHEMOTH (Big Electronic Human-Energized Machine ... Only Too Heavy), it was assembled in Silicon Valley by a team of volunteers assisting Roberts. It had almost every piece of mobile and computer technology at the time: a hacked Macintosh and other computer systems, tons of radio communications devices, GPS navigation, even a radiation monitor.

"I got so distracted by the tech stuff," Roberts admits. "I'd be reading a trade journal and go, 'Ooh, ooh, I could use that!' and then I would schmooze and get it."

The media and public were enchanted with his journeys, and wowed by the technology-laden, though impractical, bike. Roberts was interviewed by many reporters and appeared on TV talk shows. But as public interest in his project was reaching its peak, the tech-nomadic biker's passion for his original dream was dying. Ironically, to make public appearances and do speaking gigs, he traveled the country in a diesel truck that carried the BEHEMOTH in its trailer.

The Laboratory on an Island

Using the money he earned from his speaking tour for the BEHEMOTH, Roberts bought property on Camano Island, choosing the region for its variety of surrounding waterways. He put up the 3,000-squarefoot workshop to facilitate the research, construction, and testing of small watercraft that would utilize tech-nomadic technologies. He brought over the volunteer-community ethic of BEHEMOTH by inviting engineers and other specialists to take part.

The "Microship Project" began with the idea of outfitting a basic kayak with communications devices, but evolved into a pair of specially modified boats, Songline and Wordplay. These water vehicles expanded upon the embedded systems technologies that Roberts and his BEHEMOTH team developed, and which allowed the pilot to control almost every aspect of the craft through a Palm-OS PDA.

In his workshop, Roberts shows me the inside of a thick plastic project box that's sitting at one end of a worktable. It looks large enough to hold several hundred sheets of 8.5x11 paper. This is the power supply he designed for the Wordplay. It uses seven microprocessors to regulate and distribute 600 watts. While the boat's solar panels alone can provide 5 knots to move the craft, this power unit was also designed to enable the pilot to divert all power to the thrust for emergencies (like quickly veering away from another boat).

Roberts took the Wordplay on a 132-mile ride through Puget Sound in September 2001.

"Unlike the other systems [Winnebiko, et al.], which were really lifestyle choices, this is much more 'grab and go,'" he says. "With Shacktopus, I'm avoiding any more multi-year projects that tie me to a specific bike or boat."

He shows me the prototype for the Shacktopus, sitting on the same worktable where he has the power supply box for the Wordplay. The Shacktopus is essentially a clear plastic project box with a medley of off-the-shelf communication and other computer components inside,

> "It was like I reached around the back of my head and hit the reset button," he says of the day his journey began.

Though he still considers it to be in development, the Microship Project has gone through extended inactive periods over the years, as personal priorities for him shifted.

"When I started this ten years ago, I was perfectly pleased with the idea of taking out on a canoe-sized hull, spending two years sleeping in the bilge — it's the size of a coffin. Now I'm 52, and I'm like, 'I don't want to be that uncomfortable for that long!'" he says, chuckling.

Project Shacktopus

Along with a weakening tolerance for physical discomfort, Roberts has been wondering lately if maybe he has spent too much time over the past 20 years designing machines and not enough of it going out on actual adventures. He calls this consequence the "BEHEMOTH effect."

Thus, last year he decided to create an all-in-one mobile communications pack, which he named the Shacktopus. He plans to make it the size of a notebook computer, or small enough to fit into a messenger bag. It's the first project of his that he hopes to turn into a commercial product.

bashed and interfaced together through a unified control system designed by Roberts. There's an HF/VHF/UHF transceiver; a GPS system; environmental telemetry; internet access with wi-fi bridging; a lithium-ion battery power system that can be charged by a solar panel, automobile cigarette lighter, or AC outlet; speech synthesis; audio recording; Bluetooth interfacing to a notebook computer or PDA; and a deployable antenna array from HF to 2.4GHz.

For something that's supposed to help simplify Roberts' life, to free him from dealing with such technically complicated projects as the Microships, this early version of the Shacktopus itself already looks to be ... complex.

While it stems from the restlessness that has been growing within its creator over the past few years, the Shacktopus also seems to represent the same conundrum. Roberts came to the woods of Camano Island with what sounded like a simple enough plan: build a boat and move on to the next big adventure. This didn't happen fast enough. Apparently, it takes a long time to build a boat, at least the way Roberts likes to build one, thanks to fancy things like retractable wheels. Things became complicated.

"I want to get moving again, and I see Shacktopus as the way of making that happen — kind of short-cutting that whole process," he says, perhaps hopefully. "I'm looking for a big boat that I can live on and do some world traveling."

The Allure of Human Power

When I started talking to Roberts for this story last summer, I immediately noted that his tech-nomadic vehicles — the bikes and the boats — shared the distinction of being small vehicles that relied mainly on people power. I wondered what the appeal in that was for him.

"I like the human scale of it. I find that when you cruise on a motorcycle or car, you're really anonymous. You're just somebody passing through on the freeway. Whereas, when I was on a bicycle, I was completely non-threatening," he said to me back then. "Back in the early 80s when I was [biking through] small towns in the South, people would take me home. They weren't worried about me. Also, it's a lot more satisfying. Kayaking to an island is more exciting. There's an old saying: the smaller the boat, the bigger the adventure."

Afterwards, I mused for several months: what could be the thematic connection of a human-powered vehicle, or wind- and solar-powered one (as in the case of the Microships), to mobile communications technology? What was the appeal of the two brought together?

When I meet the tech-nomadic pioneer in person on a sunny afternoon in February 2006, I ask him about this. Beyond the fact that the two subjects have always interested him personally, he cannot come up with some profound, satisfying explanation for it all.

I climb aboard the Wordplay — or climb into it, to be more precise. At 5'8", I'm much shorter and thinner than Roberts, so the inside doesn't feel "coffin-like" to me at all. There's a lot of legroom. I fiddle with some of the levers and try to relax myself into the hard seat. I look ahead, out the canopy. A compass is affixed to the top of the dashboard, and beyond that I see a large marker board with technical-looking diagrams drawn on it, hanging from the wall in front of the Wordplay. If I were on Puget Sound, my view instead would be of the water rippling out beyond and into a backdrop of the mountain ranges of northwestern Washington, I imagine. But it also feels like I'm in a starfighter.

It then gradually dawns on me: there's a unique feeling about being inside such a small craft that relies upon your own physical skills and wits to control. It becomes like an extension of your own skin, your own body. It becomes personal.

Maybe that's the connection: mobile communications, and vehicles like Roberts' bikes and this boat, both evoke a personal, emotional bond between themselves and the user. The more physically invested the rider is in the direct powering of the vehicle, the more personal the journey becomes.

"My goal is not to spend my life in the lab building electronics," Roberts says. "I got this beautiful place in the woods. It looks like it ought to be paradise, but I'm just itchin' to get moving again."

The last time he felt that itch, he was living in a three-bedroom, ranch-style house somewhere in the suburbs of Columbus. There was no eBay, so he was stuck with a bunch of unwanted stuff that he couldn't easily get rid of.

But one day, he left it all behind — there were still dirty dishes in the kitchen sink — and didn't even bother to lock the front door of his house. He just pedaled away on an odd-looking bike that he had slapped computers and mobile communications gadgets onto. "It was like I reached around the back of my head and hit the reset button," he says of that day when his journey began.

The microship Wordplay:
1. Retractable wheels facilitate transport.
2. The cockpit is covered with a bright red canopy made of heavy waterproof fabric.
3. Solar panels can provide thrust in an emergency, otherwise it's wind and sail.

1

2

3

Sky Cutters

Pulling the strings at the Berkeley Kite Festival.

By Arwen O'Reilly

IF YOU'RE IN BERKELEY IN JULY, YOU CAN'T miss the kites. After almost 20 years, the Berkeley Kite Festival is a time-honored tradition and one of the largest kite festivals on the West Coast, attracting kite-flying professionals, families with children, and casual passersby. Held on parkland surrounding the Berkeley Marina, you can see the kites — swooping, billowing, and soaring — from almost anywhere in the surrounding hills.

Tom McAlister, in love with the beauty and accessibility of kites, founded the festival in a desire to give back to his community. "At the time," he points out, "most festivals were primarily either children's events or exhibitions for professionals. We wanted to offer all of the best that modern kiting had to offer."

McAlister, whose stepfather built gyrocopters, grew up loving anything that flew. He discovered kites in college, and started selling them out of the back of his Honda Civic in 1985. "You can be a kid or an 80-year-old in a wheelchair," he says, "and for a modest investment be cutting up the sky like the Red Baron."

Lucky for him, he was in kite-flying heaven: Berkeley, just opposite the Golden Gate, is directly in the path of some truly gorgeous and reliable wind. Over time, the weekend gig turned into a full-time job, and the festival blossomed, too. The timing was perfect: a park was being constructed on top of landfill when the festival began, and as the park grew, the festival swelled to fill it, now attracting crowds of up to 25,000 people every summer.

And diverse crowds they are. Your first sight once you actually get to the festival is a hilltop covered in picnic blankets and tangled kite strings, as little kids run amok trailing kites behind them. The variety is beautiful to see: there are elaborate, store-bought dragons clashing joyfully with colorful, homemade box kites and simple, hand-painted classic diamonds. The ecstasy is palpable. In the afternoon, everyone streams over to the demonstration area, where a kite showers down candy, tiding kids over until Halloween. "We had the first candy drop anywhere in the world," McAlister delights in mentioning. "It's kind of silly, but hey — we're proud of it."

He's just as proud of the festival's other acts. Over the hill from the families are the professionals: champion kite fliers mix with giant three-dimensional kittens dancing in the wind, team kite demonstrations share airspace with exhibition kites, and kite manufacturers like Prism, Revolution Kites, and Ozone let kite buffs test out their latest designs.

The bronzed Ray Bethell, a seemingly ageless multiple-kite-flying world champion, draws gasps from even the most seasoned kite watchers with his nimble fingers and whizzing stacks of kites. Kite aerial ballet groups perform with 96 kites at once, and 20,000 square feet of Peter Lynn's giant creature kites gallop, slither, and leap in the wind.

All in all, to say the event is "fun for the whole family" is understating it, but there is more in the appeal of kites than mere child's play. They have long been a symbol of freedom and hope (check out national newspaper archives for images of kite flying, and you'll often see articles about emerging democracy), and it seems hard to find someone who doesn't like them. McAlister, who has been designing his own kites for 15 years, is a firm believer in kites as an art form: "That doesn't mean every kite is art — and most aren't — but there's the potential, whether it's the design or the act of flying." When asked why he thought the festival had been so successful, he says: "A wise old man once told me, 'Kites are holding the wind in your hand.'"

Berkeley Kite Festival, July 29–30, 2006
www.highlinekites.com/Berkeley_Kite_Festival

Arwen O'Reilly is staff editor at MAKE.

Photography courtesy Highline Kites

Peter Lynn, who's been designing kites in New Zealand for well over 30 years, came up with his huge and magical trilobyte designs in the mid-1980s, and snagged the world record for largest kite in 1995. His creature kites (top) now represent all parts of the animal kingdom.

Ready for takeoff, dual-line stunt kite trains make a colorful ground display (left).

What Steam-Up would be complete without steam-powered personal transporters?

Boiler Room

A visit to the annual Yankee Steam-Up.

By Brian Jepson

Dozens of steam engines meet in Rhode Island every year, connect to a massive boiler, and crank away for all to see. Everything from the smallest tabletop engines to large industrial models share the steam and show off what they can do with it. This is the annual Yankee Steam-Up, held on the grounds of the New England Wireless and Steam Museum (users.ids.net/~newsm), an organization dedicated to preserving the beginnings of wireless and steam technologies.

Throughout the day, volunteers kept feeding the boiler as the steam engines kept cranking. And there are a few steam-powered vehicles prowling around as well: a steam-powered motorcycle and a vintage Stanley Steamer Bus were on hand.

In addition to this impressive display of steam power, the museum operates the oldest working wireless station in the world, Massie Wireless Station PJ. Because the spark gap transmitter's bandwidth is so wide, it can't be connected to an antenna without creating interference that would draw in the FCC's vans and black helicopters. The museum has a collection of Morse code keys, radios, television equipment, and transmitters — all vintage and antique equipment from throughout the history of wireless communications.

Whether handmade or rescued and restored from an earlier time, all of the engines and wireless equipment have felt a maker's touch. Museum members meet every Thursday to tackle the work-in-progress. The Yankee Steam-Up is the museum's major public event. To see the museum at other times, visitors should contact the museum in advance at (401) 885-0545.

Brian Jepson is an O'Reilly editor, programmer, and co-author of three O'Reilly Media books.

Photograph by Brian Jepson

Ship of Cards

Popular card games become an exercise in miniature construction. By Alex Handy

THE WORLD OF COLLECTIBLE CARD games has long been dominated by Magic: The Gathering and Yu Gi Oh. But the 2D nature of these games has recently been tossed aside and the world of card games has a new caveat: *to play, you must first build*.

Up-and-coming games company WizKids, founded by Jordan Weissman, created Pirates of the Spanish Main in 2004. It's a collectible *constructible* card game based around the idea of using desks and tables as a substitute for a board, and tiny punch-out boats as the pieces in a game of high-seas combat and plunder.

Since then, WizKids and its competitors have taken this concept to new extremes, with RocketMen, a 1950s serial-style space game; a NASCAR-based racing game; a driving/jousting game called Racer Knights; and the catchall game, Z Cardz.

No matter who makes them, constructible card games all have one thing in common: you have to build stuff to play. Game pieces come in the form of perforated polystyrene cards, the parts punched out and assembled into 3D in-game characters, ships, and weapons. First came pirate ships; now there are pterodactyls, fleets of spaceships, and even race cars that roll like their Matchbox equivalents.

And it all started with Mike Mulvihill, a member of the game design team at WizKids and a grizzled veteran of the game design world. "There was such a resurgence in pirates, but we couldn't afford to make ship miniatures," he says.

"When the styrene thing fell into our laps, we knew we could do ships with the plastic. But then the question was, if you do this, how do you make everything in that pack of cards become something that is useful to play the game with? It was like Apollo 13."

Mulvihill and his team eventually built Pirates of the Spanish Main into a game that can be played by two people with only two packs of cards. Total price to play is about $8. Each game pack comes with a tiny die, a punch-out island, two ships, a card filled with gold punch-out pieces, and a card full of crew.

Since playing the game typically requires more than two islands, the instructions suggest using the frame left over from the island card as an island itself. More islands can be made out of the leftover pieces of ships, and even the card pack wrappers. During a game, ships move set distances that are measured by lines drawn along the edges of the empty cards that once housed masts, hulls, and deck plates. The game is truly a wonder of resourcefulness.

Alex Handy is a freelance journogeek, and his office is now adorned with little styrene pirate ships. He blogs at gism.net.

Photograph by Alex Handy

Space Cases

The balloon men at NASA's Jet Propulsion Laboratory.

By David Pescovitz

In 1835, an article appeared in the *Southern Literary Messenger* telling the remarkable story of one Hans Pfaall, a bellows-mender from Rotterdam who escaped his creditors by flying a homemade hot air balloon to the moon. The story turned out to be a prototypical bit of science fiction dreamt up by a young Edgar Allen Poe to dupe his readers. The idea of ballooning to the moon or elsewhere in the solar system never took off, probably because it's impossible. But for a certain group of engineers, flying balloons on other planets isn't far-fetched at all. In fact, the Russians have already done it. And NASA's Jet Propulsion Laboratory (JPL) is keeping the tradition alive.

A sprawling campus just north of Los Angeles, NASA's Jet Propulsion Laboratory was established by the California Institute of Technology in the 1930s. Emerging from the DIY mindset that launched the science of rocketry in this country, JPL was where the first U.S. satellite was designed, the Mariner and Voyager probes were born, and the long-running Mars rovers were built.

In a small cluster of bunkers at the heart of the campus lies JPL's Mobility and Robotic Systems laboratory. About 100 engineers develop all of the robotic components, from the autonomy software to the mechanical appendages that enable us to explore other worlds while staying safely on terra firma.

In a back room of the laboratory, a white blimp, about 15 feet long, floats above a table, secured by tethers. The blimp doesn't seem particularly unique, that is until you hear that it's a prototype for an autonomous nuclear-powered aerobot that could someday explore Saturn's moon Titan. The blimp would fly below the dense clouds that hide the terrain from orbiting spacecraft, snapping high-res photos and possibly even scooping up surface samples for onboard analysis.

"Balloons provide a unique observation platform for doing planetary science that you simply can't do any other way," says Jeffery Hall, the senior engineer who leads JPL's aerobot research.

An aeronautical engineer by training, Hall joined JPL in 1997 after graduating from CalTech down the road. At the time, he worked on cryogenic technologies for satellites. Then, he says, the funding dried up and he was forced to expand his horizons.

"I've always been motivated to figure out how things work, and that inevitably leads you to building stuff for experiments," he says. "One of the great attractions of balloons is that it's fairly easy to make them. Then you get to take them outside and fly them around. From an experimentalist's point of view, they're fun things to work with."

The Titan balloon is just one of several very different aerial vehicles that Hall and his colleagues are designing. Each vehicle's technology is dictated by its ultimate destination. For example, the very thin atmosphere of Mars requires a spherical balloon at least ten meters in diameter to carry just a couple kilograms of scientific instruments.

In one scheme, the balloon package would be released from a spacecraft upon entry into the Mars atmosphere. As the canopy drifts down beneath a parachute, onboard helium tanks inflate the "envelope," the actual balloon fabric. Once the envelope is filled, the parachute and tanks are cut loose and the balloon settles at an altitude a few miles above the surface. According to Hall, a wind-blown helium superpressure balloon like this could float around the planet for up to a year, all the while transmitting data back to mission control on Earth.

Photography by Chris Leschinsky

Viktor Kerzhanovich stands beside the JPL prototype aerobot. During high-altitude flight tests, internal combustion engines provide enough propeller power to overcome Earth winds.

Following page: On Titan, the juice would come from an onboard nuclear source that would "effectively last forever," says Jeffery Hall (pictured).

To put the technology through its paces, the researchers have flown their balloons in the environment closest to the Martian atmosphere that's readily accessible — well, relatively accessible anyway: 100,000 feet up in Earth's stratosphere.

"It's a difficult place to do experiments, but that's where you have to go If you want something similar to what you'd see on Mars," Hall says.

A commercial scientific balloon tows the rolled-up test balloons to the right altitude. From the ground, the researchers activate a pyrotechnic cutter that releases the packaged prototype with accompanying helium tanks. Then, they cross their fingers and watch the live video feed.

"We've had a lot of failures and a couple of successes under our belt," Hall says.

The Jet Propulsion Laboratory didn't invent the idea of flying balloons on other planets. Early in the summer of 1985, a pair of superpressure helium balloons floated through the upper atmosphere of Venus. They were part of the Soviet Union's Vega mission to study the planet's atmosphere and surface.

The windblown balloons, 12 feet in diameter and sewn from a Teflon-like fabric, carried a gondola outfitted with thermometers, velocity sensors, light and pressure sensors, and a nephelometer to measure cloud particles. The high winds quickly swept the balloons across the dark side of the planet at an altitude of 35 miles while their progress was tracked by radio telescopes on Earth. After two days, the batteries died as expected, contact was lost, and the balloon mission was declared a resounding success.

The JPL Robotics researchers have inside knowledge of the Vega mission. That's because one of the lead engineers on that project is now on their team. As part of his lifelong career in the Russian space program, Viktor Kerzhanovich worked on 15 of the USSR's Mars and Venera missions, including Vega.

"In the Soviet Union of the 1980s, the space program tried several noble things and the balloons were one of them," he says.

While a follow-up balloon mission to Mars was planned by the USSR and the French Centre National d'Etudes Spatiales (CNES), the collapse of the Soviet Union essentially wiped out Russia's pioneering space science program before the technical challenges could be met. Kerzhanovich knew it was time to take his talents elsewhere.

In 1994, Kerzhanovich emigrated to California where he worked briefly at an aerospace engineering consultancy before joining JPL. While he felt fortunate to have the opportunity to follow his passion, the difference in culture was like, well, visiting another planet.

"The (space program) here is much more bureaucratic than it was in Russia back then," he says. "There, the governmental role was not great. Most of the decisions were made by the projects' chief designers. Here everything goes from the top to the bottom. The process is much more regulated and it's directed to avoid as much risk as possible."

Kerzhanovich may once again have the chance to fly balloons on Venus. He and Hall are applying to NASA's Discovery Program, an effort to launch many small missions with short development cycles, for a flight mission to Venus. (JPL's Stardust mission that brought comet samples back to Earth was the most recent Discovery Program mission.) The Venus trip would be a long-awaited encore of the pioneering Russian mission, but the scientific payload would be far more elaborate, including a mass spectrometer to analyze the planet's atmosphere.

The balloon material is also much more advanced. Developed by JPL with private-sector collaboration, the fabric is built to withstand the Venusian clouds of sulfuric acid. It's a multilayer laminate material consisting of Mylar, Teflon, and Vectran, the high-strength fabric used for the airbags that cushioned the fall of the Mars rovers. Right now, engineers are constructing an 18-foot spherical balloon from the material to help sell the proposal.

"It's not common to make prototypes as part of these proposals," Hall says. "But JPL ponied up money to make one so we can demonstrate the technical maturity of this technology and convince skeptical reviewers. After all, they'll get 15 or 20 proposals submitted for just one or two flight missions. Not everyone gets to go."

And if this particular design doesn't fly, there will certainly be other shots. After all, the space race may have ended, but the makers at JPL still have their eyes on the sky.

"I don't know if I could work like I do if I didn't think one of these projects would fly in the relatively near future," Kerzhanovich says.

David Pescovitz, MAKE's editor at large, is also co-editor of BoingBoing.net and a research affiliate with the Institute for the Future.

The blimp doesn't seem particularly unique, that is until you hear that it's a prototype for an autonomous nuclear-powered aerobot that could someday explore Saturn's moon Titan.

Distributist Technologies

At last, Tolkien meets capitalism.

By Tom Owad

IN *THE LORD OF THE RINGS*, J.R.R. TOLKIEN contrasts the bucolic "hobbit sense" of the Shire with the noxious industrialism of Mordor: "The one small garden of a free gardener was all his need and due, not a garden swollen to a realm; his own hands to use it, not the hands of others to command."

Communal in structure, nonmechanical, and inhabited by individual craftsmen and farmers, the Shire clearly parallels preindustrial England. Its ravaging bears an epic similarity to the transformations that have taken place in our own world since the Industrial Revolution.

As was abundantly clear in Tolkien's time, to control the means of production is to control life. Industrial capitalism placed this control in the hands of a wealthy elite; communism placed it with the state. Whether at the Carnegie Steel Company or the Lenin Steelworks, and whether the bullets came from the Pinkertons or the Red Army, the power over life and death did not rest with the individual but with a hostile external force.

The distributists of the early 20th century rejected both visions. "Too much capitalism," wrote G.K. Chesterton, "does not mean too many capitalists, but too few." Drawing upon the principles of subsidiarity and solidarity put forth by Pope Leo XIII's social encyclical *Rerum Novarum*, the distributists sought a society where each individual provided both his own labor and his own capital. The small garden of Sam Gamgee was the distributist ideal.

Distributist principles took root all over the world, but the luxuries of the Industrial Revolution triumphed over the freedom of the homestead. A similar phenomenon occurred in the 1970s, with the introduction of Schumacher's economic treatise *Small Is Beautiful* and the back-to-the-land movement. Urban idealists also found themselves ill-equipped and unprepared to deal with the hardships of living an essentially preindustrial agrarian life.

Today, freedom and technology no longer stand in juxtaposition. A CNC lathe or milling machine can be built for $500. Raw materials can be prepared in homemade foundries. Solar power and wind turbines make it possible to live comfortably off the grid.

Open source software is ubiquitous, and hardware is following the same path. The pages of MAKE illustrate the countless sophisticated devices that can be built without a factory. The internet allows for worldwide collaboration on a level never before possible. Goods can be exchanged between individual craftsmen without the need for distribution networks and middlemen.

Advancements in personal fabrication at the MIT Center for Bits and Atoms hold the greatest promise. Their aim is to produce a personal fabricator that can manufacture anything from a doll to a precise replica of itself. The current MIT Fab Lab (*see MAKE, Volume 01, page 23*), which has been deployed in nine countries, is itself an impressive feat. The equipment can be assembled for $20,000; the software is open source and free. For the price of a new car, it's already possible to establish an effective personal fabrication laboratory.

Tools such as these have the potential to personalize the Industrial Revolution — to place the means of production not in the control of select capitalists or of the state, but with every individual seeking to partake in the act of creation. Like the soil, the machine yields in response to the labor of our hands.

Tom Owad (owad@applefritter.com) is a Macintosh consultant in York, Pa., and editor of applefritter.com.

Illustration by eboy.com

RG 259 9
PL 269 9
PR 279 9
VISA
MasterCard

Don't assume he's wearing a pair of old-fashioned *pince-nez* eyeglasses. Look more closely and you'll see there's nothing pinching the bridge of his nose. In fact, they're attached by magnets on a steel rod that pierces the fleshy part above the bridge of his nose.

You Are The Platform

How hardware hackers are remaking their bodies.

By Quinn Norton

THE TERM *BODY MODIFICATION* BRINGS to mind piercing and tattooing for cosmetic reasons. But body mods can also be assistive, by changing a person's experience of their own body or even giving them a new ability. How would you like to be able to sense an electric field at a distance? Talk to Steve Haworth of Phoenix (stevehaworth.com). He has experimented on human volunteers by making a small incision in the fingertip and inserting a small neodyme magnet coated in gold, then silicone. The magnet is nestled into a bed of nerves between the epidermis and fascia layers of the ring finger, and the cut is neatly sutured up. For most magnetic implants, Haworth can have the job done in less than ten minutes.

Over the next few weeks, the finger heals, and the nerves begin to interpret electromagnetism and movement in the magnet. Some things, like other magnets, can even cause the magnet to spin freely up against the nerves. The new sensations pass to the brain, where evolution's own favorite primate hack takes over — parts of the brain become more in tune with the new signals from the hand, and the perceptual sensitivity improves. How much sensitivity

Photograph courtesy of James Sooy

you experience depends on many factors — the size and placement of the magnets, how many you have, and, probably, how well you train to use them.

Every magnetically modded person tested, from two weeks to several months, could easily pick out a live cord running with 110 volts. Many report being able to sense electrical motors running from distances of a few inches to a few feet, depending on the strength of the motor and the size and healing time of the implant.

Shannon Larratt, of the body modding publication and community BME, reported that walking through a retail security device was like "sticking my hand in an ultrasonic cleaner." Any strong inductive wire has a chance of causing a sensation in the magnetic finger.

Every magnetically modded person tested could easily pick out a live cord running with 110 volts.

(For a time, Haworth stopped implanting the magnets because, in some cases, the medical-grade silicone sheathing had broken, exposing the magnets' metal to the body. He is now using a harder type of silicone to make the implant safer.)

LOOK MA, NO BRIDGE!

When you meet James Sooy, you might initially assume he's wearing a pair of old-fashioned *pince-nez* eyeglasses. But if you look more closely, you'll notice there's nothing pinching the bridge of his nose. Each lens seems to be glued to either side of his nose. In fact, they're attached by magnets on a rod of steel that pierces the fleshy part above the bridge of his nose. They're more like *pierce-nez* glasses (piercedglasses.com).

Cheaper and less permanent than Lasik, easier to cope with than contacts, Sooy chose piercing to get rid of the earpieces of his glasses. While he plans to make a product, for now it's strictly DIY.

If you want your own pair, you'll need: a small tabletop mill, four small rare earth magnets, a bridge barbell piercing, six small screws that fit the barbell piercing, and a pair of plastic eyeglass lenses.

The first step to pierced glasses is the bridge piercing itself: a barbell that goes through the skin above the bridge of the nose. Get internally threaded jewelry — this is a good idea in general, and very important in this case. "I'd suggest to wait at least a couple months, healing time, before playing around with any piercing," says Sooy. Next, you're going to need screws that match your barbell. "We called up the piercing studio to find out exactly what threading the barbell was, and just ordered the screws."

The elbow brackets that fit on the end of the barbell were cut out of aluminum on a tabletop mill. There's some artistic license here, but make sure you leave a front-facing divot to fit in a set of rare earth magnets. It's the magnets that actually hold on the glasses — and make it safe for them to be knocked off your face. Hold the lenses up to your piercing and mark the drilling spot to mount the fitting for the other pair of magnets. Don't discard the nose pads — as with traditional glasses, it's actually the nose pads that hold the weight.

Quinn Norton is a freelance writer and co-blogger at ambiguous.org.

Health and Safety

Your body modification procedure will involve both pain and blood. The magnetic implant pain is worse than a simple injection or having your ears pierced, but not as bad as a long tattoo over a sensitive region, or many medical procedures.

For any piercing or modification, find a local professional who follows careful protocols for dealing with biohazardous waste — which is what blood is. Ask to see their autoclave, and ask how often they calibrate it; if they are reluctant to answer, leave. More info is available from the Association of Professional Piercers (safepiercing.org).

Any body modification, functional, aesthetic, sensual, or fun, needs to be carefully healed. This is an open wound, and needs to be properly cleaned and sutured, where appropriate. Make sure you get aftercare instructions and understand them.

During the healing period, don't be overly rough on the modified area or expose it to unclean conditions. Ask how long the healing period normally takes for your type of piercing. Keep an eye out for signs of infection: fever, pus discharge, excessive swelling, or redness, and talk to your modder or doctor if you notice any of these.

Saul Griffith

SIMPLY CAD

COMPUTER-AIDED DESIGN DOESN'T HAVE TO BE AWFUL. THERE'S A CAD PROGRAM OUT THERE FOR YOU.

HAVING USED VIRTUALLY EVERY CAD (computer-aided design) program out there at least once, I thought there was no chance they would ever be simple ... until I met Takeo Igarashi. Takeo was a grad student from Japan who wrote an incredible little piece of software called Teddy. It's basically CAD for 5-year-olds: if you want to design a plush toy, this is the program. It's intuitive. It's beautiful. CAD doesn't have to be awful or expensive. It doesn't have to come with 1,000-page instruction manuals.

My hope was that Teddy would influence the world of CAD and make incredible packages cheap and easy to use. That was six years ago. Things haven't changed that much, but there is still hope.

Why, as a reader of MAKE, should you care? I care because I just enjoy using CAD; it's better than playing a computer game. You can spend the same 36 sleepless hours in front of the computer, but at the end of it, you have a beautiful 3D object from your imagination, planned out and ready to be built or shared. But the real and exciting reason to care is because someday kids should be able to unleash their imaginations on a Teddy-type CAD program and then print their imaginings into the real world on a chocolate- or sugar-powder 3D printer.

You should care because the capacity to share 2D and 3D models is at least as important as sharing music and video. Making CAD representations of your "makings" makes it easier for other makers to make things too.

Before I launch into my CAD diatribe, let me sketch out the playing field for the uninitiated. Very broadly, CAD programs can be classed in a number of ways. On the hardcore engineering side, there is parametric 3D solid modeling, with exact equation-driven descriptions of objects; there are surfacing or wire-frame-based programs, which bring nice rounded surfaces and squishy shapes to life; there's a world of rendering tools for making the 3D models look cool; and there are architectural-type 2D drafting programs for laying out plans in 2D.

For completeness, I should also mention CAM, which is computer-aided manufacturing software. It takes your CAD design and converts it into a "toolpath" for a drillbit, mill, laser, inkjet head, or some other tool to follow and construct your object.

Contrary to popular belief, you don't need a laser cutter or a 3D printer or an NC mill to realize your CAD designs. On your desk, you likely already have a high-precision machine tool: an inkjet printer. These things will print designs onto paper with 50-micron, or better, accuracy. I very often use the inkjet printer to print a pattern or template, which then can be used to align the holes you will drill by hand or to cut out the fabric you will stitch together. So, in fact, the world already has the capacity to build cool CAD 3D models and share them via their desktops.

> Build your dream house and stick it in Google Earth! How cool is that?

One program taking advantage of this is Pepakura Designer — it calculates a pattern of triangles that, when folded up, can make your 3D object. It's high-tech origami. I've seen business cards that fold into toy cars made by this technique, and Japanese anime characters and dinosaurs.

In the same vein is a program called Lamina Design, which takes 3D curved surfaces and figures out how to construct them from flat sheets. Again, the nice thing here is that you don't need NC machine tools to get real-world output; you could use this program, print to paper, trace to plywood, cut with a hacksaw, and make beautiful organic furniture, or even design your own kayaks.

A great little company called SketchUp has a CAD package that allows you to build buildings and place the 3D models in Google Earth. Of all the packages

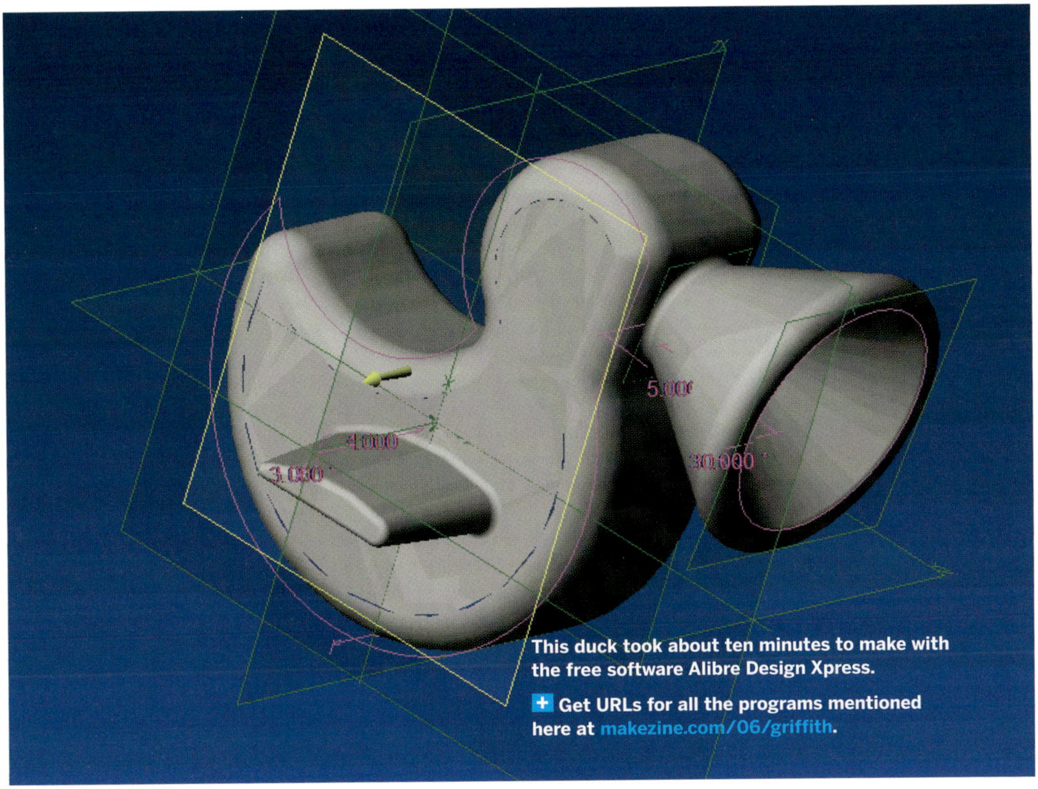

This duck took about ten minutes to make with the free software Alibre Design Xpress.

+ Get URLs for all the programs mentioned here at makezine.com/06/griffith.

out there, SketchUp is the closest in my mind to the Teddy ideal of simplicity. They have great tutorials and an intuitive interface, and if you've never used CAD before, you'll just enjoy this. Build your dream house and stick it in Google Earth! How cool is that?

The online world Second Life now has CAD packages built into it, so you can realize fantasy objects in the other world. This is cool to me because I can now see real/virtual world crossover. Make something über-cool in Second Life? Bring it into the first life!

So where should you get started? Before you spend a few thousand dollars on a pro CAD package, try some free alternatives: freeCAD.com reviews a bunch. My favorite right now is Alibre Design Xpress. It's a free parametric CAD program with most of the functionality of the pro packages.

Their business model is interesting (a professor of mine once described it as "crack for babies"): they give you the free software, you get addicted, and then if you need higher functionality (more parts, more tools), you can purchase more. Their prices are reasonable, and you only purchase the extras you need.

If you're comfortable in Illustrator or CorelDRAW, you can do a surprising amount in these programs. There are dimensioning plug-ins for Illustrator, and with these programs you can produce nice 2D plans of your wooden boat, air-powered potato cannon, or cigar box guitar. Eagle CAD is a free circuit board design and layout tool. If electronics is your gig, you can CAD it up in Eagle and split the cost of sending out for boards with your friends who also want mint-box MP3 players.

Another idea is to write your own CAD program. David Aberdeen wrote a fantastic CAD package for kite design, called Surfplan, that led a whole community of amateur kite builders to build cool kites for surfing. A lot of people have built small CAD packages within MATLAB and Mathematica with outputs for DXF or PostScript. Robert Lang uses such a thing to create his awesome computational origami models.

The world of CAD still isn't perfect. Converting between platforms is a nightmare. Learning curves are long. Things have come a long way, though, and there are a few packages to get you started with little pain. Hopefully, out there in the MAKE readership, there are hackers who will write cool new CAD programs with simple interfaces and open source extensibility. Get to work.

Saul Griffith thinks about open source hardware while working with the powernerds at Squid Labs (squid-labs.com).

CAD image by Saul Griffith

TOTCH BROWN'S PIT PAN GATOR BOAT

By Tim Anderson

START YOUR BOAT-BUILDING HOBBY BY BUILDING THIS ONE.

Loren "Totch" Brown lived in the Thousand Islands region of the Everglades his whole life. His fascinating autobiography, *Totch, A Life in the Everglades* (University of Florida Press, 1993), relates his adventures, and its cover depicts him propelling a strange little boat with a push pole.

It's called a "pit pan" and is now on display at Smallwood's store in Chokoloskee, Fla. (florida-everglades.com/chokol/smallw.htm).

When the water was too deep for the push pole, he used a paddle, and when the water was too shallow, he dragged or carried it.

He made many such boats for hunting alligators over the years. Despite its crude appearance, this is a very elegant boat. It is burdened with no unnecessary features. You can carry it and you can sleep in it. That makes it a magic carpet to freedom.

In designing boats, people get hung up about the speed of the boat rather than the speed of building the boat, and they forget that to go fast you have to work hard. If you want a boat you can take a nap in, you need something like this.

Rolled up in the bottom of the boat is Totch's mosquito net or "skeeter bar," so named because it's where the insects go to drink and socialize. It's a rectangular mosquito net just big enough to not touch the person sleeping under it. In the old days, gauze, cheesecloth, or any open weave fabric was used. The net is held up at the corners by sticks. If it rains, a tarp just big enough to keep the rain off goes over that.

A boat can be part swan, part guitar. This boat is a box with no lid. Start your boat-building hobby by building this one.

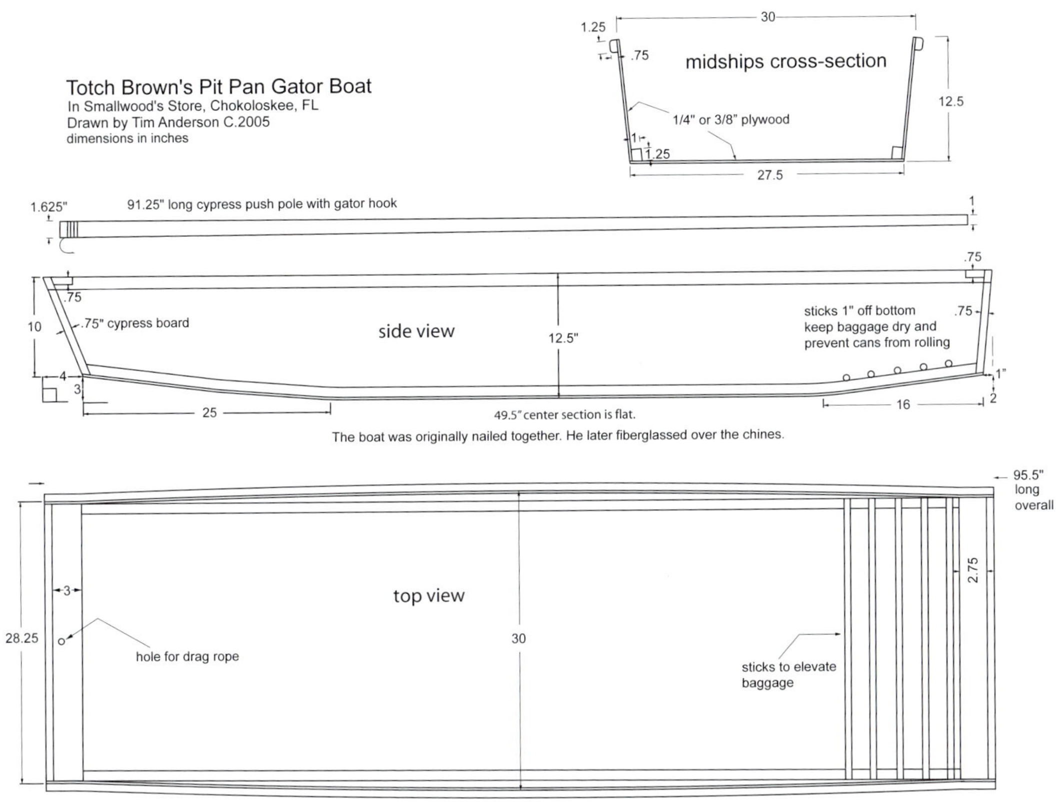

The plans for the pit pan.

Attach all the pieces together with nails and screws. To make the boat watertight, paint the mating surfaces of each joint with epoxy glue before nailing. To keep the epoxy from running out of the joints, add enough white flour to thicken it to the consistency of pancake batter. Fine wood dust from sanding works also. Whole wheat flour makes lumps and isn't as good as white flour for this. Don't use 5-minute epoxy for anything; it's not fully water resistant. To get more working time with your glue, pour it out onto a pan. That will keep it from getting too hot and curing before you're ready.

Paint the whole boat with linseed oil, and after a couple of days, paint the whole boat with whatever paint you can find, preferably oil-based. Roofing tar is also really good for sealing the joints.

Totch covered the outside edges of his boat with fiberglass, which is also a good method.

If you don't have plywood, Dan Beard's 1882 classic, *The American Boy's Handy Book* (see Figure 1), shows how to build this boat from boards. Beard, the founder of the Boy Scouts of America, suggests putting paint-soaked strips of wool between the boards to seal them.

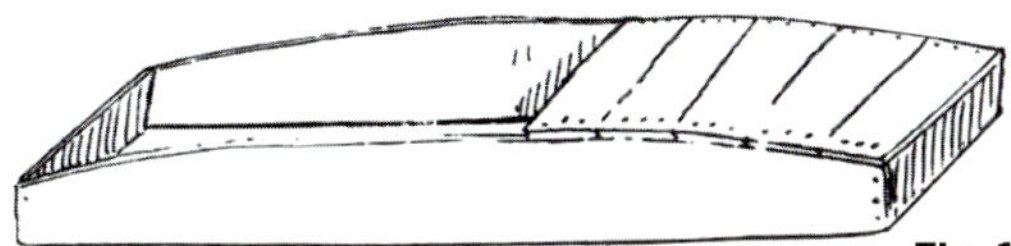

Fig. 1

EVERGLADES AND SWAMP BOAT RESOURCES

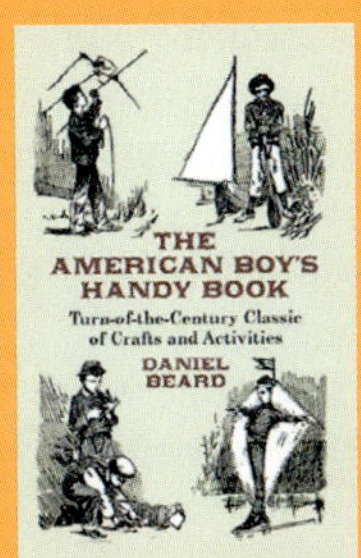

Totch died in 1996, but his website is still online at florida-everglades.com/totch/totch.htm

You can order a facsimile edition of *The American Boy's Handy Book: Turn-of-the-Century Classic of Crafts and Activities* from Dover at store.doverpublications.com/048643138x.html

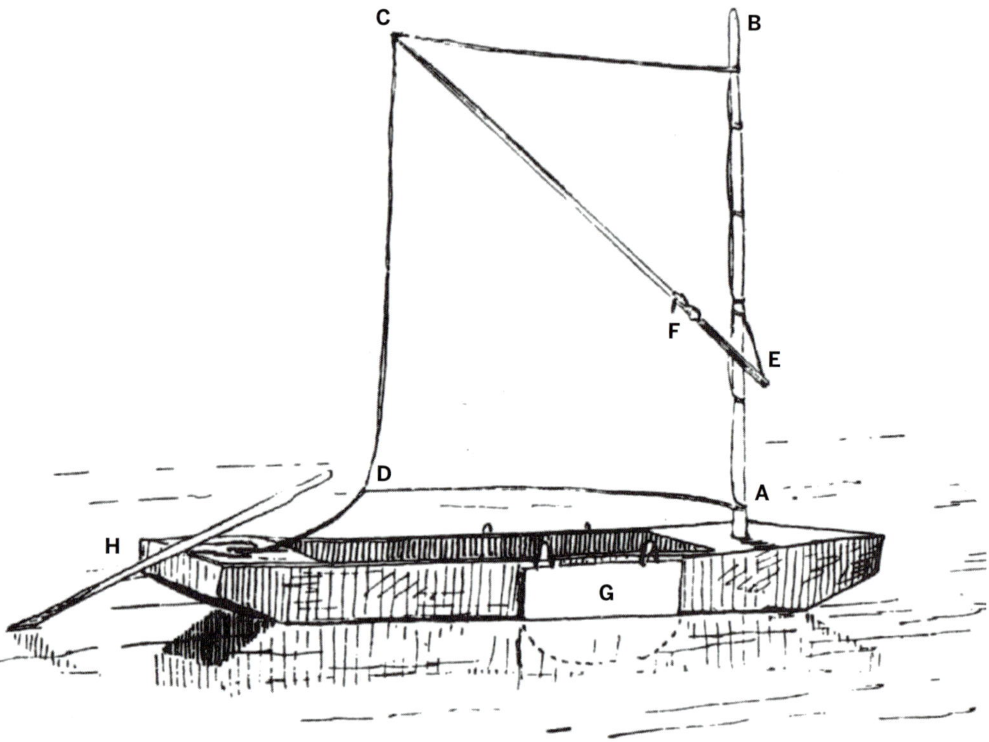

Fig. 2

SHOVING OFF

Want to try sailing your new boat? Figure 2 from the same book shows a good spritsail rig for this boat. The push pole becomes a mast, and the tarp becomes a sail.

A leeboard (G) hangs over each side to keep the boat from drifting sideways. (A "no parking" sign works well for this.) The sail is a rectangular tarp, sheet, poncho, or tablecloth tied to the mast and sprit at the corners. The mast (B) and sprit (C) are short enough to stow in the boat. The mast can be stepped off-center, if you prefer. The sprit has a notch (E) at its end for the "snotter line." The snotter is tightened at the cleat (F) to adjust the sail. In light winds the sail doesn't need to be fully laced to the mast. Just the corners is fine.

Figure 3 shows a simpler way of attaching the sprit. The snotter is a loop of cord formed into "cow hitches" on the mast and sprit. The sail is "peaked" or adjusted by sliding the cow hitches up and down. To stow the sail, lift the notched end of the sprit off the snotter. That will give you enough slack to put the mast and sprit together, so you can roll the sail up on them.

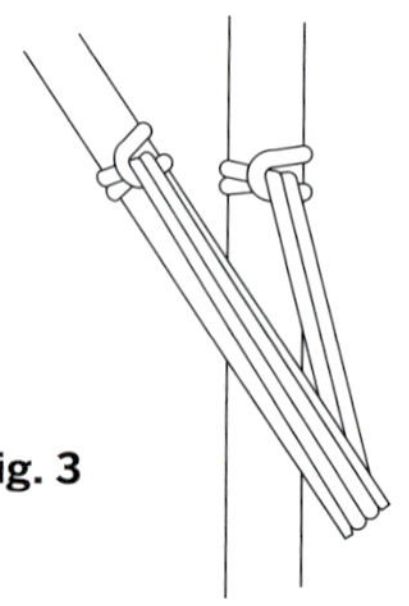

Fig. 3

Attaching the sprit to the mast: The cord attached to the sprit (diagonal) and mast (vertical) is called a "snotter," with "cow hitches" on either end.

Fig. 4

If you prefer rowing to paddling, these oarlocks are easy to build. Any scrap wood will do, but they take a fair amount of friction and pressure, so the wood should be strong. Make sure to anchor them securely.

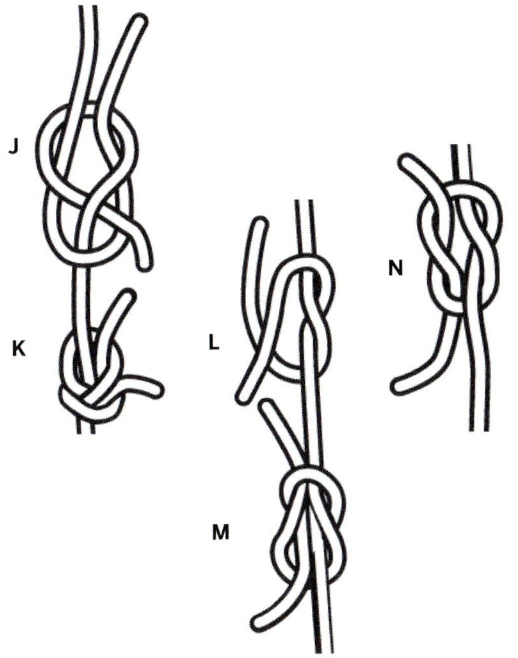

KNOTS

Useful knots for the sailor begin with the "sheet bend," a.k.a. "bowline" (J and K). Learn it well and don't use any other knot without a good reason. This knot will not jam up. It can join dissimilar ropes.

The square knot, a.k.a. "reef knot" (L and M) is easy to untie, which can be a good thing. It only works with same-size rope and must be properly formed to hold.

The "granny knot" (N) is often tied by mistake instead of the square knot. This knot slips, causing many tragedies. I'd always been told that the granny knot was only ever a mistake and sometimes a tragic one. Then an Ojibway female canoe builder showed me how to tie split spruce root with a granny knot. Other knots bend the material too much and it may break. So even the granny knot has a use, and a pretty good one at that.

For more knots, see realknots.com/knots.

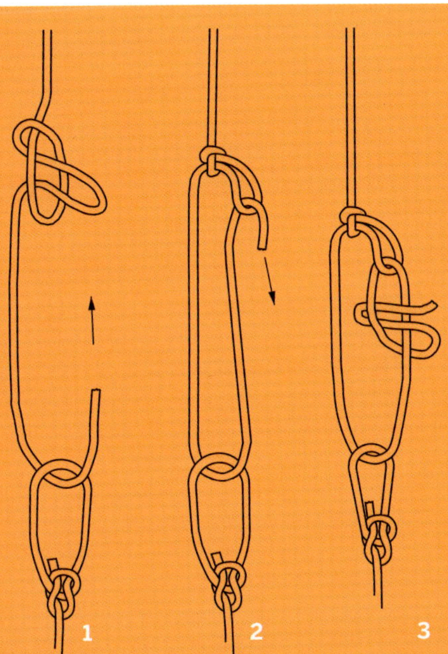

TIE YOUR BOAT TO THE ROOF OF YOUR CAR WITH A TRUCKER'S HITCH:

I learned this knot from a sailor named Jim Horn. It's a trick that really impresses people. When someone sees me cinch down a load with one of these beauties they start asking for advice on all sorts of serious matters.

1. Tie a bowline in one end of the rope, as shown.
2. Make a loop in the standing part of the rope, then pass the end through the bowline and back through the standing loop, and cinch it down.
3. Secure it with a bow, as shown, and pull that tight.

NOTE: If you don't have a roof rack, open your doors and put a rope around the roof load through the car.

Illustrations by Tim Anderson and excerpted from *The American Boy's Handy Book* 1882

Tim Anderson, founder of Z Corp., has a home at mit.edu/robot.

A Beginner's Guide to BEAM

The BEAM design approach creates nimble robots from simple components, with no programming required.

By Gareth Branwyn

EXPECTATIONS FOR CREATING THINKING and actuating machines were high in the 1960s and 70s, as universities began establishing robotics and artificial intelligence labs. But researchers soon realized that even the most basic physical tasks, such as getting a cart to sense obstacles in a space, plan a route, and execute it (called *sense-plan-act* architecture) were daunting assignments. Some wrestled with these problems by developing tweakier algorithms and throwing more, faster hardware at the problems, but MIT's Rodney Brooks took a radically different tack. Inspired by insects and other teeny-brained critters, he wondered what a *sense-act* architecture might look like, one where the bots didn't bother to create a map of their world to plan from, but rather, reacted directly to environmental stimuli.

The results were impressive: bug-like bots that could do most of what the sense-plan-act robots could do, but with scant little computing power. Many roboticists were inspired by Brooks' work, and one, Mark Tilden, took the idea even further. After seeing Brooks lecture in the early 90s, he wondered whether it was possible to create sense-act robots that used no digital computation at all. The answer was yes, and BEAM robotics was born.

The initials "B.E.A.M." stand for the four components of the BEAM design approach:

Biology: Like other sciences and technologies these days, BEAM looks to nature for design inspiration. After all, billions of years of design, fabrication, and in-field testing shouldn't be taken lightly. BEAM also encourages the human's role in robotic evolution. Olympic-style BEAM competitions are held each year, and in the spirit of hacking and open sourcing, builders share their winning design concepts for incorporation into future robotic competitors. In BEAM evolution, humans are the way for a robot to make a better robot.

Electronics: A hallmark of BEAM technology is the cyberpunky "abuse" of electronics, re-purposing components in ways that their designers never intended. Chips used in common household electronics to amplify sounds, direct data traffic, or invert signals are connected in new ways in order to control motors, process sensor input, and act as state timers. Chain some of these sub-circuits together, and you have a robot brain with no need for a programmable microcontroller.

Aesthetics: Buckminster Fuller once said: "When I'm working on a problem, I never think about beauty; I only think about how to solve the problem. But when I have finished, if the solution is not beautiful, I know it is wrong." Tilden and fellow BEAMers understand that elegance in robot design serves the evolutionary purpose (see Biology) of attracting the enthusiasm of fellow builders, which leads to replication. Many also take pains to make their creations beautiful because they see their robots as a form of kinetic sculptural art.

Mechanics: Like BEAM's electronics, its mechanics and build techniques are often quite clever in how they maximize efficiency. For instance, many BEAM bots use free-formed circuitry, where the components are soldered directly to each other rather than to a printed circuit board. This lowers cost and weight, and it shows off the engineering prowess and soldering skills of the builder.

Harold R. Ilano's Black Vermin 1.1, a replica of Mark Tilden's infamous Vbug 1.5, using hobby servos instead of Walkman motors. More images, schematics, and video at raysbeambots.solarbotics.net/harold/harold.htm

Photograph by Harold R. Ilano/haroldsbeambugs.solarbotics.net

A few other key BEAM precepts are worth detailing, as they say a lot about how BEAM differs from other robot design approaches.

Keep It Simple, Stupid (K.I.S.S.)

BEAM's Zen-like simplicity strips away assumptions to get at the essence of a problem. Look at the Robosapien, a toy humanoid robot designed by Tilden that incorporates many BEAM concepts. Robosapien uses almost no digital processing, yet it performs many of the behaviors of corporate research robots like Sony Qrio and Honda ASIMO. A Robosapien costs less than $100, while these high-end industrial fussybots aren't even on the market — and would cost a fortune if they were.

> 1. Protect thy ass. 2. Feed thy ass. 3. Get thy ass to better real estate.

Junkbots and Appropriate Technology

BEAM also emphasizes recycling. One of Tilden's early BEAMbots, Vbug 1.5, was made from little more than a few cheap analog ICs, parts lifted from several Sony Pro Walkmans, an oven timer, and some wire. BEAM enthusiasts do use microcontrollers and other high-end components, but only if cheaper and simpler solutions are not possible.

My Many Dumb Bots Beat Your One Smart Bot

Implied within the BEAM philosophy is Brooks' idea that a bunch of working dumb bots is better than one broken smart bot. BEAM enthusiasts envision a future in which swarms of cheap robots could do such things as clean skyscrapers, toxic waste sites, and even your house. If a few of them break or run out of juice, that's fine, but if a single high-priced fussybot breaks so much as one end effector, you're out of luck.

Real Robots Fend for Themselves

Whenever possible, BEAM builders design bots that can take care of themselves. Many older designs are solar-powered. Some more recent BEAM and BEAM-inspired designs do require batteries, but ingenious engineering maximizes their efficiency.

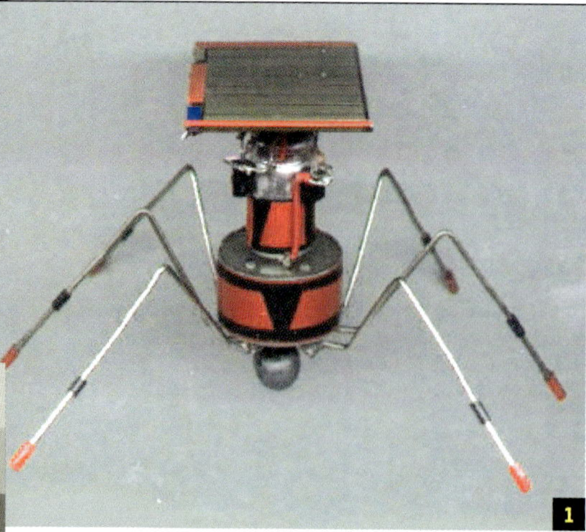

1. Jim Mullins' Symet3-Virus, which is basically a Type 1 (voltage-triggered) Solarengine with an added ball-tip "motivator" on the motor shaft and a set of six legs, upon which this robotic top can skitter around.

2. A swarm of Robosapiens, a commercially available toy robot designed by Tilden that incorporates many BEAM principles.

Robosapien, for example, contains seven motors, but its gearboxes are designed to allow powered movement along one joint to transfer over to neighboring joints without engaging the additional motor. So the Robosapien's torso will passively rotate back and forth in response to its powered waving of an arm.

Thanks to such optimizations, BEAM bots will likely be among the first to achieve power autonomy with the arrival of cheaper, more efficient solar cells and power storage technologies. This will fulfill one of Tilden's Laws of Robotics (laid down in response to Asimov's Laws): 1. A robot must protect its existence at all costs. 2. A robot must obtain and maintain access to its own power source. 3. A robot must continually search for better power sources. (Or, as they are more widely recited: 1. Protect thy ass. 2. Feed thy ass. 3. Get thy ass to better real estate.)

Basic BEAM Circuits

BEAM prides itself on minimalist design and an anyone-can-play openness. Even so, there are dozens of basic BEAM circuits, far too many to explain here. But to give you a taste of BEAM electronics, let's look at two fundamental circuits: the Solarengine and the bicore.

The Solarengine (SE): A *Solarengine* is a power circuit built around a solar cell and a storage capacitor. The solar cell trickle-charges the cap, and when the amount of stored energy is high enough, the cap dumps the charge into the circuit to power the robot.

Different types of Solarengines are defined by the method they use to trigger the dump. Type 1 SEs use a voltage-sensing trigger, Type 2 SEs use a timer, and Type 3 SEs monitor the current that flows into the cap and trigger the dump once the capacitor's charge-rate begins to slow down as a result of getting full. A fourth and final flavor of SE is the nocturnal variety, which reads ambient light levels and turns on the power once darkness descends on the robot.

A Type 1 Solarengine provides the basis for the Trimet and Solarroller project later in this issue *(page 76, with an annotated circuit diagram on page 79 that shows how its basic Solarengine works)*. In this design, the voltage trigger is a 1381 chip, a three-pin component that opens the connection between the power and output pins when there's a 3.4 volt (or other voltage) difference between the power and ground pins. This sends the stored capacitor power into a motor.

The Bicore: Many of the most interesting BEAMbot behaviors are controlled by "nervous nets." These are simple control circuits that emulate low-level peripheral nerves in a spinal column. Each "neuron" within these circuits is typically composed of a timer sub-circuit (a resistor and capacitor) combined with one or more inverters (logical NOT gates). Because a 20-pin inverter chip can house eight gates, one inverter chip can simultaneously support several such neurons.

Photography by Jim Mullins (left) and Wow Wee Toys (right)

A one-motor walker I built from a 74HCT240 bicore and a servomotor with its control electronics removed. The bicore creates an oscillating signal that generates a back-and-forth walking gait.

Photograph by Jay Townsend/Primal Design

The most basic nervous net has two neurons, and is therefore called a *bicore*. When the two neurons feed back to one another, they can create an oscillating signal to generate a legged robot's walking gait. Bicores can also control motors for steering and LEDs for signaling, like in Dave Prochnow's Panzeroids project on page 58.

I made a simple walker using a Tilden-designed bicore built from a 74HCT240 eight-gate inverter chip. In making this two-neuron oscillator, you solder four gates of the inverter chip together. Hook this simple controller up to a gearmotor, and the back-and-forth signal makes the little critter walk and even climb over low objects.

To take it further, you can chain several bicores together in master-slave configuration, creating a multi-motor walker. The network's master bicore sets the rhythm and sends the signal down to the next bicore. Resistors chaining the series together set the time delay, determining the walker's gait.

You can also attach a touch sensor bumper to a bicore, and use it to invert the direction of power to the motors. This will make the walker back up when it runs into any objects. Combine a bump sensor bicore with a two-bicore, two-motor walker and you have a very cool little bugbot that can navigate a space and negotiate small obstacles. Not too bad for a couple of gearmotors and a handful of cheap analog parts.

Getting Started in BEAM

For more information, check out Solarbotics' *BEAM from the Ground Up* online tutorial, at solarbotics.net/bftgu. Also check out the rest of the Solarbotics site; it has massive amounts of how-to articles, circuits, component datasheets, and a BEAM encyclopedia and "bestiary." The commercial side of the site offers BEAM parts, books, kits, and other robot-related products. Any one of their inexpensive kits is a great way to get your feet wet.

Also see Dave Hrynkiw and Mark Tilden's indispensable book *Junkbots, Bugbots, and Bots on Wheels* (Osborne, 2002), and my own book, *The Absolute Beginner's Guide to Building Robots* (Que, 2004).

Gareth Branwyn writes about the intersection of technology and culture for *Wired* and other publications, and is a member of MAKE's Advisory Board. He is Cyborg-in-Chief of Streettech.com.

Panzeroids

Become a desktop general with these battlin' bot tanks.

By Dave Prochnow

WELCOME TO THE BATTLEFIELD OF the future, where robot gladiators fight to decide the political power of their human commanders. Robots have replaced human soldiers, and human deaths on the field of battle are a historical memory. Yeah, right — dream on!

Well, actually, this vision of tomorrow is here today, scaled down a bit. Behold: Panzeroids, robot tanks that battle to the death — an LED-flashing death. The field of conflict? Your kitchen floor. The scale may be small, but the combat is real. Best of all, Panzeroids won't cost you the national debt; you can easily build two dueling tanks for less than $50.

To build a Panzeroid, you start with a stock motorized model tank kit. Then you hack in a type of control circuit called a "bicore" and an infrared (IR) LED emitter flasher. What you get is a robotank that can independently locate, track, and "kill" another similarly equipped tank. Here's how I built mine.

A Robot Hacker's Delight

For tanks, I used two Leclerc French Army Main Battle Tank (MBT) kits from Academy Models. I bought them for just $16 each, and at that price, these kits represent one of the best bargains in robot hacking. You get two motors, complete with gearboxes, mounted in a neat, self-contained chassis. Even if you don't build a tracked robot, you can use the kit's motor/gearbox combination for numerous other projects.

The first step is to assemble the plastic body of your model tank, as described in the kit's instructions. But you should stop just before the part where you attach the remote control cable to the motor gearboxes. This is where the hacking begins and the Panzeroids are born. Instead of connecting to the tethered remote, each motor gearbox will connect to your robot's bicore "brain."

Photography by Dave Prochnow

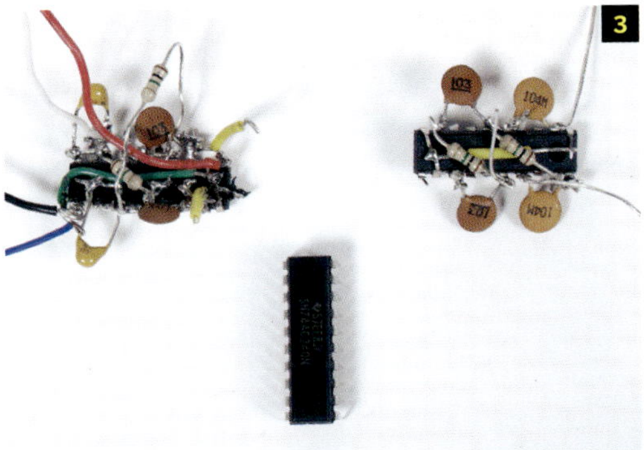

1. The Academy Models Leclerc MBT kits include motors and gearboxes, and represent a terrific bargain for robot hackers.

2. The Leclerc model tank fits together without cement, so it's easy to access the batteries inside.

3. Each Panzeroid tank has two suspended bicore "brains" based on a TI 74AC240 chip.

Two-Track Minds

The next step is to build your robots' brains. There are many different and equally valid ways for building bicores, and I won't explain how to do it here, but check the "Plans and Sources" list at the end of the sidebar on page 61 for where to find some good instructions. You'll need two suspended bicores for each tank. One will control the tank's movements by dividing and adjusting the power (0V to +5V) between each of its two motors, and the other will control the IR LED flasher "gun" that's mounted inside the tank turret.

The bicores I built are based on the 74AC240 Octal Buffer Inverter IC. In addition to these chips, you also need a few garden-variety capacitors, resistors, photodiodes, and LEDs. They're simple circuits to put together, and all told, the total cost for two tanks' worth of bicore brains was about $6.

Flasher vs. Motor Bicores

Note that for the Panzeroid flasher controller bicores, the suspended resistance is a simple resistor, as shown in most bicore schematics. But the motor control circuits have one substitution: instead of a resistor between the two cores, you use a pair of photodiodes connected in series and oriented in opposite directions. These twin photodiodes function like a resistor, and also alternately modulate the power running to each tank motor, depending on the level of IR they receive. This is what makes the tank locate and steer toward the strongest IR signal.

Before soldering together my bicore circuits, I tested them out on a breadboard, temporarily attaching two red LEDs to the control voltages. Each bicore has four connection points: power, ground, and the two control voltages. With the 74AC240 chip-based design I used, these correspond to pins 20, 10, 9, and 12, respectively.

Rebraining Procedure

Once you have your bicores built, it's time to hook them in. Each Academy Leclerc model kit comes with a tethered remote control. We'll use the remote's wires to connect the motors to the bicore brains. Open up the remote unit and desolder the four control wires (blue, yellow, red, and white) from the circuit board.

Take the four wires you just desoldered, and solder them to the four terminals on the tank's two motors. Connect the white wire to the top (ground) terminal of the tank's left motor and the blue wire to the top terminal of the tank's right motor. Solder the remaining red and yellow wires to the bottom (power) terminals of the tank's left and right motor terminals, respectively. Then cut off all but about 18 inches of the ribbon cable, and separate and strip the other ends for connecting later.

4. Basic bicore (from *Junkbots, Bugbots, and Bots on Wheels*, p. 288) assembled for testing on a breadboard. Resistor-LED pairs simulate the motors, and the black rectangles at left are photodiodes.

5. Inside the remote, four wires to desolder.

6. Wires from the remote soldered to the motors. The other ends connect to the motor bicore in the turret.

7. Master bicore brains inside tank turret (also from *Junkbots*, p. 278). One connects to two IR LEDs on the left side, and the other connects to the motors and two photodiodes on the right side.

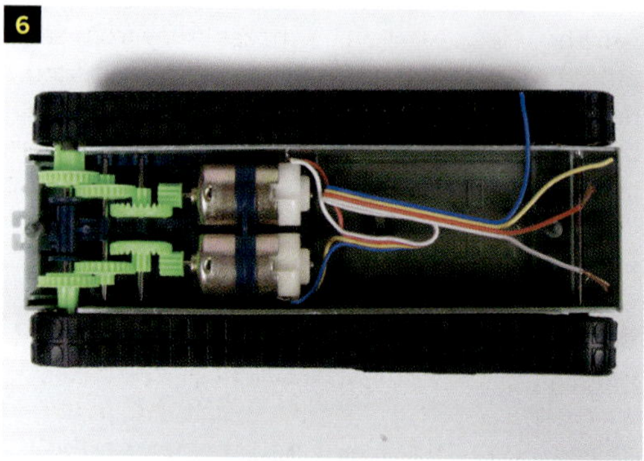

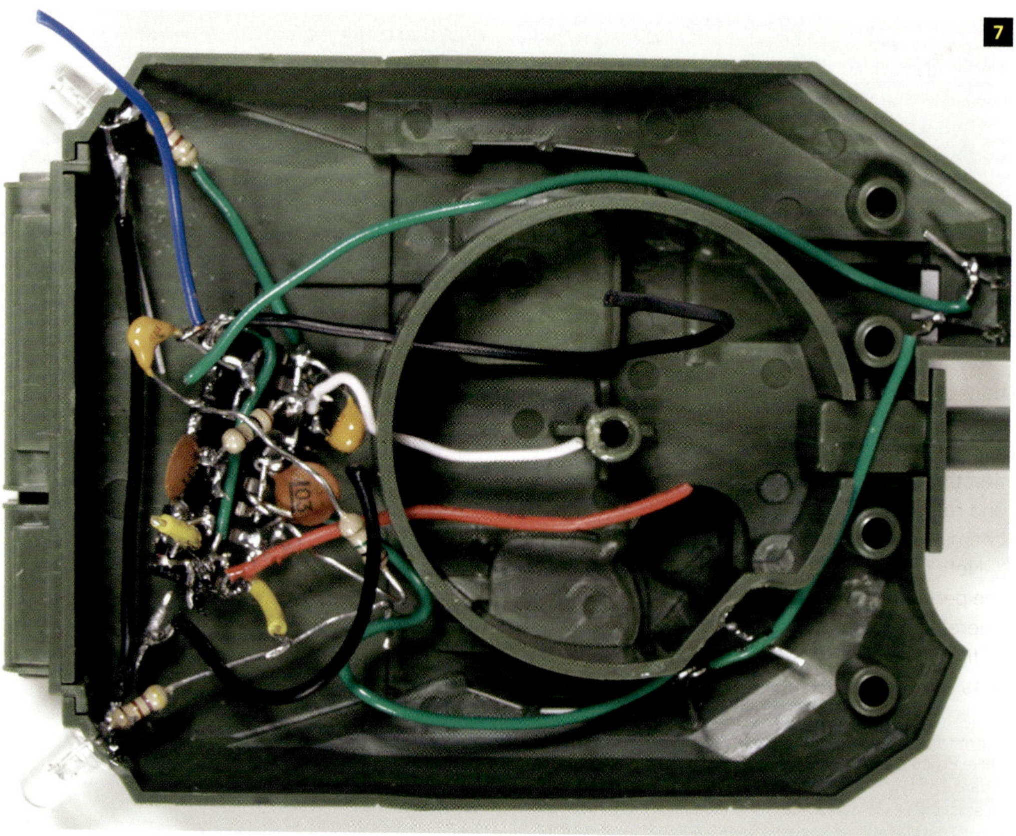

Try to squeeze your two bicore brains into the tank's turret. Connect the bicore with the resistor to the IR LEDs positioned along the left side of the turret, and put the twin photodiodes from the motor-control bicore on the opposite, right side. Then run some wires up the turret from the battery compartment in order to hook the two bicores up to power. Connect the power pin on each IC (pin 20) to the positive battery terminal, and the ground pin (pin 10) to the negative battery terminal.

Decide how you want to position the photodiodes and LED flashers. For optimal IR light sensitivity, the two photodiodes should be spaced away from each other. I installed them poking out of different port details in the turret, one next to the barrel and the other on the side. I put the LED flashers on the rear corners of the turret.

Hook up the bicores' outputs. With the IR bicore, wire its IC's output pins (pins 9 and 12) to the anodes of the two IR LED emitters. Then connect the LED cathodes to ground. For the motor control, with the twin photodiodes, solder the left motor power (red wire from the remote cable) to pin 9 and the right motor power (yellow wire) to pin 12. Connect the motor grounds (white and blue) to ground, via pin 10 or the negative battery terminal. Finally, join the turret to the tank's body, connect the battery, and let the tank run on a test drive. The motors should steer the tank around as it seeks out an IR light source.

Patton vs. Rommel

Now, let's watch two Panzeroids battle it out. Each will pulse IR from its flashers in back, while simultaneously trying to drive toward another IR light source — namely, the flashers on the other tank.

The tanks will steer around, reacting to variations in flash intensity, battery strength, ambient light, visual obstacles, and other factors. When one tank senses the other, it will steer toward it. The first tank that locks on long enough to drive up to and ram the other tank is declared the Panzeroids champ. Die, robot, die!

Note that strong sunlight can adversely affect the performance of your Panzeroids competition. Therefore, if you don't want your Panzeroids trying to fight the sun, you should confine your combatants to a dimly lit coliseum.

Dave Prochnow is a frequent contributor to MAKE and an award-winning author of 25 nonfiction books, including the bestselling *PSP Hacks, Mods, and Expansions* (McGraw-Hill, 2006). You can learn more about his projects at pco2go.com.

More on the Bicore

The bicore was invented by Mark Tilden, the father of BEAM Robotics and the Robosapien toy (*see page 54*). During early research into using electronics to simulate animal neural networks, Tilden developed the Nv neuron, a simple circuit that uses a resistor-capacitor (RC) timer and an inverter to slow electrical signals down, so that they propagate at meatware speeds. This lets electronic circuits time-slice changes in the real world at about the same rate that biological systems do, giving them a "reflex" speed that works well for controlling inertia-bound physical objects like motors and wheels.

Tilden and others combined these Nv neurons into networks, which he originally termed Very Slow Propagation Artificial Neural Systems (VSPANS). Now they're called "nervous networks," and people refer to them by the number of Nv "cores" they contain. A network with two Nv neurons is a bicore, one with three neurons is a tricore, and so on.

Our Panzeroid control circuits have two cores built using one chip, connected in series, with a resistor suspended between them. This is a "suspended bicore" design, and the resistance that separates the cores provides a "virtual ground" that swaps circuit action back and forth between the two. This creates an oscillator that can drive many types of circuits, including the Panzeroids' motors and IR flashers.

Plans and Sources

You can learn how to build bicore circuits from either of the following sources:

* **BEAM Online: makezine.com/go/beamonline**

* ***Junkbots, Bugbots, and Bots on Wheels* by Dave Hrynkiw and Mark W. Tilden (McGraw-Hill, 2002). This is the must-read bible for anyone interested in BEAM robotics.**

Or, you can buy a Bicore Experimenters PCB from Solarbotics (solarbotics.com, product #BEP, $35).

Model Tank Sources

If you have a hankering for hacking motorized models into viable robots, here are three manufacturers who produce acceptable tank kits. You can order kits online from Squadron Hobby (squadron.com) and Tower Hobbies (towerhobbies.com).

* **Academy (academyhobby.com): 1:48 scale Korean kits; best for general robot hacking.**

* **Micro-X-Tech (dragonmodelsusa.com): 1:72 scale Hong Kong kits; tiny, detailed kits that require advanced hacking skills.**

* **Tamiya (tamiyausa.com): 1:35 and 1:16 scale Japanese kits; outstanding quality for high-performance hacking.**

Bots in a Snap

The nuts and bolts of Lego robot design. By Michael Rosenblatt

WHEN I BUILD LEGO ROBOTS, I GET out the drill. I have been doing this since 2002 when I had the job of building a robotic search and rescue demonstration for the U.S. Office of Naval Research. This job was commissioned to develop an immersive activity exposing trainees to the very real challenges of robotic search and rescue. With an aggressive budget and schedule, I decided to use a combination of Lego and custom-fabricated materials to build the working search robots. The two robots had to be durable and powerful enough to climb steep inclines and simulated staircases, have working pan-tilt cameras, and, particularly challenging, carry a smaller "scout" robot onboard.

The Lego purists will cringe, but I adopted a liberal approach to construction, using any materials and fabrication methods available to achieve functionality and scale. This article shares ten tips I learned during this project that I think you will find useful when applied to your own contraptions.

1 Use vertical through-bolts for stronger structures.

Why: You can reinforce structures by adding more Legos, but bolts are stronger and use less space.

Materials: #4-40 size machine screws, nuts, and washers.

Procedure: Drill through Lego Technic studs with a ⅛" bit. With a washer, slip a 4-40 screw through the holes. Snug, but don't over-tighten, as you might crack the Lego plastic.

2 Use machine screws to mount non-Lego materials on to your creations.

Why: Add a platform as large as you want for your Lego bricks, as well as reinforcing the structure.

Materials: #6 machine screws, #6 washers, and any flat sheet material (⅛" acrylic works well).

Procedure: Drill holes in your sheet material at the appropriate places for your design. Drill through the mating Lego studs with a ⅛" bit. The larger machine screws will securely thread into the hollowed Lego studs to mount your sheet. Be careful not to strip the holes by over-tightening.

Photography and illustrations by Robin Tafel

3 Use art-store aluminum tubes to make linear slide mechanisms.

Why: Make longer Lego slide mechanisms than you can with the longest plastic axles.
Materials: Metal tube or rod 3/16" diameter, available at many art/craft stores and hobby stores.
Procedure: Cut the tube/rod to desired length. Actuate with Lego flat-gears, a string/pulley mechanism, or springs.

4 Install cheap rotary sensors (encoders, potentiometers, rotary switches).

Why: Use a multitude of inexpensive sensors that interface directly with your Lego projects.
Materials: Any 1/4"-base, 1/8"-shaft rotary sensor, and 16T Lego gear.
Procedure: Expand the hole in a Lego Technic beam with a 15/64" bit. Force-thread the sensor into the enlarged hole, being careful to keep the shaft perfectly centered in the hole. Press the 16T gear onto the shaft. If the gear slips on the shaft, use a little plastic epoxy.

5 Make your own Lego encoder with an LED and photosensor.

Why: It's a cheap, compact rotation sensor.
Materials: T-1 3/4 size LED, T-1 3/4 photo-diode or photo transistor, Lego pulley, and Technic beams.
Procedure: Press LED and sensor into the beams. Allow the pulley to spin. May require an analog microcontroller input to read, as the holes do not fully block the LED light.

6 Mount a servo with horizontal axis to drive Lego gear trains.

Why: Actuate Lego gear trains with your servo.
Materials: R/C hobby servo, #4-40 screws, nuts, washers, Lego pulley, Lego Technic beam, and two-sided foam tape.
Procedure: Tape the servo onto smooth-top Lego plates. Reinforce with vertically locking Lego structure. To drive Lego gear trains, make sure the servo's output shaft is aligned with the hole of a Lego Technic beam. Create a shaft coupling by screwing a pulley to the appropriate servo horn.

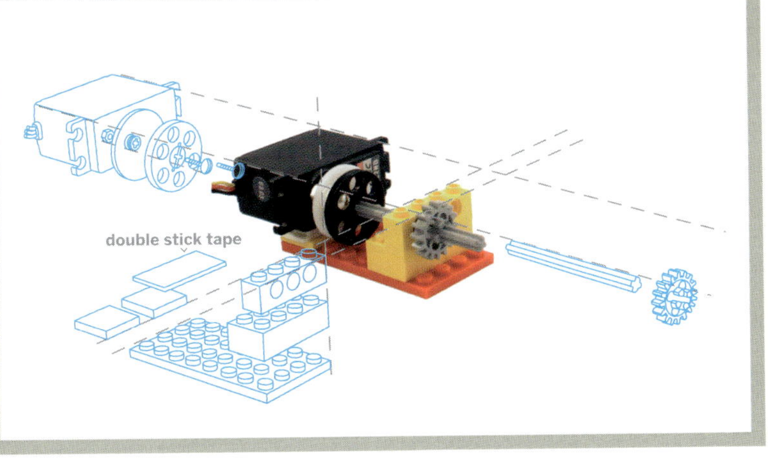

7 Mount a servo with a vertical axis, to drive mechanisms via linkages.

Why: Servo mount to actuate basically anything.
Mounting materials: #4-40 screws, nuts, washers, servo horns, #2-56 ball swivel or ball joints, and #2-56 threaded rod.
Mounting procedure: Construct a structure around your servo. Refer to tip #2 for screw mounting.
Push-rod procedure: Mount one ball swivel onto servo horn with screw. Mount the other ball swivel onto any Lego part you wish to actuate.

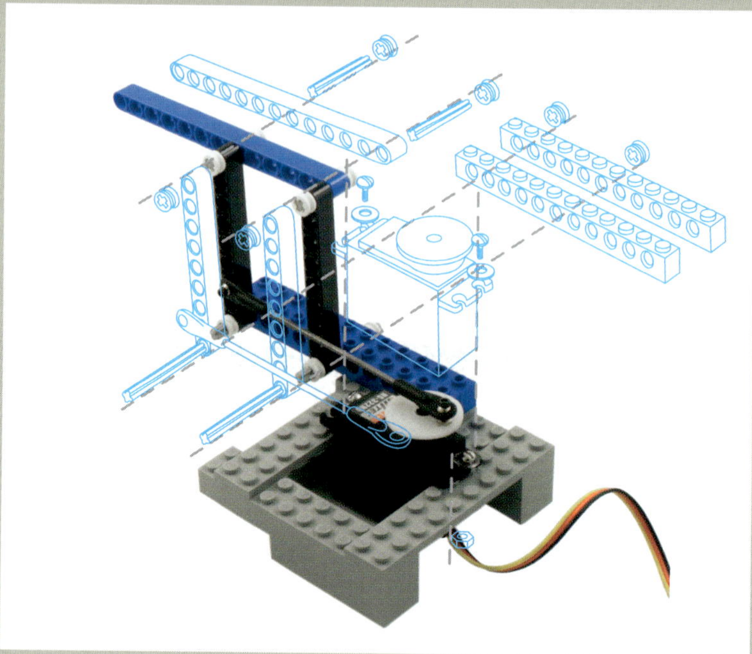

8 Mount PCBs and electronics to Lego models.

Why: Save space by removing unnecessary electronics enclosures.
Materials: Hot glue and dispenser, or 3M double-sided foam tape.
Procedure: When mounting PCBs on top of a Lego stud surface, use hot glue. When mounting on the smooth surfaces (sides and smooth plates), the 3M tape will work best.

10 Mount graphics on your Lego creations.

Why: Add graphics to the smooth sides of your Lego models.
Materials: Paper (photo paper or transparency), 3M spray mount, and inkjet printer.

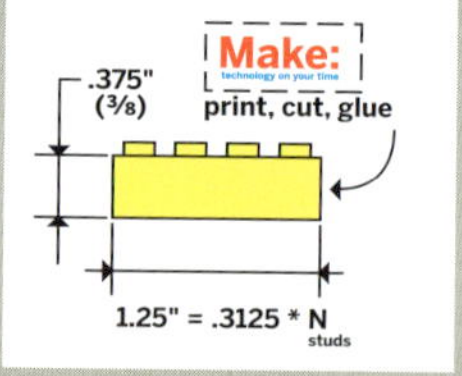

Procedure: Design your graphic to fit the dimensions of an individual Lego brick side or an entire surface. Print and cut out. Spray glue onto the back of the graphic and press onto your smooth surface.

9 Make acrylic or wood plates that fit with Lego studs.

Why: Make custom-sized plates to add functionality and structure to your Lego constructions.
Materials: ⅛" acrylic, aluminum, or basswood sheet.

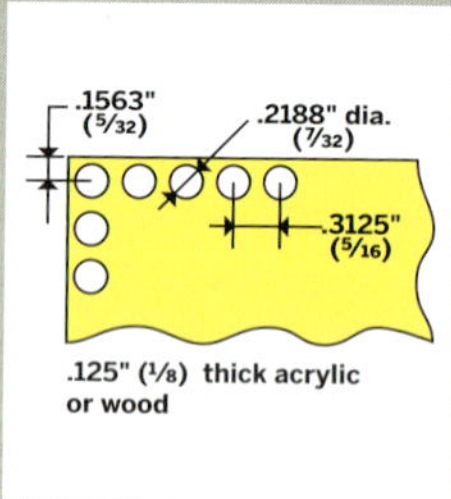

Procedure: Design your pattern in Adobe Illustrator or a 2D CAD package, based on the dimensions given in the drawing. Then fabricate your plate with a drill press, scroll saw, and/or band saw, or send the job to a fabrication shop that can make it with a laser or water-jet cutter.

For a list of resources, visit makezine.com/06/lego.

This article was inspired by *The Art of LEGO Design* by Fred G. Martin (1995): handyboard.com/techdocs/artoflego.pdf.

Michael Rosenblatt works as a program manager at Apple Computer and is a graduate of the MIT Media Lab. He is the founder of inventionDB.com, which assists designers and engineers in quickly documenting projects.

Cogs and Cocktails

Meet the drink-serving, drunk-driving droids at Roboexotica.

By Cory Doctorow

THE MOJITO-MIXING ROBOT HAS A long, steel spiral slide through which limes tumble, landing in a machine that squeezes their juice into a glass. Then, a clanking chain drive moves the glass through a series of stations: a mint dispenser, a muddler that pounds the leaves in with the lime juice, thence to rum, sugar syrup, ice, and soda dispensers. The whole thing is festooned with blobby welds, snaking cables, and exposed logic.

For three years now, Robert Martin, an ex-CNC technician, has brought his mojito robot to Vienna's annual Roboexotica convention (roboexotica.org), where he and other amateur roboticists show off their motley, semi-functional "cocktail robots" — droids that mix and pour drinks, light cigarettes, and above all, inspire conversation. Participants come from all over the world and every level of expertise, from Silicon Valley professors to a guy who simply exhibited a tank of beer with a submersible in the suds.

The mojito robot never successfully prepares a drink, but it always draws a crowd. It's so pretty, so impractical, and so teasing in its potential that watching it not work is more fun than watching it work would be. "It's about beautiful failure and interactivity," says Johannes Grenzfurthner of Monochrom, a Viennese tech/arts group that co-organizes Roboexotica along with another tech-creative outfit called Shifz.

Magnus Wurzner, the ringleader of Shifz, leads me past two 7-year-olds remotely piloting a robot to bump into people and offer them peanuts. Down the hall, a roaring slot-car race is controlled by EEG contacts — the drunker the drivers get, the faster their cars go. At a chrome bar, an animatronic drunkard vomits copiously into a cocktail glass.

Behind the bar looms Thunder One, a giant commercial drink-mixing machine with an LCD interface and a snake's nest of tubes that require regular flushing to clean them. Wurzner points and shouts over the din: "Thunder One is what cocktail culture will come to if we don't intervene. It's all about efficiency and not about aesthetics or irony. That's what cocktail robots and cocktail culture are for."

Cory Doctorow (craphound.com) wrote "A House Divided" on page 26.

Photograph by Jacob Appelbaum/jacob@appelbaum.net

The Vex Robotics Design System

Versatile, powerful design raises the bot in prefab robotics construction kits. By Gareth Branwyn

GADGET GEEKS HAVE COME TO EXPECT that anything Segway inventor Dean Kamen is associated with is going to be innovative, possibly a bit over-hyped, and undoubtedly expensive. Such is the case with the Vex Robotics Design System (vexrobotics.com). Offered exclusively at RadioShack, Vex grew out of Kamen's FIRST Robotics Competition (www.usfirst.org).

Where the Lego Mindstorms Robotics Invention System grew, in part, out of research and robotics competitions at MIT, Vex has origins in Carnegie Mellon University's Robotics Institute. The result is, like Lego, a friendly and open-ended user experience with a wealth of online resources.

Even at $299, the Vex Robotics Design System Starter Kit is reasonably priced for what you get (the add-ons, however, are another story). Its handsome carrying case comes filled with over 500 parts, including three gear motors, one servo, a variety of tires and hubs, 14 different gears, a microcontroller module, an RF receiver module with a replaceable frequency crystal, and a surprisingly decent 6-channel radio transmitter.

If Mindstorms is the robotic answer to Lego, then Vex is the Erector/Meccano analog. The main building components are stamped steel beams, rails, angle pieces, and panels, with attendant nuts, bolts, washers, and other hardware. The structural pieces are Swiss-cheesed with square holes to provide a maximum number of attachment points. The Starter Kit also includes four sensors: two limit switches and two bump sensors.

Your first Vex project: assemble the binder. The *Vex Inventor's Guide* that comes with the kit is a thick D-ring binder and a bundle of loose-leaf pages that you have to organize into it, including instructions for building a starter bot and other kit details. But each tabbed section has an excellent beginner's tutorial on mechanical engineering, electronics, or other concepts used in robot building. The guide is thoughtfully modular, and as you buy add-on subsystems, they come with hole-punched documentation that you add to the proper tabbed section of the binder. Guide sections are all color-coded with the kit parts themselves — silver=structural materials, green=motion, red=sensors, and so on — and this color-matching helps newbie builders understand the mechanical and electronic subsystems and how they interrelate. Product boxes for the add-on subsystems and components are similarly color-coded.

If Mindstorms is the robotic answer to Lego, then Vex is the Erector/Meccano analog.

The Add-Ons Add Up

It is in the add-on systems that the real cost of the Vex system reveals itself. While the Starter Kit does give you a lot of great components for the money, you don't get a full, true robot experience until you add a number of parts. Beyond using the included remote control radio to operate your creations (not very bot-ish), the Starter Kit's microcontroller can be programmed only minimally out of the box, using banks of jumper blocks and little keys that turn on pre-wired routines. To do true programming, you need to purchase the Programming Accessory Kit ($99). Also, the Starter Kit ships

Photography courtesy of Innovation First, Inc.

Here's one sophisticated design, the Vex Tomahawk from VexLabs, which originally designed the Vex system. It's a four-wheeled vehicle that uses two servos and differential power to lean into turns, up to 30 degrees, with Segway-like ease.

VexLabs' remote controlled (no microcontroller) Ping-Pong-Bot uses an accumulator arm to scoop up and carry up to seven ping-pong balls, which it can then fling into the air one by one with a spring-loaded firing mechanism.

with a battery holder, but you have to supply 14 AA NiCad rechargeables yourself, or else buy the Vex Power Pack ($50). If you want to use your bot and RC transmitter in competition against another Vex robot, you need another crystal set with a different frequency from your competitor ($50). And limit switches and bump sensors get old pretty quickly. To add other sensors, you'll need to buy them, at a cost of between $20 and $40 per sensor kit. So a complete user-programmable, competition-ready Vex system can approach $500 or more.

While the true cost of entry may be higher than RadioShack's promotional Starter Kit pricing might imply, things do start to get interesting once you add the Programming Kit and additional sensor sets. When the Lego Mindstorms line was introduced, Lego was surprised to see it embraced not only by kids as an educational toy, but also by robot hobbyists, serious robot developers, and prototypers. But these groups discovered some drawbacks in Lego-based robotics, including the relative fragility of the interlocking bricks, and the limitations of exclusively brick-based design. Vex components offer many more structural configuration options than Lego. Everything is fastened with conventional nut-washer-bolt hardware, and you can bend and cut components. This allows the Vex system to scale from kid's building set all the way up to professional development tool.

The Versatility of Vex

To demonstrate the Vex system's versatility, Grant Imahara, of *BattleBots* and *MythBusters* fame, built a working replica of iRobot's military PackBot using Vex, including the Starter Kit, Programming Kit, Tank Tread Kit ($30), and the Ultrasonic Range Finder Kit (also $30). An article describing the build can be found at makezine.com/go/vexrobo.

Imahara was impressed at how quickly he could construct the Vex PackBot, and the degree of sophistication it could achieve. The dual-tank-tread robot used a remote, semi-autonomous co-control system which allowed it to make some of its own decisions about how to negotiate obstacles.

Jamie Hyneman, also known for *BattleBots* and *MythBusters*, repurposed the Vex system's radio control system and microcontroller unit to create a series of full-sized, computer- and remote-controlled bumper cars for a corporate client — ahead of time and well under budget.

The Vex system offers a great "out of box" experience, and more robust components than Lego. With its versatile microcontroller and easy graphical programming environment (based on the EasyC language), Vex combines construction-toy ease with grown-up computer power. Add the already large user community, thanks to FIRST, and there's monstrous potential for this system. Being able to snag parts at any RadioShack doesn't hurt either.

Hacking the Vex Microcontroller?

As with its Mindstorms forebears, it didn't take long before builders were cracking open the Vex system's microcontroller unit (MCU) and trying to figure out how to make a programming cable that could talk to it. The Vex is basically a simpler, cheaper, commercial version of previous FIRST robotics building components sold by Innovation First (ifirobotics.com), and the Vex MCU uses the same Microchip PIC 18F8520 microprocessor found in the older FIRST Robotic Control (FRC) systems.

This means that if you can build a serial interface cable for the Vex MCU, you should be able to load code onto it using existing FRC software tools such as the IFI Loader (free) and Microchip's MPLAB IDE and C18 Compiler (free 60-day demo).

As of this writing, no one has done this yet, and discussions of the possibility are frowned upon on official Vex and FIRST sites. But it's only a matter of time before such programming options become available. There is a preliminary discussion on investigating the MCU's serial pin-outs at makezine.com/go/vexhack.

Gareth Branwyn wrote "A Beginner's Guide to BEAM" on page 54 and the "Two BEAMbots: Trimet and Solarroller" project on page 76.

Mini Mars Rover

A wireless remote control camera on wheels.

By Tom Zimmerman

WHEN I GO TO SCHOOLS, I TELL THE students about the Mars Rover, a remote control robot with cameras that roams around Mars. Then I tell them they're going to drive one. Well, not exactly. I bring out a remote control truck with a wireless video camera on top and magnets on the front bumper. Then I spread tin cans on the floor and have students remote-control the truck to collect them – just like the Mars Rover, except with cans instead of rocks. Sounds easy, right? Wrong. They can only see what the camera sees, on a TV screen, and they soon find out that driving is a lot harder when their field of vision is as narrow as a video camera's.

You, too, can build the Mars Rover. Here's how to do it in a few hours for under $150. The main components are a remote control (R/C) truck and an X10 wireless camera and receiver. The only tools you need are a hand drill, diagonal cutters, a saw, and scissors. And surprisingly, the hardest part of the project is arranging the magnets.

Photography by Tom Zimmerman

TEST THE CAMERA

Before you start drilling, make sure the wireless camera works. Load 4 AA batteries into the camera's battery pack, and plug the wireless receiver into a TV using the RCV audio/video cable. (If the TV lacks separate video and audio jacks, plug its RF output into the antenna input.) Use the small slide switches on both the camera and wireless receiver to set each device to channel 1, and try other channels if you don't get a clear signal. If you are near a wireless

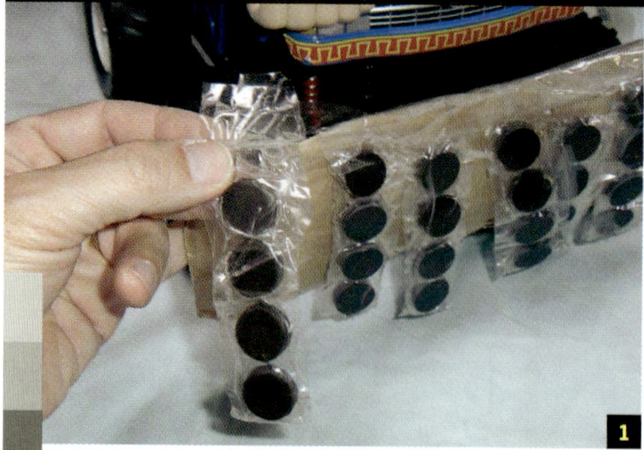

1

2

SUPPLIES

- **Wireless camera and receiver** X10 #XCAM2, $80
- **Battery pack** X10 #XCam2 Battery Pack, $20
- **RCA audio/video cable** 12' RadioShack #15-1512, $15
- **Remote control truck** Big enough to carry camera and battery pack (try RadioShack or a hobby or toy shop)
- **Magnets (28)** RadioShack #64-1888, 5 for $2.60
- **Diagonal cutters** RadioShack 64-2951, $5
- **Silicone rubber sealer** RadioShack #64-2314, $3.20
- **AA batteries (4)** For battery pack
- **Hand drill and ¼" bit** For drilling into plastic and wood
- **Cable ties (6 or more)** Less than ¼" wide and 6" long
- **Clear 2" wide tape** At least 5'
- **Cardboard** About 4"x10", enough to cover the truck's front bumper
- **Scrap wood** At least 3"x6" to mount the camera to the truck

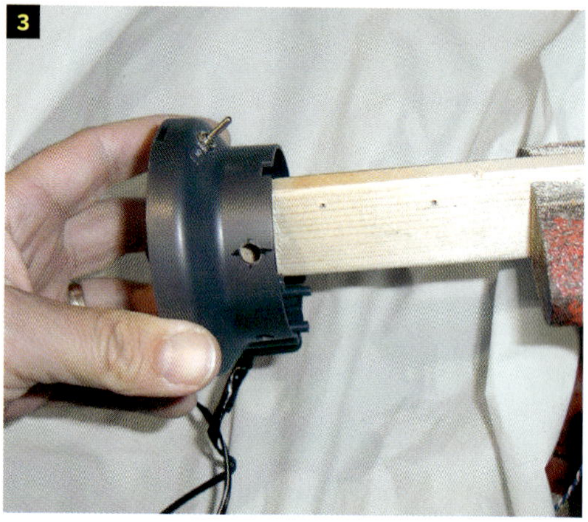

3

network (WLAN), which also operates at 2.4GHz, you may see white horizontal lines in the image. The 2.4GHz band is unlicensed, and devices operating on this band must accept interference. If this is a problem, move to another room or somewhere else.

Another thing to consider is communication range. The camera manufacturer claims a range of up to 100 feet, but as with all wireless communication, this depends on your environment. Outside gives the best range, and buildings with metal in the walls will give you the shortest. White snow in the video image, which is how analog signals like this degrade, means that you're reaching maximum range and should move closer. Once you lose the video, you'll be driving blind.

The camera's battery pack has a long cable, which you can either fold neatly and tie out of the way or shorten down. I cut mine down because I wanted to get to the wires anyway, in order to test the camera's voltage and power consumption, so I shortened the cable to about 2 feet. (Power consumption turned out to be about 1 watt, 16.5V x 65mA.)

CHOOSE THE TRUCK

When you shop around for the R/C truck, think about where you will attach the camera and battery pack. The truck I bought has railing in back where you can attach cable ties, and the cab's roof is hollow, so you can drill mounting holes into the plastic.

ADD MAGNETS TO THE BUMPER

Cut a piece of cardboard to fit the front of the truck bumper, about 4"x10", then punch pairs of holes in it and cable-tie it on. If your truck doesn't have a bumper, drill some ¼" holes in the front plastic, but make sure you don't drill into the turning mechanism or any other critical components.

The cardboard supports flexible, hanging strips of magnets that wrap against the cans to ensure a strong hold (Fig. 1). To make each strip, cut a 6" piece of tape and place 4 magnets lengthwise, starting at the bottom left side of the tape. This leaves about 2" of tape at the top, where you connect it to the cardboard. Carefully fold the right side of the tape over the left side, enclosing the magnets in the tape. I used 7 strips of magnets for my truck.

Finally, tape the top of each strip to the cardboard bumper. As the truck moves, the tape will allow the magnets to grab and hold a can. If cans slip under the bumper, move it lower to the floor.

MOUNT THE CAMERA

Before you commit to a mounting location, try the camera out at different places on the truck's roof and see what kind of image you get onscreen. Ideally, the view will be straight ahead, with the top of the bumper visible at the bottom of the image.

Mount the camera with cable ties, which are cheaper and easier than nuts and bolts, and don't require exact alignment. If the cable ties are too wide to pass through the camera's mounting holes, widen them with a ¼" drill.

After driving my truck around, I realized that I'd get a better view if I placed the camera higher and farther back. So I cut a piece of scrap wood to fit between the truck cab and the plastic railing, sanded the top corners, and then drilled holes to cable-tie the wood to the railing and the camera. I secured the block of wood to the truck bed with a bead of silicone glue. It took me a couple of tries to get the camera properly aligned — and if you look carefully, you can see a few extra holes — but the result was a better view of the floor and bumper (Fig. 2).

MOUNT THE BATTERY CASE

It's a good idea to mount the battery pack case upside down so you can replace batteries without removing the case from the truck. I put my case to one side of the truck's back railing. Make sure the power switches for both truck and camera are easily accessible. Label the camera's power switch, since it will be facing downward and there is no power indicator light to show that it's on.

Once you know where to place the case, mark and drill four ¼" holes into the plastic on the rim. Start with ⅛" pilot holes, since big drill bits tend to grab in plastic, and use a strip of wood to support the fragile lip while the case is being drilled (Fig. 3). Cable-tie the case onto the truck, and trim the excess length. If you make a mistake or change your mind, just cut the ties and try again.

ROLL IT!

Now your Mars Rover is ready to roll. Remember: The magnets will attract only tin cans, which contain iron, and not aluminum cans. The audio track will be noisy, picking up the sound of the grinding gears and the giggles of children as you drive by their feet.

Going Further

Here's a game for your Rovers and a few people: Place four cans in the middle of the floor and challenge teams to pick up as many as possible in one minute. One team member watches the screen and drives, while the other is "home base" and removes cans. The driver collects a can, drives it home, and returns for more, only viewing the screen — no peeking allowed!

For the advanced tinkerer, here are some more ideas to extend your fun:

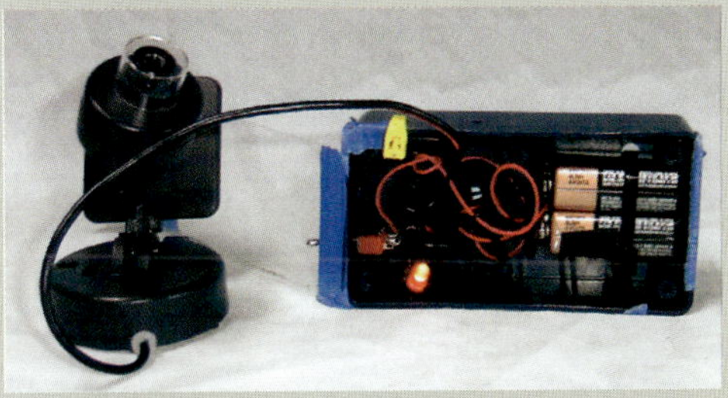

MAKE YOUR OWN BATTERY PACK The camera says 12V, but I measured the battery pack at 16.5V, 65mA. At that current, a red LED drops nearly 2V. So I connected two 9V batteries and added a red LED in series with the camera, that acts as a power indicator.

ADD A CAMERA Since the wireless camera and receiver support four channels, you can put extra cameras on the truck, to look backward or to see cans on the bumper. Set the wireless cameras on different channels and use two receivers and TVs, or else switch between the channels/cameras on the receiver.

ADD LIGHTS The camera doesn't do well in the dark, but you can add some white LEDs or bulbs on the truck for illumination. For power, tap into the truck's battery. Even cooler, open the camera and remove the infrared-blocking filter (a little circular magenta piece of glass) to boost IR sensitivity, then use infrared LEDs as headlights, and see in complete darkness.

ADD ULTRASONICS This is way advanced, but you could build an ultrasonic range finder that converts distance to audible pitch, and send it through the camera's microphone. This would allow you to drive in complete darkness, like a bat, if a bat could drive.

FLOATING CAMERAS Instead of a truck, attach the wireless camera to a 3' weather balloon filled with helium. Use 100 feet of strong fishing line or string. I attached my camera to a 4"x12" piece of plastic, fashioned like a windmill tail, to stabilize it in the wind.

Resources

X10 (wireless camera, receiver, battery pack): x10.com

RadioShack: See radioshack.com for a store near you.

Edmund Scientific: Two 3' diameter weather balloons, $26, item #3041755 scientificsonline.com

Hot Air

Build a do-everything manifold to control, dry, route, and use compressed air.

Gather the Necessary Parts

A brand-new compressor might impress your mother-in-law, but when you're ready to hit the big time, build this manifold system to get clean, dry air where you need it.

List of Materials

Most of the material you need for the Quick and Dirty Air Compressor Manifold can be purchased at a well-stocked hardware store.

[A] Compressor outlet Even small "pancake" compressors will operate a nail gun or small air tools, but don't expect to spray paint or repair a big rig. The size of the air outlet will vary by compressor size and output, so adapt it to 3/8".

[B] Flexible hook-up hose Compressors tend to jiggle around, so use a flexible hose to connect to your hard line manifold. Don't use a regular air hose! Use a special compressor connection hose that is rated at 200psi working pressure/800psi burst pressure.

[C] Ball valve In addition to turning off the air from the compressor, an inline 3/8" valve can be used as a manifold drain valve. A simple manifold won't accumulate much moisture, but when running longer lines, it's a good idea to have a drain at the lowest point.

[D] Galvanized steel lines Why use 3/8" galvanized steel pipe for air lines? PVC is a poor choice as it can shatter under pressure. Copper is ideal if you are not daunted by higher costs, increased labor, and great difficulty to make changes. I would use copper exclusively if I were a quadzillionaire with a plumber on staff. The pipe should have a slight decline down toward the drain valve to make sure moisture doesn't get trapped in the line.

[E] Regulator/moisture trap/filter Moisture is an unavoidable and damaging byproduct of compressing air. An all-in-one trap/regulator/filter is cheap insurance to protect air tools from wet, dirty air pumped out at potentially damaging pressure.

[F] T-connector Use a 3/8" galvanized steel T wherever you want a drop.

[G] 3/8" male to 1/4" female connector Maybe you will get lucky and find a 3/8" T connector with a 1/4" outlet, but I had to adapt 3/8" to 1/4".

[H] 1/4" brass nipple Rather than a galvanized nipple, use a brass nipple with a hex in the center so you can use a wrench to get a tighter fit.

[I] 3-way air connector What kind of a savage doesn't love a brass 3-way air connector? Tidy and professional, it is an absolute delight.

[J] Combo coupler The bad news is there are three different "standard" compressor plugs — Industrial Interchange, Automotive Standard, and ARO Speed. The good news is Amflo makes a combo coupler that works with all three.

[K] 3/8" elbow Use to get around corners.

[L] Drop hose The ubiquitous yellow drop hose is found in every self-respecting auto shop in the world. I wouldn't dream of rebuilding a carburetor without one.

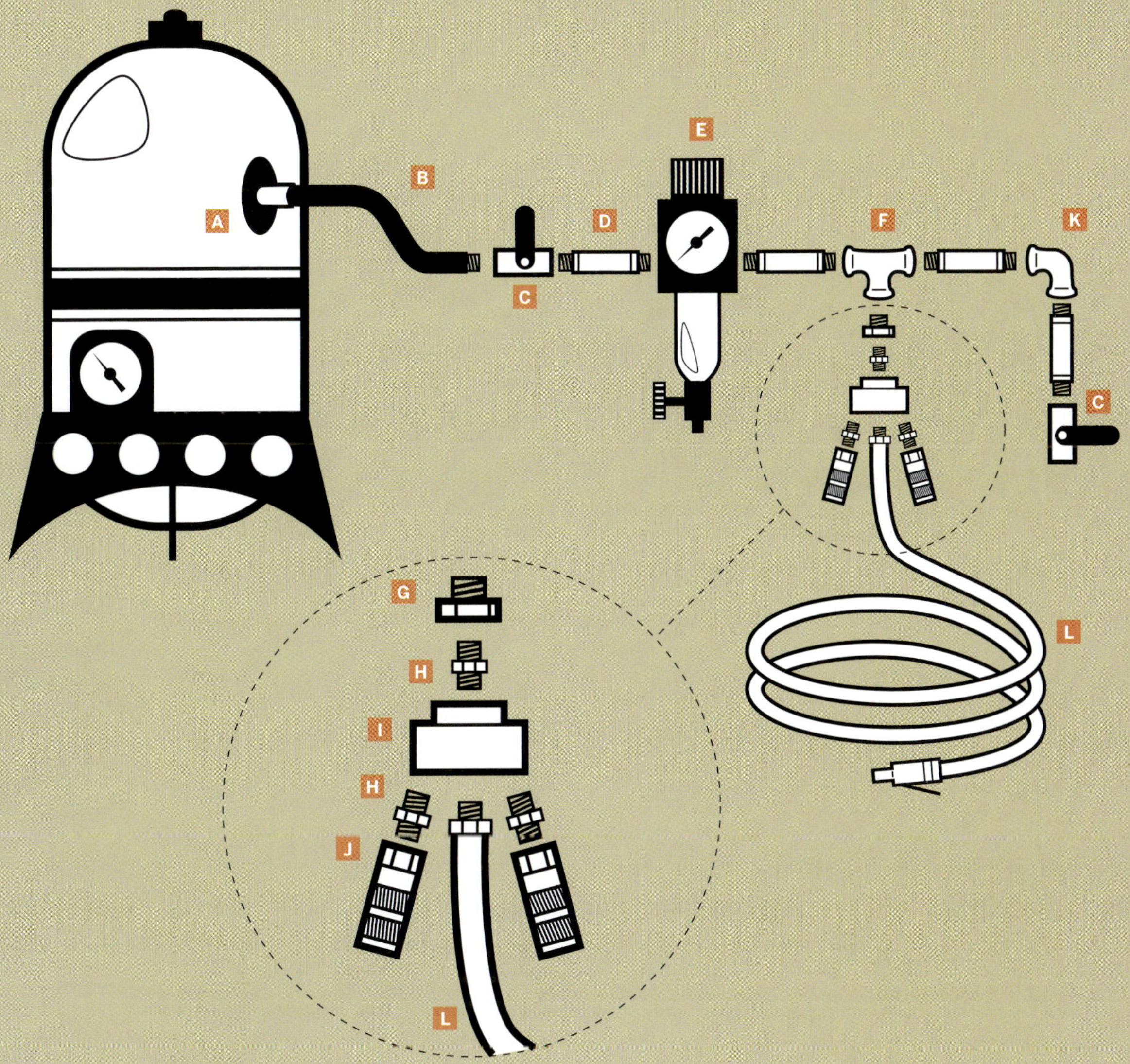

Illustration by Mister Jalopy and Mark Frauenfelder

Great Tire Gauge

Sure, there are digital tire gauges that speak, but I have enough friends and don't want to worry if I am spending enough time with my tire gauge. For $20, you can get a polished brass EZ-Air (accu-gage.com) with a standard feature list more impressive than the window sticker on a new Cadillac. Fill tires directly through the gauge and eliminate swapping between the fill hose and gauge — just watch the pressure rise as you fill!

Why not ½" or larger air lines? This ⅜" system is adequate for most tasks, but some jobs — like spraying auto paint — will require a more robust system with larger lines and specialized filters.

WARNING: **Compressed air dangers are no joke. Sounds far-fetched, but I know a fellow who is disabled due to a compressed air accident. Wear ANSI-approved safety glasses, read the manufacturer's instructions, and NEVER point a compressed air stream at friends, enemies, dogs, or even parakeets. Be safe!**

Q&D

How to Make an Airtight Threaded Connection

There are only two ways to put on Teflon tape — and one of them is wrong.

1 Holding the pipe with the open end facing you, start tape at 2 o'clock. Always wrap the tape clockwise around the threads. Otherwise, when you connect the female fitting, it will "roll up" the leading edge of the tape, resulting in an inferior seal.

What if the connections are reverse thread? OK, smarty-pants, reverse threads (a.k.a. left-hand threads) are taped counter-clockwise.

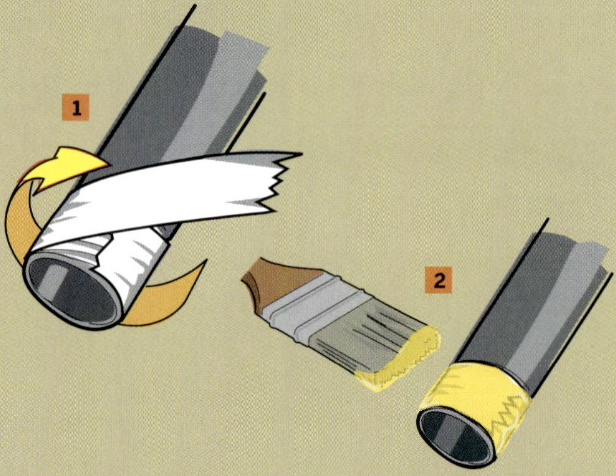

2 Why use pipe joint compound if you are already using Teflon tape? Some would say it's overkill, but the compound is a lubricant that allows you to turn the pipe an extra half turn tighter.

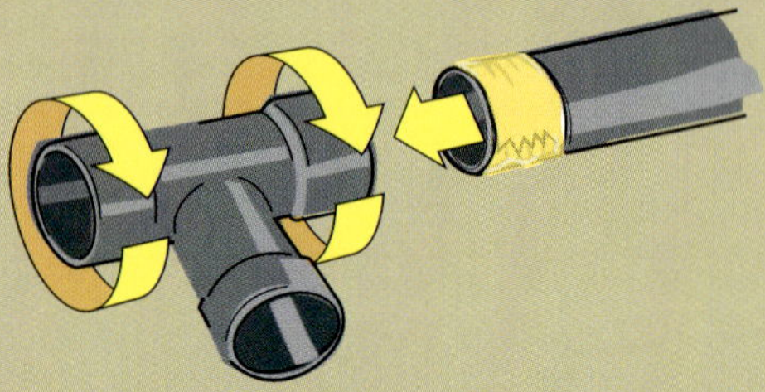

Now Make Yourself Useful

A whole world of nail guns, impact wrenches, cut-off wheels, pneumatic ratchets, die grinders, air hammers, and high-speed sanders will soon fill your dreams and make you forget about plugging power tools into electrical outlets. Behold the awesome power of compressed air!

Powerpuff Boy

If you don't require a heavy-duty air compressor, Hitachi's EC79 6-gallon pancake air compressor might fit the bill. The oil-free compressor, which weighs 60 lbs. and delivers 2.7cfm at 90psi, is perfect for finishing nail, staple, and other short burst work. hitachipowertools.com

Illustrations by Tim Lillis

Mister Jalopy breaks the unbroken, repairs the irreparable, and explores the mechanical world at Hooptyrides.com.

Make: Projects

Bask in the glow of your soldering iron and build a geeky-cool bot battalion that follows the sun. If it's gloomy outside, just start practicing for RoboCup 2050 with a soccer-playing robot. Or reach new heights by delving into tensional integrity with tabletop sculptures that soar.

BEAM Robots

76

Lego Soccer

88

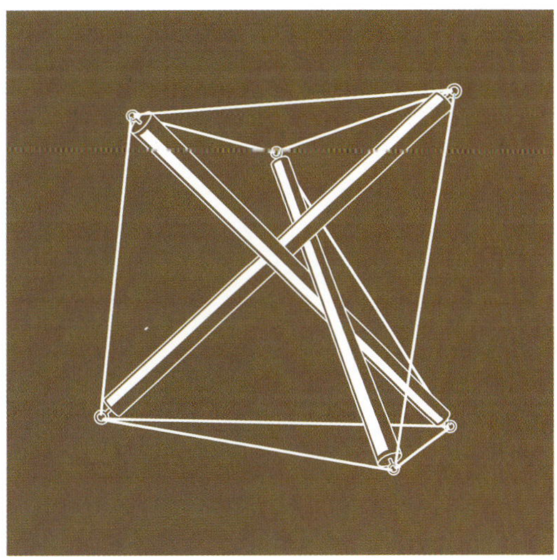

Tensegrity Tower

100

Photography by Douglas Adesko

TWO BEAMBOTS: TRIMET AND SOLARROLLER

By Gareth Branwyn

Solder together one simple circuit and use it to control two very different solar-powered robo-critters: a little satellite that scoots and bumps around, and a mini cart that just keeps a-rolling until the sun goes down. »

Set up: p.80 **Make it:** p.81 **Use it:** p.87

GO SOLARENGINE!

The low-tech, analog, dumpster-diving, and hack-friendly world of BEAM robotics (*see page 54*) has produced a bestiary of bot types, including Symets, Rollers, Walkers, Jumpers, Climbers, Swimmers, Flyers, and Crawlers. Many of these creatures can be powered and controlled by a Solarengine, a simple and popular BEAM circuit that draws energy from a solar cell and temporarily stores and dispenses it using one or more capacitors.

We'll make a couple of voltage-triggered Solarengine circuits, and then build them into two little bots: a Trimet, which looks like a satellite in orbit as it's moved around by a spinning, top-like base, and a Solarroller, which drives straight ahead in fits and starts. These light-sensitive critters will look cool and très geeky on your desk, as long as you can keep them from wandering off the edge (they're both active diurnally, and they don't have an off switch).

Gareth Branwyn wrote "A Beginner's Guide to BEAM" on page 54.

FAST REACTOR DESIGN

BEAMbots use simple circuits that interact with the world directly. Unlike control-freak robots, their brainless, reflexive reactivity is the whole point.

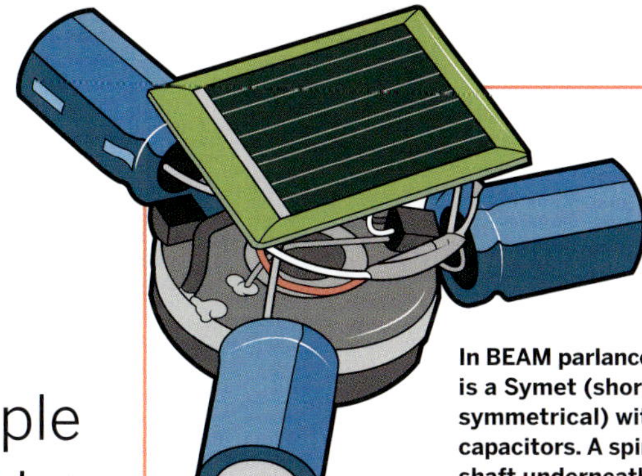

In BEAM parlance, a Trimet is a Symet (short for symmetrical) with three capacitors. A spinning drive shaft underneath pulls its top-like body, which bumps around any obstacles.

Solarrollers are little solar-powered race cars. At BEAM and other robot competitions, builders pit Solarrollers against each other in a kind of robo-mechanical Pinewood Derby.

Our Trimet and Solarroller bots are based on a voltage-triggered (Type 1) Solarengine. These circuits collect energy from a small solar cell, and periodically release it when there's enough stored up to actually do something, like run a motor.

The solar cell converts light into electrical energy, slowly juicing up the capacitor (or multiple capacitors).

The capacitor collects and stores a voltage, which discharges whenever the circuit is completed between its two terminals.

The 1381 voltage trigger measures the voltage across the capacitor, and sends a trigger signal once it's high enough (2.4 volts with a 1381-G trigger).

When the base pin of the 3904 transistor receives the trigger signal, it completes a connection that allows the capacitor's power to discharge through the motor.

The motor runs intermittently, whenever it receives a power dump from the capacitor.

During discharge, current flows to the base of the 3906 transistor. This takes the 1381 trigger offline, allowing it to reset, and routes current to the 3904 base, which keeps the motor circuit flowing until the cap is fully discharged.

The 2.2kΩ resistor reduces the voltage to the 3906 base pin, so it diverts less power away from the motor during discharge. This makes the circuit more efficient.

➕ Solarengine schematic: makezine.com/06/beambots

Illustration by Timmy Kucynda

SET UP.

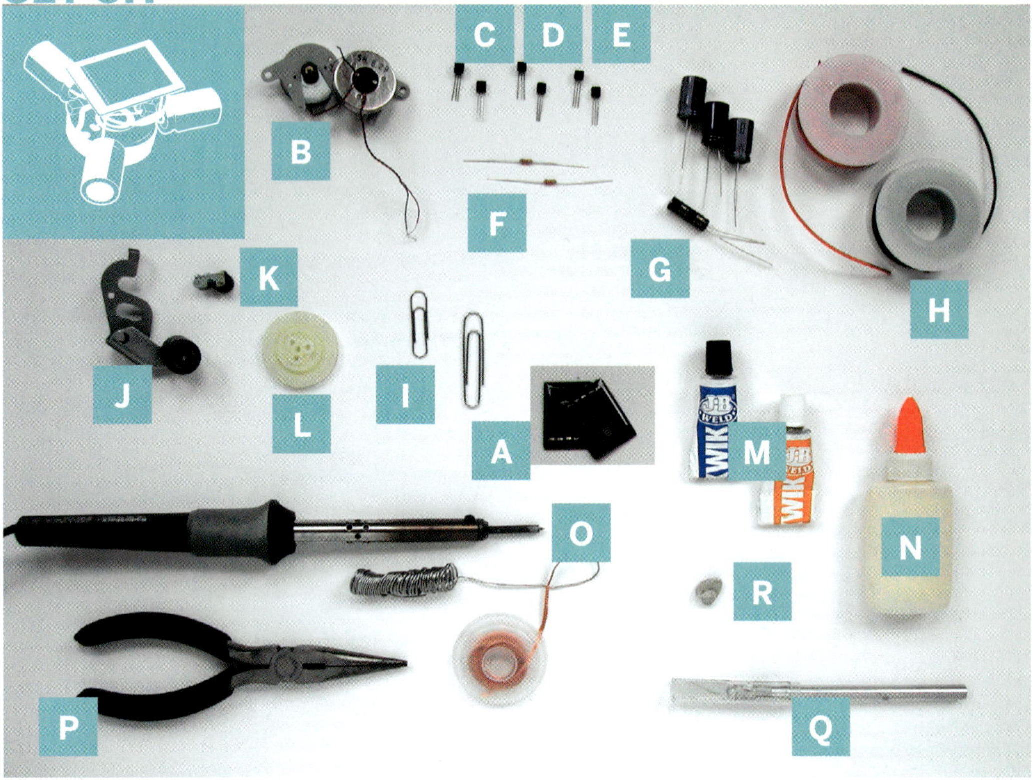

MATERIALS

The following parts will build two Solarengines. Just get one of each if you're only building the Trimet or the Solarroller, but not both. All part numbers refer to Solarbotics (solarbotics.com):

[A] 37x33mm polycrystalline solar cell, part #SCC3733 (2)

[B] Cassette motor (2) From an old Walkman or other player, part #MCM2

[C] 1381-G voltage trigger IC, part #1381-G (2)

[D] 2N3904 ("3904") NPN transistor, part #TR3904 (2)

[E] 2N3906 ("3906") PNP transistor, part #TR3906 (2)

[F] 2.2kΩ resistor (2)

[G] 4700μF capacitors (4) Or use three 4700μF capacitors (for the Trimet), and 1 "supercap" such as a 0.33F Gold Capacitor, part #CP.33F (for a higher-performance Solarroller)

[H] Hook-up wire, red and black 24-gauge stranded

[I] Paper clips (2) one small, one large

PARTS YOU'LL ONLY NEED FOR THE SOLARROLLER:

[J] Pinch roller and arm from a VCR, or similar Smooth rubber roller, about ⅝" in diameter and ⅝" wide

[K] Pinch roller and arm from a cassette player, or similar Smooth rubber roller, about ½" in diameter and ¼" wide

[L] Drive wheel of any lightweight material, with a diameter slightly greater than the motor casing Between 1½" and 1⅝" is good. An old VCR might have a suitable pulley, or try the disc that holds the control rods in a servomotor. You can also use a wheel from a toy, or any other right-sized plastic disc.

Rubber band

[M] Epoxy

[N] White glue

TOOLS

[O] Soldering equipment Iron, stand, solder, and solder-sucker, desoldering bulb, or braid

Dremel tool with grinding wheel, cut-off wheel, and router bits

"Third hand" tool with alligator clips Two are ideal

[P] Needlenose or long-nose pliers

Wire cutters

[Q] Hobby knife

Medium-grade sandpaper or metal file

Ruler

[R] Poster putty or tape

Safety glasses

MAKE IT.

BUILD YOUR BEAMBOTS

START Time: **A Day** Complexity: **Medium Low**

1. BUILD THE SOLARENGINE CONTROL CIRCUITS

We'll be freeforming these circuits, which means connecting components together directly, without a board. Normally I would breadboard and test my circuits before soldering, but this one is so simple and has so few parts that we can live dangerously. Parts are easily desoldered and resoldered if there's a problem.

1a. Face the two transistors up with their pins toward each other. Solder the base pin (middle) of the 3904 transistor to the collector pin of the 3906 (the right pin, as you read the printing).

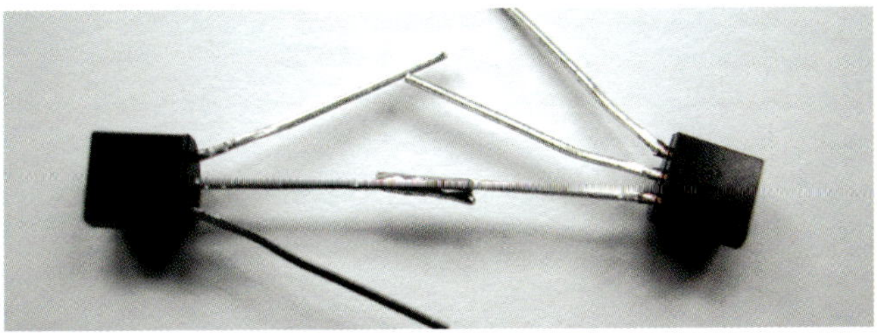

1b. Use needlenose pliers to gently bend the 3904 emitter pin (left) 90 degrees to the side and its collector (right) 90 degrees up. Bend the 3906 base pin (middle) 90 degrees up and its emitter (left) 90 degrees to the side. Solder the 2.2kΩ resistor from the 3904 collector to the 3906 base.

1c. Trim excess lead length from previous step. Place the 1381 voltage trigger to the right of the 3906, facing the same way. Solder its Pin 3 (right) to the 3904 emitter and its Pin 1 (left) to the 3906 collector. Finally, arc its Pin 2 (middle) around and solder it to the 3906 emitter (left). There's your basic circuit, ready for motor and power!

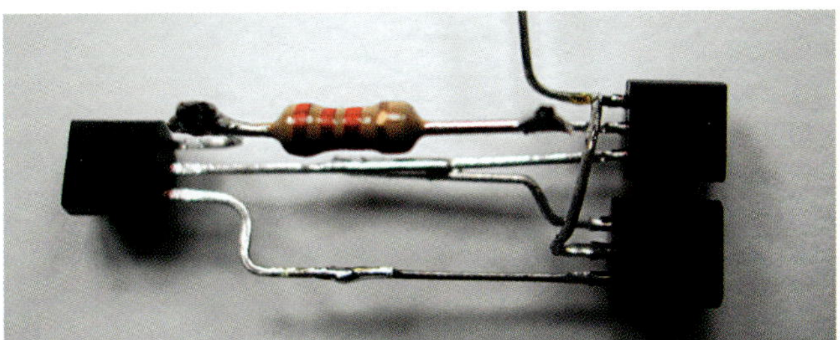

1d. If you're making both BEAMbots, build a second Solarengine circuit by repeating steps 1a-1c above. From here, you can continue on to step 2 to build a Trimet, or jump ahead to step 3 and build a Solarroller.

2. MAKE A TRIMET

2a. Prepare the motor by removing any mounting tabs with a Dremel grinding wheel. Then use sandpaper or a metal file to scuff the drive-shaft side of the case until you're down to the shiny metal underneath. Really scuff it up good; you'll be soldering capacitors directly to the case, and they'll need to hold as the Trimet drags and bumps around.

2b. Clip the negative/cathode leads on the three 4700µF capacitors so there's just enough wire to solder them to the motor casing. Bend the positive/anode leads up, making sure they comfortably clear the casing. Find 3 equidistant points at the perimeter of the motor, and solder the 3 cathodes to these points so that the capacitors form an equilateral triangle radiating out from the motor's center. Use generous gobs of solder, and use poster putty or tape to hold the caps in place while you solder.

2c. Center the circuit assembly over the motor, and solder a scrap lead from the 3904 emitter to the motor casing. This grounds the circuit, while also attaching it to the motor. For optimal balance, bend this connecting wire at 90 degrees, and try to position the circuit in the middle of the motor.

2d. The motor case is our circuit's ground (-); now let's work on the power (+) side. Take a small paper clip and bend it into a ring with the same diameter as the motor. (Conveniently, Walkman motors are the size of a quarter, so you can use one as a form to bend the clip around.) When you have a decent circle, solder it together.

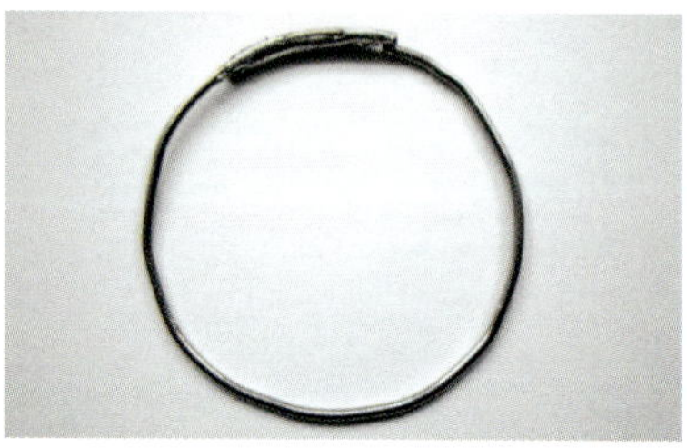

2e. Bend and trim the capacitor anode leads evenly, so that they extend just above the control circuit. Solder the "power ring" to the ends of the 3 leads, preserving the equilateral symmetry.

2f. If you can bend the 3906 emitter lead to reach the paper-clip ring, do so, and solder it on. Otherwise, connect it with a short piece of wire or scrap component lead.

2g. Now, the solar cell. If yours has pre-tinned pads but no wires (most small cells come this way), start by soldering the 2 wires onto it — but be careful, because solar cells are fragile. Then solder the positive/red wire to the ring and the negative/black wire to the motor casing. Make the wires long enough so you can still work on the circuit, but short enough so they'll stow neatly underneath when you finally glue the solar cell down onto the ring.

2h. Connect the motor. Solder the negative/black motor wire to the point where the 2.2kΩ resistor meets the 3904 collector. Solder the positive/red motor wire to the paper-clip ring.

Now, place the solar cell on top of the Symet and shine a light on it, or put it in the sun. After 10 seconds or so, it should fire and scoot along, or spin around if you're holding it by the driveshaft underneath. If so, congratulations — you're the proud parent of a BEAMbot! You can go ahead and glue the solar cell onto the paper-clip ring. Or, if the cell stays in place without glue, leave it that way so that people can peek under the hood.

3. MAKE A SOLARROLLER

Solarroller builders have used all sorts of materials, from Lego bricks to soldered paper clips to computer mouse cases. This popular approach relies on parts from an old cassette player and VCR. Your mileage may vary, depending on the parts that you use for the body and drivetrain.

3a. Cut the arms on the 2 pinch rollers with a Dremel and cut-off wheel, so that they make full, flat contact against the motor casing. The Solarroller will stand on the triangular base that's formed by these 2 idler wheels and the larger drive wheel that will go onto the motor's drive shaft.

3b. Prepare your drive wheel. First, check that it will fit on the motor's drive shaft. (The hole in the hub of the disc I used was too small, so I reamed it out using a Dremel router bit.) Then glue a rubber band around the outside of the wheel, to improve traction. Cut the band, smear a thin layer of glue onto one side, and when it gets tacky, carefully roll the wheel over this "tire" until it comes full circle. Let the join overlap, then use a hobby knife to cut away excess rubber and make sure the ends are perfectly joined.

3c. Epoxy the 2 idler wheel arms into position on the motor casing, then fit the drive wheel onto the motor shaft without gluing it (use poster putty to hold it on, if needed). It is critical that all 3 wheels run parallel to each other and make full contact with flat ground when the Solarroller is standing. If you're using the Solarbotics motor, you can affix the larger roller arm to the motor's large mounting tab, pointing toward what will be the front, and leave the two other mounting tabs and holes pointing up on top, for attaching the circuit and solar panel.

3d. Cut about 4" of wire from a large paper clip and fashion it into a U shape. For the Solarbotics motor, it can be just wide enough to run between the two upper mounting holes. Trim the remaining piece of paper-clip wire and solder it across the U as a cross-brace, about ¾" from the open end.

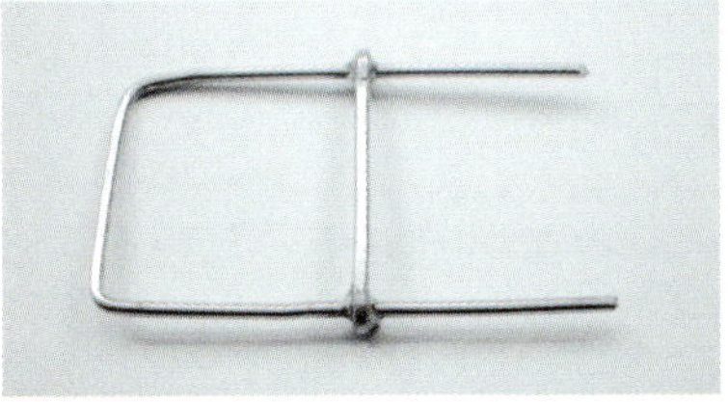

3e. Epoxy a capacitor directly to the motor casing, running horizontally, on the side opposite the drive wheel. The leads should point backward, with the cathode (-) closer to the motor.

3f. Solder (or epoxy) the paper-clip frame atop the motor casing, using the two mounting holes if present. Since we didn't glue the drive wheel on yet, you can remove it to access the top of the motor. For extra sturdiness, you can position the frame so the cross-brace rests on the capacitor, and epoxy the brace onto the cap. Glue on the drive wheel.

3g. Position the Solarengine circuit underneath the paper-clip frame, next to the cap, on the side opposite the motor. Solder the 3906 emitter (left) pin to the positive/anode lead of the capacitor. The connection should be short enough so that the cap holds that end of the circuit up in the air.

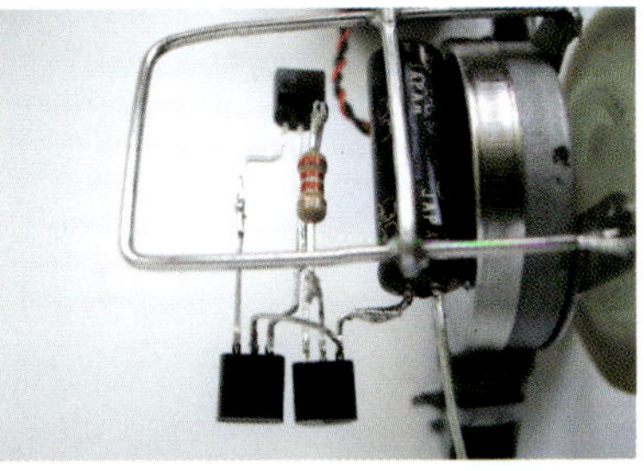

Transistor Action

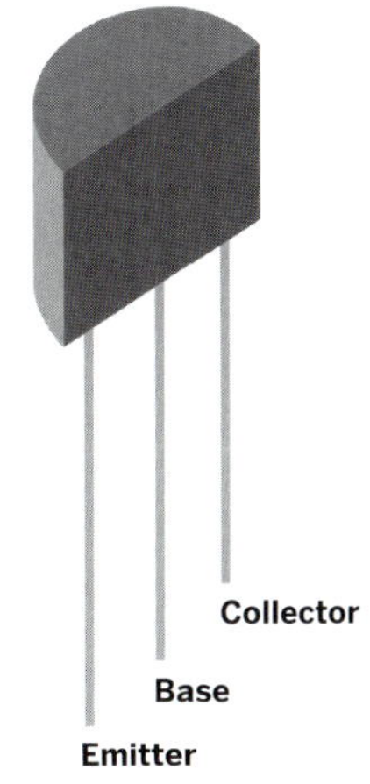

Bipolar transistors can act as switches, connecting parts of a circuit just like a mechanical switch would. In an NPN transistor, applying a voltage with the positive side to the base and the negative side to the emitter allows current to flow from emitter to collector. A PNP transistor goes the opposite way; running a negative voltage across from the base to the emitter allows current to flow from emitter to collector.

3h. Turn the Solarengine upside down and solder a scrap component lead to the 3904 emitter pin at the point where it attaches to the 1381 trigger's Pin 3. Bend the capacitor's negative/cathode lead around the undercarriage side of the cap's barrel, and solder it to the lead you just connected to the 3904. This will anchor the other end of the circuit.

3i. If your solar cell doesn't have wires, attach some to the pads marked (+) and (-). The wires only need to be long enough to reach the pins on the capacitor. Thread the solar cell's wires through the frame and epoxy the cell to the top. When the epoxy is set, solder the solar cell's positive to the cap's positive/anode and the cell's negative to the cap's negative/cathode.

3j. Finally, connect the motor. Solder the positive/red motor wire onto the 3906 emitter (left) pin and the negative/black wire to the 3904 collector.

Now, put the Solarroller on a flat surface in the sun, or shine a flashlight on the cell. After a little while, the circuit will trigger, the capacitor will dump, and your Solarroller will take off for a short run. Shine, wait, and repeat.

FINISH X

NOW GO USE IT »

USE IT.

BEAM ME UP, SCOTTY

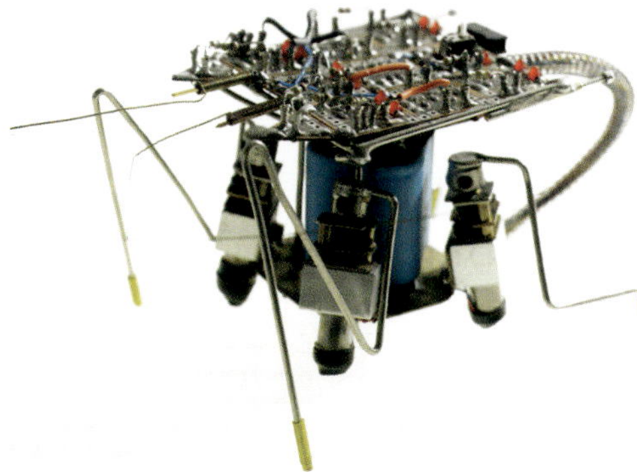

TROUBLESHOOTING

If your BEAMbot doesn't make you beam, carefully examine all connections, resolder anything that looks weak, and separate any components that might be touching (shorting). It's a simple circuit, so not much can go wrong besides incorrect connections or bad joins.

FURTHER HACKING IDEAS

On the Trimet, add an outer paper-clip ring. This creates a bumper that will help prevent the robot from getting stuck.

On the Solarroller, replace the regular 4700μF capacitor with a "supercap" like a 0.33F Gold Capacitor, as shown in the project photos. These capacitors can take several minutes to juice up, but they'll make your Solarroller take off like a bat outta hell.

You can easily convert an old Sony Walkman into a great Solarroller. Leave the motor, roller wheels, and pulleys in the original frame's base piece, and use it as the vehicle's chassis.

Try Andrew Miller's more efficient variant of the basic Solarengine, which is almost as easy to build. You need a different resistor, an additional capacitor, and a diode, but you can lose the 3906 transistor. Varying the value of the small cap, between 0.47μF and 47μF, lets you "program" different discharge times. (See schematic at: makezine.com/06/beambots.)

Once you have the basic ideas down, you can go crazy, improvising BEAMbots with greater storage capacity, better obstacle-avoidance strategies, or swankier, more attention-getting designs. Here are some Symet and Solarroller variations (pictured at right).

RESOURCES

There are many more hacks and variations on these two project types, as well as other applications for the Solarengine. For more information, see "Getting Started in BEAM" on page 57.

+ Schematic for Miller variant of Solarengine circuit: makezine.com/06/beambots

SOCCER-PLAYING ROBOT

By Matthew Russell

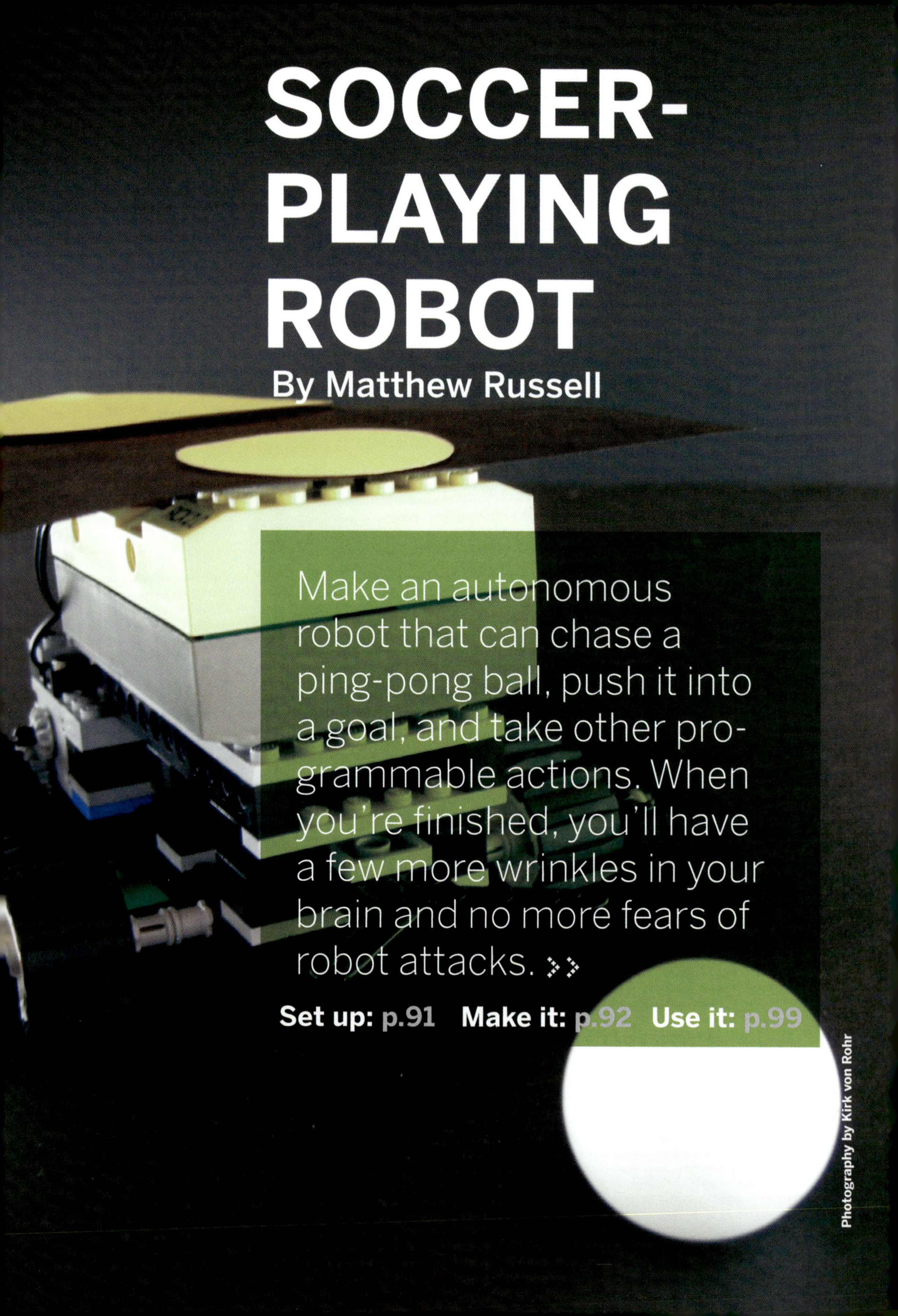

Make an autonomous robot that can chase a ping-pong ball, push it into a goal, and take other programmable actions. When you're finished, you'll have a few more wrinkles in your brain and no more fears of robot attacks. >>

Set up: p.91 **Make it:** p.92 **Use it:** p.99

Photography by Kirk von Rohr

ROBOCUP DREAMS

Robotics is one of the most multidisciplinary topics imaginable, combining artificial intelligence (AI), mechanical engineering, math, signal processing, sensor fusion, circuit design, and psychology. The resulting synergy animates applications that range from the International Space Station to kids' toys.

The most ambitious research project in robotics today may be RoboCup (robocup.org). This effort aims to develop a team of autonomous robots that will beat the human World Cup champions at soccer by the year 2050.

You can get a small taste of this complex endeavor by building your own soccer-bot, using the Lego Mindstorms Robotics Invention System. This kit includes motors, sensors, and infrared transceivers. Add a webcam and some code, and you can make an autonomous mini robot that will retrieve a ping-pong ball and bump it into a goal box. *Goooooal!*

Matthew Russell tries to live life as a renaissance man, but hasn't made much progress since becoming enthralled by the cult of Mac.

GOAL-DIRECTED DESIGN

Showing brightly against a black field, the two soccer goals, ping-pong ball, and Lego Mindstorms robot are tracked by a webcam and simple Java software. The computer sends commands to the robot via infrared transmitter, and presto — soccer-playing bot.

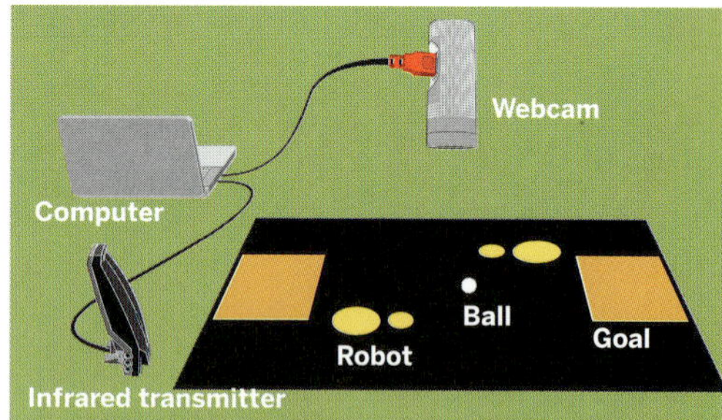

An overhead webcam above the playing field feeds an image into a computer. Software processes the image in order to identify the robot and the ball, determine the spatial relationship between the two, and calculate the robot's next moves. The software then sends the corresponding commands to the robot via infrared (IR) transmitter. The robot acts accordingly.

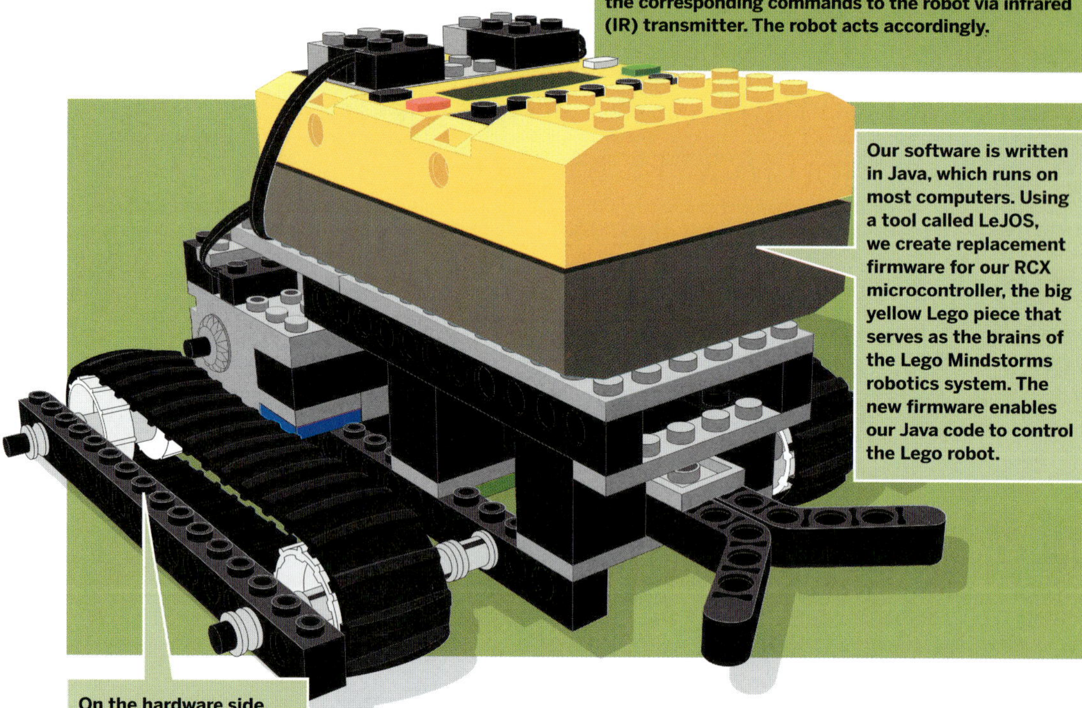

Our software is written in Java, which runs on most computers. Using a tool called LeJOS, we create replacement firmware for our RCX microcontroller, the big yellow Lego piece that serves as the brains of the Lego Mindstorms robotics system. The new firmware enables our Java code to control the Lego robot.

On the hardware side, we'll construct the playing field, mount the camera to the ceiling, and build the robot using Lego Mindstorms. The robot follows a simple design that resembles a tank. There's only so much room on the playing field, so the robot is small, and it follows the Lego design maxim of using as few pieces as possible to get the job done.

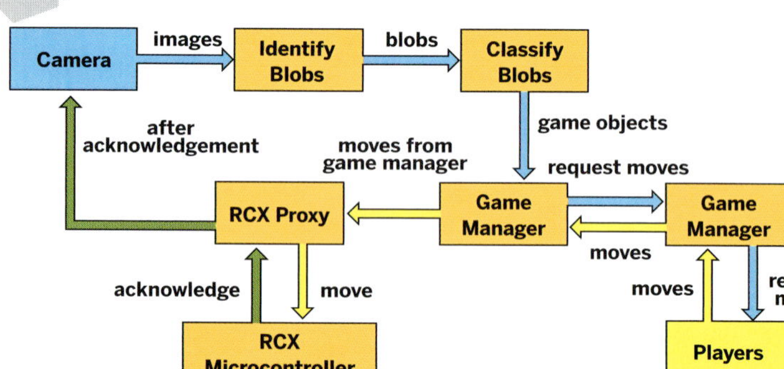

The software is where the black magic comes in. Here's the top-level loop, which continuously tracks the field and sends commands to the bot.

Illustrations by ikosy / ikosy.com

SET UP.

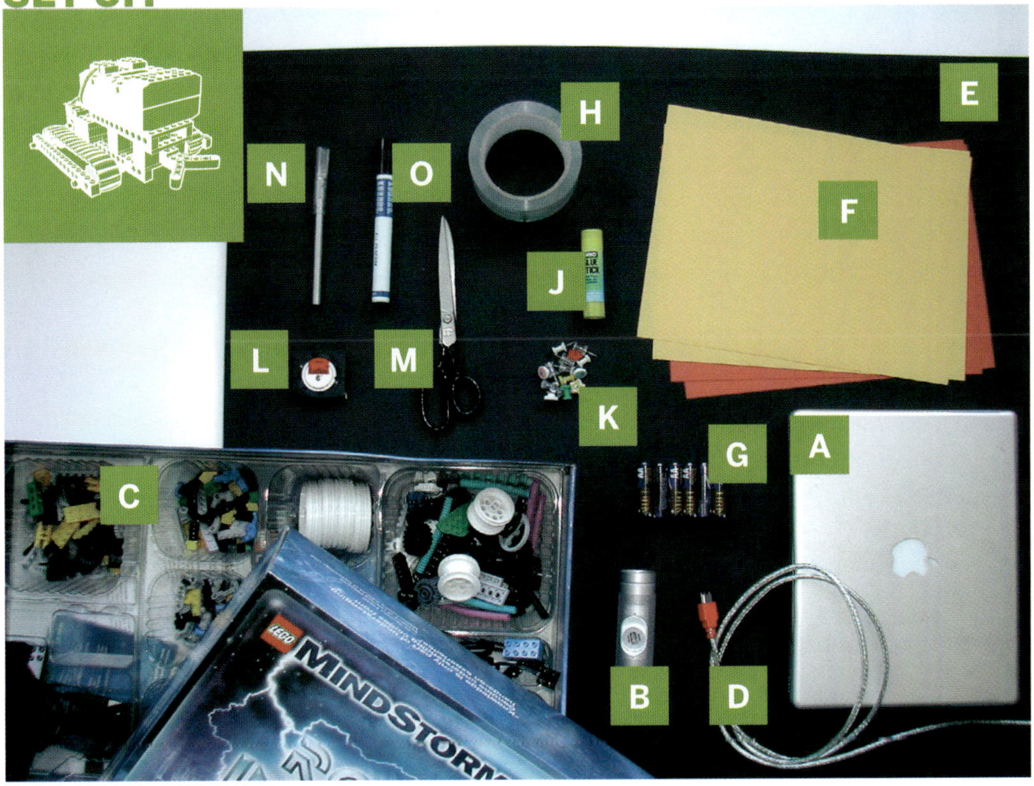

MATERIALS

[A] Computer with internet connection

[B] Apple iSight, or other webcam

[C] Lego Mindstorms Robotics Invention System (RIS) 2.0

[D] 15' FireWire cable

[E] 32"x40" black rigid-weight art boards (4), or other materials to make a flat, smooth, evenly black 3:4 ratio rectangular surface
Such as (not shown): 20"x30" white foam core boards (8), with 22"x28" black posterboards (6) to cover, or 20"x30" black foam core boards (8) — make sure it's really black, not dark gray

[F] Assorted colors of construction paper

[G] AA batteries (6)

[H] Clear packing tape

[J] Glue stick

[K] Thumbtacks

Scotch tape **(not shown)**

Ping-pong ball **(not shown)**

Velcro **(optional)**

Duct tape **(optional)**

Rubber cement **(optional, if you're covering non-black foam core)**

TOOLS

[L] Tape measure

[M] Scissors

[N] Utility knife

[O] Black marker

Drawing compass **(not shown)**

Step ladder **(not shown)**

MAKE IT.

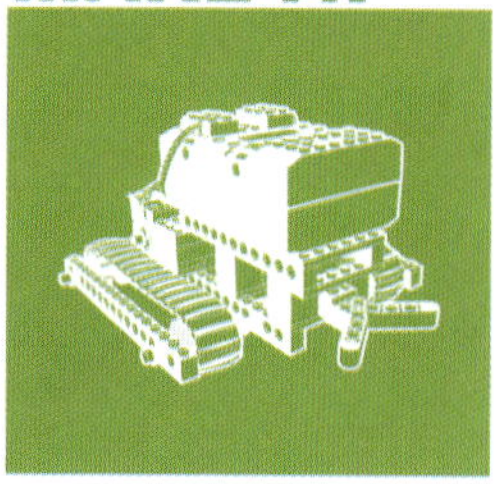

BUILD YOUR LEGO SOCCER ROBOT AND STADIUM

START **Time: 3-7 days** **Complexity: High**

1. MAKE THE PLAYING FIELD

A black field sets the stage for simple, clear image analysis by reducing shadows. Goal areas are brightly colored rectangles, and a cover for the robot uses two clearly visible circles to indicate which end is front and which is back.

1a. Cut and tape together foam core, art board, or other rigid material to make a field proportional to your webcam's aspect ratio (4:3 for the 640x480 iSight and others). This lets the camera capture maximum information. If you need to cover the field with black posterboard or paper, rubber-cement it down and touch up any non-black areas with a marker.

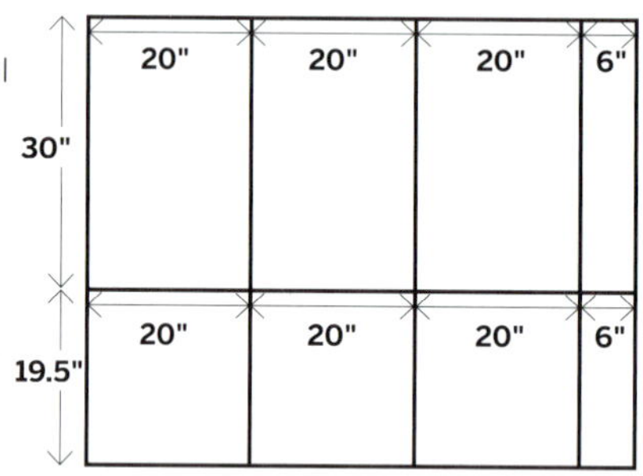

1b. Make the goal boxes. Cut two rectangles out of bright construction paper. To help the software (*see sidebar, page 97*), their width/depth ratio should be at least 1.5 or, even better, 2. Glue-stick the rectangles to opposite ends of the playing field.

1c. Make the robot cover. Pick out a different color of bright construction paper. Use a compass and scissors to cut out two circles with radii of 1" and 1.75". Glue the circles, lined up and centered, on opposite edges of a 6.5" square cut from black paper or posterboard.

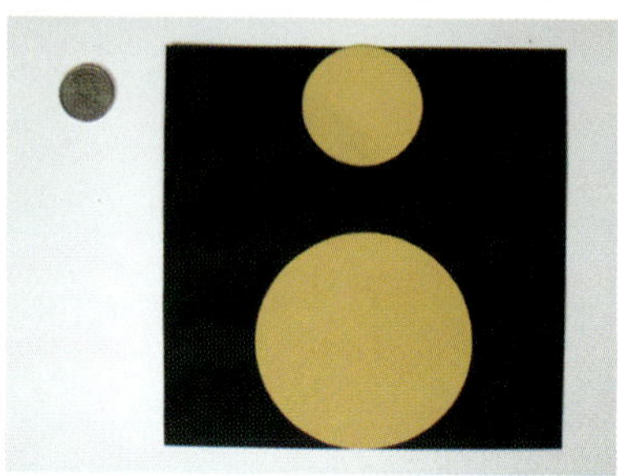

Fielding Questions

The size of field needed to fill the view of a ceiling-mounted webcam will depend on the height of your ceiling.

I used eight 20"x30" sheets of white foam core, trimmed as shown, to make a 49"x66" field. Then I covered the top with 6 pieces of 22"x28" black posterboard.

You can also use 4 sheets of 32"x40" black art board to make a 60"x80" field. Just cut 4" off the width of 2 boards.

Depending on the color of your floor, it may help to put a white border around the field, by tucking paper under the edges.

To make your field easier to store, cut it back down along the edges of the original pieces, and attach Velcro across the seams underneath so you can piece it back together.

2. INSTALL THE CAMERA

You probably don't want to drill holes in your nice ceiling to mount a webcam, so here's an easy camera bracket that combines the two greatest prototyping materials ever: Lego and tape.

2a. Figure out how to snap webcam images from your computer. I used QuickTime to control an Apple iSight. If you're not using an iSight, use another webcam that's supported by Java Media Framework (JMF) or other open source drivers. Setup details can be frustrating, but they're just one of those pains that you sometimes have to wade through.

2b. Connect 2 pairs of 6-unit and 10-unit beams.

2c. Add 2 half-bushes threaded over a 10-unit axle.

2d. Snap beams crossways on top of four 2x10 plates. Thread two 12-unit beams upright over the axle, and thread a second, 8-unit axle just above. Secure beams with 4 bushes.

2e. Thread an 8-unit axle and bushes through the 8 holes up in the beams. Place the FireWire cable plug on top, and secure it firmly in place with another axle and bushes.

2f. Thread the FireWire cable between the beams along the plates. Finish off by placing two 2x8 grey plates with holes over the low ends of the beams, and plug the camera to the cable.

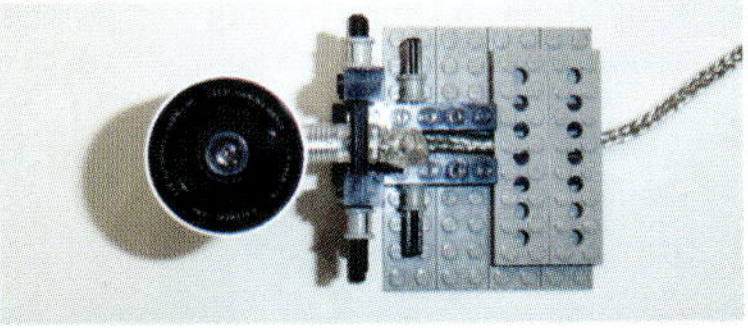

2g. Unplug the camera and secure the bracket to the ceiling with long strips of tape, reinforced with thumb-tacks so it won't peel away. Run the cable along the ceiling toward the computer, and tape it up. When you're confident your bracket is secure, connect your camera and point it directly downward.

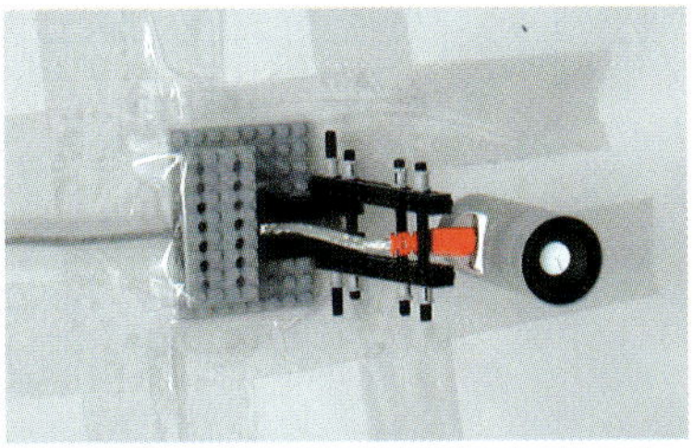

NOTE: In the legosoccer **code bundle at makezine.com/06/legosoccer, I've included a Java script named** webcampreview.sh **that uses QuickTime for previewing with the webcam. Check the *ReadMe* file for details.**

TIP: When you remove your bracket from the ceiling, use adhesive remover to clean things up.

2h. Bring up the preview pane in your webcam software. Position the field to fill the camera's field of view.

+ For advice on balancing your camera's focal length and distance, see makezine.com/06/legosoccer/focal.

3. ASSEMBLE THE ROBOT

3a. Slide 3 bushes, a wheel sprocket, and a fourth bush onto a 10-unit axle. Push them together so the fourth bush sits 1.5 bush-lengths from the end.

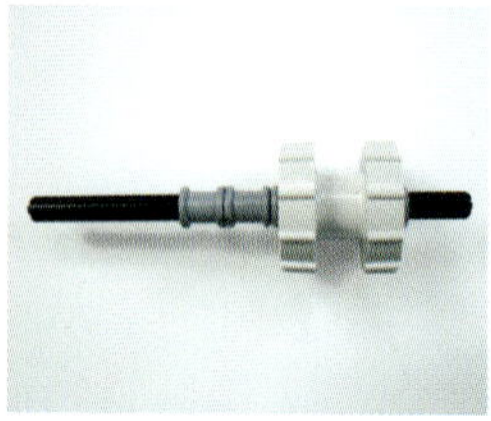

3b. Slide a bush, a wheel sprocket, a 16-tooth gear, and a 24-tooth gear onto an 8-unit axle. Push together so that the smaller gear fits inside the wheel sprocket and holds it in place.

3c. Take a 16-unit beam with the studs facing upwards. Slide the 8-unit and 10-unit axle assemblies through either end of the beam, sprocket side first, one hole in from the ends. Secure them in place with 2 half-bushes, and stretch a tank tread around the wheel sprockets.

3d. Thread a 4-unit beam and a 12-unit beam, studs pointing up, onto the other sides of the axles as shown, with the 4-unit beam on the geared axle. Secure with half-bushes.

3e. Make a mirror image of the track you just made, for the other side of the chassis. Then take four 2x8 green plates and assemble them into a square. Snap the tank tracks onto the square, with the beams staggered by 1 unit, as shown.

3f. For the robot's motor block, take two 1x2 plates with extended rails, and sandwich one corner of a 2x4 black brick, with the rails pointing outward. Place a plain 1x2 grey plate next to each, to level out the surface.

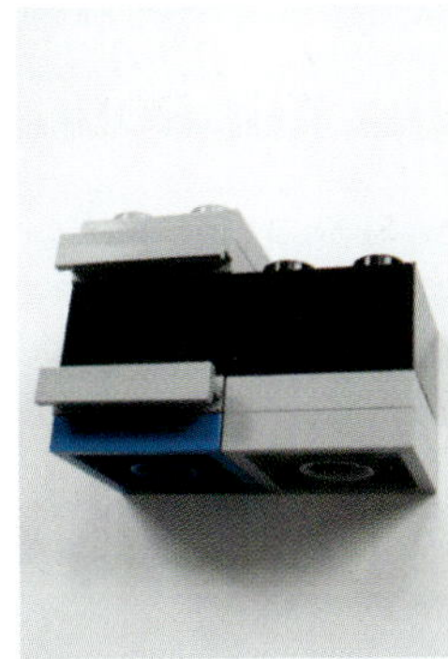

3g. Sandwich the 1x2 plates with a grey 2x2 plate above and a blue 2x2 plate below. Stack 2 more grey 2x2 plates and attach them under the black brick to level out the bottom.

3h. Slide the rail extensions into the grooves in the side of a motor piece.

3i. Repeat steps 3f to 3h, making a mirror image, for the other motor. Place the motors side by side so the gears face outward.

3j. Join the 2 motors together with a grey perforated 2x4 plate connecting the black bricks on top. Top the inside half of the perforated plate with a 1x4 gray plate stacked on a 1x4 beam, and then reinforce the join with two parallel 2x4 yellow plates, building a platform above the motors as shown. Set a 1x2 brick on each side of the yellow plates, at the rear edge.

3k. Join the motors together with two 1x4 grey plates on the underside of the block.

3l. Slip a 12-tooth bevel gear over each of the 2 axles. Set the motor block on the chassis so that the gears mesh and the back of the motor block is flush with the back of the chassis.

3m. Stack 4 pairs of 2x2 blocks onto either end of two 2x6 perforated plates. Then lay these compound beams across the chassis beams in front of the motor block, one block on the very front, and the other toward the middle. These will support the deck.

3n. Place a 1x2 plate on each of two short (6") wire connectors, and place the connectors on top of the connection spots on the motor block, with wires facing outward.

3o. Make a deck with two 12-unit beams supporting a 6x10 plate, a 1x6 plate joining the beams underneath, and 2 more 1x6 plates extending along the beams above, as shown (underside).

3p. Affix the deck onto the chassis so that it overlaps the chassis' front edge by only 1 unit. The deck's 1x6 plates should flank the yellow plates. Set a perforated 2x4 plate in the center of the yellow plates.

3q. Cover the hole with two 2x10 plates centered on top of the chassis, leaving the front 4 units on the 6x10 deck plate uncovered.

3r. For the robot's "grabber" arm, arrange two 1x9 lift arms so that they form a Y shape, and connect them with 1x2 and 2x2 grey plates.

3s. Insert AA batteries into the RCX unit. Place 2 pairs of stacked 1x2 gray plates on the RCX, just above the #1 and #3 connection slots. Attach the grabber arm to the grey 2x6 plate under the front of the robot, and secure the RCX on the chassis so that the front of the robot is flush. Finally, connect the cables from the motor block onto the RCX so that they're joined to opposite sides, with the cables wrapping around the outside as shown. The port (left) side motor connects to block C, and the starboard (right) motor connects to block A.

3t. Your robot is complete! Tape the cover on top, with the smaller, "head" circle facing forward, and make sure the grabber arm extends out past the cover, so the ping-pong ball will always be fully visible from above.

4. DOWNLOAD AND INSTALL THE SOFTWARE

4a. Install the Java SDK. If you've done any Java programming, you probably already have it installed. Confirm by typing `javac` and `java` into a terminal window. If not, download the latest from java.sun.com/j2se.

4b. Install LeJOS. Download the version specific to your platform from lejos.sourceforge.net. Read the documentation, which explains how to set up your Java classpath and other environment variables. The LeJOS site also has some good tutorials and a signup page for the lejos-discussion list, which is a great way to stay up-to-date with other LeJOS users (including me). After downloading LeJOS, install it and test your environment by opening up a prompt and typing in `lejos` and `lejosc`, to make sure they're in your path.

4c. Download, install, and run NetBeans (netbeans.org). Alternatively, you can run any other Java development environment, but NetBeans is what I used to develop the LegoSoccer code.

4d. NetBeans isn't aware of LeJOS by default, so we need to point it to the LeJOS libraries. On the Tools -> Library Manager menu item, create a new library item and add the four LeJOS JAR files located in your *$LEJOS_HOME/lib* directory.

4e. So that NetBeans can compile our RCX code, we also need to modify a couple of project files to make them aware of the LeJOS compiler. Follow the LeJOS FAQ instructions at makezine.com/go/lejos to make these changes. If you're not using NetBeans or don't feel like customizing it, you can also just compile the RCX microcontroller code manually.

4f. Download the legosoccer project source code at makezine.com/06/legosoccer.

4g. From within NetBeans, open the project files and compile each of the two project directories included, *LegoSoccer* and *RCXServer*. Be aware that by default, only the project you have set to be the "main project" will compile. So you need to select each project separately and navigate to Set Main Project for each before compiling with Build Project.

4h. Plug the Lego IR transmitter tower into your computer and position it to face onto the field.

You're done! All the pieces are in place. You've constructed a playing field for less than the cost of a few refreshments at a stadium. Your ceiling-cam is mounted and ready to provide live coverage. And a lightweight and agile player is waiting on the sidelines screaming, "Coach, put me in!" Now, let's pull out the ping-pong ball and get ready to rumble!

FINISH X

NOW GO USE IT »

LegoSoccer Software

As our software processes the playing field image, it performs three tasks: it locates the objects against the background, it identifies them and the directions they're pointing, and it figures out how the robot should move in order to get where it needs to go. Here's how the software accomplishes these steps.

Identifying Blobs in an Image

Our brains automatically translate what we see into a 3D map of the objects around us, but this is a complex process, as any AI researcher can tell you. To make it easier, we set up a black background and brightly colored objects. This lets our software scan the image until it finds a bright pixel. When it finds one, it searches the immediate area for more pixels that are above the brightness threshold, thereby discovering any large blobs of contiguous brightness. Then it continues on to the rest of the image. This is called a *depth-first search* because the program explores as far along each branch as possible before moving on.

Classifying Blobs

Now we want to know which color-blobs are the goals, which are the robot's head and tail, and which is the ball. We do this by looking at their colors and shapes. Our blob-classifying routine first determines the *bounding box* for each blob, which is the rectangular area that fully contains the blob. If the width of the bounding box equals its height, then we probably have a circle or ball, rather than one of the long rectangular goal areas. As a double-check, we then compare the area of the bounding box with the area of the blob inside. If the blob area is significantly smaller than the box, the blob is probably round, and the bounding box area includes corners that the blob doesn't fill. With a noisy source image, both of these simple tests can be fooled by a square-ish rectangle — which is why the goal box rectangles need to have width/depth ratios of at least 1.5.

Getting Where You Wanna Go

Now we need to tell the robot what to do. First we determine the robot's location as the midway point between the two circle's centers, and its heading as the angle formed by the two centers and the x-axis:

```
slope = (y.tail - y.head) / (x.tail - x.head)
```

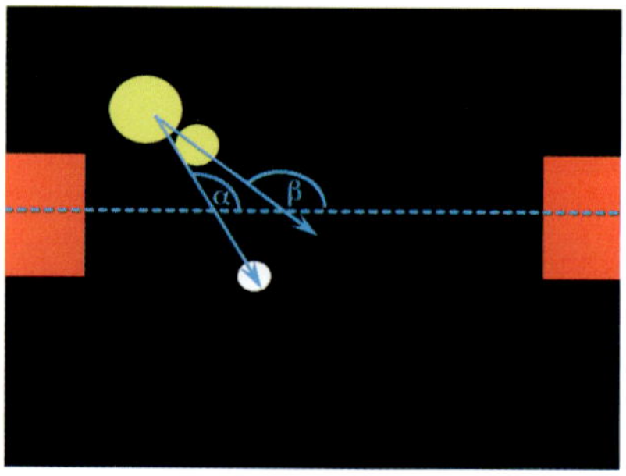

The robot's angle to the ball (α) is the arctangent of the slope between robot and ball. The difference between this angle (α) and the robot's current heading (β) determines how far it should turn.

Using calculations like this, you can get the robot to hit the ball into the goal, avoid collisions, and do other tricks. The action isn't fast-paced, since it takes time to pass the infrared and process the image. But we're stretching the limits of Lego here.

Our software defines a Java class with methods for each of these low-level behaviors. These are called, in turn, by simple shell scripts you can invoke from the command line:

gotoball.sh, gotogoal.sh, **and** goalshot.sh

USE IT.

LIGHTS, CAMERA, ROBOT

LEARN TO LET GO

You can remote-control your robot manually from the computer or run it on automatic using our software. But, this being the real world, performance can be imperfect either way. Like, when your robot receives the instruction to turn 90 degrees, it may turn 80, or 100. Be prepared for some troubleshooting. If a hardware glitch knocks out the IR tower, you'll need to restart, and if the IR tower loses contact when the robot is facing away, set up a shiny, reflective surface on the opposite side. If the ball keeps rolling away, put a dent in it and let it slide.

REMOTE-CONTROL PLAY

1. Place a ping-pong ball on the field. (For best results, scrub off any printing with a cleaning pad.)
2. In the `legosoccer` source code, run the `remote control.sh` script. The GUI controls should soon appear on your screen, or you can use the keyboard shortcuts: W=forward, A=left, S=right, Z=backward.
3. Remote-control your robot, trying to put a ball in the goal. Remember: "Slow is fast, and fast is slow." In other words, take your time, plan your moves, and you'll get a feel for how to go where you want.
4. With two robots, you can duke it out with a `legosoccer`-enabled friend. Just make sure you assign unique IDs to your robots in the source code. Otherwise, the robots will receive and process the same commands, and act like copycats.

AUTONOMOUS PLAY

Getting your robot to work autonomously will likely require some environment-tweaking, including adjusting the lighting and trying different colors for the paper dots and rectangles. Also, make sure to keep the playing field clean and uniform; scratches, glue spots, and other glitches can confuse the software. Check the *ReadMe* files in the `legosoccer` source code for some tips on easing the frustrations.

1. Run the script `servertest.sh` from a terminal window, and follow any diagnostic advice it spits out. If all goes well, you'll see a snapshot of the field (below) with bounding boxes around all of the objects. Use this to center the field in the camera's field of view.

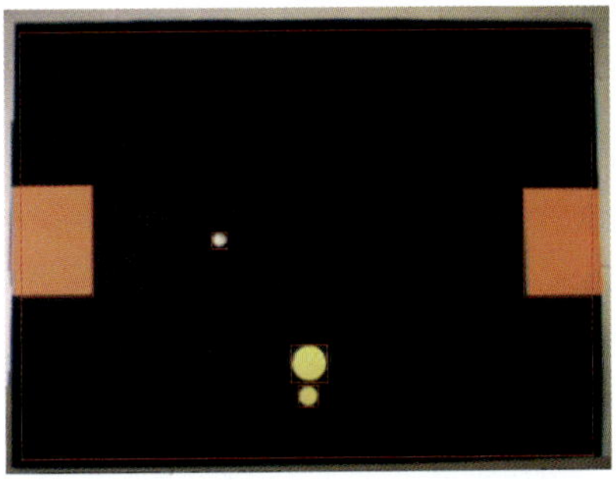

2. Run the script `gotogoal.sh`. You'll see the field again, some more diagnostics, and if all goes well, your player should turn and head into a goal box. Place your player anywhere on the field, run `gotogoal.sh`, and watch it navigate over to the goal.
3. In addition to `gotogoal.sh`, there are two more action scripts: `gotoball.sh` makes the robot hit the ball from wherever it's standing, and `goalshot.sh` makes it line up behind the ball, facing the goal, then move forward and hit it in. He shoots, he scores!

You can add as many other behavior scripts as you can write. You'll have to pay some coding dues, but the voodoo magic you'll see as a result is worth it.

RESOURCES

Core Lego Mindstorms Programming, by Brian Bagnall — Excellent introduction to Java programming, Lego Mindstorms, and LeJOS, by the creator of LeJOS.

Vehicles, by Valentino Braitenberg — Classic, thought-provoking text on designing low-level robot behavior.

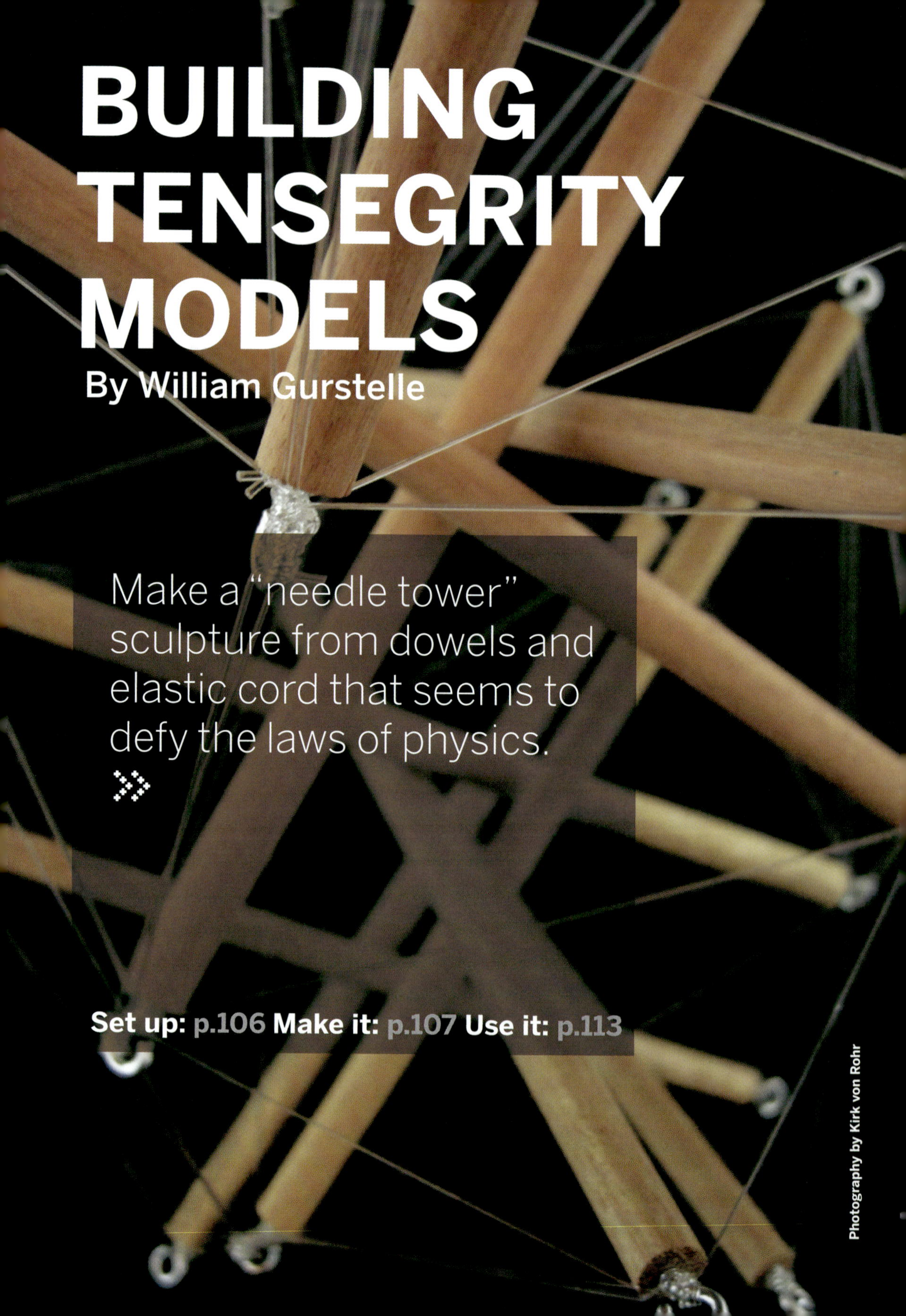

BUILDING TENSEGRITY MODELS

By William Gurstelle

Make a "needle tower" sculpture from dowels and elastic cord that seems to defy the laws of physics.

Set up: p.106 **Make it:** p.107 **Use it:** p.113

Photography by Kirk von Rohr

HOW HIGH CAN YOU GO?

One of the most striking sculptures on display at the Smithsonian's Hirshhorn Museum in Washington, D.C., is the soaring, 60-foot-tall *Needle Tower*. It's made of a latticework of aluminum tubes held together by crossed stainless steel wires under tension.

Sculptor Kenneth Snelson's creation is a fragile-looking construction. Common sense tells you it ought to fall over or implode. Yet it's been standing strong and upright since 1968, unperturbed by storms and wind or even the occasional renegade urban climber.

The rods that make up the bulk of the tower float in midair, held together only by wires that are not guyed or anchored to the ground; they are not attached to anything but the compression rods themselves. It forces you to think: just what is pushing on what? How does it stand?

William Gurstelle is a frequent MAKE contributor. His new book, *Adventures from the Technology Underground*, was published in January.

MEET THE TENDULE

Tensegrity describes structures composed of three or more rods held in place by a network of tension wires. No rod may touch another rod.

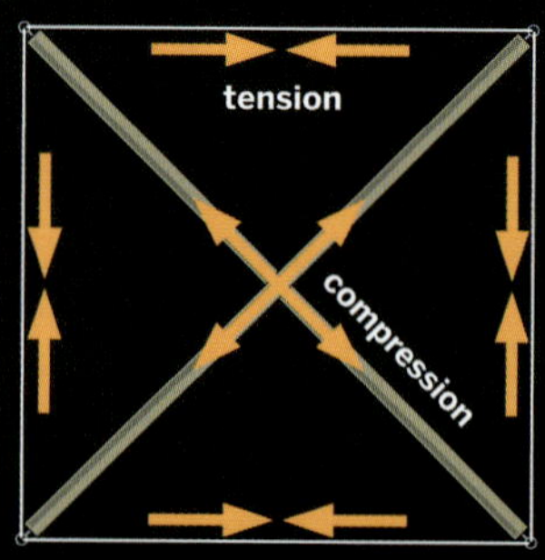

Fig. 1

In its most basic form, a tensegrity structure is simply a construction made from two crossed rods, held in place with a tightened cord around its perimeter (Figure 1). It's easy to see that if the cord is in tension, then the rods are under compression. But this kite shape is actually a quasi-tensegrity, because the crossed rods touch each other in the middle. In a true tensegrity, no compression members make contact.

It's a only a small step from the kite frame to the simplest true tensegrity structure, a basic, versatile, and elegant structure which I call a tendule (Figure 2).

Tendules have three compression members that form a stable and (unlike the kite shape) three-dimensional building block. This is the basic building unit of Snelson's *Needle Tower* and a host of other tensegrity structures.

The three essential components of a tensegrity tendule are as follows:

1. The compression members (wooden dowels)

2. The tension members (stretchy cords), which consist of:

A. The cord triangle loops: The two tension members in every tendule that connect the three compression members in the horizontal plane.

B. The vertical cords: The three tension members in every tendule that connect the top of each compression member to a corresponding bottom in an adjacent compression member.

3. The tension-compression interface (eye screws)

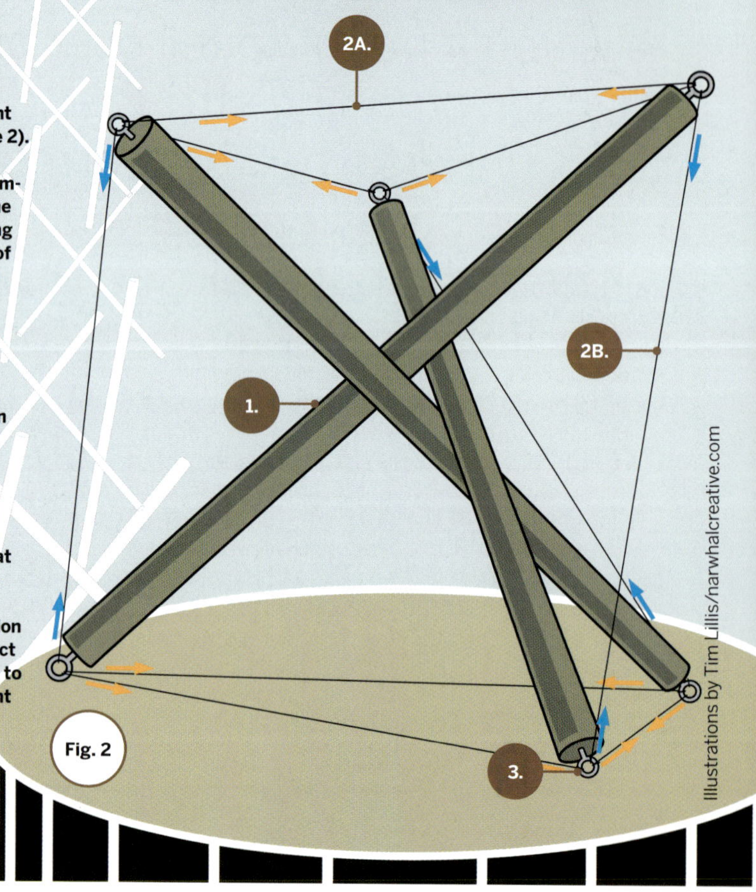

Fig. 2

Illustrations by Tim Lillis/narwhalcreative.com

A SHORT HISTORY OF FLOATING COMPRESSION STRUCTURES

The key to understanding the *Needle Tower* and a number of other Snelson works is a principle popularly referred to as *tensegrity*, a combination of *tension* and *integrity*. This is a neologism coined by Buckminster Fuller, a man who loved to create new words by truncating and fusing others: *dymaxion* (*dynamic maximum tension*), *synergetics* (*synergetic geometry*), and *sunsight* (*sun* and *first sight*).

Photography detail courtesy of the Hirshhorn Museum and Sculpture Garden, Smithsonian Institution, Gift of Joseph H. Hirshhorn, 1974. Image by Ricardo Blanc

Origins of Tensegrity

Some writers claim the notion of tensegrity was first put forward by Snelson in 1948, when he described a technique he used in building artistic constructions as "floating compression." Snelson shared his ideas with Fuller, who quickly saw both novelty and practicality in the idea. An expert at providing interesting names to concepts, he reworked and then renamed this generalized idea of structures held together by compression members, which are held in place by flexible wires in tension.

When Fuller coined and began popularizing the term *tensegrity* in 1961, apparently he did so without Snelson's consent. In fact, there is something of a controversy among scientists, artists, and architects as to who rightfully deserves more of the credit for devising the concept. Snelson claims quite vociferously that the famous and powerful Fuller more or less usurped his idea and denied him recognition, while Fullerites scoff at the idea that anybody but Bucky deserves credit.

Yet others maintain that the idea predates both of them, and belongs either to an obscure Latvian artist named Karl Loganson, who some say developed the idea in the early 1920s, or to French architect David Georges Emmerich, who began exploring and refining the basic building structures of tensegrity in the 1950s.

Tensegrity's true provenance is difficult to ascertain given the number of competing claims and the documentation on the subject. Many, if not all, of these erstwhile creators likely had a hand in developing and extending the concept. Almost certainly, each advanced the idea in important ways. Emmerich, Fuller, and Snelson each filed patent claims on differing aspects of the technology in the 1960s.

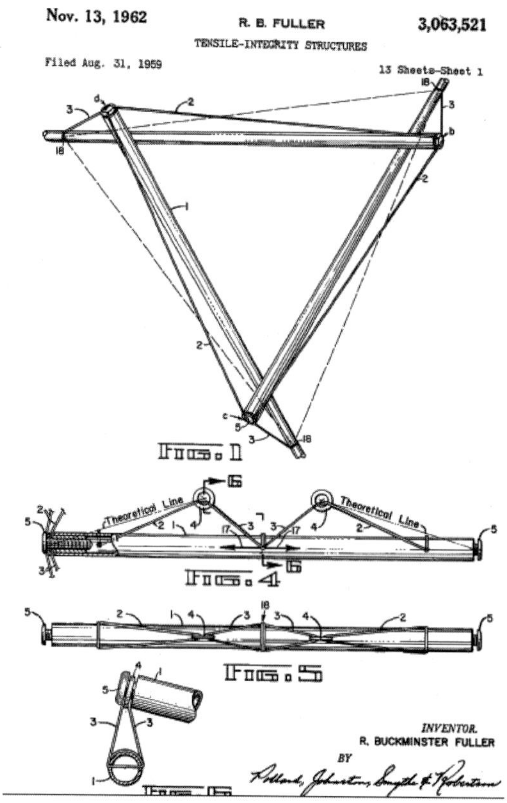

Now, it appears that each man had a much different opinion on the highest and best use of this idea. Fuller's primary interest in the concept was its application toward engineering — specifically, the spherical and dome-like enclosures (i.e., geodesic structures) for which he became famous. Emmerich, the architect, saw tensegrity's value residing primarily in its utility as a building construction technique. And Kenneth Snelson, the artist, primarily sees tensegrity's value in an aesthetic sculptural context.

Irrespective of who actually first developed the concepts and the principles behind the building technique, tensegrity structures are fun to make and interesting to look at.

Toward a Floating Tower

Ever want to make something that is cool simply for its own sake? The English might term such a project as "folly" — an eccentric sort of activity without much practical value, but valuable because of its interesting and out-of-the-ordinary nature. It's the sort of unusual, impractical, but enthralling project that makers seek out. If you're a regular reader of this magazine, you know what I'm talking about. This is one of those projects.

Be aware, the project that follows can lead to a pleasant sort of addiction. Once you master the techniques shown here, you may find yourself building a great number of tensegrity structures. Also, be forewarned that the first time you build one, it will be frustrating. Cords will break, the modules that make up the structure may be uneven and tilted, the towers will lean or list, or worse, come crashing down in a heap. Eventually, the technique will make sense. The materials are somewhat forgiving, and misalignments can often be corrected by gentle massage and your force of concentration.

A Few Background Notes

I first came across the idea of tensegrity in a 1996 edition of *The Physics Teacher* magazine. The author of that article, a University of Michigan faculty member, described his encounter with Snelson's *Needle Tower* at the Hirshhorn Museum as an experience that left him puzzled and intrigued. The author provided directions and sketches to enable the magazine's readers to replicate on a smaller scale the fragile-looking, semi-rigid column that seems to float on air.

When I tried to make my own column, I was less than successful. I followed the instructions closely, but it was apparent that the technique described in the article required either the manual dexterity of a Flying Karamazov chainsaw juggler or a prehensile tail to act as a third hand. Lacking both, the structure I tried to build invariably collapsed, sometimes slowly, sometimes quite spectacularly, and often just on the brink of success.

But, perseverance is the mother of good luck, and I eventually came up with an easy-to-replicate modeling technique that works quite well, even for people limited to just ten fingers.

Building Tensegrity Structures

In Fuller's 1962 U.S. patent application, tensegrity is described as "a plurality of discontinuous compression columns arranged in groups of three nonconjunctive columns connected by tension elements forming tension triangles." Huh? No wonder tensegrity sculptures seem hard to make.

Illustration from the U.S. Patent & Trademark Office website

Buckminster Fuller built this geodesic dome in 1962. He developed some of his ideas about tensegrity after meeting a young Kenneth Snelson in the late 1940s. A fascinating letter written by Snelson and printed in the International Journal of Space Structures in 1990 is reprinted here: grunch.net/snelson/rmoto.html.

Photograph courtesy, The Estate of R. Buckminster Fuller

Happily, a lot of other people, from artists to scientists, have developed their own definitions:

> *Tensegrity is a structure composed of three or more rods held in place by a network of tension wires. No rod may touch another rod. Rods may be supported in place solely by the tension wires to form a firm, prestressed, and semi-rigid structural unit.*

This description is somewhat more user friendly. It provides a set of conditions that can be used to determine whether a structure uses tensegrity construction techniques or not. But after you work with it a while, tensegrity becomes easy to identify, even if it remains difficult to describe in words. To paraphrase Supreme Court Justice Potter Stewart's 1964 quote, "I shall not attempt further to define this kind of material, but I know it when I see it."

Floating Compression Modeling

Tensegrity models can be made from inexpensive materials such as hardwood dowels and elastic cord, or even just rubber bands and pencils. More elaborate models occasionally are made using brass rods or steel shafts.

All tensegrity structures described in this article are built up from basic building blocks called tension modules (which, to be Fullerian about it, I call *tendules*). A tendule is composed of three rods that support compressive loads and five triangulated wires or bands that handle tension loads. When arranged in a specific style of latticework, the rods and wires form stable building blocks that can be turned into towers, bridges, and other, even more elaborate space-enclosing structures.

To take on larger shapes, the tendules rely on a building principle called *floating compression*. None of the rods touch one another, and loads are transmitted from rod to rod through a matrix of tensioned wires. Because the wires are very thin and the rods are relatively thick and widely separated in space, the overall visual effect is that of unsupported struts or poles floating in air — a most unusual and novel spatial aesthetic.

Building a Tensegrity Tower

The instructions in the following section detail the materials and techniques used in building a model of Snelson's *Needle Tower*. The model is four tiers or stories high, although fewer or more tiers can be incorporated according to the wishes of the builder. The original *Needle Tower* at the Hirshhorn Museum has more than 15 tiers.

SET UP.

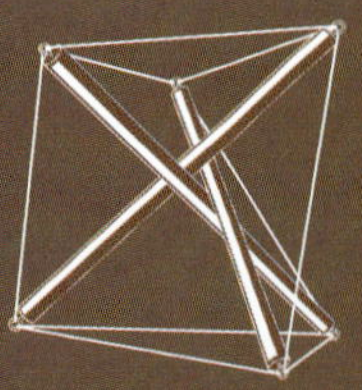

MATERIALS

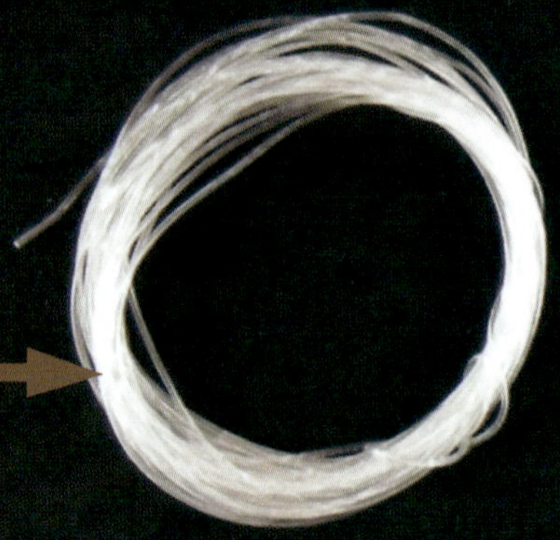

5m-long spool of 0.8mm to 1.0mm diameter clear plastic stretch cord Pepperel Braiding Company's Stretch Magic cord worked for me. Available at most crafts and beading stores.

36" long, 5⁄16" diameter wooden dowels (3) Other diameters will work satisfactorily, but I find 5⁄16" diameter to be the most aesthetically pleasing.

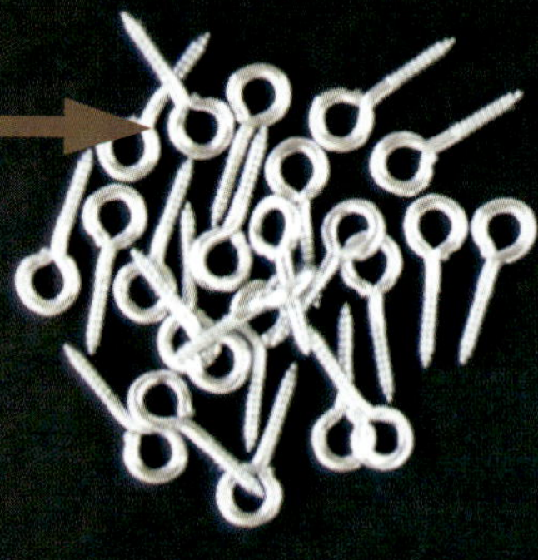

#214 eye screws (24)

Box of assorted medium-sized rubber bands

Masking or electrical tape (optional)

Spray paint (optional)

Wooden or tile base (optional)

TOOLS

Handsaw

Standard or locking pliers

Small regular screwdriver

Electric drill

5⁄64" drill bit

Scissors

MAKE IT.

CONSTRUCT A TENSEGRITY TOWER

START **Time: A Day Complexity: Low**

1. FABRICATE THE COMPRESSION MEMBERS

In tensegrity structures, the tension (pulling) stresses and the compressive (pushing) stresses are completely separated. The structure's compression loads are carried by rigid wooden dowels.

1a. Cut the wooden dowels to 8" lengths using the handsaw.

1b. Open up the end of the eye screws slightly with the pliers and a small screwdriver or needlenose pliers.

1c. Drill a 5/64" hole in the center of each end of each dowel, deep enough to accommodate the stem of the eye screw.

TIP: Roll a width of masking or electrical tape around each end of each dowel to prevent end splits.

1d. Insert eye screws into holes. There are 3 dowels required per story. A 4-story tower requires 12 dowels.

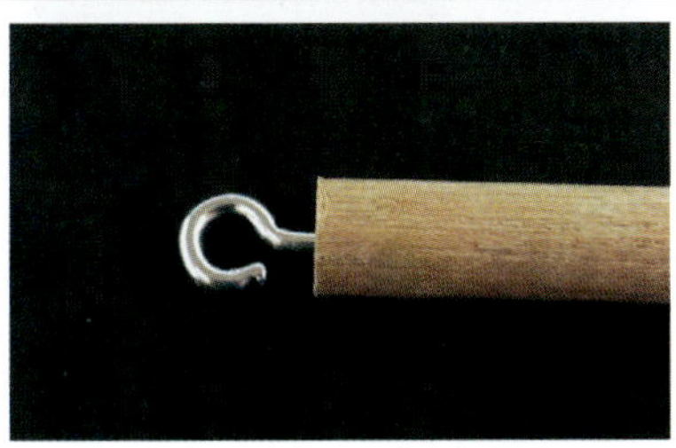

2. FABRICATE THE CORDS

Tension in the structure is provided by loops of stretchy plastic cord. The cord packaging should carry instructions on how to tie the cord into loops. If there are no instructions, try either a square knot or a double overhand knot, pulled as tight as possible.

2a. Make 8 loops from stretchy cord pieces, each piece being 15" long. Making the knot will consume an inch or so of cord, so the total length of the cord in the loop will be slightly less than 15". Because the cord is stretchy, there is some wiggle room in making the loops.

2b. Make 9 loops from cord pieces cut 8" long.

2c. Make 12 loops from cord pieces cut 10" long.

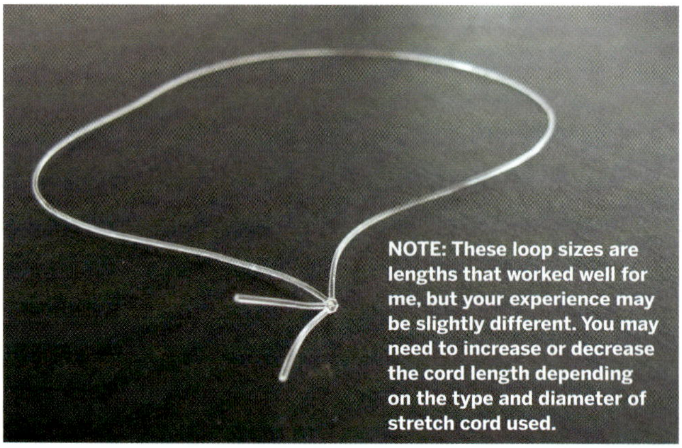

NOTE: These loop sizes are lengths that worked well for me, but your experience may be slightly different. You may need to increase or decrease the cord length depending on the type and diameter of stretch cord used.

3. CONSTRUCT THE WEAVE TRIANGLES

Tendules begin as a set of dowels arranged in groups of 3 to form preliminary weave triangles. There will be 2 left and 2 right weave triangles.

3a. Place a pencil mark at a point 3" from each end of each dowel. Align dowel pencil marks to form a left weave triangle. Tightly wrap a rubber band around the pencil marks to hold the triangle together. Repeat for the second left weave triangle.

3b. Now align the dowels to form a right weave triangle, pencil mark to pencil mark, and wrap it tightly with rubber bands. Repeat for the second right weave triangle. You now have 4 flattened weave triangles, 2 left-handed and 2 right-handed. (You can make more later, if you want to build a higher tower.)

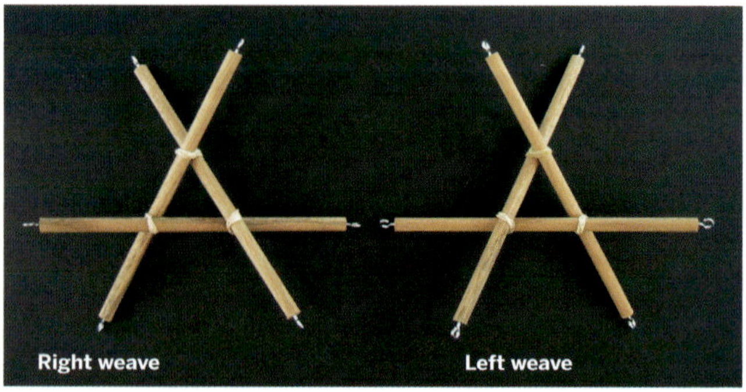

Right weave Left weave

NOTE: The rubber bands are only temporary. You'll cut them off later.

4. FORM THE CORD TRIANGLES

The cord triangles are the triangles formed by the stretchy cords. In the photo below, weave triangles are the triangles formed by the wooden dowels, and the cord triangles are the triangles colored red and blue.

4a. Attach the 15" cord to the top set of opened eye screws on a weave triangle (as shown in blue). Then attach another 15" cord to the bottom set of eye screws on the triangle (as shown in red). Note how the top and bottom stretch cords form 2 cord triangles.

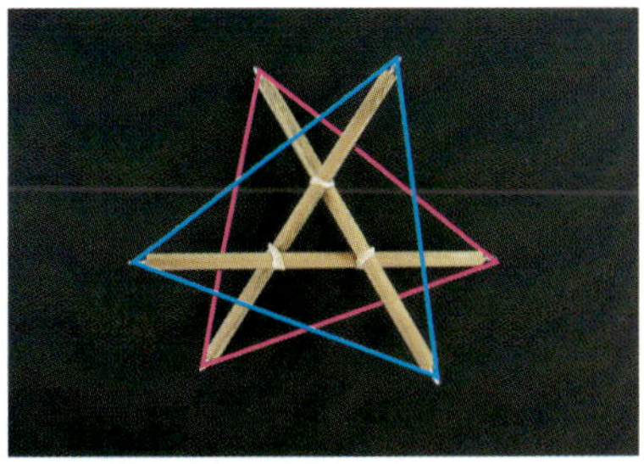

4b. The best way to form the cord triangles is to loop the cord once around the base of each eye screw. The cord should be stretched tightly but evenly in each cord triangle; that is, each third of the triangle should have roughly the same tension. You can easily check the tension by plucking each segment of the stretched cord triangle with your fingernail and listening to the resulting pitch. When the plucked pitches are all equal, the tension in each segment is equal as well.

4c. Adjust the tension by wrapping extra turns of the cord around the eye screw post until the tension in all 3 segments of each cord triangle is the same. Although the plastic cord is applying lateral tension to the weave triangle, the rubber bands should prevent it from moving. If the weave triangle shifts or is unstable, tighten the rubber bands to prevent movement.

5. ATTACH THE VERTICAL CORDS

5a. Attach one of the 10" long cord loops to each top weave triangle eye screw and to the eye screw on the bottom triangle closest to it. The vertical cord connecting the eye screws (top to bottom) will be very loose at this point. Initially, you may need to loop the vertical cord once or twice around the eye screw base to keep it from slipping off.

6. FORM THE TENDULES

Tendules are the basic tensegrity building blocks. A single tendule is a compression-tension structure that is freestanding and stable. Larger structures are made by linking one tendule to another.

6a. Use scissors to carefully cut the rubber bands. When the last band is cut, the tension within the upper and lower triangle cords will lift up the structure to form a 3D shape, while the tension in the vertical connecting cords will constrain the structure from moving too far. The result is a box-like shape. Finding the correct balance between the lateral (cord triangle) tension members and the vertical tension members aligns the compression members into semi-rigid, 3D tendules.

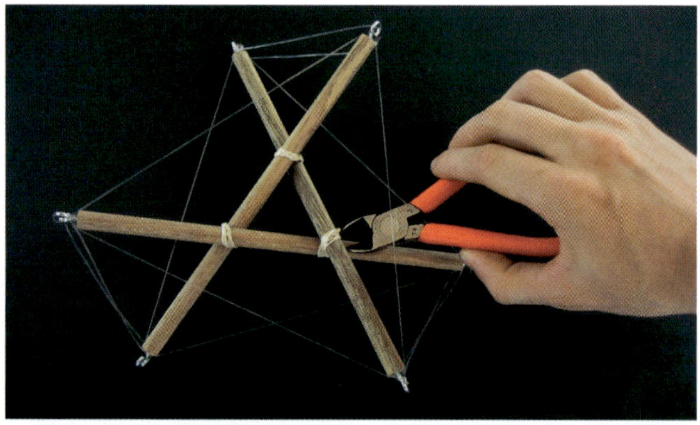

6b. Fluff and adjust the tendule by looping the plastic cords over the eye screws until it forms a box-like shape. The top and bottom triangle tension members are in blue, and the vertical connecting cords are in white.

6c. If the tendule droops, loop the top and bottom triangle cords additional times around the eye screw post while simultaneously reducing the tension in the vertical connectors. I've found that the hardest part of making a tensegrity tower is learning how to properly adjust the tension ("tuning") in the plastic cord loops. Each adjustment affects all other tension members — it's a sort of mechanical sudoku puzzle for your hands — so considerable tuning is required.

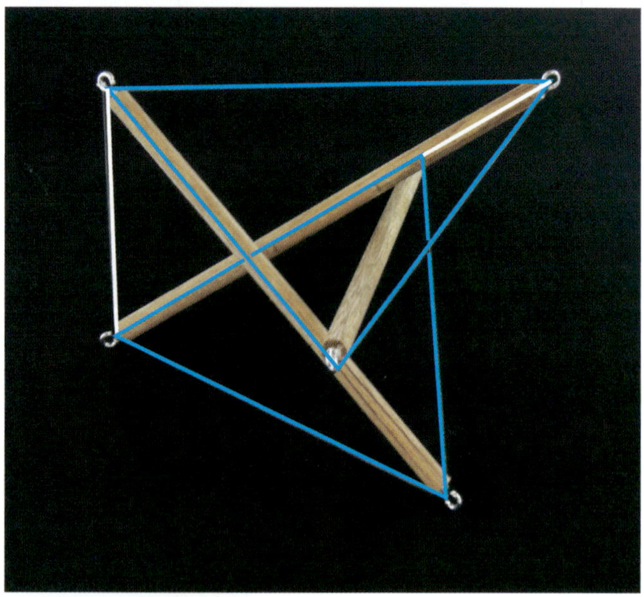

6d. Start by making single loops over the eye screws. Go easy at first on the number of loops, because making too many loops too quickly will cause over-tightening in one plane of the tendule and misalignment in the others.

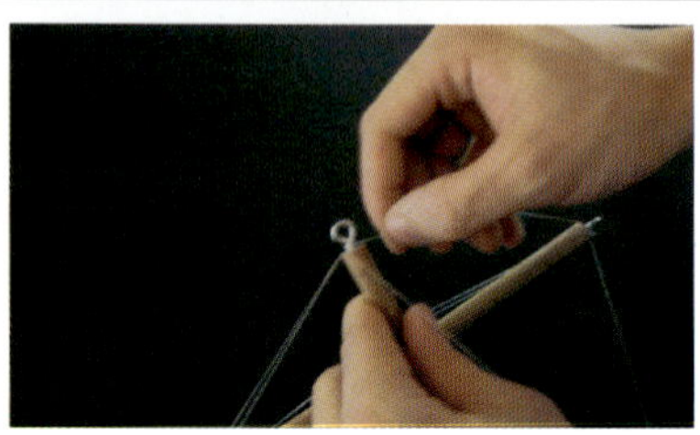

6e. Check the tension by plucking the cord after each eye screw loop adjustment until the tension is evenly distributed and the desired boxy shape is attained.

6f. Once you've figured out how to manipulate the tension in the triangle to form an evenly aligned tendule, make 3 more.

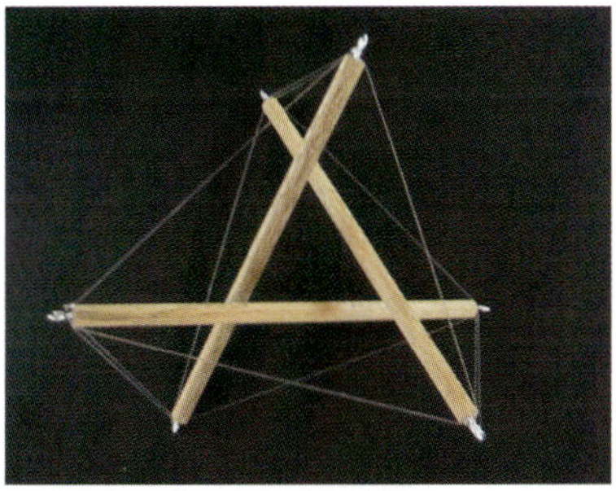

7. ERECT THE TOWER

The tensegrity tower is erected by placing one tendule on top of another, alternating left-handed and right-handed tendules as you go. Each tendule is fixed to the ones above and below by triangulating. This means that tension members (additional stretchy cords) are affixed to form stability triangles that connect stories together.

7a. Place one tendule on the base. Set a tendule of the opposite handedness on top of the base tendule. Rotate the upper tendule so that its eye screw touches the triangle cord on the base tendule halfway between eye screws. Loop the upper tendule's eye screws around the triangle cord (colored red in the photo) several times. Adjust tension in the tower cords by looping until the tension in all cords is equal.

7b. Repeat for the third and fourth tendule in the tower, alternating the handedness of the tendule each time.

7c. Once all the tendules are stacked, it is time to triangulate the tower to make it rigid. Do this by connecting the 9 remaining 8" loops to form stability triangles between each row. Complete the stability triangles by extending the 8" loop from each bottom eye screw to the bottom eye screw directly below it.

7d. The colored lines in this photo show how each stability triangle is formed between the ends of 2 compression members in the lower tendule and the end of the descending compression member in the tendule above.

7e. Adjust all cords so that the tension in each tendule is even and the cords that connect the stacked tendules are evenly distributed. Again, test by plucking and listening. Add or subtract loops around the eye screws to even up tension in the structure.

7f. After adjustments are complete, the tensegrity tower should be stable and level.

USE IT.

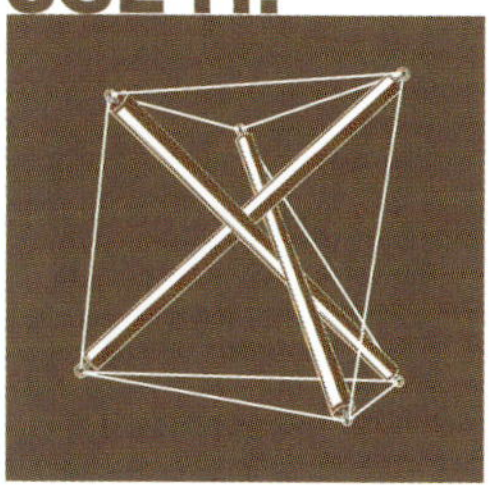

TENSEGRIFY YOUR WORLD

HINTS AND TIPS:

1. Often, much fluffing and tension adjusting is required to get the tendules to stand squarely. Take as much time as necessary to get this right.

2. Once you learn the basic technique, you can experiment with unequal-length compression members and variable tensions to obtain interesting shapes. Interesting variations include: square and five-sided weaves, using tendules whose compression members are all longer or shorter than the members in the adjoining tendule, or using compression members of varying sizes in the same tendule.

Fuller vs. Snelson on the Definition of Tensegrity

"The word 'tensegrity' is an invention: a contraction of 'tensional integrity.' Tensegrity describes a structural-relationship principle in which structural shape is guaranteed by the finitely closed, comprehensively continuous, tensional behaviors of the system and not by the discontinuous and exclusively local compressional member behaviors. Tensegrity provides the ability to yield increasingly without ultimately breaking or coming asunder."

—Buckminster Fuller
(from *Synergetics*, p. 372)

"Tensegrity describes a closed structural system composed of a set of three or more elongate compression struts within a network of tension tendons, the combined parts mutually supportive in such a way that the struts do not touch one another, but press outwardly against nodal points in the tension network to form a firm, triangulated, prestressed, tension and compression unit."

—Kenneth Snelson
(kennethsnelson.net/faqs/faq.htm)

FOR FURTHER STUDY

Tensengrity is a fascinating topic and there is much material available for further research.

Books:

An Introduction to Tensegrity by Anthony Pugh, University of California Press, 1976.

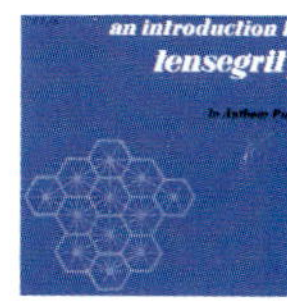

Engineering a New Architecture by Tony Robbin, Yale University Press, 1996.

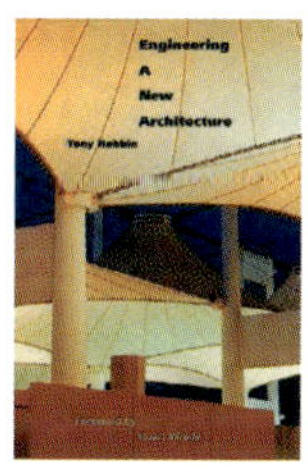

Websites:

Kenneth Snelson: kennethsnelson.net

Buckminster Fuller: bfi.org

A Practical Guide to Tensegrity Design (a PDF file): angelfire.com/ma4/bob_wb/tenseg.pdf

Other:

Images and Community: tensegrity.com

How to Build Tensegrity Structures out of Soda Straws: georgehart.com/virtual-polyhedra/straw-tensegrity.html

The Hirshhorn Museum: hirshhorn.si.edu

Tensegrity in a Cell: makezine.com/go/cell

OWN YOUR OWN

Tower PC

Photograph by Robert Bruce Thompson

The standard tower-style PC is blessedly modular, designed to be opened up and upgraded.

By Robert Bruce Thompson

Tower PCs are endlessly modifiable. Most tower PC upgrades take only a few minutes and require no more than a Phillips screwdriver. At relatively small cost, you can increase your system's performance, improve reliability, reduce noise, extend life, and add features. Here, from the recent book, *Repairing & Upgrading Your PC* (O'Reilly), are the ten most popular PC upgrades and how you make them.

A CPU Extend the useful life of your system and dramatically increase performance with a faster CPU. Check documentation to verify which CPU models are compatible with your motherboard.

• **Remove the CPU cooler fan, lift the ZIF lever to release the old CPU, and remove it from the socket. Drop the new CPU into the socket, re-latch the ZIF lever, apply thermal compound, and install a new fan. ($50 to $150)**

B CPU Cooler Replace this to reduce noise and CPU stress. Choose a premium, quiet fan from Arctic Cooling, Thermalright, or Zalman. Verify that it's compatible with your CPU and motherboard. • **Mounting methods vary, so follow the directions included with the cooler you choose. Make sure to apply thermal compound per the instructions. ($15 to $50)**

C Case Fans Add one or more fans (or replace current fans) to improve cooling, increase system stability, and reduce noise. Choose the largest fans for which your case provides mounting locations, favoring quiet models from Panaflo, Nexus, Arctic Cooling, and others. Models with variable speed settings let you balance cooling levels against noise. • **Install the fan using the four provided screws or flexible mounting studs, and connect the power lead to a motherboard fan power header. ($5 to $15)**

D Optical Drive Replace a CD/DVD-ROM drive or CD writer with a DVD±R/RW writer for fast, high-capacity backups and DVD duping. Pick a standard (parallel) ATA model that supports at least 16X DVD±R, 4X DVD±RW, and 4X DVD±R/DL dual-layer, such as the Plextor PX-716A or PX-740A, NEC ND-3540A, or BenQ DW-1640. • **Detach cables from old drive, unscrew four screws, and remove. Reverse the process to install new drive, making sure it's jumpered the same (master or slave) as the old one. ($40 to $100)**

E Hard Drive Install a second hard drive to store audio and video files, photos, and other personal data. Choose a 7,200 rpm ATA or SATA model, such as a Seagate Barracuda 7200.9, from 40GB to 500GB, depending on your needs and budget. • **First, verify that the master/slave jumper is set correctly (required for ATA models only). Then slide the drive into a drive bay and secure it with the four screws provided. Connect the power and data cables, restart the system, and format the drive. ($50 to $375)**

F Power Supply An ATX 2.0 power supply can support additional peripherals, improve cooling, reduce noise, and increase system stability. Choose a premium, noise-reduced model with a wattage rating at least 50% higher than your current unit. • **Disconnect power cables from the motherboard and drives, then unscrew four screws to remove power supply. Reverse the process to install the new unit, making sure to reconnect all of the cables. ($50 to $125)**

G Memory Add memory to improve system performance and stability. Consider 512MB a minimum for a lightly loaded Windows XP system, and 1GB for a mainstream system. Buy only name-brand memory, such as Crucial or Kingston, and use their online configurator/advisor to find modules that will match your system. If your motherboard supports dual-channel memory operation, install modules in pairs. • **Press new modules into empty memory sockets, making sure they seat completely, and that the socket retaining arms snap into place. ($25 to $100)**

H Video Card (not shown) Replace your current video card or integrated video adapter with an AGP or PCI Express video card to improve 3D gaming performance, support a second display, or add a digital interface (DVI) for a digital-only LCD display. • **Remove the old video card (if any), align the new video card with the AGP or PCI Express slot, and press the card firmly into place until it seats fully. ($25 to $500)**

I Sound Card (not shown) Replace your current sound card or integrated audio adapter with a PCI sound card to add hardware-accelerated audio for gaming and surround-sound support. Choose a SoundBlaster Audigy-series card for the best gaming audio, or an M-Audio Revolution card for both excellent gaming support and superb sound quality. • **Remove the old sound card (if any), align the new sound card with an available PCI slot, and press the card firmly into place until it seats fully. Follow the card's installation directions carefully, noting how and when to remove the old audio drivers and install the new ones. ($50 to $150)**

J Motherboard This Big Kahuna upgrade brings your system up to current standards. Choose a motherboard that fits your case and is compatible with the CPU and memory you plan to use. If you also replace the CPU and memory, you essentially have a whole new computer at a cost of $125 to $350 and an hour's work. • **Disconnect all cables from the old motherboard, remove mounting screws, and lift it out of the case. Attach stand-offs at each location required by the new board; install CPU, CPU fan, and memory on the board; then slide it into position. Install mounting screws and reconnect all cables. ($50 to $125)**

Robert Bruce Thompson is the author or co-author of several PC hardware books from O'Reilly Media, Inc., and has been building, repairing, and tinkering with PCs for more than 25 years.

LED Throwies

By Graffiti Research Lab

Make and toss a bunch of these inexpensive little lights to add color to any ferromagnetic surface in your neighborhood.

You will need:

10mm diffused LED, any color(s) 20 cents each from HB Electronic Components (hebeiltd.com.cn).

1" strapping tape One roll will make many throwies.

CR2032 3V lithium batteries 25 cents each from CheapBatteries.com.

½"x1" NdFeB disc magnet, Ni-Cu-Ni plated 25 count for $13 from Amazing Magnets (amazingmagnets.com).

Conductive epoxy (optional) Weather-resistant alternative to tape. Available from Newark InOne (newark.com).

1. Test the LED.

Pinch the LED's leads to the battery terminals, with the longer lead (the anode) touching the positive terminal (+) of the battery, and the shorter lead (the cathode) touching negative (-). Confirm that the LED lights up.

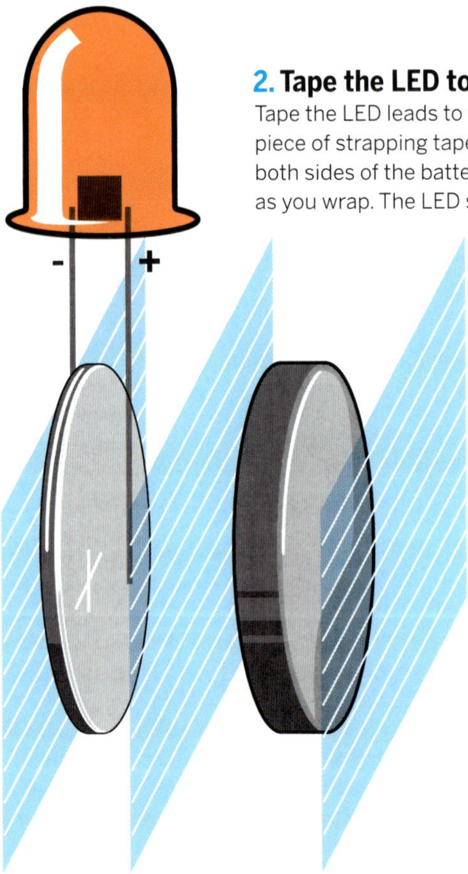

2. Tape the LED to the battery.

Tape the LED leads to the battery by cutting off a 7" piece of strapping tape and wrapping it once around both sides of the battery. Keep the tape very tight as you wrap. The LED should not flicker.

NOTE: The battery's positive contact surface extends around the sides of the battery. Don't let the LED's cathode touch the positive terminal, or you'll short the circuit.

3. Tape the magnet to the battery.

Place the magnet on the positive terminal of the battery, and continue to wrap the tape tightly until it's all done. The magnet should hold firmly to the battery. That's it — you're ready to throw (or make a few dozen more). Throw it up high and in quantity to impress your friends and city officials.

Throwies naturally chain together in your pocket, making multi-segmented throwie bugs, which will also stick to metal surfaces if they aren't too long.

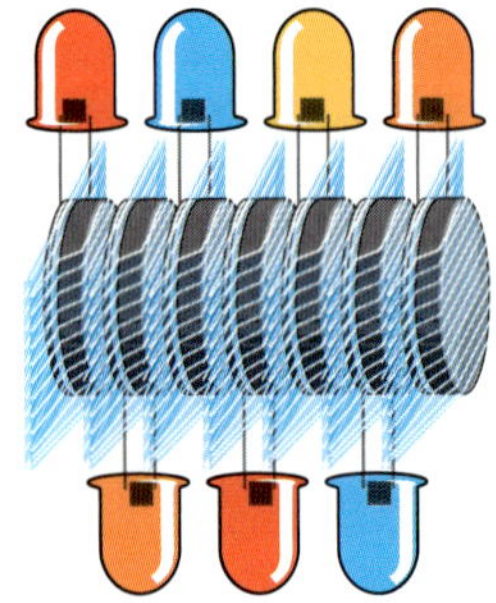

A throwie will shine for about 1-2 weeks, depending on the weather and the LED color. To get one off a ferromagnetic surface, don't pull it, or it may come apart. Instead, apply a lateral force to the magnet base, and slide it off the surface while lifting it with a fingernail or tool.

Graffiti Research Lab (graffitiresearchlab.com) is dedicated to outfitting graffiti artists with open source technologies for urban communication.

Illustrations by Kirk von Rohr

The time just seems to fly by when your unread email is piling up.

IT'S EMAIL TIME

Innocent-looking "clock" monitors the unread-message pileup in your inbox.

By Tom Igoe

I have a lot of anxiety about email. Every kilobyte in my Inbox destroys another minute of my life, but I can't stop checking it. So I decided to embody my anxiety in a device that would worry about my incoming mail for me. I've always liked clockwork mechanisms, so I made my email fetish object in the form of a clock. For each kilobyte of new mail I receive, the clock ticks relentlessly forward.

Here's the basic design I came up with. The clock itself is driven by a microcontroller, which connects to the internet and queries a program that checks my email accounts. The program reports back the number of kilobytes, and the microcontroller moves the clock forward a tick for each kilobyte. Simple!

To build this yourself, you need to know how to program a microcontroller and how to do some basic web programming. For microcontroller programming, see the Primer article in MAKE, Volume 04 (*page 158*), or my book *Physical Computing: Sensing and Controlling the Physical World with Computers*. For the web component, I used a Common Gateway Interface (CGI) script written in Perl, but you can also write CGI scripts in PHP, Python, Ruby, and other languages.

Building It

I started by looking for the simplest way to drive the clock. I carefully took the clock apart and examined the circuit board inside. There was a

Photography by Tom Igoe

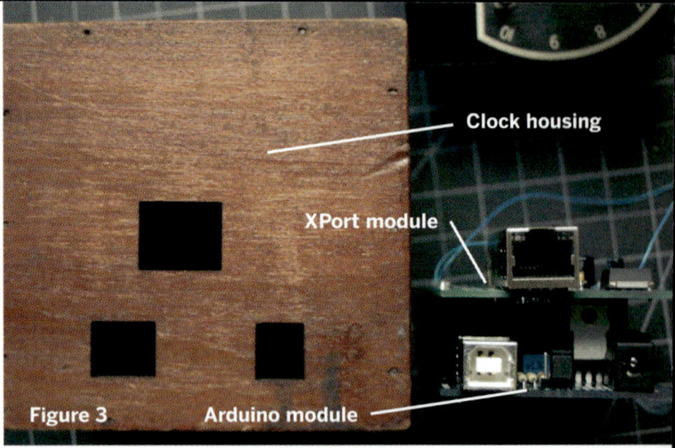

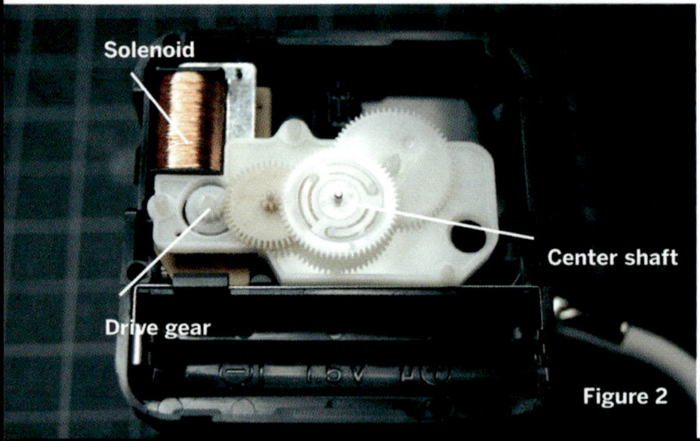

Wires soldered to solenoid contacts. Note crystal at bottom and "mystery chip" under blob in center (Figure 1). Cheap battery clocks all generally work the same way (Figure 2). Rear view of clock, showing holes drilled in housing for Ethernet, power, and the USB-B port for programming the Arduino microcontroller board (Figure 3). Front view of clock, face removed (Figure 4).

MATERIALS

Here's what I chose for the basic building blocks of my email clock.

Microcontroller I used the Arduino module (arduino.berlios.de), a small I/O board and development environment that's built around Atmel's ATMega8 microcontroller. Arduino is based on Wiring, another open and easy development environment, which uses the ATMega128 chip. I opted for Arduino because I knew I wanted to eventually make my own custom circuit board, and the ATMega8 is easier to solder than the ATMega128.

Serial-to-Ethernet Converter Instead of bothering to write my own TCP/IP stack for the Arduino board, I used some handy hardware: Lantronix's XPort serial-to-Ethernet module. These modules can route data between an Ethernet connection and a serial port that feeds a microcontroller. To avoid some soldering, I also recommend the Cobox Micro, which has the same programming interface as the XPort (you simply telnet in) but a simpler physical interface.

Clock I would have loved to build a mechanical clock but didn't have time, so I took the guts out of a cheap battery clock and connected its drive shaft into the hub of a clock-like antique test gauge.

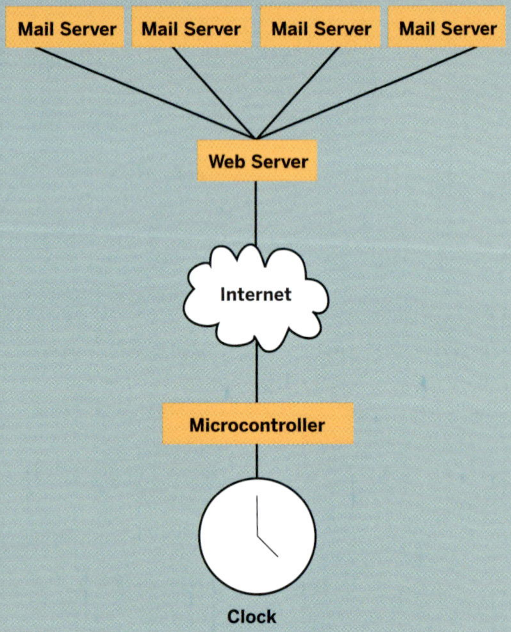

System diagram. A CGI script queries my mail accounts and returns new mail volume. A microcontroller calls the script, and moves the clock's hands forward accordingly.

mystery chip in the middle, sealed in plastic, but I found that the main gear of the clock, which drives all the other gears, is controlled by a solenoid. The two solenoid terminals were pretty easy to spot, and I figured that all I'd have to do was send a pulse through these connections, and the clock would tick happily away.

I soldered a couple of leads onto the solenoid contacts and put the clock back together. Then I pulsed the solenoid directly with 5VDC power, bypassing the mystery chip. The motor jumped, but the clock didn't tick. Clearly, there was more to learn, so I put the clock's battery back in and connected the leads to an oscilloscope to see how the voltage changed as it ticked. The pattern was more complex than I expected. Each second, the pulse would alternate: high-low-zero one second, and low-high-zero the next.

I programmed the microcontroller to duplicate this pattern, and fed its output into the clock. This took some experimentation, but eventually I got the microcontroller to control the clock pretty well. Each tick it generated moved the clock forward about two seconds. Since I wasn't concerned with keeping actual time, this was fine.

The next step was to get the microcontroller to check mail. The XPort needed only three wires connected from the microcontroller side: serial receive, serial transmit, and a reset connection to allow the microprocessor to restart the port. Making a circuit board for the XPort was a challenge because its pins don't follow a 1/10" grid, the hobbyist perfboard standard. So I used CadSoft's Eagle software to design a custom board to mount the module into. You can find the layout file at tigoe.net/emailclock.

Then I configured the XPort from my laptop through a USB-to-serial converter cable. Following Lantronix' instructions, I gave it an IP address, gateway address, and subnet mask. I also configured the serial port settings.

I opened a terminal window, telnetted into the XPort, and entered "Hello World!" to confirm that messages were passing through. Then I quit telnet and tried connecting to my web server by entering its numerical address (port 80) in the serial window: `C82.165.199.37/80`.

The XPort confirmed by returning a "C". I responded with an HTTP request for a web page on my server, *http://tigoe.net/pcomp/index.shtml*:

```
GET /pcomp/index.shtml HTTP/1.1
HOST: tigoe.net
```

The server returned the HTTP header and contents of the requested page:

```
HTTP/1.1 200 OK
Date: Tue, 13 Dec 2005 20:50:27 GMT
Server: Apache/1.3.33 (Unix)
Transfer-Encoding: chunked
Content-Type: text/html
<html>
  <head>
```

... and so forth.

Success! Seeing this exchange of HTTP (HyperText Transport Protocol), the normally hidden language of web browsers and servers, meant that I had gotten the XPort to perform as a browser. Now I just had to get the microcontroller to do the same: open a net connection, request a page, and read the results. You can find my code to do this at tigoe.net/emailclock. I uploaded the compiled firmware from my laptop via the Arduino's onboard serial-to-USB converter.

The microcontroller isn't requesting an HTML page — it's calling a common gateway interface (CGI) script, also available at the link above, that checks all of my mail accounts, queries for the volume of new messages, adds the numbers up, and sends the total back to whoever's asking. Since this script isn't expecting to be called from a browser, it doesn't format its results as HTML. To minimize the programming needed on the microcontroller side, it keeps things simple, returning just the HTTP header and one line of text: `<KB: 1234>`.

Once the whole system was working, I found a housing for the clock: a nice, antique piece of electronic test equipment with a hole behind its clock-like face that could accommodate a driveshaft. The clock shaft screwed in easily, and the Arduino module and the XPort fit snugly inside. I drilled a couple of holes in the back for the power and Ethernet cables, and the clock was done.

How well does it work? I still check my email compulsively, but for the couple of hours it took to build this clock, I didn't open my mail program at all!

Tom Igoe heads the physical computing area at the Interactive Telecommunications Program at New York University.

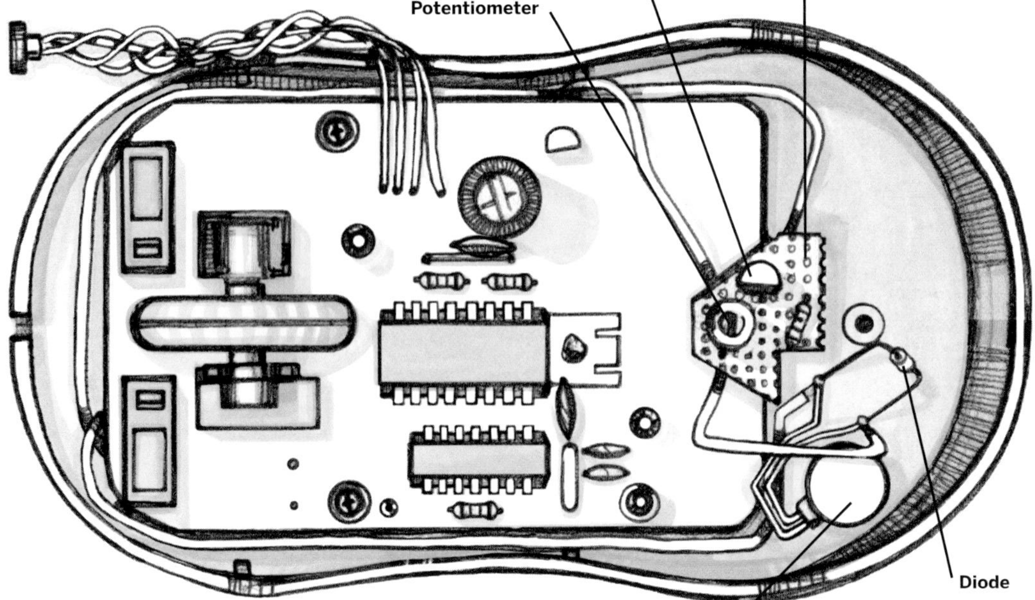

Inside the completed rumble mouse. Note the diode that is soldered in parallel with the motor.

RUMBLE MOUSE

For FPS gaming, a cellphone vibrator gives kick to your clicks.

By Greg Lipscomb

Have you ever been playing your favorite first-person shooter with someone's rumble controller and thought, "I would love to have that capability while playing on my computer"? A friend of mine had a spare rumble pack lying around, so he decided to stick it in a mouse. This inspired me to create my own version.

I determined that my rumble mouse should meet certain specifications. I wanted it to be fully enclosed, with no parts sticking out of the case. It should be an optical mouse, rather than a roller-ball one, connected and powered by USB. In play, the mouse would give a satisfying rumble-recoil when you click the left button — the trigger in most FPS (first-person shooter) games.

I found a cheap suitable mouse from a local surplus store. It had a scroll wheel that I liked, and was large enough to fit the extra components inside. For the rumble motor, I wanted something small, and my fiancée suggested that I use a cellphone vibrator. I had several old cellphones lying around, so I cut one open and located its small motor near the top left corner, which looks like a watch battery with two wires coming out of it.

Naturally, different phones are different. Nokias I've cracked open use small cylindrical motors, not as flat, and in an old flip phone I took apart, the motor was in the same piece as the speaker, close to the LCD screen.

Illustration by Damien Scogin

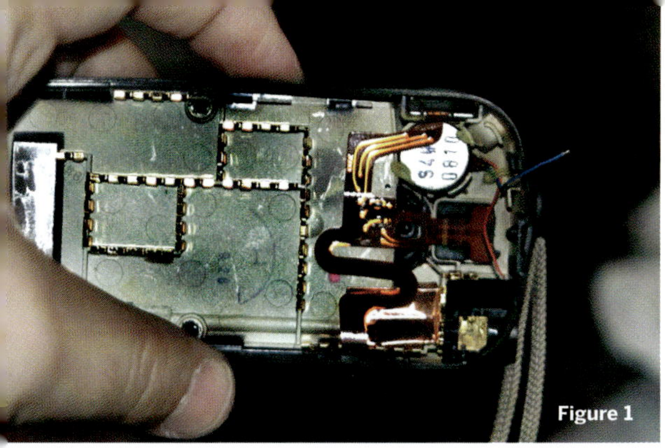

Figure 1

Figure 2

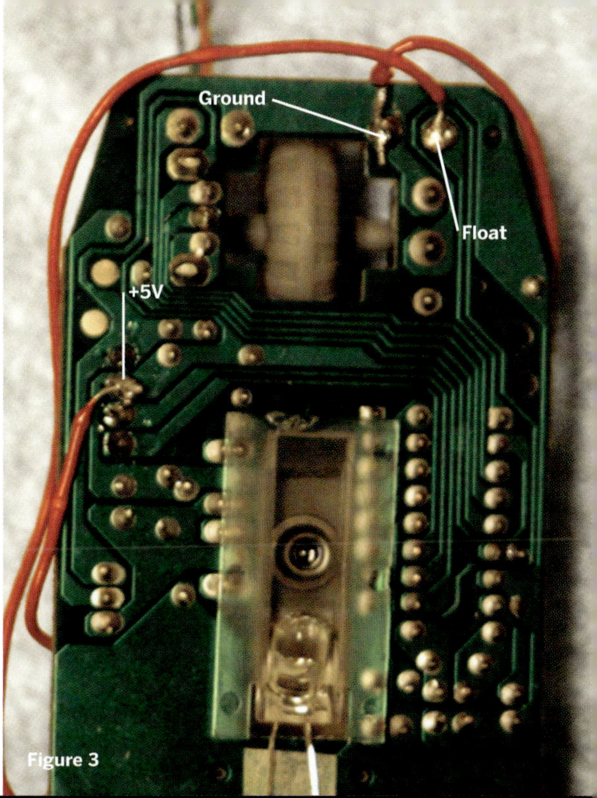

Figure 3

For the rumble motor, I opened up an old cellphone and dug out the vibrator, which looks like a thick watch battery (Figure 1). Bench test for the rumble mouse circuit on a solderless breadboard (Figure 2).

The three pins used in the circuit: +5V is direct from the USB cable, Ground is common ground for the circuit, and Float is the pin that momentarily grounds when you click the left mouse button (Figure 3).

MATERIALS

Optical USB mouse I used an off-brand, model HTM-67WT

PNP transistor, 2N3906 RadioShack #276-1604

Electric vibrating motor Salvage from cellphone, or use Sanko Electric #1E120, available at allelectronics.com, catalog #VB-1

5kΩ or 1kΩ micro trim potentiometer A 10kΩ pot, like RadioShack #271-282, works, but lower values fit the adjusting range better

Diode, 1N4001

1kΩ resistor

Mini perfboard RadioShack #276-0148

I also needed a switch to connect the USB power to the motor when the left mouse button clicks. I chose a PNP transistor for this. The final mandatory part was a 10kΩ potentiometer, which would be used to regulate the current going to the motor.

Find and Adjust the Power

The rumble circuit takes power from the USB port and directs it to the motor, so I needed to find +5V power and ground contacts both on the USB cable and inside the mouse. For the cable, I found a USB pin-out diagram of a USB cable online, and learned that the two outer pins of the USB cable are the +5V and the ground. Then I opened up the mouse and used a digital multimeter to probe for terminals on the board that connected to the USB's power and ground pins. On my mouse, I found that the green wire from the USB was ground, and the blue wire was the +5V. Then I connected the cellphone vibrator directly to the USB power, just to make sure that it could pull enough power to run. It worked perfectly.

I soldered jumper wires to the +5V and ground contacts on the phone, and started putting my circuit together on a solderless breadboard. The vibrator motor had a labeled rating of 3.6V, so I needed a resistor in series with the motor to lower the incoming voltage from 5V. I chose to use a 10k potentiometer, which would let me adjust the voltage and therefore the speed of the motor. You can use your multimeter to make sure that the voltage across the motor is less than 3.6V.

An alternative to the potentiometer is to use two diodes in series. Each diode drops the voltage 0.7V, so if you have two, you would drop 1.4V to give 3.6V over to the motor. Either way, it's important to have a separate diode in parallel with the motor. When the transistor is first activated, the

Photography and illustration by Greg Lipscomb

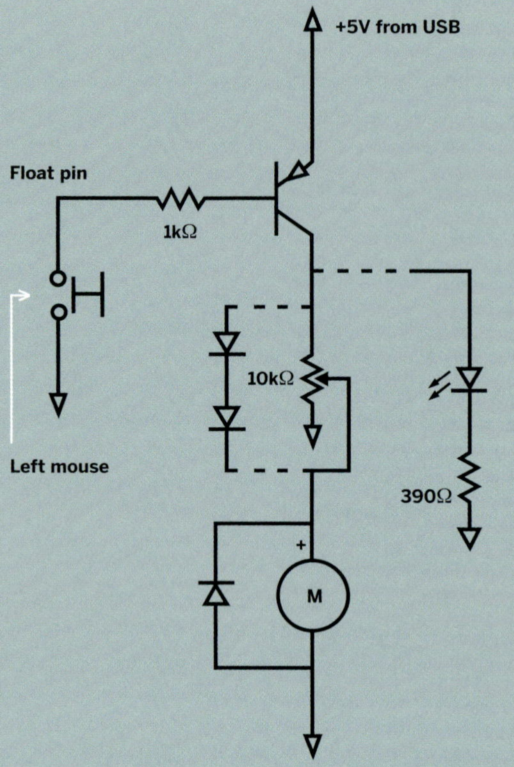

Rumble mouse schematic (optional circuitry shown by dotted lines). Note that the potentiometer can be replaced with two diodes in series, as shown. There is also an optional LED-resistor subcircuit that lights when the left mouse button is clicked.

motor voltage can spike really high, and a diode will protect the transistor from this high-voltage spike.

Intercept the Left-Click

The next step was to locate the board contacts for the mouse's left-click button. It was a single-pole momentary switch, and using my multimeter, I found which pin was ground and which pin was floating. I soldered a jumper wire to the floating pin contact, and continued breadboarding. Since the mouse button operates by grounding the floating pin, I needed an electrical switch that would be activated by a low voltage signal. A PNP transistor was perfect for this requirement. Connect the float pin contact to the base of the transistor, and low voltage from a click action completes the circuit between the emitter (connected to power) and the collector (connected to the motor). Note that this is opposite from an NPN transistor, which is open with a high base current. (Another approach would be to forget the transistor and just hook the motor up between +5V and the floating pin. Then, while the floating pin is grounded, 5V runs through the motor to ground.)

To limit the current flowing through the transistor base, I put a 1k resistor between the floating pin and the transistor base. Then I connected the emitter of the transistor to my +5V from the USB port, and connected the collector of my transistor to the motor through the 10k potentiometer. The other pole of the potentiometer was connected to ground. (When wiring a potentiometer, it is important to attach one wire to the middle pin, and place the other wire on either of the two outer pins. If you connect to both outer pins, you will always get the total 10kΩ.) This completed my circuit on the solderless test board, so I tested it and fixed the bugs.

Put It Together

The final step of my project was to solder everything together. I soldered my components to a small piece of perfboard, using small wires to connect the appropriate leads. Then I used double-sided tape to connect the motor to the casing at the back of the mouse. I also routed the wires to fit nicely into the body of the mouse.

When I completed the assembly, I had a fully functioning rumble mouse that would vibrate on every left mouse click. It is perfect for any first-person shooter game.

Some ideas for improvement would be to add a simple toggle switch between the +5V from the USB and the emitter of the transistor so that the rumble part can be turned off. All in all, this is a straightforward DIY project that is sure to enhance your gaming experience.

Greg Lipscomb is an electrical engineer (Auburn University) who is in his second year of medical school at the University of South Alabama in Mobile. You can see his work at diylive.net.

Since Version 4.9, iTunes has treated podcasts just like content from the big boys.

VIDEO PODCASTING

Producing TV shows on the cheap. By Howard Wen

Video podcasting is easy; it's just like audio podcasting, but you point your RSS feed to video files instead of audio. (*To learn about RSS feeds, see "Podcasting 101" in MAKE Volume 02, page 86.*) For example, in the `<enclosure>` tag, you point to a QuickTime MOV file and list `type="video/quicktime"` instead of pointing to an MP3 and listing it as `type="audio/mpeg"`. And that's it; that's why it's called Real Simple Syndication.

The hard-but-fun part is making the video itself. Here are my suggested tools and tips for making video for podcasts on a budget.

For shooting, I recommend MiniDV cameras, which have fast and easy FireWire download interfaces. A basic model costs less than $400.

Mac software: All you need is iMovie HD, part of Apple's iLife suite (apple.com/ilife, $79). It transfers video from the camera, lets you do simple editing, and exports as QuickTime.

Windows software: Download with WinDV (windv.mourek.cz, free), edit as AVI in VirtualDub (virtualdub.org, free), and convert to MPEG-4 with PSP Video 9 (pspvideo9.com, free). Or download, edit, and save as QuickTime using the Windows version of QuickTime Pro (quicktime.com, $30).

Apps like Final Cut Pro, Premiere Pro, Vegas, and Liquid Edition also do all of the above.

To minimize file sizes, try a few compression settings and compare the results. The most important ones are bit rate, sound quality, and image resolution. 320x240 or 240x180 are best, so shoot everything simple, clear, and high-contrast.

Save in QuickTime format if you can. In my experience, QuickTime files are smaller and better-looking than MPEG-4 and MPEG-1, the other two video formats that iTunes supports.

Whether or not you're already a podcaster, you can make your TV-producer dreams come true.

The typical Linux desktop today looks a lot like a Windows desktop.

DUAL BOOTING LINUX AND WINDOWS XP

Turn your PC into a switch-hitter the easy way, with a LiveCD. By Mark R. Brown

That voice in your head knows that you won't get real geek satisfaction unless you're running Linux, but your practical side knows that you need your favorite Windows applications (read: games). So what can you do? You dual boot. You install both Linux *and* Windows XP.

Are your palms sweating? You've probably heard horror stories about hard drive repartitioning, data loss, and configuration nightmares, but if you take a few precautions, it can be easy.

NOTE: This tutorial covers dual booting with a single hard disk. Setting up separate disks for XP and Linux requires a more convoluted procedure.

Step 1: Pick a Linux Distribution

You know there are at least a million Linux distributions, right? Red Hat. Mandrake. Debian. The names all run together, and each one comes on seven or eight CDs, and takes days to install.

Not necessarily. You don't need a full Linux distribution yet. You want a LiveCD, a distro that runs from a single CD that you can boot from and try out before committing to a full installation.

There's a list of LiveCD distros at frozentech.com/content/livecd.php, sortable by platform and function. You want one that supports x86 hardware, and is listed as Desktop and OS Replacement.

Which should you try? You can start by tracking the herd and looking at the ones with the most votes. Then visit websites, read specs, and look for the features you want. Pay attention to the hardware support. Many distros I researched didn't support my motherboard's sound and video. I eventually chose PCLinuxOS (pclinuxos.com) because it supported both.

Once you've found a few promising distros, download the *.iso* file for each and burn it onto its own CD. You need software to do this, since XP won't write *.iso* images directly. I used Nero Burning ROM, which came with my CD burner. You can also use ISO Recorder Power Toy (isorecorder.alexfeinman.com, free). Boot from each CD, and test them out. Find one that works with your hardware, internet connection, and peripherals.

You can stop here, and just boot from CD when you want to run Linux. But what kind of a geek does that? Time to repartition your hard drive.

Step 2: Partition Your Hard Drive

Here's the problem: Your Windows XP is installed on a partition formatted in NTFS. Most Linuxes can read NTFS but can't write to it. Meanwhile, XP can't read a Linux partition at all. So any data you want to be readable and writable from both XP and Linux needs to be moved to a separate partition formatted in FAT32, which all OSes support.

And yes, you're going to have to reinstall XP, so back up all of your data to CDs. Burn it *all* — there are hidden address files, configuration files, cookie files, and who knows what else on your hard drive that you will regret not having later.

Boot from your Linux CD and find the partitioning program. Mine was called `qtparted`, but your mileage may vary. Use it to subdivide your hard disk into four partitions, like this:

TYPE	FILE SYSTEM
Windows	NTFS
Linux	Linux
Linux swap	Linux swap
Data	FAT32

Make sure they're in this order. How big you make them depends on several factors. Don't give XP any more or less than you really need for XP itself (1.5GB minimum) plus all of your Windows applications. I generously allocated 7GB because I use some big programs like Visio and Paint Shop Pro.

Your Linux partition should start no further up on the drive than 1,024 cylinders (about 7.8GB). Read your LiveCD's help files to find out how much disk space your Linux distro needs, and how much to allocate for swap space. Mine needed 10GB for the Linux and 3GB of swap space. Allocate all remaining space to your data.

Step 3: Install Windows XP

Remove your Linux LiveCD and reboot with your Windows XP installation CD. Install XP. When it asks if you want to make a recovery diskette, say "Yes" and make one. Be safe.

After installation, remove your Windows CD, boot to XP, and make sure that it still works OK.

Step 4: Install Linux

Use your LiveCD to reboot to Linux, and read the help files for instructions on doing a full install.

The installation program will ask you which partitions to install to for Linux and Linux swap. Make sure you point it to the ones you allocated when you repartitioned your drive. Note that whereas Windows refers to the first hard drive as `C:`, Linux calls the same drive `hda`.

The installer will probably also ask you to configure the boot loader program (usually GRUB or LILO), which allows you to choose your OS on startup. My install ran LILO and auto-detected both my XP and Linux partitions, but you may not be so lucky; you'll find out soon. Install your boot loader in the MBR, or Master Boot Record, so it runs automatically every time you reboot.

Remove your Live CD and reboot. You should see a menu letting you choose XP or Linux. But if your computer boots to XP without seeing Linux, you'll have to find and run your boot loader and manually point it to the Linux and XP partitions.

Step 5: Restore Your Data

Finally, to restore your data from the backup CDs, just drag and drop from either Windows or Linux.

Of course, you'll also need to reinstall your Windows apps, configure, customize, and get them to see your hardware, and also learn Linux. But that's what being a computer geek is all about, right?

Once upon a time, Mark R. Brown was managing editor of *.info*, the legendary Commodore computer mag. He also wrote lots of now-out-of-print books about the internet for Que.

DA XEROX CODE

How to read the tracking dots in your color Xerox printer. By Annalee Newitz

Early in 2005, Internet Archive founder Brewster Kahle noticed something a little weird about the color copies he was making. When he looked at the pages really closely, he could see hundreds of tiny yellow dots — and they looked like they formed some kind of pattern. He recalled seeing an article in *PC World* about how the Secret Service busted counterfeiters by using marks on the fake bills that could be traced back to the printers that made them. If these dots were a method of tracking, then Kahle wanted to know exactly what information they contained.

He turned to Electronic Frontier Foundation technologist Seth Schoen for answers. Schoen, with the help of EFF intern Robert Lee, combed through hundreds of pale-yellow-speckled color copies from Kahle's machine. They contacted a Xerox research fellow, who confirmed their worst suspicions: yes, the dots were for tracking. How widespread was this practice? Was it just Xerox?

To find out, Schoen posted a call for color copies on the EFF website. He soon received hundreds of printouts from all over the world. "One was from an ice cream parlor in Australia," he recalls. "Another was from a symphony orchestra." All of them had the telltale yellow dots.

It turned out that companies like Dell, IBM, Canon, Lexmark, Toshiba, and many others were creating the dots on color laser printouts (for a list of printers that generate tracking dots, see eff.org/Privacy/printers/list.php). Schoen speculated that the nearly

Barely visible dot-matrix "fingerprints" enable any page to be traced back to its color printer.

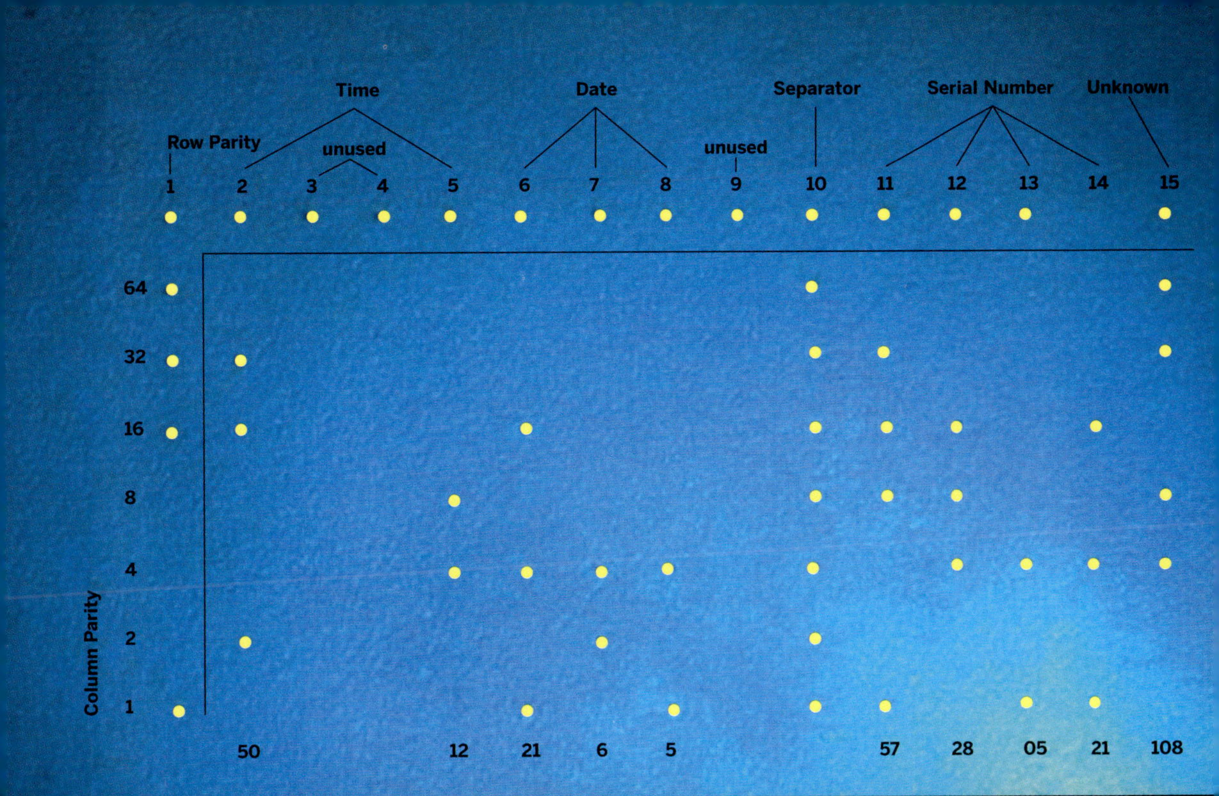

This repeating dot pattern encodes the DocuColor printer serial number, the date and time the page was printed, and other bits of info not publicly known. Parity bits along top and left keep the rows and columns odd (except top row). Rows carry binary digits, numbered up from the bottom. Columns 6-8 encode date (DMY order), 2 and 5 encode time (minute, hour), and 11-14 encode product serial number, two decimal digits per column.

invisible patterns — which seemed to repeat several times on a given page — must encode the serial number of the machine that prints them, possibly along with other information such as a time/date stamp.

Blue Light Reveals

By now, Schoen's office at EFF was awash in printouts, and he'd become adept at finding patterns in the dots by shining a blue LED on each page while it was magnified ten times under a USB microscope. The blue light made the yellow dots look almost black. When magnified, it became clear that almost every printer uses a repeating dot pattern that forms a grid of 15 columns and 8 rows. It also seemed obvious that the dots probably represented binary numbers. But which direction should the numbers be read? And how were digits being represented? Was each column one number, or part of a larger number? Schoen was stumped.

Luckily, a cryptanalyst came calling. Swiss graduate student and crypto expert Joel Alwen took one look at the dot patterns he found on Schoen's desk, and knew he had to solve the puzzle. He decided to work on dots produced by the Xerox DocuColor 12 because Andrew "Bunnie" Huang had posted some pictures of them in his blog.

"It took me about five or six hours total to decode it," Alwen says. After looking at several pages made by the same printer at different times, Alwen made his breakthrough. "The watermarks only differed in column 2, so I guessed that this was a timestamp that was arranged with low order left and high order right."

Once he knew the numbers read from right to left, he quickly discerned that the next few columns were the date, and the columns after that spelled out the serial number (*see "Read the Dots," next page*). He and Schoen also realized that the first column and first row were essentially parity bits — they speculate that this may be the case for dot grids created by most printers.

Right now, only the Xerox DocuColor 12 dot code has been cracked completely. Dell dot patterns are similar to Xerox, so Schoen speculates it will be the next code broken. That leaves dozens of other dot patterns from other printer manufacturers unsolved.

Image by Annalee Newitz and Seth Schoen

How to Read the Xerox Codes

With a few of the tools and techniques developed by Schoen, Huang, and Alwen, you'll have everything you need to find the dots, analyze them, and (hopefully) crack a few more of the encoded patterns. Before you delve in, though, you might want to play around with reading the Xerox codes.

To read the Xerox codes, you'll need a few items.
Color printout from a Xerox DocuColor 12 printer The dots look best on the white portions of the page, so make sure your printout has lots of white space.
Scanner or USB microscope Schoen recommends the $60 Digital Blue for Windows and a blue LED.

One method for finding the dots, of course, is simply to shine the blue LED on the paper and try to see them with your naked eyes. Schoen is able to see them this way, but I find it impossible. Most people use magnification or image enhancement.

Using a Scanner

If you want to go the scanner route, simply scan a chunk of the page at a slight magnification, and dump it into Gimp or Photoshop. Fiddling with the color balance and background noise will allow you to turn the yellow dots gray or black. You can see some excellent enhanced scanned images of the dots at makezine.com/go/dots1 and makezine.com/go/dots2.

If you want to add on another project, try building Huang's blue light scanner, which makes the dots look extremely clear, bunniestudios.com/wordpress/?page_id=51.

Using the USB Microscope

You don't need to see the whole page, since the dot pattern will repeat several times. A good bet is to magnify it to 10X. Once you've got the paper positioned in the microscope, shine the blue LED on it. You should see a nice example of the dot pattern fairly clearly, with the yellow dots turned a dark gray.

Read the Dots

Remember that the Xerox pattern has 15 columns and 8 rows. To find the grid, look for the only row with an even number of dots; that's the topmost row, the horizontal parity. All the other rows (and columns) are odd. Once you've located the pattern on your sheet, save it or write it out on graph paper. If you know how to read binary, you can use the EFF's guide to decode each column (read right to left):

15: Unknown (often zero; constant for each individual printer; may convey some non-user-visible fact about the printer's model or configuration)
14, 13, 12, 11: Printer serial number in binary-coded decimal, two digits per byte (constant for each individual printer)
10: Separator (typically all ones; does not appear to code information)
9: Unused
8: Year that page was printed (without century; 2005 is coded as 5)
7: Month that page was printed
6: Day that page was printed
5: Hour that page was printed (may be UTC time zone, or may be set inaccurately within printer)
4, 3: Unused
2: Minute that page was printed
1: Row parity bit

Alternatively, you can use a handy program that Schoen wrote that automatically decodes the dots for you. It's at the very bottom of the page at eff.org/Privacy/printers/docucolor.

Now you should be able to figure out the serial number of the printer that created your page, as well as the date and time it was made. (Note: Some machines print the wrong date/time because they've been set up incorrectly). The next step, for inveterate geeks and makers alike, is to start cracking codes on other printers. You can use the same procedure for seeing the dots, but keep in mind that some dot patterns could be bigger or oriented differently than the Xerox pattern. Some are probably read bottom to top, or left to right. Binary representation may vary.

The point of this project — aside from hours of amusement — is to defang a tracking system that can be quite privacy-invasive. What if you want to print political flyers anonymously? Or you don't want people to know when you printed a particular document? The more we know about these dot patterns, the better we can protect the anonymity and privacy of innocent people.

So get to decoding!

Annalee Newitz (techsploitation.com) writes a syndicated column called Techsploitation, is a contributing editor at *Wired* magazine, and is a huge fan of Upper Paleolithic technology.

DIY GAMING

Use a couple of white LEDs to illuminate the original Game Boy Advance's display.

LET THERE BE (FRONT) LIGHT

Add an LED front light to your Game Boy Advance for $5. By Dave Prochnow

One of the biggest sore points with the original Game Boy Advance (GBA) was its lack of a backlit screen. You and your GBA could while away many waking hours, but with the arrival of dusk, the fun faded as quickly as *Donkey Kong* high scores.

MATERIALS

Game Boy Advance

White T-1 LEDs (2) I got mine from goldmine-elec.com, part #G155b2.

10-ohm, ¼-watt resistors (2)

Red and black insulated wire

Nintendo rectified this failing later with the release of the Game Boy Advance SP, which features a lighted LCD.

Barring the purchase of a new GBA SP, what can a lowly GBA owner do? The Afterburner kit from Triton Labs made GBA screens glow with an ingenious, clear, front-lit LED panel, but that kit has been discontinued. Here's a simpler and rougher solution, inspired by Benjamin Heckendorn's book *Hacking Video Game Consoles* — just hack a couple of white light-emitting diodes (LEDs) that draw from the GBA's own power and illuminate the screen from inside its case.

Finding the parts and tucking them inside the

Photography by Dave Prochnow

Figure 1

Figure 3

Figure 2

Figure 4

Underneath this protective mat is the LCD. The two white LEDs will be placed in opposite corners along the bottom (Figure 1). Wiring harness and LEDs in position on GBA (Figure 2). LED cathodes connected by black wire, anodes connected to red wire through resistors (Figure 3). The toughest solders are hooking up to power and ground, S2 and GND, on the game cartridge interface board (Figure 4).

GBA is easy; the hard part is getting the case open in the first place. In what must have been an attempt to tick off hackers, Nintendo assembled the GBA with six tamperproof Triwing screws. You really need to use a Triwing screwdriver to remove these; you can't simply force 'em out with a regular screwdriver. One option is to get the Craftsman 32-piece Security Screwdriver and Bit Set from Sears (item #00947486000, $25), which includes four Triwing bits, plus Torx and others.

If buying a new, special-purpose screwdriver to work on a vintage game console seems ridiculous, take your GBA into your local electronics repair shop. If you're lucky, they'll have some Triwings, and you can replace the tamperproof screws with regular Phillips screws later.

Once inside the GBA, it's a simple two-screw (Phillips screws) process to separate the main circuit board from the display area. Carefully peel away the protective gray cushion from the lower bottom edge to gain access to the LCD itself. Then you should be able to pry the LCD up and out with a gentle push from a thin screwdriver.

The LCD is supported and protected by a black rubber frame covered with an adhesive film. To make room for the white LEDs, you need to remove the lower right and lower left corners of this frame. Do this carefully! I used a sharp knife to gently cut two rectangular pieces out of the rubber.

Place the diodes in the corners with their legs sticking out. Then assemble the wiring harness by soldering one end of a small length of red wire to both of the 10-ohm resistors, forming a T.

Connect the negative/cathode (shorter) legs of the two LEDs with a piece of black wire, and solder another piece of black wire to either one of the cathodes. Solder the LEDs' positive/anode (longer) legs to the free ends of each resistor.

Now, the tough part: carefully solder the other end of the black wire to the pad labeled GND on the GBA cartridge interface board, and solder the red wire to the pad labeled S2.

Finally, realign the main circuit board, lower the speaker back into its receptacle, tidy up your wiring, make sure all the buttons line up, and screw everything back together. You now have a new, bright-eyed GBA that will still be able to play all of your fave games. Game on!

Dave Prochnow wrote "Panzeroids" on page 58.

Young kids have more fun playing CD-ROMs using a touchscreen.

Photograph by Damien Stolarz

STICKY FINGERS

Modifying tech for pre-mouse toddlers.

By Damien Stolarz

As anyone raising a toddler knows, baby minds are hungry sponges for information. As a parent, I try never to pass up a learning opportunity. I'm a big technology enthusiast, and so naturally I want my kids to know everything about the hardware and software that runs our world.

Many parents are (understandably) overprotective of their devices at the expense of their children's education. They worry that kids will put sandwiches in the VCR or even shock themselves. What they fail to realize is that only through working with objects and technology — by making mistakes and learning from them — will their children ever gain the necessary knowledge and competence. And in my own case, as a maker and tinkerer who has broken more than my fair share of gadgets in the process of learning how they tick, it would be downright hypocritical of me to stop my children from breaking all my digital devices for the noble cause of their education.

Toddler Kiosk

As our daughter grew older, she wanted to imitate what her parents did. She liked to sit in front of our PCs, bang on the keyboard, and inform us that she was "working."

We bought her a brightly colored PS2 keyboard that has a picture next to every letter (apple for A, snake for S, dog for D, and so on). For a while, we would just pull out her keyboard and let her type on it when she wanted to work along with us.

After a while, we knew she needed her own computer — heck, she was already 2. So I dusted off an old G3 Macintosh and bought a handful of games designed for toddlers.

However, I saw how much she struggled with the so-called "intuitive" mouse user interface. She would move the mouse to the screen with no results; she would look at the mouse and move it, then look at the screen, lifting the mouse. With Mom and Dad's help, she got very interested in several of the interactive games, but we always had to be there and work the mouse for her.

I was familiar with a series of VGA touchscreen monitors, from a company called Newision.com, that were used for in-car computers. These 7" and 8" monitors are designed to stick to the dashboard of a car, but I found that they worked just as well stuck to the top of this G3 Macintosh. Since they provided a Macintosh touchscreen driver, I simply hooked up the display, and my daughter was immediately able to play the games all by herself.

I rigged up a launcher program that had big buttons, so she could even launch the games herself. And since one of the games asked her to type in her name (and I had the foresight as a parent to give her a name with only three letters), she quickly learned how to hunt and peck her name in and press Return.

Mobile Entertainment

One of my hobbies/professions is car hacking. Instead of purchasing the $1,800 factory DVD option with our new Dodge Caravan, I installed a Mac mini ($599) and a fold-down 15" touchscreen monitor ($199), so that my daughter could watch *Dora the Explorer* in hi-def resolution.

The essential feature of in-car entertainment systems these days is "multi-zone" entertainment, meaning the kids in the back have headphones so the driver and front passenger can listen to their own audio program.

I initially tried giving her headphones, but you will find that babies take a while to warm up to hats, headphones, sunglasses, or anything else you try to wrap around their heads. I wanted to find a way to create an isolated sound environment for her so we could listen to our own music.

Most toddler car seats consist of a washable, foam-filled padded cover wrapped around the plastic frame. Inspired by a noise-canceling headphones project (*MAKE, Volume 05, page 56*), I took the connecting loop out of a pair of inexpensive headphones, duct-taped them to the left and right sides of the plastic car seat, and then put the foam back on. For $5, I had upgraded my daughter's car seat with a sophisticated "surround sound for babies."

I used a mini-jack extension cord to connect this to an in-car DVD player. Now, I can have my own Mac mini-based music in the front seat, and my daughter can watch her DVDs and hear them in the back seat. Because of the perfect positioning of the speakers on either side of her head, she has her own aural experience, apparently unaffected by the music in the front seat.

Toddler Shuffle

A toddler's rabid affinity for media makes her able to listen to the same song repeatedly, apparently indefinitely. On one of our daughter's CDs, she has found a two-minute song that she likes to hear over and over.

Unfortunately, the CD player in our car does not have a "repeat song" function, and despite the integrated steering wheel controls in the minivan, it is still unnerving to have to press "back" every two minutes during long drives.

I thought about getting my daughter an old iPod, but the interface would be too complicated. Fortunately, Apple came out with the iPod shuffle, which has no breakable screen, is small, and more importantly, has a very simple three-button, one-switch interface.

We had also by now acclimated her to headphones — not earbuds, but a small pair of over-the-top earphones.

We showed her that to get her music, she needed to move the switch on the back of the iPod (to turn it on) and then press the "play" triangle. And when her favorite song comes on, she can simply press the "back" arrow to repeat it.

She's still getting the hang of the iPod shuffle — she'll overshoot, rewinding too far — but we don't have to intervene nearly as much now. And when she needs help, we can usually talk her through how to get back to her song again ("Press the left arrow two times. Is that your song? Good."), and she can do it all by herself.

Damien Stolarz is an inventor with a decade of experience making different kinds of computers talk to each other. His book *Car PC Hacks* is published by O'Reilly Media, Inc.

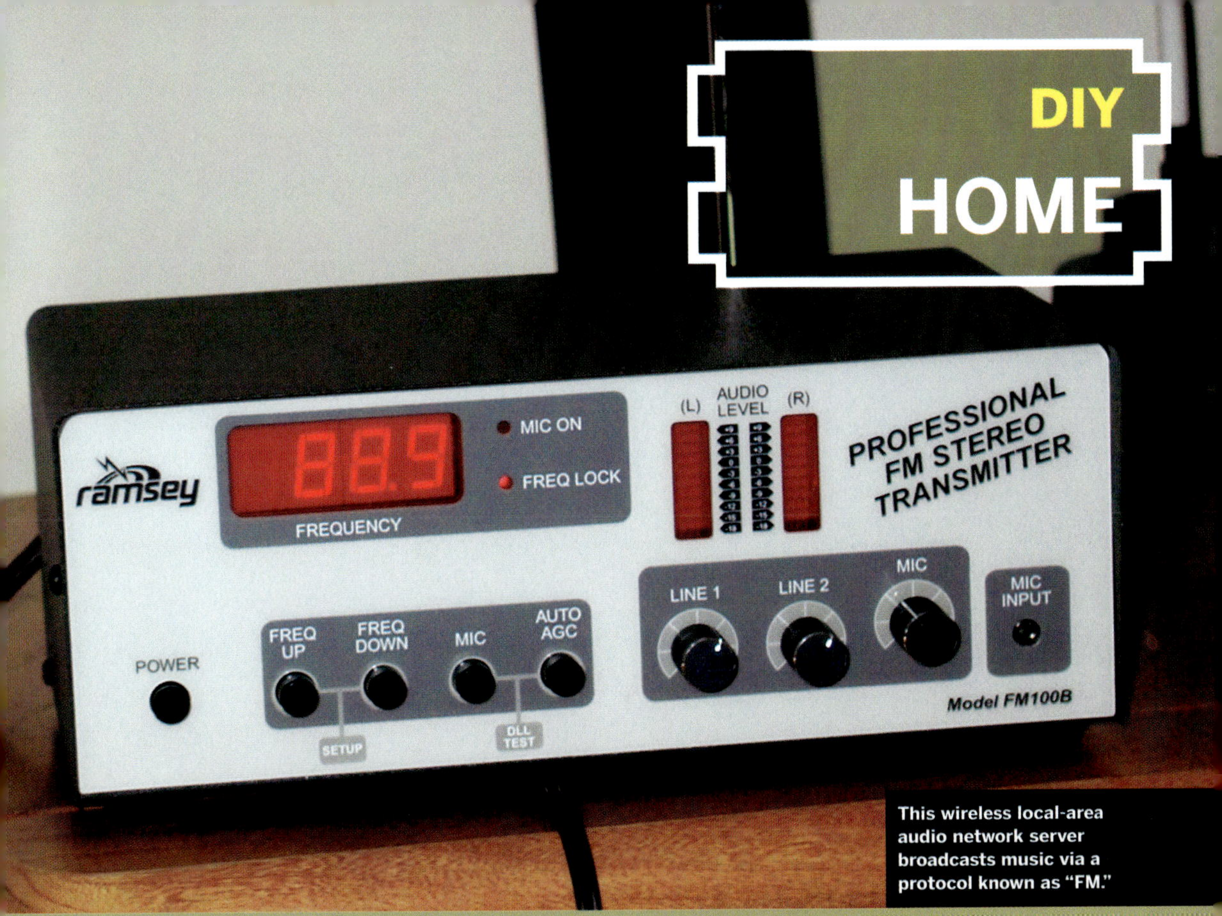

This wireless local-area audio network server broadcasts music via a protocol known as "FM."

HOMECASTING DIGITAL MUSIC

Good old FM beats wi-fi for sending streamed music around your house. By William Gurstelle

Photography by William Gurstelle

I subscribe to Real Networks' internet-based music service Rhapsody, which lets me stream a wide variety of music through to my computer's speakers. Digital music services like Rhapsody access millions of songs at the touch of a computer mouse. At ten bucks a month for unlimited use, this seems like a deal. The problem is, I don't listen to music only at my computer — I want to listen on my living room stereo, in the driveway as I work on my car, in the kitchen, in the basement, and so on. You need to untether the music from the computer's often-mediocre sound system in order to get the full benefit of your customized, on-demand music programming.

Rhapsody recommends using a multi-room digital music system call Sonos (sonos.com), but it can cost thousands of dollars. Products like Sonos distribute the music stream via Ethernet, either cabled or wireless. This makes listening outside problematic, because even if I have wi-fi set up to transmit, I still need a digital device and electrical power out where I'm listening. Using wi-fi also makes it impossible to listen using just portable headphones.

Legal FM Transmitters

A more cost-effective and flexible solution is a small FM radio transmitter. This trick is familiar

The completed power supply assembly from the Ramsey FM100B Super Pro FM Radio Station Kit.

to iPod users who play songs through their car sound systems, but broadcasting throughout your house and yard requires a bit more power.

Legal FM transmitters are a nearly perfect solution. They are relatively inexpensive, and they use the most ubiquitous of music players — FM radios.

I tried a few of these transmitters, starting with the C. Crane FM transmitter ($70, hobbytron.com or amazon.com). The Crane was plug-and-play simple, but its tiny size was matched by tiny power. Its out-of-the-box range was just a few feet. To increase this, I peeled off the metallic FCC label on the bottom, and used a screwdriver to turn the small potentiometer hidden underneath. But even at its max, the Crane's range is limited and will reach only a few adjacent rooms.

I also tried the much larger Ramsey FM100B ($250, hobbytron.com), which puts out a quarter watt of power, the maximum the FCC allows for unlicensed radio transmitters.

Some Assembly Required

The FM100B comes unassembled, as a large kit with over 800 solder points. According to the instruction sheet, it takes ten hours for an advanced hobbyist to assemble the kit, and 24 hours for a beginner. Given the number of components and the amount of soldering, these estimates seemed optimistic.

Unfortunately, the kit I received contained a bad volume control and a bad fuse holder. This added some troubleshooting time to the build, but the company was responsive and replaced the defective parts quickly.

I enlisted the aid of my friends Al and Rick, who had previous experience building Ramsey FM transmitters. With a kit this complex, they knew that we needed to set up an efficient assembly line. They prepared a work area with adequate lighting and work surfaces, and then began work in earnest, systematically identifying components (resistors, capacitors, switches, integrated circuits, etc.), soldering them into place, checking items off the list, and testing connections.

The kit directions divide the assembly process into discrete sections organized by subsystem, including the display board, mixer, filters, and transmitter. Each section clearly documents what needs to be accomplished, and builders with reasonable soldering skills should have no problem

My friends Al and Rick, who had built FM transmitters before, set up an efficient assembly line; Rick wrangled the components, and then Al soldered them into place (Figure 1). The Ramsey FM100B's main board, in the early stages of assembly (Figure 2). The completed transmitter, ready to fill the airwaves (Figure 3).

following along. As a bonus, the manual provides technical background as to what the subsystems do and how they work, turning assembly into a learning experience.

Many hours later, the kit was assembled and debugged. It was time to try it out. The FM100B has several input options. I hooked my computer into its stereo RCA jacks with a mini-to-RCA adapter cable ($8 at RadioShack).

The next step is to select a broadcast frequency, using the digital frequency selector on the front panel of the transmitter. I live in a big city, where the FM spectrum is crowded, but I found a fairly empty space toward the lower end of the FM band. I dialed it into the FM100B, launched Rhapsody on my PC, queued up several songs, and started broadcasting. Then I tuned the FM radios in my house to the broadcast frequency. The sound on several radios was pure and clean, as clear as any other FM station. A few radios did not work as well at first, but after fiddling with the placement of the FM100B's antenna, I decreased or eliminated the static and fade.

There is definitely a feng shui aspect to antenna placement, as the proximity of walls, windows, metal, and concrete have a big effect on the transmitter's ability to "hit" individual receivers.

The FM100B has a microphone input and mixer knobs, so you can do sophisticated radio station tricks. Assembling and troubleshooting the FM100B was a great learning experience, and it gave the desired result: freedom of movement indoors and out, without interrupting the music.

DON'T PUMP UP THE VOLUME

Home FM transmitters have low power because Federal Communications Commission regulations restrict the ability of any device to cause interference with legal radio signals, especially those put out by licensed stations. The maximum power an unlicensed transmitter can legally develop is 0.25 watts of signal. That's not much to work with, but it keeps the radio spectrum free from interference. Compare this with the 110,000-watt FM signal radiating from a big station such as Los Angeles' KPFK or San Francisco's KQED.

William Gurstelle's latest book, *Adventures from the Technology Underground: Catapults, Pulsejets, Rail Guns, Flamethrowers, Tesla Coils, Air Cannons, and the Garage Warriors Who Love Them*, came out in January.

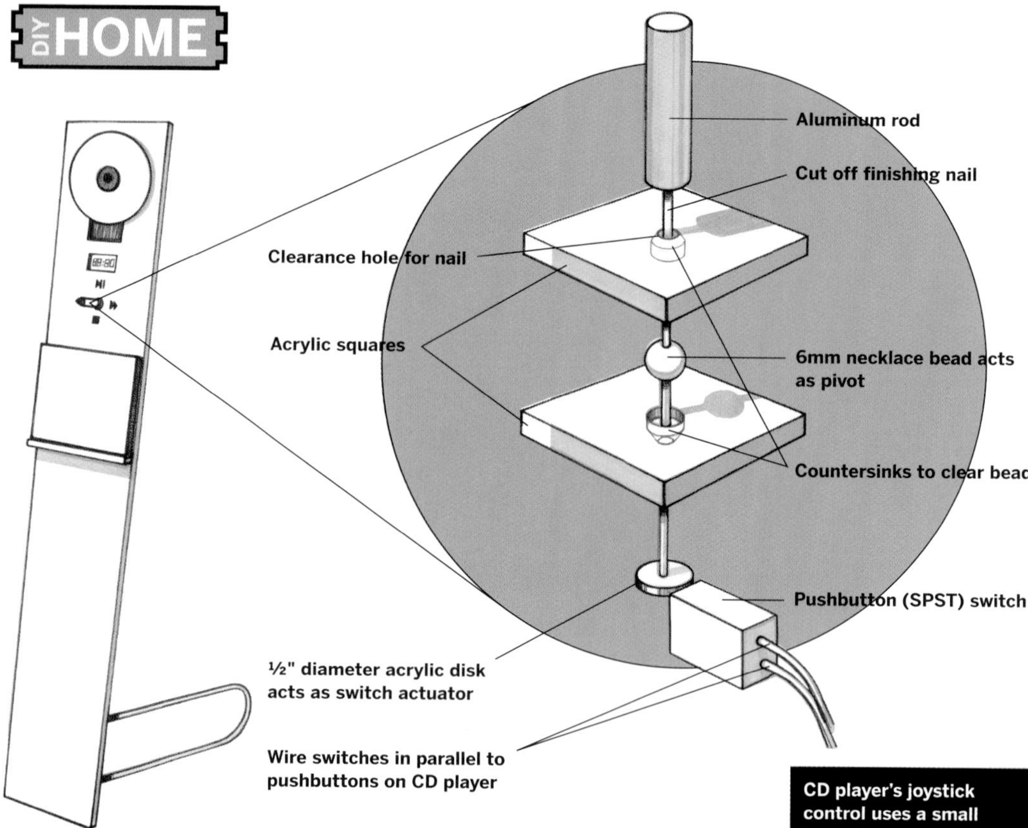

CD player's joystick control uses a small bead that leans into four surrounding buttons.

BORN-AGAIN BOOMBOXES

Up from the ghetto blaster, a new life as a functional-art stereo component. By Larry Cotton

Don't you love the fact that off-brand, bait-and-switch consumer electronics can be so cheap? I've seen portable CD players for under ten bucks, including headphones, and you can get a commodity CD boombox for about $30. Unable to resist such great values, I've bought a few of these beauties to see what makes them tick, and then reconfigured their aesthetics and turned them into attention-getting CD decks.

The procedure involves taking the boombox apart, and deciding which components will live and which you'll save for who-knows-what later. Then you arrange the keeper components onto some structure, which can range from a beautiful piece of hardwood to a coat hanger. Rather than keeping them as portable stereos, I've been turning the boxes into deconstructed CD players to use with headphones or plug into an amplifier.

This means dispensing with the speakers and wasting the FM tuner, but it's still worth starting a project like this with a boombox rather than a cheaper portable CD player, because its modules are larger, more robust, and more forgiving when you rearrange and attach them to strange objects.

I'm especially proud of a reconfigured boombox that I've named *The White CD*, a play on the Beatles album that's commonly (and erroneously) referred to by a similar title.

The White CD's innards are interconnected with relatively long wires, which lent themselves to a vertical arrangement. The CD spindle could be on top, the display would be below that, and — ta da! — a joystick below the display.

A joystick? In previous boombox mods, I kept

Illustration by Damien Scogin

Figure 1

Figure 3

Figure 2

Figure 4

Original cheap boombox on the half-shell, ready to resurrect as *The White CD* (Figure 1). Prepare the mounting board area, dug out for the display (Figure 2). Finish the surface and fill in cracks and seams with automotive body filler and glazing compound (Figure 3). Bend heated Plexiglas base over wooden form and hold until cooled and set (Figure 4).

the original push buttons for operating the box, but with *The White CD*, I wanted to do something different. I made a joystick and wired its switches in parallel with four of the switches on one of the box's circuit boards. I didn't modify any of the other circuit boards or the power transformer.

You should retain the box's headphone jack and make it accessible on the finished piece; that's how you'll plug it into a stereo amplifier, via mini-to-RCA cable. If it's destined to be a component for your stereo and you won't be using headphones, don't bother making the volume knob accessible, since you'll be using the volume on the amp. I've found that turning the CD player's volume control almost all the way up matches the output of a typical component CD deck.

A minimal FM tuner is usually included on the box's amplifier/power circuit board. You can incorporate it if you wish, and if so, be sure to make the CD/FM switch accessible.

The boomboxes I've reconfigured play standard CDs with reasonably good fidelity. For more bucks, you could choose boxes that will also play MP3-encoded discs, CD-Rs, and even audiophile SACDs and DVD audio disks.

Photography by Larry Cotton

Making *The White CD*

Here's how to "transmogrify" a boombox into *The White CD* configuration. As you work, test that the player still functions after each change you make, and then disconnect from power before moving onto the next step.

NOTE: The lasers in CD/DVD players can be dangerous. Don't stare at the laser when the device is on! Also, never rout or drill near digital circuitry, especially on the board that controls the CD stepper motors. Flying sparks can fry sensitive chips.

1. Carefully take your boombox apart without breaking or disconnecting any wires. Be especially gentle with the assembly that carries the turntable, stepper motor, and tracking mechanism. You'll likely find three circuit boards: the control electronics board, the display/switches board, and the amplifier/power board. Watch out for all of them.

I've seen two designs for securing the CD to the spindle. Some players use three spring-loaded plungers at 120-degree angles. Others use a ring magnet in the lid. If yours uses a magnet, be careful when removing it because it could be fragile.

Circuit board and transformer from donor boombox attached to mounting board with standoffs and heat-shrink tubing to prevent shorts.

2. Make a mounting board that's big enough to hold all the pieces you're using. Hardwood looks cool, but *The White CD* is made of primed masonite laminated to plywood. You'll also need auxiliary mounting posts and bosses.

3. Cut a hole from the front of the mounting board, then rout a recess in the back of the board deep enough to let the CD just clear the front surface. This lets you really see and get to the CD, which I like.

4. Mount the turntable assembly to the board using some sort of vibration-damping material in between. You could use the grommets originally attached at the corners of the turntable, or glue it down (or in) with bathroom caulking compound.

5. Mount the display/switch board in a similar manner.

6. Mount the CD control electronics board using special care, noting that certain wires must be able to flex as the CD plays.

7. Mount the other circuit boards and transformer with screws and possibly short pieces of tubing as standoffs, if necessary. Disconnect any unused subsystems, such as speakers. Make sure all 115VAC wires are well-insulated. Keep all circuit boards away from each other so that no short circuits develop.

8. If you need to switch the CD deck on and off independently of the other components in your stereo system, add a simple, single-pole, single-throw (SPST) switch that can handle a few amps, placing it in series with the power line cord and the transformer.

9. Connect the CD player to your sound system, put a CD on the spindle, add the magnet (if used), and ensure clearance for the CD. Power it up.

If there is an FM radio circuit, be sure that the CD/FM switch is in the CD position. Turn up the volume on both the player and the amplifier. By now, you should be hearing deafening sounds from the Stones or possibly John Prine.

10. Now you can build and connect the joystick, wiring its switches in parallel with the boombox's buttons (*see the diagram on page 136*). My joystick functions are: Up = Play/Pause; Down = Stop; Right = Next; and Left = Previous.

On the control PCB, find the leads that the buttons connect to, and solder in the same connections through the joystick. Be prepared to spend a fair amount of time figuring out where the joystick's switches should connect to before soldering them in.

11. Make front-side guides for the joystick, to prevent two diagonal switches from being actuated simultaneously.

12. A joystick makes the boombox's original switches redundant, so I covered them up with a scrap piece of 1⁄16" Formica. Make sure you can still clearly see the CD track number.

13. Make a stabilizing bracket for the mounting board from ¼"-thick x 4"-wide clear acrylic (Plexiglas), by carefully heating the acrylic over an electric stove burner and bending two 120-degree angles. Strategically place aluminum foil to reflect heat as necessary, and control the bend angles with a fixture.

14. Glue a jewel-box shelf to the mounting board about 7½" below the joystick centerline.

15. Unless you go with hardwood, finish your baby with Bondo body filler (available at auto stores), automotive glazing compound, and paint. Fill large flaws with body putty and smaller ones with glazing putty. I then painted the back flat black.

16. Finally, use a hobby knife to carefully cut vinyl tape (available in black and colors) into the graphic shapes that indicate the joystick functions. Precisely locate the shapes around the joystick, and press into place. You're done!

17. OK, I lied — you're not done. Reconnect your cable, admire your handiwork, put on a Kimmie Rhodes CD, power it up, and enjoy your new CD deck!

Larry Cotton is a retired engineer and part-time math teacher who lives in New Bern, N.C., and likes to listen to, write, and play anything musical. He built the player marimba featured in *MAKE, Volume 04, page 20*.

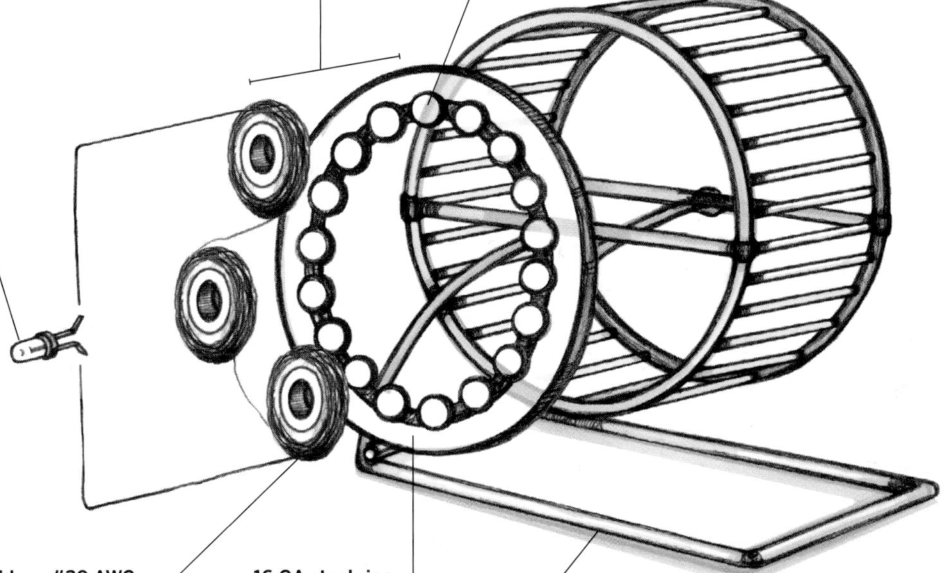

Hamster Power! Your pet's energy production at the exercise wheel will be lost no longer.

Illustration by Damien Scogin

RUN, LALA, RUN

A rodent-powered nightlight.

By William Gurstelle

Which maker hasn't thought, at some time, about putting the pet dog/cat/horse/weasel on a treadmill and actually getting something in return for all the money they've spent on the animal? Well, that time has come. Look out, dear pets, your free ride has come to an end.

The rodent-powered nightlight (RPNL) described here is a custom-made, low-rpm alternator. The original concept came from the folks at otherpower.com, who also sell magnets and wire for the insides of the alternators.

If built well, the RPNL can generate enough voltage to energize an LED when Lala the hamster merrily spins her exercise wheel in the wee hours. According to people who study these things, hamsters and similar rodents typically spin their exercise wheels at about the same rate as an LP record (which opens up the intriguing possibility of a hamster-powered phonograph as well, but that's something for a later issue.)

The Design

My Lala turns an axial flux permanent magnet alternator that has two main parts. First, an even number of strong permanent magnets are arrayed on a steel ring attached to the side of the hamster wheel. This is the Magnet Assembly.

In close proximity to the plane of the magnets goes the second main part: the Coil Assembly. The Coil Assembly consists of two or three bundles of tightly wound wire coils made from 30 AWG magnet wire. As the hamster spins the exercise wheel, the wire coils slice through the magnet's flux lines and generate voltage.

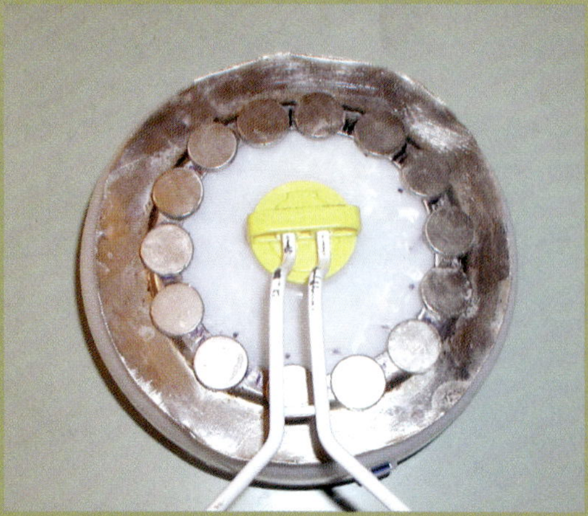

Magnet wheel assembly attached to side of rodent exercise wheel.

The diagram on the previous page shows the basics. Making this nightlight is straightforward, but care is required to make it work. Lala isn't capable of putting out a lot of energy (1 hamster-power = .00016 horsepower, by my calculations), so there isn't much tolerance for inefficiencies.

Step 1: Make the Magnet Assembly

MATERIALS
16-gauge (or so) steel ring Exact ring dimensions depend on your particular exercise wheel.
¾" diameter by ⅛" thick round magnets (18) You can use more magnets, if they will fit, but there must be an even number of magnets.

Space 14 of the magnets evenly around the perimeter of the steel ring, forming a perfect circle. The magnets *must* be placed with their north and south poles alternating. Test each magnet before gluing by holding it over its neighbor. They should repel. Then mark out the spacing and glue the magnets in place with Super Glue.

Once the magnets are glued down, use the remaining 4 magnets to attach the magnet ring to the back of the exercise wheel.

Step 2: Make the Coil Assembly

Take a large spool of #30 AWG magnet wire and wind 3 coils by hand. The coils should be tightly wound and thin — around ¼" thick. I used 2 of the plastic separators that came with the magnets as a core (any thin, non-magnetic disk will do) and simply wound 400 turns of wire around the separators, using 2 large washers on a bolt as a wire form. Once the coils are wound, squirt on Super Glue to keep the coils from unwinding.

MATERIALS:
Spool of #30 AWG magnet wire
Magnet core material
Wire winding device Consisting of one 3"-long, ¼" bolt, 2 fender washers, and 2 nuts
Wooden framework The exact shape depends on your hamster wheel.
LED
Tape

Now, mount the coils to a piece of wood so that each coil is directly over the center of one of the magnets. It doesn't matter which magnet the coil is centered over. What is important is that each coil is directly over a magnet, and the gap between the face of the coil and magnet is as small as possible.

Step 3: Connect Them Together

Once that's done, it's time to connect the wire coils. The best way to do this is experimentally, using a voltmeter. Remove the insulation from the wire ends with sandpaper and connect the wire ends to a voltmeter. Hold the first coil near the wheel and spin it. You should see a voltage reading on the dial.

Connect the second coil to the first coil in series and hold both coils near the spinning magnets. When the voltage doubles, you'll know the coils are wired correctly. Likewise, connect the third coil, and tape all coils to the wooden coil holder you've rigged up.

Connect the 2 loose ends of the wire coils to the LED leads. If you've made the RPNL correctly, a peripatetic hamster will generate enough voltage to light up an LED. For those with lazy or phlegmatic rodents, note that red LEDs require the least voltage to operate.

Now if only we could harness the power of those drinking bird toys with the top hats.

William Gurstelle wrote "Homecasting Digital Music" on page 133.

Photograph by William Gurstelle

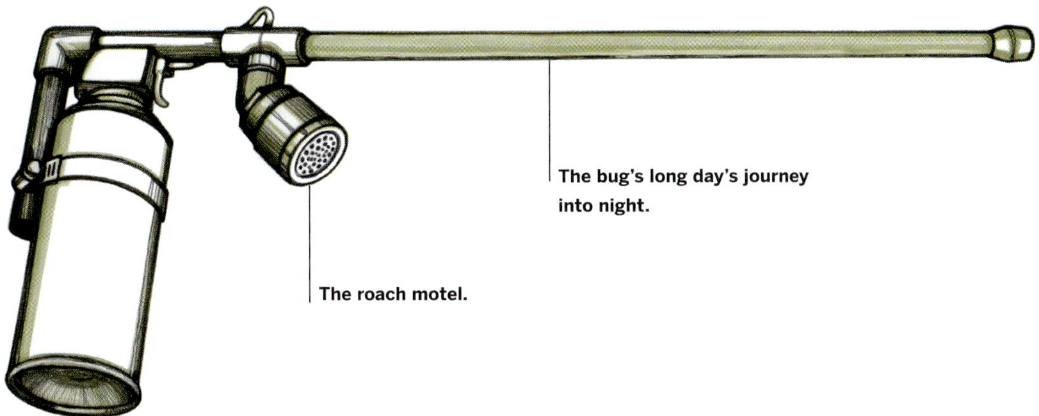

This bug suction device is so much fun to use that you'll relish home spider infestations.

SPIDER RIFLE

Humane, compressed-air-powered bug trapper.

By Matt Lind

I created the spider sucker rifle because my wife and mother-in-law are terrified of spiders in the house. I had seen "bug vacuums" on infomercials but didn't want to shell out the $30-$60 for something that runs on batteries and probably doesn't perform all that well.

What you are about to build is not technically a rifle, as it does not launch a projectile, but it does look like a rifle. It works by directing compressed air downward through the "drop tube," which draws air from the barrel, along with any hapless bug that happens to be next to the barrel end.

Illustration by Damien Scogin

Make It

Important! If you use a saw for cutting the pipe, make sure you wear proper respiratory equipment. This also applies for drilling. Dust from CPVC is toxic and can stay in your lungs. Using a pipe cutter instead of a saw will minimize dust. Check out the materials list on the next page, then begin.

1. Cut the ½" CPVC pipe into 4 lengths: 18" (1), 4" (2) , ¾" (1).

2. On one of the two 4" pieces of pipe, cut a slit lengthwise ½" from one end. The slit should be about ¾" long and go completely through the pipe. This slit will be used to clamp the can of compressed air to the Spider Rifle.

You can either drill a few pilot holes through the pipe and cut the rest out with a utility knife or use my preferred method: a Dremel with a cutting bit. Be careful with a knife, as it can slip easily, and as always, wear a dust mask or respirator.

3. Use a 3⁄32" drill bit and drill a single hole in the center of the CPVC T-fitting. This will be the

MATERIALS

To construct the Spider Rifle, you will need the following parts:

- 5' or 10' length of ½" CPVC plumbing pipe $2 for 10 (10' usually costs only 50 cents more than 5')
- ½" CPVC coupler
- ½" CPVC T-fitting
- ½" CPVC 45° elbow fitting
- ½" CPVC 90° elbow fitting
- 1"x½" PVC coupler
- 1" PVC end cap
- 70mm hose clamp
- Can of compressed air $5
- Small piece of thin, semi-rigid plastic, such as from a coffee can lid or the walls of a soda bottle

Total approximate cost is $10. This can be more or less depending on the cost of individual components in your area, as well as your choice of adhesive.

OPTIONAL ITEMS

You can use PVC cement, hot glue, or another form of adhesive for connecting parts. I use PVC cement, as it provides a permanent bond.

- Combination pack containing 1 can of PVC primer and 1 can of multipurpose PVC cement $4
- Can of spray paint in color of your choice $1
- Can of spray clear coat (protects paint) $1

hole that the straw from the can of compressed air goes into. You can use other sizes of drill bits, but you may have to ream it out a little if you use a smaller one or fill the gaps with hot glue if you use a larger one. Any type of drill should work, but a drill press is best. Since you are working with a fitting, you can connect a piece of scrap pipe into the end, to give you a handle to hold on to.

4. Drill as many holes as possible into the 1" end cap. The more the better, to let air through.

5. Trace and cut out the T-fitting stop valve. This can be cut out of pretty much any semi-rigid plastic. I prefer coffee can lids, as they are thick enough, easy to trace on, and easy to cut.

First, take a piece of ½" pipe, stand it on top of the plastic, and trace around it using a marker such as a Sharpie.

Then cut out the circle with scissors and place it into the T-fitting. If it doesn't slide in easily, cut small amounts off until it does. Be careful not to cut too much off or it will slide through. You want the stop to rest up against the ridge near the middle.

6. Push the 4" piece of pipe without the slit firmly into the side of the T-fitting that contains the plastic stop.

7. Connect the 90° elbow to the end of the 4" pipe.

8. Push the ¾" pipe into the bottom of the T-fitting.

9. Connect the 45° fitting to the ¾" pipe as shown in the illustration on the previous page (it's used to attach the "roach motel").

10. Connect the other 4" pipe to the 90° elbow fitting.

11. Connect the ½" to the 1" coupler to the end of the 45° elbow.

12. Connect the barrel to the T-fitting.

13. Connect the ½" coupler to the end of the barrel.

14. Prime and cement all the joints, or use the adhesive of your choice.

15. Cut the ½" coupler at the end of the barrel in half. There is usually a seam on the coupler that you can use a guide.

16. Insert the clamp through the handle and partially tighten.

17. Twist the 1" end cap onto the bottom of the 1" coupler.

18. At this point, you can paint the rifle if you so choose. Allow the paint to dry thoroughly.

19. Insert the plastic air tube that came with the can of compressed air into the top of the T-fitting.

20. Slide the can of compressed air up through the clamp so that the top rests firmly against the rifle. Tighten the clamp so that it is snug. Don't overtighten, as the can is under pressure. Connect the air tube to the can of air.

To use the rifle, simply point the barrel near the insect and squeeze the trigger. The air is forced through the bottom of the drilled PVC end cap. Since the T-fitting is blocked on the side connected to the handle, by forcing air downward and out the bottom a vacuum is created in the barrel. This vacuum draws the insect inward and into the "catch" (drilled 1" end cap). Build and use this at your own risk. But most importantly, enjoy your soon-to-be spider-free home!

See an animation of the Spider Rifle in action here: mslworks.com/SR1.gif

---

Matt Lind is a frustrated consumer living in Appleton, Wis., who has always enjoyed thinking of new uses for the parts in existing products.

Recursion Alert: View this image through red-cyan glasses to see how to take a 3D photo, in 3D.

3D Photography

Taking stereographs is easier and more fun than ever. By Bill Coderre

3D photography has been around for almost as long as regular photography. But the 1950s brought the fad of 3D horror movies, starting with *Bwana Devil* (1952) and *House of Wax* (1953). Many of these were so bad that 3D soon became an embarrassment. Even 50 years later, 3D still hasn't shaken its reputation as a cheap gimmick that causes headaches — which is a shame, because the medium can produce compelling images worthy of contemplation.

It's time for a 3D renaissance. In the last few years, computers have made it simple to produce color 3D images in just a few minutes. All it takes is a digital camera, some free software, and a pair of red-cyan glasses. I'll explain how to do this, and also suggest some camera projects that can make taking 3D pictures even easier.

If you don't have red-cyan glasses, you can find some at a comics shop, attached to any Ray Zone title. Or, you can buy them from American Paper Optics (*see page 145, "Further References"*).

The Stereo Shuffle: Taking Basic 3D Photos

Grab your digital camera, and head outside. Find a scene with a fun variety of distances to focus on. The closest objects should be at least 5 feet away, and perfectly still. Also, make sure that everything in the scene will be in sharp focus. Sunny conditions will help your depth of field.

Ready? Hold the camera to your eye, and lean very slightly to the left. Snap a picture. Lean very slightly to the right, and snap another picture. That's your basic "stereo shuffle" right there.

Photograph by Bill Coderre

Stereo 3D Software

A big list of 3D software is at anabuilder.free.fr/Concurrents.html

Java **(runs on Mac, Windows, Linux, and probably your toaster)**
AnaBuilder anabuilder.free.fr/indexEN.html

Windows
Anaglyph Maker stereoeye.jp/software/index_e.html
StereoPhoto Maker stereo.jpn.org/eng/stphmkr
3D Maker tabberer.com/sandyknoll/more/3dmaker/3dmaker.html

Mac
StereoPress (runs in Classic mode) makezine.com/go/stereopress
Anaglyph Maker lamarchefamily.net/nakedsoft
3D Maker tabberer.com/sandyknoll/more/3dmaker/3dmaker.html

For best results, make sure that the camera moves from left to right about 2 inches, and does not move up or down, toward or away from the scene, or tilt in any direction. Also, do not realign the frontmost object to the same position in the frame. (For the nerds: no y or z translation; no roll, pitch, or yaw; and do not "toe in.")

Now head back to your computer, import the pictures from your camera, grab some free software (see table above), and use it to load the left and right frames of the pair. I recommend AnaBuilder, which has all the functionality you'll need, once you figure out the buttons and menus.

Display the picture as a color anaglyphic image, and before you put the glasses on, correct any twist or vertical offset. The horizontal lines in both views should overlap and be parallel, and objects in each image should occupy the same vertical position in the frame.

One last step: make sure that the frontmost object in the scene has its two views completely superimposed, with no red and blue halos around it. This means that the nearest object in the picture will be at the stereo window depth, not sticking out in front of the frame. This makes the scene easier for most people to view. You can experiment with "eye-poking" stereo later.

Now, put on the glasses and admire your handiwork. Believe it or not, that's it for the basic method. Everything from here on is just enhancements and alternatives.

Near or Far Subjects: Adjust the Interaxial Distance

If your subject is a small flower or a mountain, having it 5 feet away won't work. With a close shot, 2 inches between pictures (called the "interaxial distance") is too much, and the picture will be unviewable, or at least headache-inducing. With a distant subject, 2 inches will not be enough to give any appreciable 3D effect.

For these situations, the basic rule is to divide the distance to the nearest object by 30, and move the camera that amount between shots. That means 176 feet for an object a mile away, and 0.8 inches for an object that's 2 feet away.

These approximations, however, do not take into account important factors such as lens focal length and the distance of the farthest object. For a more rigorous approach, see the Bercovitz Formula article listed under "*Further References*."

More Accurate Alignment: Make a Candy-Tin Slide Bar

A stereo slide bar will ensure that your camera moves only horizontally between pictures. You can easily make one out of a tripod, a bubble level, and a candy tin.

Use a nibbler, a drill, or some other tool to cut a slot in the bottom of the tin. The slot should be a bit wider than ¼", running the full length of the box from side to side, parallel to the hinges, centered halfway between the hinge side and the front. Place the slot onto the ¼-20 head screw (¼" diameter, 20 threads per inch) that sticks up from the top of your tripod, and align the candy tin so that it is parallel to the tripod's handle. Use poster putty (such as Fun-Tak) to attach the camera to the candy tin, and the bubble level to either the camera or the tripod. To take pictures, make sure the tripod is level, then snap one photo, slide the camera over, and snap the second photo. It also helps to add a ruler to the back of the candy tin, so you can control your interaxial distance.

Live Subjects: Link Two Cameras

You can take 3D photographs of subjects that don't hold still by using two cameras. Place them next to each other on a bar, or tack them together by their bottom surfaces and rotate or invert the images so they match later. If the lenses cannot be placed close enough together, you can rig up a mirror system. The other half of the solution is finding a way to trigger both shutters at the same time. This depends on the type of cameras you're using (*see "Further References"*).

"Found" 3D

My friend Eric von Bayer pointed out that you can create stereo pairs by selecting frames from 2D movies. Copyright considerations prevent me from showing this example, but you can download the *Star Wars Episode I* trailer, and extract two frames from the shot that pans horizontally over a landscape of Naboo. Drop them into AnaBuilder, and you end up with a pretty nice 3D scene.

Loreo Cameras

Loreo (loreo.com) makes inexpensive 3D cameras that use mirrors to put the left and right stereo images on one standard 35mm film frame. For viewing, you could either use their included viewer, or get a digital image, and then cut and paste the two sides into the software provided. The Loreo doesn't let you adjust the exposure, but at least it has a built-in flash.

Other Display Options

Red-cyan glasses take some getting used to, and you can see "ghosts" when the disparity between the images is large. Here are some other display technologies.

The **ColorCode** system substitutes blue and amber for red and cyan. This reduces ghosting, but it's also patented, and illegal to distribute software that produces ColorCode images.

Adjacent-image pairs are viewable by crossing or uncrossing your eyes, but this is a tricky skill to learn, and it limits image size. To view bigger images, you need a stereoscope, like the ones from the turn of the 20th century. These devices have two lenses mounted off-center, to allow for close-focus, and a prism that lets the eyes point straight ahead while the view-lines diverge. Fancier viewers use mirrors to accommodate larger stereo pairs. You can make your own stereoscope, but note that the brain usually interprets misaligned left and right eye views as caused by food poisoning, and reacts by the sudden expulsion of the contents of the stomach. Therefore, it's important to align the mirrors precisely.

Time-interlaced images are viewed by LCD shutter glasses, which you can get for $30 from Razor 3D (razor3donline.com). Unfortunately, you cannot use these glasses with LCD monitors, progressive scan monitors, or any monitors set to refresh faster than 90Hz. As a result, images often seem to flicker, but it helps to turn down the brightness of the monitor, and reduce room lighting. At standard television frequencies (roughly 50-60Hz), movements appear jerky, and fast-moving objects will often "tear."

Horizontally interlaced image sets are displayed by putting the interlaced image behind a lenticular screen, like with 3D postcards. Free software such as Interlace produces interlaced images that you can print on a high-resolution printer, and then display in a sleeve behind a reusable lenticular screen. The big plus of this technology is that it does not require the user to wear glasses or use a stereo viewer. The downsides include low horizontal resolution, and the difficulty of aligning the image with the screen.

Several television companies have hinted at future **3D TV** products based on lenticular screens, but the NTSC format's width of 720 isn't enough to split in half and retain a good picture. With horizontal resolutions of 1,280 and 1,920, HDTV seems like the "silver bullet" that might make it all work. We can only hope.

Further References

General 3D information: stereoscopy.com and dddesign.com/3dbydan/3dlinks

Explanation of the Bercovitz Formula: makezine.com/go/vic3d

Synchronizing twin cameras: makezine.com/go/twincam

American Paper Optics, manufacturer of 3D glasses and experimenter kits: americanpaperoptics.com

Ray Zone, "The King of 3-D Comics": ray3dzone.com

Bill Coderre dredged most of this article from class notes and memories of a seminar by 3D pioneer Stephen A. Benton, professor at the MIT Media Laboratory, to whose memory this article is dedicated. These days, Bill is a programmer for Apple.

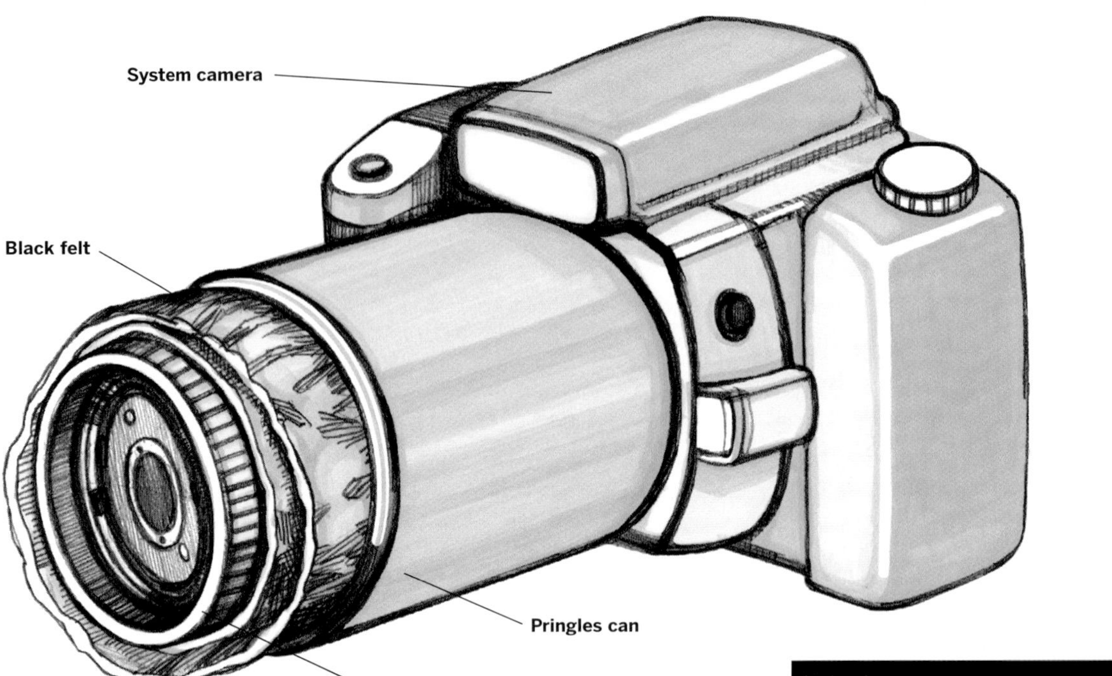

Make this macro extension tube out of $5 worth of materials.

Macro Photography On a Budget

Pringles-can lens extender produces dazzling ultra close-ups for peanuts. By Haje Jan Kamps

Photography is often seen as an expensive hobby, but here's a way to create a lot of fun for very little money: a macro extension tube you can make for less than $5 worth of materials.

Macro Tube Theory

The concept of a macro tube is as old as photography itself. Normally, a lens takes incoming parallel light and focuses it down onto the film or imaging chip. By turning the lens the wrong way around, the opposite happens: light focused in on a small area is refracted to become more parallel.

With the lens in its wrong-way configuration, either by using a reversal ring or by just holding it against your camera body, you can take photos up close. In theory, the farther away from the camera you hold your lens, the higher magnification you get. The problem is that the "air" between the camera and lens element can't have any light leaking in. Furthermore, there cannot be any lens movement when you take photos, as this will result in blurry photos. The solution? A macro extension tube, which holds the lens and shuts out light.

Extension tubes and bellows can be bought, but that's definitely more expensive and probably a lot less gratifying than building something yourself. For our tube, we'll use a Pringles can,

Illustration by Damien Scogin

which is long and lightproof. At one end, we'll attach a camera body cap with a hole cut out of the middle, which will make it easy to screw the extension tube into the camera body. At the other end, we'll put the reversed lens.

MATERIALS

- System camera with removable lenses (SLR or similar)
- Body cap for your camera
- Pringles can, or tube of similar dimensions
- Standard (i.e. non-zoom) lens
- Dremel or other cutting tool
- Black felt
- Sandpaper
- Hot glue or epoxy

WARNING: Throughout this project, the insides of the camera are exposed to dust and dirt, and cameras hate both. It may be best to do this experiment with an old, film-based SLR. If you wind up having to take your Digital SLR camera in for servicing, it will be far more expensive than just buying a macro lens in the first place.

Perforate the Cap

Use the Dremel to cut a hole in the camera body cap. Cut out as much as you can, but make sure to leave the whole bayonet mount intact. Sand the inside edge to matte the surface; this edge may be subjected to light from the lens, and you want to prevent reflections.

Prepare the Tube

Mark the hole in the body cap on the bottom metal end of a Pringles tube, and use a Dremel to cut a neat hole in the bottom of the tube. When this is completed, make sure to completely clean the inside of the tube — you don't want aluminum shavings (or salt and grease) in your camera body. Then line the inside of the tube completely in black felt. This will stop reflections, which will dramatically improve the photo quality.

Attach the Cap

Glue the body cap to the Pringles tube using hot glue or strong epoxy. This connection has to bear quite a bit of weight, so use strong glue. If the glue dries translucent, use black paint or a felt-tip marker to darken it and prevent light leaks.

Choose and Install the Lens

Most lenses, including zoom lenses, should be perfectly usable for this project, but I recommend a standard lens, as they are usually cheaper, sharper, and easier to focus. I used a Canon 50mm f/1.8 MK2. It is the cheapest lens in Canon's catalog, but also one of the best.

If you're lucky, you'll be able to slip the lens inside the open end of the Pringles tube, and it will fit snugly against the felt. Failing this, wrap another layer of felt around the lens, tape it down, and try again. Be sure to put the lens into the tube backwards, with its electrical contacts facing away from the camera body.

Take Photos!

You don't focus this lens by using the focusing rings; instead, you just move either the camera or your subject matter nearer or farther. This lens has next to no depth of field, so focusing in this way can be frustrating.

For setting exposure, your camera's internal light meter is actually a good starting point — it isn't always accurate, but it will give you an idea. Be careful, though, because any movement during the exposure will blur the image. To prevent blurring, use the self-timer or a remote shutter lead. If your camera supports mirror lock-up, use this as well, to further reduce vibration.

Taking good macro photos takes a lot of practice, and it's often a technical exercise rather than a creative one. But it's fun, and it opens up a world of photos that would otherwise have been impossible. Experiment, create, and enjoy!

Canon EF Aperture Hacking

All of Canon's newer lenses have electronically controlled aperture, so you can't normally stop down a lens unless it is attached to the camera. There is a trick you can use, however: put the lens on the camera, and set the camera to fully manual (M). Select the aperture you want. Then, press and hold the aperture preview button, and take the lens off the camera. If you have done it right, you are now holding a lens that is still stopped down at the aperture you selected.

Haje Jan Kamps is a freelance writer and photographer based in Bristol, England. He runs photocritic.org for DIY photographers and their projects.

An easy alternative to insanely overpriced digital picture frames.

$130 Digital Picture Frame

Build one in 15 minutes.

By Mike Kuniavsky

Call me a cheapskate, but $400 for a digital picture frame when a complete laptop costs $600? I won't stand for it!

Call me lazy, but a week of evenings with a Torx screwdriver, a razor blade, and an old laptop, trying to get a memory-resident Linux install to boot? I have better stuff to do!

I've wanted a digital photo frame for a long time. Unfortunately, my cheapness and laziness have stood in the way. Paying $400 for an LCD and some driver hardware (when you can get a DVD player for $85) just seems wrong. And hacking apart an old laptop is the kind of project that invariably gets put off indefinitely.

Until recently, I figured I'd end up waiting for the commercial ones to get cheap, but then I stumbled on a solution that satisfied the lazy cheapskate in me. Here's my $130, 15-minute solution.

Assemble

The fast version:

Plug everything together, and stick it all into a picture frame.

The slightly slower version:

1. Unpack everything (1 minute).
2. Plug the audio and video cables together. The yellow cable is the video cable, and is the most important. Plug the red and white ones together if you want to listen to MP3s or hear the sound of video (1 minute).
3. Copy some photos from your PC to the memory card, plug it in, and turn everything on to make

Photography by Mike Kuniavsky

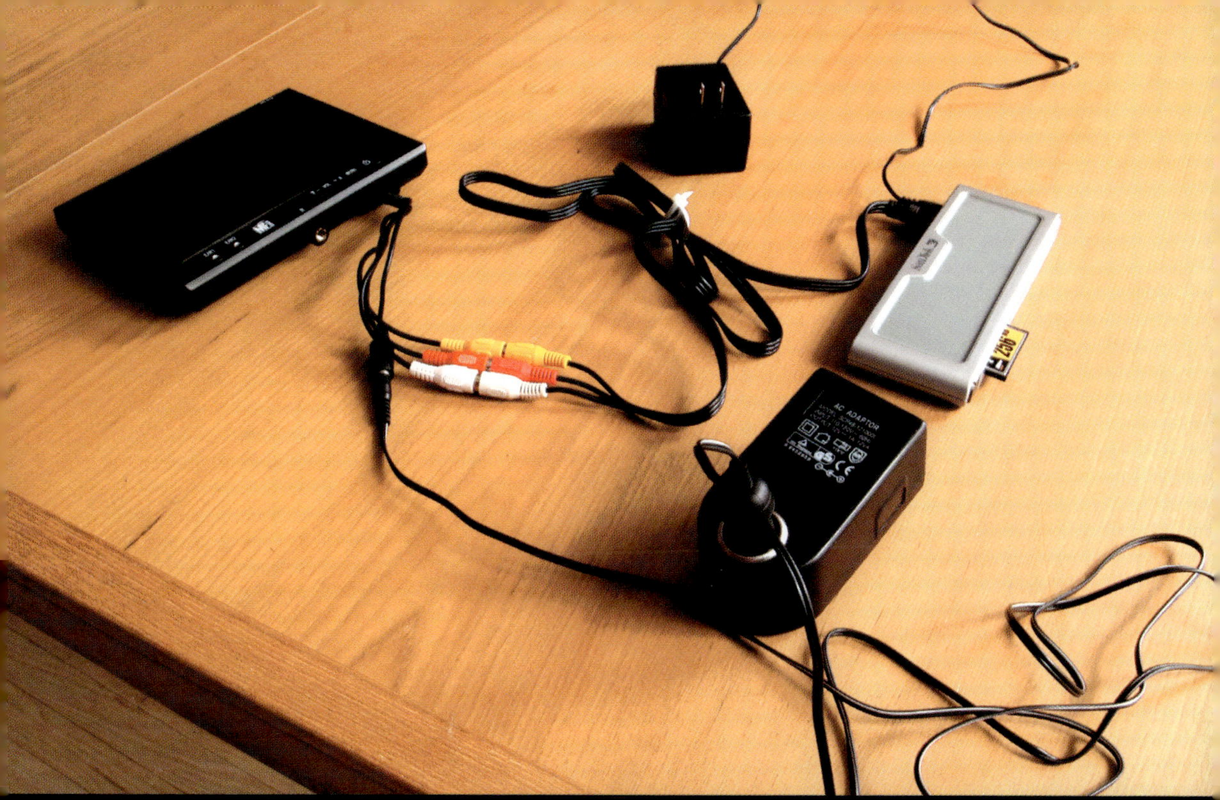

Photos are stored on a memory card that's plugged into a digital photo album/solid-state video player. These are small devices that let you watch photos and video from your digital camera on a TV screen. The player feeds a car video display of the type designed to keep kids quiet in the back seat. The display is powered by an adapter that converts house current into 12-volt, cigarette-lighter DC.

MATERIALS

Headrest video display for cars These are available on eBay for between $50 and $200, depending on the size, the brand, and whether you get them singly or in packs of two (one for each headrest). Avoid monitors that have extra hardware. Monitors already embedded in headrests or that come with dash-board/ceiling mounts have to be extracted, which, you know, takes time. Make sure it has an RCA video input jack.

Memory-based video player Solid-state video players are likely to be smaller, cheaper, and more reliable than hard-disk or DVD players. I tried the SanDisk Digital Photo Album and the Neuros MPEG-4 recorder for this one, but there are others. These run between $30 and $100.

Auto accessory wall adapter Essentially 12V transformers with a car cigar lighter plug, these let you recharge your mobile phone using your car charger, at home. I don't know why you'd want this, but someone clearly does, and these are on the market. I found one with 1 amp of output power, which is more than enough to drive a big LCD without breaking a sweat, for $3.50 at Electronic Goldmine.

Picture frame from a thrift store I was lucky to find one for about $2 that was an old perfume box/ picture frame, just the right size to fit the monitor and the player. The box-style frame is nice because the LCD screen, the player, and the data and power cords all fit inside. Otherwise, you might have to build a box to hold all the stuff neatly. And that would take energy and money.

A memory card You'll want a dedicated one, since it's going to sit in the player most of the time, and the bigger the card, the more photos you can put on it.

Tape I used clear packing tape because it was handy, though any kind of wide tape will work.

sure that all the electronics work. If everything works, take it apart to install all the stuff in the frame (2 minutes).

4. Install into the frame. This step will probably take the longest. In my case, I stuck the components into the box with zip ties and tape, but you could build an armature, cut a mat for the frame, and permanently affix the whole thing inside a frame with glass. I'm too lazy to do that, but you go for it (9 minutes).

5. Copy more photos to the memory card (2 minutes).

6. Plug it in (0 minutes)!

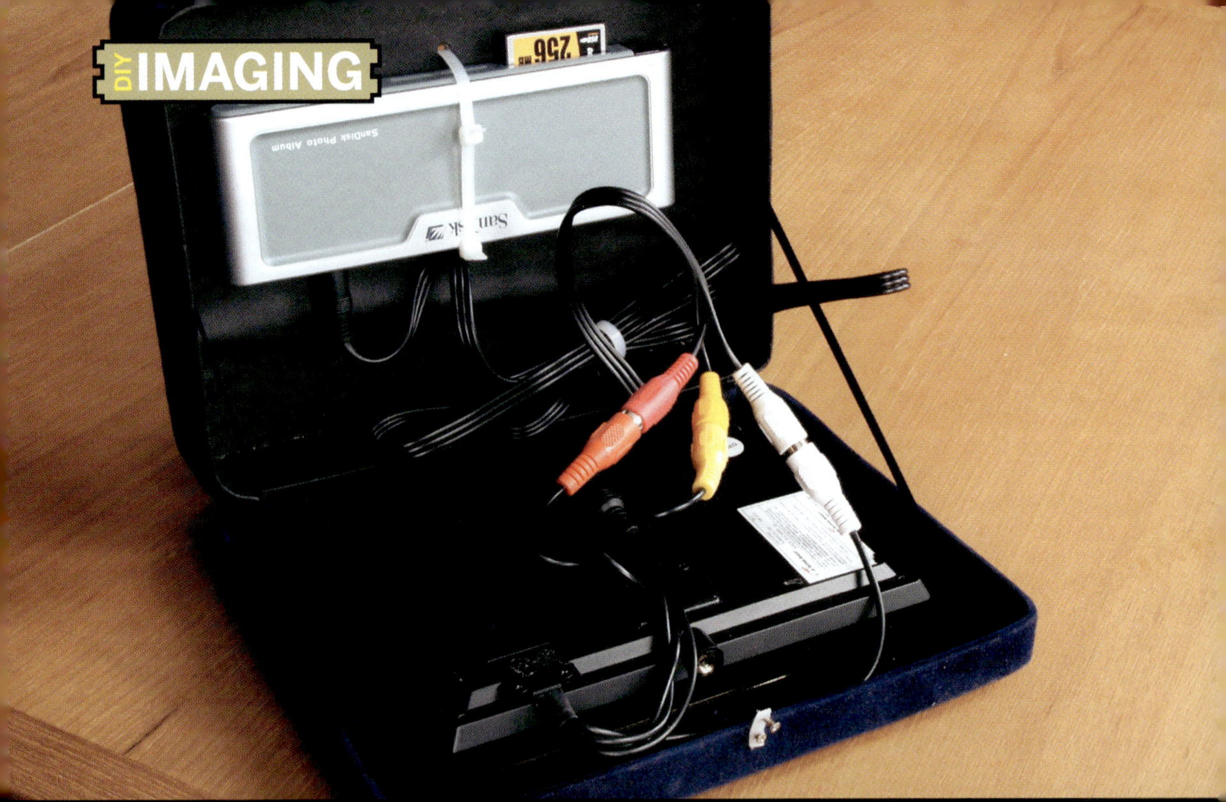

Components tucked inside the perfume box/picture frame. I zip-tied the player to the back, taped the display to the front, and zip-tied the cabling out of the way.

Things to remember:

1. Leave enough room to remove the memory card, so that you can add more photos later.

2. Don't entirely cover the infrared LEDs, so you can use the remote controls for the monitor and the player when it's assembled.

3. Adjust the screen to your liking before you cut the mat. LCD display proportions are 16:9, but most digital photos are 4:3. This creates a dilemma. Here are your three choices:

3a. Stretch pictures to fill the whole screen horizontally, like many people do with their TVs. This makes everyone in your pictures look 20% shorter and fatter.

3b. Leave black bars of unused screen to the left and right of the picture. You can mask these out with an appropriately cut mat.

3c. Clip the top and bottom of the images so they fill the full width of the screen without distortion.

I prefer not distorting or cropping my photos, so I chose 3b. Both the player firmware and the monitor setup menu let you configure this.

More Stuff to Do

These memory-based players can play back video (that's what the Neuros is best at; the SanDisk has better photo slide show functions), which adds a whole other dimension to what they can do. In addition to photos, I've been taking short video clips for the last year with my digital camera. The quality isn't as good as a video camera, and the players are finicky with the file formats they'll play, but slide shows of 10- to 30-second clips are surprisingly engaging, while still being ambient.

Of course, if you converted a full-length video to the appropriate format and copied it onto a memory card, you could watch a feature film on your picture frame. But why do that when that's what TVs are for!

My next step is to set up a script to automatically download the latest photos from my friends' Flickr streams onto a memory card, and then, maybe once a week, I'll swap cards in the frame — if I remember.

Sometimes being cheap and lazy is just so much work!

Mike Kuniavsky is a San Francisco- and Portland-based ubiquitous computing and user-experience consultant and writer who blogs at orangecone.com.

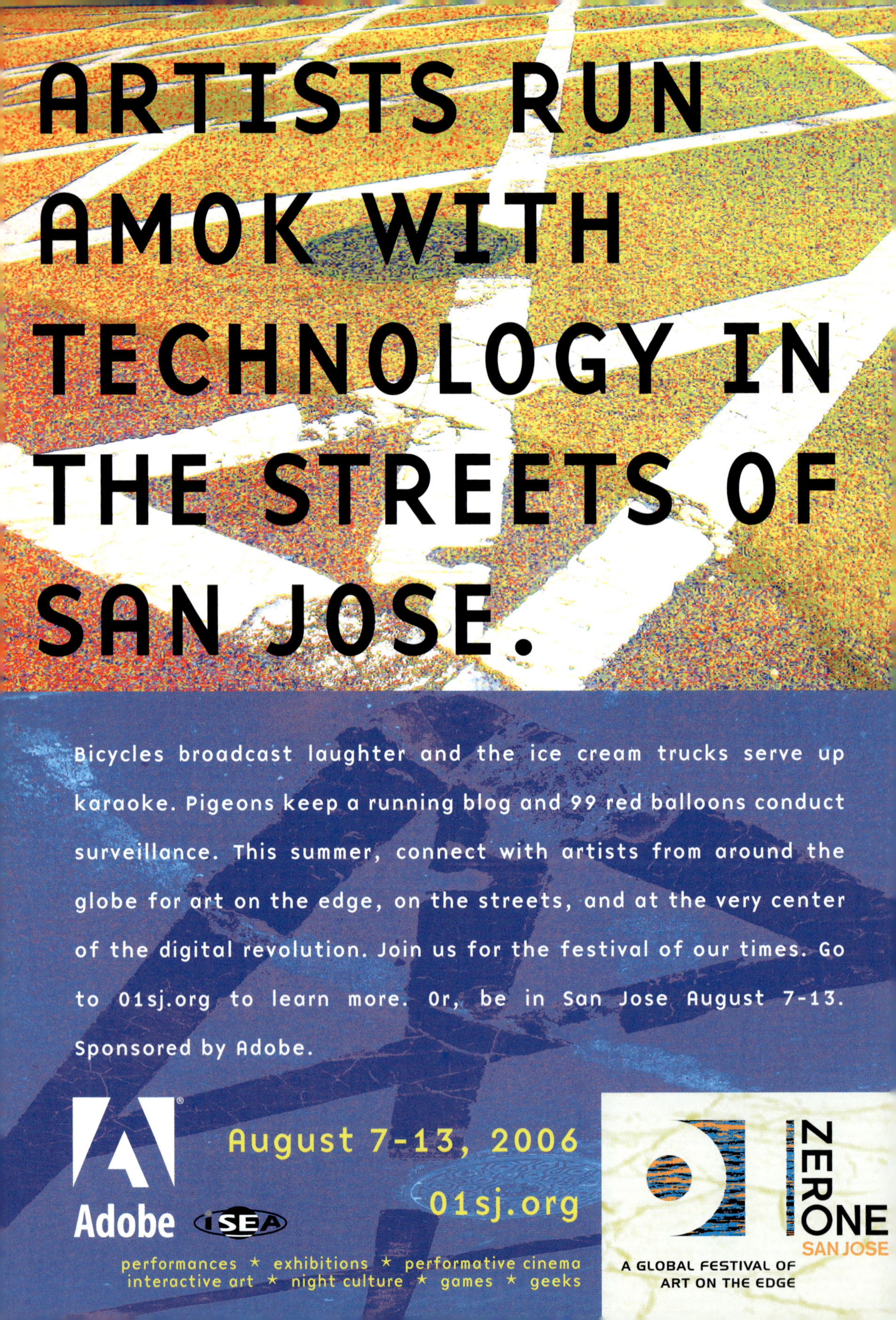
ARTISTS RUN AMOK WITH TECHNOLOGY IN THE STREETS OF SAN JOSE.
Bicycles broadcast laughter and the ice cream trucks serve up karaoke. Pigeons keep a running blog and 99 red balloons conduct surveillance. This summer, connect with artists from around the globe for art on the edge, on the streets, and at the very center of the digital revolution. Join us for the festival of our times. Go to 01sj.org to learn more. Or, be in San Jose August 7-13. Sponsored by Adobe.
Adobe
ISEA
August 7-13, 2006
01sj.org
performances ★ exhibitions ★ performative cinema
interactive art ★ night culture ★ games ★ geeks
ZERO ONE
SAN JOSE
A GLOBAL FESTIVAL OF
ART ON THE EDGE

THE SECRETS OF MONITORING ATMOSPHERIC HAZE

MAKE A MEASURING INSTRUMENT WITH AN OLD VIDEO CASE AND $20 WORTH OF PARTS.

By Dr. Shawn

EVEN ON THE CLEAREST OF DAYS, THE atmosphere over our heads buoys up thousands of tons of nature's crud — pollen, mold spores, microscopic fragments of decayed plants, insect parts, wind-borne dust, water vapor, smoke from forest fires, and volcanic ash — that we collectively call haze. Of course, we humans contribute more than our share as well. Smog is the best-known example of artificial haze.

Although many cities monitor the quality of their local atmosphere, shockingly little is known about how haze is changing globally because, believe it or not, no one is coordinating haze observations around the world. Considering how important these observations could be for the health of all life on our planet, this lack of attention is not only shameful, it could also be downright dangerous.

All this may soon change, thanks to the inventiveness of one of America's greatest living citizen scientists. Forrest M. Mims III is renowned as a science writer for mentoring more than a million professional and amateur gadgeteers through a series of wonderfully earthy, hand-drawn how-to booklets that he wrote for RadioShack. And even though he has no formal scientific training, Mims is a living legend amongst amateur scientists for his knack at besting the professionals at their own game.

In 1991, for instance, while standing in his backyard testing a $200 hand-held device of his own design, Mims discovered that an instrument costing perhaps a thousand times more than his, flying aboard the Nimbus 7 satellite, was steadily drifting out of calibration. That highly sensitive spectrophotometer, developed by NASA's elite corps of scientists and engineers, was the very same that the space agency had used to discover the "ozone hole" (really, a roughly 50% reduction in ozone concentration) that appears during the coldest winter months in the stratosphere above the Antarctic. Indeed, back then, this spectrometer was NASA's principle source of information about the health of the protective sheath of molecules that block the sun's most damaging ultraviolet rays.

NASA was understandably reluctant to admit that it had been shown up by an amateur. However, to their credit, when the red-faced researchers finally did own up to the error, they made Mims a collaborator. NASA sent this dogged citizen scientist to Brazil, Hawaii, and the western United States to study the atmospheric impacts of biomass burning and volcanic eruptions using his clever collection of low-cost homemade instruments.

MIMS' VISIBLE HAZE PHOTOMETER

One of Mims' ingenious gadgets is his Visible Haze Photometer. This remarkable device can be cobbled together from an old videocassette case and about $20 worth of RadioShack parts. Moreover, it's so simple to make that even the most hardened technophobe can put it together in an afternoon. This instrument could revolutionize haze research by opening the field to all comers: citizen scientists, science fair students, and weather watchers of all stripes.

The Visible Haze Photometer exploits the fact that the sun is the perfect probe for measuring haze. The intensity of sunlight striking the top of the Earth's atmosphere is essentially constant. So, by measuring the intensity of the sunlight at the ground and knowing the thickness of the atmosphere it has passed through, one can determine how much light has been either scattered or absorbed and then determine how much gunk is floating about overhead.

There is a complication, however. Air molecules themselves scatter light, a phenomenon known as Rayleigh scattering. The effect may create beautiful blue skies and brilliant red sunsets, but the fact that air molecules scatter blue light more strongly than red is a pain when it comes to measuring haze. To compensate, one must look only at a narrow sliver of the spectrum over which the scattering is constant.

To achieve this selectivity, professionally designed instruments typically started with a detector that could see the entire visible spectrum, and then added an expensive optical filter that permitted only a narrow band of frequencies to pass. But with no government grants to rely on, Mims needed to find a cheaper way.

LEDS: A CHEAP AND SIMPLE SOLUTION

His solution was to use a miracle of modern technology, called a light-emitting diode (LED), that costs just pennies. Electronically, LEDs couldn't be simpler. Just dump current into the bottom, and light of a single color flows out the top. Mims knew that this, like most natural processes, could run backward — light falling on an LED could create a current that could then be detected. Furthermore, just as LEDs produce a single color, an LED generates a current only when light of just the right color strikes it.

Mims' photometer uses a green LED that emits light at around 555 nanometers and detects light at around 525 nanometers. RadioShack sells them for under a dollar. The circuit contains just two other active elements: an operational amplifier and a resistor. Mims houses his instrument inside a black plastic VHS videocassette case with a hole drilled in one end to admit sunlight.

An LED can only detect light that has a shorter wavelength than the light it emits. Photons are emitted in an LED from a very special set of energy states that atomic physicists call the "conduction band." These are basically states in which an electron is free to wander about the substance. When emitting light, the energy to pump the electrons into the conduction band is supplied by current flowing in through the leads. It turns out that, to be detected, a photon has to supply this extra energy itself. And since photons with shorter wavelengths have more energy than photons with longer wavelengths, an LED's detectible photons always have shorter wavelengths than its emitted photons.

MAKE A PHOTOMETER

1 Prepare the LED

The body of the LED is epoxy. The top is shaped like a lens to concentrate the LED's relatively dim light into a narrow cone that is turned toward the viewer. It turns out that more of the sun's light reaches the sensor when the lens and most of the LED's body is removed. Coarse sandpaper and a little elbow grease do the job fast. File the case down vigorously until you've whittled it to within about 4mm of the light-producing element at the bottom.

2 Wire and Install the Circuit

Now, build the circuit and install the solderless breadboard in the VHS case. Keep a big gap between the LED and the hole that you will eventually drill in the top of the case to make sure that only the sun's light — and not the ambient light in your surroundings — reaches your detector. (You can use a hot glue gun if you're making this instrument for a short-term project, like a science fair. If you intend to get serious, then use tiny ¼" machine bolts to lock the breadboard into place.) Next, install the terminals, the battery holders, and the indicator light into the case where shown.

3 Test the Circuit

It's time to test your circuit. Indoor electric lights are too feeble to produce much signal, so take your instrument outside. With a multimeter, record the so-called "dark voltage" — the voltage you measure when the case is closed. With Mims' choice of components, it's likely to be rather large — perhaps several hundred millivolts. Next, open the case and let the sun's light shine down directly onto the circuit board. If the signal jumps several volts when you open and shut the case, your circuit is working.

4 Secure the LED

Use one corner brace to keep your LED sensor from shifting about over time. Slip the LED through the

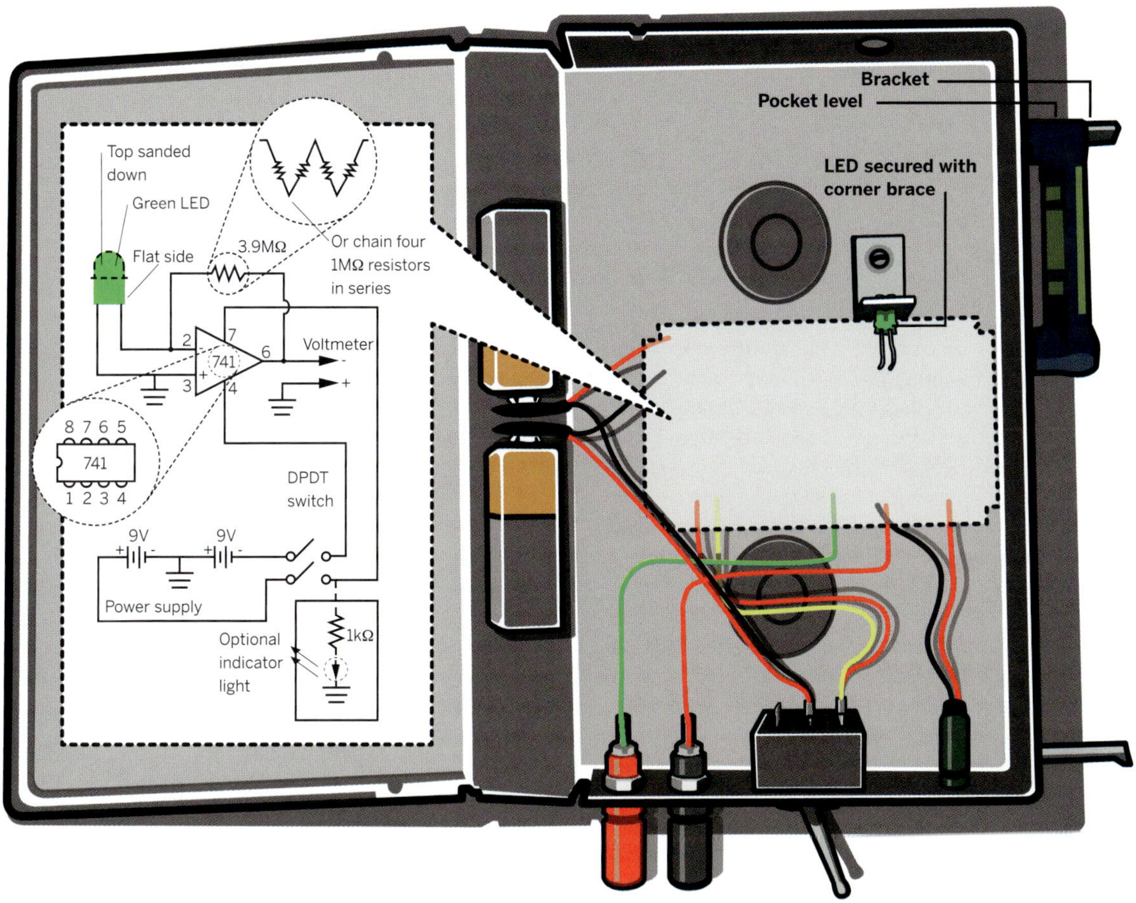

Illustration by Damien Scogin

Make Your Own Haze Photometer

Here are the parts you'll need:

741 op-amp (RadioShack 276-007)

3.9M ohm resistor (RadioShack does not happen to sell resistors with this value), **or four 1M ohm resistors connected together end-to-end to form a chain** (271-1130). Mims' circuit calls for a 3.9M ohm resistor because this is a standard value. But linking four 1M ohm resistors together to create a 4M ohm resistor will work just fine.

Small solderless breadboard (276-175). This device eliminates the need to solder wires together and makes building the circuit a snap.

Green LED (276-022A)

9V batteries (2)

9V battery caps (2) (270-0325)

Hookup wire or, much better, a wire kit (276-173), which is a small plastic case containing pre-stripped wires of various lengths. Wire kits are a great time and aggravation saver.

Double-pole double-throw switch (DPDT), since it can turn off both the positive and negative power to the op-amp. RadioShack's model 275-652 is perfect.

1" (25mm) L-shaped corner braces (3) from your local hardware store.

Black (and therefore, opaque) VHS case

Pocket level

Optional, but strongly recommended:

Battery clips or holders (2) (270-326B). These hold the batteries in place inside the case. You could always use duct tape, but holders make fieldwork easier.

Banana plug terminals (2) (274-662). You could feed the signal to your voltmeter by passing two wires through the case, but terminals make data taking more convenient.

If you decide to use banana plug terminals, then you'll also need **banana plug connectors (2)** (274-721) to go with them.

I've added an **LED indicator light** (276-309) to Mims' original design so I can tell at a glance when the circuit is on. If you want to add it to your instrument, you'll also need a 5kOhm resistor (271-1124) to limit the current in the LED.

hole in one side of the brace as shown, butt it firmly against the breadboard, and either hot glue, bolt down, or epoxy the brace in place.

Next, you'll need to drill the hole in the top of the case that lets in the sun's light. Carefully position it *directly above* the LED detector. The hole should be roughly the same size as the hole in the corner brace; 6" is just about perfect.

5 Mount Corner Braces

The other two corner braces, when attached to the outside of the case, will enable you to both align the instrument with the sun and measure the all-important sun angle — the angle between the sun and the horizon. If you'd like to be able to make measurements even when the sun is low on the horizon, you'll want to install the braces as far apart as possible. But make sure the vertical tabs face each other so that a paper ruler can be pasted flat between them. Hot glue, screw down, or epoxy the tabs firmly in place.

6 Make Alignment and Sun Angle Gauges

Next, place a small piece of white tape over the vertical tab on the bracket nearer the middle of the case. Then open the case and point the instrument toward the sun. Move the case around until the bright spot of sunlight that falls through the hole you drilled is centered on the LED sensor. If the brackets are approximately aligned, a second image of the sun, this one caused by the sunlight passing through the hole in the front brace, will appear somewhere on the tape. Lightly mark its center in pencil. The case may flex slightly when closed, so you may need to adjust your mark. Close the case and align the instrument to the sun using the outside brackets. Then tilt the instrument slightly in different directions while watching the voltmeter until you find the orientation that produces the largest voltage. Make a permanent mark at the center of the sun's image. From then on, whenever you center the bright spot on that mark, you'll know that your sensor inside the case is fully bathed in sunlight.

Next, paste a paper ruler on the case between the two brackets. You can photocopy a metric ruler. Finally, glue a pocket level to the case. Pocket levels work well, or you can purchase a small line level and, using a pair of pliers, break off the plastic parts used to connect the level to a string. With the level attached, you're ready to go!

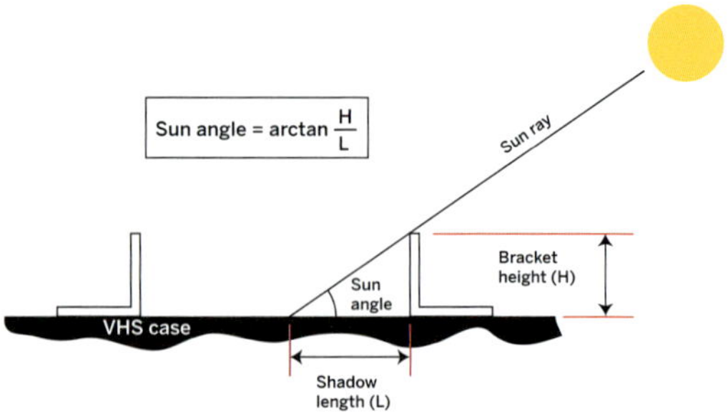

Calculate the sun angle by dividing the height of the bracket by the shadow length of the bracket. The arctangent of this value is the sun angle.

TAKE A HAZE MEASUREMENT

To make a measurement, you first record the dark voltage, that is, the voltage produced when no light strikes the detector. Cover the hole in the top of the case and record the voltage. Now, center the image of the sun on the mark you made on the lower brace and record the resulting voltage. The signal is this voltage minus the dark voltage.

Next, you'll need to find the sun angle. To do so, hold the case level, with the corner braces up. When you aim the case at the spot on the horizon directly below the sun, the front brace will cast a shadow along your rule. By simple trigonometry, the sun angle is the arctangent (use the `atan()` in your spreadsheet program) of the height of the brace divided by the length of the shadow. Don't forget to also record the time and the sky condition near the sun.

Although you can begin collecting data right away, you will want to calibrate your photometer as soon as possible. Remember that the signal you measure only tells you how much sunlight is reaching your detector. But to measure haze, you need to know how much light isn't making it because it's getting scattered or absorbed. To calculate that, you have to know what signal your instrument would read if you could position it above the Earth's atmosphere. That is, you need to know what signal voltage corresponds to the solar extraterrestrial constant (*ET*). While a bit tedious, it's not hard to find out.

You'll need about half a day, either early morning to solar noon or solar noon to early evening, when the sky is clear blue and there are few or

Aerosol Optical Thickness and the Langley Plot

Aerosol optical thickness (AOT) is the quantity normally used to quantify haze. To calculate it, you must subtract out your photometer's dark voltage and compensate for Rayleigh scattering and sun angle. The equation is:

$$AOT = \frac{1}{M}\left[\text{nl}\left(\frac{ET}{V_s - V_d}\right) - \frac{MP}{8660} \right]$$

where ET is the "extraterrestrial constant" (the signal that would be measured above the atmosphere), V_s is the measured voltage, V_d is the dark voltage, M is the air mass (as calculated from the sun angle), and P is the atmospheric pressure in millibars. The constant 8660 is an empirical factor for Rayleigh scattering at the wavelength detected by the LED.

Langley Plot

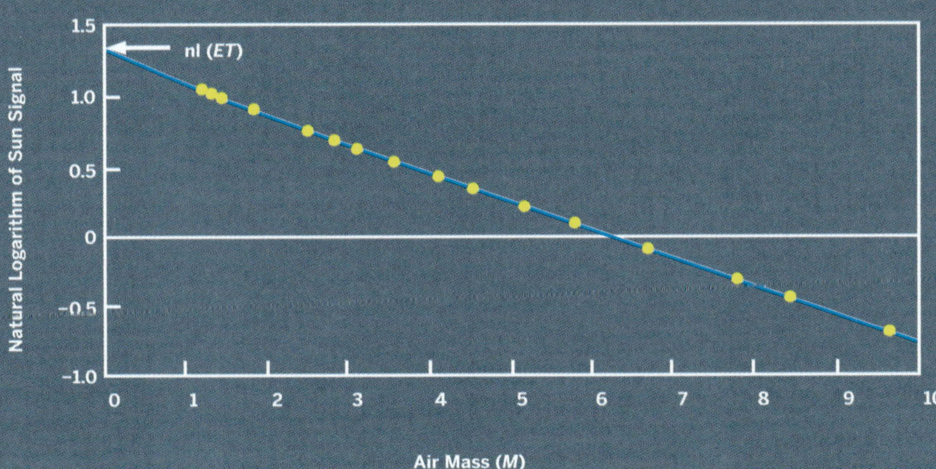

no clouds. First, determine the signal voltage and sun angle every 20 minutes when the sun is high overhead and every 10 minutes when it is low on the horizon. Next, calculate the air mass for each measurement. The air mass, *M*, is a measure of the depth of the air column between you and the sun; by convention, the air mass is 1.0 when the sun is straight overhead (sun angle = 90 degrees) and infinite as the sun sets (sun angle = 0). The formula is $M = 1.0 / \sin(\textit{sun angle})$.

To find *ET* for your instrument, simply plot the natural logarithm of each signal voltage against the corresponding air mass. The result is called a Langley plot (see above), and it should be a straight line out to an air mass value of about 10. On this plot, an air mass of 0 corresponds to the top of the Earth's atmosphere. So simply extrapolate the line back to the y-axis. The intercept is the natural logarithm of *ET*. Once you know this number, you can use the equation above to measure something called the aerosol optical thickness (AOT), the critical number scientists use to quantify how much crud there is in the atmosphere. Now it's up to you to publish your data on the web for others to use. Please visit the MAKE site and post a comment at makezine.com/06/citizenscientist if you do.

Dr. Shawn (Shawn Carlson, Ph.D.) is the founder and executive director of the Society for Amateur Scientists (sas.org). He won a MacArthur Fellowship for his work helping ordinary people do extraordinary science.

MAKIN' MAN

Ross Shafer is a busy man. When he's not working in his machine shop, he's working in his woodshop. When he's not woodworking, he's practicing guitar or playing in his band, South Bend (named after a machine in his shop). Or he's riding one of his bicycles (that *he* made), or riding his Ducati through the winding roads of Sonoma County in northern California. Or taking his Ducati apart to replace the lifters. Or refurbishing old amplifiers and old guitars. Or taking metal scraps and making birdbaths, candlesticks, and nightstands for his wife.

"I'm always making something," he says. "I'd rather make something that I need than buy it. If I don't know how to make it, then I learn how."

Right now, he's just finished a steel staircase for a custom house, and is remodeling his own home, a small farmhouse on a gorgeous piece of land that actually *is* a working farm. (There are chickens, sheep, a huge garden with a greenhouse built from salvaged windows, and, at one time, a pig named Homer.) Besides these accoutrements of rural life, there are two refurbished chicken coops and two barns. They contain a music studio, a woodshop, an office, a sewing room, a proper car garage, and a machine shop.

Once the owner of Salsa Cycles, a successful high-end bicycle company known for its hand-built frames and accessories, Ross sold it and began his next career, as a metal fabricator. Or as a rock band guitarist. Or as a sought-after woodworker. Or as an accomplished artist. You can check out this, and his other work, at six-ninedesign.com. —*Shawn Connally*

1. Ross' fixed-gear bicycle (his favorite). He made it. 2. Bedside table he made as a present for his wife, Mari. "It's soooo beauuuuutiful," she says. 3. Tube guitar amps, built by Ross. 4. Side table with attached candelabra. Another gift for his wife. 5. Ross' first self-built tube amp. 6. Big-ass milling machine. 7. The sweet, sweet sound of the first guitar Ross ever built. 8. Ross at one time raced motorbikes; these are his racing leathers, emblazoned with the Salsa Cycles logo. 9. This Ronnie poster was stolen by his wife off a wall in Scotland. 10. Another big-ass milling machine. 11. Ross built this bicycle for his son Max's third Christmas. Max is now 20.

Photograph by Christopher Lucas

1
6
7
8
10

THEORY & PRACTICE

RFID for Makers

Build this kit to read radio frequency ID tags.

By Joe Grand

THEORY

Radio frequency identification (RFID) technologies are rapidly becoming a part of life. RFID tags receive or transmit data (usually a unique serial number) over short distances when scanned by a reader. These tags have worked their way into robotics navigation, inventory tracking, access control, automatic identification, payment systems, and dozens of other applications. Casinos are beginning to embed RFID tags into their gambling chips in order to monitor gambling activity, detect counterfeit chips, and catch cheaters who try to surreptitiously add or remove chips from a wager.

Needless to say, we've seen only the tip of the iceberg for the possibilities of RFID applications, deployments, and projects. The technologies are out there and they're cheap enough for anyone to start experimenting with. Now's your chance to create that access-control-based pet feeder you've always wanted to make. Or maybe you can finally track your kids throughout the house.

In this article, I'll briefly explain the concepts of RFID technologies, introduce the Parallax RFID Reader Module, and show you how to put together a simple, portable RFID "scanner" so you can experiment with RFID and snoop around on your own to discover any hidden or embedded RFID tags.

How RFID Works

(*Material in this section is based on information provided by rfidjournal.com.*)

There are two major types of RFID tag technologies. "Passive tags" do not contain their own power source or transmitter. When radio waves from the reader reach the chip's antenna, the energy is converted by the antenna into electricity that can power up the microchip in the tag (known as "parasitic power"). The tag can then send back any information stored on the tag by reflecting the electromagnetic waves (think of your access card at work).

"Active tags" have their own power source and transmitter. The power source, usually a battery,

Needless to say, we've seen only the tip of the iceberg for the possibilities of RFID applications, deployments, and projects.

is used to run the microchip's circuitry and to broadcast a signal to a reader. Since active tags have their own transmitter, the reading distance is much longer than with passive tags. Active tags are typically larger and more expensive, and require occasional service. They're commonly used in automatic toll collection systems like FasTrak and E-ZPass. Within both passive and active tag types, there exist read-only, read/write, and cryptographic variations.

Just as you tune your radio to different frequencies in order to hear different radio stations, RFID tags and readers have to be properly tuned in order to

Illustration by Dustin Amery Hostetler/UPSO.org

communicate effectively. RFID systems typically use one of the following ranges: low frequency (or LF, around 125 and 134.2kHz), high frequency (or HF, around 13.56MHz), ultra-high frequency (or UHF, around 868 and 928MHz), or microwave (around 2.45 and 5.8GHz). Generally a higher frequency equates to a faster data transfer rate and longer read ranges, but also means more sensitivity to environmental factors such as liquid and metal that can interfere with radio waves.

There really is no such thing as a "typical" RFID tag. The range of a tag ultimately depends on many factors: the frequency of RFID system operation, the power of the reader, and interference from other RF devices.

RFID for Fun and Profit

There are a few inherent problems with RFID that generally make it not so secure. If you have a reader designed for the correct tag family and frequency, you can communicate with the tag. There is no security between most tag and reader transmissions. This means that you can easily create an RFID scanner, as we'll do in this article, and can snoop around to search for RFID tags (and retrieve the information right off them). If the RFID tags are used for any sort of financial transactions (electronic commerce, stored value), or access control or authentication (storing a unique ID or personal information to prove your identity), you should use them with appropriate caution.

Figure 1. The Parallax RFID Reader Module next to a basic tag.

Photography by Joe Grand

Cracking an RFID Code

In January 2005, the challenge/response scheme of the "secure" Texas Instruments Digital Signature Transponder (DST) tag was cracked. The proprietary cryptographic cipher used in the challenge/response scheme was successfully reverse engineered by some folks at RSA Laboratories and Johns Hopkins University based on a single PowerPoint slide inadvertently discovered on the web. The tag is used around the world in such applications as vehicle immobilizers and the Mobil Speedpass gas pay system. Details of the attack can be found on the "Analysis of the Texas Instruments DST RFID" website at rfidanalysis.org.

Not only did these researchers crack open the secret egg with their knowledge of the algorithm and a distributed, brute-force key search using 100MHz field-programmable gate arrays (FPGAs), but they also created a system to allow them to eavesdrop on the communications between a DST transponder and reader, or to participate actively in the protocol by emulating either device. What exactly does that mean?

Over 150 million deployed DST devices are now at risk and could be cloned or spoofed. Free gas for everyone!

Mark of the Beast

Meet Amal Graafstra, a maker who embedded an RFID tag into his hand. Now, Amal can unlock his house door without reaching for his keys and log into his Windows box without having to remember a password. (His interview with MAKE can be found at makezine.com/go/amal.)

The "Tagged" RFID implant forum (tagged.kaos.gen.nz) has been created to allow folks to share information and their experiences regarding implanting RFID technologies into their own bodies.

Certainly not for the faint of heart, this is a far cry from the civil liberties and personal privacy communities who consistently push back against legislation enabling embedded RFID tags in such personal items as cash and passports. What better way to have your privacy invaded than by doing it yourself?

This will give us a portable RFID scanner that you can use as a "bug finder" to snoop through your house and scan yourself and your friends, family, and pets to check for RFID tags.

The Parallax RFID Reader Module

I designed the Parallax RFID Reader Module (Figure 1, previous page) as a low-cost solution to read passive, low frequency (125kHz) RFID tags from up to 3 inches away. The module requires a single +5VDC supply and it interfaces to a PC, BASIC Stamp, Microchip PIC, or other processor via a 1-wire, 2400 baud serial TTL interface. It costs $39 from Parallax (parallax.com, part #28140) and can be integrated into any design using only four connections (VCC, /ENABLE, SOUT, GND). It just can't get much easier.

An LED shows the state of the reader module. When the module is successfully powered up and is in an idle state, the LED will be green. When the module is in an active state and the antenna is transmitting, the LED will be red.

The RFID Reader Module is controlled with a single TTL-level active-low /ENABLE pin. When the /ENABLE pin is pulled low, the module will enter its active state and enable the antenna to interrogate for tags. Keep in mind that when the antenna is active, the device is broadcasting and consuming about 100 to 200mA from the power supply. If you're going to build a project that involves batteries, you may want to add a physical button to activate the reader only when a card is actually present, or use a timeout in your controlling software/firmware to disable the reader periodically and reduce the load on the power supply.

This particular product is designed to detect common, passive read-only tags within the EM Microelectronics-Marin SA EM4100 family. When a valid RFID tag is placed within range of the reader, the tag's unique ID will be transmitted as a 12-byte ASCII string via the TTL-level SOUT (Serial Output) pin in the format shown in Table 1.

The start byte and stop byte are used to easily identify that a correct string has been received from the reader (they correspond to a line feed and carriage return characters, respectively). The middle ten bytes are the actual tag's unique ID.

All serial communication is non-inverted at 8 data bits, no parity, 1 stop bit (8N1). The baud rate is configured for 2400bps, a standard communications speed supported by almost any microprocessor or PC, and cannot be changed. The RFID Reader Module initiates all communication and can connect directly to any TTL-compatible UART.

If you plan to connect the RFID Reader Module directly to a PC's serial port, you'll need a common RS232 line driver (also known as a "level shifter"), such as the Maxim MAX232 or Linear Technology LT1780, to bring the module's TTL-level outputs to the PC's expected range (Figure 2).

Making Something Fun

Armed with our knowledge of RFID and the Parallax RFID Reader Module, let's make something. Because the RFID Reader Module simply sends a 12-byte ASCII string containing the tag's unique ID, we can connect the output of the reader to the input of a serial LCD module, such as Parallax's 2x16 Serial LCD regular (part #27976) or backlit model (part #27977), which will display whatever ASCII characters it receives from its input pin onto the LCD.

Combining the RFID Reader Module and the serial LCD will give us an instant, portable RFID scanner that you can use as a "bug finder" to snoop through your house and scan yourself and your friends, family, and pets to check for RFID tags. The best thing about this project is that no software is required! The RFID Reader Module itself handles all of the underlying RFID-reading functionality, and the serial LCD is already expecting to receive serial data.

To build our project, simply hook up the two pieces as shown in Figure 3. If you're using a 9V battery or any other power source above 5V, you'll need to use a voltage regulator (like the standard 7805) to bring the voltage down to 5V. Don't even think about connecting an unregulated power source directly to the RFID Reader Module or LCD!

Your final product will look something like Figure 4. I built mine on a simple breadboard to show off at the Make Fest at O'Reilly's Emerging Technology Conference in San Diego (March 2005) and still haven't taken it apart. The text string shown on the LCD, "04129C1A89," corresponds to the unique ID of the most recent tag held up to the reader.

With your RFID scanner ready to go, you can begin your hunt for RFID tags — hidden or otherwise.

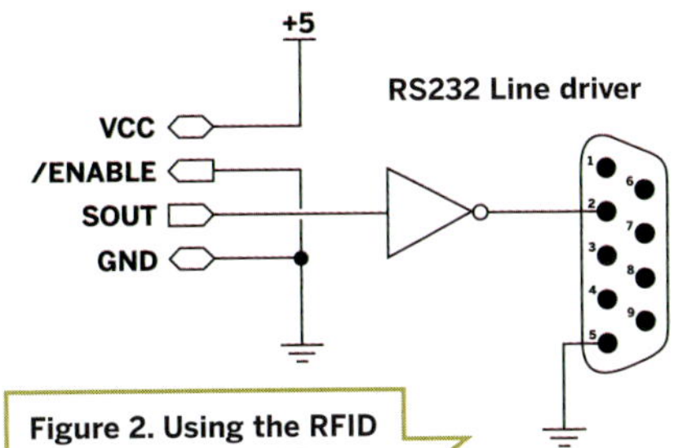

Figure 2. Using the RFID Reader Module with a PC.

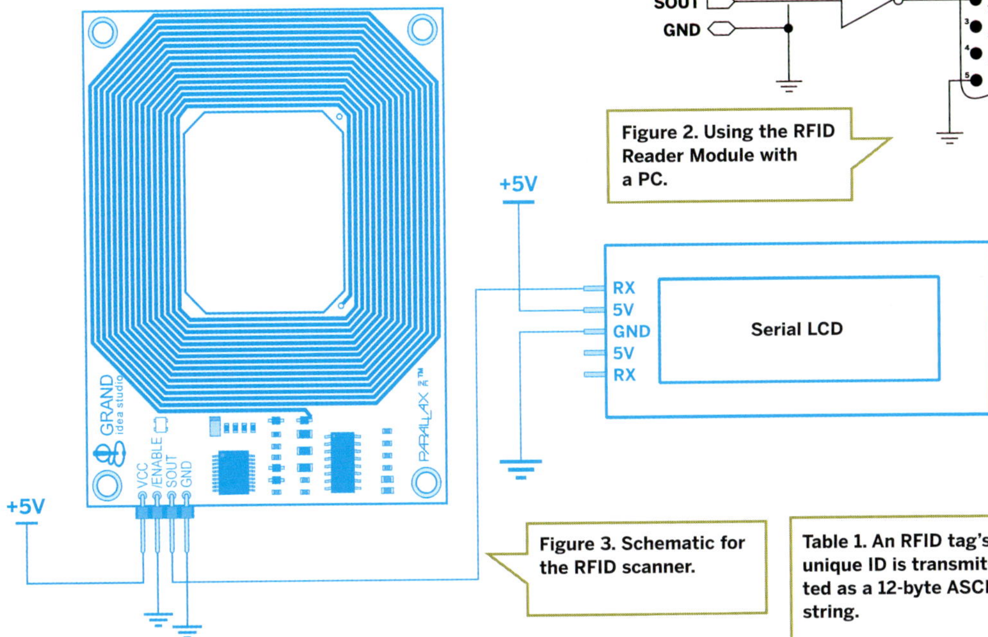

Figure 3. Schematic for the RFID scanner.

Table 1. An RFID tag's unique ID is transmitted as a 12-byte ASCII string.

MSB											LSB
Start Byte (0x0A)	Unique ID Digit 1	Unique ID Digit 2	Unique ID Digit 3	Unique ID Digit 4	Unique ID Digit 5	Unique ID Digit 6	Unique ID Digit 7	Unique ID Digit 8	Unique ID Digit 9	Unique ID Digit 10	Stop Byte (0x0D)

Simply walk around to find any tags within range, or modify the design to work with your specific project or application.

NOTE: Spurious "reads" can be caused by RF interference from nearby electronic devices. Also, text on the LCD can eventually get jumbled, due to the way the LCD handles the start and stop bytes — if so, just reapply power.

For information on interfacing the RFID Reader Module with a Parallax BASIC Stamp to make some interesting access control projects, check out Jon Williams' article, "You Can't Touch That: Non-Contact Access Control," in the April 2005 issue of *Nuts & Volts* magazine.

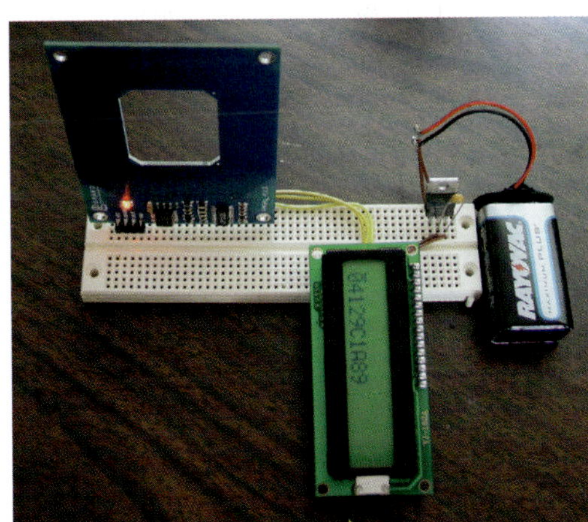

Figure 4. RFID reader module in action.

Joe Grand (joe@grandideastudio.com) is the president and principal electrical engineer of Grand Idea Studio, Inc., a product research, development, and licensing firm. He specializes in the invention and creation of consumer electronics, video game accessories, and toys.

Illustrations by Joe Grand and courtesy of Parallax, Inc.

www.makezine.com/06/makecontroller

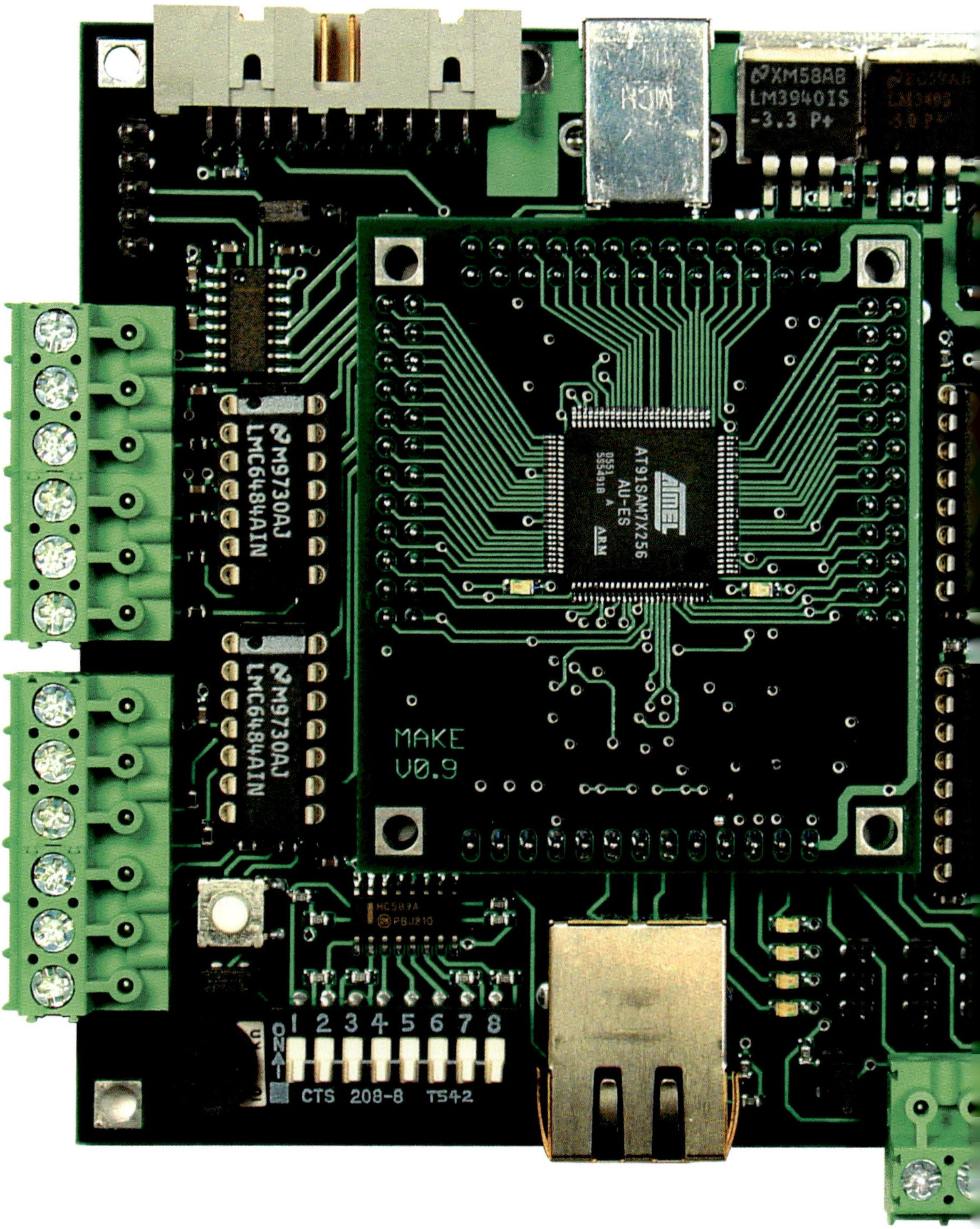

The MAKE Controller

Announcing a just-maybe-revolutionary microcontroller for all things DIY.

By David Williams and Liam Staskawicz

Photograph by David Williams

Microcontroller chips and the tools required to use them were once so specialized that few people could explore their potential without a degree in electronics and software engineering.

From BASIC Stamp kits to Wiring and Arduino boards, manufacturers and open source communities now offer somewhat easier ways to program microcontrollers, enabling cross-pollination between the engineering world and hobbyists, do-it-yourselfers, artists, and students.

But the effort needed to program a chip and design a circuit is still a barrier to countless creative people, even as the latest 32-bit chip designs offer new features like built-in networking.

All the elements are in place for the next revolution in DIY microcontrollers: a friendly, open platform that can do it all, launch a thousand projects, and empower both novices and pros.

Open Source Hardware

It's always exciting when powerful technologies become accessible to ordinary people. This is happening now with microcontrollers. After revolutionizing many industries (and requiring industrial-sized development budgets), now these do-anything chips are becoming easier to program and use with free, open source tools.

We realized, as did MAKE, that a general-purpose controller kit served by a free and open development environment would offer limitless potential to makers. A smart building block like this would shortcut (and even enable) diverse projects, collapsing pages of complicated instructions into, "connect these components to the controller board, upload this code, and *voilá* — you've got an amazing new device for your kitchen" (or garden, or living room, or backyard, or ... *see sidebar, page 169*). This platform would also inspire more advanced makers to develop and share their own original projects. Ideally, such a controller kit would be visible enough to attract a healthy developer community, and rewarding enough to see it grow.

Last year, MAKE magazine approached MakingThings to create this platform, the MAKE Controller. We were delighted, and we're even more delighted to announce that it is now available. In this article, we explain how the kit's two PC boards work, and why we designed them the way we did.

Basic Principles

Right from the outset, it was imperative that this be an open project. We knew from our experience with similar products, and from our customers (MakingThings has been producing and selling a range of controller boards since 1998), that anything short of freely available hardware and software would frustrate the creativity of the experimenter and the hacker — precisely the audience we're aiming for. We will publish all of our schematics, to encourage modifications and improvements to the hardware. The software environment(s) had to be free and easily accessible as well, to promote development. By supporting all aspects of the project with cross-platform, free, and open tools, we created an open hardware platform that many disparate communities can support, contribute to, and enjoy. For further accessibility, we also wanted the kit to be relatively inexpensive, and we managed to keep it down to less than $150.

To serve both novice users and experienced engineers, we decided to base our design around two boards rather than one. The Controller Board contains the microcontroller itself, plus the delicate circuitry needed to run it it, and all essential peripherals. This Controller Board can then plug into a larger Application Board that provides circuitry to interface with real-world electronics like sensors and actuators. Users of an Application Board can just plug a stepper motor or a pneumatic valve right in, rather than having to design the circuitry to drive them from the chip. We designed one general-purpose Application Board with a nice mix of control and communication capabilities, and over time we will add others with different features, such as CMOS cameras and audio input and output.

> We decided on the newer, faster class of chip, to create a new generation of controller boards.

With the two-board approach, novices can experiment with an Application Board with little fear of breaking anything, while experts can either use an Application Board or plug the Controller Board into a board of their own design.

For the most advanced programmers, the Controller Board makes it easier to get at the microcontroller's densely packed pins. Everything is accessible. They can read the 700-page processor manual, dig deep into the chip's architecture, choose or write their own software and device drivers, and have full control over every other aspect of the board's operation.

Novices also have options. The Application Board includes onboard switches that select and control useful and fun software that's pre-installed on the Controller Board. This makes the kit enjoyable and configurable right out of the box, without any programming. After exhausting the switches, a novice's next step can be downloading different code and uploading it into the boards using USB.

Liam and David with the final MAKE Controller Kit and several prototypes.

Choosing the Microcontroller

The first and most difficult task was choosing the microcontroller. Most microcontrollers that address the hobby/artist/DIY markets are 8-bit systems, in which most operations in the processor work on 8 or 16 bits at a time. All of the predecessors of the MAKE Controller are 8-bit, including our own products and the Wiring and Arduino boards.

32-bit microcontrollers have recently become available, some of which are ten times faster than 8-bit chips, with far more onboard memory and sophisticated on-chip resources. They can comfortably multitask and communicate over bridges and networks, in addition to running simple control operations. On the downside, they're also a bit more expensive (about $9 rather than $7 or less) and they use more power.

We decided on the newer, faster class of chip, to create a new generation of controller boards. Then we had to choose a specific one, since Philips, STMicroelectronics, Texas Instruments, and Atmel all have excellent 32-bit products. Several meetings and a masochistic one-day trip from San Francisco to Boston later, we had enough information to select our chip: the Atmel AT91SAM7X256 — or "SAM7X." This is a brand-new part with a staggering list of features that hobbyists could only dream of before its release. In addition to quite exciting basic stats (a 32-bit 55MHz ARM core, 256KB of flash memory, and 32KB of RAM), the part has Ethernet addressability, a USB port, a CAN controller (CAN = Controller Area Network, a simple, flexible, two-wire protocol originally developed for cars), several serial ports, and more, making it one of the most versatile microcontrollers on the market.

The SAM7X's Ethernet port opens up entire realms of possible applications — for example, it can host a web server or telnet session, or retrieve data from the internet all by itself. And its USB port lets the system interface with a computer, to offload any computational heavy lifting. With either port, you can have a computer perform specialized processing while the connected microcontroller exchanges information with the rest of the world.

Finally, because of early excitement around the introduction of the SAM7X, many software providers have adapted their development environments and other tools to this chip, resulting in strong software support even before we got started.

Selecting the Development Environment

The next thing we did was confirm that there was a full set of free tools that would allow people to develop with the SAM7X. Fortunately, GNU's incredible open source compiler, gcc, will compile code for most ARM processors. With every release, gcc becomes more capable and more efficient, although at the time of writing, the debugging environment that gcc provides for ARM parts is somewhat limited.

A few expensive professional tools also target the SAM7X processor and use gcc, including Rowley and Associates' CrossWorks for ARM. This means you can use either a high-end environment or open source tools to develop for the MAKE Controller.

Designing the Hardware

By far, the hardest part of the process is designing the hardware. Here are the steps we followed:

1. **Decide on basic functionality.** What will each board include? One such decision was to squeeze all Ethernet support (except for the connector itself) onto the Controller Board, making it available to all users, instead of putting it on the Application Board.

2. **Decide which microcontroller pins do what.** The SAM7X has 100 pins, but that's not much, considering what the chip can do. Manufacturers rely on

Photograph by Susan Williams

MAKE CONTROLLER KIT: CONTROLLER AND APPLICATION BOARD LAYOUTS

Layout of Controller Board plugged into Application Board. **Dashed connectors are detachable.**

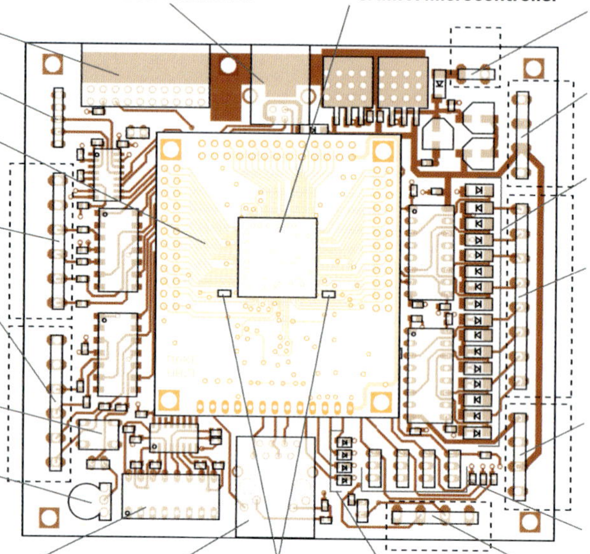

pages worth of usage modes and dependencies in order to squeeze more functionality onto their chips, so the task of allocating all the signals becomes a monumental logic puzzle.

3. **Work out power issues.** What power will be needed by what parts? What is the maximum voltage the board should withstand?

4. **Draw schematic.** Wire all components per manufacturer's notes and common design practices.

5. **Select packages for all devices.** We favored surface mount (SMT) components since they're smaller, cheaper, and cheaper to assemble — although you can't solder them easily by hand.

6. **Map schematic onto a PCB board.** Each trace connects two points, and they can't overlap or run too close. Ground and power can go on the internal layers of a four-layer board, as a wiring convenience and to reduce interference from electrical noise.

7. **Check and double-check everything.**

8. **Generate Gerber files.** PCB fabricating machines use these to manufacture boards. Most PCB layout software has a Gerber export option.

9. **Send Gerber files to a board shop that can do the job.** The traces on the Controller Board, and even the pads of the SAM7X itself, are very close together. With a design this fine, many cheaper shops will send boards back with shorts between contacts.

10. **Assemble the boards.** Order all the parts, and send them with the boards to a manufacturing facility.

11. **Test.** With our first boards, we found mistakes in the complex Ethernet area. Correcting these required some real white-knuckle soldering tricks.

12. **Celebrate!** Designing a new board is tense, and failure means a loss of money as well as pride. When it all finally works, everyone breathes a sigh of relief.

Illustration by Liam Staskawicz

Designing the Software Environment

One of the difficulties of programming a microcontroller directly is having to write for the device's specific inner workings. This means structuring code to permit other functions to operate, and deciphering arcane header files and the bit-by-bit operation of complex peripherals. We wanted none of that.

To liberate our users (and ourselves), we wrote a series of code layers that access all the devices on the boards at a more abstract level. Instead of needing to understand every aspect of each device, developers can include these files with their projects, and then call simple methods that encapsulate lots of fiddly hardware settings. For example, with one simple call, our `AdcInit()` routine initializes the SAM7X's Analog to Digital Converter, a process which otherwise requires a hideous chunk of code.

Another difficulty with bare-bones microcontroller programming is having to explicitly manage all control-passing, sharing the processor among all running tasks. The cure for this pain is to use a Real Time Operating System (RTOS). With complex projects where functions run simultaneously or timing needs to be managed closely, an RTOS makes coding applications quicker, simpler, and more elegant.

RTOSs have traditionally been expensive, and they consume too much overhead to be practical with 8-bit microprocessors, but there is an excellent, free RTOS, helpfully called FreeRTOS.org, that runs on the SAM7X. When connected to lwIP, a small library of code designed to provide internet capabilities, we have a full RTOS with internet capability.

Programming with the benefit of an RTOS means you can start tasks upon initialization, and run them concurrently as permanent loops that poll for input and communicate with each other. There's no need for spaghetti code that breaks out of each ongoing task to check on everything else.

Conclusion and Future

We are very pleased with the MAKE Controller Kit. It is an absolute delight to program, and connecting real devices to it is very simple. We look forward to adding more and more functionality, and seeing what others think of doing with it. Try one and tell us! Email us at info@makingthings.com.

Project Possibilities

Some applications of the MAKE Controller that you may be seeing in the near future.

Drink-O-Mat
Timed control of valves and motorized swizzle stick can dispense and mix measured amounts of different beverages from inverted bottles.

Plant Cyborg
Light sensor and motor controllers position houseplant for optimal growth, while moisture sensor triggers valve to ensure perfect watering.

Glue Gun 3D Imaging
3-axis stepper motors and trigger controller render stable 3D forms by building them up layer-by-layer using hot glue.

Automated Pet Feeder
Motor device feeds pet remotely via web-based interface or timer program, and dispenses treats when pet stands in a specific position.

Liquid Temperature Controller
Thermostat feedback loop keeps liquid within a specified temperature range — for brewing, aquariums, candy-making, photo developing, etc.

Global Sensor Network
Large-scale survey system registers sensor input from hundreds of physical nodes connected to a common server, which publishes sensor info.

Get Your MAKE Controller Kit!
On sale now at makezine.com/controller, where you can also learn more about the kit, including full schematics, and share your experiences in our online forum.

David Williams enjoys the connection between machines and intelligent software, and has designed and built professional tools for software, electronic, and mechanical engineering. Liam Staskawicz has explored sensor-driven hardware and software systems from the perspectives of both engineer and musician. See makingthings.com.

HowToons.com

BY SAUL GRIFFITH, NICK DRAGOTTA & JOOST BONSEN '06

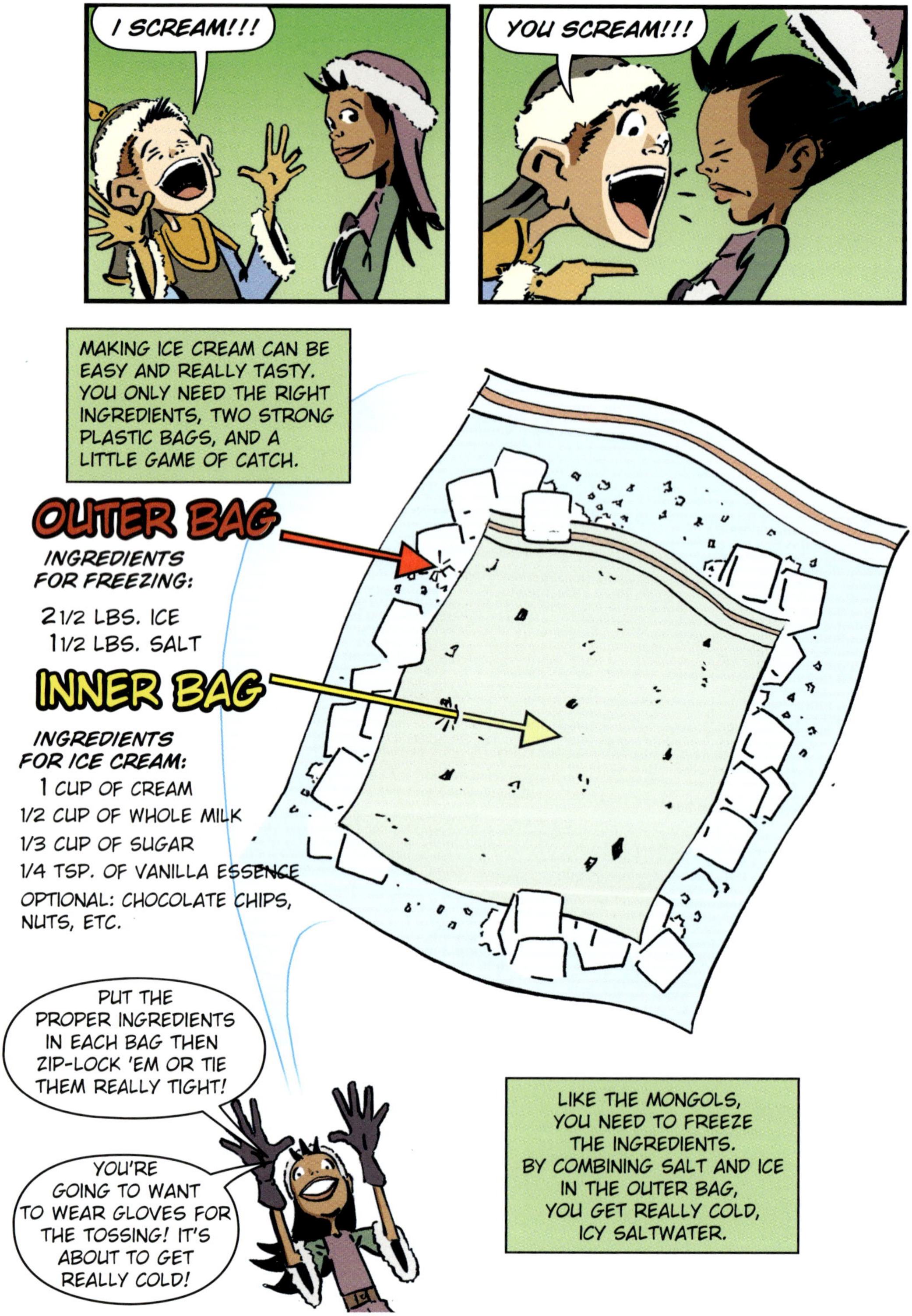

From the forthcoming book *HOWTOONS Book 1* by Saul Griffith, Nick Dragotta, and Joost Bonsen. Published by arrangement with ReganBooks, an imprint of HarperCollins Publishers, Inc.

MakeShift

By Lee D. Zlotoff

The Scenario: You live with your wife and two teenage children in a rustic, one-story house next to a roaring mountain stream in a steep, narrow canyon. Early one morning, you're awakened by a serious earthquake. The family is OK, but the power is out. The portable radio tells you that the quake has hit the region hard, and power will probably be out for days. The land and cellular phone systems are out as well, but you don't get cell reception up here anyway.

You look outside and see that a rockslide just below you has not only cut off your driveway, but also dammed the stream, which is rising fast. Your house will be mostly underwater in 15 minutes.

You have two cars: A Toyota Prius and a Chevy Suburban. Both have full tanks of gas and cellphone chargers in the dash, but you have only enough open space up behind the house to keep one of them out of the water, above the level of the rockslide. Your home has typical belongings and a basic set of tools, but all your camping equipment is lent out to your brother — and it's been getting cold at night.

The Challenge: You have 15 minutes to move one of your cars up to dry ground, and tell your family what to collect from the house to ensure that you can all survive up here for at least five days. You also need to devise a way to let the outside world know that you're here, and need to be rescued.

There's one other wrinkle: Your only neighbor, a single guy named Dave, probably partied hard last night and slept through the quake. But his house will flood just as quickly as yours. You can send someone to rouse Dave, but then you're down one collector, and if you do rescue him, he'll join you with nothing more than the clothes he has on. Should you factor in Dave, or figure that in a disaster like this, it's every man for himself?

Now, look at your watch again. You need to make some key decisions immediately, and figure out who should do and grab what. Because in 15 minutes, the party's over, and you're completely isolated and homeless ... Go!

Send a detailed description of your MakeShift solution with sketches and/or photos to *makeshift@makezine.com* **by July 17, 2006. If duplicate solutions are submitted, the winner will be determined by the quality of the explanation and presentation. The most plausible and most creative solutions will each win MAKE sweatshirts. Think positive and include your shirt size and contact information with your description. Good luck! For readers' solutions to previous MakeShift challenges, visit** *makezine.com/makeshift*.

Lee D. Zlotoff is a writer/producer/director among whose numerous credits is creator of *MacGyver*. He is also president of Custom Image Concepts.

Photograph by Topher Lucas

GASOLINE

The best tools, software, gadgets, books, magazines, and websites.

TOOLBOX

Turn your living room into a high-tech pool of primordial ooze.

Roboraptor Rules the Roost

$100 roboraptoronline.com

Guided by the design brilliance of Mark W. Tilden, WowWee Robotics created the best-selling robot of all time, Robosapien. Dubbed the only robot fluent in "caveman talk," Robosapien has now "evolved" into the dinosaur-like Roboraptor.

Bristling with IR, touch, and sound sensors, the 32-inch-long Roboraptor is fully equipped to interact with its environment. A remote controller, fondly reminiscent of a PlayStation 2 game controller, controls body movements, personality modes, and dino moods. Forget the Prozac: mood swings are encouraged with Roboraptor. Hunting, playful, and cautious moods are simple two-button commands. Unlike Robosapien, Roboraptor has a special "laser" targeting feature that can be accessed with the remote controller. Just point the green "targeting assist light" on any surface, like a wall, and Roboraptor will hunt down the target.

Is Roboraptor as hackable as Robosapien? I asked Mark Tilden, and discovered that you can saddle up Roboraptor with a Barbie or a G.I. Joe, add a microcontroller for responding to the dino's impressive sensor array (a higher brain function hack), and design a Ni-Cd wall charger for Roboraptor to stalk. Now get out there and make it.

—Dave Prochnow

The Übergeek Wondertool

Eklind 6-Blade Screwdriver $12 eklindtool.com

While it looks like something you would put in your glove box and forget, I've kept my original Eklind 6-blade Screwdriver in my pants pocket for more than 15 years and have used it almost every day.

I've adjusted headlights on a dark road with the T-15 Torx, cracked thousands of computer cases with the #1 Phillips, fixed eyeglasses with the small slot, and mounted routers using the awl to drill starter holes in mounting boards. While it doesn't have pliers like some wannabe geek tools, it is far superior in terms of weight, choice of blades, and feel in your hand. —*Steve Johnson*

Maker Heaven

Ax-Man Surplus Stores ax-man.com

▪ I moved from Minneapolis a few years ago and I miss this store so much. They resell industrial parts, wheels, motors, speakers, fabric, foam, biohazard suits, and just about anything you can use to make stuff. As you walk through Ax-Man, you'll notice large motorized moving displays with baby doll parts and humorous descriptions accompanying their products. I bought some beautiful, large Hollywood-style metal film canisters there and sanded them down to make a portfolio case. Twin City residents, take note.

—*Joel Raedeke*

Government Leftovers

govliquidation.com

▪ Looking for "Space Survival Equipment" or "Nuclear Reactors"? Government Liquidation is the place to go. Just be prepared for a bit of paperwork if you bid on anything from the "Combat Ships & Landing Vessels" category.

The selection is wide: lathes, presses, steam turbines, saws, tractors, cranes, tents, boats, jeeps, bulk computers, solenoids, amplifiers, sirens, motors, teletypes, cameras, oscilloscopes, gauges ... in short, everything a geek could dream of.

Beware of misleading picture sizes. I once purchased a Hughes FACT PC (a giant DEC PDP-11 system used for testing aircraft circuitry) that included a box of cables. The box turned out to be 6 feet square and 4 feet high. Some items are labeled "bring your own crane." Also, the government doesn't ship, so make certain the item you're bidding on isn't in Guam. Most of the equipment is stored at military bases. Be sure you bring ID and be prepared to have your vehicle searched.

It can be a hassle to deal with, but Government Liquidation is an incredible source. It's the sort of site that makes you wonder, "What would I do with 300 snowshoes, a Gould Instruments Thermal Array Recorder, and a 1970 AM General Cargo Truck?"

—*Tom Owad*

MAKE LOOKS AT KITS

Editor Phillip Torrone talks about kits that caught his eye on makezine.com.

9V USB Charger Kit

$9 aarondunlap.com/blog/1130885615

Aaron Dunlap made a great little kit for folks who'd like to make their own 9V USB charger. And, you guessed it, it's all inside an Altoids tin. It's pretty simple to construct one of these, but Aaron put everything together in kit form if you're not a hardcore maker yet.

Sun SPOT Kit

$499 sunspotworld.com

This Java-based wireless-sensor development kit, an ongoing project from Sun Research Labs, is all about sensors. It's pretty techie, but looks like a lot of fun. Release is scheduled for May 2006, and people are using the Sun SPOT forum already.

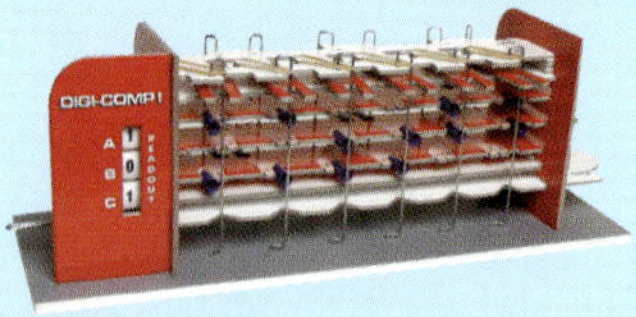

Build Your Own Digi-Comp!

$55 mindsontoys.com

▪ Tim Walker missed the classic 3-bit Digi-Comp mechanical computer from the 1960s, so he figured out how to make his own. Now he's reissued it as a kit.

Wooden Clock Kits

wooden-gear-clocks.com **and** krazydad.com/blog/2006/01/24/

▪ One of our readers wrote in and suggested a few wooden clock websites for kit builders. He started out with the Ascent, which is precut and ready to build. (Plans are also available for those who want to get down and dirty with their jigsaws.) The finished product looks gorgeous.

Carnivorous Plant Kit

$25 scientificsonline.com

▪ Edmund Scientific has a deal on Audrey II meat-eating plants. Their kit lets you create an "authentic" bog for growing various carnivorous plants from seed in a specially designed terrarium. Feed them, Seymour!

Arty Electronics for Kids

$250 playfulinvention.com

▪ I saw an early demo of these tiny computers at the Exploratorium in San Francisco, and they're amazing. The PicoCricket Kit is designed for fusing art and robotics using lights, sound, music, and motion.

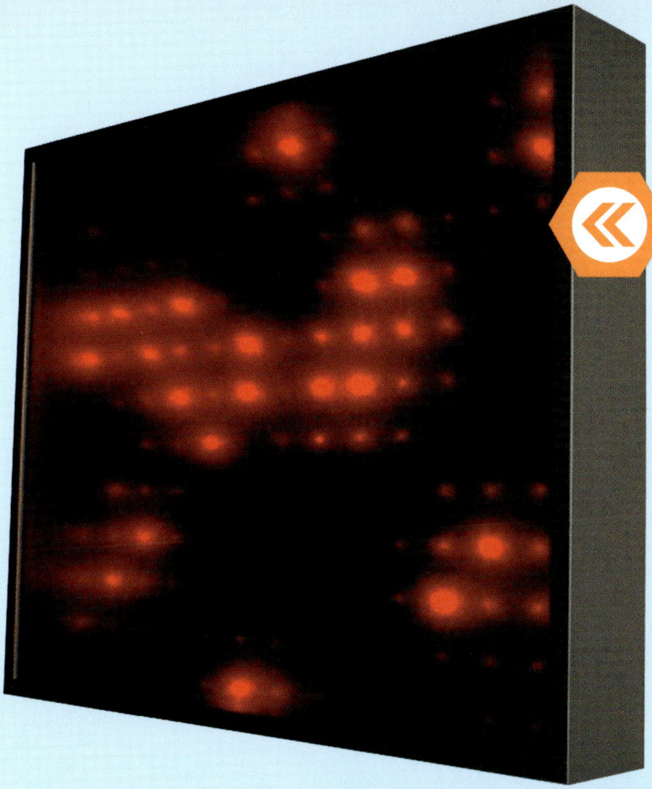

DIY LED Panel Kits for Artists

$235–$359 biteditions.com/diy.html

Send off your own animations to Bit Editions, and then assemble the kit to display your LED-ified art. The displays are 12x14 inches, and contain one LED per square inch, with each LED capable of eight levels of brightness. Driven by a microcontroller and 128K of programmable memory, the panel gives artists up to 1,984 frames for animation. The new version, scheduled to be on sale in May, will be reprogrammable from a PC through a serial port.

Lie Detector Kit

$11 apogeekits.com/lie_detector.htm

We've had a few emails and posts from folks who wanted to build their own lie detector, and now you can. The website explains how lie detectors work (by measuring the change in skin resistance of the subject — you sweat when you're lying) and also has a circuit diagram.

Blow Your Mind

Lego Mindstorms: Robotics Invention System 2.0 $199 lego.com
Contrary to what you might have thought, you can do a lot more with Legos than build another useless static display. You can pick up a set that includes a bunch of gears, motors, sensors, and a fairly sophisticated microcontroller that runs all of this stuff for under a couple of Bennies. But don't just settle for what comes out of the box. If you really want to impress the girls, be sure to reprogram the firmware with an alternative like LeJOS to really squeeze out that last bit of juice. Also, keep an eye out for some fat price cuts coming up, since Lego announced a newer version of this set coming out later in the year.

—Matthew Russell

FAVORITE THINGS

Chosen by the editors of *Ambidextrous* magazine.

« Abrasive Cut-off Wheel: No hacker's toolbox is complete without a rotary cutter. And no rotary cutter is complete without fiberglass-reinforced abrasive cut-off wheels. The bionic cousins of the wimpy wheels that usually come with your tool, a set of these in your bag moves you from the world of plastic and wood into the realm of aluminum and stainless steel. Just mangled the slot on that screw head? No problem. With this, you can cut your own.

« Digital Calipers: Whether you're gauging the width of gaps, the diameter of pegs, or the depth of holes, calipers are immeasurably more useful than rulers. It's true that the digital variety require batteries, but the ease of use and accuracy of digital calipers will save you a lot of heartache when you've been working too late, too long, and too hard on a project.

« Gaffer's Tape: Move over, duct tape — gaffer's tape is as versatile as its Space-Age cousin, but looks nicer and doesn't leave behind a sticky, messy residue when it's pulled up. Absolutely indispensable when you're attaching your inventions to something valuable, like a brass instrument or someone else's carpet. Gaffer's tape also rips easily but secures firmly. The catch? It's pretty pricey, about $20 a roll.

« Spray Can Trigger Handle: No more tired index fingers! With the Spray Can Trigger Handle, avoid the frustration of keeping even pressure applied to the plastic nub on the top of the spray paint can. This handle transforms any cheap can of paint into a much-easier-to-use, airbrush-like system. Your projects can now be professionally primed and painted. And at less than $3, the Spray Can Trigger Handle is a cheap must-have for any paint shed or toolbox.

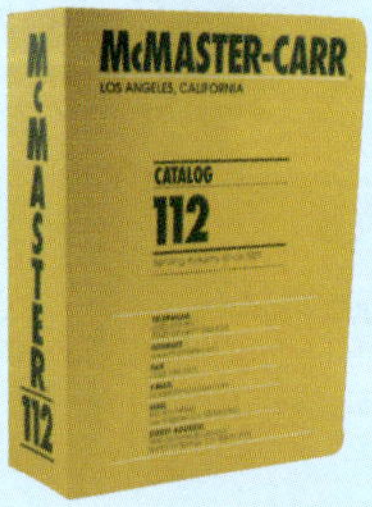

« McMaster-Carr Catalog: In our book (a big yellow and green one, in fact), McMaster-Carr trumps all other industrial supply companies. You can get anything from McMaster. Standard shipping is surprisingly speedy, with orders usually magically appearing on our doorstep in a day or two. And, there is no better source of brainstorming fodder than their well-organized catalog chock full of extra-descriptive info and masterful line drawings. Can't get your hands on one of their coveted catalogs? Their website is excellent as well: mcmaster.com.

« Safety Glasses: Actually, safety glasses are more of a favorite thing by association. What we are truly in love with is our eyes. They've been invaluable in most every project we've done to date, and given their all-around utility and the fact that they're amazingly hard to replace, we're committed to protecting the ones we've got.

When not distracted by their advisers' urgings to finish their Ph.D.s (all in mechanical engineering at Stanford University), these design-obsessed makers are devoted to the publication of *Ambidextrous* magazine, a quarterly that celebrates the people, process, and products of design. Check it out and/or subscribe at ambidextrousmag.org.

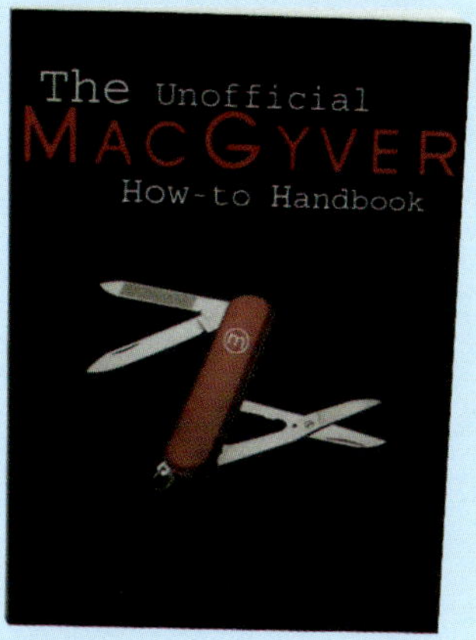

MacGyver Handbook

***The Unofficial MacGyver How-To Handbook: Revised 2nd Edition*,**
by Bret Terrill and Greg Dierkers $20, American International Press

MacGyver always seemed like one of those "kids, don't try this at home" shows. Well, that was up until this spiffy book came out that includes a complete listing of every trick ever done on the show with step-by-step instructions for many of them. Checking in at about $20, you can get it for a reasonable price, and it actually makes for pretty good weekend project fodder (although disarming a missile with a paper clip is an academic exercise for most of us). Each Macgyverism includes a list of materials and a psuedo-scientific explanation of why it works; so much of it could be adapted without too much effort as a good middle-school science-class supplement.

—Matthew Russell

Design Yourself

***D.I.Y.: Design It Yourself*, edited by Ellen Lupton**
$20, Princeton Architectural Press

Finally — good design for the masses! This fabulously designed (of course) book was produced collaboratively by the self-described bastard children of the fine arts: MICA graphic design students and faculty. Based on the premise that "design is art people use" and should be accessible to everyone, the authors begin with basic design principles, in a writing style that feels like a good conversation.

Featured project ideas run the whole gamut from blogs, books, brands (fight the corporate power by making your own), and business cards (even chocolate brownie cards) to embroidery, flyers, housewares, logos, and yes, T-shirts. There's a great section on kids and design that highlights how design can serve as the best defense for kids against our corporate-sponsored landscape. Each page is an open invitation to reinvent your world, as well as yourself.

—Goli Mohammadi

Hitchhiker's Guide to the Workshop

Machinery's Handbook, 27th Edition $53, Industrial Press

Machinery's Handbook is the ultimate reference tool for makers and hardware hackers. In just the past few weeks, I've used it to look up a nice patina mix for copper artwork, calculate what thickness of steel I need to build a long-span desk, and figure out how many tons of sand my godchildren require in their sandbox. From A (ABEC bearings) to Z (zinc-based, die-casting alloys), the 70-page index alone will keep a hands-on person dreaming up new projects for a good long while. It's so much a part of hands-on culture that most machinist's toolboxes are manufactured with a special drawer that fits just one thing: this book.

—Travis J.I. Corcoran

Big Little Sharpie

Sharpie Mini **$1 at office and drug stores,** sharpie.com

I, like most women I know, have used the venerable black Sharpie for many wardrobe fixes — scuff marks and bleach splatters are no match for the queen of permanent markers. Then the colored Sharpies arrived, and oh, the possibilities. An eight-pack of various colors means I can hide the chips on the rim of my expensive platter and mask the gouges in my not-so-expensive kitchen cabinets. Pastels come out at Easter; red and green come out in December.

But none of them hold a candle to the new Sharpie Minis. Complete with a ring on the end so you can attach them to your keychain or backpack, they are maker nirvana in miniature. Now I always have a powerful writing utensil with me — I can copy something off a bulletin board; make adjustments to myriad items, from handbags to posters; or get an autograph on my son's T-shirt from his favorite musician. Permanent, colorful, stylish, non-toxic, utilitarian. Dreamy!

—Shawn Connally

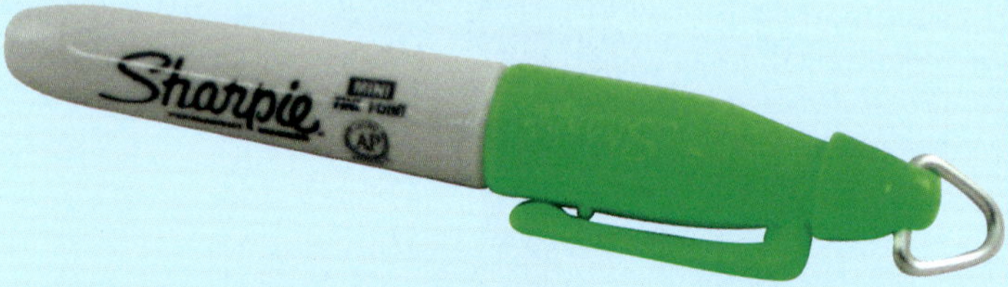

Shawn Connally is managing editor of MAKE.

Travis J.I. Corcoran hacks software, wood, and metal. He is founder and president of TechnicalVideoRental.com.

Steve Johnson (mojocode.com) is an IT consultant with a point.

Terrie Miller is online manager of MAKE.

Goli Mohammadi is copy chief of MAKE.

Tom Owad is the author of *Apple I Replica Creation* and editor of applefritter.com.

Dave Prochnow is a frequent contributor to MAKE. You can learn more about his projects at pco2go.com.

Joel Raedeke misses Ax-Man Surplus.

Matthew Russell tries hard to live life as a renaissance man, but is distracted by the cult of Mac.

Dan Strunk and his scratch-free iPod live in Minneapolis, Minn.

Phillip Torrone is senior editor of MAKE.

Have you used something worth keeping in your toolbox? Let us know at toolbox@makezine.com.

30 Amazing Minutes

***The Way Things Go*, by Peter Fischli and David Weiss** $20, amazon.com

You may have seen the Honda commercial where parts from an Accord are used to create a fantastic Rube Goldberg-like contraption. That footage pays homage to artists Fischli and Weiss' 1987 film, *The Way Things Go*, which opens with a large garbage bag suspended from a twisted rope. As the rope unwinds, the bag lowers until it bumps into a tire, which begins to roll, and the chain reaction begins its way through the warehouse.

Common objects are employed: tires, ladders, chairs, and bottles — even a humorous old pair of shoes. And there's more than just kinetic interaction: chemical reactions, fire, and even crude rockets and explosions are used to make things go. Sometimes events happen fast; other times, you have to endure the anticipation, perhaps while a container slowly fills with enough water to make a can float and tip a chair.

—Terrie Miller

See the Honda commercial at makezine.com/06/toolbox

Always Use Protection

Vinyl Film

Prices vary, available at craft stores

Five years ago, I wanted a clear protector for my iPod that worked like the sticker that Quickie-Lube puts on my car windshield. So I went to my local craft store to see what they had. I was initially intimidated by the different thicknesses of vinyl, but they let you decide if you want light protection from minor scratches or something that might absorb the impact of a drop onto a hard floor.

The thin stuff doesn't even require a cutout to let the click wheel move — it flexes just the right amount. I cut the material to size and wrapped it around my iPod like a burrito. It sticks really well to a clean iPod (and itself) without any adhesive. When my iPod died after five years of use, it didn't have a scratch on it.

—Dan Strunk

Microsoft
I am.
I think,
therefore, I am!
Tools for Your Imagination
Add brains with Microsoft® Visual Studio® Express Editions at:
http://msdn.microsoft.com/express/
Learn how at:
http://msdn.microsoft.com/coding4fun
Microsoft Visual Studio
Express Editions
Coding4Fun

AHA!

Puzzle This

By Michael H. Pryor

MAKE's favorite puzzles. (When you're ready to check your answers, visit makezine.com/06/aha.)

Coins on the Table

Your roommate challenges you to a game with the quarters from each of your sock drawers. Whoever wins the game gets to keep all the money. (Assume you each have an unlimited supply of quarters.)

You sit down at the perfectly round kitchen table and each person takes a turn, placing a quarter down anywhere on the table. No quarters can overlap and the entire quarter must rest on the table surface. The first person that can't put a quarter down on the table loses. You each have plenty of quarters and won't run out during the game.

Your roommate wants to go first. But you think about it for a while and realize you would much rather go first because you've figured out a surefire method to win (without cheating). After some convincing, your roommate allows you to place the first quarter.

Where do you place it, and what is your winning strategy?

In-Your-Head Math

How many trailing zeroes are there in 100 factorial (100!)? (100 factorial is 100 * 99 * 98 * etc., down to 1.) For example, 5,030,499,400,000 has five trailing zeroes.

One Mile South

How many points are there on Earth where a person can walk one mile south, one mile east, and then one mile north and end up in the same spot they started in? To be precise, let's assume the Earth is a solid smooth sphere, without oceans or mountains.

The North Pole is one such place where you can walk one mile south, one mile east, one mile north and end up where you started. But is there another starting point that works?

If you think you've figured it out, I'll give you a hint. There is more than one point. In fact, there are more than two points.

Calendar Cubes

Each morning I come into work and arrange two normal cubes on my desk to display the current day of the month. For example, on the 1st of the month, the cubes have a 0 and a 1 on them. On the 16th day of the month, one cube has a 1 and the other a 6. On the 31st day of the month, a 3 and a 1 appear on the front of the cubes. Each cube has six sides and one number painted on each side. What are the numbers painted on each cube?

Michael Pryor is the co-founder and president of Fog Creek Software. He runs a technical interview site at techinterview.org.

Illustrations by Roy Doty

READER INPUT

Where makers tell their tales and offer praise, brickbats, and swell ideas.

I've had to rewrite this letter in my mind several times to keep it from sounding like a fan letter penned by a silly schoolgirl to Elvis.

Wow! I just got my first issue of MAKE and I'm giddy with excitement. I'm a not-very-closeted geek, and have attempted (sometimes successfully) some fairly strange projects, but your people have shattered any illusions of grandeur I might have held. I read it cover-to-cover-to-cover and now I'm re-reading it again!

I'm reminded of the stories they tell about Nikola Tesla, how he had ideas, visions, of fully fleshed-out inventions flying through his mind so fast that he couldn't even snag them all as they raced by. Now after reading your magazine I think I may know just a little bit how he felt!

I've turned some like-minded friends on to this incredible ride, who now feel like they *need* to subscribe, simply by describing some of what I read in Volume 05. *—Dave Sica*

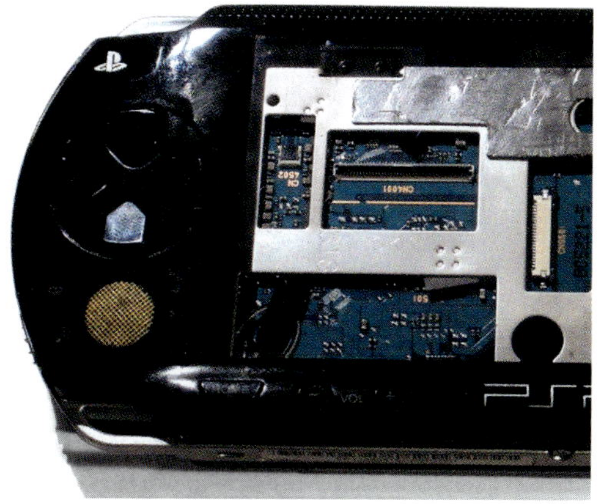

Hey, I love the magazine, keep it up. I was not all that happy with my PSP; I also attained an extreme distaste for Sony due to events following my purchase of the PSP. And so I was compelled to disassemble it. As soon as I broke the seal I knew that I had voided the warranty and so, I thought, why not proceed? In the process of putting it back together I decided to put the case back on and take this picture, and wanted to share it with you. I hope this makes you feel as good as it did me.

—SpotDart

MAKE AMENDS

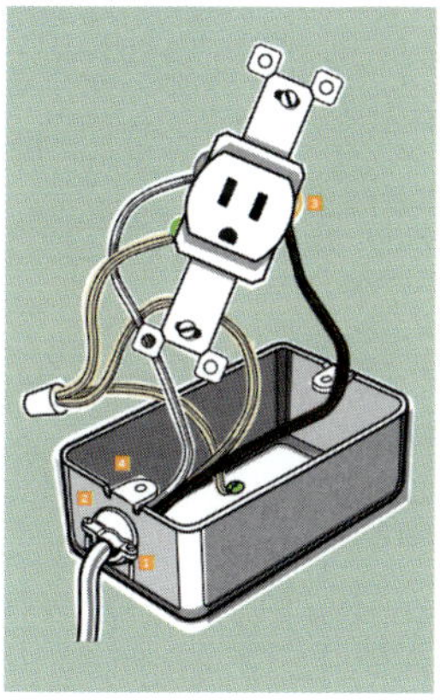

We heard quite a lot from our readers about the misdrawn illustration on page 112 of Volume 05. In the piece, "How to Build a Power Tap," the illustration can cause some confusion as well as a tripped circuit breaker. The drawing shows a dual plug receptacle, which, when wired improperly, can cause a short circuit. Do NOT attach the wires from the extension cord to the screws on the same side of the receptacle, because it will create a short circuit.

Many eagle-eyed online readers of our digital edition tipped us to our error, even before the print edition was mailed to subscribers or showed up on newsstands. This quick communication let us notify tens of thousands of subscribers via email before they even received their copies. It also allowed us to put a warning sticker on another 15,000 copies that hadn't yet shipped to retailers.

Thank you for the help, and please know that we've added an electrical proofing step in our editorial process.

The corrected illustration is at left. You can also download it from makezine.com/images/05/p112revision.pdf. The PDF is designed so you can paste it right over page 112 in MAKE Volume 05.

OTHER ERRATA:

In Volume 05, page 58, "No More Scratches!" suggested using toothpaste to remove scratches from glass. To clear up confusion and avoid disasters, please remember that this ONLY works on glass (and should be tested in an out-of-the-way area first). It should never be used on plastics or on glass with coatings, and it will damage many modern eyeglasses, which, despite the name, are not made from glass.

In Volume 05, page 138, Fig. 2, the distance between the left edge of the base and the center of the leftmost mirror block was incorrectly labeled as 5⁄16" rather than 15⁄16". You can find the correct dimensions on author Rob Hartmann's blog, at pseudoscope.blogspot.com.

In Volume 05, page 175, the price of the Poqet PC was wrong. The Poqet PC actually costs $159 plus shipping.

BLAST FROM THE PAST

The Boy Mechanical

By Mister Jalopy

A look back at the glory years of engineering for common folk.

Being an insufferable snob is an awful lot of work. It can be heartbreaking, wallet-emptying, and full of regret — often all at the same time. In the world of snobs, the most pure would have to be the book snob. Sure, you can spend more money on Tiffany lamps and art from the New York School, but it is difficult to match the mania of a true-blue book snob. To winnow that group of individuals to an even narrower class of snobbery, consider the perfectionist bibliophile who buys books on eBay. While intoxicated. A surer recipe for regret has never been devised. My own disappointments are manifold — limitless potential and hope that nearly always end in disaster.

As a result of mixing eBay and Scotch, I own the original four-volume set of *The Boy Mechanic*, which was printed from 1913-1925. Naturally, I paid too much and was terribly disappointed with the condition of the books, but not with the content. Slack-jawed wonder lies within those tattered covers.

The projects range from the utterly pedestrian, like "How to Weave a Shoestring Watch Fob," to the ridiculously ambitious, as illustrated by the young lad who is gracefully flying his homemade glider off a cliff and over a locomotive only to land in his hometown. I bet the girls at his school were mighty impressed.

These were the glory years of engineering and mechanics for common folk. The Wright brothers had flown their airplane a few years earlier, the Model T was changing the world, and unlimited opportunity was waiting in the workshop. The insanely complicated projects, like the "Homemade Steam Turbine," are immensely entertaining, but are they any more enjoyable than the 44 parlor tricks, like "Removing 36 Cannon Balls From a Hand Bag"? Why, the toboggan made of homemade skis and an old ladder, in "Making Skis and Toboggans," looks to be every bit as good at knocking out front teeth as today's modern version!

Luckily, Volume 1 is available for free from Project Gutenberg, so you can dodge the high-price-and-poor-condition bullet that nearly killed me. Furthermore, if you are OK with having a PDF instead of a crumbly, yellowed, chipped volume from 1913, then you will have avoided the dreaded fate of the book snob.

The airborne boy.

Published in the early 20th century, *The Boy Mechanic* provides instructions on making "wireless outfits, boats, camp equipment, aerial gliders, kites, self-propelled vehicles, engines, motors, electrical apparatus, cameras and hundreds of other things which delight every boy."

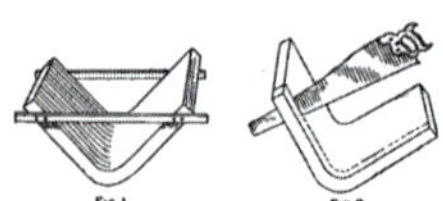

Making a boomerang from a "plank of well-seasoned hickory." It's noted that "any worker in wood can turn out a great number of boomerangs cheaply."

Images excerpted from *The Boy Mechanic*, 1913

Mister Jalopy breaks the unbroken, repairs the irreparable, and explores the mechanical world at Hooptyrides.com.

RETROCOMPUTING

Old Hardware: Anything but Garbage

By Tom Owad

Laser printers and photocopiers are full of especially interesting mechanical parts, with myriad gears and a variety of sensors and electromechanical components.

When my Laserwriter 12/640 PS tragically passed away, I ended up with drawers full of useful and fascinating components.

Photograph by Tom Owad

Old computers vary greatly in their desirability. A handful of systems are in great demand. No Altair 8800 or DEC PDP-8 is going into the trash, but more common computers, especially ones that were popular in schools, are still headed for the dump by the truckload. What's to happen to all those Apple IIe classics, PC clones, and printers that can't even be given away?

I now tear apart anything I can get my hands on before passing it along to the garbage man. Rugged old computers are sure to have some quality parts within them, perfect for use in future projects. Any circuit board from the 80s is likely to have TTL or CMOS chips, and if you're lucky, they'll be socketed and take a second each to remove.

Many boards will also have EPROMs, which you can quickly identify by the sticker covering the chip's window. Remove the sticker, expose the circuit to ultraviolet light, and they're as good as new. Pick up an old EPROM burner at a swap meet or on eBay, and you can reprogram it.

Mechanical devices are even more likely to contain interesting parts. A typical scanner includes a stepper motor and a handful of belts and gears, plus a long stabilizing bar that guides the scanner head (and could be used to guide anything). The one I tore apart last week also had a nice transformer and an EPROM on its circuit board. Invariably, they'll have a fluorescent or xenon lamp, too, which can be made into a bench lamp, if nothing else. I haven't actually done this, but it would be fun to be able to point around to various objects in your house and say, "That lamp used to be my scanner; that night stand was my server; that digital picture frame is my old PowerBook ..."

Hard drives are such a frequent point of failure that I've yet to scrap a working one, but I've encountered plenty of dying drives — most of them have two extremely strong magnets in them. When I bought a magnetic CB antenna for my car, I was disappointed to find that it wouldn't stick to my trunk lid. I placed four magnets scavenged from dead hard drives on the inside of the lid, and they've successfully held the antenna in place at more than 80 mph.

5.25" floppy drives almost always contain a stepper motor to adjust the head and a DC motor to spin the disk. Scrap out a pallet of old PCs, and you'll have enough motors to build some pretty complex devices. The drives also typically have two stabilizer bars, just like in the scanners but smaller. I haven't yet figured out what I can make with these, but I've got a drawer full of them. 3.5" floppy drives tend to have junkier parts (especially the newer ones), but the motors are much smaller, which can be useful.

As systems get more compact, it gets harder to find decent parts, but it's still not impossible. The death of my wireless router yielded a nice collection of LED arrays, two crystal oscillators, an 802.11b card, and two antennas, complete with internal cabling.

Tom Owad (owad@applefritter.com) is a Macintosh consultant in York, Pa., and the editor of applefritter.com. He is the author of *Apple I Replica Creation* (Syngress, 2005).

Recycling & Remarketing High Technology
WEIRDSTUFF®
DOOHICKIE CENTRAL
WAREHOUSE
Silicon Valley's Source for Great Surplus Deals!
Computers Electronics Equipment Software
Sign up for our SPAM FREE newsletter!
Computer Parts
Jetway 64-BIT AMD MotherBoard $35
CPU's 486 / Pentium / PII / PIII *call / web
Video Cards PCI / AGP *call / web
Memory SIMM/ SDRAM / DDR *call / web
Extensive Intentory of Hard Drives *call / web
Weird Stuff
USB Powered Tranquility Ball $10
Pigskin Toss Football $10
Authentic 8-Way Arcade Joystick $8
Customizable Arcade Cabinet In Store Only $100
8" Silicon Wafers Set of 3 $10
Cables
HP 2m (6.6ft.) HPIB/GPIB Cable $19
3Com Ethernet 10/100 PCMCIA Dongle $10
Belkin Pro Series 10ft. VGA Cable $10
RJ45 CAT5 STP 25 Meter Cable $7
6 Outlet Power Strip with Motion Sensor $15
Accesories & Peripherals
2.5" External USB2.0 Harddrive Enclosure $20
Light-Up Mini Optical Mouse $10
Backlit Low-Profile Keyboard $20
Magic Link Phone Line Video Distributor $15
HP Laser Printers In Store Only $50
Servers and Stuff
Sun Ultra Sparc 1 / 5 / 10 In Store Only $50/$99/$160
Sun Ultra Enterprise 250 / 450 / 220R In Store Only $389
4U P4 1.5GHz Server In Store Only $290
2U Dual PIII 600 Mhz Server In Store Only $150
Battery Back-Up 19" Rack Mount In Store Only $300
Industrial Stuff
Tygon 50FT General Lab Tubing $35
Optrex 2-Line Back-Lit LCD Module $15
Erie Controls Motortrol 2-Way Valve $23
Motor - 230V 60HZ 35RPM $5
Semtronics Grounding Monitor $50
* = for the latest, lowest prices, please call our store or visit www.weirdstuff.com
Hard to Find Products, Novelty Toys, Closeout Items, Obsolete Stuff, & Much More!
WE BUY & SELL - NEW/USED Visit www.weirdstuff.com
SUNNYVALE STORE - Open 7 Days 384 West Caribbean Dr. S'vale, CA 94089 408-743-5650 - Store x324
SEALED BID AUCTION SALE - Weekly Local Auction of High Tech Items
WEB-ECOMMERCE www.weirdstuff.com Email - orders@weirdstuff.com
WEIRDSTUFF® Web Cam! view online!
The majority of our products are surplus and are available in limited quantity. First come first serve - while supplies last! Don't delay!
© 2005 WeirdStuff, Inc.

Hone your skills.
Black Hat®
Briefings & Training USA 2006
July 29 - August 3 • Caesars Palace • Las Vegas
sponsors
platinum
ERNST & YOUNG
Quality In Everything We Do
Microsoft
gold
CITADEL
Configuresoft
QUALYS
silver
ArcSight
BARRACUDA
CORE
GRANITE EDGE
Juniper
Lancope
nCircle
IOActive
SPI DYNAMICS
StillSecure
TippingPoint
radware
WEBSENSE
SAINT
netForensics
watchfire
Defeat the attackers roaming your cyber streets!
Black Hat USA will once again gather the world's information and computer security elite to share their knowledge and experience with you.
Six days. Twenty nine classes. Sixty presentations.
www.blackhat.com

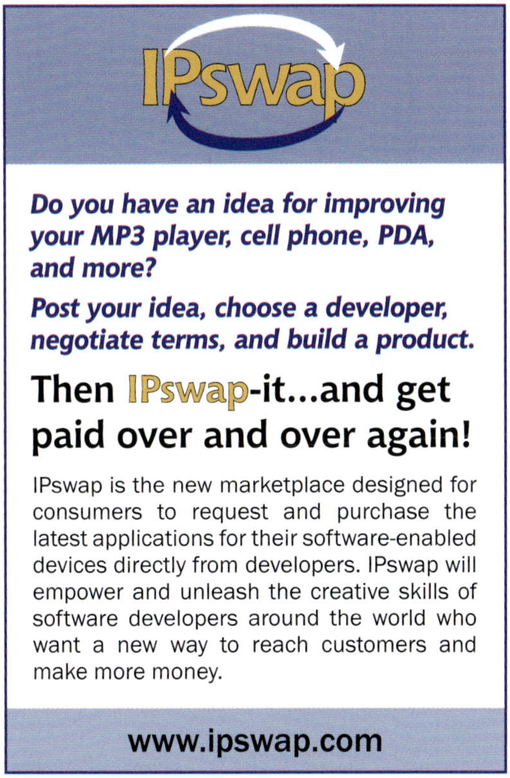
IPswap
Do you have an idea for improving your MP3 player, cell phone, PDA, and more?
Post your idea, choose a developer, negotiate terms, and build a product.
Then IPswap-it...and get paid over and over again!
IPswap is the new marketplace designed for consumers to request and purchase the latest applications for their software-enabled devices directly from developers. IPswap will empower and unleash the creative skills of software developers around the world who want a new way to reach customers and make more money.
www.ipswap.com

HOBBY ENGINEERING
Robotics
Electronics
Science
Tech Hobbies
Metal Working
More ...
www.HobbyEngineering.com
1-866-ROBOT-50
1-650-552-9925
180 El Camino Real, Millbrae, CA 94030
Visit our store near SFO!

Maker Faire
Thank You to Our Sponsors
Microsoft
YAHOO!
Sun microsystems
Geek Squad
SQUIDOO
c|net
HOBBY ENGINEERING
BEST BUY
LEGO
PARALLAX
ZAP!
ROBOT
MOZES
GRIFFIN
simplified building concepts

TALES FROM MAKE: BLOG

Roomba Tronic

By Phillip Torrone

Take a tour into the underground robotic relentlessness of Roomba hacks, robots in the streets of Austin, and robot cockfighting.

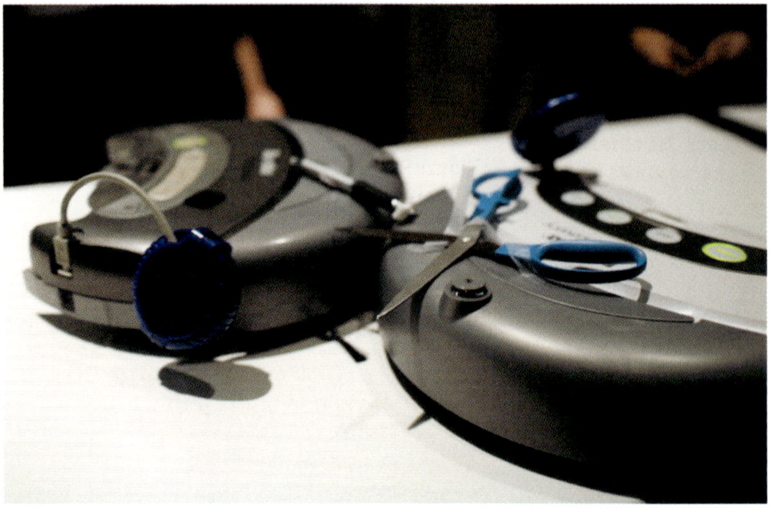

Roomba wars!

It all started out innocently enough. iRobot recently opened up their Roomba vacuum for hackers, tinkerers, and makers to send control signals via a little hidden port (irobot.com/hacker). On Makezine.com (makezine.com/blog), we posted how to control your Roomba with a serial cable or a Bluetooth computer, and all was well in the world. But you simply cannot have robots under human control; they're better off sticking to Isaac Asimov's "Three Laws" than listening to us meat-based guardians. With great robot power comes great robot responsibility.

At O'Reilly's yearly Emerging Technology conference, alpha geeks and technologists from around the world gather to tune in, turn on, and trade RSS feeds. One of the nightly events, the MAKE Fest, showcased a dozen makers dutifully explaining their creations, from wired clothing to retro arcade mods.

This is when it happened.

The ambient noise of conference chatter stopped, cheers were heard, money was thrown on the table. An underground Roomba cockfighting competition spontaneously erupted. Maker Tod Kurt (todbot.com) just happened to have his Roomba with him; we happened to have a couple extra with us (you just gotta love a conference where people have extra Roombas with them).

Bets were taken; weapons were formed from what we had available (mostly conference schwag). The crowd frenzied; mild-mannered AJAX programmers were pushing their way in to get a piece of the robot action. Oh yes, it was on.

A few days later the MAKE team was off to Austin's South by Southwest Interactive Festival. Our little MAKE blog picked up two awards for Best Educational Resource and Best Craft Blog, and in celebration we continued our relentless pursuit of robot gaming. The details are hazy, but basically

> With great robot power comes great robot responsibility.

we hustled a bar full of festival goers with some Roomba pool. And still later, with the help of some Austin residents, we played the first game of "Real Frogger." This in celebration of Frogger's 25th anniversary, of course.

The future of robotic games has arrived, it just hasn't been evenly vacuumed. Complementing this issue of MAKE (our robotics theme), check the makezine.com/blog each day; we'll have new projects, updates, and more things you should (and shouldn't) do with robots.

Phillip Torrone is associate editor of MAKE.

Photograph by Scott Beale, Laughing Squid / laughingsquid.com

MAKER'S CALENDAR Compiled by William Gurstelle

Our favorite events from around the world — June - July - August 2006

June

»The Power of DC June 3, Hagerstown, Md. Fast, powerful, and silent electric vehicles race for ecologically sound glory. nedra.com

»Combine Demolition Derby June 10, Lind, Wash. Lind's yearly rodeo weekend kicks off with a "Combine Demolition Derby" showcasing kamikaze farm machinery. lindwa.com

EDITOR'S CHOICE
»Made in America Factory Tour Weekend June 14-17, York, Pa. Visitors are invited onto the production floors at companies such as Harley-Davidson, Pfaltgraff, Utz, and nearly 30 others. factorytours.org

»RoboGames San Francisco State University, June 16-18, San Francisco. Robotics galore: FIRA, BEAM, Mindstorms, and Robotic Combat. robogames.net

EDITOR'S CHOICE
»LDRS 25 Silver Anniversary June 28-July 4, Amarillo, Texas. The largest amateur rocketry event in the world, featuring hundreds of high-powered rocket launches. ldrs.org

July

»Montréal International Fireworks Competition Weekends, June through July, Montreal. Teams from around the world compete in what may be the largest and most prestigious pyrotechnical competition anywhere. internationaldesfeuxloto-quebec.com/en

»Warship Weekend 2006 July 8-9, The Museum of Naval Firepower, Gosport, UK. Display features old and modern warships. explosion.org.uk

»Da Vinci Days July 14-16, Corvallis, Ore. A multidimensional festival of Da Vinci-esque activities, including a kinetic sculpture race, robots, speakers, and more. davinci-days.org/index.php

»The Royal International Air Tattoo July 15-16, Fairford Air Field, Fairford, UK. RIAT is the biggest military air show held in Europe and attracts participants, airplanes, and technology from around the world. airtattoo.com

»ASME's Human Powered Submarine Contest July 19-23, Escondido, Calif. Engineering students will test their skills by designing, constructing, and racing their own submarines. sections.asme.org/sandiego/HPS2006.htm

»World Freefall Convention July 21-30, Quincy, Ill. The USA's largest skydiving convention occurs every summer at this central Illinois location. freefall.com

»Farnborough International Airshow July 22-24, Farnborough, UK. This huge airshow features a 4+ hour flying display, static displays, exhibitions, seminars, rides, and other attractions. farnborough.com

»The 2006 Fire Arts Festival July 12-17, The Crucible, Oakland, Calif. Pyro-artistic festival includes lectures, classes, sculptures, demonstrations, and fire-related performances. thecrucible.org/calendar

»EAA AirVenture July 24-30, Oshkosh, Wis. More than 750,000 people attend AirVenture, the annual convention of the Experimental Aircraft Association with thousands of aircraft-related events. airventure.org

August

»DEFCON August 4-6, Riviera Hotel, Las Vegas. DEFCON is one of the largest and oldest computer hacker conventions in the world. defcon.org

»6th Annual Telluride Tech Festival August 12-14, Telluride, Colo. The festival looks at the past and future of technology, through speakers and discussion groups. telluridetechfestival.com

EDITOR'S CHOICE
»Bonneville Speed Week August 12-18, Bonneville Salt Flats, Utah. The fastest racers on Earth streak across the saltpan attempting new speed records. scta-bni.org

Important: All times, dates, locations, and events are unconfirmed and subject to change. Verify all information before making plans to attend.

*Do you know of an event that should be included? Please use our online form at **makezine.com/events**. Sorry, it is not possible to include listings for all events in the magazine, but they will be listed online.*

*Did you attend an interesting event? If so, please tell us about it in our Forums. (**forums.makezine.com**).*

GEORGE DYSON

Off-the-Shelf Parts, Off-the-Wall Ideas

Looking for the dawn of the digital universe? Check in the basement, next to the lavatory.

Sixty years ago, in the spring of 1946, a handful of electronic engineers showed up at the Institute for Advanced Study in Princeton, N.J., and requested a place in the basement to plug in their tools. They were out of luck. "The only really useable space in our basement is that adjoining the men's lavatory, to which you are most heartily welcome," Electronic Computer Project director John von Neumann was told. Despite this inauspicious welcome, the engineers moved in, and in April, the project's accountants recorded the first $4 for "electrical work."

The Institute for Advanced Study was established in 1930 as a refuge for scholars in mathematics, physics, history, and art. There were no laboratory facilities of any kind. "What could be wiser than to give people who can think the leisure in which to do it?" the founding director Abraham Flexner had been advised in 1939. Not a word about workbenches or tools.

John von Neumann was a pure mathematician who, along with many of his colleagues, had been drawn over to the applications side of things during World War II. He liked it there, and refused to leave. Having been exposed to the powers of digital computing through his involvement with the nuclear weapons program at Los Alamos, he was impatient to build the next generation of computers, and was well aware of the implications of Alan Turing's 1936 results on Universal Machines. "We are building one," he told his arriving engineers, and the copy of Turing's paper in the IAS library has been so heavily consulted that its binding has come unglued.

Other groups were engaged in similar projects, but none quickly and boldly enough for von Neumann. "If he really wanted a computer," explained Arthur Burks, "the thing to do was to build it."

Burks and Herman H. Goldstine, an Army colonel transplanted from the ENIAC project at the Moore School in Philadelphia, were installed on the second floor of Fuld Hall in a small office belonging to logician Kurt Gödel, where they set to work, under the guidance of von Neumann, outlining the logical architecture of the machine.

"The heart of the system is a central clock, carrying an enormous load," say the minutes of the first meeting of the Electronic Computer Project. The circuitry would be modular, because "this sort of design is favorable for mass production," added the engineers. "Words coding the orders are handled in the memory just like numbers," explained von Neumann, breaking the distinction between numbers that *mean* things and numbers that *do* things, ensuring that all hell would break loose.

Instead of importing a few mathematical logicians into a lab full of engineers, von Neumann imported a handful of engineers into a place run by mathematicians. Although they had to start out by building their own workbenches at the Institute, and scrounge for war-surplus components as best they could, there was no one around to say, "Well, Dr. von Neumann, this looks interesting, but we build computers like this ..."

The lead engineer was Julian Bigelow, former colleague of Norbert Wiener and co-author of the 1943 paper "Behavior, Purpose, and Teleology," around which the beginnings of the cybernetics movement coalesced. His job was to take the logical design as laid out in the abstract by Burks, Goldstine, and von Neumann, and coax it to life as a machine. "

> Every single bit of the digital universe we now inhabit can be traced back directly to those first binary digits that flickered to life in the summer of 1951.

Julian would have the ideas, and Ralph Slutz would kind of detail the ideas, and then James Pomerene and I would go try and make the electrons do their thing," explains Willis Ware, one of the original engineers. "Julian was the architect of that machine." Bigelow's gang soon outgrew their room in the basement of Fuld Hall, and were pushed outside to a hastily constructed building at the edge of the Institute's 600-acre woods.

At the Institute for Advanced Study, it was easier to find an expert in quantum mechanics than someone who worked on their own car. Bigelow was the exception — the only permanent member of the

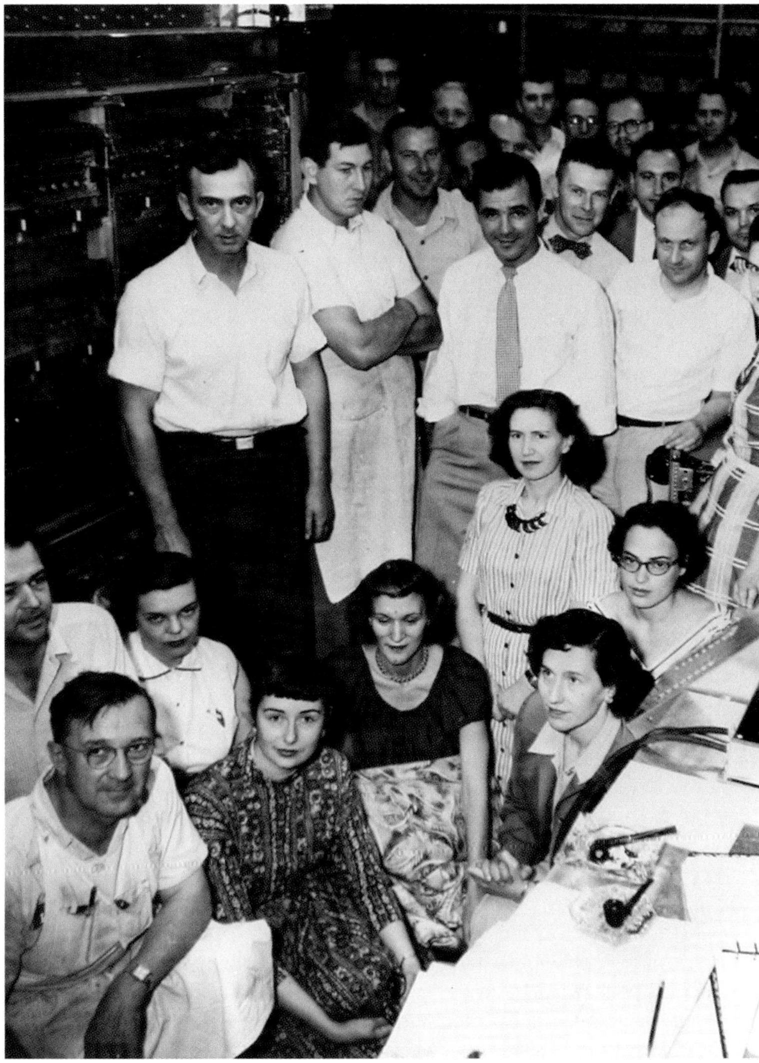

Making a Revolution: Staff of the Institute for Advanced Study Electronic Computer Project, ca. 1952. After John von Neumann decided to construct a modern, stored-program, high-speed electronic digital computer, others had to design and build the machine, and write the code to make it work. Some of these people, incompletely identified, are pictured here.

The entire computer was built in the project's primitive machine shop, and all code was assembled by hand, using absolute addressing, bit by bit. The finished computer, comprising a total of 3,474 vacuum tubes, had an internal memory of 40,960 bits, stored as 32x32 arrays of charged spots on the faces of 40 cathode-ray "Williams Tubes." It entered service during the summer of 1951 and was duplicated widely around the world.

Photograph by Alan Richards, courtesy of the Archives of the Institute for Advanced Study

Institute who routinely dismantled engines with his own hands. He approached the construction of the new computer as a true engineer. The guts of the new machine (the first to have a substantial, fully random-access memory available at megacycle speeds) were 40 "Williams Tubes," 20 on each side of the main chassis, so that the computer not only resembled, but functioned, as an enormous V-40 engine with overhead cams and valves.

Every single bit of the digital universe we now inhabit can be traced back directly to those first 40,960 binary digits that flickered to life, in the summer of 1951, as a 32x32x40 matrix of charged spots on the face of some delicately modified cathode ray tubes.

The first serious computation, a thermonuclear problem for Los Alamos that was coded by a crew including Hedy Selberg and Klari von Neumann, ran for six weeks nonstop, and led to the success of the hydrogen bomb. Not bad for a shakedown run.

Cosmologists may never know what existed before the big bang, but at the dawn of the digital universe, we know that it was Julian Bigelow, James Pomerene, Herman Goldstine, Ralph Slutz, Jack Rosenberg, Willis Ware, and a small group of helpers who worked to bring the dreams of Leibniz, Babbage, Turing, Atanassoff, Zuse, von Neumann, and many others to life with their bare hands.

George Dyson, a kayak designer and historian of technology, lives in Bellingham, Wash., and is the author of *Baidarka*, *Project Orion*, and *Darwin Among the Machines*.

HOMEBREW

My Version of Einstein's Amplifier

By Tyler Rourke

After World War II, many of the best physicists and mathematicians in the world could be found at the Institute for Advanced Study (IAS) in Princeton, N.J. In 1949, a young engineer named Jack Rosenberg was part of a team building one of the world's first digital computers. Having knowledge of electronics, a keen ear, and an interest in classical music, Rosenberg's home-built hi-fi was far superior to commercially available audio systems, and before long, his system was widely talked about at the IAS.

Erwin Panofsky, an art historian and close friend of Albert Einstein's, heard about Rosenberg's amazing hi-fi and asked him if he would build a system for "The Professor" as a 70th birthday gift from everyone at the IAS. Rosenberg agreed, and on March 14 he and J. Robert Oppenheimer personally delivered the system to Einstein at his home.

Last summer, I found myself speaking with George Dyson, who had spent time during his youth at the IAS. He told me that he had recently interviewed Rosenberg about the hi-fi. He offered a copy of the original hand-drawn schematic of the amplifier portion of Einstein's hi-fi system and encouraged me to attempt a recreation of it. I immediately began figuring out how I was going to build the amplifier.

I started the project by purchasing a chassis that was similar in size to the one specified in the parts list. Although I was unable to obtain the original transformers, I found suitable substitutes and positioned them on the chassis. I then laid out the positions of the tubes and tried to arrange all of the parts (switches, jacks, pilot light, etc.) according to function, so that high voltages would be isolated from the signal path whenever possible. I printed templates to mark the positions of all holes, drilled the chassis, installed the hardware, and began laying out the amplifier's internal components. Resistors and capacitors were connected directly to tube sockets, switches, and jacks wherever possible. The remaining components were soldered to terminal strips, which were then connected according to the schematic. Friends at CDI Signs in Bellingham, Wash. made the beautiful faceplate for me.

Visually, my version is not identical to the original, and I've had to substitute some parts that were no longer available, but I'm confident that this amplifier sounds nearly the same as Einstein's, which is a wonderful thing.

Tyler Rourke has a penchant for electronics and good food.

Photograph by Tyler Rourke (thanks to Jack Rosenberg)

Make:
technology on your time

Volume 05

VOYAGE TO THE BOTTOM OF THE POOL
Students from universities and trade schools in the U.S., Canada, and India entered the 8th International Autonomous Underwater Vehicle Competition. At stake was $20,000 in prize money, which considering the cost and time to create a submarine, isn't the motivating factor. It's all about the bragging rights.

INTEL'S IN-HOUSE MAKER
Eric Paulos is a computer scientist at Intel Research Berkeley, a corporate "lablet" in the penthouse of a downtown office building where a dozen or so scientists explore the edges of computing and networking technology. For Paulos, every city is an ideal laboratory to test what he calls "objects of wonderment."

Vol. 05, February 2006. MAKE (ISSN 1556-2336) is published quarterly by O'Reilly Media, Inc. in the months of February, May, August, and November. O'Reilly Media is located at 1005 Gravenstein Hwy North, Sebastopol, CA 95472, (707) 827-7000. SUBSCRIPTIONS: Send all subscription requests to MAKE, P.O. Box 17046, North Hollywood, CA 91615-9588 or subscribe online at makezine.com/offer or via phone at 866-289-8847 (U.S. and Canada), all other countries call (818) 487-2037. Subscriptions are available for $34.95 for 1 year (4 quarterly issues) in the U.S. Canada: $39.95 USD; all other countries: $49.95 USD. Application to Mail at Periodicals Postage Rates is Pending at Sebastopol, CA and at additional mailing office POSTMASTER: Send address changes to MAKE, P.O. Box 17046, North Hollywood, CA 91615-9588.

Make: Projects

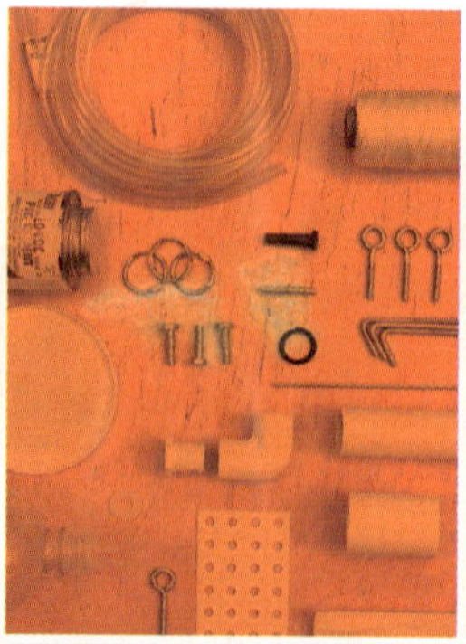

Volume 05

ON THE COVER
Most people think of water rockets as being those cheap red and white plastic toys. But a thriving community of water rocket enthusiasts have launched their hand-built creations to heights of nearly 1,000 feet. Our project, which starts on page 78, comes with a parachute for a soft landing. Photograph by Topher Lucas

Make:
technology on your time™

EDITOR AND PUBLISHER
Dale Dougherty
dale@oreilly.com

EDITOR-IN-CHIEF
Mark Frauenfelder
markf@oreilly.com

MANAGING EDITOR
Shawn Connally
shawn@oreilly.com

ASSOCIATE EDITOR
Phillip Torrone
pt@makezine.com

PROJECTS EDITOR
Paul Spinrad
pspinrad@makezine.com

SPECIAL PROJECTS EDITOR
David Pescovitz

ASSISTANT EDITOR
Arwen O'Reilly

COPY CHIEF
Goli Mohammadi

COPY EDITORS/RESEARCH
Terry Bronson
Keith Hammond

CREATIVE DIRECTOR
David Albertson
david@albertsondesign.com

ART DIRECTOR
Kirk von Rohr

DESIGNERS
Sarah Hart
Kristy McKoy

ASSOCIATE PUBLISHER
Dan Woods
dan@oreilly.com

CIRCULATION DIRECTOR
Heather Harmon

ADVERTISING COORDINATOR
Jessica Boyd

MARKETING & EVENTS COORDINATOR
Rob Bullington

ONLINE MANAGER
Terrie Miller

MAKE TECHNICAL ADVISORY BOARD:
Gareth Branwyn, Joe Grand, Saul Griffith, William Gurstelle, Bunnie Huang, Tom Igoe, Mister Jalopy, Steve Lodefink

PUBLISHED BY O'REILLY MEDIA, INC.
Tim O'Reilly, CEO
Laura Baldwin, COO

Visit us online at makezine.com
Comments may be sent to editor@makezine.com

For advertising and sponsorship inquiries, contact:
Dan Woods, 707-827-7068, dan@oreilly.com

Printed in the USA by Schumann Printers, Inc.

Contributing Artists:
Howard Cao, Roy Doty, Paul Hansen, Chad Holder, Dustin Amory Hostetler, Christopher Hujanen, Tim Lillis, Topher Lucas, Jenny Pfeiffer, N55, Emily Nathan, Eric Rife, Nik Schulz, Anne Sigler, Damien Scogin, David Sobo

Contributing Writers:
Tim Anderson, Thomas Arey, Dave Battino, Joost Bonsen, Gareth Branwyn, Abe Connally, Cory Doctorow, Nick Dragotta, George Dyson, Jonathan Foote, Joe Grand, Saul Griffith, William Gurstelle, Keith Hammond, Larry Harmon, Rob Hartman, Christopher Holt, Bunnie Huang, Tom Igoe, Andy Ihnatko, Mister Jalopy, Stefan Jones, Chris Kentworthy, Jason Kohrs, Mark Lengowski, William Lidwell, Steve Lodefink, Robert Luhn, Merlin Mann, Dave Mathews, Josie Moores, Danny O'Brien, Tim O'Reilly, Ross Orr, Tom Owad, Bob Parks, David Pescovitz, Jenna Phillips, Dan Picard, Charles Platt, Julie Polito, Michael Pryor, Douglas Repetto, Matthew Russell, Brian Sawyer, Bob Scott, H.B. Siegel, Ewan Spence, Bruce Sterling, Bruce Stewart, Cy Tymony, Steve Vigneau, Howard Wen

Interns: Adrienne Foreman (web), Jake McKenzie (eng.), Ty Nowotny (eng.), Aly Rasmussen (design)

MAKE is printed on recycled paper with 10% post-consumer waste and is acid-free. Subscriber copies of MAKE Volume 05 were shipped in recyclable plastic bags.

Please note: Technology, the laws, and limitations imposed by manufacturers and content owners, are constantly changing. Thus, some of the projects described may not work, may be inconsistent with current laws or user agreements, or may damage or adversely affect some equipment.

Your safety is your own responsibility, including proper use of equipment and safety gear, and determining whether you have adequate skill and experience. Power tools, electricity, and other resources used for these projects are dangerous, unless used properly and with adequate precautions, including safety gear. Some illustrative photos do not depict safety precautions or equipment, in order to show the project steps more clearly. These projects are not intended for use by children.

Use of the instructions and suggestions in MAKE is at your own risk. O'Reilly Media, Inc., disclaims all responsibility for any resulting damage, injury, or expense. It is your responsibility to make sure that your activities comply with applicable laws, including copyright.

Contributors

Tim Lillis (*Wind Powered Generator* illustrations) is still waiting for the hand-drawn blueprints that he sent to Hasbro as a young lad to be made into Transformers, but in the meantime, he works as a freelance designer and illustrator. A recent arrival to San Francisco, Lillis once worked with Kaiju Big Battel, the world's only Live Monster Fighting Tournament, making monster costumes, miniature cityscapes, and giant cartoonish props out of foam, and acting as a producer for their live events. Tim's hobbies include ruining karaoke, making way too many puns, and designing whatever he can at narwhalcreative.com. He loves AMC Eagles, the best car ever.

Cy Tymony (*Sneaky Uses*) reads six newspapers a day. "I am inquisitive about all aspects of how life works," the Los Angeles author/inventor says. His career started as a child, when he made a sleeve-mounted shocker device to defend himself from schoolyard bullies. His *Sneaky Uses* books explore the delight of finding. He also roller-skates, practices aikido and judo, plays with cardboard boomerangs, and eats tuna salad when he is not working on finding still more sneaky uses for everyday things.

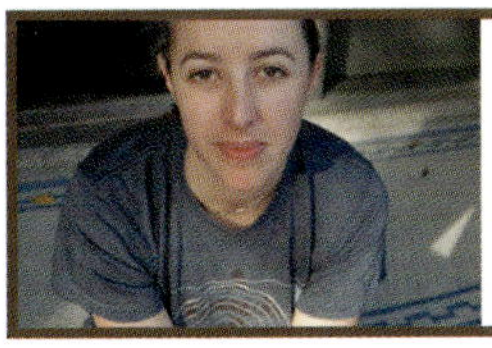

Jenny Elia Pfeiffer (*Woody's World* photos) is a San Francisco-based artist and photographer. Her portraits and travel photography have appeared in several local and national publications. She also collaborates with her sister, Lisa Pfeiffer, on multimedia works that have been featured in galleries in San Francisco and Prague. Her favorite food is ice cream (Oreo cookie).

George Dyson (*Treehouse*) is a historian of technology whose interests range from the development (and redevelopment) of the Aleut kayak to the evolution of digital computing, telecommunications, and nuclear bomb-propelled space exploration. Dyson, who lives in Bellingham, Wash., divides his time between building boats and writing books. He considers himself a follower of Nathaniel Bishop, who, while paddling a paper kayak to the Gulf of Mexico from Québec in 1874, urged his audience at Princeton University to "seek in his friendly canoe that relief which nature offers to the tired brain. Let him go into the wilderness and live close to his Creator by studying his works."

An inveterate tinkerer and "broad-spectrum hobbyist," **Steve Lodefink** (*Soda Bottle Rocket*) just can't say no to a cool project. At 3, he was already reverse-engineering the peanut butter and jelly sandwich: "I figured out where all of the parts were, found a good tool, and built one. I've been doing it ever since." He lives in Seattle with his wife and two sons, two cats, five tarantulas, and 24 African cichlids, and thinks that one of life's great pleasures is a really sharp aged cheddar cheese. "I'm a simple man," he says. He looks at life's debris at finkbuilt.com.

"Tall and skinny, with an affinity for firecrackers and vinyl records," **Larry Harmon** (*Underwater Vehicles*) has been exploring San Diego's underbelly for years. His fanzine, *Genetic Disorder*, is an ode to anything "that would embarrass our Chamber of Commerce." When not at the library doing research for a book about the history of San Diego's punk rock scene, he plays guitar in the Dissimilars (myspace.com/thedissimilars), "a trashy garage rock band." He spends his days as an editor at a community newspaper, three blocks from the beach.

If **Tom Owad** (*Retrocomputing*) could be any animal, he'd be something from the Galapagos. His interest in extraordinary and incomparable things extends to his human existence as well. He spends his days tinkering and learning, and is the owner and webmaster of applefritter.com, an online Macintosh community of artists and engineers dedicated to the "obscure, unusual, and exceptional." He serves on the board of directors and is webmaster and archivist of the Apple I Owners Club, and is also the author of *Apple I Replica Creation*.

Maker Faire
Saturday & Sunday
April 22 – 23, 2006
San Mateo Fairgrounds
San Mateo, California
Meet the Makers
makezine.com/faire
Two-day family-friendly event celebrating science and technology projects and the DIY mindset. It's for creative, resourceful people who like to tinker and love to make things.
We call them Makers.
MAKE Readers get 50% off
on advance purchases of weekend passes. »» Visit
makezine.com/faire and use access code MFWinter*
2 days
HANDS-ON WORKSHOPS
100,000 square feet of indoor exposition
5-acre outdoor midway with stage
DIY competitions
6 EXPOSITION & WORKSHOP PAVILIONS
WEIRD SCIENCE
ULTIMATE GARAGE
ROBOTICS
DIGITAL ENTERTAINMENT/GAMING
GREEN TECH
ULTIMATE WORKSHOP
Latest in tools, kits, and DIY resources
SPONSORED BY
Microsoft
intel
Geek Squad
*Discount tickets for Maker Faire must be purchased by March 31, 2006.

ON TO YEAR TWO

Dale Dougherty and Mark Frauenfelder look back on MAKE's first year.

Q: Now that MAKE is a year old, what have you learned?

Dale Dougherty: Someone once told me that you don't create a magazine — you discover it. You discover an audience, if you're lucky, and then you have to work hard to keep up with them.

Mark Frauenfelder: We hoped there would be an audience for MAKE. The surprising thing is how big that audience is. The world has far more makers than I would have dreamed. It makes me feel good about the human race.

DD: I had an interviewer ask me, "Don't you have to be pretty smart to do the projects in MAKE?" I answered, "I think most people are pretty smart. It's a matter of having the time to make things, and that's true of cooking, photography, electronics, and lots of areas." Often, there's some technical knowledge you need to acquire, but you learn as you go, and you can have lots of fun. It's like playing a game.

MF: The best games are ones that are easy to learn but challenging to master, like chess. I think the same thing can be said for great projects. The kite aerial photography project in our first issue was really simple, giving you a lot of bang for your buck. But more importantly, it pointed the way to a community of photographers who are improving on one another's designs, and building remarkably sophisticated systems.

DD: I guess it's my move now! So one of the big surprises for me in MAKE's first year is hearing from kids or from their parents. I got a note recently from a father who said he was saving back issues of MAKE to do projects with his son when he was old enough. I hope MAKE's projects engage kids of all ages in science and technology.

MF: I agree. There's a reason why we don't label any of the projects as being for kids or for adults. If kids are interested in a complex project, they'll rope in an adult to help them. I'm sure just as many adults will ask their kids to help them with the projects. My 8-year-old daughter is already teaching me how to work the user interfaces on the technology around our house.

DD: When you're a kid and you see something cool, such as the soda-bottle rocket (which is a project in this issue, page 79), you don't just want to watch someone else launch it. You want to do it yourself. You want to know how to build it yourself. That's the maker spirit. It's fun to meet makers and see what they do. It's inspiring. I'm looking forward to our Maker Faire in April.

MF: So am I. The best part of my job is meeting with makers and hearing their stories. One of the things we'll be focusing on more in MAKE this year is the stories behind the projects we feature, and the community of makers involved in them.

DD: Finally, I'd like to thank all those readers who have written us saying how much they love the magazine. I'd also like to thank the team we've put together to make the first year of MAKE such a success. We're looking forward to another exciting year.

Mark Frauenfelder is editor-in-chief and Dale Dougherty is editor and publisher of MAKE.

Tim O'Reilly

NEWS FROM THE FUTURE

"THE FUTURE IS HERE. IT'S JUST NOT EVENLY DISTRIBUTED YET"— WILLIAM GIBSON

I HAD LASER EYE SURGERY THE OTHER DAY, and after more than 40 years of wearing glasses so strong that I was legally blind without them, I can see clearly on my own. I had a perfect outcome: 20/20 for far vision, yet still able to read and do other close work as well. I keep saying to myself: I'm seeing with my own eyes!

But in order to remove my need for prosthetic vision, the surgeon ended up relying on prosthetics of her own, performing the surgery with the aid of high-tech equipment and specialized technicians.

First, they mapped my eyes with a device called a corneal topographer, and came up with a modification plan. Then they used a laser to blister the surface of my cornea, and 20 minutes later, the surgeon used a microkeratome to lift the flap of the blister so another laser could do the real mods to the deeper layers of the cornea. During the actual surgery, apart from lifting the flap and smoothing it back into place after the laser was done, her job was to clamp open my eyes, hold my head, utter reassuring words, and tell me, sometimes with urgency, to keep looking at the red light. Afterwards, I asked what would happen if my eyes drifted and I didn't stay focused on the light. "Oh, the laser would stop. It only works when your eyes are tracking."

In short, surgery this sophisticated could never be done by an unaugmented human being. The human touch of my superb doctor was paired with the inhuman accuracy of complex machines, a 21st-century hybrid freeing me from the tyranny of assistive devices first invented in 13th-century Italy.

Whether or not we're heading for a Kurzweil-style singularity, in which humans merge with machines, an increasing number of our activities are only possible with the aid of computers and other complex devices. My eye surgery is only one example.

The revolution in sensors, computers, and control technologies is going to make many of the daily activities of the 20th century seem quaint as, one by one, they are reinvented in the 21st. It's all over the news — remote drone warfare, powered exoskeletons, automatic language translators, cellphones that know their owners by the way they walk — but these high-profile examples tend to obscure the ways that computer technology is remaking the everyday world.

Take the automobile: it's a short step from high-tech vehicles that can only be debugged with the aid of computer diagnostics to vehicles that can only be driven by a computer. "The long run" — the last road rally of those diehards who insist on driving their own vehicles — imagined in the 1989 novel of the same name by Daniel Keys Moran, may not be that far off.

As we appreciate the wonders of technology, let's not forget the dangers of taking them for granted.

For everything that is gained, something is lost. The classic 1958 essay "I, Pencil," by Leonard Read, compellingly makes the point that in a connected world, even simple objects like the humble lead pencil require the expertise and activity of machinery, business processes, and thousands of coordinated people. How true that is today of our complex devices, composed of throwaway components themselves made in multi-billion-dollar factories.

In this world, the role of the maker is especially important, lest we fall into the future foretold nearly a century ago by E.M. Forster in "The Machine Stops," in which people cocooned in their high-tech homes are helpless when their technological assistants go silent.

Knowing how things work, being able to fix them when they break, and knowing how to create acceptable substitutes when they can't be repaired are not just hobbies but essential skills in a world growing ever more complex.

Check makezine.com/05/nff for related stories.

Tim O'Reilly (tim.oreilly.com) is founder and CEO of O'Reilly Media, Inc. See what's on the O'Reilly Radar at radar.oreilly.com.

BUILDING A SMARTER TO-DO LIST

PLANNING YOUR PROJECTS ONE BITE-SIZED PIECE AT A TIME.

By Merlin Mann and Danny O'Brien

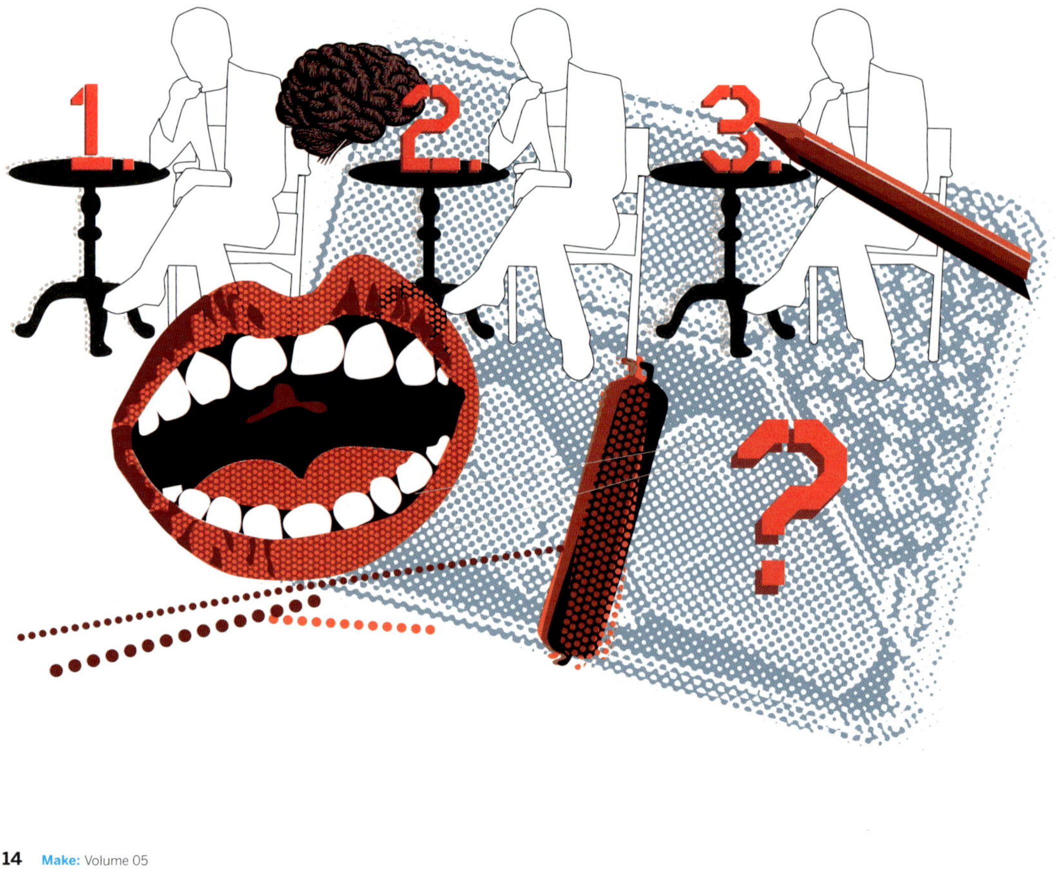

Illustration by Aly Rasmussen

WHILE YOU CAN ARGUE FOR THE flavor and approach to task management that best suits your style, it's hard to disparage the benefits that come from getting task commitments out of your brain and captured in a consistent location. Think of the to-do list as the evolving strategy for focusing your effort and attention in the immediate future.

Anatomy of a To-Do

The primary idea of a to-do is that it's a task that can and should be done — a point that might seem obvious until you start uncovering how many of the items on your to-do list may not belong there (or, conversely, how many uncaptured items do). The best and most useful to-dos share some common qualities: It's a physical action. It can be accomplished at a sitting. It supports valuable progress toward a recognized goal.

Glancing at your own to-do list, do you see any potential troublemakers? Notice any items that make you squeamish? Any "mystery meat" tasks that seem "un-doable" as is? Don't sweat it. We're going to have you shaped up in no time.

Let's Get Physical

Articulating your to-dos in terms of physical activity — even when they require only modest amounts of actual exertion — has a variety of benefits.

Most importantly, it ensures that you've thought through your task to a point where you can envision how it will need to be undertaken and what it will actually feel like once you're doing it; it's not just a bunch of words you've written on a page.

Framing your work in the physical world is easiest when you imagine what's being done, and the best trick here is to simply phrase your task in a form like: "verb the noun with the object." That means instead of reminding yourself with the mystery meat of "year-end report," you'd more accurately first start with "download Q3 spreadsheet from work server." And, instead of "get with Anil," you'd probably want to "email Anil on Monday to schedule monthly disco funk party." Get specific in whittling the task down to one activity that you can accomplish in ten minutes or less.

Now, Now, Now

Avoid the trap of littering your horizon with piles of crufty pseudo-tasks that can't actually be addressed (or, almost as often, can't be addressed *yet*). While you want to always stay aware of future obligations and the work that they are likely to generate, the to-do list is absolutely not the place to do it. Keep your to-do list a sacred tabernacle for current activity, and maintain longer-term task and support materials as well as appointments where they belong — in a project support folder and your calendar, respectively.

The trick is that most jobs can be made easier long before they're undertaken, simply by framing and naming them properly and in the right-sized units. As early as the capture and planning phases of building your to-do list, you hold the power and responsibility for defining your work properly. The last thing you want is to wonder whether you're doing the right thing at the right time.

> "Get specific in whittling a task down to one activity that you can accomplish completely at a sitting."

Why Am I Doing This Task?

When compiling a list of everything on your mind (and plate), it's crucial to unpack how each task you accept or assign to yourself will support your projects, your roles, and your goals. Before adding a new item, reflect on the value that each chunk of work brings to your world. Remember: Every minute you work on one task is necessarily a minute you can't work on another. Don't accept or assign crap. If it goes on your list, it's a commitment to seeing that task through to completion. Respect your time and attention enough to take your work seriously and plan it with care and forethought. Alpha geek status awaits once you can get your to-dos framed, flipped, and retired as efficiently as possible.

Note: This column originally appeared in an expanded version on Merlin Mann's site, 43Folders.com.

Learn how to reel in your mind at Danny O'Brien's lifehacks.com and Merlin Mann's 43folders.com.

Cory Doctorow

TRAITORS TO HISTORY

A COPYRIGHT-CONTROLLED MUSEUM IS A CRIME AGAINST HUMANKIND.

THIS PAST AUGUST, MY PARENTS CAME to visit me in London, and we had a grand sightseeing day that culminated in a visit to the Royal Greenwich Observatory, where clever makers in the 17th and 18th centuries made homebrew instruments to measure their physical environments, sensor arrays so clever that they enabled sailors anywhere in the world to divine their longitude. This was the scientific and military problem of the day, because sailors who can't calculate their longitude can't reliably cross oceans.

Today, the Greenwich Observatory is given over to a brilliant museum that tracks the history of this paleo-maker initiative. There are the most cunning brass clocks, handwritten logbooks in the crabbed script of Royal Astronomers centuries dead, and towering, proud telescopes built of wood and hand-polished optics.

And the first sign you see when you go through the door says "NO PICTURES." There's even a picture of a camera in a red circle with a slash through it.

The temple of measurement demands that it not be measured. The museum that is supposed to inspire a generation of scientists demands that the most basic tools of science — recording devices — be kept in our pockets.

I went and asked a curator why this was: was it because they're worried about flash pictures fading the exhibits? No — photons don't hurt old brass, no matter how precisely machined. The curator said that there's no photography allowed in the Greenwich Observatory because of copyright!

Copyright? How can that be? There's no such thing as a copyright on a clock. There's certainly no copyright in centuries-old logbooks; facts aren't copyrightable, and copyright runs out in substantially less than a couple hundred years — besides, the photographing of a single page out of a logbook falls firmly in the realm of fair dealing, Britain's equivalent of the United States' fair use.

All right, the curator admitted. It's not really copyrighted per se, but we want to be the exclusive purveyors of photos and picture-postcards and so forth. What's more, some of the exhibits are on loan from third parties who made us promise to prohibit their being photographed.

There's no nice way to say this: a museum curator who takes this attitude to the exhibits in her charge is a traitor to history, to heritage, and to science. The point of a museum is to spread culture, not restrict it in order to run a penny-ante picture-postcard racket. A museum curator should not accept an unphotographable exhibit of historic interest any more than she should accept an exhibit that comes with the condition that patrons who wish to see it must swear a loyalty oath, view a propaganda film, stand on one foot, or accept any other abridgment of their personal and cognitive liberty.

> "A wise steward of *David* would take every conceivable measure to see that this piece of world heritage was spread to every corner of the globe."

However, this attitude toward curatorship is fast becoming the norm, not the exception. For example, when the Florentine curators who control access to Michelangelo's *David* allowed Stanford to undertake a high-resolution 3D scan of the famous sculpture, it was on the condition that the resulting 3D files be restricted with Digital Rights Management technology to prevent their use in making copies of the *David* that were outside of Florentine control.

The *David* is not copyrighted. It is a centuries-old artifact that is part of the world's heritage. It does not "belong" to a museum any more than the

Mark Neel took this photo of Michelangelo's *David* and uploaded it to his Flickr.com account under a Creative Commons license.

Mona Lisa or the Elgin Marbles do; it is merely in the temporary stewardship of a museum that cares for it so that all of us may someday see this magnificent sculpture and understand a little more about what it means to be a member of the human race.

A wise steward of *David* would make it his business to see to it that every conceivable measure was taken so that this piece of world heritage is spread to every corner of the globe. The surest way to do that would be to use Stanford's high-resolution scan to print millions of these at the highest quality possible, distributing them to every school, park, museum, city hall, and public bath in the world. In an era of cheap, accurate 3D scanning and reproduction, there should be no person alive who lacks the opportunity to confront Michelangelo's masterpiece in living physical reality.

Instead, the stewards of *David* and innumerable other works of art take it as their duty to restrict access to their charges, all to sell a few tawdry postcards, to preserve their station and power as gatekeepers.

I've been to Florence. I've seen the *David.* Not only that, I've seen about a squillion copies of the *David.* There's a David on practically every street-corner in Florence, because for the past 500 years, the way that budding Florentine sculptors learn their craft is by hacking out copies of the *David.*

Every creator stands on the shoulders of giants: so said Sir Isaac Newton. The sciences and the arts are built on copying, on observing, on measuring, on the public disclosure of facts and discoveries. The difference between alchemists (who didn't publish their results) and chemists (who did) is that every alchemist had to learn the hard way that drinking mercury was a bad idea, while chemists read their colleagues' papers on mercury, skirted danger, and devoted their energies to making genuinely useful new discoveries.

No curator of human knowledge has the moral right to restrict the recording of our shared history. The next time you find yourself at a museum — generally funded by your own tax dollars — where the rules say you have to keep your camera in your pocket, seek out the curator and ask him why he's turned traitor to history.

Cory Doctorow (craphound.com) works for the Electronic Frontier Foundation (eff.org) and co-edits boingboing.net. Doctorow's most recent novel is *Someone Comes to Town, Someone Leaves Town*, from Tor Books.

MADEONEARTH

Report from the world of backyard technology

Graffiti Temple

Artist **David Best** is known for his gigantic temple installations at Burning Man in Nevada's Black Rock Desert; his latest was 115 feet tall, a quarter mile long, and visited by thousands. The temples — largely made out of scrap and the plywood cutouts left over from a local toy factory — are intended to be places of solace for people of all faiths. They're also a blank slate for everyone who comes to experience them. Visitors are encouraged to write or draw on exposed surfaces, so that the end result is a stunning collaborative statement of mourning.

This last summer, though, San Francisco Mayor Gavin Newsom invited Best to temporarily install a temple on Hayes Green in San Francisco, not far from City Hall and the opera house. While some might consider a plywood temple that actually *invites* graffiti to be a risky bet, Best points out that temporary art allows artists to bypass the red tape usually necessary to approve public art. This public work, however, was so well loved that it was up for almost six months, far longer than originally planned. By September, the surprisingly elegant plywood structure — part Thai temple, part Brussels lace — was covered with names, drawings, messages, stickers, photographs, and flowers as high as the arm could reach.

Best builds his structures with a part-volunteer crew, which includes everyone from a master architect to a person paralyzed except for the fingertips. He comes up with a drawing, and then he and his experienced crew try to turn it into reality. "There are no mistakes," says Best. "It's important to me to empower other people to work with the material." He has worked with thousands of volunteers over the years.

The Hayes Green Temple took 30 people three days to build. Still, despite the intricacy of the design and the shapes involved, the process is nothing complicated. According to Best, "It's real caveman stuff."

— Arwen O'Reilly

>> **David Best Hayes Green Project:** blackrockarts.org/david_best.html

Photography by Arwen O'Reilly

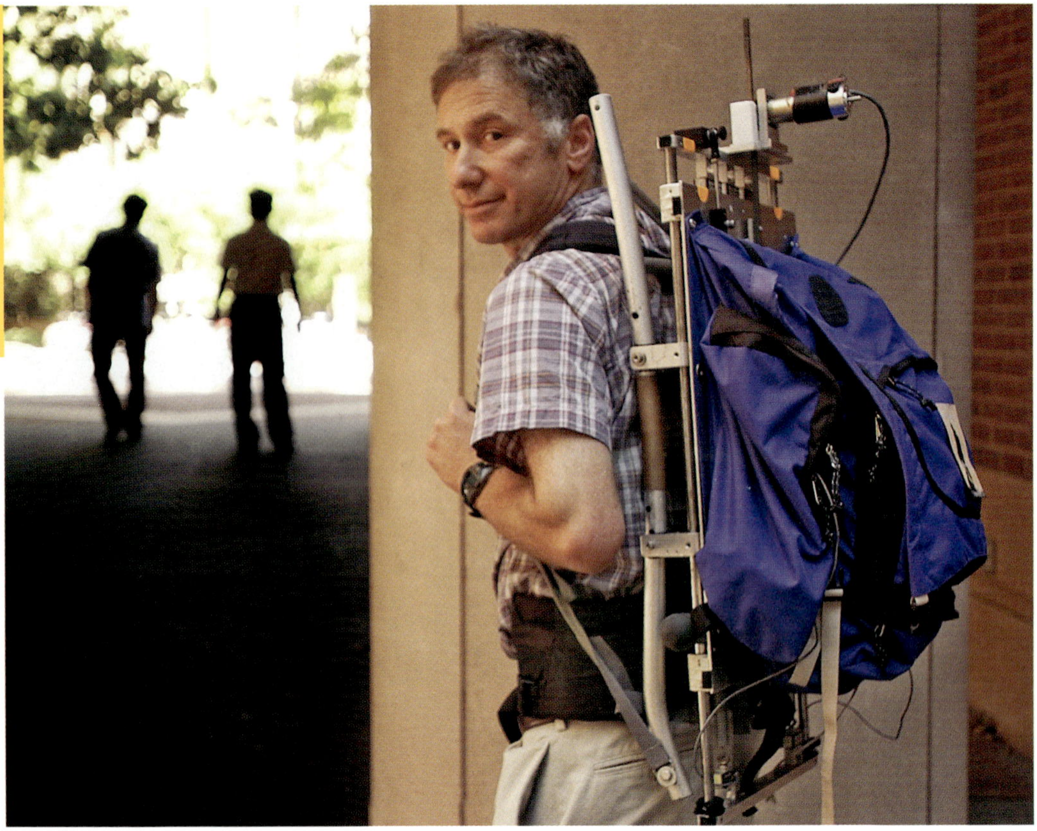

Power Pack

Finally, you can recharge your iPod with Clif bars.

When the military needed to recharge batteries on the move, they turned to University of Pennsylvania professor **Larry Rome**, an expert in muscle power and, it turns out, a capable inventor. His solution: the world's first electricity-generating backpack.

Rome, who studies fish muscles, says the idea struck him in a Navy meeting. U.S. troops were lugging 80-pound packs, including 20 pounds of batteries for high-tech gadgets. The brass wanted to use muscle power to generate electric power, but the best existing technology was shoe generators, straight out of *Get Smart*.

"I said, 'That's a terrible idea,'" recalls Rome. "The force of the heel strike is only over a couple millimeters. The right way became obvious: with every step, these guys are lifting 80 pounds 5 to 7 centimeters — that's potentially 36 watts of mechanical energy."

To turn his brainstorm into hardware, Rome grabbed an old external-frame backpack from college days and called his lab's "very fine machinist" Fred Letterio. In their basement shop full of mills and lathes, the two added springs to suspend the cargo compartment from the pack frame. As the wearer's stride raises and lowers the pack, the load slides up and down, driving vertical rods to spin a geared DC servomotor up to 5,000 rpm to generate electricity.

With a 40-80 pound load, Rome's pack generates 7 watts, plenty of juice to simultaneously power a two-way radio, GPS receiver, and night vision goggles (or cellphone, PDA, digital camera, and iPod). The load can be locked for stability on sketchy terrain, and then unlocked to generate power again.

Ultimately, the generator pack (patent pending) will weigh just a couple pounds more than a regular backpack. Carrying it burns 3% more energy, but wearers say it's more comfortable, and the extra work costs only a couple of extra candy bars. ("Food is 100 times more efficient than batteries.") Green bonus: the technology could keep tons of toxic batteries out of landfills. —*Keith Hammond*

>> **Lightning Packs:** lightningpacks.com

Photograph courtesy of Larry Rome

Flying Fruitcake

Photograph courtesy of the Manitou Springs Chamber of Commerce & Visitors Bureau

Back as far as 400 BC, catapults were used to hurl the vilest of projectiles: rocks, arrows, cow manure, diseased horse carcasses, barrels of venomous snakes, wasps' nests, the severed heads of captured soldiers. But perhaps the most notorious missile of all may be ... the fruitcake. At least that's what the citizenry of Manitou Springs, Colo., seems to think.

Each January, Manitou Springs residents gather for the city's annual Fruitcake Toss. The town's best and brightest combine their initiative and tinkering ability to build an array of mechanical and pneumatic-powered contrivances — including air cannons, catapults, and trebuchets — all designed with one thought in mind: to hurl unwanted fruitcakes out of town.

Toy designer **Dave Meyers** is the architect of the M-63 Fruitcake Remover. It's an industrial-grade, slingshot-style catapult powered by the ample potential energy stored in 36 feet of overextended ⅝-inch surgical rubber tubing. The machine has a yoke 10 feet across, and two large men are required to cock the spring. The trigger is similar to a sailplane hitch release, sailplaning being another of Meyers' interests.

Meyers built this machine large and strong because it has a big job to do. See, fruitcakes are hard to get rid of. The specific density of fruitcake is equivalent to mahogany. It's the one gift that the U.S. Postal Service has not found a way to damage in transit. And you simply cannot flush a fruitcake. Hence, the M-63.

"When I started, I essentially began doodling in AutoCAD," says Meyers. "I had a vision in my head for a slingshot, and this design just sort of developed as I stared at my computer screen."

One year, the M-63 launched more than 300 feet. Some of the larger air cannons present at the Fruitcake Toss hurl even farther. As fruitcake-slinging technology improves, the bar (and the fruitcake) gets ever higher.

—William Gurstelle

>> **Fruitcake Toss:** makezine.com/go/fruitcake

LoCost Sports Car

Now midlife crises don't have to be quite as expensive. Ron Champion's cult classic, *Build Your Own Sports Car for as Little as £250*, has spawned a host of clubs devoted to doing just that. Slightly more complicated than your traditional kit car, the LoCost is assembled entirely from secondhand parts and raw materials (metal bar, aluminum sheeting), and requires quite a bit of welding.

You also need a donor car. "Basically you need an engine, a gearbox, and either a live rear axle or an entire independent rear axle unit — the rest is made up," says **Adam Streatfeild-Jones**, of Britain's LoCost Car Club. The chassis design is generally built from Champion's original drawings, but even a casual look at the images posted on enthusiasts' websites makes it clear that there's a range of vehicles. Some build road cars, for tooling around the countryside, and others concentrate their energy on racing cars. All the cars seem to be a labor of love.

In fact, LoCost car building became so popular that in 1999 the LoCost Formula race was started, run by the 750 Motor Club, with tight construction regulations to make sure that the winners are not just the ones with the deepest pockets. In fact, the founders intended it to be a way to encourage younger drivers and builders, and entries include cars submitted by colleges and groups as well as thrifty individuals.

According to one website, a car has been built for as little as £47.50, but Streatfeild-Jones admits that "most builders want their car to look good, and thus they go for expensive paint jobs, nice wheels, and so on." Even with all the trimmings, it's still a far cry from the cash needed to buy a sports car hot off the assembly line. So don't let that $360 burn a hole in your pocket.

—*Arwen O'Reilly*

>> **The LoCost Car Club:** locostcarclub.co.uk

Photograph courtesy of Adam Streatfeild-Jones

Photography by Monica Mills (left) and Alyce Santoro (right)

Sonic Fabric

Throughout history, music has played an important role in our social, political, and religious rituals. It's been used to express the most elusive of emotions and situations: love and loss, conquest and oppression, joy and mourning. You name it, there's a piece of music that will fit it. And now, the music can fit you: **Alyce Santoro** has created a series of products that allow you to wear your music and listen to your clothes.

As a child, Santoro would watch the tell tales, made from cassette tape, of her family's sailboat, and imagine them playing their song to the breeze. Later, she learned about Tibetan prayer flags, whose sacred mantras are released by the wind and carried out into the world. The two seemed to fit together.

Using cassette tape recorded with music and sounds significant in her life, the New York-based artist started making fabric. She had planned to produce enough material to make only a few prayer flags, but when her friends saw her knitting and crocheting tape, they too became inspired. With the added input, the project took on a life, or lives, of its own.

Her friends suggested various techniques of weaving and decorating the cloth, but friend Joe Ball made one of the most significant contributions. He suggested they might be able to listen to the tape. Until then, the music was purely symbolic. Sure, Santoro knew it was there and what it represented to her, but her creations remained silent to others.

"It hadn't occurred to me that it might actually be possible to hear them," she says. "And sure enough, by simply retrofitting an old Walkman by mounting the head on the outside, Ball created the first Sonic Fabric reading apparatus."

Santoro even crafted a dress for Jon Fishman, drummer from the band Phish. The Sonic Fabric was woven using tape culled from Fishman's extensive collection, and the dress was "played" onstage in Las Vegas, using specially designed tape-head gloves during a 2004 Phish show in.

—Josie Moores

» **Sonic Fabric:** sonicfabric.com

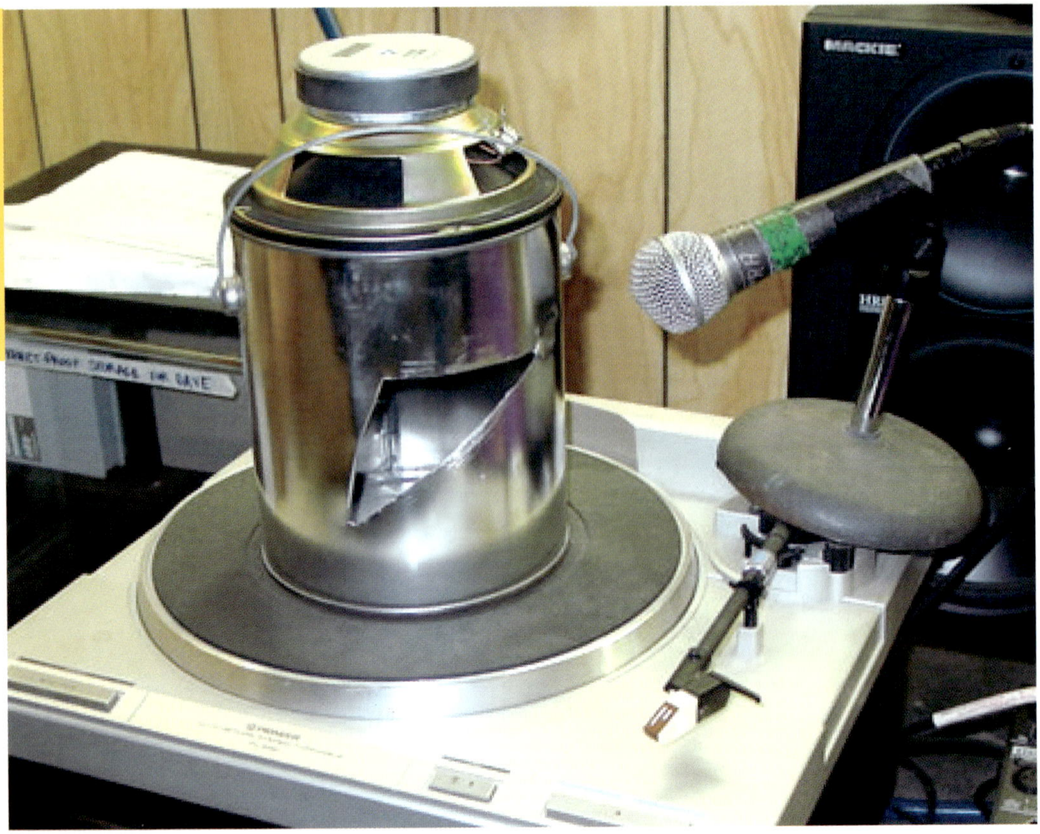

Spatial Education

Video game composers have to be crafty, because every project is a new world with new rules. **George** "the Fat Man" **Sanger**, who's been at the top of the craft since the Mattel Intellivision ruled the Earth, is crafty indeed. His homemade Leslie speaker — which he first built out of a punctured paint can and record player — taps a polyethylene drainage device to get a sound he describes as "one of the most organic, spacious, and beautiful there is."

Invented by Don Leslie (1911–2004), the Leslie rotating speaker defined the sound of the Hammond organ, becoming a staple of rock, jazz, gospel, country, and pop music. It's frequently used on guitars as well. As the speaker spins toward and away from the listener, the Doppler effect alters the sound's pitch and timbre, producing a rich vibrato. And the way the projected sound sweeps around the room? "Man, that is good medicine for the soul," says Sanger.

To make his desktop Leslie, Sanger used a garage-sale turntable ($5), a 4x1-foot wire shelf ($4), and a double-outlet catch basin ($7). He then added plastic cups to extend and focus the sound. Inserting a CD-R spindle loaded with unwanted discs into the basin's bottom provided a solid base and allowed the cups to clear the tone arm.

By increasing the diameter of the rotor, the cups also heighten the Doppler effect, because the output ports travel faster. Sanger found that one foot is the optimal diameter at standard turntable speeds. The amp he used is a Vox Brian May Special (about $149 new), which has a big sound and a sealed back, making it very directional. Its controls are on the top rear, which puts them in perfect operating position when the amp is face down.

Next up: Sanger is currently plotting how to make a Leslie speaker out of a ceiling fan.

—David Battino

The Fat Man: fatman.com

Hear audio examples at makezine.com/05/made.

Photograph by George Sanger

Huggable Tec

Making something out of T-shirt material can mean different things to different people. For **Phoneticontrol** and **Shawnimal**, it meant designing Pulse Width (PW), a "happy little 'bot" filled with "stuffing and love. Mostly stuffing."

The bug-eyed, blue collaboration was created as an entry in I Love Your T-Shirt's (iloveyourtshirt.com) competition for *something* made out of T-shirt material. Named after a term used in analog synthesizers, Pulse Width was designed in two dimensions by the pizza-loving **Eric Broers** (Phoneticontrol), an illustrator and designer. "We both needed to change our thinking a little to bring him to life," Broers says. "It was an interesting challenge. I designed the PW, and Shawnimal sculpted the crap out of it."

Shawn Smith (Shawnimal), a multi-talented artist who paints, sculpts, draws, and creates plush creatures, brought the vision into the third dimension. "The challenge was getting Eric's style to come through in plush form," he explains. "Somehow, maybe the planets aligned or something, but I think we nailed it."

The two have worked together in the past and had a blast, so when Phoneticontrol found out about the contest, he immediately thought of working with Shawnimal to create a plush. Once it was created, the duo decided that Pulse Width needed to get airborne. After a little old-fashioned tossing into the air and rapid-fire photography, PW was captured airborne on film above Logan Square in Chicago, where they both live.

Smith, who sells both big and pocket-size plush ("because small equals cute"), ends on a philosophical note: "All I can say about Pulse Width is frumpy, huggable robots rule."

—Shawn Connally

>> View photos: phoneticontrol.com
shawnimals.com/projects.php
studio606.com

Photograph courtesy of Phoneticontrol and Shawnimal

Bruce Sterling

HANDS ON: MAKING A MOBILE

STAYING POISED AT THE FERTILE EDGE OF CHAOS, WHILE THE CLOCK TICKS.

I'M IN RESIDENCY AT ART CENTER COLLEGE of Design in Pasadena, Calif., and it didn't take me long to figure out that most everybody here makes stuff.

Here at design school, it's a demo-or-die situation. For instance, everybody in my "Ecology of Things" class is busily making demos and prototypes for highly distributed, embeddable computer chips. These microchips are supplied to the students courtesy of the research and design wing of Sun Microsystems.

Since I'm an author and have spent my entire career blabbing, I decided my design students would surely benefit by seeing me build something myself.

The prospect of ubiquitous, ad-hoc'ed, networked, pervasive, teensy, radio-frequency, spooky, action-at-a-distance, internet-of-things-ish, Java-swapping, motorizable chips is not just talky, but a little daunting. Luckily, there's already a sweet, artsy, familiar, and highly popular version of a working "ecology of things." It's from the pre-digital age. It's called a "mobile."

Alexander Calder (1898-1976), that Modernist artist with a mechanical engineering degree, invented the mobile in the early 1930s. The materials for mobiles — steel wire, scrap aluminum, and paint — are cheap and easy to find. A vintage Calder mobile goes for upwards of half a million bucks at art auctions now, but Sandy Calder used to be able to whip one out in a day. So how tough could it be to just make one?

Equipped with hardware-store wire and cheap Chinese hand tools from the local Target store, I soon discovered that it's, in fact, dead easy to make a mobile. A mobile is simply a cascade of levers, a multilevel, energy-trophic ecology of levered elements, if you will. The key to success is the relationship between elements. If the links are made too tight and rigid, then the mobile has a lifeless, clockwork, phony feel, much like a Soviet bureaucracy. If the links are too loose, then it gets all corrupt and herky-jerky; its extreme fringes start clattering around in violent incompetence, much like the Bush administration.

A successful mobile artfully combines central control and local initiative, artfully poised at the fertile edge of chaos. Ideally, the thing should dance.

My own model mobile (after many returns to the drawing board) dances just fine now. Unfortunately, since I lack physical skill at artfully working steel wire,

"A vintage Calder mobile goes for upwards of half a million bucks but Sandy Calder used to be able to whip one out in a day."

it's ugly; it looks like it was dropped off a ten-story building. That doesn't matter, though, since the original model, made from paper plates, was merely the working prototype for a far more ambitious effort. My completed mobile is supposed to have a motor on board, and an embedded microchip to assure interaction with passing humans. It's going to be about 40 feet across.

This entails scaling up my model by a factor of ten. Scaling up chaos by a factor of ten is a rather nonlinear effort. The sweet and subtle balancing forces within my tiny mobile maquette are packing some serious heft now. In my misspent youth, I used to stretch barbed wire and string Texan cattle

Photography by Bruce Sterling

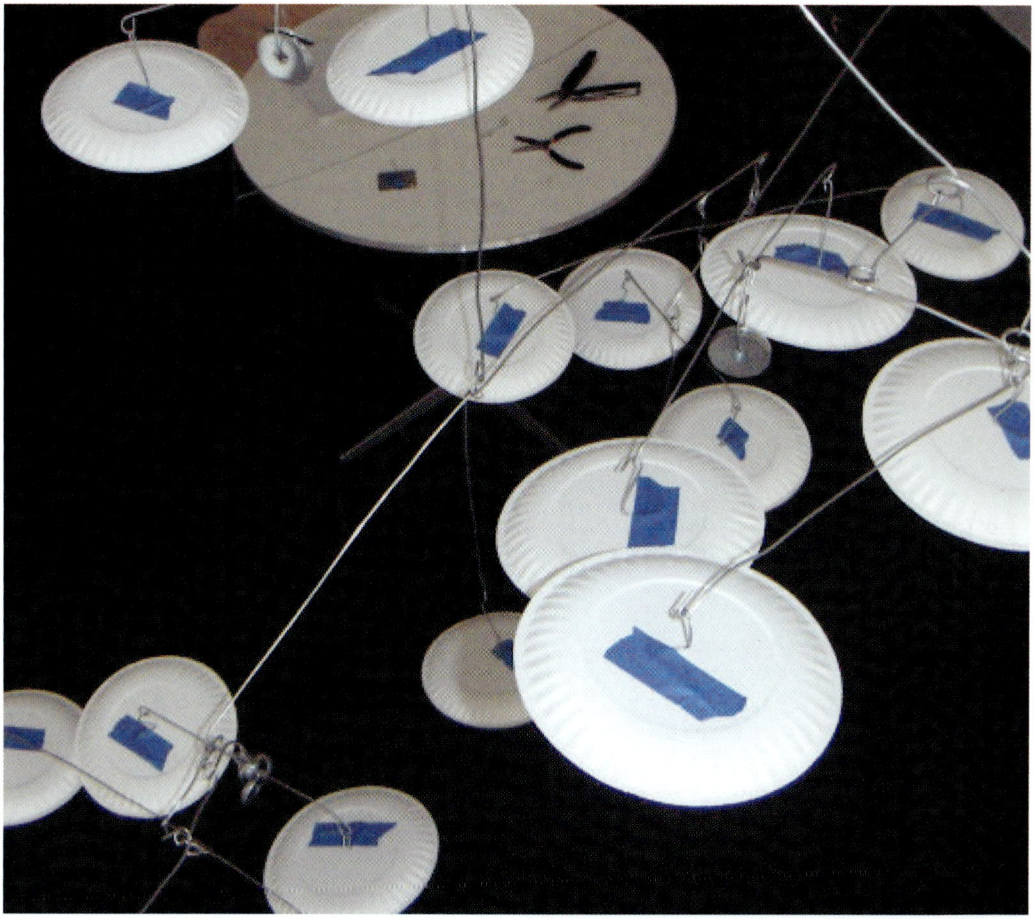

fences. Every once in a while, under pulley tension, a strand would pop and lash free, wickedly eager to scar and/or blind. Then you'd learn instantly that barbed wire is a torture device with extensive military applications.

As I wrestle with my wobbling polyvinyl pipes and dangling chunks of illustration board, that early lesson is returning with redoubled force. I now finally and fully understand the wisdom of a good design-school education. You can lie to the board of directors, but folks, you just can't lie to steel wire.

I'd love to rattle on about this subject, because the many weird, sexy applications for digitally interactive mobiles are so intensely novelistic and science-fictional. But I've got a deadline and a budget to meet. The clock is ticking and I'm running out of cash. I'm learning a lot really fast. So I'd like to keep talking here, but, well, I gotta go.

Bruce Sterling (bruce@well.com) is a science fiction writer and part-time design professor.

Maker

Woody's World

A Q&A with the inventor of a pirate-scaring noisemaker, a helicopter for every garage, and a way to hack gravity itself.

INTERVIEW BY WILLIAM LIDWELL
PHOTOGRAPHY BY JENNY PFEIFFER

Woody Norris is standing 50 feet away from me, pointing a menacing-looking sci-fi contraption at my head. We're in the back lot of his American Technology Corporation in San Diego, Calif. Woody's device is called the LRAD (for Long Range Acoustic Device). It looks something like a big spotlight, the main difference being that it projects sound instead of light. Woody yells to me, "We are going to start with some music. This is really cool!"

Now hear this: Elwood "Woody" Norris' Long Range Acoustic Device was developed for the U.S. military to send a painful blast of noise at small boat crews attempting to get near U.S. vessels.

He fiddles with the controls and I hear classical music as clear as if I were next to the speaker. I smile and nod back. Woody grabs a microphone and speaks through the LRAD, "Now I'll play some pre-recorded commands being used by our troops in Iraq." I can hear him clearly. I begin to hear a series of commands like "Halt" and "Put your arms over your head" quickly followed by their Arabic equivalents. Again, I nod. Then I see Woody and his prototyping guru, Jeff Belka, chatting back and forth, nodding, and smiling. They are up to something. Woody looks my direction and all I see is a big set of teeth slowly revealed by a widening smile.

He turns a control on the LRAD and a 150db siren screams out of the device — my hands instinctively shoot up to cover my ears. The siren is so loud I can't think or do anything. I am paralyzed. The siren lasts for only a couple of seconds before Woody turns it down. I start to put my hands down and he cranks it. My hands shoot back up and I stand there at his mercy. We do this dance a few times until Woody is convinced that I get it. The LRAD is cool technology. It has all manner of communication, crowd control, and acoustic weapon potential. Before I can clear my head and convert my thoughts to words, Woody starts to reach for the LRAD again. "It is cool! It is cool!" I shout. He gives me an approving look and motions me in. "Yeah, cool," he says. "Now we can continue."

A recent winner of the Lemelson-MIT Award for Invention and holder of more than 40 patents, Woody Norris is the classic American success story. Born of modest means and largely self-educated, Woody is the founder of a number of companies based on his inventions, ranging from personal flying aircraft to a variety of revolutionary acoustic technologies. As inventors go, he is a rare breed that enjoys selling as much as inventing — an unusual blend of P.T. Barnum and Thomas Edison.

Woody's workshop is within eyeshot of his huge and opulent home on an adjacent hilltop. As we go inside, my eyes are immediately drawn to the notes and diagrams on his whiteboard. Intriguing terms like "plasma antenna" and "optical cobweb" fill the board. On the workbench, I spot some very strange-looking speakers, a gas chromatograph, and circuit boards that he appears to be reverse engineering. Seeing this array of gadgetry leads to my first question ...

When did you discover that you wanted to be an inventor?

My whole life I have been dragged along by circumstance. I became an inventor by accident. Around 1960, I was reading a magazine that had an article about a new kind of electric shaver that ionized off your whiskers — it had drawings and everything. The article intrigued me. The further I read, the more incredulous I became, and sure enough, at the end it said, "April fool." Here is the part that changed the rest of my life. To the side of the article, there was a small box, outlined in black and titled, "Editor's Note: Submissions Wanted." It was a contest announcement inviting readers to write next year's article. The winner would receive $250. I was only making about $400 a month at the time, so that was a lot of money to me. I dropped what I was doing, dreamed up a new invention, and wrote the article.

My invention idea was a record player tone arm that made a straight line across the record instead of pivoting at an arc. Since that is how they cut records, it seemed reasonable to me that a straight-line playback arm would be superior. I was literally in the process of licking the envelope to submit my article when I stopped myself. I decided that I should call some hi-fi stores pretending that this was the real deal and see if they would believe it — the thought being that if it could fool them, then it could definitely fool the readers of the magazine. So I called up every retailer in the Seattle area and not only did they believe it, they were excited about it. At that point, I decided I better build the thing instead of submitting the article. And that set me on the path to inventing things.

Of your many inventions, I am most intrigued by the HyperSonic Sound technology. How does it work?

HyperSonic Sound is my landmark invention. It allows you to project sound like a very focused beam of light. With a normal speaker, sound propagates outward as waves in the air. If I point a normal speaker away from you, you still can hear the sound. With a HyperSonic Sound speaker, if I point the speaker away from you, you won't hear anything. How does it work? Ultrasonic frequencies are very directional because their wavelengths are so short — the higher the frequency, the more like a beam they become. So I figured out a way to increase the frequency of standard audio content to ultrasonic levels. If you do this in a particular way, the ultrasonic signal demodulates in the air and turns into audible sound. So instead of emitting audible sound from a single source like a speaker, the sound is actually made in the air at an infinite number of points along this ultrasonic column. It's like having a billion little speakers lined up and pointing in the same direction. It is very cool — like magic.

Can you describe your process of invention?

I largely invent by analogy. I find things that are successful in one environment and then think about parallels in other environments where people have not yet discovered or exploited the concept. For example, I came up with the idea for HyperSonic Sound by analogy. I was watching television one day and I noticed that the logo on the TV had three colors: red, green, and blue. These are the colors that are squirting onto the screen from the back of the picture tube from which millions of colors are created. By mixing just a few colors, you can get millions of colors. I thought to myself that this was pretty cool. And then I thought about electronic circuits, where there are components called mixers that allow you to mix a couple of frequencies together to get new ones. So it occurred to me that if you can mix energy optically and electronically, why couldn't you mix sound energy in a similar way? Then I started thinking about the possibility of mixing sound that is inaudible to people — in this case, ultrasonic sound — in a way that would result in audible frequencies. And that is where the HyperSonic Sound idea came from. I solved the artificial hip problem in a similar way.

The artificial hip problem?

When someone gets an artificial hip, it can fall out of socket quite easily. It takes many months of conscious effort for patients to learn how to do things in a way that prevents the artificial hip from disengaging. When one disengages, it is very painful and costs thousands of dollars to reset. Anyway, I was approached by a group of doctors who wanted a way to detect when an artificial hip started to separate and then set off an alarm. In this way, people could stop what they were doing and tense up their muscles to prevent the hip from disengaging. Additionally, they didn't want any active electronics or power sources to be inside the body.

Now, around the time they contacted me, I had been playing around with a guitar. I noticed that when one string was tuned to the same note as another string, plucking one vibrated them both. This is sympathetic vibration. So I started thinking about the string I plucked as a radio transmitter sending out a signal and the receiver as the other string that is tuned to the same frequency. The second string did not start vibrating for free — it takes energy to get it to move. It then occurred to me that I could make a transmitter that sends a signal into the body, and if there was a thing inside the body that was tuned to the same frequency, then we could use the change in frequency as a trigger. So I embedded a resonant circuit — which is simply a coil of wire — in the artificial hip that was tuned to the same frequency as an externally worn transmitter. If the hip started separating, the

1. American Technology, Norris' company, donated several MRAD "sonic lasers" for Hurricane Katrina crowd control and rescue operations.
2. A component of an artificial hip alarm, which warns patients that their implant is about to dislocate.
3. A HyperSonic Sound speaker.

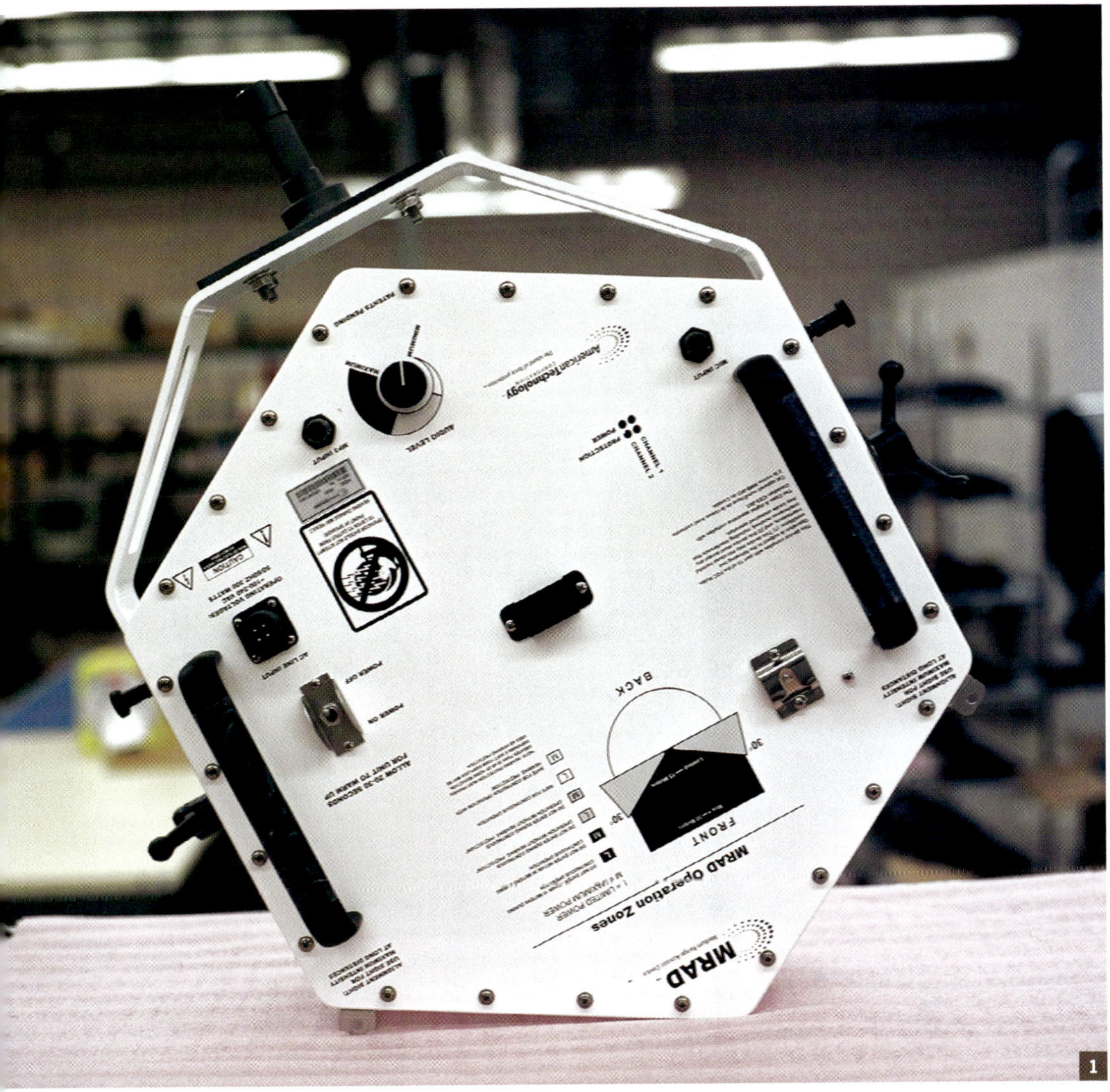
MRAD
MRAD Operation Zones
American Technology
FRONT
BACK
AUDIO LEVEL
MP3 INPUT
MIC INPUT
POWER ON
POWER OFF
AC LINE INPUT
ALLOW 20-30 SECONDS FOR UNIT TO WARM UP
CAUTION

1

2

3

frequency would shift, and that shift would trigger an alarm. Sure enough, it worked.

The AirScooter is a bit of a departure from your other inventions. What is the story behind it?

My business partner and I began discussing how people love to fly. So we decided to invent a flying machine. I then made a list of all the things I didn't like about airplanes: I don't like to have to be going 100 mph to stay in the air, I don't like having to drive to the airport, etc. I then made a list of all the things I didn't like about helicopters: I don't like how hard they are to fly, I don't like the cost to get licensed, etc. Then we simply went about coming up with a product that met all of these criteria.

We knew people needed to be able to take off and land anywhere, so we went with a rotor-based craft. The tail rotor on helicopters drains about 20% of the power without adding lift, so we replaced it with a second rotor that counter-rotates. This eliminated the wasted power and the gyroscopic effects. We knew the controls had to be easy to use, so we eliminated the foot pedals and yoke and went with a handlebar. Push, pull, or turn it the direction you want to go and you go there.

We knew it had to be low-maintenance. The ratio of fly time to maintenance for a helicopter is about 1:1, which is not acceptable to most people. If we could make it light enough, we could eliminate a lot of the rotor complexity (and corresponding maintenance) and control lift through rotor speed. Additionally, if we could get the craft under 254 pounds, it would qualify as an ultralight and be exempt from licensing requirements. So we reduced the weight, primarily by inventing a new high-power, lightweight, four-stroke engine. This allowed us to get the weight down and reduce the rotor complexity. Now a pilot can use a simple handlebar throttle to go up and down. It has taken us several years to work out all of the kinks, but for the price of an expensive car ($47,000), anyone can now buy and safely fly their own AirScooter out of their backyards. Pretty cool!

Rumor has it that you are working on some inventions that involve gravity. Anything you can share?

I am fortunate that I am now in a financial position that allows me to pursue big problems. So for the last several years, I have tried to deliberately invent over my head. When I can't get the answers I need for problems I'm working on, I find professors who specialize in the areas that I am working on and then pay them to conduct the research or write me papers. This allows me to get experts in niche areas to evaluate my ideas and tell me if they are smart or stupid.

So with that context, let me say that nobody really understands gravity. What we do know is that it has something to do with mass, and it is the result of something going on in the nucleus of the atom. I think I know what that "something" is. I might be the only person in the world who knows what that "something" is. I know that sounds brazen, but I believe it's true. I have been working on a variety of experiments, over the course of years, to test my theory in a way that yields clear, unambiguous results so that it can be independently reproduced — I don't want a Pons and Fleischmann situation (*see MAKE, Volume 03, "A Fusion Reactor for the Rest of Us," page 25*).

Based on the current rate of progress, I am confident that within ten years we will understand the mechanism of gravity and be able to influence it in specific ways. I realize that most scientists reading this will think I am nuts, but I am not risking anybody's money or reputation but my own. So if I am wrong, no big deal. If I am right, the rewards for everyone will be terrific.

Read more of William Lidwell's interview with Woody Norris at makezine.com/05/interview.

Racks of HyperSonic Sound speakers ready to ship. Customers include Disney World and supermarkets, who use them as "audio spotlights."

What Woody's Wrought

Norris' key inventions, from audio spotlights to flying cars.

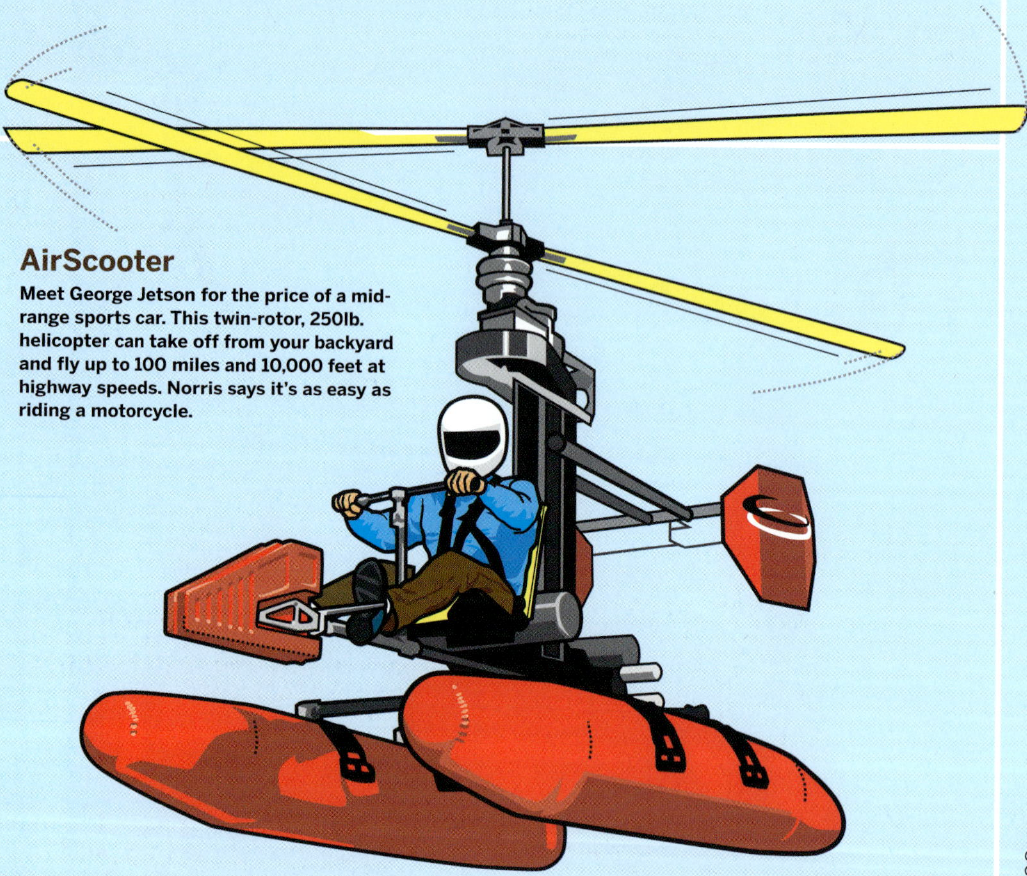

AirScooter

Meet George Jetson for the price of a mid-range sports car. This twin-rotor, 250lb. helicopter can take off from your backyard and fly up to 100 miles and 10,000 feet at highway speeds. Norris says it's as easy as riding a motorcycle.

HyperSonic Sound System

The first major breakthrough in acoustics since the loudspeaker was invented 80 years ago. HyperSonic Sound speakers send out two different ultrasonic signals. When they arrive at their intended destination (any object within 300 feet), they combine to form audible sound.

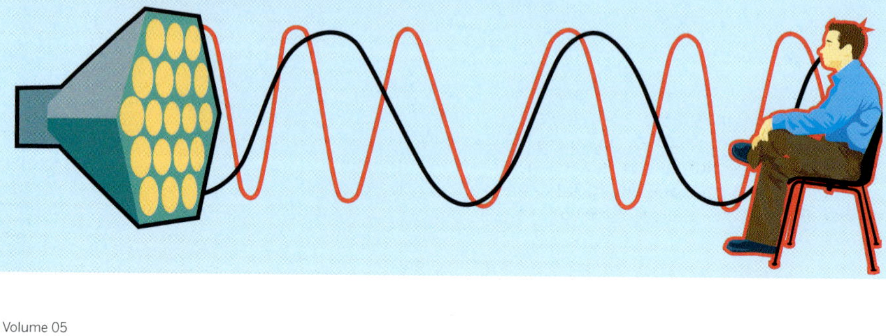

Illustrations by Dustin Amery Hostetler / UPSO.org

Long Range Acoustic Device

Related to HSS, the LRAD can project audio in a tightly focused sound beam over 500 yards at insane volumes. Good for any situation requiring long-distance communication. Want to disperse a drunken riot after a football game? Point one of these at the mob, crank it to 150db, and watch them scatter. Norris' LRAD was recently used by the crew of a cruise ship in Africa to scare off pirates trying to board the vessel.

Gravity Machine

Gravity shield? Gravity distortion device? Woody has been conducting experiments and working with experts. He is not talking details, but stay tuned — he says something big is just around the corner.

Ear Radio

At less than ¼" thick and ¼ ounce, this tiny FM radio caught the attention of NASA engineers and ultimately led to the first hands-free earset, which integrated the microphone and speaker in the earpiece and used bones in the head to conduct sound.

Art in Living

N55, a conceptual group from Denmark, creates functional art for living, breathing, and growing.

By Bruce Stewart

MAKING PUBLIC SPACES INHABITABLE and useful to anybody is the goal of N55, an artistic effort based in Denmark that has created functional art for over a decade. N55 creates inexpensive, modular, portable structures. Installations have included floating platforms and living structures, urban-farming plant modules and floating fish farms, and even a mobile bar. At the core of N55's values is the idea of art as part of everyday life.

Originally a collective of a dozen artists living, working, and exhibiting together, N55 was later pared down to the husband-wife team of Ion Sørvin and Ingvil Hareide Aarbakke. With the passing away of his wife last year, N55 is now carried on solely by Sørvin.

He is currently working on the first module for a new system called *Micro Dwellings*, a low-cost, easy-to-build, amphibious modular house designed to function on both land and water (floating or submerged). The modular *Spaceframe* is a reconfigurable, easy-to-build, low-maintenance living space structure. Other projects include a *Floating Platform* that can support these *Spaceframe* buildings, and a *Snail Shell System* that resembles a giant plastic shell-like wheel and lets people "move around, change their whereabouts, and live in various environments." Various cramped environments anyway. Many of N55's projects are about demonstrating the minimum space and materials needed for basic survival and producing them easily and cheaply.

Photography by N55

N55's projects range from the practical to the political. Besides dwellings, food-production systems, and portable soil factories, there is a language education kit for illegal immigrants in the works, and a "protest" rocket that can distribute various things over a vast area from high altitudes (for example, printed matter or seeds).

One of N55's most popular and long-standing projects is *Land*, which encourages people to donate property for public use. Parcels of land have been contributed to this effort in the United States, Switzerland, Denmark, the Netherlands, France, Norway, and Sweden. The land is maintained by whoever is currently using and inhabiting it.

N55 started out as an artist's group in 1994, when they were 12 members living together in an apartment in the center of Copenhagen, trying to "rebuild the city from within." The group took its name from its address at Nørre Farimagsgade 55. Sørvin and Aarbakke continued to make use of this space as the launching pad for their functional art projects. Financed primarily by exhibitions, grants, and educational work, Sørvin considers N55 noncommercial, and creates and documents all the projects expressly so they can be freely duplicated and refined by others around the world. You could say N55 makes open source domestic technology.

The Snail Shell System (previous page) is a low-cost way for people to move around and live in various environments. Pictured is the late Ingvil Hareide Aarbakke, one of the founders of N55. The Spaceframe (above) is low-cost and movable; N55 envisions it as a living space for three to four people. "It demands practically no maintenance."

N55 publishes manuals for all of its projects online (N55.dk), including detailed instructions and photos, so others can easily implement the ideas. There is also an N55 book that includes most of these manuals (available as a free download).

Bruce Stewart has covered technology issues for various publications for over a decade and is the editorial director of O'Reilly Media's online publishing group.

C.S. and the City

Intel's Eric Paulos and the edges of urban computing.

By David Pescovitz

There's a bomb inside Eric Paulos' storage space. It's not your typical homebrew explosive, though. This is an information bomb. If you happen to walk by when Paulos fires it off, you may not even notice right away. But your wristwatch will probably stop. Your credit cards certainly will no longer be readable. Your cellphone will instantly become a paperweight. Carrying a laptop computer? Hopefully, you have a recent backup. With the press of a single button, the I-Bomb unleashes a powerful electromagnetic pulse that kills electronic devices and corrupts all storage media within several meters.

Paulos has demonstrated the I-Bomb at places like the San Francisco Museum of Modern Art, but he'd be ill-advised to show it off at work. Not to say that his colleagues wouldn't appreciate it. In fact, projects as unusual and provocative as the I-Bomb landed Paulos his job in the first place. Still, he works for a company where fried chips are particularly distasteful.

A BIZARRE MACHINE LABLET

Paulos is a computer scientist at Intel Research Berkeley, a corporate "lablet" in the penthouse of a downtown office building where a dozen or so scientists explore the edges of computing and networking technology. When Paulos joined the lab several years ago, the first thing he did was set up a machine shop with lathes, scopes, and assorted prototyping tools. The laser cutter took a bit more wrangling but is finally due to arrive this week. A computer science Ph.D., Paulos can hack C++ with the best of them, but he's really a lifelong maker who made a career out of building bizarre machines.

As a graduate student, he designed groundbreaking telerobots that enabled people to physically explore remote spaces over the internet. His work at UC Berkeley bled into years of collaboration with infamous San Francisco machine performance group Survival Research Laboratories. In 1997, Paulos and SRL director Mark Pauline invited anonymous web users from around the globe to remotely aim and fire a massive air launcher loaded with concrete-filled soda cans. It was the first time lethal machinery was operated over the internet. Paulos' own tech-art collective, Experimental Interaction Unit, has exhibited around the world.

At Intel Research, Paulos directs the Urban Atmospheres program, a group exploring technology in the cityscape, from Bluetooth-enabled phones to ad hoc networks of tiny sensors. While many companies are developing urban computing applications — from location-enabled restaurant recommendation systems to real-world buddy lists — Paulos says he prefers to look between the cracks in the asphalt for research topics. After all, entire conferences are already devoted to the likes of location-enabled services and Geoweb technology.

"There are certainly moments in life for productivity and efficiency, but there are also moments for wonderment and reflection," he says. "I'd like to use technology to celebrate non-places, non-events, non-activities that actually matter a great deal to the emotional experience of urban life."

ONE MAN'S TRASH IS ANOTHER MAN'S RESEARCH PROJECT

Urban Atmospheres aims to illuminate the most subtle characteristics of city living, like the relationships you have with people you see every morning at the train station but never acknowledge, the invisible cellular infrastructure you move across, even the

For Paulos, every city is an ideal laboratory to test what he calls "objects of wonderment."

Photography by Emily Nathan

Left: Jabberwocky is a mobile phone application that detects "familiar strangers" nearby. To avoid privacy issues, the color and motion of the blocks on the display provide information about crowds rather than specific individuals.

Right: One prototype for the Connexus project is a "friendship bracelet" containing superbright LEDs that display a range of color and tactile outputs based on input from another bracelet wearer. Another version mimics the form factor of a wristwatch. Surrounding the Connexus devices are several "motes," coin-sized wireless sensors. The matchbooks were part of an experiment about anonymous text messaging.

public trash cans you walk by on the way to the office.

Last year, Paulos and Tom Jenkins, a student at the Royal College of Art, spent countless hours stalking a single garbage can in San Francisco's Financial District. After their field study, they built what Paulos half-jokingly refers to as "the most expensive trash can in the world" and set it on the corner outside the lab. Called Jetsam, the can is augmented with a variety of hidden sensors, processors, and a video projector. An infrared switch detects when refuse has been tossed into the can (or grabbed out of it), and snaps a digital picture. A highly sensitive scale then weighs the item and a laptop PC categorizes each bit of refuse by time and size. Meanwhile, an evolving visualization of the garbage photos and data is projected out of the can's opening, creating a rotating galaxy of garbage on the sidewalk.

"Archaeologists dig through trash to learn history, so we looked at what trash can tell us about urban living," Paulos says. "Is there a lunch binge at noon or a coffee craze at 3:15? If there is a continuous stream of lottery ticket stubs, does that mean the local denizens are risk takers? Instead of hiding trash, we changed it into something that people interact with."

And they did. A young graffiti artist tagged a piece of paper and then tossed it into the can, projecting his name onto the sidewalk. Later, a businessman exiting a photo developing shop tossed his unwanted double prints into Jetsam's mouth, creating an instant photo album on the pavement.

"It's not as if we think Intel should be developing trash cans," Paulos explains. "But we want to propose research ideas that might feel a little awkward, because it's from those unusual vantage points that we may see other interesting possibilities."

> "Making things is much harder than programming because there's no quick Control-Z undo."

KEEPING TABS ON YOUR FAMILIAR STRANGERS

Indeed, Paulos charts urban territories and dynamics that are invisible, ignored, or taken for granted. Sashay is a mobile phone application he wrote that maps your path through the various cells in a city's mobile phone network. The point of Sashay, Paulos says, "isn't to help you get places, but rather cause you to reflect on how you move." Another project for mobile phones, called Jabberwocky, extended the Familiar Stranger experiments conducted in the early 1970s by social psychologist Stanley Milgram. A familiar stranger is someone who you may see repeatedly, at the bus stop or cafe, for instance, but never interact with. You mutually agree to ignore each other. The Jabberwocky application grabs the

unique Bluetooth identifier of the mobile phones around you, creating a log of people you've previously encountered, even if you've never talked with them. Later, a quick glance at your handset reveals if any of your Familiar Strangers are nearby.

"You might not want to make friends with everyone you see regularly, but these people color your city," Paulos says. "Jabberwocky enables you to get a sense of place based on who is there with you."

A previous project, Connexus, highlights another form of human interaction that's rarely discussed: nonverbal communication like the friendly smile across the room or a reassuring pat on the back. Connexus is a bracelet meant to contain a wireless radio, sensors, and actuators to facilitate nonverbal communication at a distance. Imagine a couple both wearing Connexus devices and perhaps miles apart. As the woman taps on her bracelet, her husband notices his bracelet glowing softly, not unlike a vintage mood ring. He responds by caressing the bracelet, an input that manifests itself as a gentle heating sensation on the other end. Paulos never managed to stuff all of the Connexus components into as small a wearable device as he'd like, but he says that prototyping is essential to communicating his research goals to others.

ENVIRONMENTAL RINGTONE SOUVENIRS

Right now, he's fabricating a device that he hopes will add a layer of audio intrigue to the urban atmosphere. Installed, say, in a public commons area, the small object will grab the Bluetooth identification numbers from the mobile devices of passersby. That data will then be translated into a unique ringtone, creating a souvenir of that location tied to a moment in time and the specific group of people nearby.

The device itself is only one part of the project. Paulos also plans to release a free developer's toolkit to inspire others to create and deploy their own interactive objects into the urban atmosphere. For Paulos, every city is an ideal laboratory to test what he calls "objects of wonderment."

"Making things is much harder than programming because there's no quick Control-Z undo," he says. "But if you're explaining your research to someone and can't present them with an object, they're likely going to default to talking about things they know. On the other hand, if they can interact with an object outside their realm of understanding, they're usually more willing to come with you into the conversation."

David Pescovitz, MAKE's special projects editor, is also co-editor of BoingBoing.net and an affiliate researcher with the Institute for the Future.

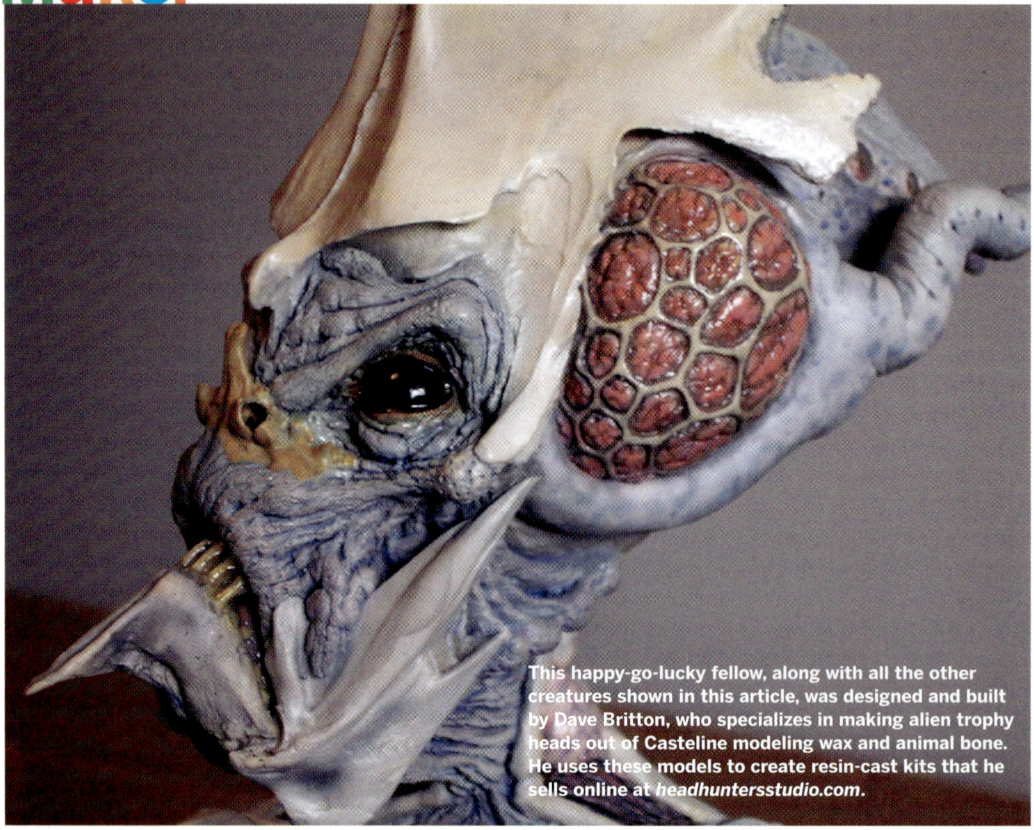

This happy-go-lucky fellow, along with all the other creatures shown in this article, was designed and built by Dave Britton, who specializes in making alien trophy heads out of Casteline modeling wax and animal bone. He uses these models to create resin-cast kits that he sells online at *headhuntersstudio.com.*

It Came from My Garage!

Model kit makers bring B-movie monsters to your home.

By Gareth Branwyn

MANY MAKERS LIKELY GOT THEIR first bug for building by putting together styrene plastic vehicle and monster model kits as kids. Until the late 1980s, the barriers to entry for actually fabricating — as opposed to assembling — such kits meant that only commercial concerns could afford to do so. But as new, easy-to-use materials for sculpting, molding, and casting became cheaper and more readily available, kit creation became a possibility for just about anyone with a lick of sculpting talent, a few bucks in the old pocket, and a world of patience. Combine all this with the power of desktop publishing, the internet, and e-commerce, and the "garage kit" modeling movement was born.

Until the emergence of garage modeling, builders were limited to commercially available model kits, mainly military hardware, cars, planes, spaceships, and mainstream movie monsters. But DIY kit creators cast their nets much wider. Today, a model

Photography by Dave Britton

enthusiast can buy kits of characters from David Lynch films, low-budget horror flicks, underground comic books, and spaceships from just about every cult sci-fi classic. And then there are the crazed, whimsical kits, models of aliens, monsters, pinup babes, and mad machines, rendered directly from the fevered imaginations of their sculptors.

Terry J. Webb, editor of *Amazing Figure Modeler* (amazingmodeler.com), one of the premier garage kit magazines, is also the author of three books that chronicle the development of the do-it-yourself kit scene, from its beginning in Japan in the 1980s to the present. With titles like *The Garage Kit That Ate My Wallet*, *Son of the Garage Kit That Ate My Wallet*, and *Revenge of the Garage Kit That Ate My Wallet*, these books say a lot about the humor that underlies much of this movement, the keen interest in B-movie subject matter, and the expense of these small-run resin- and vinyl-cast kits, which can cost from $60 to several hundred dollars. "A lot of what drives this hobby," says Webb, "is a love of sci-fi and fantasy movies and television. Many people get into kit making because they love a movie or show so much, they want to create a little 3D monument to it that other people can enjoy."

Many of the model makers behind garage kits are involved in the movie business, often special F/X techs, make-up artists, and prop masters. It is through these creative professions that they get their sculpting talent and their familiarity with molding and casting techniques. Dave Britton of Headhunters Studio (headhuntersstudio.com) is cut from this mold. He worked at several Hollywood F/X shops in the 90s, but soon found himself in Portland, Ore., where he started to make and sell latex monster masks. It was through his mask business that he discovered the kit-making scene and soon began producing figure-modeling kits.

Unlike many sculptors and casters who make direct copies of movie, TV, or sci-fi characters, Britton has developed a line of mounted alien heads cast in resin, creating a whole sci-fi backstory about the hunted aliens and those who hunt them for sport (think: DIY *Predator*). Besides playing off the idea of intergalactic trophy hunters, there's a practical purpose to the design of the line. "I heard many people say they were running out of shelf space for their model collections, so I figured, how about making heads that hang on the wall?" For his designs, Britton riffs off of creature forms found here on Earth, even incorporating organic materials, such as animal bones, into his original sculptures. His latest series is of the alien hunters themselves. Dubbed "Boneheads," the first bust, Admiral Enob, is built around part of a bird skeleton, a fish jaw, and a crab shell. While this is a sophisticated sculpt, it's also a good example of how found objects can be assembled and amended with sculpting materials to create impressive-looking models, even if you don't have Britton's sculpting chops.

> "A model enthusiast can buy kits of characters from low-budget horror flicks and underground comic books, and spaceships from just about every cult sci-fi classic."

The Shiflett Brothers (shiflettbrothers.com), Brandon and Jarrod, are another example of kit makers who are also involved in the entertainment business. These Texas brothers got their start doing sculpture work for the video game series *Oddworld*. They've also just teamed up to work with Nimba Creations, a U.K. F/X company that, among other things, worked on Peter Jackson's *King Kong*. To satisfy their kit-creator jones, the brothers have produced a line of resin kits, including the very Tank Girl-esque model Chloe: Aviator for Hire. While many sculptors in the hobby do everything: sculpting, molding, casting, and sales, the Shifletts (and others) do only the sculpting and then sub out the molding and casting to other hobbyists or small-scale commercial casting operations. "Casting is an art in itself," says Brandon, "and for us, finding the right person to mold and cast our work is extremely important."

Most people involved in the garage kit community want to give something back, often in an effort to

get new hobbyists involved and to form alliances with fellow kit makers. The Shiflett Brothers do this by maintaining "Sculpting with the Shiflett Brothers," the largest sculpting online forum of its type. "Our forum attracts pros and amateurs alike," says Brandon. "Everyone from 13-year-old newbies to pro sculptors working on the *King Kong* film."

Other modelers like Dan Perez include workshops on their websites that teach sculpting, molding, and casting. And then there's the collective policing of recasters. While most garage kit creators don't bother with licensing the subject matter of their kits, they don't take kindly to those who buy one of their kits, create a second-gen mold, and then cast knock-off kits at a third of the price (called "recasts").

"We're a close-knit community," says Terry Webb. "We look out for each other, and when recasters rip one of us off, it impacts us all." He adds, "This is an art form, and those of us involved in it take great pride in what we do."

Gareth Branwyn writes about the intersection of technology and culture and is "cyborg-in-chief" of Streettech.com.

Inspired by books about Bigfoot and the Loch Ness Monster, Dave Britton began creating models at age 15, eventually moving to Hollywood to work as a make-up artist and lab technician. Most of his models are "full size" — the Aqualarian head (right) is 23" tall.

Resources:

Dan Perez Studios Modeling Workshop

danperezstudios.com/workshop.htm

Perez's workshops take you step-by-step through beginning sculpting, mold making, casting, and other skills used in garage kit building. Highly recommended.

The Clubhouse

theclubhouse1.net/

Popular online watering hole for kit creators and builders.

Amazing Figure Modeler

amazingmodeler.com, $28/4 issues

Premier magazine for figure modeling, with how-tos on everything from sculpting your own models to assembling, painting, and basing garage model kits.

+ For more resources go to makezine.com/05/monsters.

MAKING YOUR OWN MODELS

While garage kit making can be expensive and require serious dedication to the art of sculpting, beginners can get their hands dirty with minimal effort. The easiest way to get started is to buy a casting kit, such as Bare-Metal's (bare-metal.com) Mold Making and Resin Casting Starter Kit. It comes with all of the materials you need to mold and cast a few small models.

Here are the basic steps to model making:

1. Sculpting/Master Modeling

This part of the process obviously requires some talent, but beginners can use "kit bashing" (cannibalizing existing models) to cobble together designs of their own mutant creations. By combining existing kits with plastic stock (found at hobby stores and e-tailers like plastruct.com), you can create a unique model that can then be molded and cast. Figure modeling can also benefit from bashing existing kits as well as using armatures (wire forms that can be bent into desired poses). Sculpting materials vary widely, from Super Sculpey clay, to two-part epoxy putties, to Casteline modeling wax. Every sculptor has a favorite, and different materials are used for different models.

2. Mold Making

Once your master model is complete, you need to create a mold from which the successive kits will be cast. There are a number of methods and materials for doing this, but the most common approach is to create a two-part mold around your model, one half at a time (*see MAKE, Volume 03, page 135, "Squeezable Nightlight"*).

The bottom half of the model is surrounded by a soft, non-sulfur-based modeling clay bed, leaving the top half of the model to be molded in rubber. Next, RTV (room temperature vulcanizing) silicone rubber is mixed and poured over the top to create the first mold. The two mold halves are later joined up securely, thanks to "keys," a series of dimples that are pressed into the clay bed around the model before pouring in the rubber. Air vents and a pour spout are created by pressing brass rods into the modeling clay on top of the model. A foam-core molding box is built to surround the model and contain the rubber molding material.

You will not be able to make anything this cool on your first model-making attempt. Try for a chubby penguin instead.

Once the rubber has been poured into the box to create the top mold, and is allowed to cure, the mold is flipped over, the clay bed is removed, and the first mold is treated with a releasing compound, which can be as simple as brushed-on petroleum jelly. This prevents the two mold halves from sticking together. Once both mold halves have been created in silicone rubber, casting of your model can begin.

3. Casting

Before casting, the mold halves are treated with a spray-on releasing agent (such as Parafilm mold release). With the two-part mold held securely together with rubber bands, the resin is poured into the spout until the mold is filled up. The most common casting material is a two-part polyurethane resin, such as Vagabond Corporation's Model-Cast (vagabondcorp.com).

Once the casting material has cured, the mold is carefully removed and the cast model is cleaned up, ready for painting.

A crowd gathers as jostling robots struggle their way through Peter O'Kennedy's infinite loop, titled *Escape*.

Re-Booting Art

A new art venue for the robotic age.

By Douglas Repetto

WHAT IS A ROBOT? WHAT IS ART? Those aren't easy questions to answer. Now try defining *robotic art*, or decide what exactly an art-making robot is.

In 2001, I was learning the difficulties of trying to show robotic work in traditional art venues. (It took decades for galleries and museums to embrace video art; robotic and kinetic works, with their sometimes-formidable technical and logistical complexities, have quite a way to go.) But at the same time, robots were all over the popular media. Most of the attention was focused on violent applications, like BattleBots and Patriot Missiles, yet I knew of many artists working with robotics whose strange, subtle, or whimsical ideas had little in common with the remote-controlled chainsaws on TV.

With those two issues in mind, I started a robot talent show email list. Over the next several months, many people contributed ideas for how to create such a show. As those discussions quickly made evident, robotic art is hardly a well-defined genre, and it's not clear what sorts of works qualify. Is an inkjet printer an art-making robot? Is a purely mechanical sculpture robotic art? Our response to these kinds of questions was to dodge them entirely; our open call for entries says: "If you think it's a robot and you think it's art, then send it in." That strategy of inclusiveness has paid off, and some of my favorite pieces have been ones that most liberally interpret both "art" and "robot."

Photography by Douglas Repetto

1

ArtBots: The Robot Talent Show is now an annual art exhibition (see artbots.org) where an international group of artists gather to share their work and their trade techniques, and give a general audience a chance to see what kinds of answers (and new questions!) are being cooked up in labs, garages, and studios around the world. Each year, our location changes and I invite a couple of new people to co-curate the show with me. (Philip Galanter, an artist and teacher, joined me in co-curating and producing the first two iterations of the show.)

We hope that by tweaking the specifics, we can keep the show open and accessible to a diverse range of people, works, and ideas. There have been robots that draw, robots that dance, and everything from a self-preservation machine to a robot race in an infinite loop. ArtBots is low-budget and hands-on (all of the artists attend and help mount the show), and is dedicated to the idea that technology is ours to use and create, and that making art and asking questions is an essential use of any technology.

Douglas Irving Repetto is an artist and teacher involved in a number of art/community groups including dorkbot, ArtBots, organism, and music-dsp.

2

1 Wildflower Meadow Glacier by James Powderly, Michelle Kempner, Tom Kennedy, Todd Polenberg, Brendan Fitzgerald, and Paul Bartlett. Scale model of an enormous artificial glacier moving through Central Park, leaving a path of wildflowers in its wake. The density of flower planting is determined by continuous monitoring of the CO_2 levels in the local environment.

2 Robots Like H_2O: Photosynthesis Perpetual Motion Machine by Futurefarmers: Amy Franceschini and Michael Swaine.

Calculate This!

By Robert Luhn

In the Pre-Digital Epoch, people building bridges, launching rockets, or trying to fill an inside straight used manual calculators. It might've been a snazzy slide rule, or a paper wheel with dials that (when properly aligned) computed a sales commission, the proper valve fitting, or how big a hole a 20-megaton H-bomb would make.

As one observer put it, manual calculators like the slide rule and its kin "radically increased our capacity to perform complex mathematical computations. They literally enabled us to develop our modern world." Here are six classic calculators, ranging from the mind-blowing to the arcane.

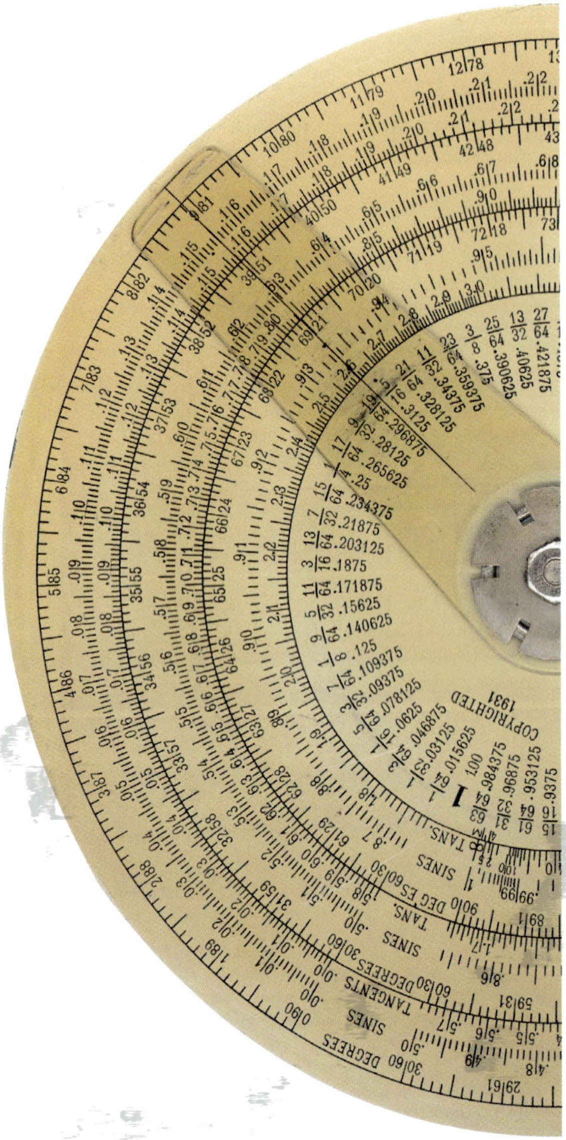

Gilson Binary Slide Rule | 1940

When people think of slide rules, they think of the classic rectangular slipstick. But many engineers, scientists, and merchants turned to this Frisbee-sized binary slide rule from Gilson (a whopping $11 with case). This was the HP calculator of its era, sporting nearly two dozen scales, from trig functions to logarithms to multiplication. But this calculator was rooted in the real world — for merchants, the Gilson could quickly compute retail pricing and add and subtract fractions. For engineers, the Gilson could compute an answer to five figures — not bad for a pre-Pentium device. As with all slide rules, your ability to estimate numbers and keep track of decimal points was key to getting the right answer. (For more calculating, go to the Slide Rule Universe at sphere.bc.ca/test/sruniverse.html.)

Images courtesy of Robert Luhn

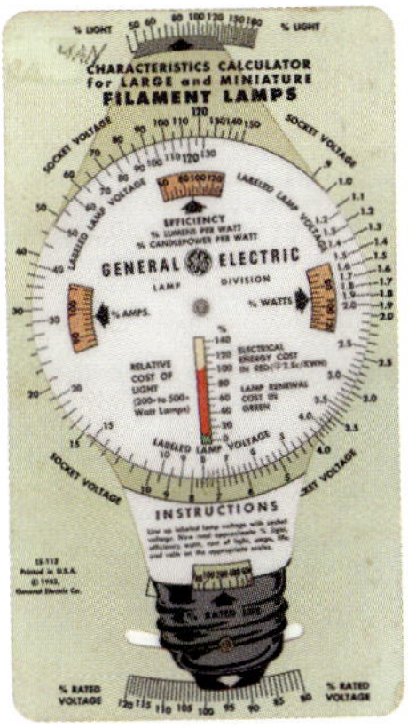

Characteristics Calculator for Large and Miniature Filament Lamps | 1952

How long before that incandescent bulb blows? This handy General Electric calculator — probably created for GE salespeople pitching the company's wares — has the answer. Turn the bulb-cum-pointer until it alights on the socket voltage and bulb voltage, and you can quickly learn how much energy the bulb consumes, the relative cost of lighting your hovel with this particular bulb, and how long the bulb will last before it goes nova. (To shed further light on this topic, seek out the Hall of Electrical History at schenectadymuseum.org.)

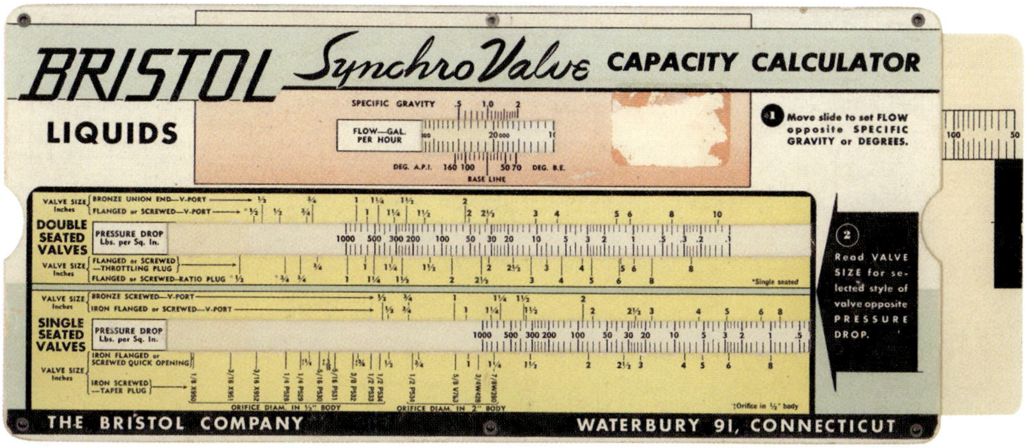

Bristol Liquids SynchroValve Capacity Calculator | 1943

No, this isn't for tuning up your Flux Capacitor. Bristol manufactured sensors, recorders, and other instruments for companies working with liquids and gases, from water to steam to pancake batter. This handy multicolored slide let engineers quickly figure out the right valve type and size for a given fluidic situation. Bristol closed its doors years ago; according to the EPA, the company site is a heady stew of lubricants, solvents, PCBs, and other toxic byproducts. Calculate the cost of *that*.

Poker Pal | 1955

When the chips are down — and you wonder if you'll ever fill that inside straight — reach for Perrygraf's Poker Pal. With a quick yank of the slide, you can determine what hand you need to win (in draw or stud) — and the odds of getting it. You can "what-if" with ease by setting the number of players, number of cards wanted, what's showing on the table, and so on. In business since 1934, Perrygraf is still "THE source for innovative Slide-Charts, Wheel-Charts, Pop-Ups and other dimensional marketing products." (To see its modern-day equivalent, slide over to poker-pal.com.)

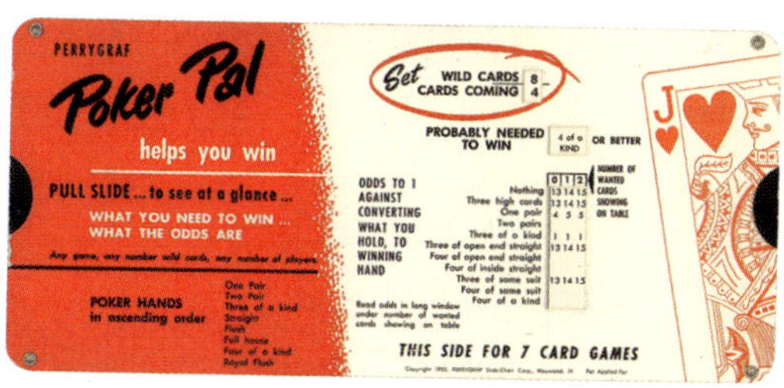

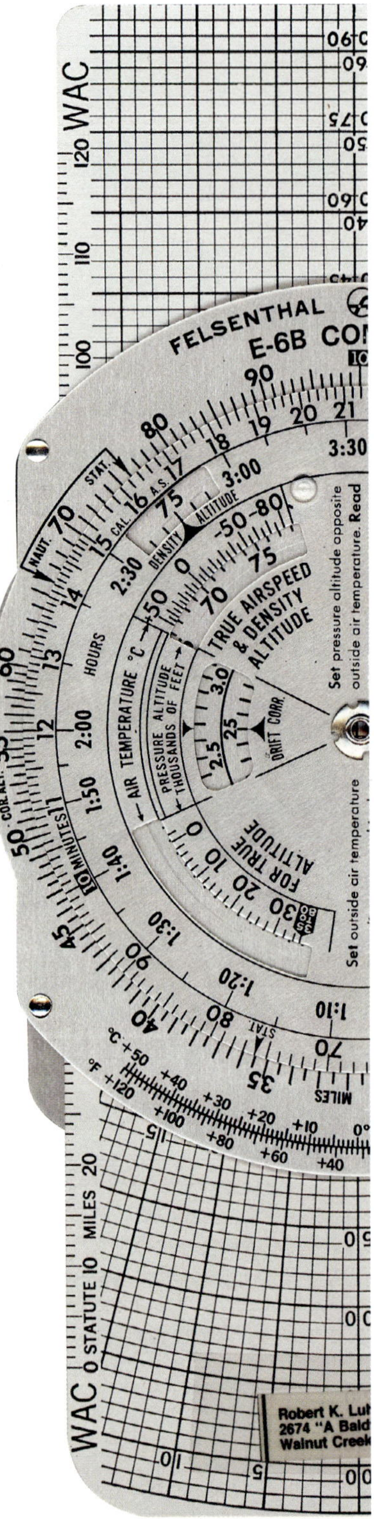

E-6B Flight Computer | 1966

Time, speed, distance, fuel. These are the mantras of the pilot. In the old days (pre-1990s), pilots relied on this device to solve such in-flight quandaries as "If I fly 40 minutes and consume 75 gallons, when will I run out of gas and plow into that mountain?" One side of the E-6B is devoted to such issues, and the other, to figuring out your true direction and speed while correcting for the nefarious effects of wind. These days, you can pick up a digital version of the E-6B for $60 that does a heck of a lot more — but nothing says "veteran aviator" like this baby tucked under your arm. (To see how modern fliers plot their flights, check out the DUATS system at duats.com.)

Today, when everything is programmable, it's reassuring to discover dedicated calculators that are as straightforward to use as a hammer.

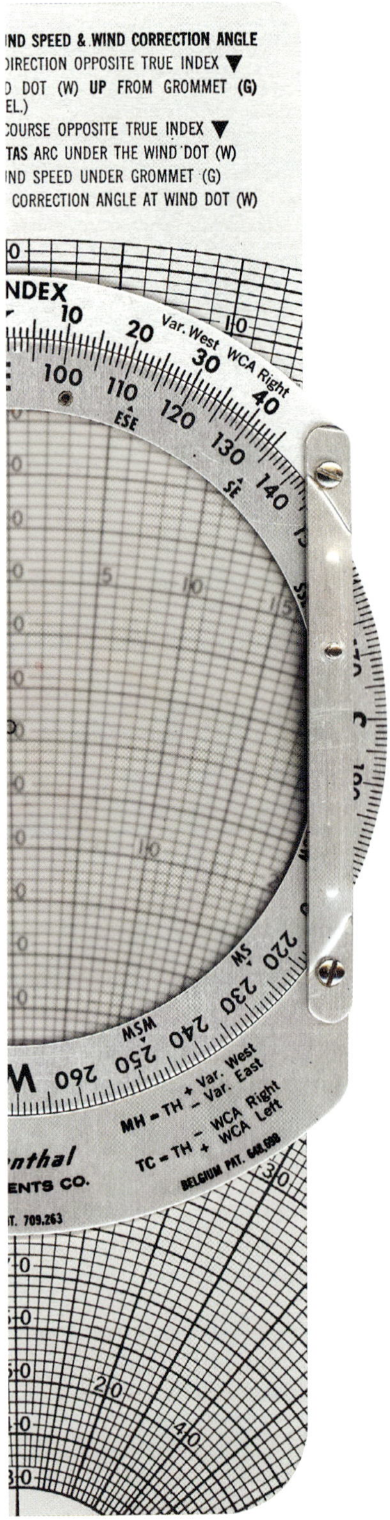

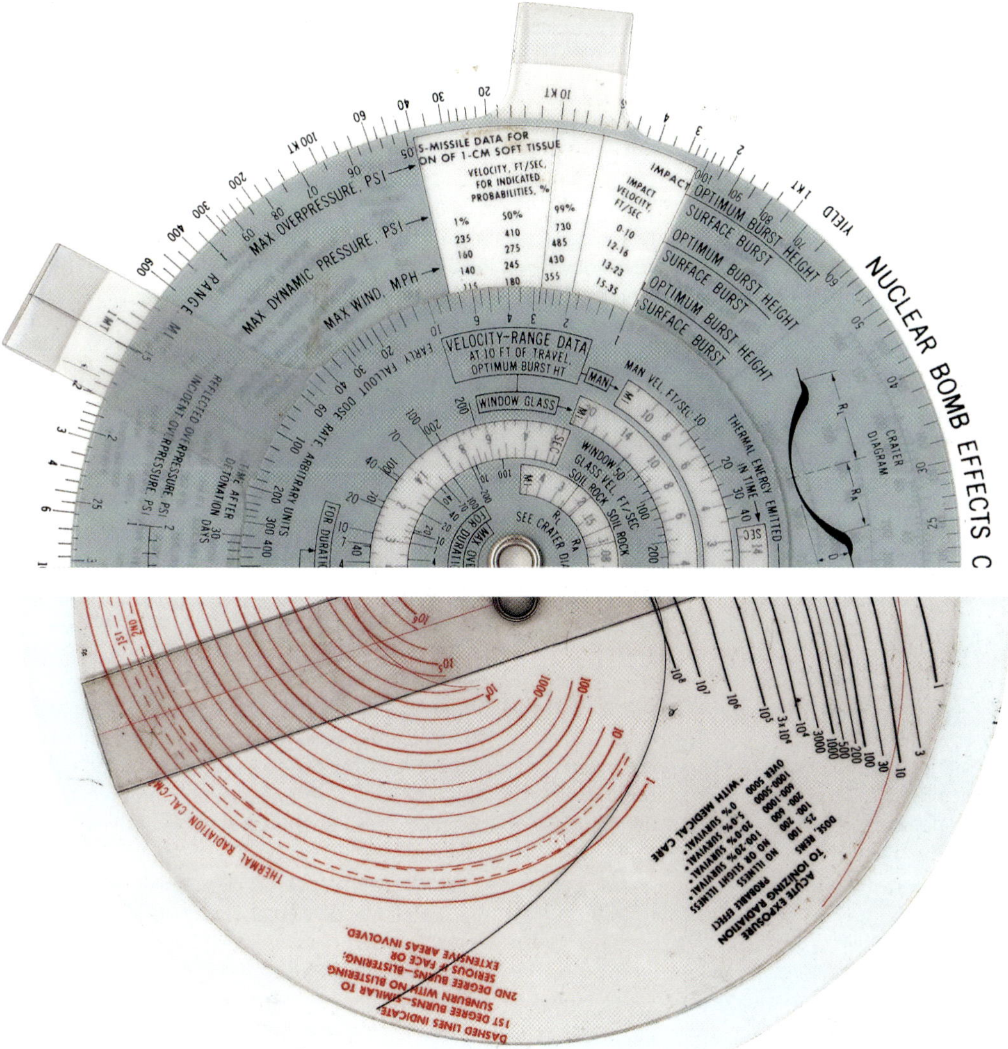

Nuclear Bomb Effects Computer | 1962

What every nuclear family needed in the 1960s. In an era when building a bomb shelter was more fashionable than building a swimming pool, the Atomic Energy Commission contracted with the Lovelace Foundation to create this handy computer. Know the megatonnage of that incoming Soviet ICBM and where it hit? Turn the wheels to those values and you can learn the maximum radius of the fireball, the velocity of burst window glass, your chances of surviving 5,000 rems, and more. Fun for the whole family! (For more history and the scoop on building your own bomb effects computer, skulk over to fourmilab.ch/bombcalc).

Robert Luhn is an executive editor at O'Reilly Media. He collects antique machines and small, unattractive animals.

Saul Griffith

PRE LOVED

OR, WHY USED CAN BE SO MUCH BETTER THAN NEW.

MY GIRLFRIEND RECENTLY SAID, "LET'S get a dirt bike."

"Woohoo!" I said. Then, "Wait, why do you want a dirt bike?"

The motives, according to her, were a string of recent events. 1. *Deep Impact*, the movie: remember the scene where the two kids escape the tsunami caused by the meteor impact with the Earth? 2. The newspaper coverage of traffic jams of people trying to get out of the way of Hurricanes Katrina and Rita. 3. We live in the East Bay of San Francisco in a low-lying area only a few miles from a bunch of major fault lines, with the potential for still more post-disaster traffic jams.

Of course, it may have been none of these reasons. She is an aesthete, after all, and her choice to Netflix the fabulous film *On Any Sunday* was probably the major influence, however subconscious. It's a 1971 documentary on dirt bike racing bankrolled by, and starring, Steve McQueen, produced by the same filmmaker who made the cult surf classic *Endless Summer*. We both loved the movie, particularly the simple beauty of dirt biking before advertising and logos made every amateur look like a lime green advertisement for extreme sports drinks.

One of the most beautiful aspects of *On Any Sunday* is the footage of trials-riding legend Malcolm Smith participating in an amazing motorcycling marathon. A multiday event where the rider was expected to ride the same bike, all components included, across a long and tortuous route including the mountains and roads and mud and dirt and deserts of Spain — sort of a dirt-bike Tour de France. The only person allowed to fix the bike during the ride is the rider himself, right down to changing his own tire. As the ride is a time trial, one must complete each section within a given time period, including repairs, so it pays equally to be an awesome rider and a talented and intuitive repairman. What a refreshing format: a competition for renaissance men, people who understood their machinery as much as having the testosterone to ride it fastest. I would love to have seen the fencing-wire and duct-tape solutions that got the winning machines over the finish line. I say winning machines because the race was not for first place, since anyone who finished with original bike within the time limit got a medal.

I could see why my girlfriend might have been seduced, but I was still suspicious. I broached the subject with one of my dear friends. "A dirt bike?" he said. "Your girlfriend is suggesting you buy a dirt bike?! Don't worry about the motives, just say yes, yes, yes!"

I started looking at dirt bikes online. I had to filter my search by cubic engine capacity as much as by space for two riders and the aesthetic requirements of my lady, because after all, if you are going to escape the apocalypse, you must escape it in style. I wanted something I could fix with minimal tools and equipment — this ruled out an awful lot of the newer bikes, which were too expensive anyway, and many of the older bikes, which were terribly trouble-some looking. Eventually, we found the Bultaco, which, it so happens, is a Spanish machine whose heyday was in the 1970s. I quickly found one nearby on Craig's List (craigslist.org). On the web, I also found a plethora of sites about passionate Bultaco hobbyists, a good sign I'd find community support and hints on keeping the machine alive. Although the last manual was printed 20 years ago, I'd still be able to train myself in the finer arts of Bultaco-owning.

Buying anything secondhand, particularly a 30-year-old machine *designed* to be abused, is always risky. I fully expected to be buying it from a 30-year-old hooligan who'd thrashed and beaten it and was now disposing of it. I was delighted, therefore, upon approaching the Cupertino driveway to see

The Bultaco's glory days were also the heyday of engineering renaissance-men.

a middle-aged man mounted atop this machine with the confidence and man-machine respect of Malcom Smith; he was having a little "goodbye ride," likely reminiscing, before we arrived. I was even more delighted when I entered his workshop. It was the epitome of the phrase "a place for every tool, and every tool in its place." This was a bike that had been nurtured and nourished and loved by an artisan.

Curiously, an old ambulance sat in the driveway alongside a new Mustang and vintage kit car. While I discussed the fine points of oil changing and the reserve fuel tank of my girlfriend's new Bultaco with Tom, she found out from Tom's wife that he was a keen competitive amateur motorcyclist, and that he had converted the ambulance so it was both bike transport and mobile workshop. Legend has it he was known as the only rider to *arrive* at races in an ambulance.

Tom's expertise with bikes extended to less tricked-out transports as well: I was happy to learn the two-tie-down trick for loading a dirt bike in the back of a pickup (one to each side of the handlebar, compressing the front suspension and locking down the front wheel — the rear then doesn't need tying). While adding this arcane detail to my repertoire, I asked knowingly, "You're an engineer?" To which he acknowledged, "Yup." Sometimes a word has curiously high bandwidth; this one signaled that he knew I too was an engineer and that I'd care for his machine as he had, not polishing it to a shine, but rather developing a symbiosis, understanding the rattles and what they mean before they develop into more terminal mechanical problems.

So now I have another old machine in my care. It will require periodic tuning (and yay for me, periodic "test" riding) to make sure it is ready should the big wave or crack in the ground ever come. I choose to believe the disasters are unlikely and that now I'll get to spend the odd "any Sunday" tuning the carburetor and riding around the block imagining myself in Spain in a man-machine marathon.

"This was a bike that had been nurtured and nourished and loved by an artisan."

Saul Griffith thinks about open source hardware while working with the power-nerds at Squid Labs (squid-labs.com).

Photograph by David Sobo

JACKHAMMER HEADPHONES

By Tim Anderson

AND A DISCOURSE ON BLIND MEN, CHIPMUNKS, WHALES, AND THE FUTURE.

I made my first pair of jackhammer headphones about 15 years ago while doing volunteer work for the blind. Many organizations are devoted to helping the blind, including the Library of Congress (LOC) and the Minnesota Braille Press, where I was a volunteer.

The LOC mailed blind people audio books on half-speed cassettes that played on special orange tape recorders. I used to repair these machines along with a group of retired telephone company workers. Other volunteers, mostly actors, read books onto tape in many sound booths. We volunteers were allowed to check out audio books ourselves from the huge collection. The first one I got was *Moby Dick*. Rather than using one of the clunky LOC machines, I used my (no-name brand) walkman. I reduced my walkman's motor speed until it seemed about right and the narrator sounded like a chipmunk. Even so, it took a long time for Mister Chipmunk to recite the book on about 30 cassettes. (In my memory, the book is like this: "Oh boy! Here's a bunch more facts about whales! Blah blah blah blah!")

At that time, I was driving a VW bus with no muffler and hardly any body left. The body had rusted away from Minnesota road salt. On the highway, it was so loud in the cab that I wore jackhammer earmuffs. I wanted to hear what happened to Captain Ahab despite the noise, so I shoved headphone speakers into my ear protectors, making the first pair of the headphones that are the subject of this article.

FUNCTION IN A NOISY WORLD

This project is the easiest and probably most useful thing I've ever designed. I'd have trouble functioning in this noisy world without them. These homemade hi-fi headphones work as well or better than Bose or Sony noise-canceling headphones, according to friends who have tried both. Unlike the commercial models, these block outside noise instead of canceling it. They save your hearing by blocking outside noise and allowing you to play your music at a lower volume.

I've been making these for many years now and using them daily. People sometimes ask, "Isn't it dangerous not being able to hear?" Maybe. Ask a deaf person. I've never had a problem. Lots of my friends use them and no harm has come to anyone.

When driving cross-country, I check out 5 gallons of audio books from the local library to listen to on the road. It's the only thing that works to keep me awake and (somewhat) sane on a long drive. When I get to my destination, I mail the tapes back to my home library.

MAKE YOUR OWN HEADPHONES

Cost: $20 Time to Make: One Minute Complexity: None

The Three Ingredients

1) Industrial ear-protection earmuffs from McMaster-Carr (mcmaster.com) or another vendor. These are Peltor model H10A, the best I've found. They are the quietest earmuffs available (30dB) and have a lip inside that holds the speaker in place.
2) Any airline or walkman headphones of the one-wire-per-ear variety.
3) A cutting tool.

Step 1: Cut off the head loop.

Clip, cut, or chew off the plastic loop that connects the headphone earpieces. Note the right-angle-style phono jack. This type is good. It is less likely to damage your walkman than the straight kind. The straight ones tend to pry the audio-out socket loose from its internal solder connections. A limp cable is also desirable.

Step 2: Shove a speaker into an earpiece.

Let me say this again, Peltor model H10A earmuffs are perfect for this. They have a rim inside that holds the speaker in place.

Finished!

Repeat with the other speaker, and you're done! Enjoy!

Tim Anderson, founder of Z Corp., has a home at mit.edu/robot.

Photograph by Tim Anderson

Tips and Tricks

Helmet-Mounted Headphones

You can see my bike and kitesurfing helmet at www.makezine.com/05/heirloom. In the photos, I'm trying it out with Peltor helmet-mounted headphones. These hangers are slightly more comfortable than wearing the headband under the helmet, but I like being able to take the helmet off and keep the headphones on. So I went back to using the headband-style headphones under the helmet.

The "Double Diaper" Trick

In a really noisy environment, use foam earplugs in addition to these headphones and turn up the volume to a comfortable level. Good foam earplugs can block 27dB. Worn correctly, these headphones block about 30dB of outside noise. In a perfect world, you'd put them together and block 57dB, but I'll leave the numbers to the scientists. It's impressively quiet.

What to Listen to?

My preference is audio books. After listening to 500 or so books on tape, I find myself with quite a store of useful knowledge. Most public libraries have a good selection of audio books. They'll also purchase items you request if they don't already have them.

TIPS AND TRICKS

FROM MAKER TO MAKER

Who doesn't appreciate a really good tip now and then? Especially the kind, as one reader put it, "that changes your life." Whether it's something as practical as revealing a way to remove scratches from glass, or something more creative, like how to make an origami envelope, we all rely on our friends and neighbors to tip us off to the new and the good. *—Arwen O'Reilly*

Tip Your Glass

Tom Bridge saves you from embarassment: "When you're knocking back a pint at the local bar, you look like an absolute idiot when that little cocktail napkin is stuck to the bottom of your glass. Grab the salt shaker from the guy next to you and dash out a bit on your napkin. It will soak up beer and condensation, but not stick to your glass. Now you can drink a beer without looking like an unwashed cretin." (That's why some of us drink at home.)

Inner Tube Tie-Downs

What won't MAKE columnist **Saul Griffith** ever be found without? Far more useful than a bungee cord, he uses bicycle inner tubes for everything, from strapping broken parts together when something fails away from the workshop, to lashing kitesurfing gear to the back of his bike, to fixing the fan belt of a broken-down Chinatown bus. Most bicycle repair shops throw out used tubes and will be happy to pass them along to you.

No More Scratches!

Scratched CRT monitor? Can't see through your glasses? Rather than replace the glass or buy an expensive scratch-removal product, just use a little bit of toothpaste (make sure it's paste, not gel, and one with baking powder is best). Rub over the scratch with some lint-free cloth "in a small circular motion with moderate pressure for a minute or two" and wipe off, says MAKE reader **Paul Short**.

Better Shoelace Tying

James Arlen wants to make sure you never trip over your shoelaces again. Tie your shoes the normal way, but after you make the first loop, wrap twice around the first loop before pushing through the second loop. "Yes, it's that simple," he says. "The knot won't come undone, there is no need to 'double knot,' and you can still undo the laces just by pulling the free ends. This works with bike shoes (notorious for coming undone), running shoes, dress shoes, boots, and especially little kids' shoes!"

HANDY WEBSITES

DIY Laptop Feet

If your poor laptop sheds feet like cherry blossoms in a light spring rain, here's a thought for you. This page has simple DIY instructions for making even sturdier replacement feet at less than one-tenth the price of commercially available ones. All you need are a package of self-adhesive surface protectors (the kind you put under lamps and such) and a sturdy hole punch. the-wabe.com/notebook/laptop-feet/

Tricks of the Trade

A whole website of tips offers "professional secrets from those in the know." Each blog entry takes the form of a reader-submitted trick of the trade. It has advice by everyone from window washers (wash each side in a different direction so you know which side streaks are on) to soldiers in Iraq, who are using Silly String to check for bomb tripwires (spray in a dark room and it clings to even the tiniest of wires). Practical and beautiful, the site is yet another example of why we are greater than the sum of our parts. tradetricks.org

The Shelf Life Index

Ever wonder just how long your mayonnaise will really keep? How about that old can of soda? What's the difference between something just tasting bad and actually making you sick? *Real Simple* has a great list of 77 common foods and their shelf lives to solve your last-minute panics and late-night musings. makezine.com/go/shelflife

Give Away Unwanted Stuff

Freecycle is an international network of local groups who use email listservs to post about things they want to give away — used couches, exercise equipment, broken-but-fixable electronics, film canisters, cardboard boxes, plant cuttings. Everyone wins here: givers get more garage space, takers get something for free, and landfills are conserved. Plus, you get to meet your neighbors. freecycle.org

GPS Stamping for Photos

The Motorola i860 on the Nextel network works with an application that allows you to upload geo-stamped camera phone pictures to a website. "There's even an Easter egg that lets you upload directly to flickr.com," says Allen Smith. "The phone adds tags of the city, state, and zip code where the picture was taken, as well as a Mapquest link. To enter your Flickr email, press # *your pin* 8 7 5 0 # from the first start-up screen, and a pop-up screen lets you enter your Flickr email and password." geosnapper.com/mobile.php

Have a tip for MAKE readers? Send to tips@makezine.com.

Origami Envelopes

Letter writing may be quickly becoming a lost art, but **Mark Brown** has found just the way to spice it up again. Flying Pig, a website of whimsical paper animation kits, also has a free downloadable PDF of instructions for making a paper envelope. "I've seen a lot of different ways to fold an origami envelope," says Brown, "but the simplest and most elegant, bar none, is this one. Too cool." flying-pig.com/pagesv/envelope.html

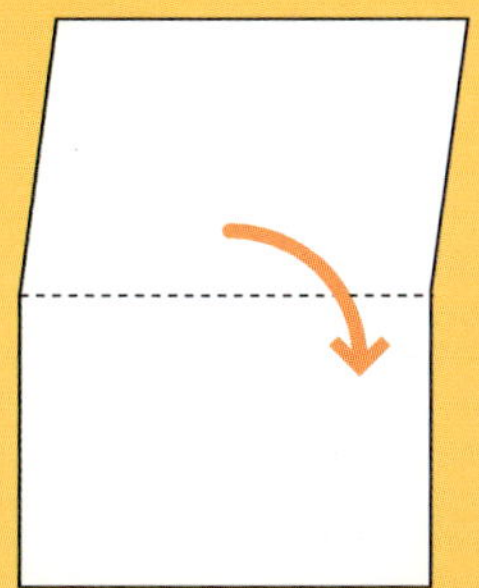

Start with a piece of 8.5x11 or A4 paper. Fold the paper in half horizontally and make a crisp crease. This is your center line. Open up the paper again.

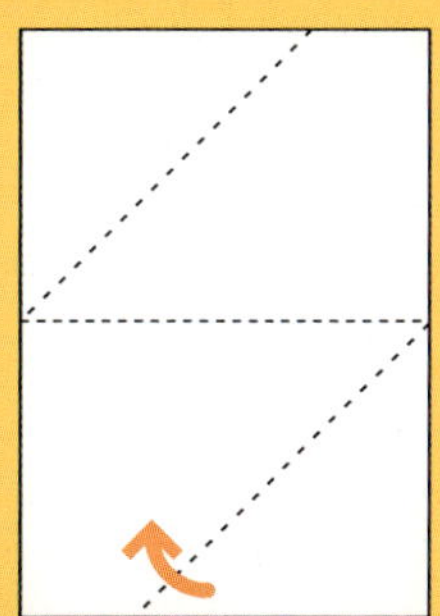

Fold over the two diagonals and carefully line up the edges with the center line.

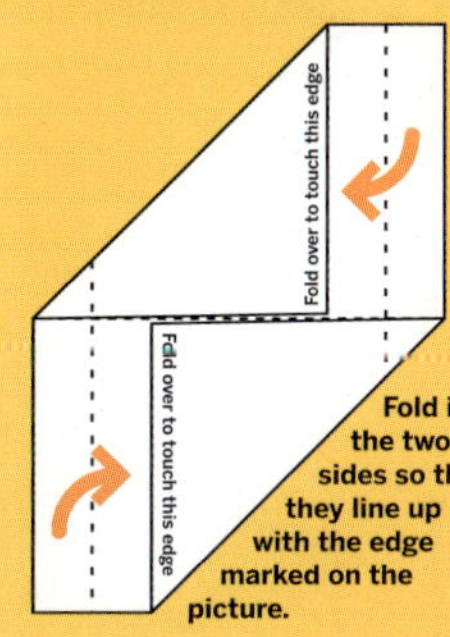

Fold in the two sides so that they line up with the edge marked on the picture.

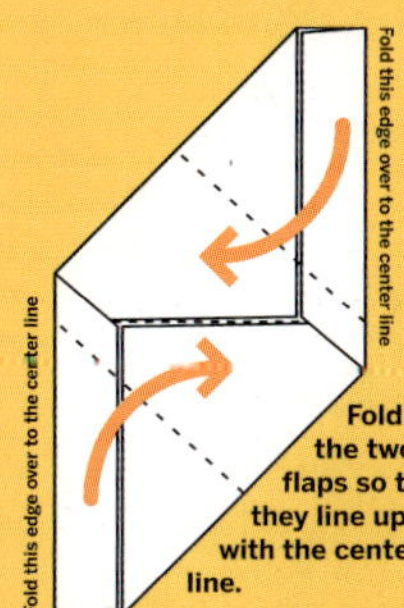

Fold in the two flaps so that they line up with the center line.

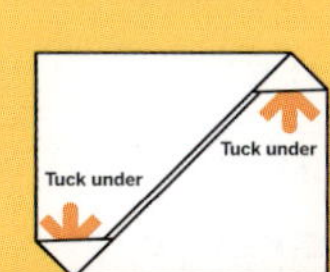

Tuck the ends of the two flaps into the two small pockets. That's it. Turn your envelope over and address it.

Be recursive! Print out these instructions, make them into an envelope, and mail them to a friend.

www.flying-pig.co.uk
From a design sent to me by Phil Warrington. Thank you!

ELECTRIC AVENUE

THE STREET-LEGAL ELECTRIC CAR OF THE FUTURE IS COMING — NOT FROM JAPAN OR DETROIT — BUT FROM YOUR NEIGHBOR'S GARAGE.

By Charles Platt

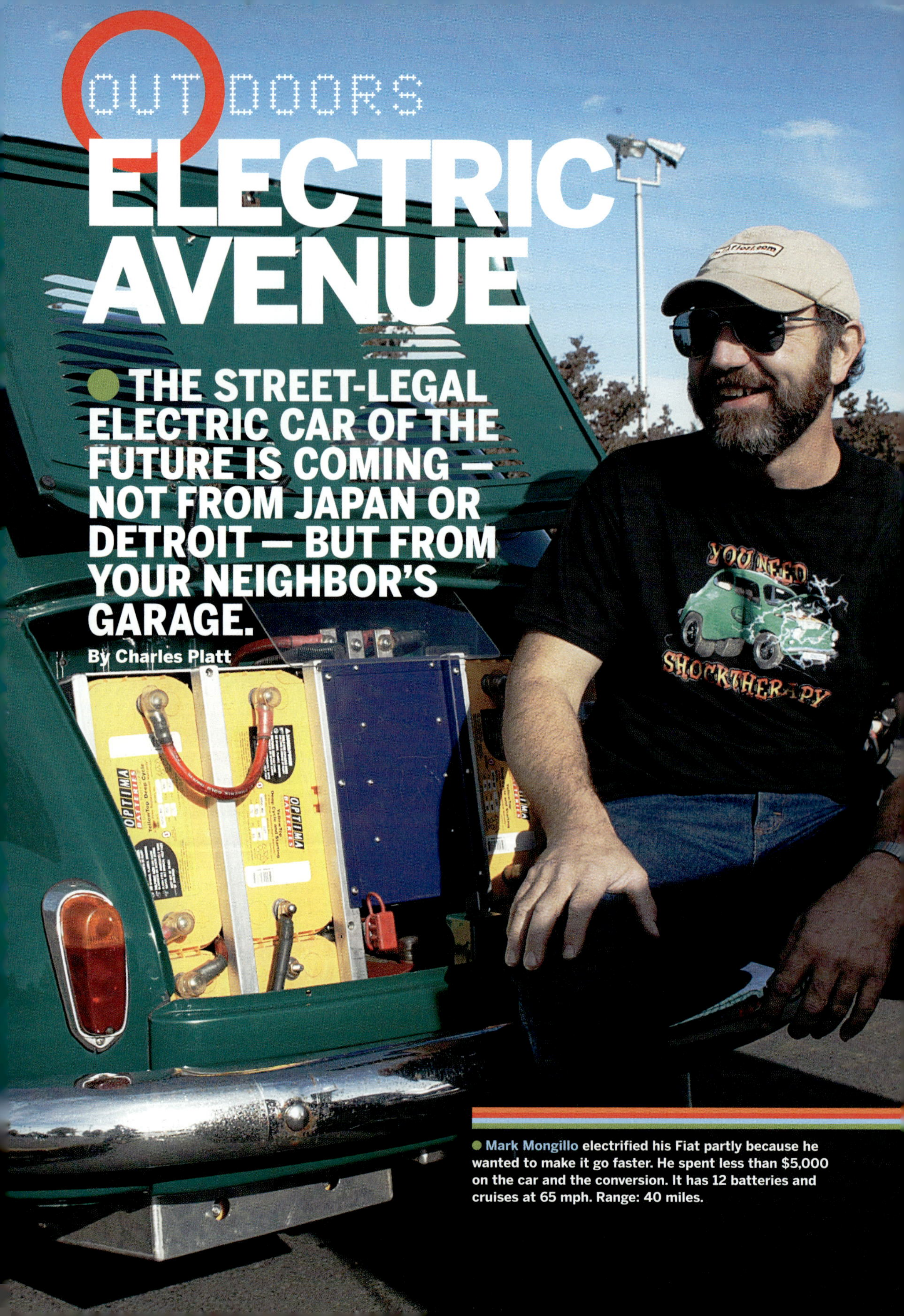

Mark Mongillo electrified his Fiat partly because he wanted to make it go faster. He spent less than $5,000 on the car and the conversion. It has 12 batteries and cruises at 65 mph. Range: 40 miles.

ON A FRIDAY NIGHT AT THE Portland International Raceway, teenage greasers in tricked-out muscle cars gather to strut their stuff. Out in the staging area, purple smoke drifts from burnouts under mercury-vapor spotlights while loudspeaker announcements echo above the rumble of blown V8s. Here in this all-American ritual fueled by gasoline and testosterone, I'm sitting in the passenger seat of a totally incongruous interloper — a cube-shaped postal delivery van powered entirely by batteries. The owner, Roderick "Wild Man" Wilde, did the conversion for a one-hour show on the Discovery Channel. His motive was simple: to prove that electric vehicles don't have to perform like golf carts.

As I perch in the passenger seat beside Wilde, I am uncomfortably aware that if he presses the pedal to the metal, the trio of electric motors driving all four foot-wide dragster radials will be juiced with 2,000 amps at 240 volts from 40 Exide XCD batteries. That's almost half-a-megawatt, enough electricity to sustain 50 average homes.

"When I was a kid," Wilde remarks, "my father was a service manager for a Chrysler dealership, and he told me that gasoline motors are stupid. The piston has to go up, stop, and come back down — it's very inefficient." He smiles reflectively. "If I can take this postal van, weighing more than two tons, and hurl it down the quarter-mile, maybe people will revise some of their anti-electric prejudices."

A few minutes later, after I step out of the van, he peels out with tires smoking and accelerates to 80 miles per hour. He's disappointed; he hoped to break 100.

In an era where alarmists speak of "peak oil" and gas prices have jumped 50% within a year, EV advocates see an unprecedented opportunity to deliver a message that has been persistently misrepresented or misunderstood. Electric vehicles do not have to be overpriced and overweight, like GM's EV-1 or Toyota's RAV4. Nor do they have to be flimsy absurdities such as the Gizmo or the Twike, which look like refugees from reruns of *The Jetsons*.

Modern batteries and high-amperage controllers can satisfy any reasonable need for speed — in a car that looks like a normal car. In fact, if you have minimal mechanical skills, you can do your own gasoline-to-electric makeover for a mere $7,000, acquiring an automobile that has zero emissions yet still boasts enough low-end torque to burn rubber or (more sensibly) merge with traffic doing 75 on the freeway.

"If you have minimal mechanical skills, you can do your own gasline-to-electric makeover for a mere $7,000, acquiring an automobile that has zero emissions yet still boasts enough torque to burn rubber."

This seemingly impossible dream of a high-performance ride within the constraints of ecotopian frugality raises an obvious question: if EVs can be so easy to build, cheap to run, and fun to own, why aren't we all driving them? Even more perplexing, why have today's auto manufacturers abandoned EVs in favor of hybrids, which deliver less performance with greater complexity at a higher price, while still emitting greenhouse gases from their tailpipes?

ON THE SATURDAY MORNING after Wilde's racetrack adventure, he tows his mail truck into downtown Portland, where a miscellaneous assortment of vehicles is gathering at the parking lot outside a Village Inn diner. This annual show of street-legal electrics is being sponsored informally by local resident John Wayland, whose innovative backyard engineering and persistent proselytizing have earned him guru status in the EV world. Wayland is displaying three of his own creations: a commuter-friendly renovated 1972 Datsun, which looks sufficiently immaculate to be a show car; another 1972 Datsun dubbed the "White Zombie," which is the world's quickest street-legal electric car, running a quarter-mile in just over 12 seconds; and — as if the man needed to prove that he has a sense of humor — an electrified lawn tractor enhanced with a thumping stereo system.

Photography by Charles Platt

Wayland is cheerful, amiable, and relentless, even when a skeptical elderly visitor utters the inevitable complaint: "Do you have to plug these things in?"

Well, duh, yes, you have to recharge an EV, just as you would recharge a laptop, an iPod, or a cellphone. And so we get to the inevitable follow-up: "How far will it take you?"

Almost all the vehicles being shown here share the same limit: around 40 miles. Inevitably, this rouses doubt and dismay, since a tank of unleaded typically lasts for 200 miles or more. The fact is, though, most of us seldom use this capacity. Half the homes in the United States contain two or more cars, and the number-two car performs local errands such as picking up the kids from school, commuting to work, or visiting the super-market. For this kind of duty cycle, an EV with a 40-mile range is ideal.

When big companies such as Ford and GM offered EVs, they didn't even try to pursue the short-range paradigm. Instead, they attempted to guarantee as many miles as possible by loading their vehicles with massive battery stacks that devoured interior space and crippled performance. The result was an unhappy compromise that pleased no one.

Mindful of this foolishness, backyard builders acknowledge the limits of today's battery technology and sacrifice some range in order to save weight, mimimize expense, and maximize drivability. Here's a classic example: a cute little green vintage Fiat, the size of an old VW Beetle. "I've always had a thing about Fiats," comments its owner, Mark Mongillo, "but they weren't fast enough as gasoline cars." After cramming a couple of battery stacks into the rear engine compartment, he has a livelier vehicle that still seats four adults. "I bought it for about $2,500," he says, "and put about $2,000 into the conversion, with a lot of used parts bought over the internet and through friends. It took me about a year of tinkering — a few hours here and there. I mostly did it because it's fun."

Mongillo drives me around some back streets, and I notice that the motor is so tolerant of speed variations that he doesn't have to touch the gearshift. "Second gear is good for up to about 45 miles per hour, and third gear takes me up to 75," he explains. Nor does he need to slip the clutch when starting from a red light, because an electric doesn't have to rev up in preparation for taking off.

Back in the parking lot, I inspect Rick Barnes' 1986 Chevy Sprint, the dowdiest vehicle on display. "I was going to convert a Yugo," he says, "but I found this already converted, for sale in Seattle." He says there is an active secondhand market for EVs, because no one wants to throw them away. He didn't bother to pimp his ride for this car show, because the Sprint is not a show car. He commutes in it every day to his job at a local Intel chip factory.

"Cruise past gas stations with the same feeling of superiority as an ex-smoker bypassing a display of cigarettes in a convenience store. 'Middle Eastern oil? Sure, I used to burn it — but I quit.'"

At the other extreme, Russian-born electronics engineer Victor Tikhonov displays an ongoing research project: a 1991 Honda CRX powered with state-of-the-art lithium-ion (Li-Ion) batteries, which offer greater range but have a higher internal resistance that limits amperage and acceleration. To compensate, Tikhonov has added a massive array of capacitors, visible beneath the Honda's fastback rear window. "There are 160 of them at 2,700 farads each," he tells me.

Another lithium-ion advocate is Jukka Jarvinen, who flew here from his native Finland. Jarvinen is tall, thin, intense, and dogmatic in his passion for all things electric. He built a lithium-ion-powered motorcycle, and recently obtained government funding for his own EV battery development company. "You get about 25,000 miles from one lead-acid battery pack before it must be replaced, but lithium-ions easily give you 150,000 miles," he claims. "That's about 10 years from one set of batteries, with 160 miles per charge." He believes that lithium-ion prices will plummet, since the Chinese have invested $2 billion in battery

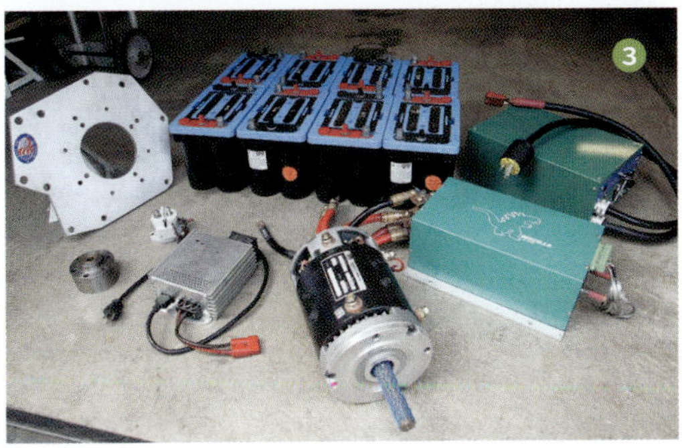

1. Roderick Wilde's 500,000-watt postal van arrives on its trailer to compete with street-legal dragsters at the Portland International Raceway.
2. Tom True (left) enlisted assistance from EV veteran Don Crabtree (right) when he wanted to electrify his Datsun Z. While all the original accessories still work, including power windows and radio, the gasoline engine has been replaced with a cluster of three electric motors.
3. On the concrete floor of John Wayland's garage are the components for a typical EV conversion: eight 12-volt lead-acid batteries, a forklift truck motor and controller, recharging module, DC-DC converter, mounting hardware, and drivetrain coupler. Total cost including batteries: $7,000.

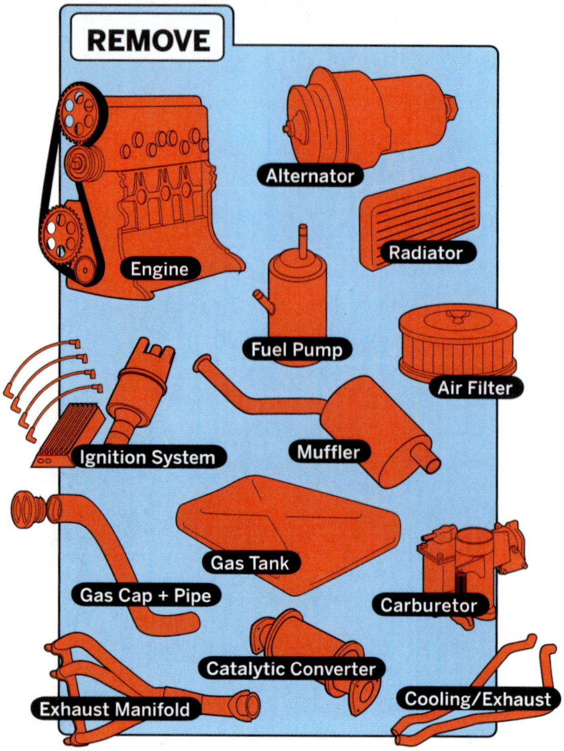

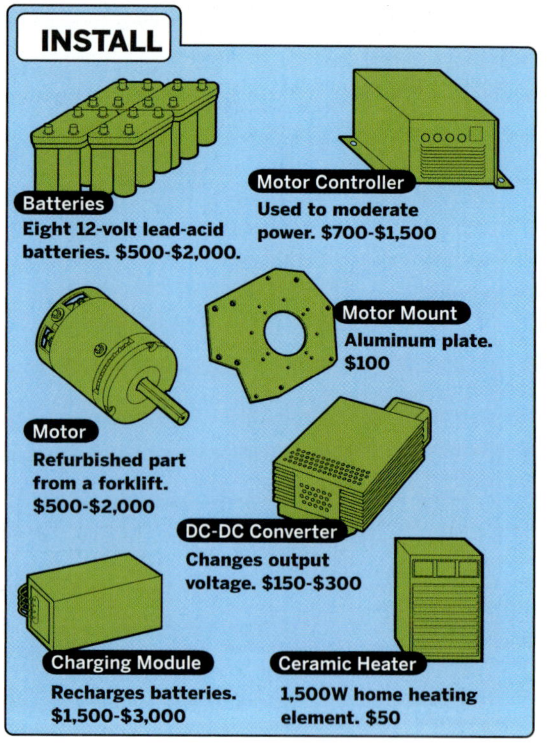

research and have lithium-rich ore deposits in Tibet.

Still, most EV builders prefer the reliable simplicity of lead-acid batteries and refurbished DC forklift truck motors. Pioneers such as Wayland and Wilde figured out 10 years ago that you can safely force 96 volts through a 48-volt motor if you turn the commutator ring to change the timing. Jim Husted, who ran a business refurbishing forklifts for 24 years, was stunned when he discovered what the EVers were doing. He is now a true believer, hanging out with the nerds and spreading the gospel with wide-eyed intensity.

"This is very exciting for me," he says. "I want to get more of these cars out on the road. I'll work within any EV builder's budget to make his car better. Send me a motor and I'll send it back with variable timing. I want to make a retrofit kit, a user-friendly way to do what these guys have been doing on a much more technical basis."

He sounds giddy with the possibilities. "There are thousands of different models of DC motors that have been built in the past 40 years. Some would be ideal for motorcycles. And I'm looking to wire a pair of motors so they can go from series, to series-parallel, to parallel to make a three-speed transmission. I'm proud to be a part of all this."

AFTER THE SHOW, I VISIT THE two-car garage behind Wayland's suburban home and ask him to show me the parts that are needed for a conversion. It takes him only 10 minutes to gather and array eight 12-volt batteries, a motor (refurbished from an electric forklift truck), a controller to moderate the power, some heavy-duty copper cables, and a recharging unit. With minor additional items such as a 5,000-ohm potentiometer that fits under the gas pedal and costs less than $10, this is all you need. The lesson is obvious: EVs are ideal for backyard builders because they can be really, really simple.

The first step in a conversion is to find a "donor vehicle," usually a lightweight compact gasoline-powered automobile from a scrap yard. You throw away almost all of its mechanical components: the engine, radiator, hoses, carburetor, air cleaner, ignition system, starter motor, exhaust pipe, muffler, catalytic converter, fuel pump, and (of course) the gas tank. An electric motor needs none of these peripheral devices. It is also one-quarter the size of a four-cylinder gasoline engine, about four times as efficient, and far more reliable. An electric vehicle needs no external cooling system, filtering systems, or oil supply.

Illustration by Tim Lillis

John Wayland's immaculate 1972 "Blue Meanie" Datsun includes protective plastic covers over additional batteries under the hood, alongside a green "Godzilla" controller.

With one massive motor under the hood driving the front wheels, and two more for the rear wheels, the postal van can smoke all four of its foot-wide radials.

After you get someone such as Jim Husted to tweak your motor (or buy a pre-tweaked motor from a source such as EV Parts (evparts.com), you'll need to attach it securely to the frame of the car. This should be simple enough, since you don't need rubber mounts. To attach the motor to your car's transmission, you can get a local metal fabrication shop to create a suitably drilled quarter-inch aluminum plate for maybe $100. You'll also need to get a hub adapter to connect the motor shaft with your existing flywheel and clutch. These are available for many types of cars on various EV parts retailers on the web.

You'll wire your batteries with caution and respect, since their high amperage at 96 volts is hazardous. It's a good idea to invest in some EV conversion how-to books such as *Convert It!* by Michael Brown or *Build Your Own Electric Vehicle* by Bob Brant, or go online to discussion sites where technical knowledge is exchanged freely.

"You get about 25,000 miles from one lead-acid battery pack but lithium-ions easily give you 150,000 miles."

Expect to spend between $5,000 and $10,000, or more if you want extras such as separate motors to power an air-conditioning compressor, power steering, or power brakes. Most EVers choose small, light cars in which power assists are not necessary.

During the winter months, the efficiency of an electric motor actually counts against it, because it generates virtually no waste heat to warm the car interior. Still, you can use the existing car heater if you insert a ceramic element from a 1,500-watt (minimum) heater of the type sold for home use. When you run it from your DC batteries, it will deduct about 10 to 15 percent of your range.

Cold weather will also affect the batteries themselves. At 32 degrees F, lead-acid electrochemical reactions create only half as much power as at 80 degrees. Still, you can line your battery box with styrofoam, and keep your car in a garage overnight. Alternatively, according to Wayland, "If you can afford a set of NiCads, they have a cycle life in the thousands, and they're on the secondhand market now. At 20 degrees, they still have 98% of their power. Nickel-metal hydrides and lithium-ions also are not significantly affected by cold."

No matter which batteries you use, they'll require some nurturing. You'll try to avoid discharging them completely, and you'll use a properly engineered unit to recharge them. With a small amount of care, they'll last several years, and you'll have the pleasure of driving a truly unique, silent, nonpolluting vehicle. You'll cruise past gas stations with the same feeling of superiority as an ex-smoker bypassing a display of cigarettes in a convenience store. "Middle Eastern oil? Sure, I used to burn it — but I quit."

Of course, electricity has to be generated somewhere, but ideally it may come from hydroelectric installations or wind turbines, and even if it originates from fossil fuels, a large power plant is more efficient and cleaner than a small internal-combustion engine. Also, it may use domestically mined coal, of which we have copious reserves.

Your EV may not have the megawattage to burn rubber on a drag strip in the style of Roderick Wilde's demented postal van, but it will be more spirited than a hybrid while also being simpler, more reliable, cheaper, and more efficient. "The only reason that hybrids are catching on," says Wayland, "is they get people over their fear of not having enough battery power. People can still go to the gas pump. They have gas to rescue them." He shakes his head sadly. "It's a crutch! Hybrids are training wheels for the masses!"

From an EV perspective, hybrids are not only lame but crippled, since their batteries can only be recharged via the car's gasoline engine. Perhaps in an effort to avoid the "EV stigma," manufacturers omitted any option to recharge them from an external source. If this feature were added, Wayland points out, "You could drive a hybrid in the city all week long and never turn on its gasoline engine. But you'd still have the engine when you need greater range."

How long will it take for large auto companies to stop thinking that consumers will be afraid of plugging a car into a wall outlet?

Fortunately, we don't have to ponder this question because the components on Wayland's garage floor offer an easy option. If your daily driving habits fall within a 40-mile limit, you can have your own rationally engineered EV right now. All you have to do is make it yourself.

HYBRID OWNERS PLUG IN

Why don't hybrid cars have charging plugs as well as gas tanks? The auto industry's stated rationale is that the larger batteries needed to store a useful "full tank" of electricity are too heavy and expensive. But many hybrid owners are doing plug-in electric hybrid (PHEV) conversions themselves, and a nonprofit organization, CalCars, has formed to support them and to lobby for PHEV development by automakers.

To convert the popular Toyota Prius, you can replace its 1.3kWh NiMH battery with a 9kWh battery pack, add a power inverter and a plug, and tweak the control system so that the gas engine kicks in at a higher speed (or can be disabled entirely). The result is a pluggable Prius that can make trips around town without using any gas at all, although it won't reach highway speed in all-electric mode.

—Paul Spinrad

RESOURCES

EV discussion and advice:
madkatz.com/ev/evlist.html

Photo album (more than 600 home-built EVs):
austinev.org/evalbum

Parts for EVs:
evparts.com, ev-america.com, manzanitamicro.com, cloudelectric.com

More EV links:
pages.prodigy.net/noela/new_page_2.htm

Victor Tikhonov's "high-end AC drive systems":
www.metricmind.com

North American Electric Drag Racing Association:
www.nedra.com

Charles Platt has been a senior writer for *Wired* magazine and has written science fiction novels such as *The Silicon Man*.

DIVE, DARN IT, DIVE!

IT'S A MATTER OF SINK AND SWIM AT THE 8TH INTERNATIONAL AUTONOMOUS UNDERWATER VEHICLE COMPETITION.

By Larry Harmon

A Navy diver unhooks Georgia Tech's autonomous robotic submarine for a trial course run at the Space and Naval Warfare Systems Center training pond in San Diego. The team placed 12th in a field of 19.

IT'S 11:30 P.M. ON A FRIDAY NIGHT in San Diego and a group of college guys from a handful of different universities are hanging out in bathrobes and swimsuits by the pool at a resort hotel on the harbor.

But it's not what you think. Instead of chasing girls after beer chasers, these guys are laid out on recliners, writing last-minute programming changes on their laptops, or they're in the pool tinkering with large submarines that look like they were built from erector sets, PVC pipe, and spare PC components.

The students — a mix of 19 teams representing universities and trade schools across the U.S. and Canada, along with a Bay Area high school team and even one from India — are in town for the 8th International Autonomous Underwater Vehicle Competition, held at the Space and Naval Warfare Systems Command (SPAWAR) training pond overlooking the Pacific Ocean.

> **"Each vehicle sports plenty of zip ties and duct tape, evidence of quick fixes after going through damage-inducing practice runs in the pool."**

At stake is $20,000 in prize money, of which $5,500 will be given to the first place team. Considering the cost and time to create a submarine, money isn't the motivating factor. It's all about the bragging rights.

The AUV competition involves three tasks. Once a vehicle is lowered by a small, portable crane into SPAWAR's murky pond, it must submerge by itself and proceed to travel approximately five yards to an underwater gate, which it must enter. This will prove the vehicle is capable of traveling in a straight line.

Next, each robotic submarine must locate a flashing beacon and bump into it to simulate an underwater docking.

Finally, the vehicle must locate a length of zig-zagged pipe laid out on the bottom of the pond and follow it until it finds a large gap. Once it locates the gap, it must drop a marker identifying the missing section. The final segment requires the submarine to listen and locate an acoustic pinger in a marked target zone and surface to be recovered.

At the resort pool, the teams are doing trial runs of the pipeline simulation and tinkering with their machines on the pool deck.

Andy Linnenkohl, an electrical engineering major from Southern Polytechnic State University, located just outside of Atlanta, says his team's chances of winning are about as likely as a high school football team beating the Atlanta Falcons.

His football reference is way out of place among the talk of vector thrust, pressure sensors, and sonar pulses, and it's quickly apparent why he stands out from the other students at the pool.

"I used to be a police officer. We had the big remote-controlled robots for the bomb squad, which was the first time I ever messed with any type of robots before, and it captured me instantly. We had four guys on the bomb squad, all about my age, and when we were alone with that control panel, it looked like a bunch of guys sitting around playing a PlayStation game. We'd be sitting there jerking the joystick back and forth, and the only thing missing was a fire button."

All of the teams are at the pond by 7 a.m. the next morning for the qualifying heats. There are 19 tents forming an L-shape surrounding the pond, organized by last year's placing. The top five tents house the odds-on favorites: MIT, followed by Cornell University, Québec's École de Technologie Supérieure, University of Rhode Island, and Duke University.

At the bottom of the list, mostly because it's their first time competing, are North Carolina State University, Indian Underwater Robotics Society, Georgia Tech's Marine Robotic Society, and DeVry.

Like artists asked to paint the same portrait, but with widely varying skills, experience, and budgets, each of the 19 teams approached the challenge uniquely.

"That's the beauty of working in the water," says Daryl Davidson, executive director of the Association for Unmanned Vehicle Systems International (AUVSI), which organized the competition.

"When you think about the aircraft industry, it's very specific and very elegant," he says. "That's not the case here. Some of them do have very elegant things to them, but it's not so much. When they get into the propulsion systems or battery systems,

Photography by Eric Rife and Paul Hansen

some of them are using very small computers, so they've been able to pack everything down into a small case. One team has a boxy design that looks like Rhode Island. If you have a long, streamlined system, it takes less energy to push it through the water, but once it gets going, it's going to have a harder time adjusting and correcting. Some of these round vehicles can pretty much turn on a dime."

The bigger the budget, the more intricate the vehicle, and it was obvious that the Indian team didn't have much of one.

In fact, team member Anuj Sehgal said they had a total budget of $1,000, which included travel expenses to San Diego. To save money, the entire team was staying at a relative's apartment in the low-rent suburb of El Cajon, approximately 20 miles east of San Diego.

They bought a majority of the vehicle parts after arriving, including the main body, which was a black, squarish, waterproof carrying case complete with handle. Bilge pump motors powered four thrusters — one facing up and one facing down on each side of the case, giving it the appearance of a gyroscopic batwing.

There wasn't enough money in the budget for any type of camera or listening device, which means the vehicle really has no way to complete any of its missions.

No matter what the teams spent, each vehicle sports plenty of zip ties and duct tape, evidence of quick fixes after going through a couple of damage-inducing practice runs in the pond. If a vehicle was having a problem with buoyancy, a handful of washers on a bolt or a plastic Coke bottle filled with hardened insulation foam would balance the craft.

But buoyancy is the least of their problems. There are fried boards, water leaks, software issues, and a host of other problems. Murphy's Law is in full effect today.

During the qualifying heats, most of the vehicles have problems finishing any tasks, and only a handful complete the entire course.

Come the finals on Sunday, the favored MIT vehicle strays off course by following the reflection of the sun, taking third, while University of Florida's *Subjugator* — a clear cylinder with a rear spoiler and foam strapped on for buoyancy — surprises everyone by taking first, with École de Technologie Supérieure's lunar-lander-looking sub winning the second place slot.

"One reason for the competition is to bring together these bright engineers to create robotic vehicles that can remove human beings from dull, dirty, and dangerous work, such as bomb disposal or de-mining a shipping lane."

Despite the friendly competition and camaraderie between the teams, the potential for heavy-duty military and industrial applications is ever present in the minds of both the students and the competition organizers.

Davidson is clear that one reason for the competition is to bring together these bright engineers to create robotic vehicles that can remove human beings from "dull, dirty, and dangerous work," such as bomb disposal or de-mining a shipping lane.

And he sees the possibilities as endless. "You can design these things to go as deep as the ocean is," he says. "They don't have that human factor of designing a very huge, robust, and very tight system to protect a person. You just pack it all in and you're not constrained by having to protect human life. The same thing goes with the unmanned airplanes. When you take out the cockpit and safety elements, you can pull all sorts of Gs and do things that no manned aircraft could ever do."

Davidson adds, "If you can take people out of the missions that have a risk of life and put a machine in that same mission and something goes wrong, who cares? You blew up a piece of machinery."

Larry Harmon has been covering San Diego music, strangeness, and crime in his fanzine, *Genetic Disorder*, and a number of other publications for more than 10 years.

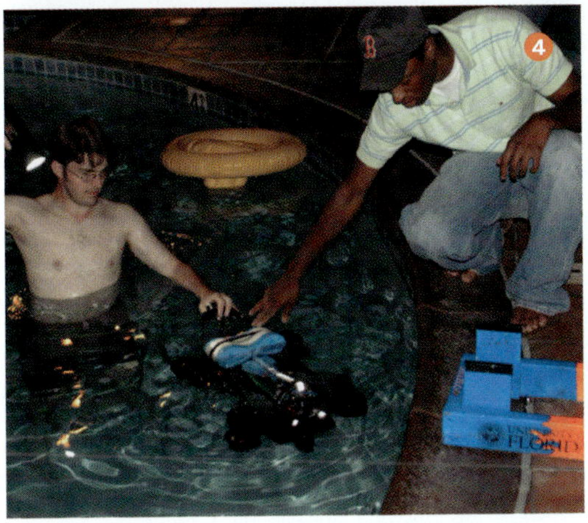

1. The MIT team's *Orca*, which was favored to win this year's International Autonomous Underwater Vehicle Competition in San Diego, came in third, losing out to the University of Florida's *Subjugator* and École de Technologie Supérieure's *S.O.N.I.A.*
2. Southern Polytechnic State University's Auvrobot landed squarely in the middle of the pack, coming in at 10th place.
3. Because of a last-minute operational mishap, North Carolina State University's delta-shaped *SeaWolf* came in 18th out of 19 entrants.
4. Best in Show: University of Florida's *Subjugator* brought home the gold by successfully completing one of three tasks, and for weighing in at only 30lbs. (far under the 70lb. competition limit).

OUTDOORS

BACKYARD ZIP LINE

BE THE HIT OF THE NEIGHBORHOOD WITH A HIGH-FLYING TREE-TO-TREE TRANSPORTER.

By Dave Mabe

The author's wife, Joan Nesbit Mabe, tries out the zip line.

YOU COULD BUY A DINKY, READY-MADE kit with a short zip line for kids, but why not make your own industrial-strength zip line that will support the heaviest of neighbors? It's a fun project you can tackle in a weekend. You can order all the parts on the web for less than $300.

First, you'll need to find a suitable location for your zip line. Depending on the lay of the land, you'll be choosing between two basic types of zip line. If you would like to put your zip line on a steep hill, you'll need to use a braked zip line, which has a brake block attached to bungee cords that slow you down as you approach the end of the line. If your site has a more gradual incline, you can use a gravity-stop zip line that simply uses gravity to slow you down. I knew small children would be riding mine, so I chose a gravity stop because it is tamer, and our property layout made this option ideal.

> "Instead of risking life and limb, I took the cable halfway around the tree and then chose the base of another tree to secure the cable."

FIRST STEPS

I surveyed my property and identified two large oak trees that were far enough apart and only had a few very small trees between them. I measured the distance between the trees and ordered the following parts from Starlight Outdoor Education (starlightoutdoored.com).

For a permanent zip line, you'll need long eyebolts, drilled through the entire tree trunk, to attach the cable to. I knew I might move the zip line in the future, so I chose a more temporary technique to attach the cable.

Depending on the run you've selected, you'll need to determine how much cable to purchase. The cable comes in a spool in multiples of 250 feet. You should get more than you need because you'll need a bit of play on both ends.

While I was waiting for the supplies to ship, I started preparing the site. I had to clear some small trees and underbrush to create a path between the two trees. I also started visualizing how far up each tree I was going to need to attach the cable and how steep the slope should be. I knew there would probably be some trial and error, but taking some time to plan ahead of time definitely minimized this once the supplies arrived.

You'll need a buffer between the cable and the tree it is attached to — otherwise, the tree will actually grow around the cable and, over time, completely engulf it. For this purpose, I bought three 1x6 boards of pressure-treated deck flooring and cut them into several 1-foot lengths. I went ahead and drove the nails partway into the boards so I wouldn't have to do so once I was at the top of the ladder. I nailed the boards into the tree vertically so that they encircled the trunk with small gaps in between them. This allows the cable to attach securely to the tree without actually touching it. On a couple of the boards, I drove a nail about halfway in so the cable would be supported by it and wouldn't slide down the tree.

LAYING CABLE

When the cable arrived, I unwound some from the spool, climbed the ladder and circled the tree one full time, and then secured it with two cable clamps. Depending on the diameter of the tree, you may need a second person and ladder to help, as the cable tends to be unwieldy, especially on top of a ladder.

I unwound the rest of the cable toward the other tree and nailed the boards to the second tree. Because the cable on the second tree is so high, I knew it would be difficult and dangerous to tighten the cable and clamp it at that height. Instead of risking life and limb, I took the cable halfway around the tree and then chose the base of another tree to secure and tighten the cable to. This allowed me to use the come-along (a.k.a. the hand winch) to tighten the cable standing on the ground rather than at the top of an extension ladder.

To tighten the cable, I took it around the tree and put two clamps on the cable without tightening them. I then clamped a loop on the end of the cable so I could attach the come-along to it. Because there was so much tightening that needed to occur

Photography by Terri Bennett

MATERIALS:	COST:
7x19 galvanized aircraft-quality cable (250 feet)	$100
Two-wheel zip pulley	$110
Drop-forged cable clamps (8)	$1.75 each
Come-along (chain puller)	$40
Climbing straps (2-3)	$5
Steel carabiners (2-3)	Around $10 each
1X6 decking boards (2)	$10 each
3" nails (several)	Around $5
1 extension ladder	Try to borrow one!

to raise the cable to the proper height, I captured the gain by tightening the cable clamps and then repositioned the come-along and the loop at the end of the cable. This took several repetitions before the cable was tightened (and therefore raised) to a useful riding height.

TEST RUNS AND SAFETY CONSIDERATIONS

For the first couple of test rides, I started well shy of the top to make sure I wouldn't get a mouthful of bark if I hit the tree at the bottom of the run. I used a wooden handle from an old ax as a hanging bar and drilled an eyebolt through it, using a carabiner to attach the pulley to the hanging bar. I made more test runs by starting closer and closer to the top tree, carefully observing ground clearance and whether height adjustments needed to be made on the bottom tree to create more or less "gravity" at the end of the line.

Depending on the age, strength, and confidence of the rider, there are a variety of options for attaching the rider to the zip line. A climbing harness is the safest and easiest for younger riders, but it is time-consuming to transfer and adjust the harness between different riders. Older and stronger riders can use a couple of climbing straps as a seat by simply attaching the straps to a carabiner and sitting in the loop. The strongest and most confident riders can simply hang from the wooden bar, although because of the height from the ground at the end of the line, this is discouraged due to safety concerns.

Dave Mabe is the author of *BlackBerry Hacks* (O'Reilly Media) and lives in Chapel Hill, N.C.

1. Protect the tree using the 1X6 boards (otherwise the tree will grow around the cable).

2. Tighten the cable using the come-along.

3. Attach the pulley to the cable.

Make: Projects

Get your sugar rush setting off a soda-bottle rocket powered by air and water. Then harness the wind with a PVC windmill generator made from pure junk. And if that isn't enough excitement for you, raise your pulse by spending an afternoon igniting a high-octane jam jar jet engine.

Water Rocket

78

Wind Turbine

90

Jar Jet

102

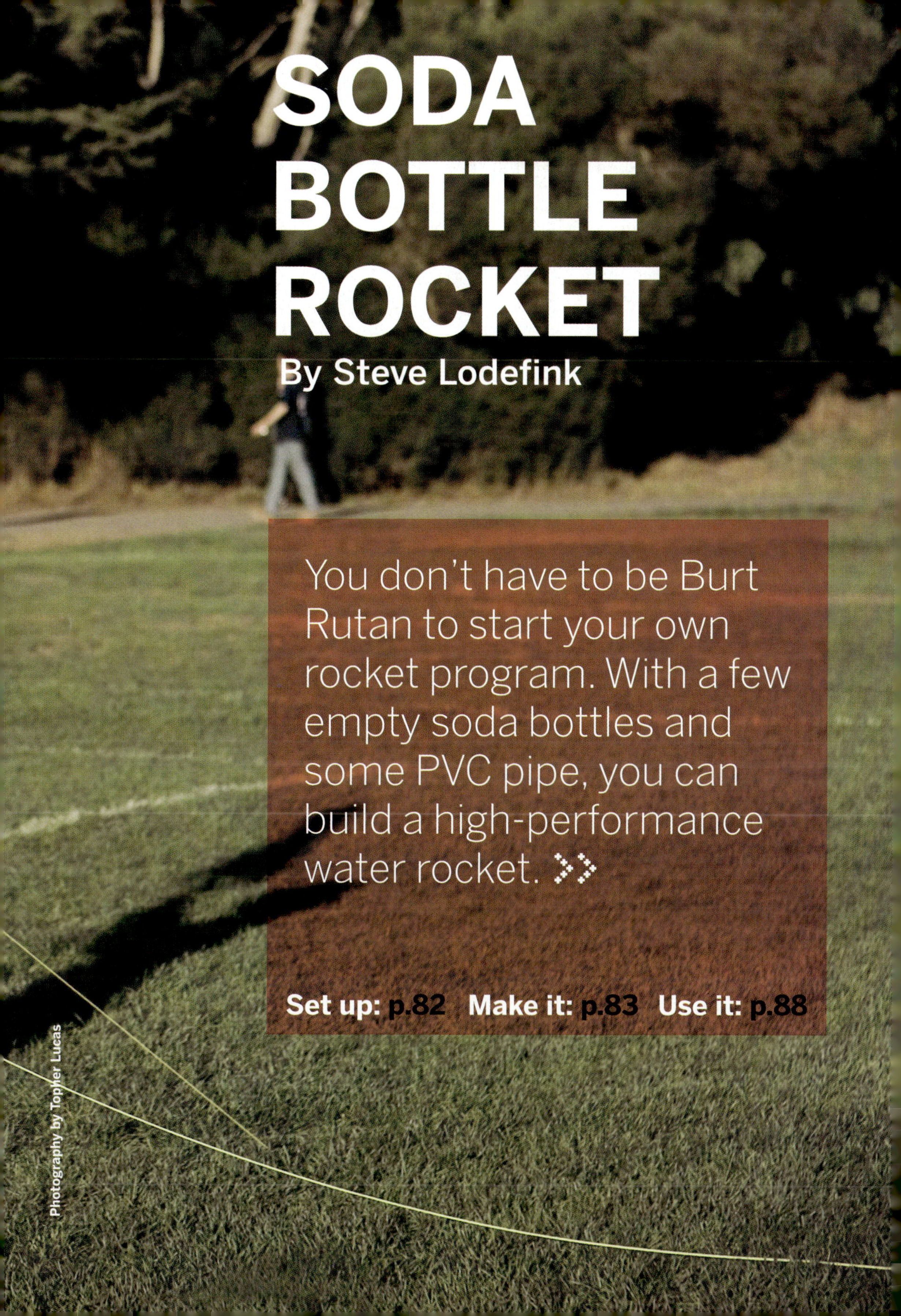

SODA BOTTLE ROCKET

By Steve Lodefink

You don't have to be Burt Rutan to start your own rocket program. With a few empty soda bottles and some PVC pipe, you can build a high-performance water rocket. >>

Set up: p.82 Make it: p.83 Use it: p.88

Photography by Topher Lucas

LIQUID FUEL ALTERNATIVE

I've been a big fan of model rocketry since I built my first Estes Alpha back in third grade. Nothing is more exciting to a 9-year-old proto-geek than launching a homemade rocket. But flying those one-shot solid-fuel rockets can burn a hole through a young hobbyist's wallet faster than they burn through the atmosphere, and with today's larger, high-powered rockets, locating and traveling to a safe and suitable launch site can require substantial planning and effort.

Instead, you can use 2-liter carbonated drink bottles to build an inexpensive, reusable water rocket. The thrill factor is surprisingly high, and you can fly them all day long for the cost of a little air and water. It's the perfect thing for those times when you just want to head down to the local soccer field and shoot off some rockets!

Steve Lodefink works as an interactive designer and web producer for The Walt Disney Internet Group in Seattle.

(WATER) ROCKET SCIENCE

The soda-bottle rocket works the same way as those little red and white plastic rocket toys you had as a kid.

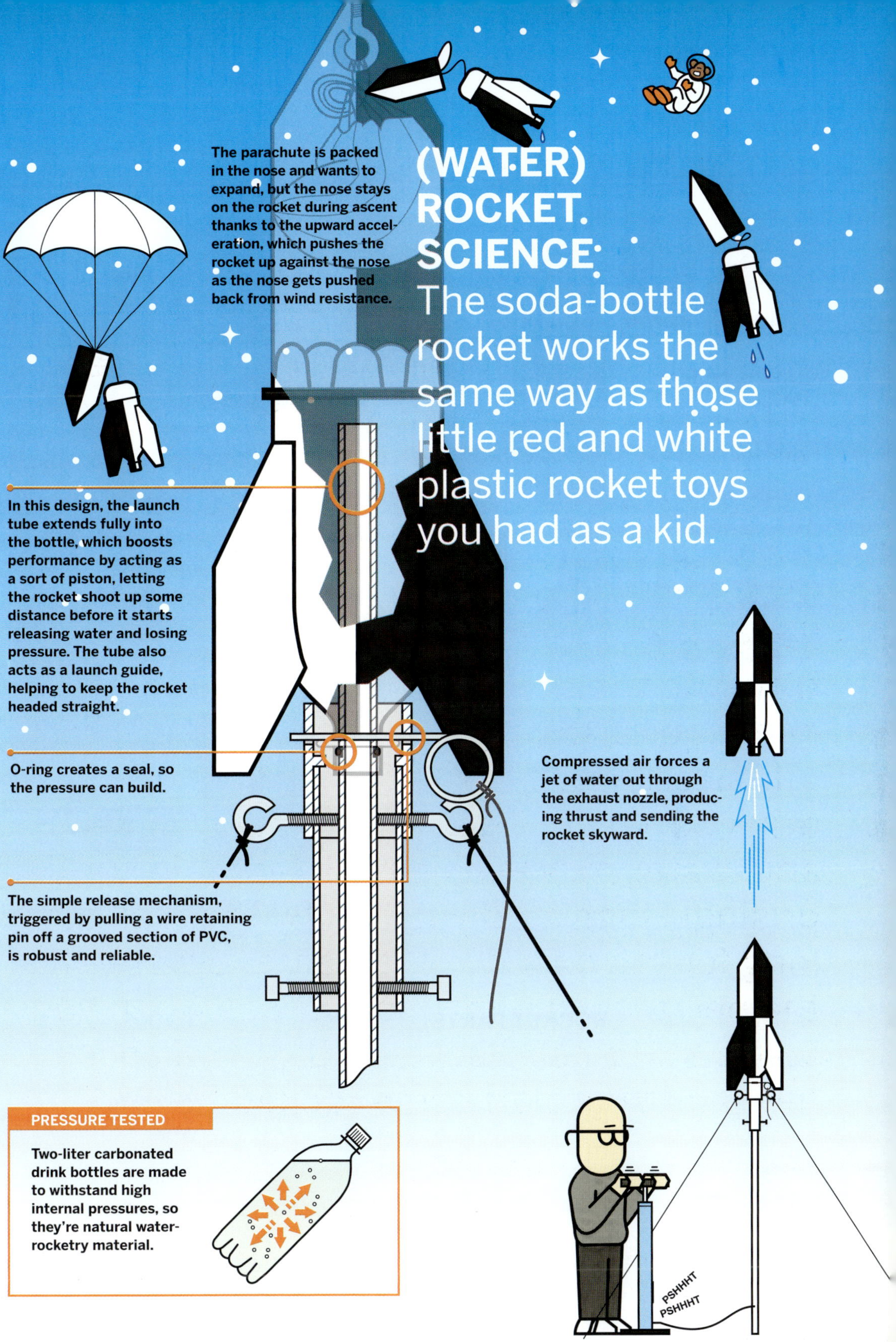

The parachute is packed in the nose and wants to expand, but the nose stays on the rocket during ascent thanks to the upward acceleration, which pushes the rocket up against the nose as the nose gets pushed back from wind resistance.

In this design, the launch tube extends fully into the bottle, which boosts performance by acting as a sort of piston, letting the rocket shoot up some distance before it starts releasing water and losing pressure. The tube also acts as a launch guide, helping to keep the rocket headed straight.

O-ring creates a seal, so the pressure can build.

The simple release mechanism, triggered by pulling a wire retaining pin off a grooved section of PVC, is robust and reliable.

Compressed air forces a jet of water out through the exhaust nozzle, producing thrust and sending the rocket skyward.

PRESSURE TESTED

Two-liter carbonated drink bottles are made to withstand high internal pressures, so they're natural water-rocketry material.

Illustration by Nik Schulz/L-Dopa.com

SET UP.

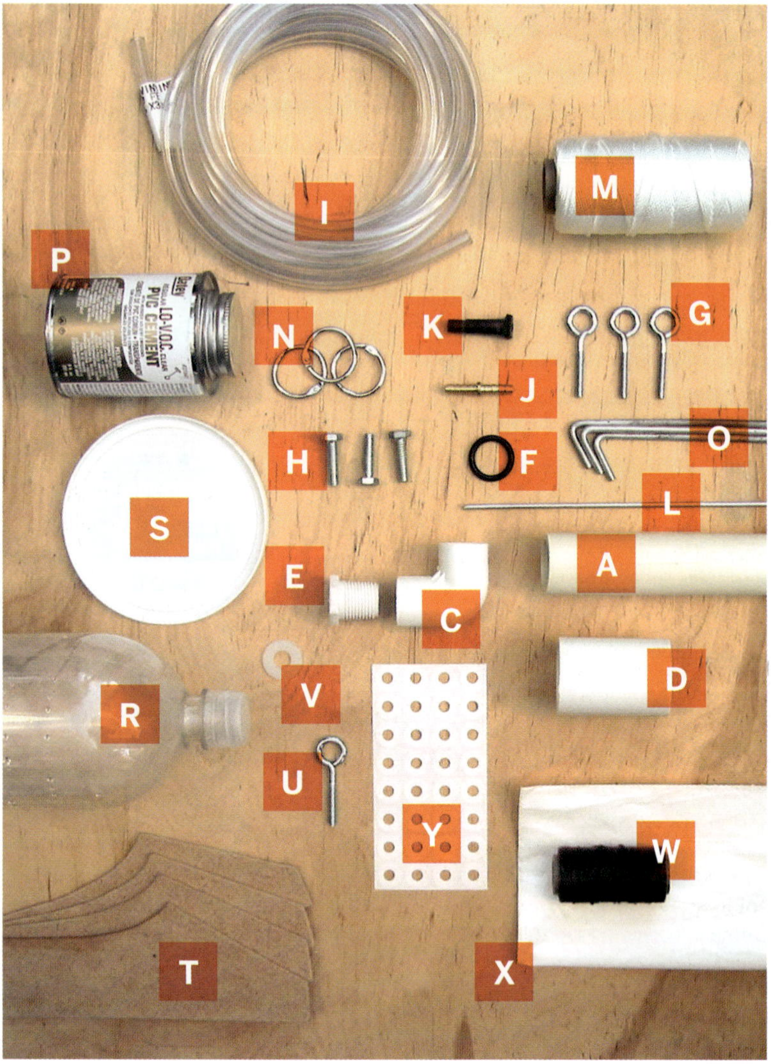

Photograph by Kirk von Rohr

LAUNCHER PARTS

[A] 4" length of 1" PVC pipe For the release body

[B] 50" length of ½" Schedule 40 PVC pipe For the launch tube (not shown)

[C] ½" PVC elbow pipe For the end cap

[D] 1" PVC pipe coupler For the release collar

[E] ½" PVC plug cap

[F] Rubber O-ring, 22mm outside diameter (OD)

[G] Eyebolts (3)

[H] Hex bolts (3)

[I] 15' length of 5/16" OD x 3/16" inside diameter (ID) flexible vinyl tubing

[J] 3/16" hose barb

[K] Tire air valve

[L] ⅛" music wire For the release spring

[M] Nylon cord

[N] Small binder rings (3) For stay clips

[O] Small tent stakes (3) For stays

[P] PVC cement

[Q] Bicycle pump with pressure gauge (not shown)

ROCKET PARTS

[R] 2-liter carbonated drink bottles (3)

[S] 4" deli cup lid

[T] Fin material, such as balsa, thin plywood, or Plastruct sheeting

[U] 2" eyebolt

[V] Medium nylon washer

[W] Kite string

[X] Large garbage bag For parachute material collar

[Y] Round hole reinforcement labels

[Z] Quick-set epoxy (not shown)

TOOLS

Hacksaw

Utility knife

⅛" file

Drill

Locking pliers

120-grit sandpaper

Thread-cutting taps and dies (optional)

MAKE IT.

BUILD YOUR SODA-BOTTLE ROCKET

START **Time: An Afternoon** **Complexity: Low**

1. BUILD THE LAUNCH TUBE

1a. Cut the launch tube. Use a hacksaw to cut the ½" PVC pipe to length. A 50" tube will make a launcher that's a convenient height for most adults to load from a standing position. The ½" Schedule 40 PVC pipe fits perfectly into the neck of a standard 2-liter soda bottle.

1b. Install the O-ring. Mark the O-ring position by fully inserting the launch tube into the type of bottle that you plan to use for your rockets. Locate the O-ring roughly in the middle of the bottle's neck. Use the edge of a file to cut a channel for the O-ring to occupy. Rotate the launch tube often while you work to maintain an even depth of cut, and be careful not to go too deep. Then slip the O-ring over the launch tube and seat it in the groove.

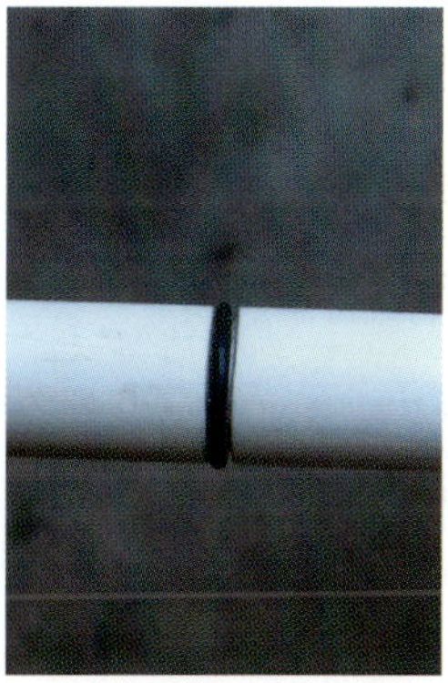

2. BUILD THE RELEASE MECHANISM

2a. Assemble the release body. Cut a 4" length of 1" PVC pipe and press-fit it into the 1" coupler. Cut squarely and deburr all PVC cuts with 120-grit sandpaper.

Photography by Steve Lodefink

2b. Cut the release spring slots. Insert your bottle's neck into the release assembly and determine the distance of the bottle's neck flange from the end of the bottle. Mark the flange location on the 1" pipe coupler and use the hacksaw to cut a 3/16" long slot on each side. These slots will hold the retainer/release spring.

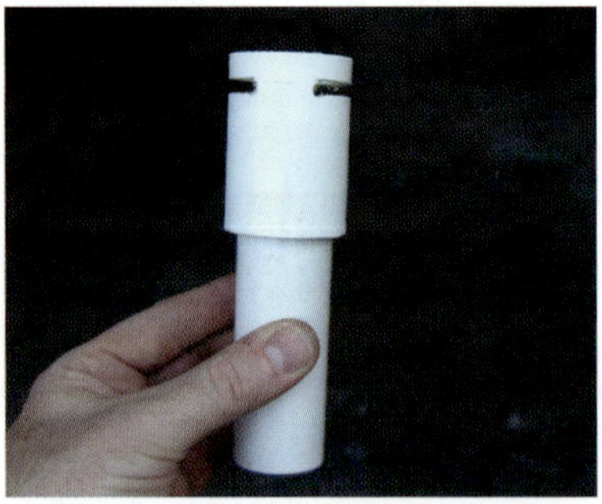

2c. Attach bolts. Drill three evenly spaced holes through the release collar and release body together, and thread the three eyebolts into these holes. Similarly, drill three holes in the lower release body tube to accept the three hex bolts.

Optional: Cut threads in these holes with a tap to accept the hex/eyebolts. If you won't be tapping them, drill the holes just undersized, and the bolts will cut through the PVC just fine. Be careful not to strip these holes.

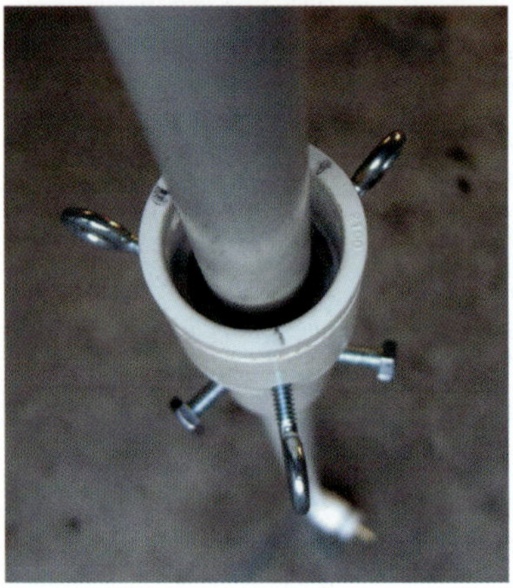

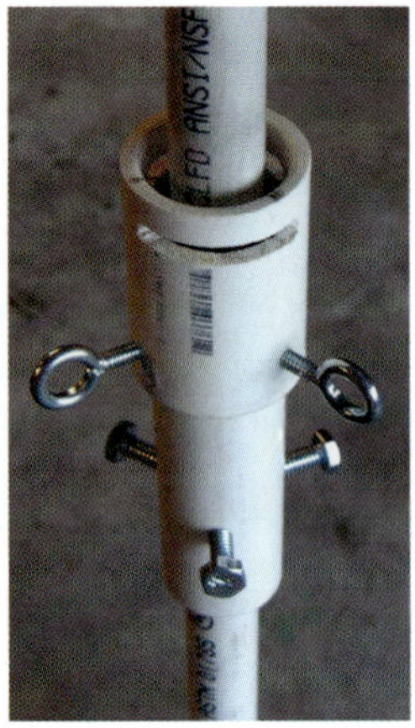

2d. Make the retainer/release spring. Bend a piece of 1/8" music wire 1½ turns around a piece of scrap ½" pipe clamped into a vise. The spring should be roughly V-shaped.

Make a retainer clip for the spring by drilling two holes in a scrap of ½" pipe. The ends of the compressed spring will fit into these holes. This keeps the spring closed above the bottle's neck flange, holding the bottle in place.

Tie a 15' trigger line to the clip. At launch time, you pull the clip off with this trigger line, which allows the spring to open and the rocket to take off.

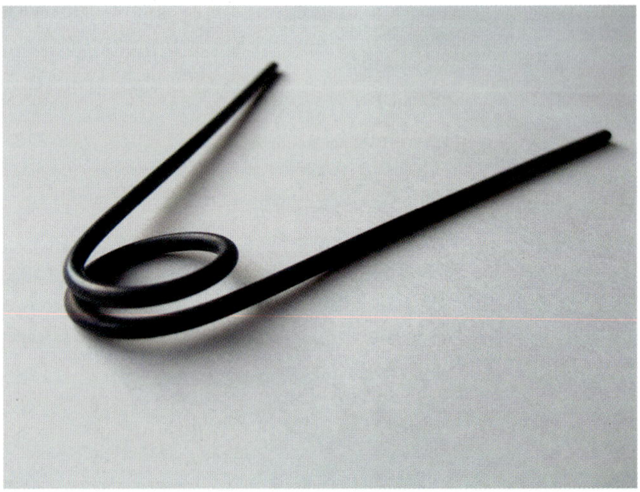

3. MAKE THE AIR HOSE

3a. Drill a 3⁄16" hole in the center of the threaded ½" end cap, and press in the 3⁄16" barb fitting.

3b. Thread the end cap into the elbow fitting and tighten it with a wrench. Using PVC cement, solvent-weld the elbow to the bottom end of the launch tube. The end cap is tapered, so it should require no Teflon tape or adhesive.

3c. Use a utility knife to strip the rubber from the tire valve to one inch from the end. Insert the valve into one end of the 3⁄16" flexible tubing.

Optional: You can use a die to cut threads into the plain end of the valve stem, and then twist it into the tube.

3d. Push the other end of the air tube onto the barb fitting.

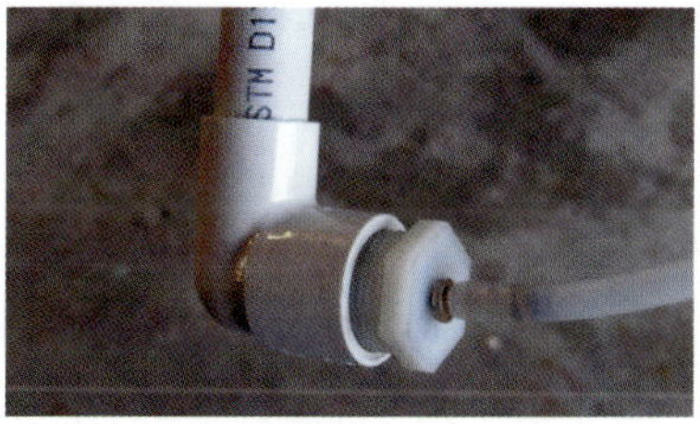

4. SET UP AND TEST THE LAUNCHER

4a. Stake down the stays. The launcher is installed in the field using three stays, each consisting of a 72" length of light nylon cord. Stake one end of each line to the ground, and clip the other end of each stay to the eyebolts on the launcher.

Photograph by Topher Lucas (step 4a)

4b. Pressure-test the launcher. Now is a good time to ensure that all the launcher's connections are airtight. Fill a bottle to the top with water (this way, if the bottle fails this pressure test, it will not explode). Quickly invert the bottle and slip it onto the launcher. A little Vaseline inside the neck will help the bottle make a seal against the O-ring. Squeeze the release spring into the slots in the release collar and clip it in place. Use the bicycle pump to pressurize the system to 70psi. If the pressure holds steady, all is well. Otherwise, fix any leaks and test again.

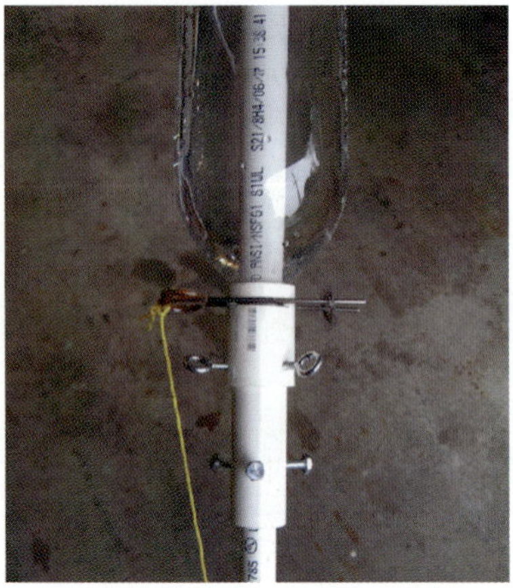

5. ASSEMBLE THE ROCKET

Water rocket designs range from a simple finned bottle to elaborate six-stage systems with rocket-deployed parachute recovery and on-board video cameras. Ours is a painted single bottle affair with wood fins and parachute recovery. Chute deployment is by the passive "nose cone falls off at apogee" method.

5a. Cut 3 or 4 fins from a light, stiff material such as balsa, thin plywood, or Plastruct sheeting. Roughen the surface of the bottle, where the fins will attach, with some sandpaper and then glue the fins to the bottle with epoxy, or a polyurethane adhesive such as PL Premium. Sand the leading edges smooth.

TIP: Gluing on the fins at a slight angle will cause the rocket to spiral as it flies, adding stability to the flight.

5b. Make the nose section by cutting off the neck and base of another bottle. Cut a 6" circle of material from a third bottle. Make a radial slit on the circle, fashion it into a nose cone, and cement it in place atop the nose section.

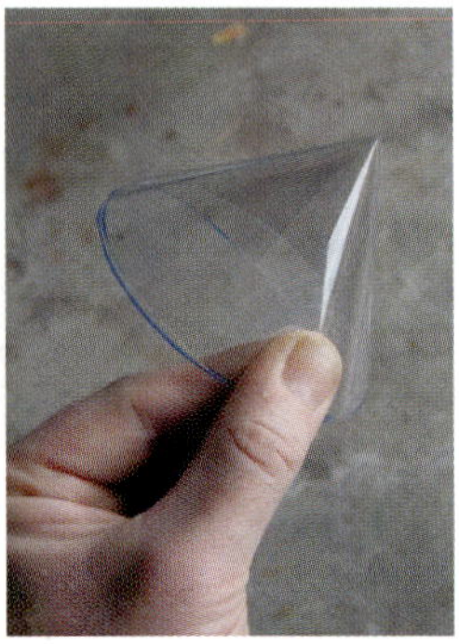

5c. Outfit the nose cone. When the cement is dry, turn the nose over and epoxy the 2" eyebolt to the inside tip of the nose cone. This bolt serves as a place to anchor the parachute shock cord. It also adds extra mass to the nose section, which will help to pull this section off as the rocket decelerates, exposing the parachute.

5d. Make the nose-stop. Cut the center from a 4" deli container lid, leaving only the outer rim. Cement the rim onto the rocket's lower "motor" section such that it allows the nose to sit loosely and straight on the rocket. This "nose-stop" will prevent the nose from being jammed on too tightly by the force of the launch, which ensures that the nose will separate off and deploy the parachute during descent.

5e. Make a parachute canopy from a 36" or so circle cut from a large trash can liner. For best results, use 12 or more shrouds made from kite string. Apply paper reinforcement labels to both sides of the chute, where the shrouds attach, to keep the chute from tearing. Tie the loose ends of the shrouds to a nylon washer or ring to make the chute easy to manage.

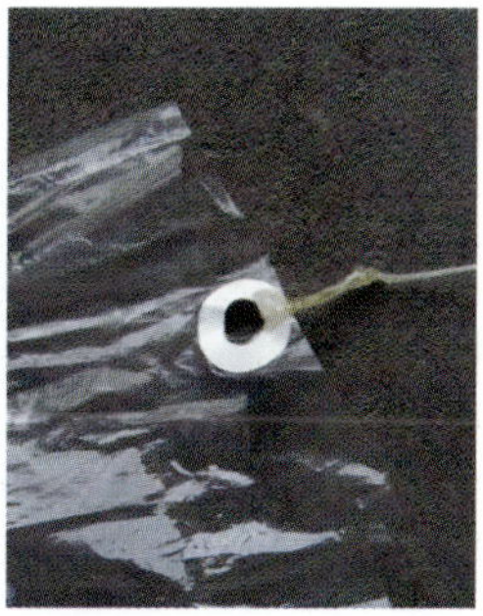

TIP: Ideally, a parachute's shrouds should be a bit longer than the diameter of the chute canopy.

5f. Epoxy a parachute-anchoring ring to the top of the rocket base and tie the parachute to the ring with a short cord. Cut a 4' connecting cord and tie it between the nose cone eyebolt and the parachute-anchoring ring. This cord will keep both halves of the rocket together during descent.

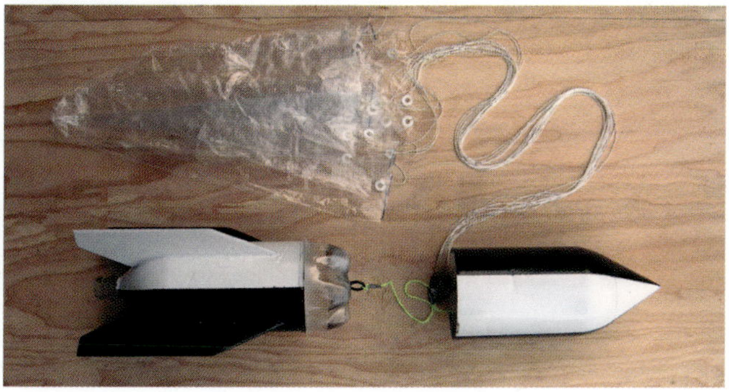

TIP: Make sure the connecting cord is long enough to allow the parachute to completely pull out from the nose cone.

USE IT.

THREE, TWO, ONE, LIFTOFF!

SAFETY

Water rockets produce a considerable amount of thrust, and getting in the way of one could cause severe injury. Take the same common-sense precautions that you would when launching any type of rocket. Make sure that everyone in the area is clear of the rocket and aware that it is about to launch. Do a verbal countdown, or yell something alarming, such as "Fire in the hole!" just before you launch the rocket.

SELECTING A SITE

A well-built, single-stage water rocket is capable of flying several hundred feet into the air and drifting a considerable distance during descent. Less well-built rockets may choose to travel several hundred feet to the side. In any case, you need to choose a launch site that is large and open enough to allow your rocket to wander a bit without getting lost in a tree, or on the roof of some Rottweiler's doghouse.

Big sports fields are the logical site choice for most of us, but if you are in a rural area, any wide-open space will work as a rocketry range. Be sure to take wind direction into consideration when deciding on which side of the field to set up the launcher.

OPERATION

1. Set up the launcher by clipping the three support stays to the launcher's eyebolts. Take up any slack in the lines and stake the other ends to the ground, evenly spaced. Uncoil the air hose and tuck it under one of the tent stakes to keep it from coiling. Attach the bicycle pump to the air hose.

2. Pack the parachute. Grab the center of the parachute canopy between your thumb and forefinger, and let it hang. Draw the chute through your closed hand to gather it, and then fold it into thirds, zigzag style. Lay down the parachute shrouds on the ground, and accordion-fold them back on themselves. Don't wrap the shrouds around the canopy; just slide the whole thing into the nose section and bring the two halves of the rocket together.

TIP: Line the nose section with parchment paper or Teflon baking sheet liner to help the parachute deploy smoothly. A light dusting of talcum powder will also help keep the chute from sticking.

3. Fill and set up the rocket. While holding the nose in place, turn the rocket over and fill it one-third full of water. Apply petroleum jelly to both the launch tube O-ring and the inside of the bottle's mouth to help it slip onto the O-ring. Hold the mouth of the rocket up to the launch tube. In one smooth motion, pivot the rocket up, slide it down

onto the launch tube, and twist it back and forth, if necessary, to help it engage the O-ring seal.

4. Compress the release spring into the slots of the release collar, locking the bottle flange in place. Install the spring-retaining clip on the ends of the spring, and carefully run the trigger line back to your "ground control" area.

TIP: Tie the release spring to the launcher with some string to keep it from flying across the field and getting lost every time you launch.

5. Launch! Jump over to the bicycle pump and bring the pressure up to about 70psi. When you are ready, clear the area, count down to zero, and pull the trigger line, releasing the spring and freeing the rocket.

If all goes as planned, your rocket will shoot upward, dispensing with its entire fuel load in less than half a second. Then it will begin to decelerate, and the nose will want to separate. As the rocket reaches apogee, the two halves will come apart, deploying the recovery chute and bringing the craft gently back to Earth, much to the excitement of the assembled crowd.

Experiment with different amounts of water and air pressure until you find the sweet spot that sends your rocket the highest. Don't exceed the amount of air pressure that your bottle is designed to withstand; 70psi seems to be about right for a standard 2-liter soda bottle.

Rocket Boy

Watch a video clip of author Steve Lodefink and his 4-year-old son Ivan launching their soda-bottle rocket at makezine.com/05/rocket.

The author kindly (and we'd say somewhat naively) allowed the MAKE team to borrow his rocket. After making the rounds through numerous photo shoots, kids' birthday parties, and a high-decibel, code-violating Thanksgiving affair, the rocket sustained the damage you see here. We deeply regret the mishap and promise not to do it to the next rocket we've asked Steve to build for us.

ADVANCED DEVELOPMENT

Once you've tasted the joys of basic water rocketry, you will inevitably want to improve and refine your rocket designs. If you want your rockets to fly higher, the best improvement you can make is to increase the volume of the rocket "motor." This is usually done by splicing or otherwise coupling two or more bottles into a single pressure chamber. There are also various schemes for building multistage rockets, as well as more elaborate parachute deployment setups.

There is an abundance of water rocket information available online. Here are a few sources to get you started:

Antigravity Research Corporation – ready-made water rocket components: antigravityresearch.com

Water rocket links:ourworld.compuserve.com/homepages/pagrosse/h2orocketlinksi.htm

The Martinet Launcher, the basis for this project's launcher design: martinet.nl/wp-site/water-rockets

Photography by Topher Lucas

WIND POWERED GENERATOR

By Abe and Josie Connally

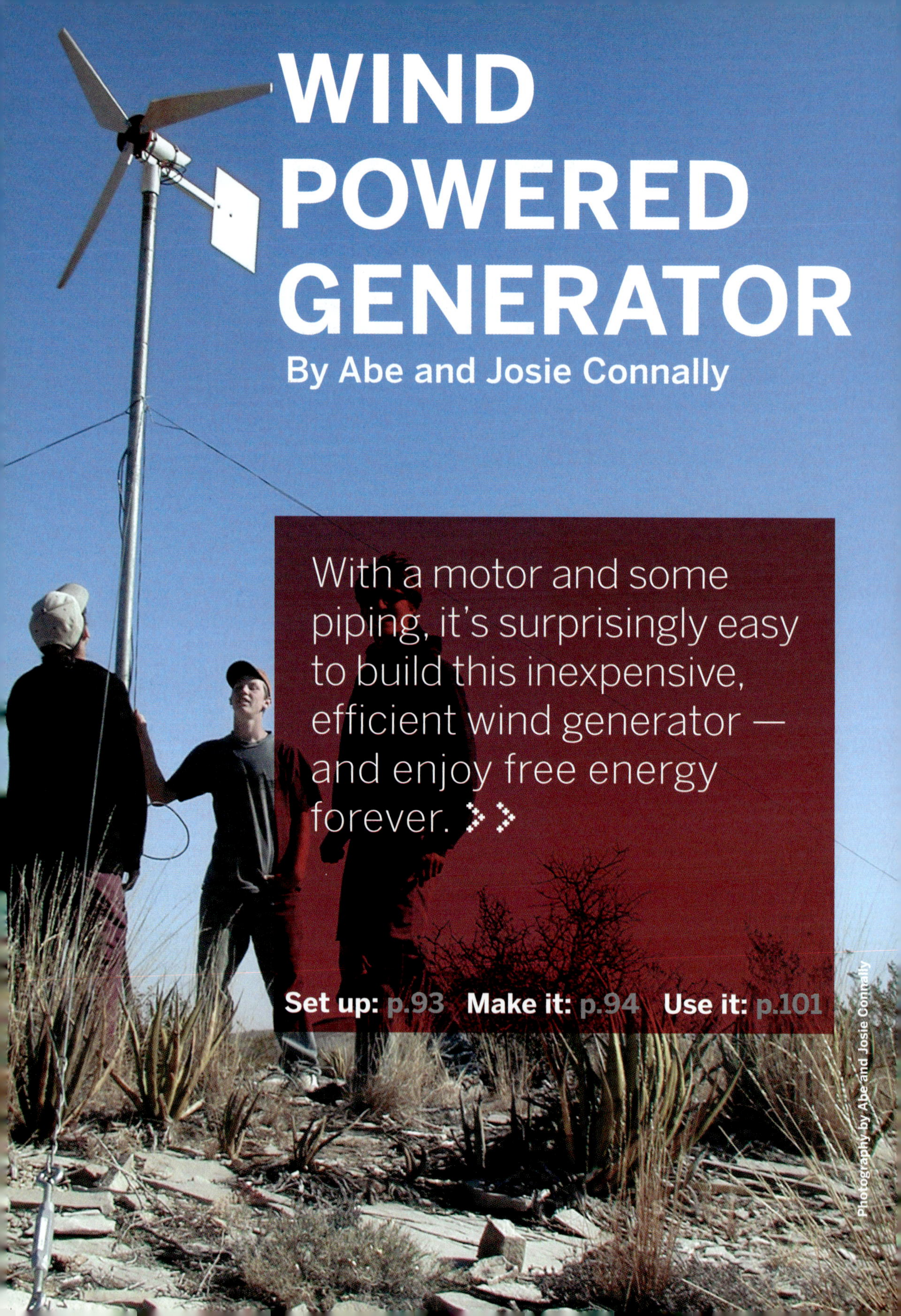

With a motor and some piping, it's surprisingly easy to build this inexpensive, efficient wind generator — and enjoy free energy forever. >>

Set up: p.93 **Make it:** p.94 **Use it:** p.101

Photography by Abe and Josie Connally

CURRENT FROM CURRENTS

There are no limits to what you can do with wind power. It's abundant, clean, cheap, and easy to harness. We designed this Chispito Wind Generator (that's Spanish for "little spark") for fast and easy construction. Most of the tools and materials you need to build it can be found in your local hardware shop or junk pile. We recommend that you search your local dump or junkyards for the pieces required. Or, if you live in a city, search freecycle.com for salvaged parts, and see if you can install one on your roof.

We believe that anyone can be in control of where his or her electricity comes from, and there is nothing more rewarding and empowering than making a wind-powered generator from scrap materials. Remember: puro yonke (pure junk) is best!

Abe and Josie Connally are off-grid adventurists based in the remote Big Bend region of Texas, where they experiment and live with sustainable technologies built from puro yonke (*velacreations.com*).

WIND GENERATOR BLOW-BY-BLOW

The Chispito Wind Generator is a simple little machine that's great for getting started with wind power. In a 30 mph wind, ours gives us about 84 watts, 7 amps at 12 volts.

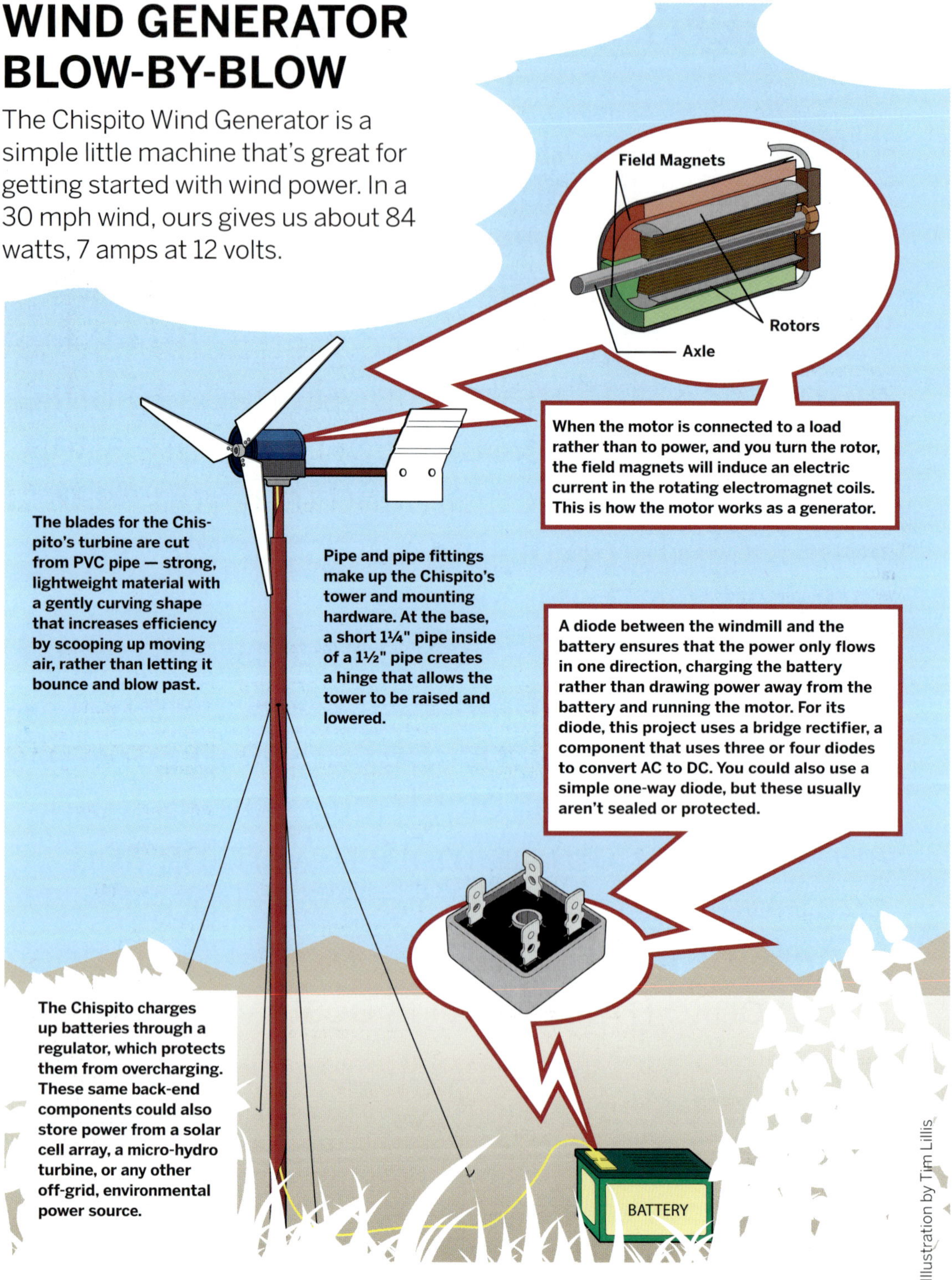

When the motor is connected to a load rather than to power, and you turn the rotor, the field magnets will induce an electric current in the rotating electromagnet coils. This is how the motor works as a generator.

The blades for the Chispito's turbine are cut from PVC pipe — strong, lightweight material with a gently curving shape that increases efficiency by scooping up moving air, rather than letting it bounce and blow past.

Pipe and pipe fittings make up the Chispito's tower and mounting hardware. At the base, a short 1¼" pipe inside of a 1½" pipe creates a hinge that allows the tower to be raised and lowered.

A diode between the windmill and the battery ensures that the power only flows in one direction, charging the battery rather than drawing power away from the battery and running the motor. For its diode, this project uses a bridge rectifier, a component that uses three or four diodes to convert AC to DC. You could also use a simple one-way diode, but these usually aren't sealed or protected.

The Chispito charges up batteries through a regulator, which protects them from overcharging. These same back-end components could also store power from a solar cell array, a micro-hydro turbine, or any other off-grid, environmental power source.

Illustration by Tim Lillis

SET UP.

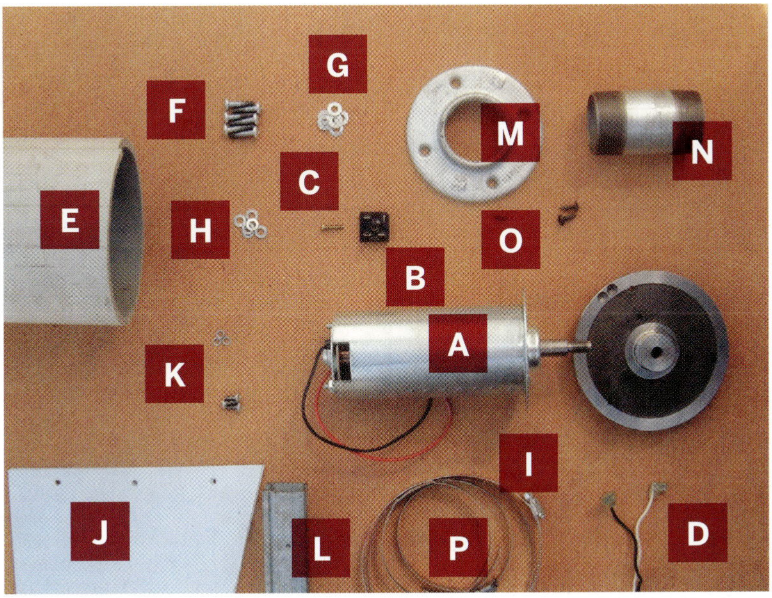

MATERIALS

MOTOR AND ELECTRIC
[A] 260 volts DC, 5 amps, treadmill motor with a 6" threaded flywheel You may use any other simple, permanent-magnet DC motor that returns at least 1V for every 25 rpm and can handle upwards of 10 amps. Our motor is rated at 5A, no load, and we've found that the coils can withstand 15A going through them without heating up.

If you use another motor, change this supply list to match. For example, if the motor lacks a flywheel, you will have to find a hub for it. A circular saw blade with a ⅝" shaft adaptor will work.

[B] 30-50A bridge rectifier with center hole mount surpluscenter.com, item #22-1180

[C] Mounting screw (1)

#8 (or larger) copper wire, red and black, both stranded Enough length for both a red and black piece to run from the top of the tower, down through length of pole, to batteries. We recommend at least #8 wire, but if your tower will be sited a long distance away from your batteries, you may need a heavier gauge.

[D] Spade connectors for wires (4) For bridge rectifier

Heat-shrink tubing or electrical tape

Battery bank We recommend deep-cycle lead-acid storage batteries, and a total battery bank capacity of at least 200 amp-hours.

Ammeter

Regulator or charge controller

Fuse

BLADES
[E] 2' length of 8" Schedule 80 PVC pipe If PVC is UV resistant, you will not need to paint it.

[F] ¼" #20 bolts, ¾" long (6)

[G] #20 washers (9)

[H] Lock washers (6)

[I] Hose clamp (1)

VANE
[J] 1 sq. ft. (approx.) of sheet metal

[K] Mounting screws and lock washers (approx. 9)

MOUNT AT TOP OF TOWER
[L] 36" of 1" square metal tubing or 1" angle iron

[M] 2" floor flange pipe fitting

[N] 2" steel pipe nipple, at least 4" long

[O] Mounting screws (2)

MOTOR MOUNT
[P] #72 hose clamps (2)

TOWER POLE
10'-30' length of 1½" steel pipe, threaded at both ends

TOWER BASE
2'x1¼" steel pipe nipple (2)

6"x1¼" steel pipe nipple

1¼" 90-degree steel pipe elbows (2)

1½" steel pipe T

10 lb. bags of quick-mix concrete (2-3)

¾" #10 sheet metal screws (4)

TOWER STABILITY
Guy wire, galvanized steel With a working load of 200 pounds

1½" U-bolt

Stakes (4)

Turnbuckles (4)

TOOLS

Drill and drill bits (5/32", 7/32", ¼"), jigsaw, thread-tapping set, pipe wrench, crescent wrench, flathead screwdriver, vise and/or clamp, wire strippers, metal punch or awl, tape measure, level, marker, tape, compass and protractor, shovel, wheelbarrow, several ropes (each at least twice the length of the guy wires), and an extra person or two to help.

MAKE IT.

BUILD YOUR WIND-POWERED GENERATOR

START >> **Time:** A Couple Weekends **Complexity:** Medium

1. CUT THE BLADES

Let's begin by cutting.

1a. Place the 24" length of PVC pipe and square tubing (or other straight edge) side by side on a flat surface. Push the pipe tight against the tubing and mark the line along the length of the tube. This is Line A (see Fig. 1).

1b. Starting from Line A, draw parallel lines at 75-degree intervals along the length of the pipe. You should have a total of five lines on your pipe as shown in left Figure 1. Note that one strip will have an arc width of only 60 degrees. That's OK.

1c. Use a jigsaw to cut along the lines, splitting the tube into five strips. Four will be wider than the fifth (60°) strip. Set the 60° strip aside for now.

1d. Place the four 75° strips concave-side-down. For each one, make a mark 20% of the width of the strip from one corner along the diagonally opposite side as shown (see Fig. 2).

1e. Mark a diagonal line between the two marks you just made on each piece, and use the jigsaw to cut along these lines (see Fig. 3). You should wind up with eight identically shaped trapezoidal blades. You can trim a ninth blade out of the 60° strip left over. You now have enough blades for three generators, or plenty of spares for one generator.

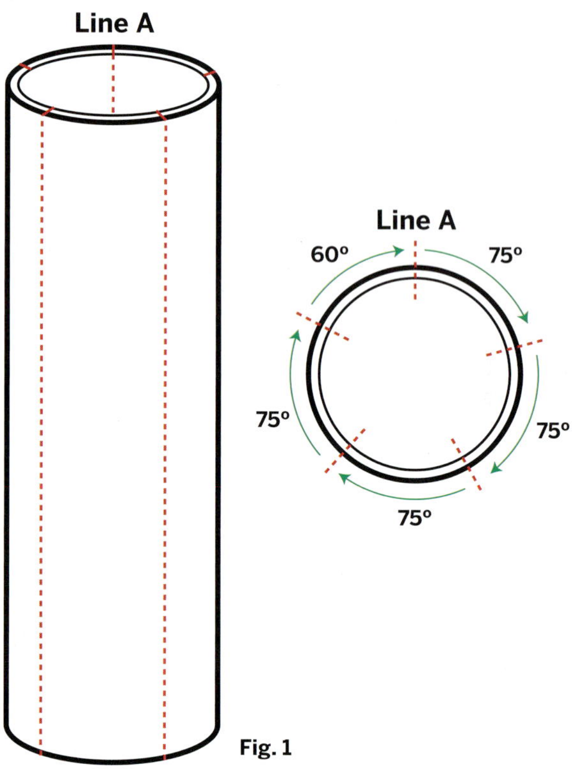

Fig. 1

1f. Now you are going to cut one corner from each blade. First, measure the width of the blade (if you are using an 8" diameter PVC pipe as your stock, it should be about 5.75" wide). Call this value W. Then make a mark along the diagonal edge of the blade, a distance of W/2 from the wide end (3" is good enough if you are using 8" PVC). Make another mark on the wide end of the blade at 15% of W from the long straight edge (1" with 8" PVC) (see Fig. 4).

1g. Connect these two marks and cut along the line. Removing this corner prevents the blades from interfering with each other's wind.

1h. The blades should look like the ones shown in Fig. 5. Pick the three best ones of the batch and let's move to the next step, making the tail.

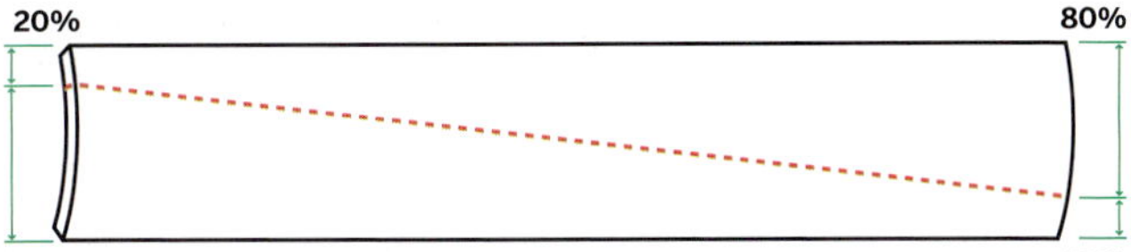

Fig. 2

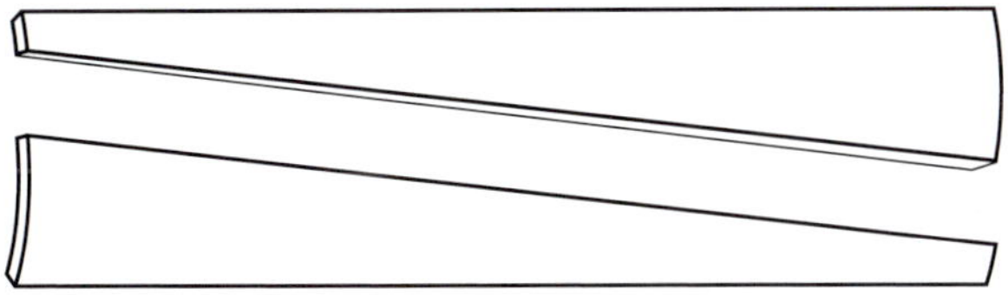
Fig. 3

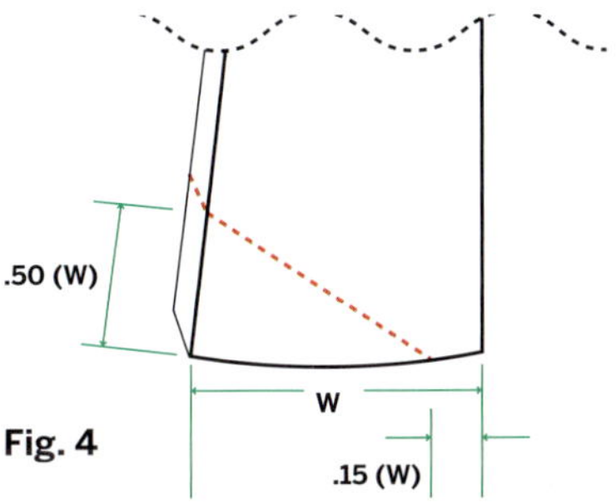

Fig. 4

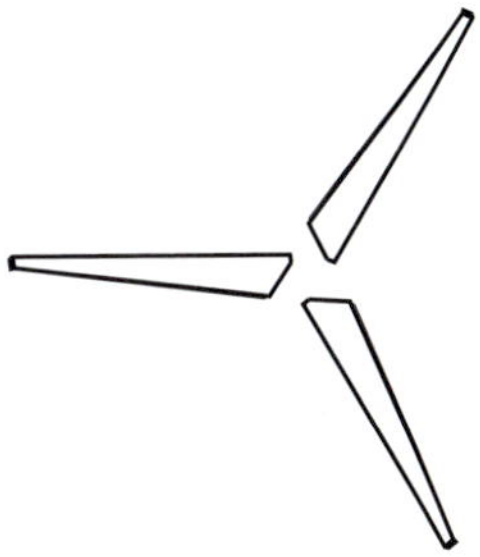
Fig. 5

2. MAKE THE TAIL

2a. Cut the tail. You can make it any shape you want, as long as the end result is stiff rather than floppy. The exact dimensions of the tail are not important, but you'll want to use about one square foot of lightweight material, preferably metal.

2b. Using the 5/32" drill bit, drill two or three holes, spaced evenly, in the front end of the tail. Then place the tail on one end of the square tubing, noting that it will attach to what will become either the right or left side of the tubing, as the generator sits upright. Mark the tubing through the tail holes.

2c. Drill holes in the square tubing at the marks you just made.

2d. Attach the tail to the tube with sheet metal screws. (Or you can do this later, so it doesn't get in the way.)

3. ATTACH THE BLADES

3a. Take three blades. For each blade, mark two holes along the long, right-angle side of the blade (as opposed to the long diagonal side), at the wide end, next to the cut-off corner. The first hole should be ⅜" from the long side and ½" from the end, and the second hole should be ⅜" from the straight edge and 1¼" from the end.

3b. Using the ¼" drill bit, drill these six holes for the three blades.

3c. Detach the hub from the motor shaft. With our motor, we removed the hub by holding the end of the shaft firmly with pliers and turning the hub clockwise. This hub unscrews clockwise, counter to the usual direction, which is why the blades turn counterclockwise.

3d. Using a compass and protractor, make a template of the hub on a piece of paper. Then mark three holes, each of which is 2⅜" from the center of the circle, 120 degrees apart, equidistant from each other.

3e. Place this template over the hub and use a metal punch or awl to punch a starter hole through the paper and onto the hub at each hole.

3f. Drill the holes with the 7⁄32" drill bit, then tap them with the ¼" tap.

3g. Attach the blades to the hub using ¼" bolts, running them through the holes closest to the ends of the blades. At this point, the three outer holes on the hub have not been drilled.

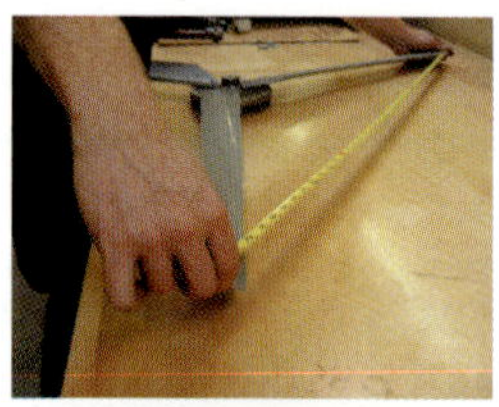

3h. Measure the distances between the tips of each blade, and adjust them so that they are all equidistant. Then mark and punch starter holes for the three outer holes on the hub through the empty holes in each blade.

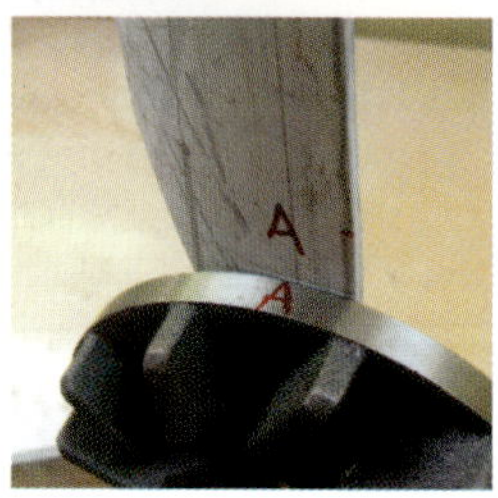

3i. Label the blades and hub so that you can match which blade goes where.

3j. Remove the blades, and drill and tap the three outer holes on the hub.

3k. Position each blade in its place on the hub, so that all the holes line up. Using the ¼" bolts and washers, bolt the blades back onto the hub. For the inner three holes, use two washers per bolt, one on each side of the blade. For the outer holes, just use one washer next to the head of the bolt. Tighten.

4. ASSEMBLE THE GENERATOR

4a. Drill a 5/32" hole in the tubing, about 5 inches from the front end of the tube, opposite the tail holes end, on any side. Place the bridge rectifier over the hole, and screw it to the tubing using a #10 sheet metal screw.

4b. Using hose clamps, mount the motor on the end opposite the tail. Do not tighten the clamps, because you will make a balance adjustment later.

4c. Crimp spade connectors onto the black and red wires from the motor, and connect them to the two AC voltages in terminals on the bridge rectifier, L1 and L2. Insulate connections with heat-shrink tubing or electrical tape.

4d. If you haven't already, attach the tail.

4e. Re-attach the blade assembly on the motor.

4f. Now we'll attach the tower mount. Using a pipe wrench, screw the nipple tightly into the floor flange. Clamp the nipple in a vice so the floor flange faces up and is level.

4g. Set the generator on the flange/nipple and balance it by adjusting the position of the motor, then tighten the hose clamps down. Mark spots in the square tubing that match up with the flange holes.

4h. Drill these two holes using a 5/32" drill bit. (You will probably have to take off the hub and tail to do this.)

4i. Attach the square tubing to the floor flange with two sheet metal screws.

MAKE THE TOWER

5. PLANT THE TOWER BASE

The tower is one of the most important components in your wind generator system. It must be strong, stable, easily raised and lowered, and well anchored.

5a. Dig a round hole about 1 foot in diameter and 2 feet deep.

5b. Feed the 6"x1¼" steel pipe nipple through the horizontal part of the 1½" steel pipe T.

5c. Screw the pipe elbows onto each end of the nipple, one on either side of the T, so that they both point in the same direction.

5d. Screw the two 2'x1¼" pipe nipples into the free ends of the elbows.

5e. Set this hinged base assembly in the hole, so that the T just clears the ground. Dig around, adjust, and position things so that the 2' nipples point straight down and the horizontal part of the T is perfectly level.

5f. With the base properly positioned, mix some concrete and pour it into the hole.

6. ERECT AND STAY THE TOWER

The higher your tower is, the more wind your generator will catch, and the more power it will produce.

6a. Drill a large hole about 1 foot from the bottom of the 10'-30' pipe, for the copper wires to exit.

6b. Screw the pipe into the vertical part of the base's hinged T.

6c. Make four strong, flexible rings out of guy wire, about 5 inches in diameter. For each ring, loop the wire around several turns, and twist it closed.

6d. Place the 1½" U-bolt around the pipe, 3 feet from the top of the pipe. Thread the four wire loops around the U-bolt, and space them evenly around the pipe. Then tighten the nuts of the U-bolt.

6e. Secure a guy wire to each of the loops on the U-bolt. Also loop the ropes (safety ropes) through loops on opposite sides of the pole.

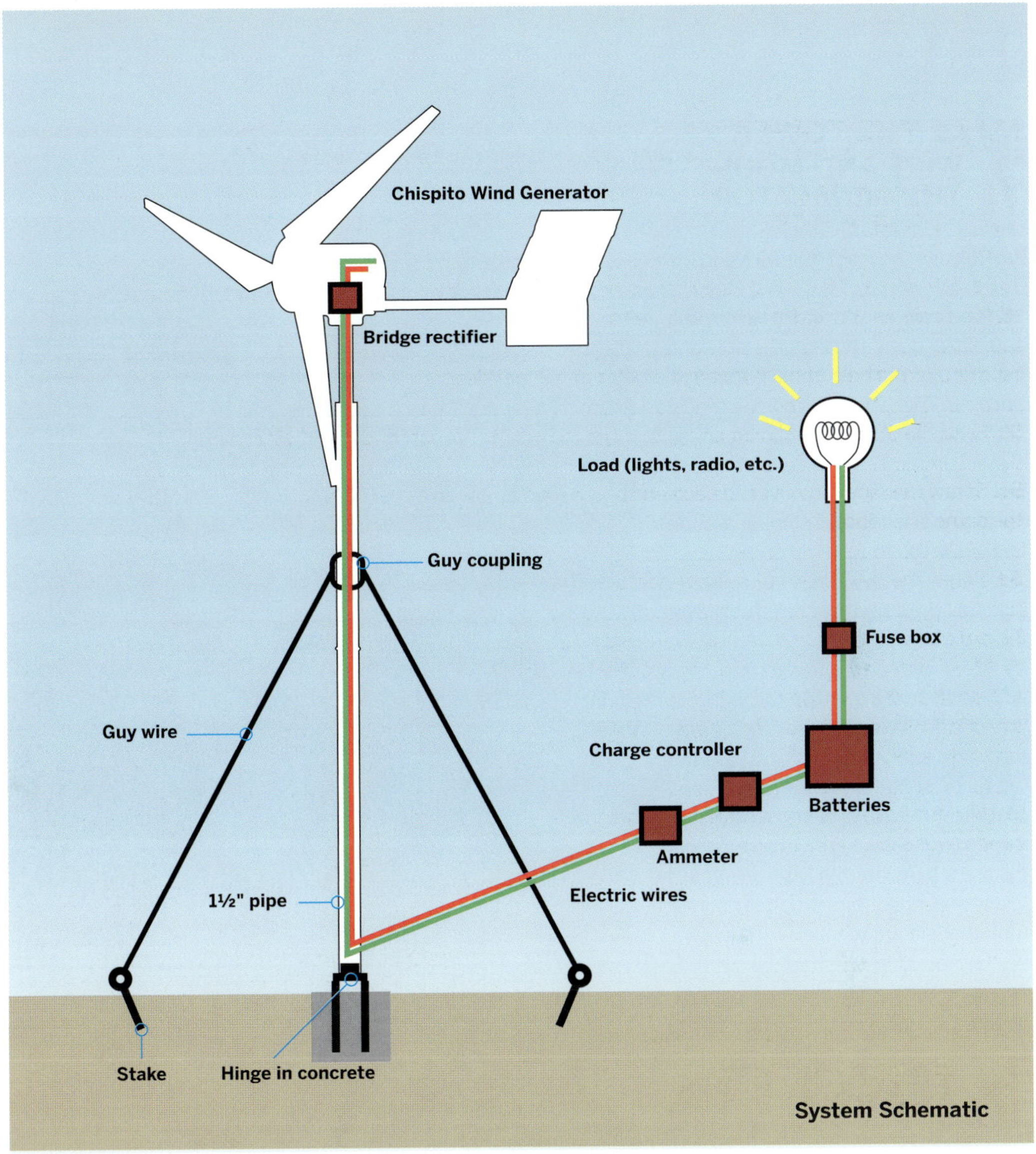

6f. Position the four stakes, spacing them evenly apart at a distance away from the base that's at least 50% of the tower's height. For our 15-foot-tall pole, we positioned the stakes 12 feet away from the base. Then drive the stakes firmly into the ground, slightly angling them away from the base. Or, for greater strength and permanence, dig holes 2 feet into the ground, and set the stakes in concrete.

6g. Wire a turnbuckle to each stake, using several strands of guy wire.

6h. Raise the pole up and tie each of the safety ropes to something solid, like a truck or a building (this is where having another person or two really helps). Attach the guy wires to the turnbuckles.

6i. Hold the pole straight upright, and tighten all turnbuckles to ensure a secure fit.

6j. Mark the front turnbuckle for future reference, so you know how far you need to screw it back in when you're re-raising the pole.

7. WIRE AND MOUNT THE GENERATOR

7a. Release the front guy wire and lower the pole to the ground.

7b. Feed two lengths of #8 wire, red and black, down through the pole and out through the hole in the bottom of the pipe. Then wrap the bottom ends of the two wires together, to create a closed circuit. This is a safety precaution; it puts a load on the wind generator to prevent it from spinning around fast while you're working on it.

7c. Slide the generator assembly over the top of the pole.

7d. Pull the pole wires up through the mount, strip the ends, and crimp them into spade connectors. Plug the red wire into the DC+ terminal of the spade connector (which will probably be perpendicular to the others), and the black wire into the DC– terminal. Insulate connections with heat-shrink tubing or electrical tape.

7e. Raise the pole by pulling the front guy wire into place, and tighten the turnbuckle to the mark made earlier.

7f. Unwrap the ends of the wires and wire up your system as shown in the schematic (previous page). Connect a regulator, an ammeter, a fuse, and a stop switch on the positive line coming from the generator, between the generator and the battery bank. Refer to the manufacturer's instructions. Then hook up the battery bank, and watch it fill up with free power!

Teach a Man to Make

Give a man a fish, or teach a man to fish? Abe and Josie Connally prefer to teach people how to make their wind generators, rather than providing the finished product. They take their wind generators to some of the poorest rural communities in northern Mexico, where rough terrain and long distances make grid power not a viable option. They also hold workshops in their construction for entire villages and have also set up solar water pumps in a few communities.

"Basically, we focus on getting communities to help themselves," Abe says. "So far, we've been met with outstanding enthusiasm."

For more information about their work, and to get complete how-tos of their projects, visit velacreations.com.

"The simplicity and cost are the driving factors. Self-repair and low-maintenance options are essential. Our goal is simple: help those who want to know how to do it."

USE IT.

FREE WIND POWER IS YOURS

SAFETY

Safety should be your highest priority. Human life is more important than electricity, so please follow any and every safety guideline you come across. Wind generators can be dangerous, with fast-moving parts and high voltage, especially in violent weather conditions. Some things to consider:

» Ground and fuse your electrical system, as well as each component within it.

» Always stand upwind when inspecting the wind generator, to avoid debris in case of failure.

» Always attach safety ropes and/or cables when erecting your tower and/or wind generator.

» Always wire connections securely, with proper insulation such as heat-shrink tubing and/or electrical tape.

» Never touch the positive and negative wires at the same time while they are connected to the battery.

» Never leave your wind generator unconnected to anything, unless it is on the ground. Always connect it to a battery or some other load, or else short it out by crossing the positive and negative wires that lead from the generator, to create a closed circuit. Without one of these precautions, the generator can spin freely and attain dangerous speeds.

» Never expose batteries to heat, sparks, or flames, and do not smoke near them. Batteries can ignite and explode easily, resulting in injury.

SYSTEM CARE

To give your wind generator a longer life span, you should paint the blades, and tape the motor with something like aluminum tape.

Batteries are usually the most expensive component in your system, and it pays to take good care of them. The Chispito Wind Generator is rated at 12V, so your system should be configured to match. We recommend deep-cycle, lead-acid storage batteries, and a total bank capacity of at least 200Ah. Check your batteries' water levels regularly to ensure the longest possible life span. Don't use car or marine-type batteries; they will wear out fast and can be dangerous if used in a battery bank.

For proper system health, you should include a regulator or charge controller, and an ammeter. An ammeter lets you view your wind generator's current-charging rate, and the regulator prevents your batteries from overcharging.

SPECIFIC USES

We use the Chispito Wind Generator, along with a 100-watt solar panel, for all of our electricity needs in our home. This includes a laptop, satellite receiver, TV, VCR, DVD player, printer, stereo, power tools, and lights. Our system consists of a 450Ah battery bank, a 1,000-watt inverter, and the proper fuses and breakers for a home supply.

The wind generator has a wide range of applications depending on your particular needs, from powering your entire house, to running your computer and printer, to recharging the battery packs in your RV. The really nice thing about our model is that you can literally build ten of them for the price of one pre-built wind generator sold in stores.

RESOURCES

Wind generator information and inspiration: otherpower.com

U.S. Wind Energy Resource Atlas wind maps, for assessing the suitability of a region for wind energy: rredc.nrel.gov/wind/pubs/atlas

Edwin Lenz's wind generator made from an old microwave: windstuffnow.com/main/microwave_wind_generator.htm

THE JAM JAR JET

By William Gurstelle

Don't think you can build a jet engine at home? Here's a simple jet engine — a pulsejet — that you can make out of a jam jar in an afternoon. All it takes is bending some wire and punching a few holes. >>

Set up: p.105 **Make it:** p.106 **Use it:** p.109

Photography by William Gurstelle

JOIN THE JET SET

Turbojets and fanjets contain hundreds of rotating parts. But the ancestors of these designs, called pulsejets, convert fuel and air into propulsive force by using a fixed geometry of chambers and ducts, with no moving parts. The simplest pulsejet is the Reynst combustor, which uses one opening for both air intake and exhaust.

The pioneering Swiss jet engineer Francois Reynst discovered this combustor as a pyromaniac child. He perforated the lid of a glass jar, put a small amount of alcohol inside, and lit the top. Flames shot out of the hole and then were sucked back into the bottle before being ejected again. This almost-magical process repeated until all of the fuel was expended. Reynst had discovered a jar that literally breathed fire, like St. George's dragon. Our jam jar jet is based on Reynst's discovery.

William Gurstelle enjoys making interesting things that go whoosh then splat. He is the author of *Backyard Ballistics* (2001), *Building Bots* (2002), and *The Art of the Catapult* (2004). Visit *backyard-ballistics.com* for more information.

HOW IT WORKS

When the fuel and air inside the jar first ignite, the jam jar jet generates a burst of hot gas, raising the internal pressure and pushing the gas out. The exiting gas leaves a slight vacuum behind, and fresh air rushes back into the jar to fill the void. More methanol mixes with the fresh air in the still-hot jar, triggering another combustion. Scientists call the resulting cycle relaxation oscillation.

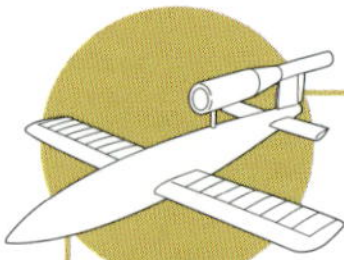

The V-1

The pulsejet engine is simple, cheap, and powerful, but isn't used in commercial aviation because large versions are incredibly noisy, and they vibrate like gigantic, unbalanced chain saws. Invented in the early 20th century, pulsejet engines had no practical use until German scientist Paul Schmidt developed a no-frills but dependable pulsejet-powered cruise missile. This was the notorious V-1 rocket, a.k.a. the "buzz bomb," which terrorized Britain during World War II.

The V-1 was a 25-foot tube with two stubby wings. Its simple engine gave the missile a range of about 150 miles with an explosive payload of nearly a ton. The first V-1 hit London on June 12, 1943. At the height of their use, 190 were launched daily. The V-1 attacks ended only when the Allies marched back through Europe, and seized the missiles' launch sites, which were located across the English Channel.

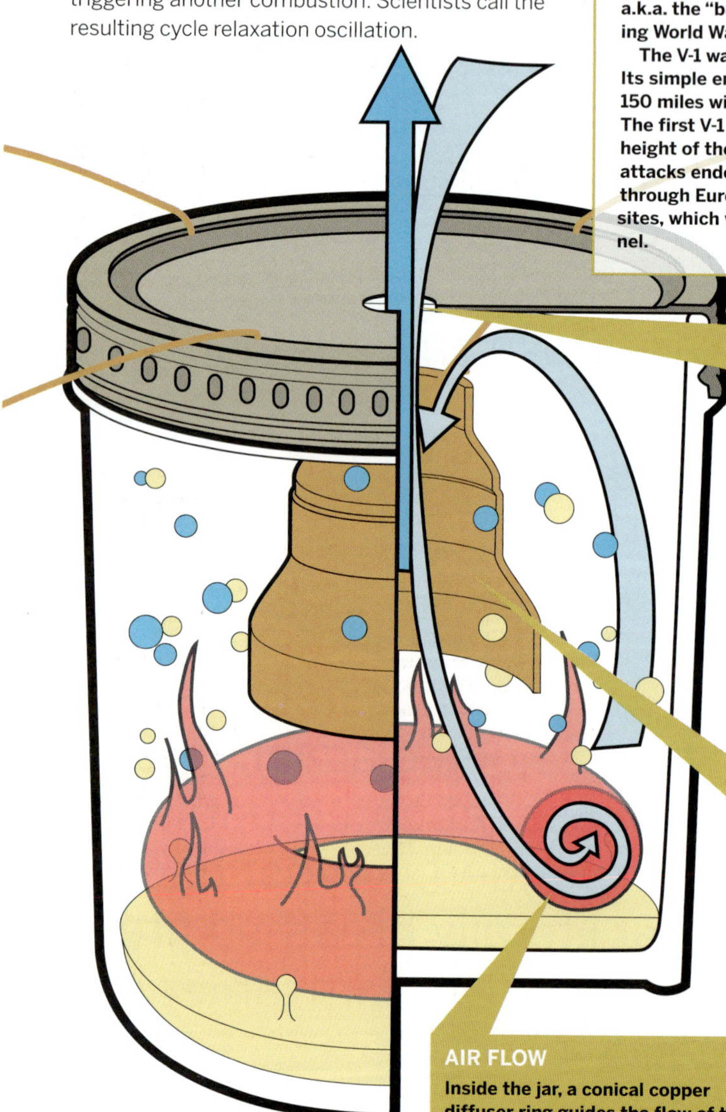

INTAKE AND EXHAUST

A ½" diameter hole drilled in the center of the Mason jar lid serves as both the air intake and exhaust port. Most functional pulsejet engine designs use two separate ports, but because the combustion cycle's intake and exhaust stages are not simultaneous, pulsejets can also use a single port. The continuous combustion cycles of more advanced jet engines, such as turbojets, require separate intake and exhaust ports.

HEAT DIFFUSION

The copper diffuser also conducts heat and transfers it out to the four wires that it hangs from. The long wires radiate heat to the air outside, which takes some thermal expansion strain off the jar, reducing the risk of cracking the glass.

AIR FLOW

Inside the jar, a conical copper diffuser ring guides the flow of the gases so that they follow a simple whirl pattern. This improves the efficiency of the combustion cycle.

Illustration by Tim Lillis

SET UP.

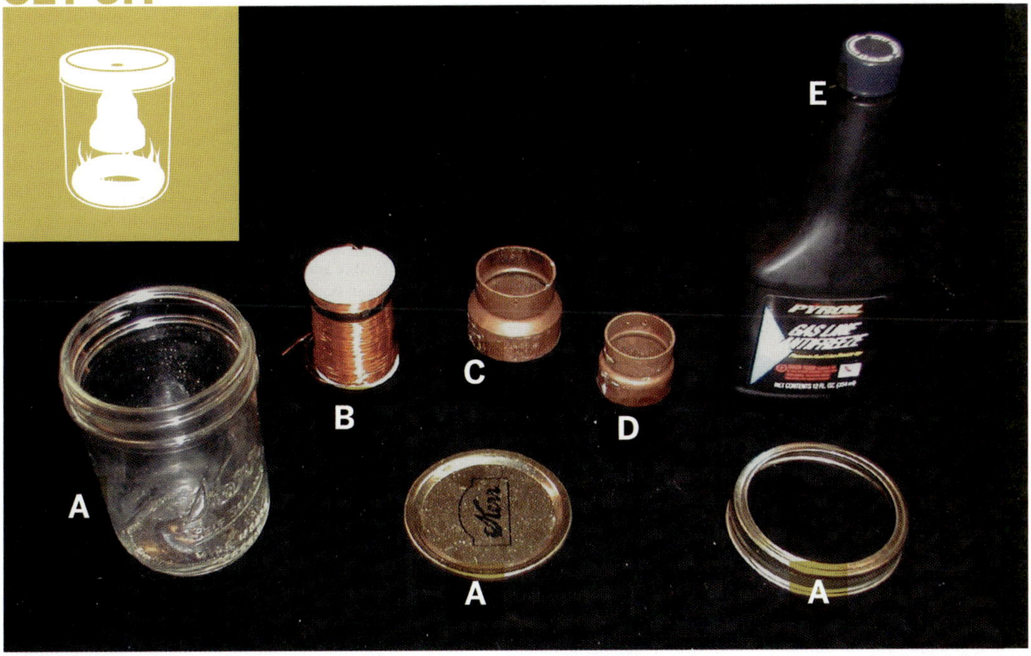

MATERIALS

[A] Pint-sized Mason jar with extra screw caps and lids

[B] 22- to 26-gauge magnet wire (thin enamel-coated copper)

[C] 1½" to 1¼" copper drain/waste/vent (DWV) reducing fitting

[D] 1¼" to 1" copper DWV reducing fitting Available at home centers or hardware stores. These two pipe fittings are for the conical air diffuser.

[E] Small bottle of methanol Available at auto supply stores as gas-line antifreeze; common brands include Heet and Pyroil. Methanol absorbs water readily, which is why it works well as gas-line antifreeze. But this property also causes it to go bad quickly, so you should always use fresh methanol.

[NOT SHOWN]
Package of long fireplace matches, or a long-handled barbeque lighter

Table salt (optional)

Boric acid crystals (optional)

1' long, 1" diameter plastic or metal pipe (optional)
Optional items are for experimental variations.

TOOLS

Electric drill with ½" and ⅛" drill bits

Wire cutters

File or sandpaper

Teaspoon measure

Cookie sheet

Refrigerator/freezer

Safety glasses

Gloves

SAFETY GUIDELINES

This is a jet engine you're building, a tempest in a teapot. I've never had any problems with this design, but no one — not me, not this magazine — can guarantee your safety. If you do choose to go forward with this project, here are some important safety measures.
1. Do not experiment with different sized jars and openings. A too-large jar with a too-small opening might result in an explosion of glass shards.
2. Use no more and no less fuel than directed. Wipe up any spilled fuel immediately.
3. Use only the parts listed in the directions. These are proven to work safely, and I haven't tried or analyzed all the substitutions that people might think of.
4. Wear gloves and safety glasses or goggles.
5. Do not handle the jar for 5 minutes after a successful run, and then be sure to tap it first to make sure it is cool enough. The Reynst combustor is an extremely efficient heating device, and it gets hot enough to burn skin after just a few seconds of run time.
6. After a long run, the glass jar may crack. If so, carefully sweep the entire assembly into a bag without touching it. Seal the bag and throw it in the trash. Jars are cheap enough, so just get another one.
7. Keep spectators at a safe distance.
8. Always ignite the engine with a long-handled match or barbeque lighter, to avoid getting burned by the pulse of hot gas that immediately follows ignition.
9. Examine all parts for wear before and after use. Discard any worn parts.
10. Always use common sense before, during, and after running the jam jar jet.

MAKE IT.

CONSTRUCT YOUR JAM JAR JET

START >> **Time:** An Afternoon **Complexity:** Low

1. DRILL THE PORT

Drill a ½" diameter hole in the lid of the jar. Use a file or sandpaper to completely remove the burr. If the hole is so jagged that it cannot be made smooth and round, discard the lid and re-drill another one.

2. DRILL THE DIFFUSER HOLES

Drill four ⅛" diameter holes in the small copper adapter. The holes should be located about ¼" down from the smaller, 1" diameter end. Space the holes evenly around the perimeter at 90, 180, and 270 degrees from the first hole.

3. ASSEMBLE THE DIFFUSER

Insert the large end of the small copper adapter into the small end of the large copper adapter. Press-fit them together firmly. This forms the conically shaped jet diffuser and heat sink.

4. CUT THE DIFFUSER WIRES

Cut four 4" long wires from the spool of magnet wire.

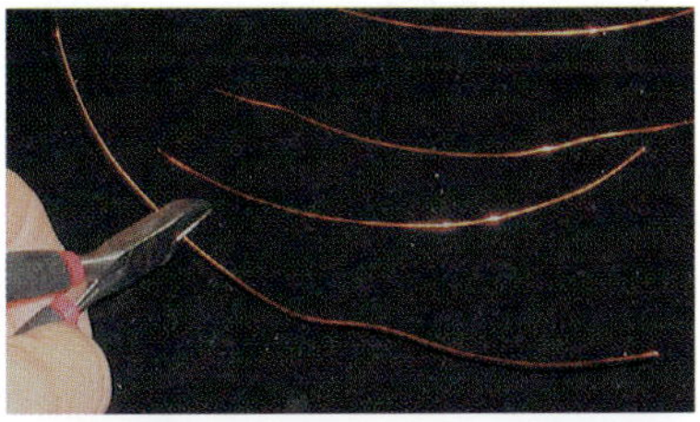

5. ATTACH THE WIRES

Loop one wire through each one of the holes you just drilled, and tie a knot. Extend the other end of the wires outward, radially, from the diffuser cone.

6. SUSPEND THE DIFFUSER

Center the copper diffuser in the middle of the jar. Crimp the wires over the edge of the jar so that the cone hangs suspended close to the top of the jar, with a gap of about ¼" between the diffuser and the top.

7. ADD THE FUEL

Carefully pour or use an eyedropper to measure and add 5 to 10ml (roughly 1-2 teaspoons) of methanol into the bottom of the jar. You can vary the amount of methanol by a small amount to improve performance. At most, the methanol should just cover the bottom of the jar.

8. CLOSE THE JAR

Screw the Mason jar lid down onto the jar and over the copper wires. The lid will hold the diffuser cone securely in place at the top of the jar.

9. VAPORIZE SOME OF THE FUEL

Prepare the jar by letting it sit in the freezer for two minutes. Hold your thumb over the opening in the lid. Vigorously swirl and shake the methanol inside the jar. Place the jam jar jet on a cookie sheet and place the cookie sheet on a secure surface, away from any flammable objects.

NOTE: When you remove your finger from the hole, you should notice a slight pressure release, and the jar should make a very faint "pffft" sound. If you feel no slight pressure and hear no sound, shake the jar again. If there is still no pressure, there is a leak in the seal of the jar that you'll need to fix.

10. FIRE IT UP

Wearing safety glasses and gloves, hold a flame over the opening in the jar's lid.

The fuel will ignite, and for the next 5 to 15 seconds, the jam jar jet will cycle, pulse, and buzz, running at a low but audible frequency of about 20Hz, depending on conditions in the jar and in the surrounding air. With the lights down low, you'll enjoy a noisy, deep blue pulse of flame that grows and shrinks under the lid as the jar breathes fire. It's an amazing effect.

Pint-Sized Fireworks

During the air-intake part of the cycle, the bottom of the jam jar jet glows brightly. The photo on page 102 shows the blue flame you'll get from burning straight methanol, and this photo (at right) shows the yellow variant that comes from adding a little salt to the fuel. By adding salt or boric acid crystals, you can color your flames in a variety of attractive, retina-burning hues, as described on the next page.

USE IT.

TIPS AND TRICKS FOR YOUR JAM JAR JET

FLAMBÉ RECIPES

Here are some interesting variations on the jam jar jet that you can experiment with:

1. For a bright yellow flame instead of the blue, add a pinch of table salt to the methanol.

2. For green-colored flame, add a pinch of boric acid crystals to the methanol.

3. To amplify the sound of the jet, hold a tube a half inch or so above the hole. You can use a metal or plastic pipe, and even the cardboard from a roll of paper towels will last a little while.

Use pliers or a gloved hand to hold the tube in position. Experiment with the length and diameter of the tube. When the size is right, you'll be rewarded with an unmistakably loud, deep, resonant buzz.

4. Some enthusiasts make Reynst combustors with metal jars instead of glass, and outfit them with resonator tubes permanently attached above the hole. These are sometimes termed "snorkelers."

The most advanced snorkelers also have fuel-feed systems that drip methanol into the combustion chamber, which allows them to sustain combustion for long periods of time.

TROUBLESHOOTING

If the methanol burns with a single big whoosh instead of pulsing:

» Check the size of the hole and make sure it is accurately drilled to a ½" diameter.

» Be sure to place the jar in the freezer for two minutes before lighting. In my experience, slightly cooling the fuel and the jar improves performance.

» Make certain the jar is charged with the recommended amount and type of fuel.

If you hold the long match over the opening and it doesn't ignite, or it does ignite but the pulse is weak:

» Make sure the methanol is fresh.

» Cool down the jar in the freezer for two minutes.

» Start with just one teaspoonful of fuel in the bottom, and vary the amount slightly until you get better performance results.

» Check the seal by listening for the "pffft" when you remove your finger from the hole. If necessary, rejigger the lid to get a good seal.

» Reposition the diffuser by adjusting the support wires, or try shortening the diffuser by removing the bottom section.

If the jar cracks:

» Carefully dispose of the broken jar and replace it with another one of the same size. The Reynst combustor/pulsejet is a very efficient burner and therefore extracts a lot of heat from the fuel very quickly. If the jar you're using cannot handle the rapid expansion, it will crack.

RESOURCES

Pulsating Combustion: The Collected Works of F.H. Reynst, Pergamon Press, 1961

Homemade pulsejets webpage and discussion forum: pulse-jets.com

Larry Cottrill's *jetZILLA*, an online magazine of amateur jet propulsion: jetzilla.com

Q&D

THE QUICK AND DIRTY
Workshop tips from Mister Jalopy's garage.

Holes, Rivets, and Bent Metal

Learn three essential shop fabrication techniques and reward yourself by making a wi-fi signal deflector.

How to Drill Equidistant Holes in a Straight Line

Pegboard is underrated. Handy as can be. Think you are too good for pegboard? Julia Child's kitchen at the Smithsonian is practically a pegboard temple.

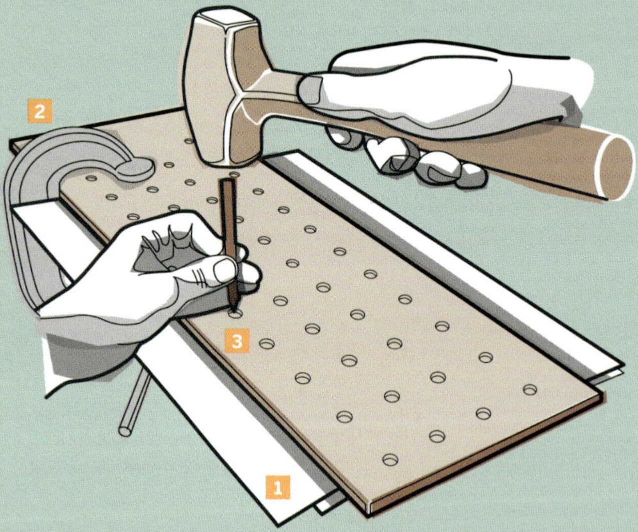

1. Clamp two sheets of sheet metal or plywood to extend over the edge of the bench.
2. Align, square, and clamp a strip of pegboard over the sheets.
3. Use a pointy center punch to mark each hole. Every pegboard hole too many? Skip every other hole. Want staggered holes? Then stagger them!
4. Remove pegboard and drill holes.

WARNING: Sheet metal wants to turn with the drill bit. Keep your fingers by always clamping the metal steadfast!

Drillin' and Snippin'

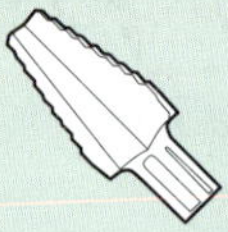

Drill Big Holes
The Irwin High Speed Metal Unibit step drill bit is the best way to drill holes up to about 1½".

Drill Bigger Holes
What about big holes? Doorknob-size holes? Hole saws can drill holes up to 6". Buy a mandrel and the appropriate hole saws to Swiss-cheese your world.

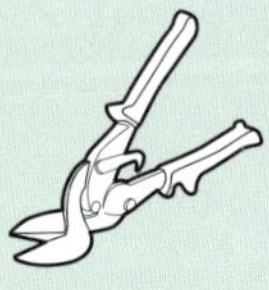

Best Sheet Metal Shears for the Money
According to legendary custom car builder Gene Winfield, there is no better tin snip for the money than the Irwin Aviation Snip from Harbor Freight. Either the right or left snips will cut a straight line, but if you want to cut a left-leaning radius, you need the left snips. The same goes for a right-leaning radius.

What about all that leftover pegboard?

Hang it on the wall and get organized! A pegboard hanging hardware kit is under $2 and is adequate if you are planning to hang spatulas and wooden spoons. If you want hammers with outlines showing which goes where, screw a pegboard-size 1"x2" pine frame to the wall studs and attach the pegboard.

Illustrations by Damien Scogin

How to Pop-Rivet

There is nothing that looks faster than pop-riveted aluminum. From race cars to airplanes, the pop rivet is the fastener of choice for joining sheet metal. It is a "blind fastener," meaning you don't need access to the backside, as you would if you were using a nut and bolt.

1 Load the right pop rivet for the job (i.e. hole size and sheet metal thickness). Squeeze the riveter handle just until you feel resistance.

2 Put the fat end of the rivet in the hole and squeeze the handles several times with great pressure. The rivet will pop when the pin has collapsed the fat end and finally snaps off due to your awesome brute strength.

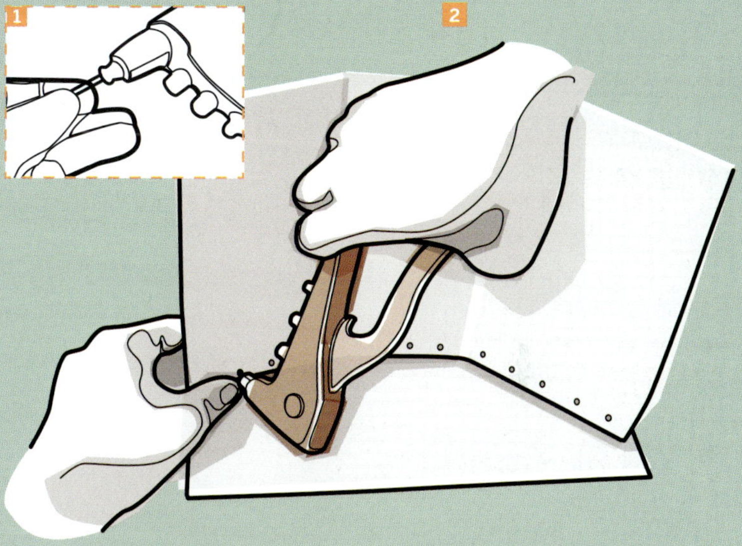

HOW A POP RIVET WORKS

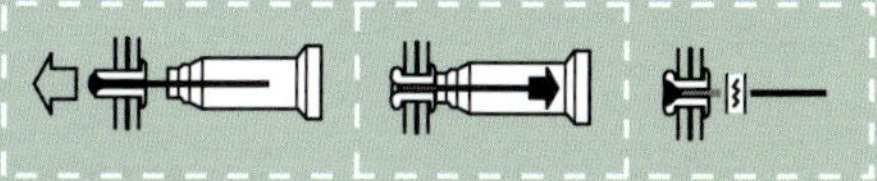

How to Bend Sheet Metal

The best way to bend sheet metal is on a sheet metal brake, but you can get decent results with a couple pieces of angle iron clamped to the edge of a sturdy table.

1 Sandwich the piece of material between the angle iron, and clamp tight.

2 Bend to the desired angle!

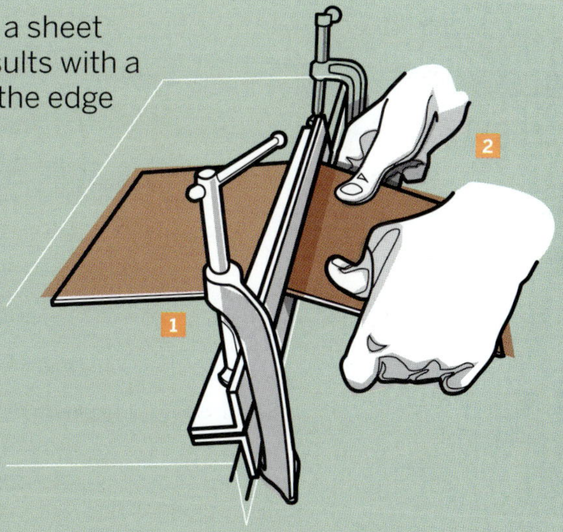

Tip: **Aluminum is easier to bend than steel, but even aluminum can be tough. If the grade is prefaced with a T (e.g., T6, T8, etc.), it is tempered for strength and can crack when bent. The .025" thick 3031 aluminum is swell for most home projects.**

How to Build a Power Tap

Build a power tap and put the power where you need it. Electrical junction box systems are the Lego of electrical. All sorts of adapters, connectors, and doodads are available to fit any situation.

MATERIALS
Electrical outlet and face plate
Single-gang junction box
Combination connector
Cheap extension cord

1. Connect a combination connector to the side of the junction box.
2. Feed the cord through the combination connector and tighten so the cord won't inadvertently pull out.
3. Separate and strip the wires. Twist the copper strands and tighten one wire to each side of the electrical outlet — not to the green ground screw!
4. The outlet comes with the correct screws to install it to the junction box.

WARNING: This is safe as an extention cord, but don't wire your whole house like this.

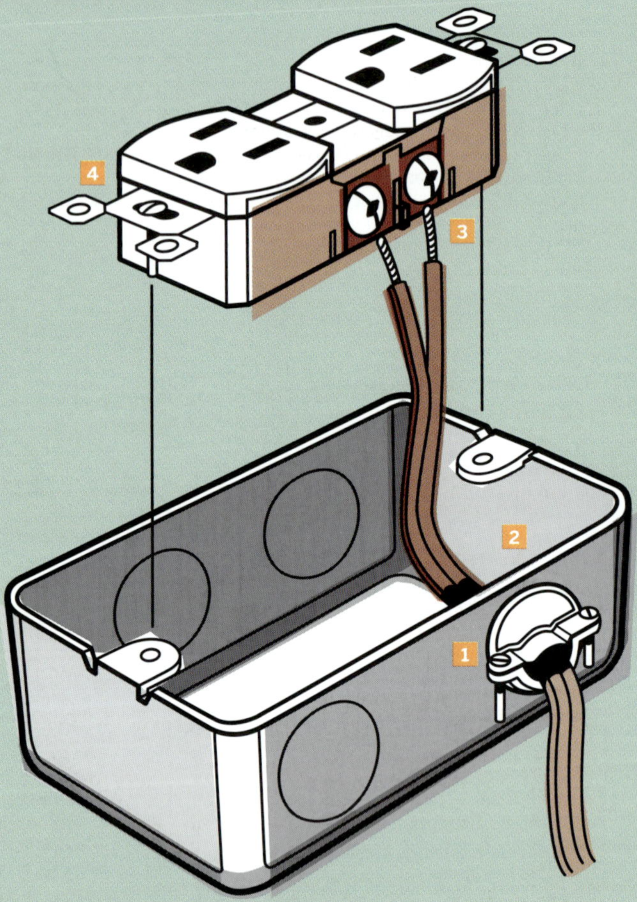

Now Make Yourself Useful

You can use all these tips to make this wi-fi hotspot signal deflector. Put it all together and you end up with a pretty decent wi-fi extender that actually works! In addition to the reflector gain, it really helps to get the Airport higher than a typical wall outlet or a floorbound surge protector. The all-important office-to-garage test netted another bar or two of wi-fi signal strength. Not too shabby considering it can be built for bubblegum money!

Mister Jalopy breaks the unbroken, repairs the irreparable, and explores the mechanical world at Hooptyrides.com.

Sneaky Uses for Everyday Things

1+2+3 By Cy Tymony

Wireless remote alarm sensor: Make different types of alarms and sensors with a cheap remote control car.

You will need: RC car, watch batteries, buzzer, wire, LEDs, seltzer tablet, universal AC adapter.

Inexpensive radio-controlled cars have many sneaky adaptation possibilities. We'll use the cheap single-function type of radio-controlled toy car — the type that continuously travels forward once it's turned on until you actuate its remote control, causing it to back up and turn. You can modify the transmitter to a more compact size, and for use as an alarm trigger. You'll also modify the receiver to activate other devices, such as lights and buzzers.

1. The transmitter is always in the Off mode until its activator button is pressed or until wires connected across the push-button leads are connected together.

The first step is to remove the circuit board from the case. Then, substitute lightweight watch batteries for its normal battery type. For each AA or AAA battery, use one small 1½-volt watch battery to supply the same voltage output.

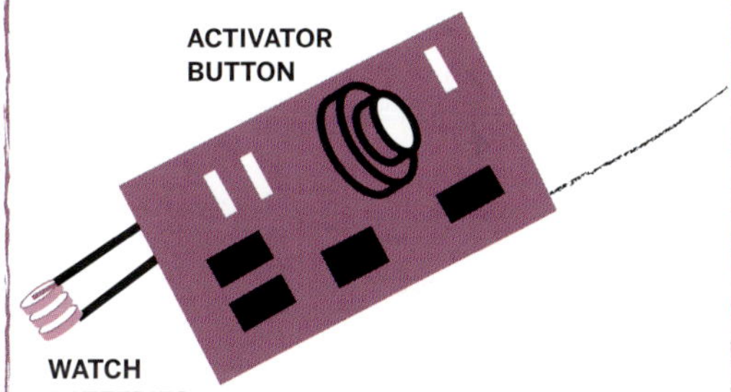

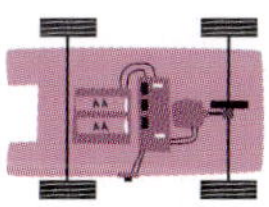

2. Remove the radio receiver from the car's chassis. Unlike the transmitter, the receiver must stay On in order to operate. You can use an AC adapter (with the correct output voltage) instead.

The car motor is attached to the receiver with two wires. If you disconnect the motor wires, the receiver can be used for more practical purposes to indicate that a remote sensor is activated. LEDs and buzzers are perfect.

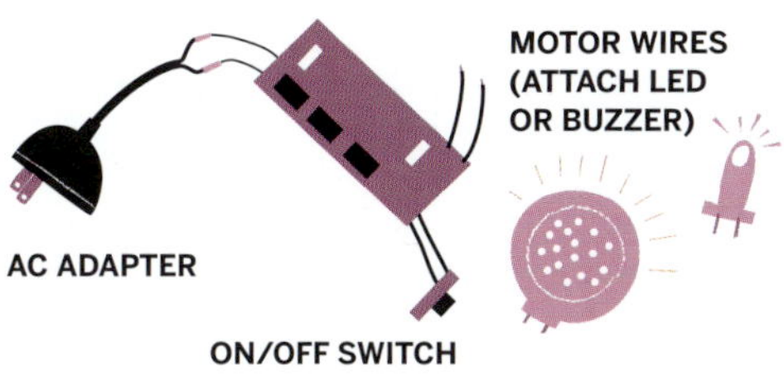

3. If you connect two wires across the transmitter's activator button, you can have another sensor or switch activate the transmitter to alert you of an entry breach or that your valuables are being removed, or that your basement is flooding.

Here, two paper clips are attached to the wires. Then, a seltzer tablet is placed between the paper clips with a rubber band. If water gets near the wires, the tablet will dissolve and the rubber band will cause the paper clips to make contact and activate the transmitter.

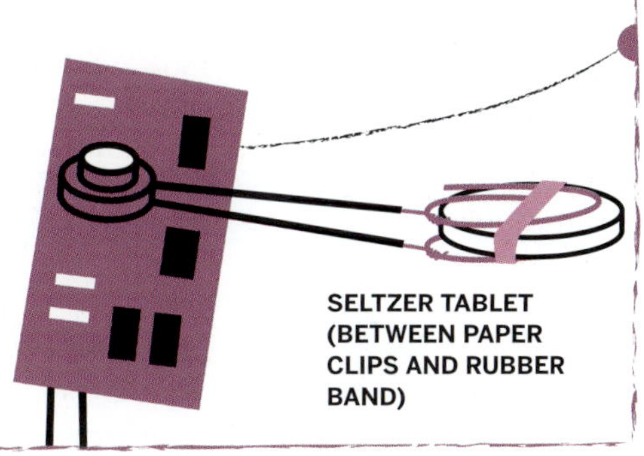

Illustrations by Mark Frauenfelder

Cy Tymony (sneakyuses.com) is a Los Angeles-based writer and is the author of *Sneakier Uses for Everyday Things*.

Give your keeper magazines the durability of a hardcover book.

OLDE-SCHOOL BOOKBINDING

Pages last longer, lie flatter, and look better inside a handsome, durable hardcover.

By Brian Sawyer

Photography by Brian Sawyer

Magazines aren't really built to last, but here's how you can turn your copy of MAKE (or any other magazine or printout) into a durable hardcover that will withstand the test of time. Your hardcover MAKE will also lie flat on your workbench, making it easier to follow instructions for other projects.

Create the Signatures

Peel away the existing cover, and use a utility knife and a heavy ruler to cut all the pages out ⅛" from the spine, freeing them from the glue. Divide your loose pages into consecutive 32-page sections (signatures). Binding loose pages as joined signatures will strengthen the spine and keep pages from falling out. For MAKE Vol. 01, without the ads, I got 192 pages, or six groups. If your total page count isn't a multiple of 32, you can fudge the signature sizes a bit, but each signature must have a page count divisible by 4.

Open the first group in half, such that pages 16 and 17 are facing. Pair these pages and set them aside, doing the same for the next facing pages (14 and 19) and the rest in the group. Just keep

MATERIALS:
- Magazine **Or pages to be bound**
- Acid-free nontoxic adhesive **I'm a Yes man, myself**
- Bookbinding tape, ¼" linen
- Binder's thread **Durable, acid-free linen**
- Decorative paper **For cover and end sheets**
- Binder's board, ⅛" thick **Or use chipboard or illustration board**
- Loose-weave, white linen fabric

TOOLS:
- Heavy-duty ruler or carpenter's square
- Utility knife
- Pen
- Binder's needle **Or tapestry or other heavy needle**
- Brush
- Scrap paper
- Wax paper
- Medium-grit sandpaper
- Medium duty awl

subtracting 2 from the left side and adding 2 to the right to determine which sheets to pair up. If you did it correctly, the last two pages you pair up will be 2 and 31.

Draw a vertical line ¼" from the right edge of page 16 (and all up-facing even pages). Cover the area to the left of the line with scrap paper. Brush a thin coat of acid-free, nontoxic adhesive (such as Yes) into the exposed ¼" gulley. Press the corresponding up-facing odd pages into the gully to glue together the pairs.

Return the pairs to the order they were in before being split in half. Align the edges of the pages and fold them along the spine. Repeat for each group and collate the finished signatures in their original order.

Stitch the Signatures

Sewing your signatures together around bookbinding tape creates the added durability of a hand-bound book. Measure and make two marks along the fold of the first signature, ½" in from each edge of the signature. These two marks represent the kettle stitches, the stitches that connect one signature to the next.

Now measure and mark six more points on the fold: three pairs of points, ¼" apart, spaced evenly between the two kettle stitches. These represent the in- and out-point for sewing the signatures around three tapes, which will run behind the signatures.

Stack the remaining signatures and make the same marks, at the same measurements. Pierce the marks with an awl, making holes just wide enough to allow a needle to pass through snugly.

Using a heavy needle, enter the spine and pull about 30" of thread through the foot (bottom edge) kettle stitch of the last signature. Exit the spine at the next hole and re-enter around the first tape.

Keep stitching around the tapes, and exit the spine at the head (top edge) kettle stitch. Using the same thread, enter the next-to-last signature at its head kettle stitch. Stitch around the tapes, and knot the thread around the foot kettle stitch of the first signature.

Continue in this fashion to stitch the remaining signatures together. If you run out of thread, knot a new 30" length to the existing thread. The best place to do this is just before re-entering the spine around a tape. When you come to the last kettle stitch, knot the thread.

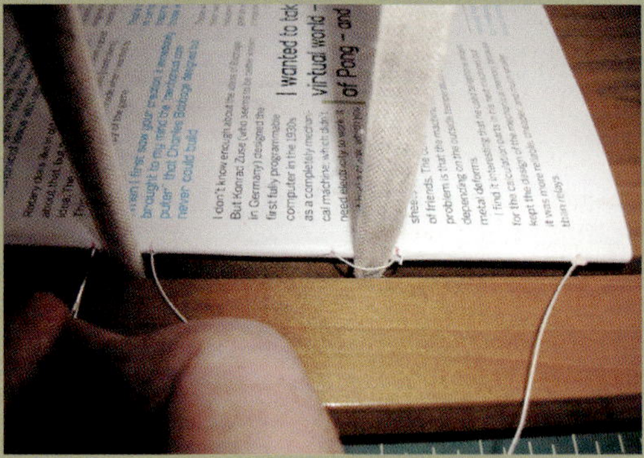

Use a heavy needle to penetrate the paper. The three tapes (one is shown here) run behind the signatures.

At this point, it's a good idea to apply a bit of glue (about ¼") to the inside of the first and last signatures (use a piece of scrap paper to protect the portion of the page you don't intend to glue), and put the work under heavy weights overnight.

Glue the Spine

Next, attach the mull (a strong strip of cloth with a loose weave that allows paste through it) to the spine and tapes. (Connecting the cover boards to the mull, rather than directly to the signatures, allows for a flexible backbone. This is the key to lay-flat binding.) Keeping the pages aligned on all sides, sandwich your work in a press or vise.

Cut a piece of mull that's tall enough to cover your kettle stitches, and 3 inches wider than the width of the spine. Brush a generous amount of

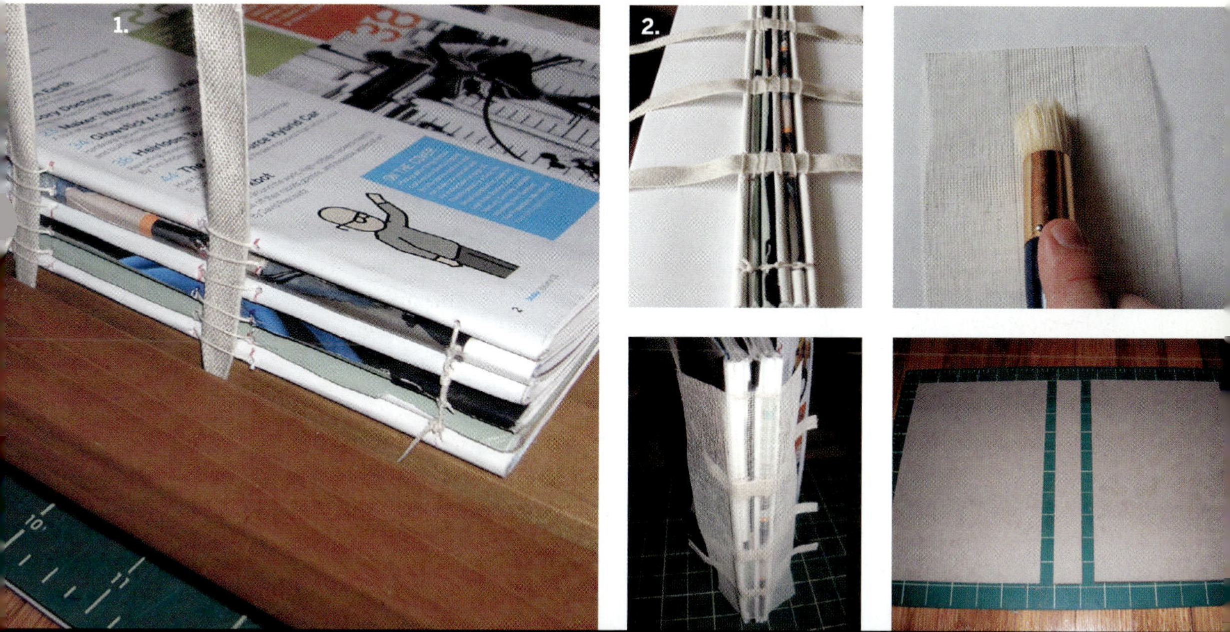

From magazine to book: Attaching the signatures with three strips of bookbinding tape (1), gluing the mull (linen fabric) to the spine (2), gluing on the cover boards (3), and smoothing out cover wrinkles (4).

To see a complete set of annotated photos of the author's bookbinding process, you can visit flickr.com/photos/olivepress/sets/14697/.

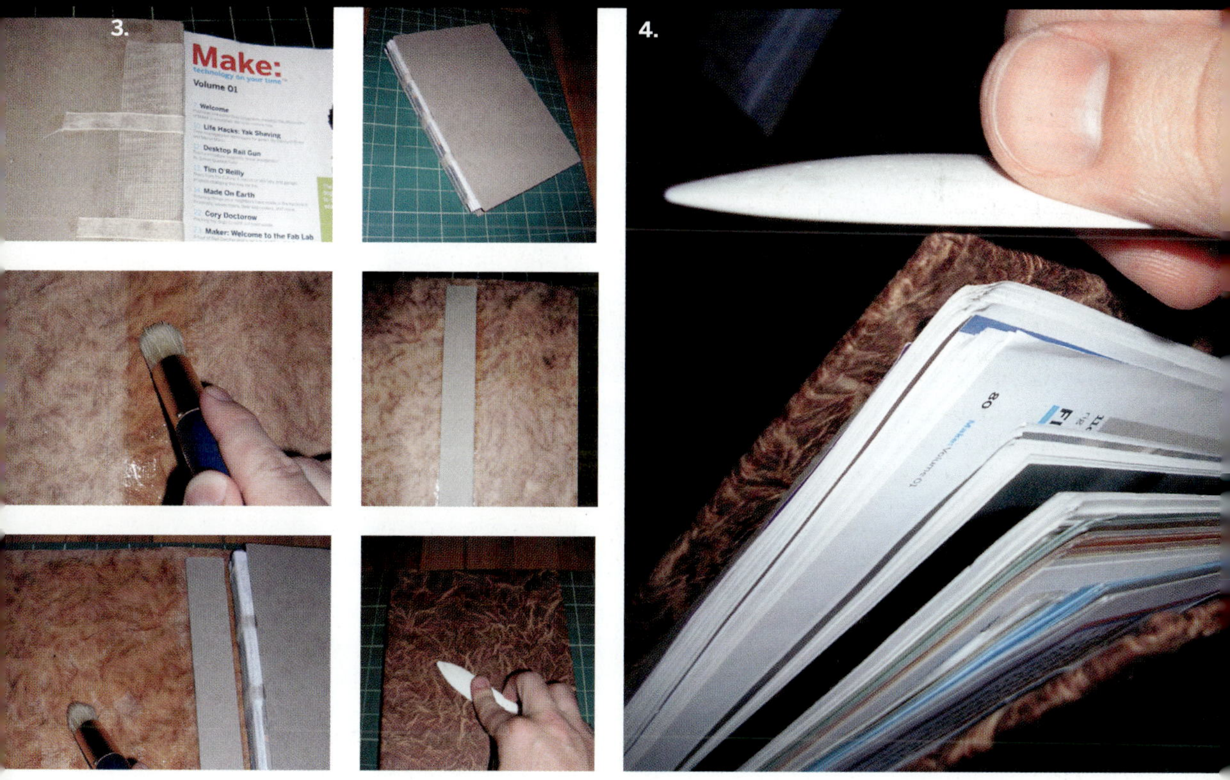

Ready to sew. Though using a stitching post is not absolutely necessary (as long as you keep a steady hand, make sure the tapes remain taut, and ensure that the signatures stay even while you sew), it does keep the work organized and easier to manipulate with the only two hands you have. If you have one available, set it up with the tapes stretched tight and spaced to match your holes.

glue on the spine, from kettle stitch to kettle stitch, across the full width. Lay the mull flat and mark the spine area on it, 1" in from either side. Brush this area generously with glue.

Position the mull symmetrically and rub it into the spine. Leave the book overnight under heavy weights or in a press.

Attach Cover Boards

Cut three boards (⅛" binder's board) for the front cover, back cover, and spine. Allow ⅛" additional clearance on the head, foot, and fore edges. Reduce the edge on the spine side of the front and back covers by ¼" (the thickness of two boards), to accommodate the hinge. Altogether, the covers should be cut to ¼" taller than the height of the book (⅛" added to the head and foot) and ⅛" narrower than the width of the book (adding ⅛" and subtracting ¼").

Cut the spine to the same height as the covers and the same width as the signatures. Sand down the rough edges.

Place one piece of wax paper between the mull and the free ends of the tapes and another beneath the tapes. Brush the free mull edge with glue. Remove the top piece of wax paper. Press the front board against the mull, extending ⅛" of the board over the head, foot, and fore edges.

Open the cover and rest it against a board for support. Rub the mull with cloth or paper, working the glue into the board. Brush the tapes with glue and press them to the cover board. Discard the second piece of wax paper, and place another clean piece between the cover board and the first signature. Repeat for the back cover.

Cover the Cover

You'll now cover this skeleton with a single piece of decorative paper, which wraps around the front cover, back cover, and spine. To create room to slide the paper over the edges of the cover and spine, use your utility knife to slit the mull by ½" where the covers meet the spine, at both the head and foot edges.

Lay your cover paper face down and mark the placement of your boards. Allow a ½" (thickness of four boards) turnover width for all edges, and ¼" (thickness of two boards) for each hinge.

Brush a generous amount of glue on the spine, from the head kettle stitches to the foot kettle stitches and the full width of the spine, including the portions of the tapes that rest over the spine (don't paste the free ends of the tapes). Then, attach the mull to the spine, covering the head and foot kettle stitches, and work the cloth into the spine with your fingers and a clean, dry rag until the glue is set.

Brush the spine area of the paper with glue, position the board, and press firmly. Turn the paper over and rub to secure the spine and mold the paper over the edges of the board.

Brush the area you've marked for the front cover with glue, brush slightly into the turnovers and hinge, and press the board between your marks. Turn the book over and rub the cover to remove air bubbles or wrinkles.

Brush glue onto the area for the back cover. Lay the back cover board's fore edge down on the paper, meeting your mark for that edge. Pinch the paper to the board on that edge, and press the remaining paper to the back of the book.

Rub the back cover, working the paper on into the hinge to seal the paper to your book block around the spine. Repeat for the hinge of the front cover.

Lay the book open, and brush glue across the length of the head turnover. Stand the book up on its foot edge and roll the edges of the head turnover over the top of the board. Repeat for the foot edge. Brush fore edges with glue, fold the turnovers, and smooth out wrinkles.

Lay sheets of wax paper between the cover boards and your block of signatures, and press under heavy weights overnight.

Finish Up

Pasting end sheets to the inside front and back covers reinforces the spine and finishes your book's appearance. I used the original MAKE wraparound cover as my front end sheet. Trim it to leave an equal distance around each edge, and paste it to the board. Allow glue to run into the spine, covering the point where the cover meets the spine and extending into the first signature by a ¼" gutter. Repeat for the back cover board.

Put fresh pieces of wax paper between the covers and the book block, and then set under heavy weights to dry overnight. You'll wake up to a long-lasting volume that will look unique, lie flat, and serve you well.

Brian Sawyer (bsawyer@oreilly.com) is lead editor for O'Reilly Media's Hacks Series. When he's not under the hood of a paid project or working pro bono for his wife, he can often be found fiddling with books in various capacities.

OUT DAMNED SPOT!

The chemistry of stain removal.

By Arwen O'Reilly

Both absent-minded and a klutz, I have had my fair share of debilitating stains over the years, and cultivated every stain tip I could get my hands on, even trying tricks from old, dusty, out-of-print books. Removing stains doesn't have to be hard; usually it's just a question of knowing the right chemistry (and practice makes perfect).

Different types of stains are more soluble in different temperatures: hot water, for example, dissolves sugars more easily, while it will set many proteins, like blood. Enzyme detergents will break up long-chain molecules, and are more successful at getting rid of protein stains like grass. In a pinch, dishwashing soaps can help, too; they are designed to break down food proteins and grease, after all. At the very worst, you may have to try a solvent. The following are a few of my favorite tricks, old and new. (Always test on a spot that isn't easily noticed first, and wash stains from the back, forcing them off fabric.)

Berries/Jam/Honey

Boiling water, especially when poured from a distance. This sounds like madness, but it is absolutely miraculous. Makes berry picking fun again.

Grease/Oil/Butter

Baby powder applied thickly, left overnight. Brush off powder in the morning (I use a toothbrush to clear the caked powder off the stain). If there's still a mark, reapply powder, and if not, launder.

Red Wine

Pour white wine onto the stain! Another miraculous save. If you don't have white wine, soda water will do. After, pour salt onto the stain to absorb the liquid. Try to get to it as quickly as possible.

Coffee/Tea

Borax and water mixture 3-to-1 (for carpets, use a non-gel shaving cream and scrub with a toothbrush). Glycerin also works magic.

Gum/Wax/Tar

Rub with ice and scrape off once hardened. With wax, if any excess remains, iron a paper bag over it, which will absorb the wax.

Blood

Cold saltwater and soap — not hot!

Avocado/Tomato/Ketchup

Enzyme detergent and a mild bleach (lemon, white vinegar, hydrogen peroxide).

Grass

Rub the back of the stain with rubbing alcohol. Scrub with non-gel baking soda toothpaste.

Sweat

Ammonia or white vinegar, then soak with laundry detergent.

White-out/Crayon/Lipstick

Use WD-40 and rinse, or try acetone (nail polish remover). Dishwashing liquid and hot water help, too. For lipstick, try bar soap first.

Ink

Use rubbing alcohol, turpentine, or acetone. Soak in milk or rub with toothpaste.

Stain-Removal Links

Cornell's *Removing Stains at Home* PDF
makezine.com/go/cornellpdf

FabricLink's stain removal site
makezine.com/go/stain2

Butler's Guild stain removal site
makezine.com/go/stain3

Visit makezine.com/05/diy_stains **for more tips.**

Arwen O'Reilly is assistant editor of MAKE.

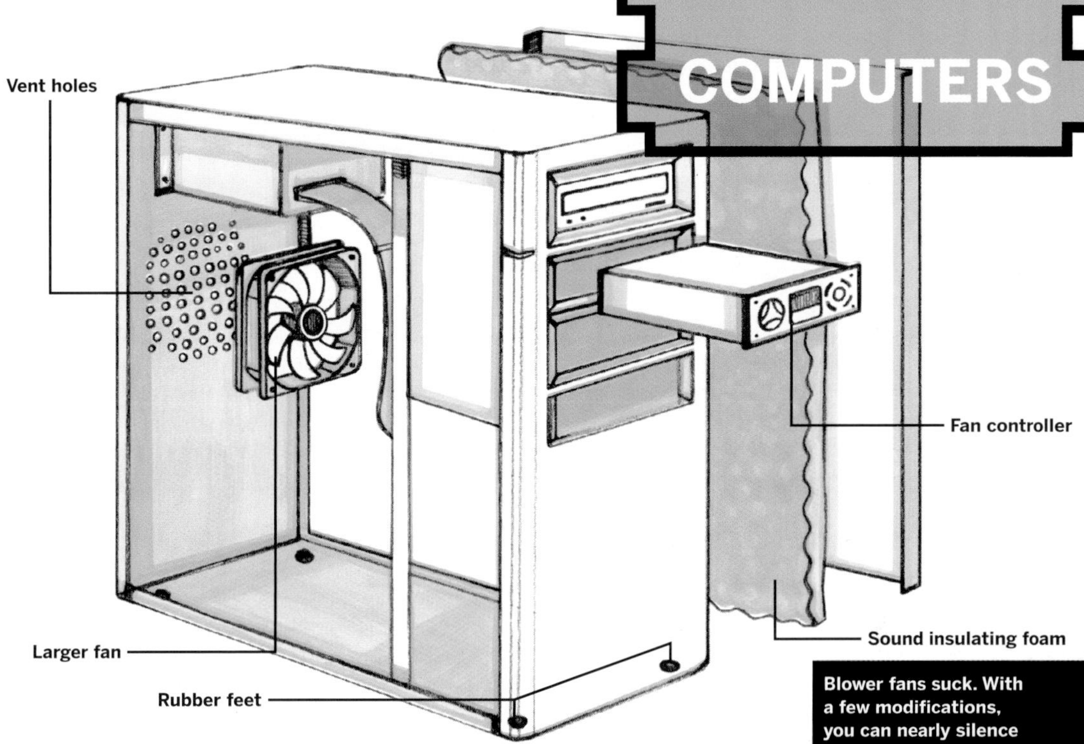

Blower fans suck. With a few modifications, you can nearly silence a noisy PC.

PUMP DOWN THE VOLUME

Five simple steps to a quieter PC.

By Jason Kohrs

Illustration by Damien Scogin

No one wants their computer to be loud, but in order to keep components running at safe temperatures, cooling fans are often needed. Unfortunately, they can make the system sound like a blow dryer. In a busy office environment, some noise may go unnoticed, but at home, the whir of a fan is annoying.

Silencing a computer can be a costly endeavor, but taking a few relatively inexpensive steps can have a drastic impact on the noise produced by the common computer system. Before starting on any sound reduction upgrades, analyzing a system to pinpoint the areas in need of the most attention will help determine the best course of action and the best way to spend any money.

Here are a few ways to quiet existing systems on a minimal budget.

Cooling Fans

The bulk of all noise in a computer system is going to come from the cooling fans mounted on the case and from any heat-generating components such as the processor. Cases generally employ 80mm fans with ball bearings to keep cool air flowing. Two steps to reduce noise include increasing the fan size and choosing a fan with fluid or sleeve bearings. If a 120mm fan can be installed where the 80mm fan presently resides, a noise reduction can be achieved because the larger fan can move the same amount of air at a lower rotational speed. In general, the slower a fan moves, the less noise it will make.

The ball bearings on many fans are a source of noisy vibration. Selecting a fan with fluid or sleeve bearings will greatly reduce the noise created, which is generally a good thing, except for one instance. Ball bearing fans can be counted on to get even noisier just before failure, letting you know when replacement is necessary. Fluid or sleeve bearings will just fail without such a warning, which could jeopardize other system components.

A quality "processor cooler" (a special add-on fan designed to blow air on a PC's microprocessor) is essential to keep a high-powered system running cool, but it isn't always necessary to run the fan installed at full speed. Some coolers, such as the Cooler Master Aero 4, include a simple speed dial that can be mounted either in the back or the front of the case for convenient adjustment. For those bold enough to run plumbing inside a computer, water cooling kits such as the Cooler Master Aquagate can take cooling performance and quiet operation to a whole new level.

Cases

A case with ample ventilation is required to keep the components cool, and a few things can be done to achieve this without adding to the noise level. Of most interest is the availability of multiple fan-mounting locations in a case, as well as the open area provided for the fans to move air.

Often, the perforations provided for air to pass through are somewhat restrictive, which could add to the noise level as the wind whistles through the small openings. This is nothing that someone handy with a Dremel couldn't remedy.

Experimenting with the size, speed, and placement of case fans can lead to a setup with adequate cooling and low noise levels. It is possible for some cases to be cooled well with a single 120mm exhaust fan, while leaving the other various fan locations empty. There is no need to use all of the fans just because they are there.

Fan Controllers

Fan controllers are available in numerous configurations, but they all serve the same function — to allow a fan to run at something other than full speed. Just reducing a fan's speed by 5-10% can have a noticeable impact on noise, but zero impact on cooling performance.

Some fan controllers operate automatically, using a thermal sensor to vary the speed of the fan in direct proportion to the temperature sensed. This type is convenient as it requires no user interaction but eliminates any possibility of custom control.

Manual speed controllers put all of the power in the user's hands, generally with a dial that adjusts the fan's speed by varying the resistance on the line powering it. The Cooler Master CoolDrive 4 is primarily a hard drive cooler, but it also serves the function of a four-channel, manual fan-speed controller. From one digital control panel, up to four temperatures can be monitored, and the corresponding fans can be monitored and controlled to maintain a healthy balance between noise and temperature.

Power Supplies

The typical computer power supply features two 80mm fans to keep it cool, which will obviously also generate some noise. Fanless power supplies are now available that generate zero noise, and there are other ways to quiet a power supply without removing the fans altogether. The MGE Vigor 450W Power Supply reduces noise by using a larger 120mm fan to move more air with less speed and a fan speed control knob to allow the user to reduce the speed even more, if they desire. Some other power supplies, such as this 500-watt unit from Clever Power, include a variable speed fan that automatically increases and decreases the spin of the fan, depending on the system's power draw.

Noise/Vibration Isolators

Products are available to reduce the vibration caused by system components as well as to insulate the case to keep noise from escaping. For the bottom of the computer case, rubber feet are available to replace the hard plastic ones generally found. Silicone gaskets can be installed between a power supply or case fan and the case to reduce the transmission of vibrations and the amplification of noise. If you want to keep the noise inside your case, there is even adhesive-backed sound insulation that can be applied to the inside walls of a computer case.

Get tech and computer tips online at Geeks.com.

Jason Kohrs is a technology enthusiast with a background in mechanical engineering. He spends his free time with his wife and daughter or at his online hideout, bigbruin.com.

DIY MOBILE

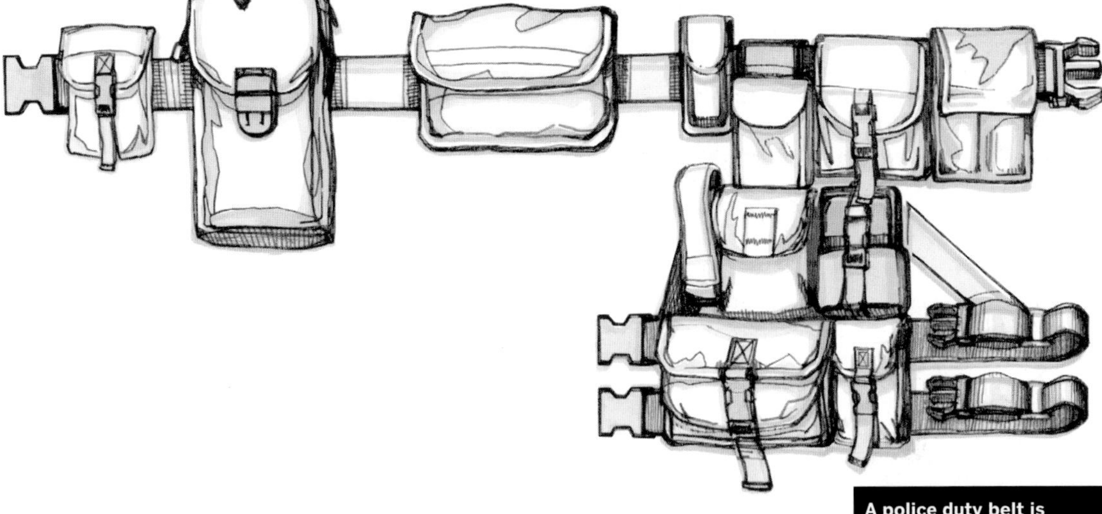

A police duty belt is great for holding cameras, phones, PDAs, multitools, and more.

GEARED UP

Use a gunbelt and leg holster to hold your gear.

By Andy Ihnatko

Illustrations by Damien Scogin

We geeks have a crackhead-like dependence on personal electronics, gizmos, tools, and other modern fetish objects. Things like pocket computers, smartphones, LED flashlights, USB thumbdrives, multitools, ZipLinq cables, notepads, digital cameras, spare batteries, and GPS units enhance our lives in obvious and inexplicable ways, but we can't deny that living in the Push-Button World of Tomorrow greatly complicates the otherwise straightforward task of changing one's pants.

Every night, you have to empty all of your pockets. Every morning, you have to fill 'em up again. And portable pockets (in the form of belt pouches) are a mere Band-Aid solution. Unless your electronics are machine-washable, you still need to unthread them from your belt and reinstall them over and over again, morning after morning. Decent men and women change their pants every day, so what else can you do?

Well, you can head off to your nearest police supply store and buy yourself a genuine, professional duty belt. With your pouches suspended off of that wide band of stiff, thick nylon or leather (structurally speaking, it acts more like a supporting frame than a belt), a simple click of the buckle leaves 6 pounds and $1,200 worth of personal electronics hanging off your bedpost until you get dressed again in the morning. And no, whistling the Clint Eastwood theme from *The Good, the Bad and the Ugly* as you put it on isn't at all inappropriate.

My day-to-day gunbelt configuration consists of a medium-size pouch for my cellphone, iPod,

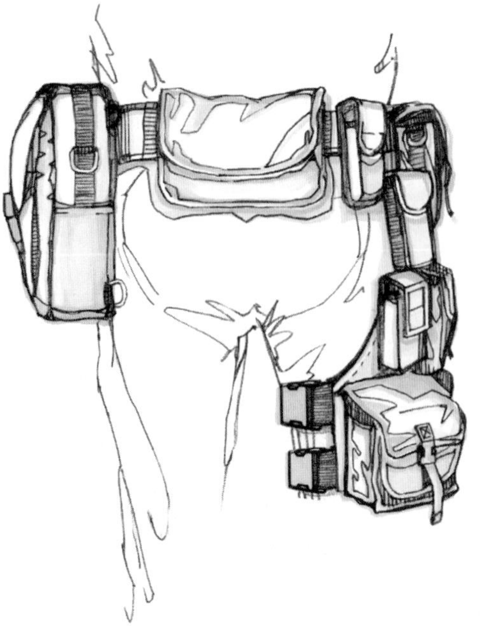

How flexible is a gunbelt system for carrying your stuff? Flexible enough that you'll no longer bristle at a concert event's "No Bags or Backpacks" policy.

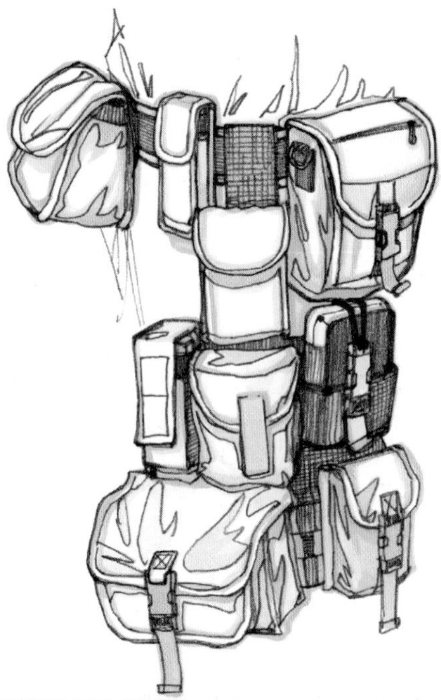

I keep a pouch containing iPod speakers, a canister of Pringles, and a thermos of frozen daiquiris ... or as I like to call it, the "Date-In-A-Bag."

and PDA, plus a holder for my Leatherman tool. But with a drawer full of pouches purchased over the years at various camping and photo stores, I can easily add capacity to suit the situation.

For the ultimate in added capacity and convenience, buy yourself a leg holster, which allows you to quick-draw your smartphone, even when you're sitting down or wearing a jacket. Tactical Tailor (tacticaltailor.com) manufactures equipment for urban SWAT units and Army Rangers. They make a "Modular Leg Rig" that can be custom-configured to your specific needs, along with a wide array of pouches that can easily be perverted to nonlethal, geeky needs and will hold everything but your PowerBook.

When I attend trade shows and conferences, my usual gunbelt is supplemented by TT's small leg rig. I've configured it with their adjustable Small Radio Pouch (which is perfect for a PDA or a chunky smartphone), a Small Utility Pouch for my camera, plus the real superstar of their line: the compact, compartmented Multi-Purpose Pouch, flexible enough to hold anything from a folding PDA keyboard to a palmcorder. You can even mount most third-party belt pouches to the leg rig, using Tactical Tailor's "Malice Clip" system.

Gunbelts are a perfect answer to the blight of personal electronics. I've been wearing one for years, and its value has only increased with recent tightening of airport and building security.

Yes, indeed: I routinely walk through airport security while wearing a police gunbelt and a SWAT tactical leg holster, and I haven't been held in a windowless room without charge even once. To the contrary, screeners and passengers are relieved to encounter a geek who can get all of his personal gear on the conveyor and walk through the archway after just 2 seconds of fiddling with a buckle, instead of holding up the line for 5 minutes while he desperately curses and pats himself down.

Just, um, be sure to refer to your gunbelt as a "utility belt" while you're in the facility.

Andy Ihnatko (andyi.com) is *The Chicago Sun-Times'* technology columnist and the author of a best-selling series of Mac books for Wiley Publishing.

BOOST YOUR SIGNAL

Improving MyFi XM satellite radio reception.

By Bob Scott

Although I love my XM MyFi portable satellite radio, I'm not as smitten with its lipstick-sized portable antenna. It's supposed to provide reception when the internal antenna is shielded from satellite goodness by a backpack, or by your water-filled self. That's the theory, anyway. In my neck of the woods, the portable antenna rarely improves reception.

In contrast, the tiny mag-mount patch antenna for car use pulls in a solid signal despite my dodgy shelf-package installation. It looked like an ideal portable replacement antenna.

My voltmeter provided the good news that the SSMB jack supplies the 4.5-volt bias needed to power the amplifier in the patch antenna.

Fooling around a bit, I found that the car cradle would work without external power, allowing me to do some side-by-side testing of the patch, via the cradle, and the portable antenna in problem portable-reception areas. As expected, the patch not only produced a stronger signal, it also was significantly less fussy about pointing.

Convinced that I now had probable cause to start violating warranties, I cut the cable on a spare patch antenna down to five feet and replaced the SMB plug with an SSMB. Connecting the hacked antenna to the portable antenna jack on the MyFi, I steeled myself for the worst and switched it on. Three bars on the signal meter!

My new portable patch has worked great in cars, buildings, or attached to my backpack (via a steel washer on the inside of the fabric) with no apparent ill effects on the MyFi. Don't feel like soldering? I just found an equivalent pre-made antenna setup available for about 46 bucks. A little over $15 will get you an SMB/SSMB adapter cable that lets you use the stock patch as a portable antenna, sans the car cradle.

Bob Scott is a statistical construct of various consumer electronics marketing departments.

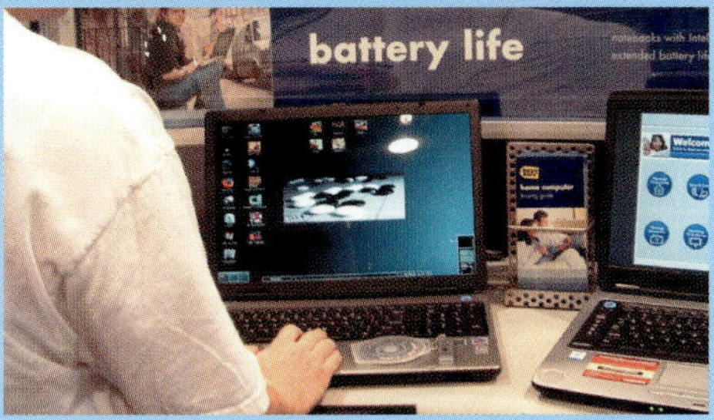

Photograph by Anne Sigler

Go on the Go

It's ironic that a security expert is helping people commandeer public PCs to play games. But Karl Sigler, who trains executives in computer security, created a miniCD and USB drive that will override a computer's normal operating system and substitute a tiny version of the Linux system. It sets you up to play Go, the 4,000-year-old Chinese game beloved by programmers and geeks around the United States. "You just pop it in the CD tray and flip the computer on," says Sigler with a certain glee. "I have yet to find a machine it won't boot on."

With a background in building self-contained CDs for computer forensics, Sigler created a Live Linux CD for his favorite new game. The technology allows users to run a whole operating system and software on a computer's RAM rather than a hard drive. Sigler modified a version of the operating system called DamnSmallLinux, added dozens of Go applications in Linux, and recompiled it back to a CD so that it's executable. Using the disc, absolute beginners can play against the machine or get an interactive lesson; fanatics can enter online games or pull up descriptions of 4,000 historical games to hone strategy. And you can run the disc in most locked-down PCs in cafes, training stations, and libraries. "When you pull out the CD, the computer works normally and there are no fingerprints that you were there," says Sigler.

—Bob Parks

>> Live Linux CD: hikarunix.org

WEIRD USB

Plug into a world of novelty.

By Phillip Torrone

USB is the "cable access" of modern computing in more than just a literal sense. Here's a look at a variety of USB gizmos, from the outlandish to the outright useful.

Ghost Detector: You know Scooby and gang need more than just to store photos of the mean old men who run the amusement parks on some type of USB drive — this one detects ghosts, too. Zoinks! makezine.com/go/ghostUSB

Ashtray: For that smoker who has everything, a USB ashtray. We're waiting for a version that emails your insurance company. makezine.com/go/ashtrayUSB

Self-Destruction Box: When the feds bust in looking for that Bit Torrent collection you've been squirreling away, the USB destruction box is the way to go — quickly and discreetly hit the nuclear option. makezine.com/go/selfdestructUSB

Light-Up Hub: What's the sense of transferring all this data unless you can see it? Routers get the blinky; why not a USB hub? makezine.com/go/lighthubUSB

Light-Up USB Cables: For the modder in all of us. That new blinged-out PC is drab with IBM-grey cables. These light up! makezine.com/go/lightupUSB

Laptop Cooler: If you're male and own an Apple PowerBook, you need this. Enough said. makezine.com/go/coolerUSB

Hot Plate: Why go all the way to the kitchen to keep that Hot Pocket warm and toasty? With the awesome 5V cooking action of USB, you can stay parked at the desk. makezine.com/go/hotplateUSB

USB for Rock Stars: USB fobs are handy, but not something you wear rocking out Bo Bice style. Now you can. makezine.com/go/rockUSB

Heating Pad: Nothing cramps up your back like staring at an RSS reader all day, but the sweet warmth of the USB port makes it go away — at least until the next refresh. makezine.com/go/padUSB

Disco Ball: Get on down with the USB badness, if you can dig it. Also makes the cat act weird. makezine.com/go/discoUSB

Eye Massager: Eyes hurt from staring at the screen all day, but don't want to leave the umbilical cord? We see a USB Eye Massager in your future. makezine.com/go/massageUSB

Battery Recharger: Why in the world is this available only in Japan? We'd actually use this. makezine.com/go/chargeUSB

Phillip Torrone is associate editor of MAKE.

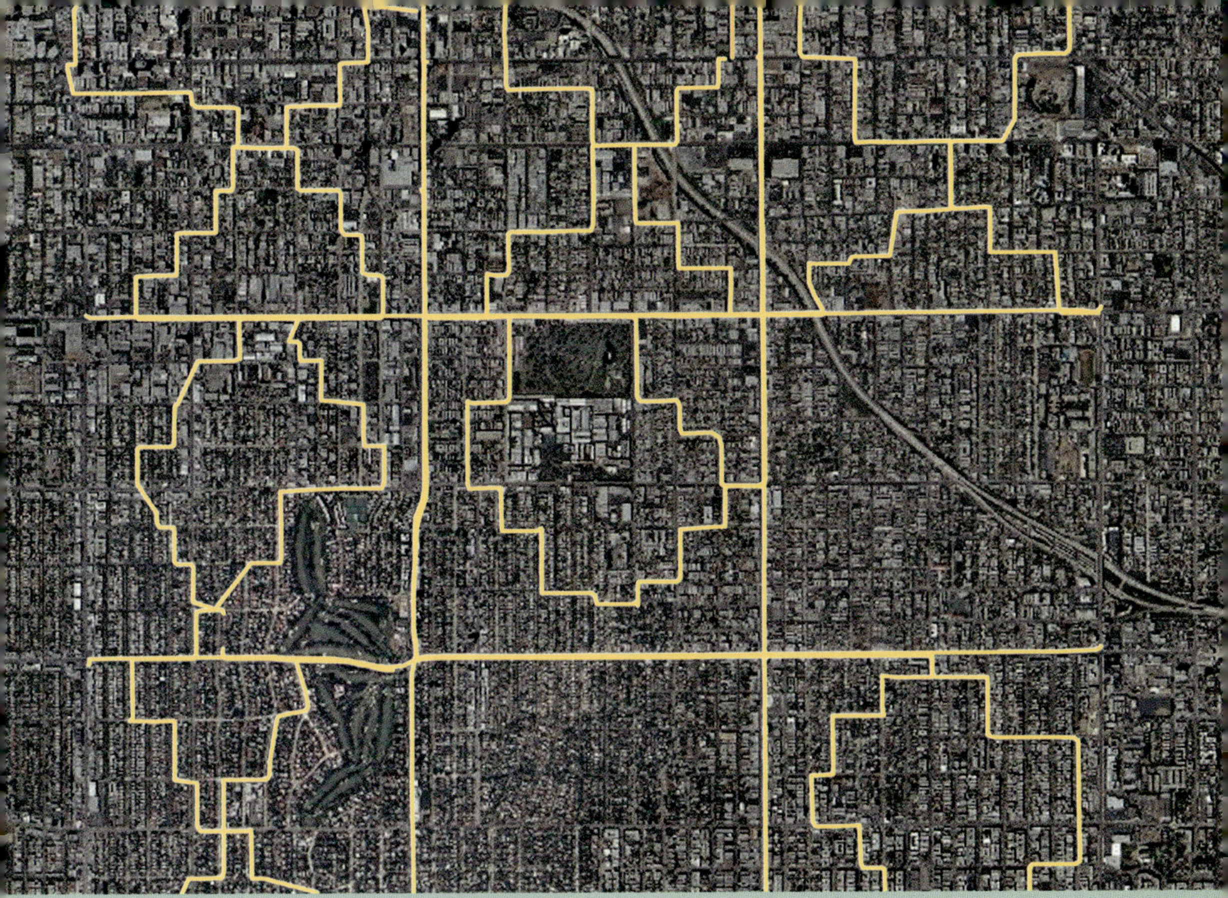

THE WORLD AS YOUR CANVAS

Use GPS to create giant-sized works of art.

By Julie Polito

Sometimes you need something bigger than a piece of paper to express your artistic self. Like, the whole world. A GPS device, a little planning, and a lot of free time can help you leave a giant footprint — or any other shape — on the planet. Think of it as crop circles for the 21st century.

Aaron Roller of Sausalito, Calif., wanted to give his girlfriend something extra special for Valentine's Day. He'd been toiling at a start-up company all year and she'd been an extra good sport. Roller is the chief technology officer for Motion Based, Inc., a company that created a web-based application that turns GPS data into functional analysis and online mapping for training athletes. He realized that those patterns on the map could also translate into pictures on a giant Valentine card.

After a little visualization, Roller set off into San Francisco Bay with a twitchy river kayak, a GPS, and a dream. "I don't think I'd been out on the bay more than a couple of times in this kayak, but I had a mission," he says. Half an hour and two miles of paddling later, Roller had traveled the shape of a large heart and uploaded the image of his path onto the Motion Based map-

Tic-Tac-Toe: A GPS unit and a mapping app can turn the world into a giant Etch A Sketch.

Photography © Jeremy Wood

The GPS unit on your car's dashboard serves as a "stylus" as you drive your car around the urban canvas. You can also use a bike or take to the water in a boat.

ping site (motionbased.com). "I had to be careful not to turn too sharply," he says. "And there is a little crook toward one end of the path where I had to navigate around a big boat in my way." His girlfriend loved the heart and the magnitude of the message.

While Roller says he is a fan of GPS drawing, he considers himself a dilettante. He points to GPSDrawing.com as an example of how elaborate you can get with your ideas. The site has several examples of impressive GPS art, including a large dollar sign in the Las Vegas desert and a larger-than-life tic-tac-toe game played on the streets of Hollywood (an exercise that took four hours of driving over a 23.6-square-mile pattern). Jeremy Wood, one of the founders of GPSdrawing.com, mapped out the grid the night before with his brother, and the two of them played the giant tic-tac-toe and mapped out their moves as they drove.

Wood has been drawing with his GPS unit since the day he bought it, and says he is merely building on an idea that is as old as humankind. "There is evidence of the concept all around us," says Wood. "Traces of trajectories can be found throughout history. They grow from footpaths to roads when we move."

If you're inspired and planning to create a little GPS art of your own, here are a few pointers to get you started.

First, grab a GPS device. Any one will do, really, as long as it maps your every move and uploads to a software program. Aaron Roller used a Garmin eTrex Vista C GPS, which costs around $428 and gives you a 256-color readable map display to keep you on track during your journey. You can also opt for a more portable GPS like the Garmin Foretrex 201 if you're traveling light. The Foretrex 201 costs around $150 and fits sportily on your wrist, but lacks the easy navigational screen of the eTrex Vista C.

Design your image, and keep it simple. Save your detailed likeness of the Virgin Mary for your sketchpad or grilled cheese sandwich. The more complex the design, the more time and editing you'll need to execute it.

For your canvas, choose an open space with very few obstacles. If you're planning a large route by car, remember that you will be restricted to usable roads. Walking in a field or paddling on an open body of water will give you flexibility.

Map out your route with a mapping program. Roller recommends tools such as Expert GPS, MapSource, or National Geographic Topo to help you plan your path. "You can download the route into your GPS and just follow the arrows," he says.

When you're following your route, think of an Etch A Sketch. "Start at a point, and draw a single continuous line," Roller recommends. You may need to backtrack to stay in your pattern.

After you've finished your design, upload your data into a software mapping program such as Motion Based (motionbased.com). Jeremy Wood uses a custom-designed program called the GPSograph, designed by fellow GPS artist Hugh Pryor. It's available on the GPSdrawing.com site and enables 3D rendering and map/aerial image alignment. Motion Based has editing tools you can use to clean up the drawing and erase start and finish points and any glitches.

Julie Polito is a writer whose pieces have appeared in *Salon* as well as *Self*, *Mobile*, and *San Francisco* magazines.

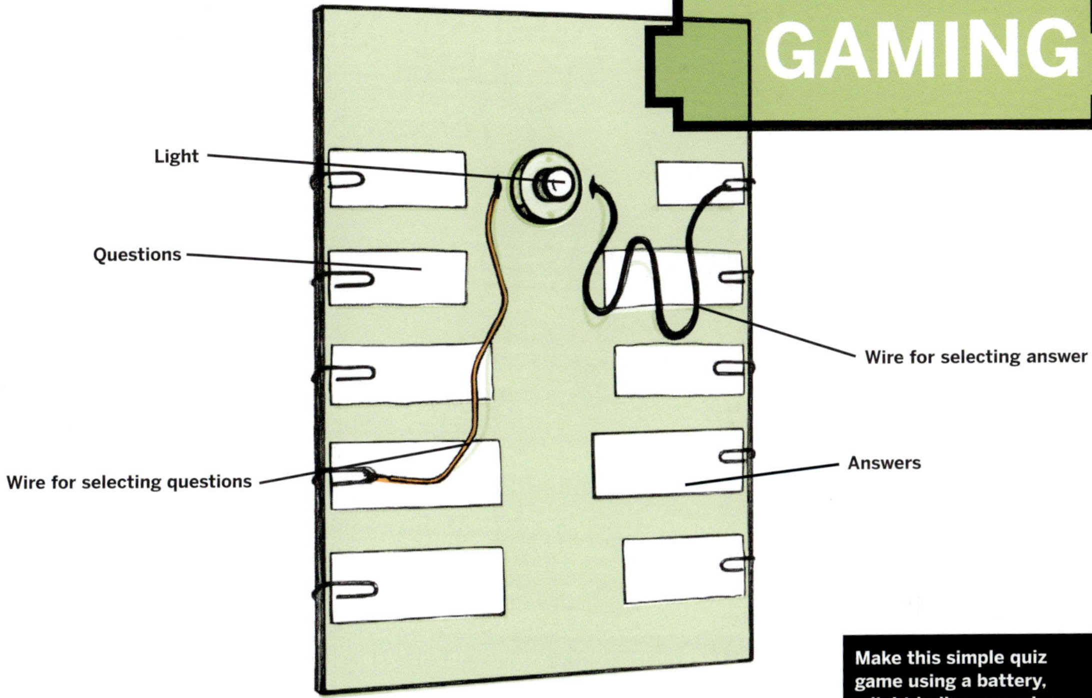

Make this simple quiz game using a battery, a light bulb, some wire, and paper clips.

CIRCUIT QUIZ GAME

Teach kids about circuitry in 30 minutes.

By Ben Wheeler

Arthur C. Clarke's Third Law states that "any sufficiently advanced technology is indistinguishable from magic." Clarke's law could use a corollary: sometimes it doesn't take much for something to be advanced enough to seem mystical. Today's kids are used to every teddy bear and softball being Turing-complete, so hidden complexity doesn't impress them. To really dazzle them, you need older technologies, like circuitry, which feel magical by their very openness and simplicity.

Here's a classroom-tested example: a simple project that introduces kids to circuitry. It functions as a quiz game, where a light bulb signals that the correct answer has been selected. The circuit itself is very basic, but if you provide enough sections of quiz material, it's surprising and satisfying to see it work. You can quickly rearrange the questions and swap them in and out, which makes for fast competition, impromptu topic-creation, and ample opportunities for ridiculing siblings.

Questions and answers are attached to a board by paper clips, with matching question-answer pairs connected in back. Two wires that dangle on the front of the board connect to the light bulb and the battery. When a player touches one of the wires to a question and the other to its correct answer, the circuit completes across the hidden wire behind the board, and the bulb lights.

Circuitry projects like this are also a great way to introduce kids to circuit diagrams, and show how they map to physical reality. Here's a simple circuit diagram for this project:

Illustrations by Damien Scogin

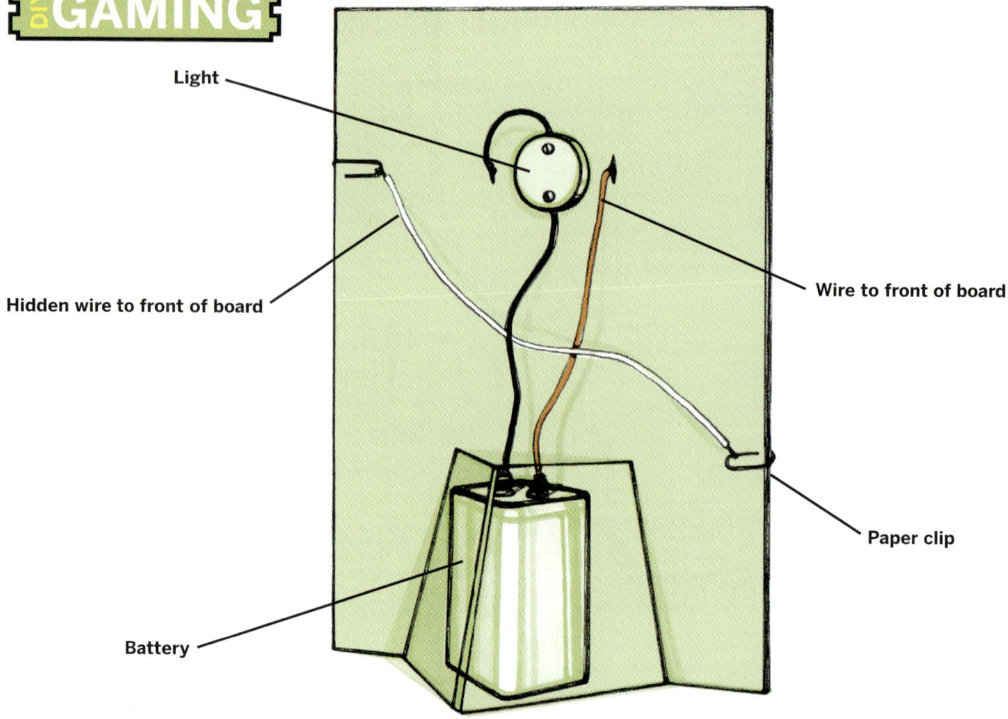

Behind the curtain: Compare the illustration on the previous page with the one shown here. The two wires on the front of the board (emanating from holes on either side of the light bulb) form a complete circuit and light the bulb only when they touch the paper clips that are connected by the wire behind the board. (For purposes of clarity, the paper clips for the wrong answers are not shown in this view.)

All of the circuit components in this project are available at RadioShack.

MATERIALS

- 6V light bulb (RadioShack #272-1128)
- Light bulb base with screw terminals (RadioShack #272-357)
- 6V lantern battery (RadioShack #23-016)
- Wire spool: 20-gauge, insulated (RadioShack #278-1225)
- Two large pieces of cardboard
- Paper clips
- Duct tape
- Paper and pens or printer, for labels

Total cost: $15

TOOLS

- Hobby knife
- Small screwdriver
- Wire cutters/stripper

Make the Board

Cut two identical cardboard trapezoids, about 6" on the top and 9" on the bottom, with sloping sides. Tape them together along one side and to the bottom of the quiz board to form a stand.

Install the Light

Cut a small circle in the center of the quiz board for the light bulb base. Cut two small slits on either side of the circle. Attach two wires to the light base. Send one wire through a slit to the front of the quiz board. Screw in a bulb and fit the base into the circular hole.

Install the Battery

Put the battery inside the quiz stand. Attach the second wire from the bulb to a battery terminal. Connect a long wire from the other terminal through the other slit to the front of the board.

Wire the Answers

Make each question-answer wire by wrapping both ends of a wire around paper clips. Make sure the connections are snug. Attach the clips on alternate sides of the back of the quiz board.

Label the Quiz

Make paper labels for the quiz questions and answers and place them next to the corresponding paper clips. Place dummy paper clips next to all the non-matching questions and answers, so they all look identical from the front.

Ben Wheeler is a math teacher living in Brooklyn, N.Y.

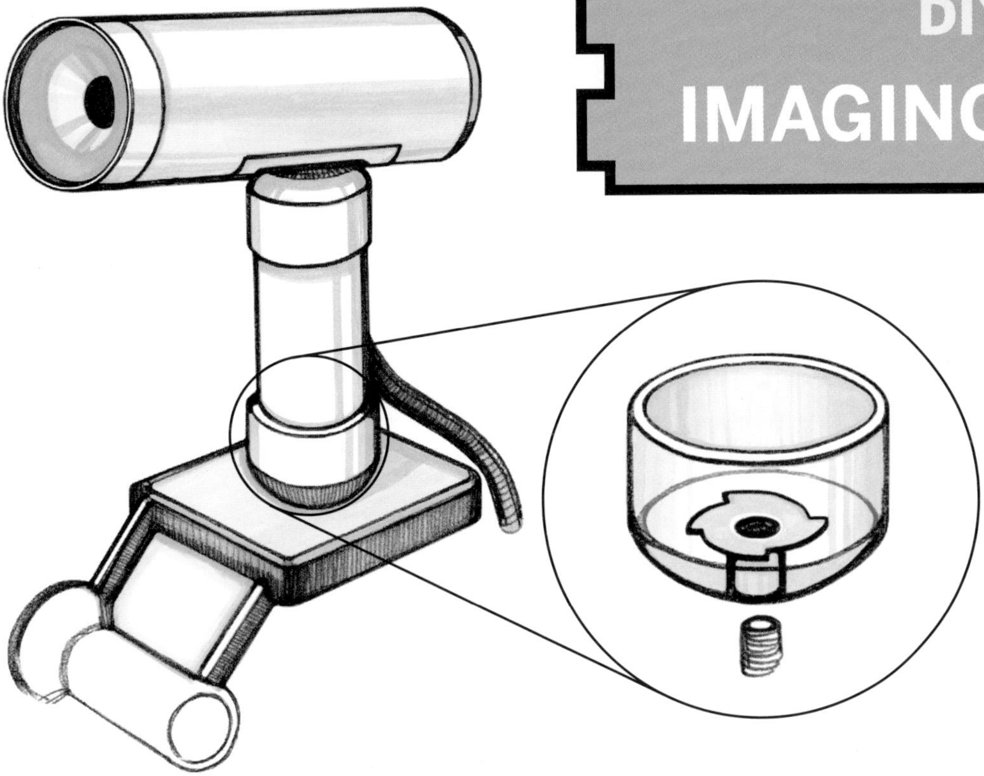

DIY
IMAGING

iSIGHT TRIPOD MOUNT

Point your webcam anywhere you want.

By Steve Vigneau

Illustration by Damien Scogin

In mid-2003, Apple announced a FireWire camera called the iSight. With true autofocus and the ability to capture 640x480 video at 30 frames per second, these cameras have become quite popular with Apple users. However, one of the camera's most innovative features — the mounting hardware — is also its most irritating quirk.

The iSight uses a bracket that supports one end of an Apple FireWire cable, and this cable snaps into a FireWire port on the bottom of the camera. Because the iSight is lightweight, this configuration works well in certain instances. Apple ships a number of attractive brackets designed to mount the iSight on a CRT, laptop, or flat panel monitor. What these brackets do not allow for is mounting the iSight on a tripod.

As I often want to point the iSight out a window or someplace other than at what is sitting in front of my monitor, I needed a way to mount my iSight on a tripod. There seems to be a lack of commercially available mounting options (Kaidan makes an adapter for $20, but it requires an Apple-branded FireWire cable and part of an iSight bracket), so I headed to the basement to build my own. Here's how you can do the same:

Beyond some basic tools, you will need about three inches of ¾" PVC pipe, two end caps to fit the pipe, a FireWire cable, a ¼-20 tee nut, hot glue, and some PVC cement.

First, drill one of the PVC end caps with five holes so that the tee nut and its wings will fit inside, points down.

A tripod bracket for your iSight gives you the freedom to place it anywhere you want.

Fig. 1

Fig. 3

Fig. 2

Fig. 4

Making the iSight tripod mount, step-by-step: Drill holes in one of the PVC end caps. The four smaller holes are for securing the tee nut, while the larger hole is for the tripod's mounting screw (Fig. 1). Squirt hot glue into the end cap to secure the tee nut (Fig. 2). Cut a rectangular hole in the other end cap, making sure the FireWire connector protrudes enough for the iSight to plug all the way in (Fig. 3). Make a notch for the cable (Fig. 4).

I used a 5⁄16" bit for the center hole, and a 3⁄32" bit for the four surrounding holes. Press the tee nut into the holes inside the end cap and secure with hot glue.

Cut a rectangular hole in the center of the other end cap, through which the cable's FireWire connector can fit snugly.

Next, cut a slot up one side of the pipe. This slot should be as wide as the cable, and as long as an end cap is deep, plus the thickness of the cable. Using PVC cement, fasten the end cap with the FireWire connector hole to the opposite end of the pipe from the slot, aligning the rectangular hole with the slot.

Orient the keyed side of the FireWire connector away from the slot in the pipe and fit it up through the pipe and into the rectangular hole in the end cap. Allow it to stick out a bit so the iSight will have room to tilt freely. Tack the FireWire connector in place, inside the tube, with a bit of hot glue.

Bend the cable so it runs out the slot in the side of the tube, and fill the body of the tube with hot glue. This secures the connector and provides strain relief for the cable.

Quick Tip: USB Cam and iSight

Apple's iChat AV application requires a FireWire camera, but if you already own a USB webcam (which can be purchased for as little as $15 online) and don't feel like shelling out $130 for an iSight, you can buy a utility to fool iChat into thinking you're using a FireWire camera.

Called iChatUSBCam (ecamm.com), this $10 program allowed my sister in Colorado to use her el-cheapo Logitech webcam to videochat with me for a fraction of the cost of a new iSight.

—Mark Frauenfelder

After the glue has cooled, cement the end cap containing the tee nut onto the other end of the pipe, and you're done. Screw your tripod's connector into the bottom of your new iSight Tripod Adapter and enjoy the myriad of places where your iSight can now be pointed.

Steve Vigneau makes, bakes, and un-breaks things around Southeast Michigan.

Photograph by Steve Vigneau

Study Andy's attire carefully. Imitation, like flattery, will get you everywhere.

The Fauxlance Photographer

How to get VIP treatment by dressing the part of a pro photographer. By Andy Ihnatko

Being a columnist for a Great Metropolitan Newspaper has its perks, first among them being the fact that the *Sun-Times* and I have worked out a little agreement by which they pay me money for my columns.

But there's another benefit. My idea of a fun time is to walk around with an SLR and make some pictures. Sometimes, getting Just The Right Shot means climbing up on a balcony or asking a street performer to turn toward the sunlight or standing rock-still with my SLR against my face, waiting for the scene to finish composing itself. And being able to flash a business card or a recent clipping is often the difference between being told "I'm sorry, but I'm really going to have to ask you to leave" and being told "You'll get much better shots of the parade from the top of the reviewing stand ... just follow me."

If you're not a columnist, well, simply dressing the part of a working journalist can go a long way. The way you appear and the way you behave subconsciously communicate to people that you're not some creep with a camera or just a tourist getting in the way. It's not about misrepresenting yourself; you're just subtly encouraging people to treat you with the same amount of respect that you invest in your photography.

Here's how you can get treated like a pro:

Overall Dress

Don't look like a bum, unless you actually are a photographer for *Sports Illustrated*.

As a freelance journo, you don't have the money for expensive clothes and they'd just get ruined

Photograph by Andy Ihnatko

as you travel from event to event anyway. So the effect to aim for is "as comfortable as you can get away with while still looking professional and presentable." Meaning: buttoned work shirt (no tees or pullovers), casual slacks, and sneakers that can pass for shoes.

Camera (general)

Yup, it's bigotry, but civilians instantly respect the Big Black SLR. A line of servers at a Chowderfest will not patiently pose for you if you're packing an $80 point-and-shoot.

Pros almost always wear two cameras (one with a short lens, one with a tele). Finally, a reason to get your old film Nikon out of mothballs! Always wear them across the shoulder on a long and comfortable strap.

Camera Lens

Put a lens shade on your camera, like every working photographer — not to combat sun flares, but to protect the glass and filter threads against bumps.

Camera Flash

An external flash is a must, even when shooting outdoors. For extra credibility, rubber-band a white index card around the head as shown. It's a cheap bounce-diffuser for better lighting.

Bag and Waistpack

Photographers need to get their hands on another lens, memory card, or filter without any fiddling around. So: A waistpack is essential, and if you wear a bag, it shouldn't be a backpack.

Necktie (in bag)

Handy to have, in case you discover that most of the other media are wearing 'em. A rare problem, but I've gotten into at least one event because I was able to put a tie on.

Microphone (in bag)

If you're recording audio for a podcast, get a mic that is, or looks, impressive. And, the shabbier the windscreen, the better.

Old Press Pass (in bag)

OK, it's a cheap trick, but I might have a press pass in my bag, too. If I'm talking to the media coordinator of an event and I can sense that they're on the fence about giving me a pass, I'll rummage through my bag for something and whoops, a genuine laminated badge to a previous event fell out! Sorry about that. Hmm? Oh, yes, that's from a space launch that I covered, thanks.

More important than a pass is a batch of business cards. No need to lie: just use your name and put the simple word "Photographer" underneath. Include your address, phone number, and email, and a link to your photo site. Photo subjects and media coordinators want to be reassured that you're for real and not some creep; 500 cards for $10 goes a very, very long way.

Breast Pocket

Keep it filled with random crap at all times.

Notepad

Hands-down the most important prop in the picture. All working media need to have a notepad handy at all times. Not a PDA — a notepad. You're always scribbling down names of people you've spoken to or photographed and millions of other notes, and it also serves as documentation that you actually were where you said you were, in case lawyers get involved later on.

Plastic Bag

If you're still shooting film, you gotta keep it in Ziploc baggies. It's just the rule.

Important Notes:

So what about getting an actual press pass? Well, it can't hurt to ask (call the event office and ask for Media Relations). Two big warnings, though: First, do not, do not, do not lie. If I'm not attending an event with the express purpose of covering it for my paper, I'm crystal-clear about my intentions. Trust should never be abused. And of course you want to get into the media tent, with its air conditioning and water and hookups. But stay out. For members of the actual media, it's essential. If I have 30 minutes to file my story but you're using the last network connection to play some MMORPG ... well, look, it'll get ugly. Finally, give away your photos and story to your local paper. Remember there's just one difference between an amateur photographer and a pro: a single published clipping.

Andy Ihnatko (andyi.com) is the *Chicago Sun-Times'* technology columnist and the author of a best-selling series of Mac books for Wiley Publishing.

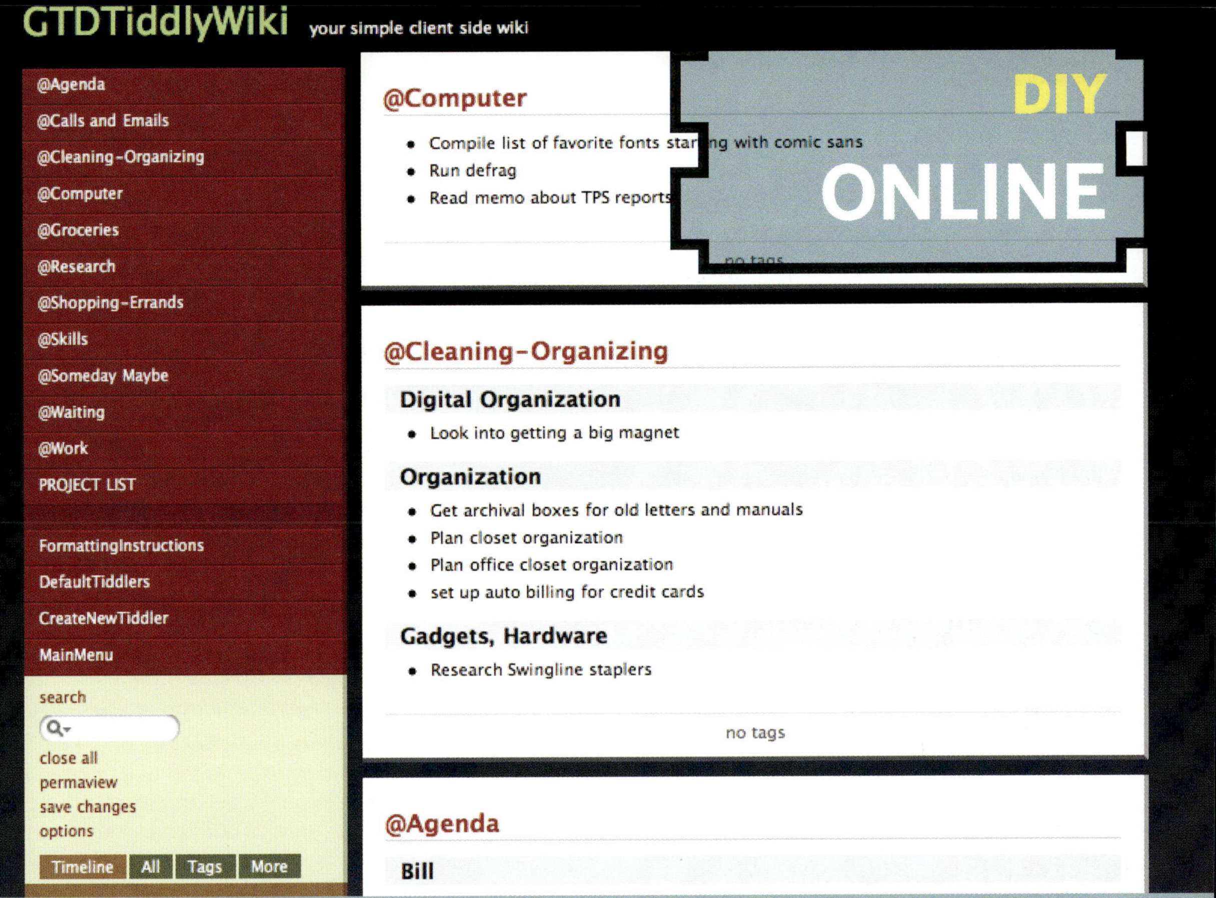

WIKI YOUR WORLD

Getting things done with GTD TiddlyWiki.

By Gareth Branwyn

It's a sad commentary on hardware and software design that few products deliver on their promises to make a meaningful difference in how we live and work. When something does rise above the clutter of flashy, but ultimately distracting, widgets and truly changes and inspires you, it's a genuine thrill. When such a discovery is free, the thrill is even headier. Such is the case with Nathan Bowers' browser-based, cross-platform application GTD TiddlyWiki.

Wikis Writ Small

You've probably noticed that wikis, open source collaborative web publishing tools, have been gobbling up a lot of mindshare recently. High-profile projects such as Wikipedia are bringing the wonders of wiki to the world. And the collaborative, open nature of wikis has spawned numerous mutations on the basic theme of quick ("wiki-wiki" is Hawaiian slang for "quick") web publishing.

One of those variants is TiddlyWiki, a client-side app by Jeremy Ruston that exists as a single HTML document on your web browser. Users can create a series of hyperlinked Post-it Note-like entries, called Tiddlers, using an extremely simplified wiki markup language. One of the most popular TiddlyWikis is Bowers' GTD TiddlyWiki.

GTD refers to *Getting Things Done*, David Allen's best-selling, geek-friendly book and personal task management system. Its appeal to a lot of geeky GTD users is in breaking down and tackling (or deferring)

Use this personal wiki to manage your tasks and plan your strategies.

actions as discreet components, and using a reliable storage system to hold and retrieve these "packetized" tasks. This fits in perfectly with the heart of TiddlyWiki, the Tiddler.

TiddlyWiki started out life as an experiment in hyperlinked microcontent. In the early days of hypermedia, there was Bill Atkinson's breakthough HyperCard, Apple's multimedia-capable, hyperlinkable application. HyperCards were discreet, non-linear ... well, cards. The web supplanted HyperCard and offered a globally hyperlinked universe of media, and the microcontent of HyperCard went macro. Then blogging came along and laid a dated journal/log book metaphor over the web. Both the web in general and blogs in specific are hyperdocuments, but they tend to be used in a rather linear manner. The idea of TiddlyWiki is to reintroduce the small, hyperlinked "card" metaphor. And, as many people have learned, it turns out to be ideal for managing tasks.

Getting Things Done with Your Wiki

To set up a GTD TiddlyWiki (or GTDTW for short) on your desktop, all you have to do is go to the app's website (shared.snapgrid.com/gtd_tiddlywiki.html) and save a copy of the HTML page to your hard drive. The entire contents of your TiddlyWiki — oodles of JavaScript and Cascading Style Sheet code, along with your content — is saved as this single document.

GTDTW comes pre-wired with a series of categories based on the system described in Allen's book, but you can easily tweak away on your own. It takes only a few minutes to set up the app, learn the simplified wiki formatting (a built-in Tiddler teaches by example), and customize the interface and content to your tastes. Once you get the hang of it, you'll be hyperlinking Tiddlers like crazy, launching apps from your wiki, hyperlinking to web content, pasting images into Tiddlers, making sortable tables, and all sorts of other "sophisticated" functions.

The power and simplicity of GTDTW are darnnear awe inspiring. It looks sharp, runs with a responsive snap you don't find in most applications, and is really fun and easy to use. It does what it's supposed to do, reliably, and it's got a rapidly growing user community that's dreaming up new hacks on a seemingly daily basis.

Getting Even More Done

If you want to really customize GTDTW, you'll need to roll up your sleeves. GTDTW is client-side only, so unlike other wiki apps, it is not collaborative. There are server-side TiddlyWikis that you probably could set up as a collaborative GTD tool (such as TiddlyWiki Remote, phiffer.org/tiddly).

Another problem, more of an annoyance really, concerns the printer function. GTDTW lets you print Tiddlers onto 3x5 cards, creating literal, analog microcontent cards. This allows GTDTW to integrate perfectly with the HipsterPDA (hipsterpda.com), another popular and refreshingly retro implementation of GTD concepts. But on browsers where you can't easily turn off headers and footers (such as Firefox on the Mac), your lovely Tiddler cards have unsightly text on the top and bottom. This is fixable via various hacks. For workarounds on this and other GTDTW bugs, and to get the latest hacks and variants, check out the forums at groups-beta.google.com/group/GTD-TiddlyWiki.

After reading *Getting Things Done*, I tried a number of GTD apps and hacks for using existing personal management tools (such as Gmail). Nothing (except the HipsterPDA) stuck, and they all seemed to get in the way of, well, getting things done. After several months of using GTDTW, I'm amazed at how much it's improved my organizational life and my day-to-day productivity. It's also a delight to use, and the whole microcontent concept underlying it has inspired a possible book idea. When's the last time Entourage did that?

Note: The GTD TiddlyWiki I actually use is one called GTD TiddlyWiki – KnipSter Enhanced (knipster.net/projects/gtd_tiddlywiki.html). The main enhancements are sortable grid tables and a dated journal entry feature.

Gareth Branwyn writes about the intersection of technology and culture for *Wired* and other publications, and is a member of MAKE's Advisory Board. He is also "cyborg-in-chief" of Streettech.com.

Build this simple eye-tricking instrument on the cheap, for about 10 bucks.

TEN DOLLAR PSEUDOSCOPE

See everything inside out through this classic optical instrument. By Rob Hartmann

Photograph courtesy of Rob Hartmann

The pseudoscope is a device that plays a trick on the eyes, switching the perception of near and far by reversing stereoscopic vision. It was invented by the great Victorian-era scientist Charles Wheatstone, and M.C. Escher used one to help create some of his famous perspective-bending illustrations.

You can buy one of these unusual devices for about $800 from Grand Illusions, so I decided to build a simple one of my own. Actually, I've built a couple of them now, with varying degrees of complexity, but here I'll show you how to make the easiest one for much less money: about $10.

Assemble It

I built the pseudoscope the way you'd guess just by looking at it. First, I measured and drilled the holes in the board. Then I drilled the cubes, glued the mirrors to the cubes, and screwed the cubes onto the board. I didn't glue the cubes to the baseboard because you need to be able to rotate the mirrors to align them depending on the distance to the subject matter.

Countersink the holes in the base so that the screws won't scratch table surfaces. Also, drill the holes in the cubes before you attach the mirrors.

Wooden cubes from craft stores, like the ones I used, aren't all perfectly square. For each cube,

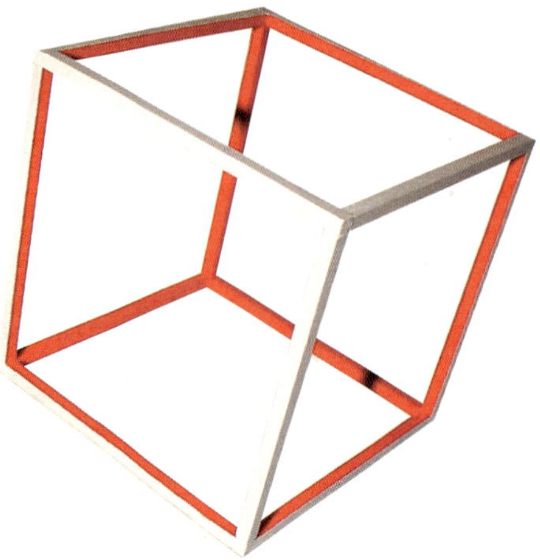

Fig. 1

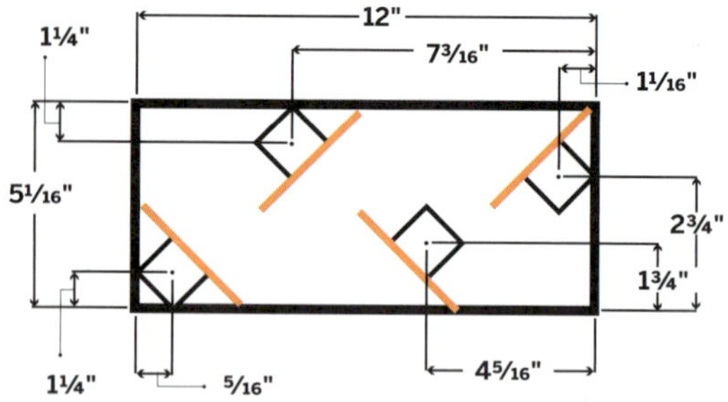

Fig. 2

MATERIALS
- Two 3"x3" mirrors
- Two 3"x4" mirrors
- Four 1¼" wood cubes
- One piece of wood, 5"x12"x¼" thick (make sure it's perfectly flat)
- Four ¾" flathead screws
- Glue

TOOLS:
- Drill and drill bits
- Screwdriver

M.C. Escher did it: With four mirrors mounted at exact right angles, you can easily create powerful illusions. Here are the dimensions I used for the screw holes (see Fig. 2). I don't think it's critical to use these exact dimensions; being a fraction of an inch off here or there shouldn't matter much.

you need to find two adjacent faces at exact right angles from each other, one side to serve as the base of the cube, where you drill the hole, and the other side to glue the mirror onto.

To find the good sides, check with a right triangle or push cubes together on a flat surface at different orientations until you find two that sit perfectly flush against each other. Just hope that the two cubes aren't off-angle by perfectly matching amounts.

To keep the mirrors at exact right angles, use a thin layer of glue, and glue them in pairs, pressing the blocks and mirrors together with the mirrors face-to-face.

Experiments and Enhancements

Your brain uses multiple cues for depth perception rather than relying entirely on comparing left eye versus right eye. As a result, inverted perceptions from looking through a pseudoscope can appear and disappear, depending on what you're looking at. The effect generally works better when you're sitting still rather than moving, and it's fun to experiment with what works and what does not.

One vivid pseudoscope experiment is to watch framework solids rotate, especially if their inside and outside faces are of different colors. I built a framework cube using twelve 6"x¼"x¼" pieces of basswood, and painted the inside faces red and the outside faces white (see Fig. 1). I hung the cube up and slowly spun it; through the pseudoscope, the cube looked red on the outside and like it was spinning in the opposite direction. When I closed one eye, the cube reverted back to its normal appearance. This illusion is best seen with strong lighting.

Always clean the mirrors before using the scope — fingerprints and smudges will greatly diminish the illusions. To take things up a notch, you can use front-surface mirrors, which eliminate the slight ghosting effect of conventional mirrors. Prices for these vary widely, so you should query different glass shops and look online.

I found some slightly larger front-surface mirrors for a good price at C and H Sales (aaaim.com/CandH), and used them to build a fancier pseudoscope with a custom-shaped walnut base.

Rob Hartmann is an electrical designer who lives in Fairfield, Ohio and enjoys mechanical puzzles and science projects.

Illustration by Damien Scogin

The Davis Vantage Pro2 collects weather data and sends it back to the console for web upload.

CITIZEN WEATHER STATION

Collect meteorological data on your roof and donate it to science. By Terrie Miller

Photography courtesy of Davis Instruments

I love the idea of citizen science, where amateur enthusiasts collect and combine data in the spirit of shared scientific discovery. After learning about citizen weather observer projects, I got the itch to plant my own weather station and start contributing data. I decided to go with a commercially manufactured station, and a local group of weather geeks recommended the Vantage Pro2 from Davis.

The Vantage Pro2 measures precipitation, temperature, humidity, wind speed, and wind direction (and you can add additional sensors to measure solar radiation and leaf and soil moisture). There are two pieces: the station itself, which sits outside and collects the data, and the console/receiver, which hooks up to your computer via USB. The station is solar powered and transmits weather data to the console/receiver wirelessly, so you don't need to worry about getting AC power cords or data cables up to your roof. Davis' WeatherLink software, sold separately, downloads data from the console, and lets you upload it to your own website and contribute it to the Citizen Weather Observer Project (CWOP), a grassroots, private-public partnership that collects and centralizes meteorological data from stations like mine.

It was easy to assemble; it took about an hour to put together and set up the communication to the console. Depending on your circumstances, the optional mounting tripod might make installation easier. We installed the weather station at the O'Reilly offices, but we had to attach it to an outside wall instead of putting it directly on the roof. We needed some extra height in order to get the station up into unobstructed wind, so we got

The Davis Vantage Pro2 Weather Station includes the sensor unit you install outside (above, left), and a console unit that you keep indoors (above, right). The sensor unit is easy to assemble.

a 10-foot conduit pipe from the hardware store, attached that to the wall, and then mounted the station at the top. When you install the station, you'll need to know which direction is north, so you can align the anemometer to correctly record the wind direction.

The console/receiver lets you view current conditions recorded by the station, as well as some past conditions. You configure the console with your latitude, longitude, altitude, and other settings that relate to the station itself. The console also displays current sunrise and sunset times and a rudimentary weather forecast. I liked that it displays the rainfall rate as well as the cumulative rainfall, and that during a downpour the console reports, "It's raining cats and dogs!"

According to the station's specifications, it can transmit up to 1,000 feet to the console line-of-sight, or 200 to 400 feet through walls. We were concerned about our installation at O'Reilly, where the building's steel frame interferes with wireless, but we had no problem transmitting data from the roof high above the third floor down to the second floor.

The Weatherlink software makes uploading data to the web or CWOP quite easy (it supports FTP uploads, but sadly not SFTP). The web template system generates a number of interesting charts and graphs from your data. The default HTML templates could use a designer's touch, but I found them easy to customize.

When I signed up for CWOP, our station got its own call letters: CW3724. Now anyone can access its data, and it's interesting to compare our station's readings to those of other citizen stations nearby. Seeing a storm roll in has never been so fun!

Davis Wireless Vantage Pro2 Weather Station, $595, davisnet.com/weather/products/weather_product.asp?pnum=6152
Current data from our station: makezine.com/weatherlink
Current station data, reported through CWOP: findu.com/cgi-bin/wxpage.cgi?cw3724
NOAA Citizen Weather Observer Project: wxqa.com

Terrie Miller works at O'Reilly Media, Inc., and publishes PointReyes.net and CitizenSci.com.

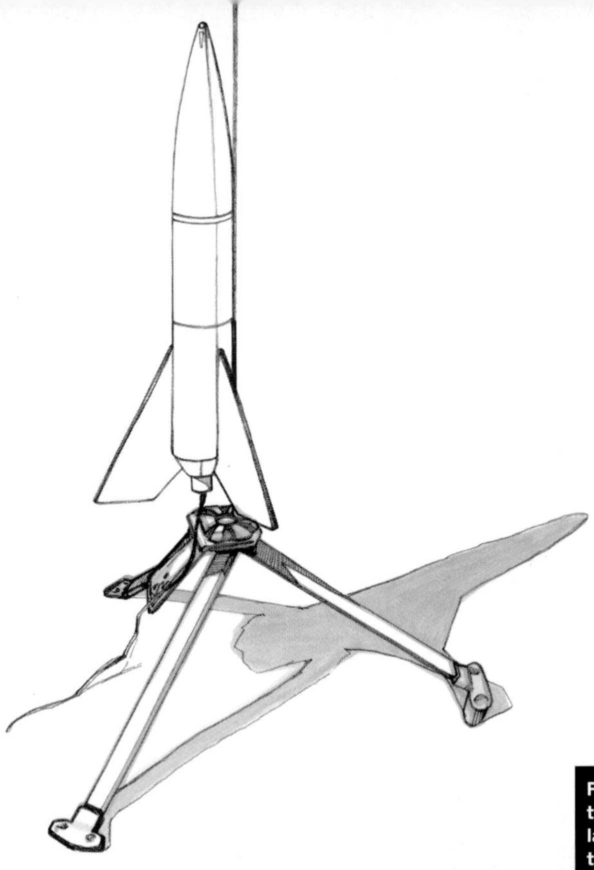

For rocketeers on the go, this easy PVC launcher is just the ticket.

LAUNCHING LIGHT

Portable, collapsible model rocket launch pad.

By Stefan Jones

My new dog is terrified of fireworks. This wouldn't be a problem outside of July 4th — if I didn't fly model rockets. But she can't tell the difference between an Estes B6-4 and a Joyous Family Whistling Fountain, so rockets terrify her too. This makes weekend trips to the organized launches in central Oregon impractical, so I've been flying locally. Launching in suburbia can involve hikes to unused fields, so I designed this sturdy, portable launch stand made of PVC pipes and fittings.

MATERIALS

- 7/8" non-plumbing-grade PVC pipe, 3' long
- 7/8" PVC T-junctions (3)
- 7/8" PVC corner junctions (2)
- Epoxy putty
- Galvanized metal shingle, 5x8"
- 1/8" steel rod, 3' long
- 3/16" steel rod, 3' long
- Skewer-type tent stakes (3)

TOOLS

- Hacksaw
- Vise-grip pliers
- Sandpaper
- Hand drill or drill press (preferred)
- 1/8", 3/16", and 1/4" drill bits

Illustration by Damien Scogin

Preparation

Legs: Cut the PVC tube into 1' lengths. Sand the ends smooth.

Feet: Drill a 1/4" hole through the side of each T-junction.

Hubs: The corner junctions form the hub of the launcher's tripod and support the launch rod. Stuff the center of each corner junction with epoxy putty, taking care not to get any in the socket areas. The hardened putty provides additional weight and support for the launch rod.

After the epoxy has cured, drill a hole straight

Drill a hole down the center of the corner junction's rounded "elbow" for different launch rods. At the launch site, thread a tent stake through the holes in the T-junctions and press into the ground.

down into the center of the junction's rounded "elbow" and into the putty, taking care not to punch through the other side. Drill different-sized holes in each hub for different launch rods, one ⅛" and the other 3/16".

Blast Deflector: Fold over the ends of the metal shingle and crimp flat. Drill a ¼" hole about 2" from one end.

Assembly

You can string the pad's legs and fittings into a tube that is easy to carry, stows in a car trunk, and doubles as storage for your launch rods. Toss the blast deflector, tent stakes, and spare hub can in a tote bag, along with your launch panel, motors, and other gear.

When you've reached the launch area, take the tube apart and assemble your tripod:

1. Choose a center hub and insert the legs into it.
2. Put a T-junction on each leg, turned so a side is flush against the ground.
3. Thread a tent stake through the holes in the feet and press into the ground.
4. Insert one end of the launch rod through the hole in the blast deflector and into the hub.

To angle the pad, pull up a stake and place an object under one of the feet.

Wherever and whatever you fly, be considerate and follow the NAR safety code (nar.org/NARmrsc.html) that keeps the hobby legal!

Stefan Jones dabbles in net-futurism, model rocketry, and gaming.

Don't DIY

Ethan Zuckerman isn't afraid to admit he's wrong. One recent project posted on his blog, a wood-fired hot tub built almost solely from materials he had lying around, fizzled dramatically. Photos show his partner-in-crime, Nate, sitting in the smoky, lukewarm hot tub during a snowstorm, looking stoic at best. Multiple reader responses point out just how dangerous such a project is, prompting Ethan to respond: "Let me be very explicit: ***Don't do this.***" makezine.com/go/hottub

Photography by Stefan Jones

Insert firefly in plastic bubble and a $5 photodiode chip converts light to frequency.

FIREFLY METER

Bioluminescence detector lights way toward insect-cyborg pollution sensor. By Christopher Holt

For me, the memory of catching fireflies in a jar brings me back to the dog days of my childhood summers. Now I'm a part-time biology graduate student, and one of my research projects focuses on the impact of air pollution on firefly flash duration, intensity, and period. Understanding this link could enable us to use fireflies as environmental contaminant sensors, providing an indirect method of monitoring air quality.

If you couple the fireflies with a low-cost way of measuring their bioluminescent properties, you have a sensor. Fusing whole organisms with hardware in this fashion is an exciting new research area, which has produced sensors capable of measuring biological and chemical agents at part-per-trillion concentrations.

Fireflies, or lightning bugs as they are commonly called, are technically neither flies nor bugs, but beetles of the order *Coleoptera*. There are more than 2,000 species worldwide, with 170 species in the United States alone. Fireflies use their light to find mates, which makes them among the few nocturnal insect types that discriminate mates visually. Numerous other insects are luminescent, but fireflies have the rare ability to turn their bioluminescence on and off. Typically, the males will signal with a specific flash pattern, and then wait for a female to respond with another flash pattern.

Most instruments for measuring bioluminescence are bulky, require advanced signal conditioning circuitry, and are far too expensive for a self-funded grad student. I needed an easy-to-use instrument that could measure firefly flash

Photograph by Kirk von Rohr

duration, intensity, and period. I surveyed my electronics workbench (yes, I'm probably the only biology grad student with one of these), and my eyes wandered to a Basic Stamp 2 (BS2) Board of Education (parallax.com). I pondered the possibilities of using this simple, easy-to-use microcontroller kit for my bioluminescence meter.

On a few robotics projects in the past, I had already interfaced cadmium sulfide (CdS) photoresistors with the BS2 to measure light intensity. So first I thought I would just plug a photoresistor into a breadboard, build a chamber to house the firefly, and collect data until the cows came calling.

I soon discovered a big problem with my CdS photoresistors: long response and recovery times. In some cases, it took the photoresistor up to half a second to recover its value after an LED flash.

I figured out that Texas Advanced Optoelectronic Solutions (TAOS, www.taosinc.com) had basically solved my problem with their $5 TSL230 light-to-frequency converter. This 8-pin chip has a photodiode and built-in signal conditioning circuitry that does all of the analog-to-digital conversion on board. Also, the chip can be programmed to measure light for a specified duration, and transmit serially to a microcontroller or PC. With this key piece, I had what I needed.

Let's Build It

The hardware components needed for making the firefly flash meter are simple: a BS2 microcontroller, 0.1µF capacitor (to reduce noise), the TSL230 light-to-frequency converter chip, firefly test chamber, and a computer.

For software, I used StampDAQ (Stamp data acquisition) to translate the data coming via serial cable from the BASIC Stamp microcontroller into a format that's usable by Microsoft Excel on the computer. StampDAQ software is available as a free download from parallax.com.

For my delicate measurements, I set the TSL230 pins S0, S1, S2, and S3 high (+5V), which tells the chip to run at maximum sensitivity and resolution. I assembled the circuitry right on the Board of Education, which is an easy way to prototype.

I built the firefly test chamber out of a plastic pipette (VWR #14670-149). I found that cutting the bulb of the pipette approximately ½" from the top provided adequate space for the firefly.

To make sure the firefly had adequate air supply, I cut an X at the top of the pipette bulb chamber. To maximize system sensitivity, I placed the firefly chamber directly on top of the TSL230 chip.

I used a styrofoam cup to remove residual background light sources. The hardest part of this project was collecting the fireflies. I enlisted the help of my 4-year-old daughter, Jordan. We learned that it's easiest to collect fireflies right at dusk, when it's still light enough to resolve the flying males from the background during periods of no flashing.

I used the BASIC command COUNT to count the number of pulses from the TSL230 (see code above). I set the COUNT command to measure for 10ms, but I determined that the actual resolution was approximately 31ms. The delay comes from the time it takes to execute the rest of the program loop and transmit the data into Excel.

Code to interface the BS2 to StampDAQ:

```
'{$STAMP BS2}
'Define variabless
light  var  word                         'stores light intensity
sPin   con  16                           'serial transmit pin
Baud   con  84                           '9600 baud, 8-bit
S0     con  0                            'S0 pin
S1     con  1                            'S1 pin

'--------------Main routine------------------------------------

Initialize:
pause 1000
serout sPin,Baud,[cr]                    'Prepares StampDaq
                                         buffer
serout sPin,Baud,[cr,"label,light",cr]   'label Excel column
                                         with light
serout sPin,Baud,["cleardata",cr]        'Clear all data
dirs = %00000011                         'make pins 0 and 1
                                         outputs
High S0                                  'Set S0 and S1 to 5V to
                                         divide frequency by 100
High S1

again:
count 2, 10, light                       'Count on Pin 2 for 10
ms
serout sPin,Baud,["DATA,light,", DEC light,cr]
                                         'Send data to
                                            StampDaq

goto again
```

Christopher Holt is a research scientist at NexTech Materials in Columbus, Ohio.

DIY TELEVISION

LEGAL AND FREE DIGITAL SATELLITE TV

Use a long-obsolete Primestar dish to pull in a wide world of programming. By Dave Mathews

In 1999, the home satellite TV provider DirecTV acquired one of its competitors, Primestar. The DirecTV and Primestar systems relied on different microwave receiver hardware, so when DirecTV upgraded the 2.3 million former Primestar subscribers to its own mini-dish antennas, it made obsolete all of those Primestar dishes, leaving them out there, unused. This article explains how you can modify one of these old antennas to receive dozens of channels of free television. You won't be able to pull in "premium" channels like HBO, which are encrypted, but for about $50 in additional equipment, you'll be able to receive regional television stations from throughout the U.S. as well as international programming from around the world.

Much of the information we'll need in this project comes from the Free TV webpage at lyngsat.com/freetv/United-States.html. To see all of the channels available, browse this page for the DVB channels. (DVB refers to the Digital Video Broadcasting standard, based on the MPEG2 format.) For the channels that interest you, click the satellite name and then find the channel on the satellite page that comes up. If it's listed on a white background

Pick up a used Primestar dish for under $50 (check eBay) and use it to watch free satellite television.

Photography by Dave Mathews

PARTS LIST:

1. Primestar or DirecPC/DIRECWAY antenna
If you can't find one of these relics in a neighbor's yard, look for them at a marina, where you might see dozens. A DIRECWAY or DirecPC antenna will also work (DIRECWAY, formerly DirecPC, is a satellite broadband service). Attached to the antenna will be one or two filtering modules, Ku-band Low Noise Block (LNB) filters with horizontal/vertical polarity.

2. Free-to-Air MPEG2 receiver
Searching eBay for "FTA MPEG2" will reveal hundreds of new and used units starting under $40. A basic unit is the VIStar 2000, which will support a couple hundred channels carried over ten different bouquets. The Cadillac of receivers is the Pansat 2700a, which supports 5,000 channels. Assuming you buy a cheap receiver, your total hardware cost for this project should be under $50, even using high-quality coaxial cable.

3. 2x4 or 3x4 satellite multiswitch (usually necessary)
You'll need one of these if you're using the older and more prevalent model of Primestar dish, which has two LNB outputs, for horizontal and vertical. If you're using a DIRECWAY/DirecPC or a Primestar dish with just one output, you're good to go as-is.

4. RG6 coaxial cables
You'll need one long cable to go from the antenna to the receiver, plus two short ones if you're using a multiswitch. Get the high-quality stuff.

5. Stable mounting platform for the dish

6. Compass

and its frequency, shown in the first column, is above 11700 (11,700MHz), you can receive this channel — provided your view to the satellite is unobstructed.

Aiming a dish takes a while, so you need to choose one satellite from which to receive your signals (or install multiple dishes, like I have). But each "bird" transmits hundreds of channels of video and audio, some encrypted, some clear. Each satellite broadcasts at dozens of frequencies, and each frequency can pack multiple digital channels simultaneously, which is why the frequencies are referred to as "bouquets." So you can get a lot from one satellite. These days, plenty of good English-language programming comes off the satellites Intelsat Americas 5 and Galaxy 10R, so these are good ones to explore.

Procedure:

1. "Correct" the polarity of the dish.

For two LNB outputs: Using two short coax cables, connect the vertical output from the dish to the 14V port on the multiswitch, and the horizontal output to the 18V port. Then connect the multiswitch's output port to your receiver. This hack lets you switch your receiver between horizontal- and vertical-polarity channels by telling it to receive either 14V or 18V input.

For one LNB output: Rotate the LNB filter so that its sticker aims to the left when you face the focal point of the dish.

2. Choose the satellite you wish to tune in and note its orbit, which is listed on its lyngsat.com page. Intelsat Americas 5 is a good one to start with because it has a very strong signal and it orbits at 97.0° W, positioned centrally to the U.S. (but geostationary over the equator, of course).

Determine the local azimuth and elevation of your target satellite by entering its orbit, along with your own latitude/longitude, into the satellite finder website at satsig.net/ssazelm.htm. Mount the dish to a stable and unobstructed surface outside, and then use your compass to pre-aim the dish towards the satellite. If the bird you want sits below your horizon, you need to pick another one.

3. Temporarily hook up a TV to the receiver using a short coaxial cable, and position the screen so you can see it while you aim the antenna.

4. Set the receiver's Local Oscillator (LO) frequency to 10750, for Ku band.

5. Refer back to your satellite's lyngsat.com page to find a channel bouquet that carries at a frequency of 11,700MHz or higher. On Intelsat Americas 5, a good one to try is at 12,177MHz, run by Pittsburgh International Telecommunications. This bouquet includes a nice variety of channels.

5a. In the bouquet's satellite page listing, note whether it transmits with horizontal or vertical polarity by looking for the "H" or "V" in the first column. Then note the signal rate and forward error correction values, listed in the column labeled "SR-FEC/SID-VPID." Our example, 12177 on satellite IA5, transmits with vertical polarity, and the SR and FEC are 23,000Mbps and 3/4, respectively.

5b. On your receiver, program the frequency of your chosen bouquet, and select its polarity.

5c. Set the bit rate value for the bouquet.

5d. If your receiver requires it, set the forward error correction value for the bouquet.

When you buy a used satellite dish, it should have a filtering module attached to it, typically a Ku-band Low Noise Block (LNB) filter with horizontal/vertical polarity. If you have an older dish, you'll also need a 2x4 or 3x4 satellite multiswitch, but if you are using a DIRECWAY/DirecPC or a Primestar dish (with just one output) you won't need the multiswitch.

6. Save these values and exit. Open the receiver's signal strength and quality screen, and choose the frequency that you just programmed in.

7. Watching the TV screen and your receiver's display, fine-tune the east/west aiming of the dish to get the maximum signal and quality level. Since this is a digital signal, you need to make small adjustments and then wait a moment after each change. This is a tedious process; it's much more difficult than with analog receivers, which react quickly to your movements. An inline analog signal strength meter can help, but an inexpensive one won't be able to identify the satellite name, and you might wind up zeroing in on a bird that's two degrees away from the one you wanted. If your elevation is off, you can adjust this as well, but note that moving in two axes makes aiming much more difficult.

8. Exit the signal strength screen and initiate a channel search or scan. If the channels you were aiming for show up, then you've aimed the dish correctly.

9. Once you know that you have a good signal from the right bird, program in all of the bouquets that are available on the satellite. Each of these frequencies is carried by a different transponder, which is why they're listed in a column labeled "Freq. Tp" (which stands for "frequency" and "transponder").

10. Initiate a full scan to find all of the channels on the satellite.

That's all there is to enjoying free digital MPEG2 satellite TV. If you get into it, you can buy a smaller, motorized dish to pick up other satellites. Sometimes you'll find surprise "wild feeds" that networks use to distribute prime-time programming to their affiliates for later broadcast. You'll get to see the shows before your friends! A few MPEG2 receivers even support hard drives for TiVo-like operation. The audio feeds you can find are also interesting, but don't expect a professional program guide. This is the Wild West of home entertainment, and it's every programmer for themselves!

Dave Mathews has been tinkering with TVROs since they were first affordable and has a very small array (VSA) of dishes in his backyard. More stories and video clips can be found on his website at davemathews.com.

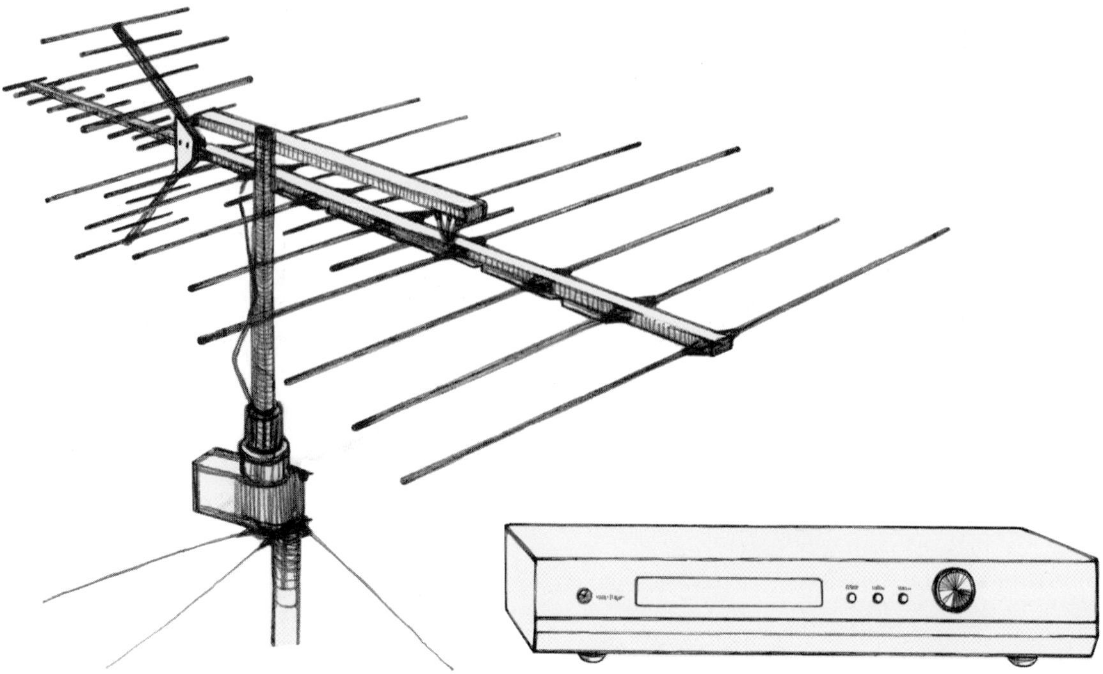

DIGITAL TV ON THE CHEAP

You can get many of the benefits of Digital TV (DTV) for as little as 20 bucks if you're willing to do a little tinkering. By Bob Scott

In most cities, some or all TV stations now broadcast free over-the-air (OTA) DTV signals (see checkhd.com). This signal, and some surplus gear, is the ticket to cheap DTV.

Despite what you might have heard, you do not need an HDTV set to benefit from DTV signals. An external DTV tuner will allow your faithful Philco to receive even high definition (HD) programming, just not at the resolution you see on the pricey plasmas down at the mall.

So why bother? First, even at the lowest DTV resolution (called SDTV, equivalent to current analog TV), the picture will be completely free of snow, ghosts, and other artifacts. If you can receive a digital signal it will be, by definition, perfect. Error-correction protocols in the transmission assure that the picture you see will be an exact replica of what left the studio.

Second, broadcasters can transmit additional channels called "multicasts" that are only available to digital receivers. These channels typically include news loops, weather radar, and children's programming, in addition to the normal network programming.

Finally, you'll get high-fidelity digital audio in either regular stereo or 5.1 Dolby Digital surround.

I got a refurbished Samsung SIR-T451 DTV tuner to see how hard it would be to make all this work (a used model sells for about $225; new ones are

You don't need a high definition television to get at least some of the benefits of DTV.

about $50 more). The VCR-sized unit has multiple outputs, ranging from radio frequency (RF) all the way up to digital video interface (DVI), making it compatible with a wide range of TVs.

A real hacker, though, goes for a surplus HDTV satellite receiver that incorporates an OTA DTV tuner. These can be had for a couple of bucks at a garage sale, but do your homework. Know exactly what you're looking for, as only certain hardware, firmware, and receiver status configurations will pick up OTA signals when not subscribed to a pay service.

The newsgroup alt.video.digital-tv is a good place to do your research. I've seen old Voom satellite boxes going for $20 on eBay. Although the company went under, the boxes contain a perfectly usable over-the-air HDTV tuner.

(If you're lazy, you can just sit back and wait until March 1, 2007. By that date, all TVs sold in the United States must have a built-in OTA tuner, as decreed by the Federal Communications Commission.)

DTV is generally broadcast on UHF, so you'll also need a good UHF antenna. A set-top loop might work if you're lucky, or you might need a roof-girdling Yagi on a tower if you're out in the sticks.

DTV signals are particularly sensitive to multipath (ghosting), so reception in urban areas can be problematic. If you can get a ghost-free UHF signal on analog TV with your antenna, you'll probably be in good shape for digital, but the only way to be sure is to try.

In my high-rise apartment, I could get only one of the six digital signals in my area, but it was gorgeous. I tried everything from a hunk of wire to an ultra-sophisticated amplified directional antenna, with the same results.

The next experiment was loaning the tuner to a friend in the suburbs of Houston. He connected it to his vintage outside antenna and instantly got every digital channel in the area — much to my annoyance.

I let him keep the tuner while I looked for a better antenna ... or a new apartment.

A little fiddling in my apartment resulted in an interesting discovery: my cable company was pumping HD down the cable, for free. With the proper tuner, I could get six HD channels for just the price of a basic cable hookup, no box or premium service required. Apparently this is getting more common, but the cable companies aren't exactly taking out any full-page ads about it, preferring to rent you an HD cable box.

Bob Scott is a statistical construct of various consumer electronics marketing departments.

Satellite TV History

Once upon a time, satellite TV was for regional cable companies, not homes. Geostationary satellites 22,300 miles high beamed unencrypted, free-to-air (FTA) broadcasts to Earth. Cable companies tuned them and rebroadcast the content to their subscribers. Home satellite TV debuted in the 1980s with 12-foot dishes that eavesdropped in on these signals. Private backyards started sprouting "BUDs" (Big Ugly Dishes) that moved from horizon to horizon in order to pick up the 24 C-Band analog channels that were spread across 30 different satellites. For a while, anyone who could afford tens of thousands of dollars for receiving equipment would be rewarded with free, unlimited programming, which included premium cable channels and commercial-free feeds from national television networks.

Before long, this equipment dipped below $3,000, and cable networks began losing significant revenue from satellite owners who had stopped subscribing because they were going directly to the source. Something had to be done. So in 1985, HBO and Showtime began scrambling their signals with an encryption scheme called Videocipher II. Under this regime, satellite users had to add a descrambler module in order to watch these premium channels. Of course, people soon figured out how to build their own decoders, or bought black-market versions. It was estimated then that only 10% of satellite dish owners were actually paying for their signals, and nearly all satellite dealers were also selling the illegal decoders.

The mid-1990s saw a closed and fully encrypted system that received digitally compressed signals through a fixed 18-inch microwave mini-dish. Today, DirecTV and EchoStar's Dish Network use this technology. It's still not uncrackable: in December, two men from Tennessee were charged with selling a $250,000 device that made illegal DirecTV decrypting cards. ***—Dave Mathews***

Illustrations by Damien Scogin

FULL MAST RECEPTION

Build your own satellite dish mast in three easy steps. By Joe Grand

If you live in an apartment complex and you've read the fine print of your lease, you might know that you're not allowed to attach anything to the railings, walls, or other physical property of the apartment. So, if you plan on setting up satellite TV (for example, DirecTV, Dish Network, or other services), you'll need to use a floor-mounted mast or base to hold your dish in place. The mast must be solid enough to stay put in rain and strong winds and if you bump into it. That way, the dish remains fixed on the network's geosynchronous satellites, orbiting at 22,300 miles above the Earth. The satellites remain in the same location in the sky, so little or no maintenance is required once the dish is aligned.

Some people use off-the-shelf tripod stands for this, but they're hard to secure unless you bolt them down or place a bunch of sand bags around them. This is not very aesthetically pleasing out on a patio, so let's build our own!

Step 1: Obtain the Parts

Visit your local home or hardware store and obtain these two items:

1. A heavy concrete umbrella base like the ones you commonly see at sidewalk cafes. I used the "Umbrella Base White 7-inch" from Home Depot, UPC #96968-60699 (which, despite the name, is 14" in diameter).
2. A 6' length of 1½" diameter fence piping.

For me, the total cost of the materials, including tax, was just $22 — well worth the money!

Step 2: Place the Base

Drag, lift, or otherwise move the umbrella base onto your apartment's patio, deck, or wherever else you desire. Be sure to wear shoes during this step, as the base can easily crush toes. Place the base in a spot where your antenna will have an unobstructed view of the southern sky, where geosynchronous satellites ring the equator.

Step 3: Host the Post

Conveniently, the 1½" diameter of the fence piping fits snugly into the umbrella base and also matches the mounting clamp of the satellite dish. I'm using a Phase III 18x20" Triple LNB dish manufactured for DirecTV, but most other dishes have similar mounting hardware.

Simply insert the fence piping into the umbrella base and tighten the base's setscrew to secure the pipe. Tighten the screw hard with a wrench, so it doesn't wobble or become loosened or misaligned over time.

With the post in place, verify that the post is vertically true, so you can accurately aim the dish. Use a standard level and just place it vertically along the post; tweak the post and concrete base as necessary. Two possible ways to even up the post alignment are to place a shim or two underneath the base, or to rotate it left or right by a few degrees at a time. When your post is firmly attached and vertically true, you can attach your dish and align it in accordance with your satellite receiver directions.

Step 4 (Optional): Watch TV

You're done! Now that your mast is complete, your dish is aligned, and all of the cables are routed properly to your TV or DVR, it's time to sit back and melt your brain with the multitude of fantastic television stations available on the satellite networks. If you're anything like me, you'll surely take much advantage of your newly built creation.

Joe Grand (joe@grandideastudio.com) is the president of Grand Idea Studio, Inc., a product research, development, and licensing firm. He is the author of the books *Game Console Hacking* and *Hardware Hacking: Have Fun While Voiding Your Warranty*.

A crop of makers' materials is ripe for the picking inside this old radio.

DUMPSTER CORNUCOPIA

Reusing components from discarded electronics. By Thomas Arey

Surface mount components have revolutionized all aspects of the electronics industry. However, they have been somewhat of a mixed blessing for electronics hobbyists. Homebuilders, especially those involved in the radio hobby, are experiencing a shortage of some through-hole (traditional leaded) components.

This parts shortage problem has happened before. The move from vacuum-tube-based technology to transistors, and the subsequent advance of integrated circuits over discrete semi-conductors, both sent electronics hobbyists in search of rare parts to build fun and exciting things.

Closeout and surplus dealers are always a possibility, as are online auctions. But I enjoy using the time-honored technique of salvaging parts from discarded electronic devices. Most households relegate broken or obsolete electronics to the trash because the industry has advanced to the point that new devices can often be purchased for less than the cost of repairing the original unit. The politics of this throwaway economy are beyond the scope of this article, but it remains both a preference and a challenge, for me personally, to turn the parts of discarded electronics into new projects.

I do this not only with electronics discarded in my own household, but also with devices scrounged from friends and neighbors. I even resort to a bit of casual dumpster diving when I am out walking my dogs on local trash days. It is not uncommon for my walks to turn up discarded radios, televisions, VCRs, children's toys, and

Photography by Thomas Arey

Fig. 2

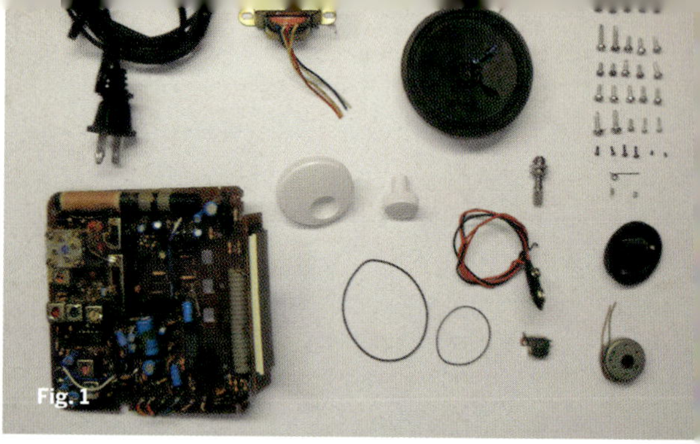
Fig. 1

Fig. 3

This simple radio took less than an hour to bring down to a nice quantity of useful parts for future projects. More complex devices, such as a VCR, can yield ten times the number of components, not to mention dozens of useful hardware pieces such as self-tapping screws.

even antique computer systems. All such devices can be treasure troves of useful and even rare parts.

Here are a couple of important warnings, though, before you join me in the quest for reusable electronics parts. First, be aware of how discarded electronics can cause you bodily harm — everything from sharp edges to imploding TV picture tubes. Handle all found items with caution and care. Also, be aware that some capacitors can hold a dangerous charge even in an unenergized, discarded device. Again, be cautious and treat every device as a live circuit, even after you have cut the line cord off and added it to the box under your workbench.

Finally, it's very unwise to pick through another person's trash without asking permission. Some people see this as a violation of their privacy.

As a test case for parts scrounging, let's take a look at this Soundesign model 3833 AM-FM clock radio/cassette player that I recently found in a trash can a few blocks from my home.

Initially, opening the case and dismantling the radio down to the circuit board level provides me with (Fig. 1): a line cord, 110-to-12-volt transformer, 3-inch speaker, a small motor from the cassette deck, 9-volt battery connector, recording head, and assorted knobs, belts, and hardware.

An examination of the circuit board shows many useful components (Fig. 2). You can strip a board such as this to whatever level you choose. I usually leave the standard capacitors and resistors but remove the semiconductors, ICs (if not proprietary), and any variable capacitors and potentiometers.

Final desoldering adds to my parts drawers (Fig. 3): an AM ferrite antenna, LED display, trimmer capacitor, potentiometer, variable capacitor, two multi-position switches, six slug-tuned inductors, voltage regulator IC, op-amp IC, two NPN transistors, and six diodes.

While not a perfect solution to the growing scarcity of some electronic parts, giving discarded electronics new life in projects can be a lot of fun.

T.J. "Skip" Arey N2EI has been a freelance writer to the radio/electronics hobby world for over 25 years and is the author of *Radio Monitoring: A How To Guide*, Paladin Press.

THE ROCKETMAN'S GARAGE

Ky Michaelson, a.k.a. The Rocketman, hails from an illustrious line of makers. On his family tree can be found inventors of the motorcycle transmission and clutch, the rotary lawn-mower blade, the flip-top aspirin box, and the oxygen mask as used in commercial aviation.

As a child dealing with dyslexia, Michaelson struggled in school, but discovered a natural affinity for building things. One day in math class his teacher discovered that he'd hollowed out his textbook and built a crystal set inside of it (the earpiece wire was concealed through his shirt sleeve). She was awestruck by his ingenuity (in addition to being pissed at him for destroying his book), and it was this ability to dazzle with his creations that sent him on a lifelong quest to build more impressive machines.

Michaelson is forever hungry for power and speed. In his Gyro Gearloose world, he's never met a moving vehicle he didn't think would go better with a big-ass rocket engine bolted to its backside. And, at this point, there aren't too many things he hasn't "rocketized," from motorcycles and go-karts to tricycles and kiddie scooters. And then there are the rocket-powered toilets and barstools. See more of Rocketman's crazed creations at the-rocketman.com.

—Gareth Branwyn

Photograph by Chad Holder

1. Replica of an early Michaelson motorcycle, originally built by Michaelson's great uncles of Michaelson Motor Company. 2. Rocket-powered go-kart. His kart holds the world record of 252 mph. 3. The payload section of Michaelson's Civilian Space Exploration Team's GoFast rocket, which, on May 17, 2004, became the first amateur rocket in space. 4. Dubbed the "SS Flusher," this rocket-powered toilet came about when somebody overheard Michaelson saying that a commode was one of the few things he hadn't made into a rocket. The bowl donor worked for the prison system and donated it on the q.t. 5. This hybrid (N_2O/plastic) trike was built for Michaelson's son, Buddy (age 6). Not surprisingly, "Rocketman" is his middle name. No, really. Legally. 6. An H_2O_2 rocketpack. Michaelson's trying to convince his wife Jodi to take it for a spin, which would make her the first woman to fly a jetpack. 7. An original 1912 Michaelson motor that the Rocketman found at a flea market in Davenport, Iowa. He'd been in search of one for 30 years. 8. A rocket-powered bicycle Michaelson built for Jodi using a Mongoose frame. 9. A twin-engine drag bike, built on a BSA frame. 10. A steering "engine bell" off an Apollo rocket. Michaelson has over 275 space artifacts in his home in the Minneapolis suburbs.

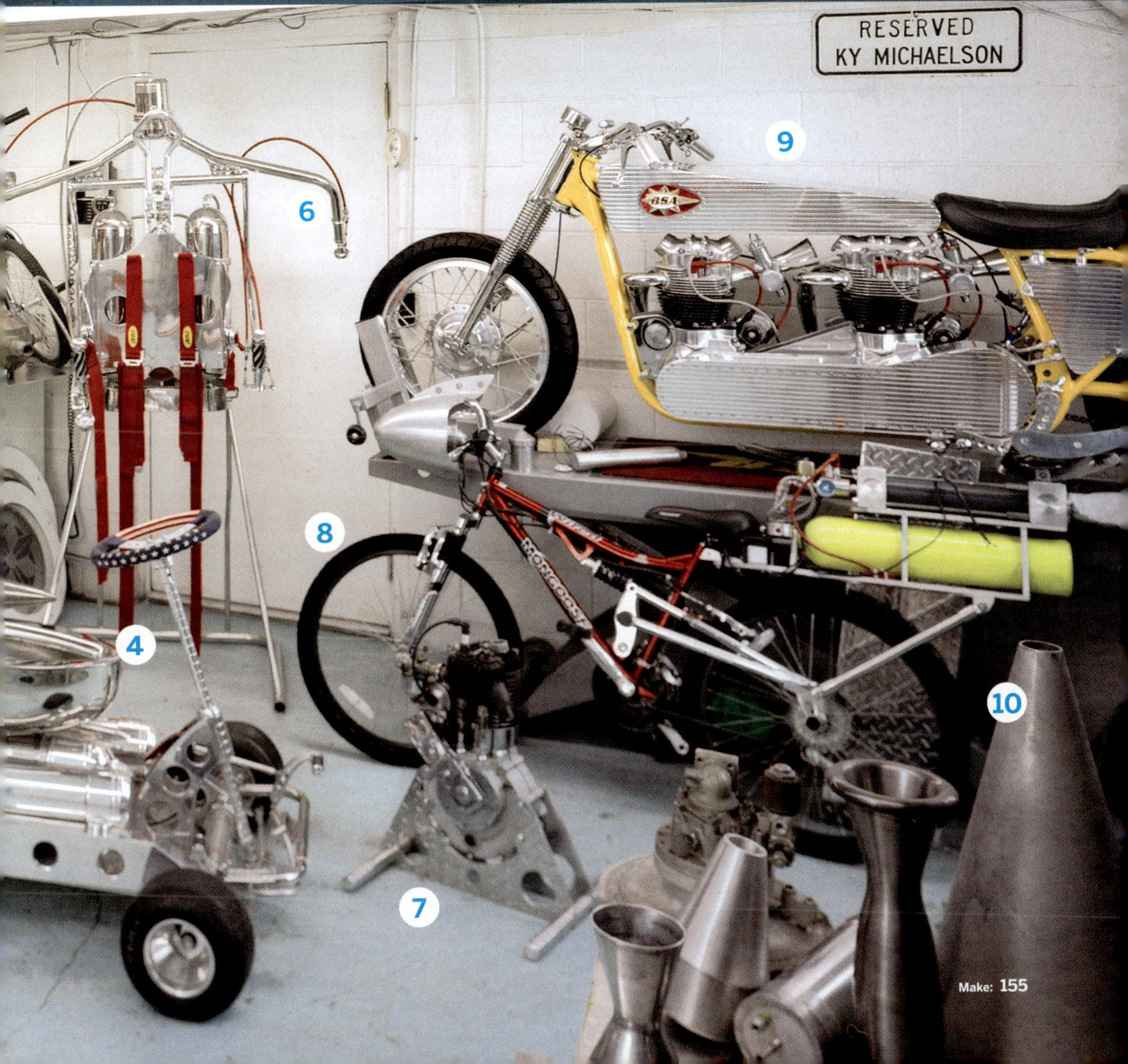

Puzzle This

By Michael H. Pryor

Welcome to the first installment of AHA!, a page of MAKE's favorite puzzles, selected by Michael Pryor. When you're ready to check your answers, visit makezine.com/05/aha.

The Rope Bridge (Easy)

Four people need to cross a rickety rope bridge to get back to their camp at night. Unfortunately, their only flashlight has just 17 minutes of juice left. The bridge is too dangerous to cross without a flashlight, and it's only strong enough to support two people at any given time.

Each of the campers walks at a different speed. One can cross the bridge in 1 minute, another in 2 minutes, the third in 5 minutes, and the slowpoke takes 10 minutes to cross. How do the campers make it across in 17 minutes?

Yarrr Maties (Sinister)

Five pirates discover a chest full of 100 gold coins. The pirates are ranked by their years of service, Pirate 5 having five years of service, Pirate 4 four years, and so on down to Pirate 1 with only one year of deck scrubbing under his belt. To divide up the loot, they agree on the following:

The most senior pirate will propose a distribution of the booty. All pirates will then vote, **including the most senior pirate**, and if at least 50% of the pirates on board accept the proposal, the gold is divided as proposed. If not, the most senior pirate is forced to walk the plank and sink to Davy Jones' locker. Then the process starts over with the next most senior pirate until a plan is approved.

These pirates are not your ordinary swashbucklers. Besides their democratic leanings, they are also perfectly rational and know exactly how the others will vote in every situation. Emotions play no part in their decisions. Their preference is first to remain alive, next to get as much gold as possible, and finally, if given a choice between otherwise equal outcomes, to have fewer pirates on the boat.

The most senior pirate thinks for a moment and then proposes a plan that maximizes his gold, and which he knows the others will accept. How does he divide up the coins? And what plan would the most senior pirate propose if the boat had 15 pirates instead of just five?

Illustrations by Roy Doty

Michael Pryor is the co-founder and president of Fog Creek Software. He runs a technical interview site at techinterview.org.

IR Remote Control Protocol

Get an infrared remote to turn your room lights on and off.

By Andrew "Bunnie" Huang

THEORY

Back when I was an undergraduate at MIT eight years ago, my fraternity had a loft in every room. This suited me fine, but there was no way to turn the light switch off from up there when it was time for some sleep, and I certainly did not want to trundle across a floor land-mined with DIP ICs, papers, and tools, then climb up the precarious ladder to my roost without the benefit of sight.

Meanwhile, I also had a Sony stereo system with a remote control that sported a TV power button, but I had no TV. And so, like peanut butter and chocolate, two worlds came together. I decided it was time to put that derelict TV button to good use: switching the lights off at night. Doing this meant making a light switch that could understand signals from an infrared remote control — specifically, one that used Sony's IR protocol, since different manufacturers do infrared in different ways.

How IR Remotes Work

At a physical level, an IR link consists of an infrared LED on the remote, which typically emits light at a wavelength of around 940-980nm, and an infrared photodiode detector on the main piece of equipment. The detector is covered by an IR-selective filter, usually a deep red piece of plastic, that helps block stray ambient light. The detector turns incident photons into electrons that can be easily sampled and amplified.

A naive implementation of an IR link would be to transmit a beam of infrared light to represent a "1" or On condition, and use the absence of signal to indicate "0" or Off. While this can yield a functional link under controlled conditions, there are several reasons why it won't work at home. Many household light sources emit energy at wavelengths that excite the IR detectors typically used in home electronics. This would make it difficult or impossible for the receiving device to distinguish intentional IR signals from stray noise in the red/infrared range — for example, sunlight reflecting off a Coke can. This problem is known as spurious pulse rejection.

> "I decided it was time to put that derelict TV button to good use."

Another problem is that remote controls need to work in a wide range of ambient light conditions, between bright sunlight and pitch dark. Without some encoding scheme for the signal, it's difficult to build a receiver that can filter out such a large possible range of ambient light and still be sensitive to the data it needs to receive.

The solution that most IR remotes use is to modulate the IR light with a carrier frequency. My Sony remote, for example, used a carrier frequency of 40kHz, which is fast enough and regular enough that it is unlikely to be confused with random fluctuations and reflections. This frequency also allows noise from ambient light hitting the IR photodiode to be largely filtered out by a relatively small coupling capacitor.

Illustration by Dustin Amery Hostetler/ UPSO.org

Many IR receivers add more complex circuitry to comb out spurious signals prior to bit detection. These include tuned filters that select (bandpass) and reject (notch) particular frequencies, and synchronous detection circuits, which multiply the signal by the carrier frequency you are looking for. Because sine waves at unrelated frequencies cancel each other out overall, this multiplication results in only the carrier frequency and its harmonics being left with a nonzero result.

Detector modules built to support the 40kHz standard also make great sensors for object detection in robotics, because of their robust handling of spurious signals.

IR Protocol

Using a carrier frequency will solve the general problem of spurious signals, but you still need a way to encode data for transmission. Sony remote controls use what's called pulse-duration modulation, where the value of a bit is communicated by the duration of a pulse of the carrier frequency — not unlike the dashes and dots of Morse code. In Sony's scheme, 1 is represented by a 1,100-microsecond (µs) burst of 40 kHz light, and 0 is represented by a 550µs burst.

Data streams over the IR link in LSB-first order (LSB = least significant bits), with an extended start bit of 2.4-millisecond (2,400µs) duration to indicate the beginning of a signal transmission. The extended start bit allows IR receivers to save power by running idle during periods of no transmissions, then have enough time to wake up and watch when there's a start signal.

Each key-press event on a remote sends an IR data packet consisting of two parts: a 7-bit command followed by a 5-bit device address. The command encodes the identity of the key that was pressed, and the device address specifies what type of appliance the key press was meant for. For example, the power button is key number 21, a TV is at device address 1, and a CD player is at device address 17.

False IR Signals

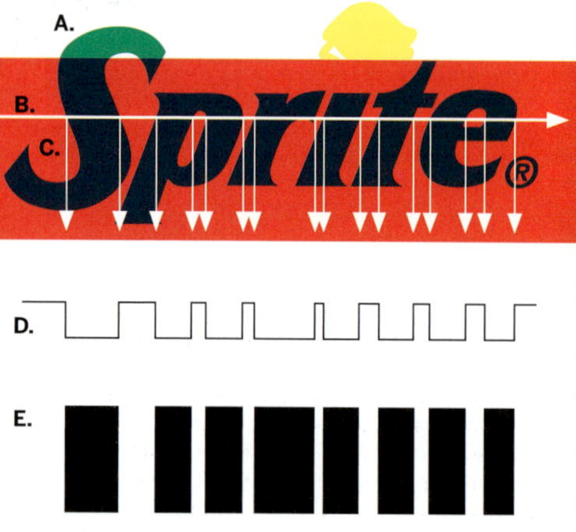

A. Bright, shiny can logo
B. Path of reflected light
C. On/off points for false binary, from reflections off can
D. False binary signal
E. Equivalent bar code

IR protocol is designed to prevent the possibility of spurious signals, but in the noisy, light-filled world, there are unlikely instances where an IR receiver might detect something that looks like binary data.

Consider this pathological case (of soda): Someone in the room moves a shiny, printed aluminum can near a window in the sun. Sunlight reflects off the can, and bounces directly into the IR receiver. The path of incidence of the beam hitting the sensor also happens to travel across a two-color logo. One of the logo's colors absorbs IR, while the other color reflects it. As a result, the sensor gets a short blast of fast-pulsing IR.

With a sensor module that combs everything it sees for binary patterns, this random reflection might appear to be a signal from a remote. Hopefully it doesn't look like the dreaded "self-destruct" button.

It's also possible to intentionally create reflective patterns that you shine light across and read as data. This is how bar codes work.

PRACTICE

Knowing this background, it's not difficult to build an IR receiver that operates a relay for switching AC devices like lamps on and off. It will be a small box with an IR detector peeking out the side, attached to the middle of an ordinary extension cord.

The Brains

The brains of our IR receiver light switch will be an 8-bit microcontroller chip (a PIC16F84A), running my adaptation of code originally written by R. Dunbar Poor that enables it to perform remote control tasks. This code is republished with his permission at web.media.mit.edu/~ayb/irx/irx2. In addition, our circuit also needs a crystal oscillator and a voltage regulator, to support the PIC.

The Eyes

For our system's "eye," we can use a standard IR demodulator such as Sharp Electronics' GP1UD28YK, available from Digi-Key for about $1. These wonderful modules perform the tasks of IR signal detection, amplification, limiting, filtering, demodulation, integration, and digitization, all in a very convenient 3-pin package. The Sharp IR module connects easily to a PIC; you just wire up its power and ground pins, then connect its output pin to one of the PIC chip's inputs. I chose to connect to the chip's port B4 (pin 10).

The Switch

To actually switch the lamp on and off, we can use any 5V-activated power relay with sufficient current rating for the lamp; one example is the Omron G6B-1114P-US-DC5. Unfortunately, the PIC's output current of 25mA isn't quite enough to activate most relay coils, which need something in the high tens or low hundreds of milliamps (the Omron relay needs 40mA). So we use a transistor to boost the current.

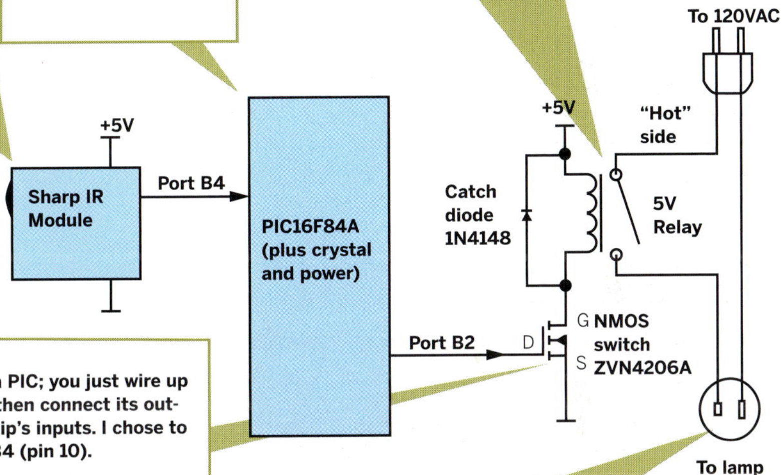

The Booster

To give the PIC output a boost, we can use a transistor, such as the Zetex ZVN4206A; it has a continuous drain current rating of 600mA and a low gate-source threshold voltage (<3V), which means the PIC output will be able to drive it. We can wire up the entire circuit on a PICProto18 prototyping board; the diagram at right shows how the main components connect.

VERY IMPORTANT: Once your switch is built, before you plug in for the first time, verify all your connections with an ohmmeter; there should be near-infinite resistance between all high-voltage lines and every other node in the circuit. If so, it's safe to plug the cord in, plug your lamp (or anything else) into the cord, and click-on, click-off. Pleasant dreams!

The Plug

After you've built the remote control switch box, you'll need to hook it into the extension cord that runs between the wall socket and the lamp. Take an extension cord, cut its "hot side" conductor (the narrower, right side of the plug), and splice the switch's relay terminals in-line into this conductor. Be careful when wiring up the 120VAC lines. As you well know, dangerous voltages will flow through these wires, so you need to use grommets to prevent the edges of your case from cutting into the wire. It's also a good idea to tie strain-relief knots in the cords, to prevent sudden tugs from pulling the wires out of your board. For good measure, seal any exposed high-voltage wires with a dollop of hot glue.

Andrew "Bunnie" Huang lives in San Diego. He likes to hack.

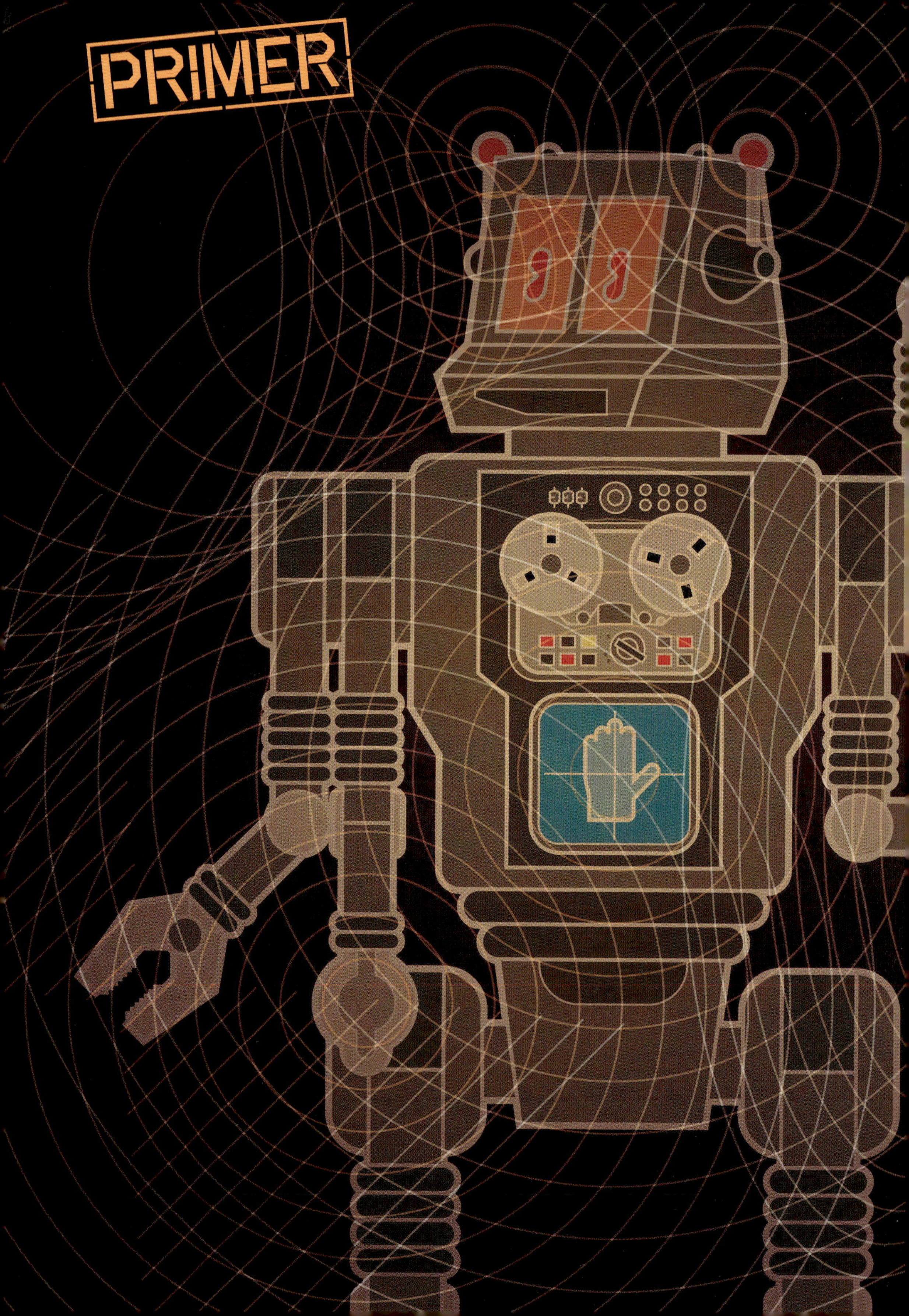
PRIMER

Illustrations by Nik Schulz

Sensor Interfaces

How circuits communicate with the outside world.

By Tom Igoe

A typical home appliance contains several sensors, from the switches and dials you interact with to the temperature sensors and limit switches that keep your coffee pot from overflowing or setting the house on fire when the last cup is gone and the heater's left on. Sensors are central to a hardware hacker's vocabulary. The more of them you know about, the more responsive your devices can be.

What follows is an introduction to the characteristics of most sensors, so you know the range of possibilities. Later, I'll show how to use a few specific sensors.

This article assumes you've got a basic understanding of how electricity works. If you don't, check out Forrest M. Mims' book, *Getting Started in Electronics*.

Types of Sensors

Electronic sensors convert other forms of physical energy into measurable electrical properties.

All sensors can be grouped in two major categories: **digital sensors**, which output one of two possible states, on or off, and **analog sensors**, whose outputs change continuously with the changing physical energy that they read.

Voltage output sensors output an electrical voltage proportional to an input signal.

Example: **A photodiode-based sensor such as the GP2D120 (page 165) can be used as a range finder for the robot to detect the distance of an obstacle.**

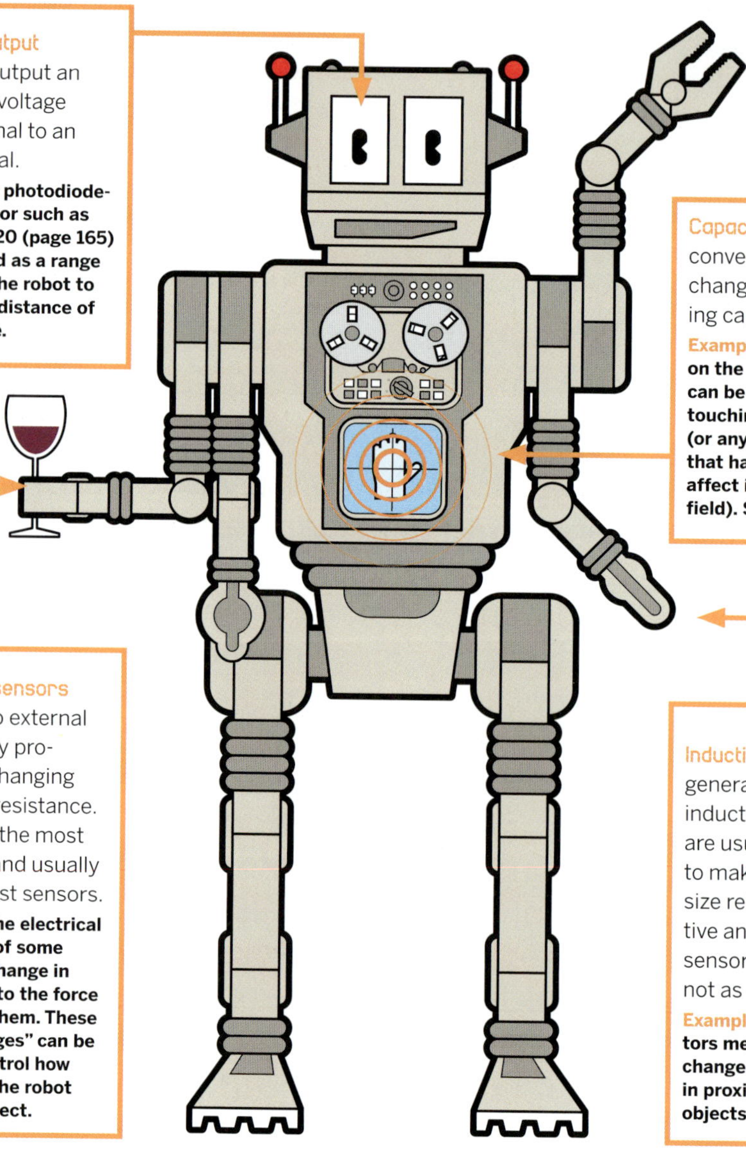

Capacitive sensors convert external changes into a changing capacitance.

Example: **The touchpad on the robot's chest can be activated by touching it with a finger (or any other material that has the ability to affect its electrical field). See page 166.**

Resistive sensors respond to external changes by producing a changing electrical resistance. These are the most common and usually the simplest sensors.

Example: **The electrical resistance of some materials change in proportion to the force applied to them. These "strain gauges" can be used to control how forcefully the robot grips an object.**

Inductive sensors generate a changing inductance. These are usually expensive to make and large in size relative to resistive and capacitive sensors, so they're not as common.

Example: **Metal detectors measure the change in inductance in proximity to metallic objects.**

Sensor Interfaces

In order to use sensors, you have to know how they interface with your computer, microcontroller, or circuit. Sensor devices typically combine a sensing component with filter circuits and amplifiers to convert the sensor's electrical changes into an electrical signal that's usable by microcontrollers and other electronic devices.

There are a handful of common types of sensor interfaces. Some can be used directly in an output circuit, and others need to be connected to a computer or microcontroller.

The simplest sensor devices output a changing DC voltage. An **analog sensor** will output a voltage that changes continuously across its output range, and a **digital sensor** will output either the top of its range or the bottom. For example, the Sharp GP2D120 analog sensor outputs a changing voltage, from 0 to about 3 volts, that corresponds to a person or object's proximity to the sensor. The Quantum QT113H sensor, a digital sensor, outputs 0 volts when a person is in contact with the sensor, and 5 volts when no one's in contact with it.

Pulse width output sensor devices will generate a series of digital pulses that change with the energy they're sensing. For example, a Taos TSL230 sensor converts the intensity of light received into a changing frequency of pulses. You'll often see the term PWM, or pulse width modulation, to refer to this form of output.

More complex sensor devices output digital data serially, either using an asynchronous serial protocol that's common to most personal computers, or using a synchronous serial protocol.

Asynchronous serial devices work just like any other device that connects to a computer's

SPI (Serial Peripheral Interface)

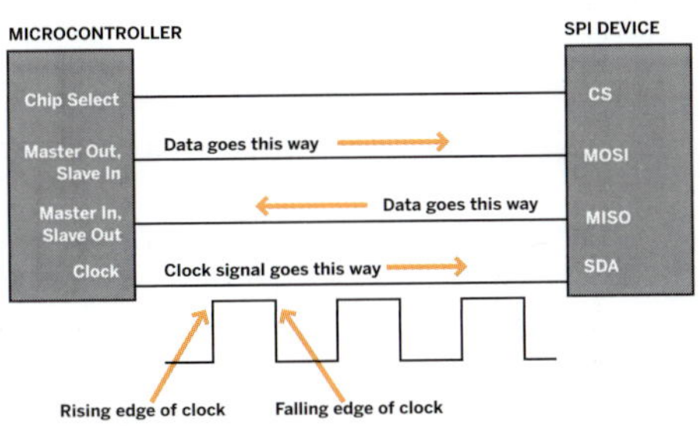

SPI devices, which have four connections and can achieve higher data exchange rates than I²C devices, are capable of exchanging data on the rising edge of the clock or the falling edge of the clock.

I²C (Inter-integrated Circuit)

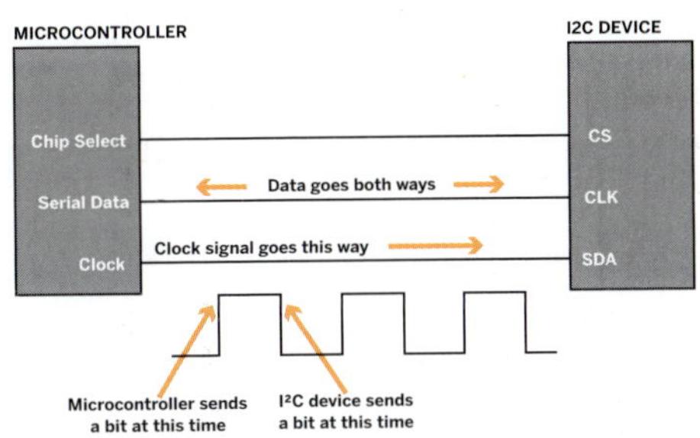

I²C devices, which have only three connections, exchange data on both edges of the clock; the microcontroller sends out on the rising edge, and reads in on the falling edge.

serial port. They have three connections: one for transmitting data, one for receiving data, and one that's a common ground. Some can be connected directly to a PC's serial port. Others can be connected through a signal converter like Maxim's MAX232 chip. Once you know the sensor device's data protocol, you can read the data right from your PC.

Synchronous serial devices don't have their own oscillators, so they need an oscillator (clock) signal from another device. The usual way to use a synchronous serial sensor is to connect it to a microcontroller, which then interfaces to an output device, desktop computer, or network. There are two common flavors of synchronous serial communication: SPI and I²C.

SPI (Serial Peripheral Interface) devices have four connections: clock, data output (sometimes called MISO, or Master-In-Slave-Out), data input or MOSI (Master-Out-Slave-In), and a chip select pin. Because it's possible to connect several SPI devices in parallel on one data bus, the chip select pin is used to tell a chip to respond to or ignore incoming data.

I²C (Inter-integrated Circuit) devices have only three connections: clock, data, and ground. The microcontroller that controls an I²C device shares the data line with the device. Typically, the controller sends a bit of data when it changes the clock pulse from low to high, and the controlled device sends data when the clock pulse goes from high to low.

Most microcontrollers have I²C and SPI interfaces built in, and software libraries for using these interfaces so you don't have to know a whole lot about SPI or I²C, as long as you know the format of the sensor device's data and which pins to connect.

Every sensor has a data sheet from its manufacturer that explains its physical and electrical characteristics, and its operating conditions. If you're lucky, sample circuits and applications will be included as well. When working with a new sensor, download the data sheet first and have it available to figure out the details.

Thermocouple: Simple Voltage Output Sensor

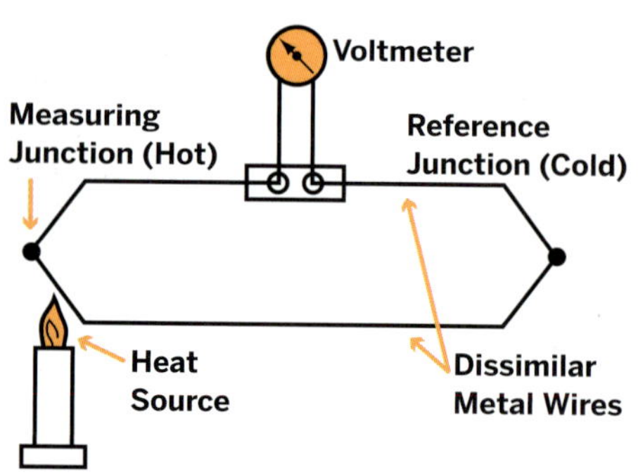

Thermocouples are inexpensive and widely used voltage-based temperature sensors. In 1831, T. J. Seebeck discovered that two dissimilar metals will produce a voltage that's proportional to the temperature difference between two junctions of the metals.

Examples: Voltage Output Sensors

Sensors that output a changing voltage are generally the simplest sensors to use, both with a microcontroller and with other circuits. Following are two examples: the Sharp GP2D120 infrared distance-ranging sensor (page 165), which detects an object and reports its distance to the sensor as a changing voltage from 0 to 3 volts, and the Quantum QT113H capacitive touch sensor (page 166), which reports the presence of a person's touch by changing its output from 5 to 0 volts. Both sensors are available from a variety of retailers. I get my Sharp sensors from Acroname (acroname.com) because they also sell a pre-made connector cable for these sensors, which can be a pain to make yourself. Digi-Key (digikey.com) sells the QT113H sensors.

If you're interested in using sensors, you should learn how to use microcontrollers as well (*see Primer in MAKE Volume 04 for a tutorial on microcontroller programming*). Microcontrollers are the small, simple computers that drive everything from your car engine to your light switches. The most common way to use a sensor is to connect it to a microcontroller that reads the sensor and controls some form of physical output device based on the sensor's readings. Parallax's (parallax.com) Basic Stamp is an excellent microprocessor development platform for sensor-based application. The company also sells a wide variety of sensors (search for "sensor" on Parallax's website).

GP2D120 Infrared Distance-Ranging Sensor

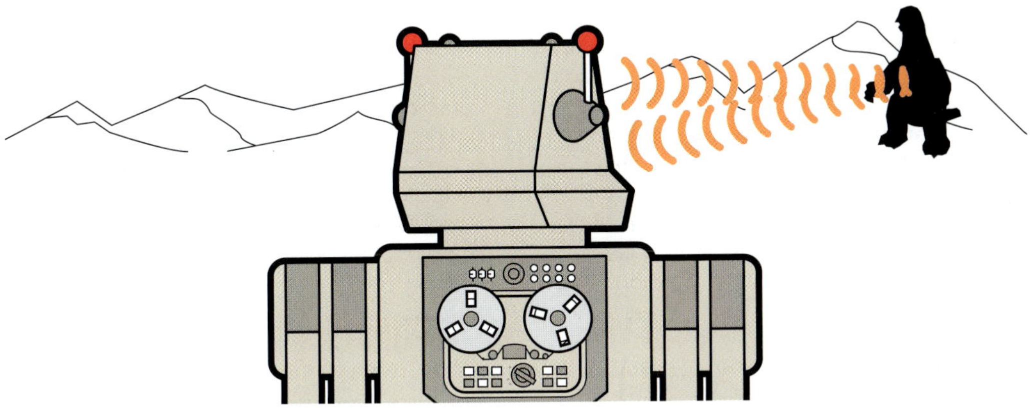

The GP2D120 measures distance by bouncing a beam of infrared light off a target and reading the reflection with an infrared photodiode that converts the incoming light to a variable voltage. The closer the target, the brighter the reflected light, and the higher the output voltage. It has a range of 1.5 to 12 inches.

This sensor comes from a family of IR ranging sensors from Sharp. Some of them have an analog voltage output, like the GP2D120, and others have a discrete output; that is, they output 0 volts until the target crosses a threshold, and then they output 5 volts. The GP2 sensors come in a range of distance sensitivities, from this one, which has the shortest range (4 to 30cm), to the GP2Y0A02YK, which has an analog range of 20 to 150cm.

The connections for the GP2D120 are about as straightforward as you'll ever see in a sensor. There are three pins: a voltage supply pin that you connect to a 4-6V DC source; a ground pin that you connect to ground; and an analog voltage-out pin that you connect to your microcontroller or sensing circuit. Place a 10-microfarad electrolytic capacitor between the voltage supply and ground as close to the sensor as possible to minimize power fluctuation when the sensor draws power.

You can build a circuit to sample and hold the sensor's voltage, or to convert it to a frequency, but I prefer to do that work using microcontrollers. Most microcontrollers have an onboard analog-to-digital converter (ADC) connected to one or more of their pins, to which you can directly connect the GP2D120's output pin. This sensor's output is not linear, but follows a curve, so using a microcontroller is an easy way to make the output linear, with a little math. See the sensor's data sheet (acroname.com/robotics/parts/R146-GP2D120.html) for the details of the output curve.

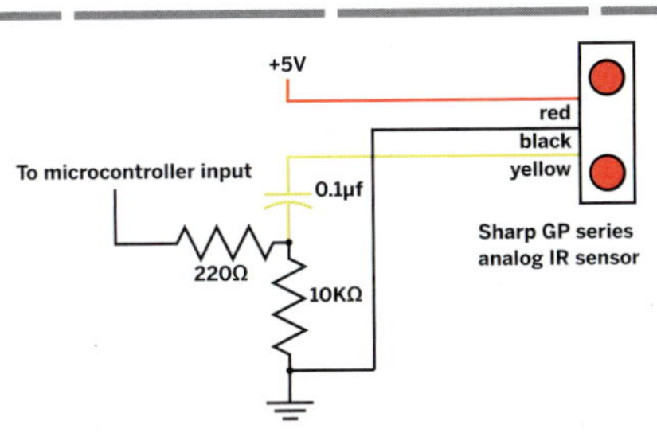

The Sharp GP2D120, attached to an AVR ATmega8 microcontroller. The schematic shows only the sensor's connections. Connect to your own circuit or microcontroller wherever it can take 0-3V DC input.

Example code to use the GP2D120 on the following microcontrollers can be found at makezine.com/go/sensor.

» NetMedia's BX-24 microcontroller
» Microengineering Lab's PicBasic pro for the PIC microcontroller
» Wiring/Arduino syntax for the AVR ATMega8

QT113H Capacitive Touch Sensor

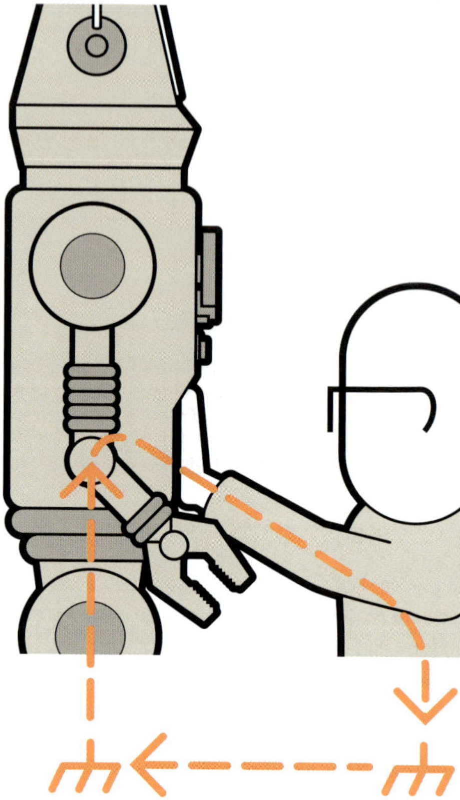

Objects and people always carry an electric charge, and every object or person has a slightly different charge. When two objects or people touch, they exchange electric charge, so that they end up with an equal charge. Capacitive sensors detect charge when we touch an object they're attached to.

Since humans and other animals are basically bags of electrolytes in water, we cause changes in the electrical fields we pass through. The Quantum Technologies QT113H touch sensor measures the change in an electrical field and outputs a voltage when the change crosses a threshold. It responds to human touch, and can read a change through several millimeters of non-conductive material, so it's a great sensor to use when you want to make a fabric switch, or a stuffed animal that responds to touch, or to make just about any surface responsive.

Capacitive sensors take advantage of the fact that objects and people carry an electric charge, and that every object or person has a slightly different charge. When two objects or people touch, they exchange electric charge, so that they end up with an equal charge. Charge-transfer sensors detect charge when we touch an object they're attached to. A conductor is attached to the object, or embedded in it, and when a person comes close to the sensor, charge is transferred, and the sensor reads it.

To use the sensor, you need to attach a piece of metal to one of its input pins. Copper mesh makes a great sensing electrode for this sensor, and it's relatively easy to solder a wire to. It can also be embedded under the surface of a fabric, wood, or plastic object fairly easily. The wire you use to connect the sensing electrode to the QT113H is also sensitive, unless you shield it. Shielded wire is easy to come by — most computer cables are shielded, and old ones can be chopped up to do the job. The shield is the metal foil or wire braid that surrounds the inner wires. Attach it to the ground of your circuit board. The end nearest the sensing electrode can hang free. Any wire that's outside the shield will sense touch, along with the electrode itself.

The sensor itself comes in two different physical packages, a SOIC chip or a DIP chip. The former is a

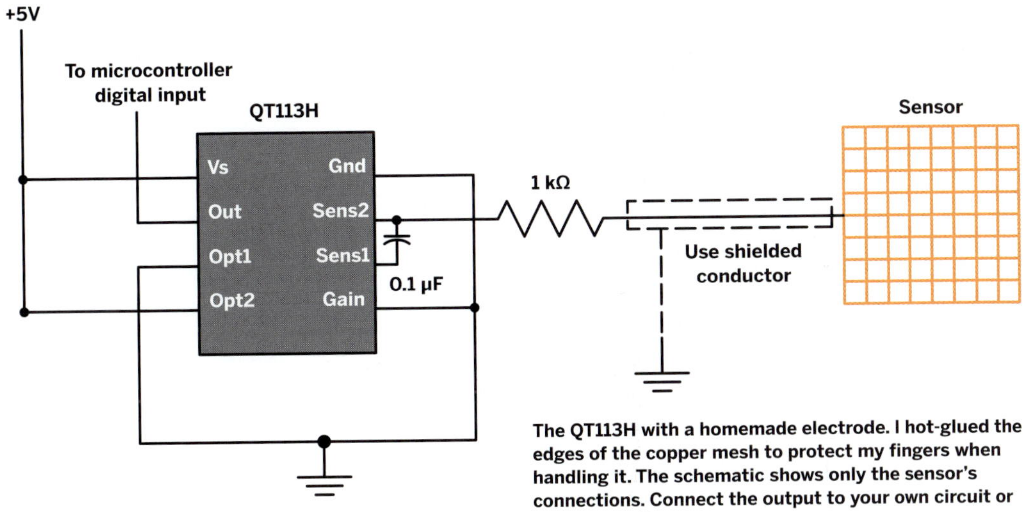

The QT113H with a homemade electrode. I hot-glued the edges of the copper mesh to protect my fingers when handling it. The schematic shows only the sensor's connections. Connect the output to your own circuit or microcontroller.

surface-mount chip, very small, difficult to solder to a printed circuit board by hand, and impossible to use directly with a solderless breadboard. I prefer the DIP models for hobby projects. The top of a DIP chip is usually marked with a notch, and Pin 1 is usually marked with a dot. The eight pins of the chip are counted in a U-shape from the top left to the top right.

The pins of the QT113H are as follows:

Pin 1: Supply voltage (V). Connect to 5V DC.
Pin 2: Output. Outputs 0V if there is contact, 5V if not.
Pin 3: Option Pin 1. The option pins control the sensor's reaction to prolonged contact. The sensor normally outputs 5V on contact, and outputs 0V after a set timeout. By changing the states of the option pins, you can set the timeout to 10 seconds, 60 seconds, infinite, or you can set the sensor to toggle every time it's touched. For details, see the sensor's data sheet (qprox.com/downloads/datasheets/qt113_105.pdf). I set the option pins for an infinite timeout by connecting Option Pin 1 to ground, and Option Pin 2 to 5V.
Pin 4: Option Pin 2.
Pin 5: Gain. Connect this to 5V for high sensitivity, or to ground for low sensitivity. In this example, I connected to ground to get more consistent results.
Pin 6: Sensing Pin 1. Connect this to Sensing Pin 2 using a capacitor. Quantum recommends 10 nanofarads, but in this example, I used a 0.1-microfarad capacitor, which worked well.
Pin 7: Sensing Pin 2. Connect this to Sensing Pin 1 using a capacitor, and to the sensing electrode. Put a 1-kilohm resistor in series with the electrode.
Pin 8: Ground. Connect this to the ground of your circuit.

With the QT113H connected like this, it will output 0 volts whenever it's touched, and 5 volts when it's not. You can connect it directly to a relay or transistor to control a motor circuit, or to a microcontroller as part of a more complex system.

If you understand the basic types of sensors and the sensor interfaces shown here, you can figure out how to use most sensors on the market. In following articles, I'll explain how to use a variety of more complex sensors. For a list of sensor-related resources, visit: tigoe.net/pcomp/resources.

Example code to use the QT11x family on the following microcontrollers can be found at makezine.com/go/sensor2.
- **» NetMedia's BX-24 microcontroller**
- **» Microengineering Lab's PicBasic pro for the PIC microcontroller**
- **» Wiring/Arduino syntax for the AVR ATMega8**

Tom Igoe teaches the sensor workshop class and heads the physical computing area at the Interactive Telecommunications Program at New York University.

HowToons.com

BY SAUL GRIFFITH, NICK DRAGOTTA & JOOST BONSEN '05

THESE LIME GREEN LEAKS INSPIRED BY SCIENCE HULK DANIEL ROSENBURG!

From the forthcoming book *HOWTOONS Book 1* by Saul Griffith, Nick Dragotta, and Joost Bonsen. Published by arrangement with ReganBooks, an imprint of HarperCollins Publishers, Inc.

MakeShift

By William Lidwell

This issue's MakeShift challenge is brought to you by Woody Norris. Woody will also participate in the analysis and winner selection.

The Scenario: You and Woody have been asked to evaluate a new bank vault with a time-release mechanism that cannot be overridden. Given the size of the vault and the ¾" vent tube in the ceiling, a person could survive no more than 24 hours before the CO_2 level becomes lethal. Woody claims that with its current contents (left inside by the construction crew), he could survive 48 hours. You feel like he is challenging you so you agree with him, even though you don't see an immediate solution. The next thing you hear is Woody saying, "OK ... see you in 48 hours!" The vault door closes and the lock is set to 48 hours before you can do anything.

The Challenge: Create a makeshift solution to stay alive until the bank vault unlocks. The lights and water cooler have electricity, but that's it. Good luck!

Supply List:

A small table with a desk telephone (line is dead)

1 water cooler with paper cups

4 emergency lights with battery backup

1 steel cable (about 10 feet)

1 garden hose

1 fire extinguisher (CO_2)

Toolbox containing hammer, pliers, screwdriver, pocket knife, electrical tape, chalk-line reel, yardstick, wire, and dirty rags.

Send a detailed description of your MakeShift solution with sketches and/or photos to *makeshift@makezine.com* by April 15, 2006. The most plausible and most creative solutions will each win MAKE T-shirts and a SwissMemory USB Victorinox 512MB. For rules and solutions to previous MakeShift challenges, visit *makezine.com/makeshift*.

William Lidwell is a consultant with Stuff Creators Design Studio and co-author of the book *Universal Principles of Design*.

Photograph by Jill Butler and Christopher Hujanen

MAKE

TOOL BOX

The best tools, software, gadgets, books, magazines, and websites.

1 The Perfectly Balanced Kitchen Knife

Wüsthof 10-Piece Classic Block Set
$400 wusthof.com

▪ Before discovering the awesome blade of the Wüsthof knife, I avoided recipes that called for cutting up fresh ingredients. Cutting things is more fun when you're doing it with a high-carbon specialty knife that was crafted with a laser beam.

Now, armed with my Wüsthof Classic knife set, I've been tossing freshly cut veggies into omelets, creating my own fruit trays, and leaving precut loaves of bread for the culinary patzers.

The knives come with a sharpening steel, and if you use it correctly, they should last longer than you will. I sharpen them every three or four uses, and it's working out fine. The entire set fits into a nice wood block for you to store away, and you get a lifetime warranty that covers defects in craftsmanship.

—Matthew Russell

2 Turn Your Records into MP3s

ART Pro USB Phono Plus $99
artproaudio.com

▪ You have "Kung Fu Fighting" on vinyl, but how do you get that into your computer to turn it into an MP3? Well, Grasshopper, try this little USB audio interface. A phono cartridge needs a pre-amp with special equalization. The USB Phono Plus has true phono inputs — along with line-level and digital audio connections — in a solid metal enclosure. It works with standard USB audio drivers on any recent Win or Mac system, and includes a wall wart so you can use it as a standalone phono pre-amp.

My favorite feature? It has a *knob*. The gain control lets you fine-tune the volume before digitizing, to record the level as high as possible without clipping. But trust your recording software's level meters; my clip LED on the front panel flashed red even when there was a bit of headroom to spare. No complaints about the sound quality, though. I can't hear any difference between my original vinyl and the digitized version.

—Ross Orr

Photography by Howard Cao

3 5-Gallon Bucket Organizer

Bucket Boss 56 $20
duluthtrading.com

▪ Who doesn't love a white 5-gallon bucket? I've personally seen 'em used as beer coolers, impromptu stools, ashtrays, boombox holder-uppers, paint mixers, clam chowder storinators, and — during one troubling night in a forest primeval — I saw one installed with a garbage bag and used as an ersatz camp toilet. Poor little bucket.

The good folks at Bucket Boss built an entire company around things you can do to mod your plain old 5-gallon bucket. Their flagship product is the Bucket Boss 56 — a super-dense, water-resistant fabric sleeve that slips over the lip of the bucket to provide a geeky embarrassment of pockets for all the tools and gadgets in your world. That's 56 pockets to be precise — 38 running around the outside, plus 18 more inside. It even has a duct tape dispenser and a big-ass holster for your drill. Rawr.

Knowing they've got a good thing on their hands here, the Bucket Bosses offer a variety of additional products, including a compartmentalized lid for storing screws and nuts, a wheeled dolly for pushing the bucket around, a cushioned grip for the handle — even a model with a little cushion that turns your bucket into a comfy-looking seat. They also sell something they call the Mug Hugger, a Bucket Boss scaled down to fit on a mug for organizing your pens, pencils, and scissors, and, "in the cause of modesty and good taste," the Longtail T, designed to make plumber's butt a thing of the past.

While some of this aftermarket frippery may turn off your stauncher bucket puritans, MAKE enthusiasts can appreciate the extreme range of happy hacks gladly accepted by our humble 5-gallon friend. If containers were software, the bucket would be GNU/Linux, and the Bucket Boss 56 would be its cleverest value-added reseller.

—Merlin Mann

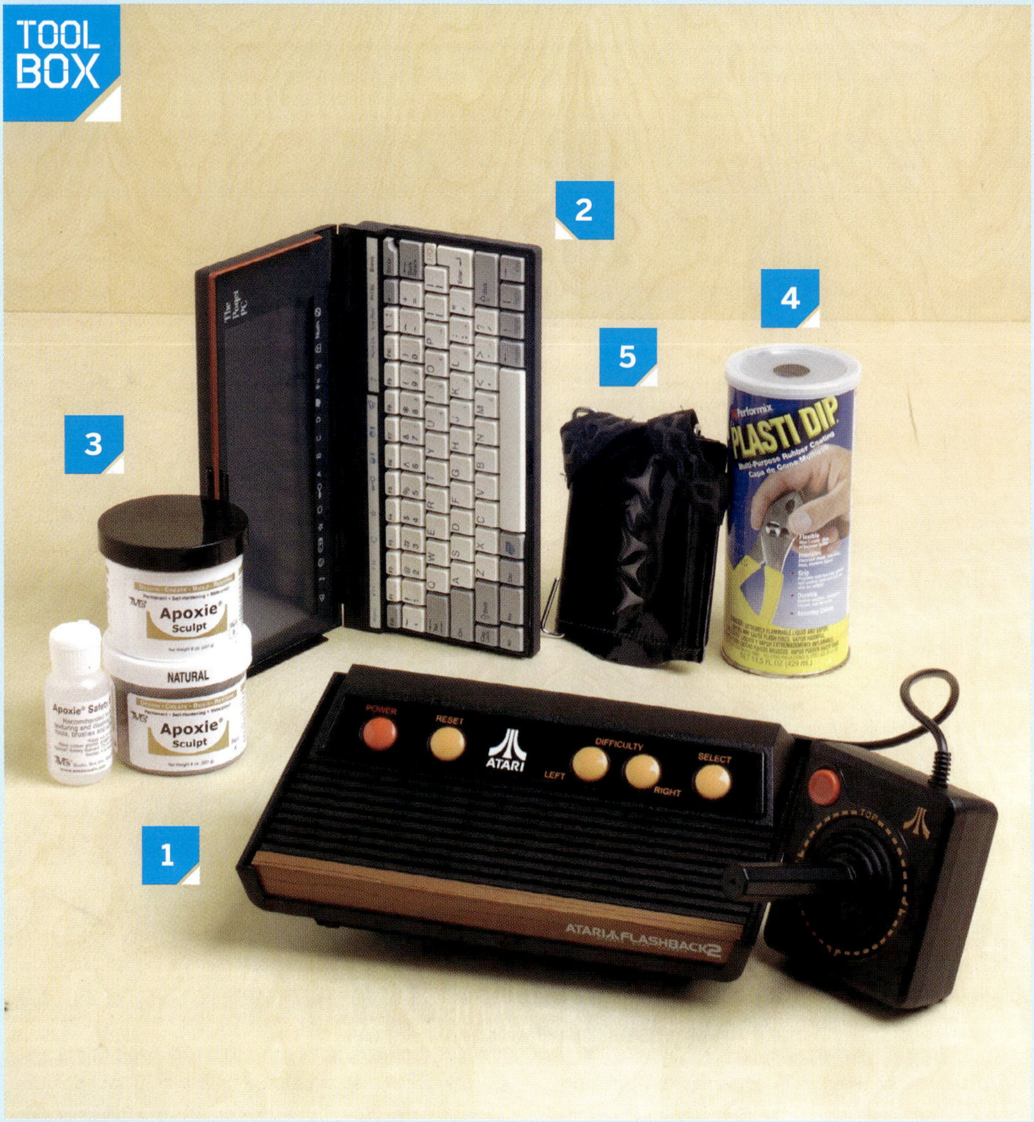

1 A Better (and Hackable) Retro Atari

Atari Flashback 2 $30

atari.com

The first Atari Flashback retro game unit was a disappointment. Worst of all, its processor was similar to (blasphemy!) the Nintendo Entertainment System's. Thank the gaming gods, the Atari Flashback 2 gets it right. Not only does it look like a smaller version of the 2600, its internals recreate the 2600 processor, and its classic joysticks actually work better than the originals (using modern-day circuitry).

The Flashback 2 comes with 40 built-in games, including old-time favorites *Adventure, Missile Command*, and *Yar's Revenge*, a few Activision titles, and brand-new games made specifically for it.

The Flashback 2 was even purposely designed to tempt one to hack it. Though it doesn't come with a cartridge slot, plans detailing how to add one have been released on the web along with some other hacks by an engineer who developed the hardware (atarimuseum. com/fb2hacks). The fact that you can mod the Flashback 2 to reuse your old Atari 2600 cartridges is its coolest feature of all.

—*Howard Wen*

2 An Inexpensive New/Old Word Processor

Poqet PC $30
cadigital.com/poqetpc.htm

▪ Some things in the computer industry were so far ahead of their time that they did not gather a following until years after the parent company quit doing business. First sold in 1991, the Poqet "pocket" computer measured just 9"x5"x1" when closed and sported near-state-of-the-art technology for its time. The Poqet PC ran DOS 3.3 (on ROM), had a 25-line, 80-character display and a keyboard large enough for touch typing, and could run for up to 8 hours on two AA cells. It stored programs and files on a pair of 1MB RAM disks, and accessories included a 3.5" floppy drive and an RS232 serial port.

Originally retailing for over $2,000, it still has a following among folks looking for a low-cost, lightweight, basic word processing tool. Refurbished units can be found on the internet, marketed by California Digital, Inc.

—Thomas J. Arey

3 Epoxy for Big Projects

Apoxy Sculpt $11
avesstudio.com

▪ Most people are familiar with plumber's epoxy putty — mix two parts together, cures rock hard in 20 minutes. Unfortunately, it's usually sold in small quantities at a relatively high price and has a short working time. Recently, I was thrilled to discover a world of epoxy putties I never knew existed; you can buy large quantities at a reasonable price and in more flavors than a Baskin-Robbins. Apparently, this is the same stuff used by professional modelers for Disney displays, museums, and movie props.

I originally bought it for modeling, but I am constantly amazed at how useful it is for other tasks, from using it as insulating material for a hot wire foam cutter to replacing large chunks of painted wood trim eaten by our dog. The epoxy putties from Aves Studio are nontoxic, stick to almost anything, won't shrink, and dry rock hard. They can be sanded, drilled, tapped, frozen, heated ... you name it.

Now, I can finally finish my Flying Spaghetti Monster idol and do some serious worshipping.

—Mark Lengowski

4 Multipurpose Rubber Coating

Plasti Dip $8
plastidip.com

▪ Plasti Dip International has a range of maker-friendly rubber coating products (like Super Grip Fabric Non-Skid Fabric Coating and Liquid Tape Electrical Brush-On Insulation), but my favorite is the classic Plasti Dip, a gooey substance in a can.

To rubber-coat an object, you lower it into the can and then remove it and allow it to dry (you can also paint the rubber coating onto a surface). The flexible rubber coating works great to add a better gripping surface to a tool, or to insulate an object from electrical shock or vibration. To create a great non-skid coating, you can mix three parts Plasti Dip to one part pumice grit.

Plasti Dip is easy to use, and easy to clean up if you goof; let it dry, and you can just peel it off. (The bit I spilled on the floor cleaned right up.)

—Terrie Miller

5 Bond Girl

Stylish Leg Purse $79
tsaya.com

▪ Doesn't everybody want a secret pocket on their person to store useful gadgets? Having just watched a sleep-inducing cigarette procured from a garter case in *The Spy Who Loved Me*, I have been hankering after a similar accessory. To my delight, the Tsaya, a sleek, patent-leather purse that straps inconspicuously around your thigh, seems to fit the bill.

My first foray is at a dinner party, where I have impetuously fastened it over jeans. Aside from having a poor grip on denim, the strap-over-pants look seems to alarm dinner guests. Worn under a black skirt, it's more successful. At a San Francisco nightspot, fellow clubbers are intrigued; one girl asks if you can use it like an iPod armband, and someone else wants to know if it comes in lace.

Rock on! While its appearance could be enhanced for a broader audience (its thick straps and shiny patent leather are more bondage gear and less Coach accessory), the appeal of carrying cellphones and other vitals hands-free more than makes up for it. Now I just have to find those 007 cigarettes — if anyone has a source, let me know.

—Jenna Phillips

MAKE LOOKS AT BOOKS

1 Ready to Make

***ReadyMade How to Make {Almost} Everything: A Do-It-Yourself Primer*, by Shoshana Berger and Grace Hawthorne**
$25, Clarkson Potter

▪ This whimsical instructional guide from the folks at *ReadyMade* magazine (a sort of sister publication to MAKE) shows you how to make useful objects from things formerly destined for the dustbin: a ceiling fan constructed from an old bike wheel and a plastic bowl, a wine rack rigged from coat hangers, a dog bed fashioned from an old pair of blue jeans, a rug assembled from shopping bags, a picture frame made out of a hardback book. Half the joy of this book came from seeing how the authors breathed new life into unwanted junk. The other half came from the way it inspired me to start thinking of novel uses for the neglected stuff in my own garage.

—Mark Frauenfelder

2 The Hippie Handbook

How to Tie-Dye a T-Shirt, Flash a Peace Sign, and Other Essential Skills for the Carefree Life, by Chelsea Cain; illustrations by Lia Miternique $12.95, Chronicle Books

▪ Offering up *The Hippie Handbook* to your friends quickly becomes a sociological study. The younger hippies — say, 35 and under who still live more or less by the hippie constructs — will most likely love this paperback (with a tie-dye cover, natch).

But the older hippies — or even worse, those fortunate (?) few who were raised by hippies — seem to dislike the book, or are at least likely to rip its tips to shreds. To this group, being a hippie is no laughing matter; a childhood full of homemade yogurt and granola can clearly make a person angry later in life.

Regardless of who your friends are, the book is a blast. You'll read about everything from how to make "oregano" brownies to how to care for a fern, how to milk a goat, and the proper way to do tree sitting (bring a laptop). The illustrations are charming, and really add to the Flower Child feeling of the book. Peace and love.

—Shawn Connally

Bill's Excellent Adventure

Adventures from the Technology Underground,

by William Gurstelle $25, Crown Publishing

▪ To find out what happens at Teslathon, LDRS (Large Dangerous Rocket Ships), Burning Man, or the World Championship Punkin Chunkin, you wouldn't want to rely on mainstream media. A typical reporter struggles not only to understand these technophiles but to explain how things work, trusting that few people actually care. Not Bill Gurstelle.

In his new book, *Adventures from the Technology Underground*, Gurstelle takes us inside the communities of amateur inventors and science enthusiasts. He spent two years researching what he calls the "Technology Underground," a counter-culture based on "self-directed science and technology." Gurstelle told me that "the media usually portrayed these people as either beer-drinking sods with a welding torch or uber-geek technology freaks. Of course, mostly they were just people who enjoyed technology."

Like Brian Basura, who built a very large electrostatic discharge machine, designed to shrink quarters in his garage. Gurstelle writes that high-voltage tinkerers also know "how to make a big-time, lethally dangerous high-voltage capacitor in an unfinished basement; and they can make it from a 12-pack of empty Rolling Rock beer bottles wrapped in aluminum foil and immersed in brine."

The author of *Backyard Ballistics*, a mix of history and how-to, Gurstelle not only introduces us to fascinating people and the places they gather, but offers straightforward explanations for the technologies that obsess them, such as high-power amateur rocketry, pulsejets, tesla coils, catapults, and flamethrowers. He helps us see these technophiles as they see themselves. —*Dale Dougherty*

How Your Birding Hobby Can Help Further Science

eBird Online Database Free
ebird.org

For centuries, backyard scientists have kept lists of the birds they've seen. These lists have been useful to scientists who are able to correlate the observed changes in range or migration times of species with global climate change. Patterns from shorter-term effects, like West Nile virus, can also emerge.

The Cornell Lab of Ornithology and the Audubon Society joined forces to create an easy-to-use online database, eBird, that lets you track your bird observations and pool them with others, so the data can be used by anyone.

With eBird, your personal birding list becomes more interesting and fun to browse right away. But by sharing your data, you also become a citizen scientist. The citizen scientists of the 1800s didn't realize that their observations would be important clues about global warming, and the data you collect now may be important in ways we can't even foresee. *—Terrie Miller*

Collaborative Music Playing on the Web

NINJAM Free
ninjam.com

Not only does NINJAM enable musicians to play together online, it even saves the jams as Creative Commons-licensed recordings. But the breakthrough is the way it overcomes the groove-destroying delay inherent in sending audio over the net; it delays each performer's output by an interval of several bars. Strange concept, but it actually encourages listening.

I downloaded the free NINJAM client (available for Windows, Mac, and Linux), logged in, and started playing electric piano. Suddenly, a guitarist chimed in, followed by vocals. NINJAM co-founder Justin Frankel (of Winamp, Gnutella, and Shoutcast fame) even joined on drums.

I experienced a few audio glitches and dropouts, but the mix seems to have made it to the website intact. This is one rough ninja, but it's free and fun. *—David Battino*

An iTunes Alternative for the Shuffle

Sean Shrum's Shuffler Free
shuffle-db.sourceforge.net

The iPod Shuffle is probably the biggest-selling Flash-based MP3 player in the world. Given that you can drag and drop files onto it from any decent file manager (on Windows, Mac, Linux), you would think you could just copy your MP3 files onto the Shuffle and listen to them.

Unfortunately, Apple wants you to keep using iTunes. Which is where Sean Shrum's Shuffler application comes in handy. It's a small application that will scan all the files on your Shuffle and rebuild the internal music database.

You still need to use iTunes to set up your Shuffle as a USB Memory Drive, although this is a one-time operation. After that, all you have to do is just copy your music over from any host machine, run the database builder, and away you go. *—Ewan Spence*

TOOLBOX

1 Rechargeable Batteries That Really Work

Rayovac Rechargeable Battery Charger $20 rayovac.com

▪ My whiz-bang digital camera consumes batteries at an alarming rate. Conventional wisdom is to use rechargeable AA cells, but they seemed to always be dead when I wanted to use the camera. So, I either put in disposables or waited a couple of hours while they recharged. Who knows what the Pulitzer committee has missed as a result?

The problem is the relatively high self-discharge rate of nickel metal hydride (NiMH) batteries, which is a problem for gadgets like my camera that go weeks between uses. I found a good workaround in Rayovac's NiMH batteries, which are available in AA or AAA sizes. These batteries each have a tiny internal charging circuit and sensors that allow them to be fully charged in a blistering 15 minutes. Their 2000mAh capacity matches standard NiMHs, and they are otherwise interchangeable with their not-so-smart brothers.

Now I can recharge my camera in the time it takes to find my shoes and pack my bag. The Pulitzer guys still haven't called, though.

—Bob Scott

2 A Really Good Universal Remote

Logitech Harmony 880 Universal Remote $249
logitech.com

▪ The joy of the Harmony 880 is in the setup wizard, which is part of a personal online account you create on Logitech's website. In the wizard, you select devices you want to control (like your DVD player) by manufacturer and product number, and then define activities (watch a DVD), which could include your receiver, video projector, and DVD player.

Then you install the remote software on to your computer, connect the Harmony 880 through a USB cable, and download the setup to your remote. You will want to set up the activities, even if all you want to do is turn two devices on and flip through channels. If this sounds like a lot, it really isn't. As you first use the remote, there is an onscreen assistant to fix any problems that appear. This is one seriously smart remote. *—Dane Picard*

3 Maximum Hard Drive Portability

CoolMax CD-311 Hard Drive Chassis $60
coolmaxusa.com

▪ I have a fascination with versatility — the more ports and switches a device has, the better. When looking for a hard drive chassis, then, the CoolMax CD-311 was an easy choice. As an on-site consultant, I'm often called upon to do emergency data transfers and system recoveries. The CD-311 will support whatever drive I'm working with, whether it comes out of an eMac or an Xserve.

The unit's external interfaces are also myriad, supporting FireWire, USB 2.0, and SATA (with cables included for all three). Internal cables for both IDE and SATA are included as well, not to mention a small Phillips screwdriver.

The design lends itself well to being cut in half, which would allay my fears about heat dissipation in the fanless, sealed aluminum chassis. It would be very easy to make a portable drive dock similar to WiebeTech's ComboDock — but while the ComboDock with SATA adapter costs $250, the CD-311 is under $60. Hacksaw not included. *—Tom Owad*

4 Sticky Situation

Self-Fusing Rubber Splicing Tape $11
digikey.com, part #3M130C-ND

▪ If you do any kind of electrical work or home repairs, you likely rely on heat-shrink tubing and/or electrical tape to seal and insulate wiring splices. But even the best vinyl electrical tape will eventually disintegrate into sticky black goo, and heat-shrink tubing needs a heat source. (Plus, if you are like me, you never have the right diameter handy.)

The pros have a solution: like other sticky miracles, it comes from 3M, and it's called "Linerless Rubber Splicing Tape." It fuses to itself to form a nearly perfect waterproof bond. To use it, you cut a short piece off the roll, stretch to activate it, and tightly wrap your splice. The result is a neatly insulated tube of tough black rubber. It also works great on anything that needs protective wrapping or a better grip, such as elderly extension cords or sports racquet handles. I even use it to bind rope ends so they don't unravel. *—Jonathan Foote*

5 Superior Nonstick Pans

8-Piece Calphalon One Cookware Set $479
calphalon.com

▪ Warranties are essential for expensive purchases, and the warranty on my set of Calphalon One cookware is phenomenal. Unless you submerge the set into an acid bath or melt it in lava, you're pretty much covered for life.

My favorite feature of the cookware is the incredible nonstick surface. While all humans may be created equal, all nonstick surfaces are not. I now cook eggs, pancakes, vegetables, and other items using a fraction of the oil I did with my previous so-called "nonstick" set. I'm also a fan of the "stay cool" dual handles on the large pieces. These are especially useful for the chef's skillet and the large pot, which get pretty heavy when you fill them up.

It's also really nice to be able to put this stuff in the oven — even the broiler — and see it come out unscathed. Ever melted a handle or two while preheating the oven? I won't be having that problem anymore. *—Matthew Russell*

TOOLBOX

A Table Saw for Hackers

Ryobi BT3100 10" Table Saw $299
ryobitools.com

▪ The Ryobi BT3100 10" table saw is accurate, inexpensive, and cleverly engineered. Its accuracy is due to a number of well-designed features, although the design may make some traditionalists a bit uncomfortable. There are no miter slots (though most people find the sliding miter table more than sufficient for all tasks), and the rip fence and sliding miter table are aligned to the blade.

The rip fence is rock-solid; it locks down and stays locked down. This allows for such a feat as ripping 1⁄64-inch strips from a 6-foot work piece. Blade run-out is negligible due to an efficient, belt-driven arbor design, which allows the saw to cut a standard 4x4 in one pass.

The BT3100 is a unique saw and sure to keep anyone with a tight budget and a small workspace very satisfied. It also has a loyal following — big enough for not one, but two, online forums.

For more information, see bt3central.com and ryobitools.com.

—Mark Lengowski

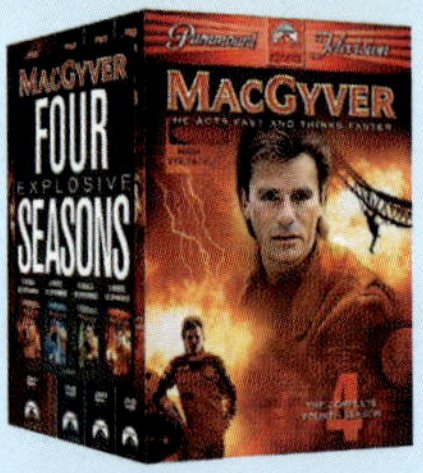

The Alpha Maker

Four Seasons of *MacGyver* on DVD
$113 amazon.com

▪ Creating an impromptu smokescreen, plugging an engine hole with egg whites, and dismantling a missile with a paperclip were all in a typical day's work for television's fictional alpha maker, MacGyver. With the recent release of the complete first four seasons on DVD, we now get to relive all of these amazing and innovative MacGyverisms.

As a kid, the show inspired me to invest several weeks unsuccessfully trying to create explosives out of pine cones, but I feel it was time well spent. MacGyver is one of the few shows that doesn't make me feel dumber after watching it, and that's the main reason I love it so much.

—Matthew Russell

T.J. "SKIP" Arey N2EI is a freelance writer and author of *Radio Monitoring: The How-To Guide*.

Dave Battino plays Mac, PC, and keyboards. More at batmosphere.com.

Jonathan Foote is a research scientist in Silicon Valley.

Mark Lengowski spends way too much time making things instead of earning an honest living.

Merlin Mann helps people make interesting things for the Global Interweb.

Ross Orr keeps the analog alive in Ann Arbor, Mich.

Tom Owad is the author of *Apple I Replica Creation* and editor of applefritter.com.

Jenna Phillips is currently launching Formula Magic, a high-tech shoe company.

Dane Picard is a video artist and filmmaker based in Los Angeles.

Matthew Russell tries hard to live life as a renaissance man, but is distracted by the cult of Mac.

Bob Scott is a statistical construct of various consumer electronics marketing departments.

Ewan Spence keeps it real in Scotland. His blog is ewanspence.com.

Howard Wen (howardwen.com) is a frequent contributor to MAKE.

Have you used something worth keeping in your toolbox? Let us know at toolbox@makezine.com.

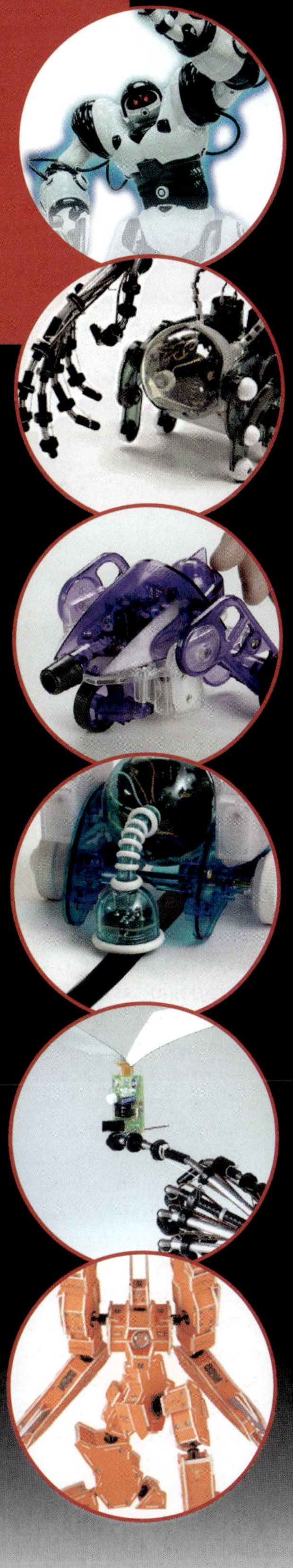

BLAST FROM THE PAST

Prewar Maker's Primer

By H.B. Siegel

Amateur Craftsman's Cyclopedia of Things to Make

It would be hard to overestimate the number of hours I pored over this book as a kid. Published in 1937 by Popular Science Publishing Company, Inc., it was obsolete decades before I was born, yet it is captivating in the glimpse it provides into the pre-WWII maker's world. At 330-plus pages and one to five projects per page, it is filled with projects that range from simple to elaborate, from relevant to bizarre. A few are even politically insensitive in our era.

I did not come from a "maker" household, so I was restricted to window-shopping. I was fascinated by the hand-drawn illustrations and black-and-white photographs. No Illustrator or Photoshop for another 50 years, thank you! The gorgeous "Rakish Privateer Clipper of 1812" was intricate beyond all hope, requiring milling, woodwork, and fittings. But the "Mayan Throwing Sticks" looked possible, and the kids in the illustration were having a great time.

While several of the projects make little sense today, such as "Comical Coconut Bird Serves as an Ash-tray" or "Eight Scoop Popcorn Server," there are hundreds that are still appropriate and approachable, e.g., "Lathe Gear Cutting," "Glass Cutting Jig Removes Bottle Tops," "Homemade Furnace Melts Aluminum," "Homemade Tool Shapes Metal Tubing," and "An Easy Way to Turn True Wooden Balls."

It was the games and toys that drew me in again and again — the "Walking Top Game," the "Magic Knives for Trick Surgery," the "Rubber Bulb Toy Submarine," and especially the "Ticket-Dropping Airplane," which offered assurances to "start a party off with a bang." I had no lack of toys as a kid, but it was mind-blowing that someone could *make* these.

The scary ones are fun as well. No dust masks, gloves, safety goggles, saw guards, or warnings in sight. Were some lives shortened while "Making a Mercury Barometer," or a "Simple Hygrometer Made with Carbon Tetrachloride," or the double-whammy of "Smoker's Set Made of Thin Sheet Lead"? What are we blithely making now that will horrify our grandchildren?

There's something for everyone in this book. It demonstrates recycling and reuse way before it was cool. Recommended.

Homemade Furnace Melts Aluminum

WITH ORDINARY ILLUMINATING GAS

The burner, asbestos baffle, crucible, and cover are shown at the left. The body of the burner is an old 2 by 1¼ by 1¼ in. pipe tee. The gas and air enter through the side outlet

FURNACES for melting brass are often constructed by amateur mechanics with a vacuum-cleaner blower for the forced draft. However, aluminum can be melted without a forced draft, and as aluminum will answer the purpose in most cases where castings are required by the amateur mechanic, this method will appeal to many who have hesitated to attempt making their own castings. The average amateur with a small lathe will find working with aluminum every bit as easy as brass.

Brass requires 1,8000 deg. F. to melt, while aluminum only requires 1,200 deg. The burner shown in the photographs melts aluminum with ordinary illuminating gas. It was made from an old 2 by 1¼ by 1¼ in. pipe tee. A pipe plug was screwed securely into the bottom and machined off. Then a shoulder was cut on the outside of that end of the tee and fitted into an opening in an old lamp base, which had been machined to match. The edge was peened over to hold it in place.

Reducing bushings were used in the side outlet to reduce it to ¾ in., and a piece of ¾-in. brass pipe, about 3 in. long, was screwed in. This piece of pipe had two openings, each about ¾ in. square, one on each side, about 1 in. from the outer end.

Castings for models and for such attractive novelties as this lamp are made with the aid of the furnace

The mixture of gas and air is adjusted by means of a sliding sleeve on the inlet pipe

The lamp ready for assembly. This project is a good example of what can be done with homemade castings

HOW MELTING FURNACE IS CONSTRUCTED

Working drawings of the furnace. Aluminum melted in it is cast in sand molds in which plenty of risers must be made, as shown by the rough casting at the right

WORKSHOP HINTS 743

Instructions for making a small furnace that can melt aluminum, good for making castings of "attractive novelties."

Copies of *Amateur Craftsman's Cyclopedia of Things to Make* appear frequently on Amazon and eBay for about $10.

Make sure to recite an "appropriate airplane verse" before setting this tiny payload-delivering aircraft aloft over the heads of a "gathering of young folks ... The resulting confusion and laughter will break the ice and start the party off in the right spirit of hilarity."

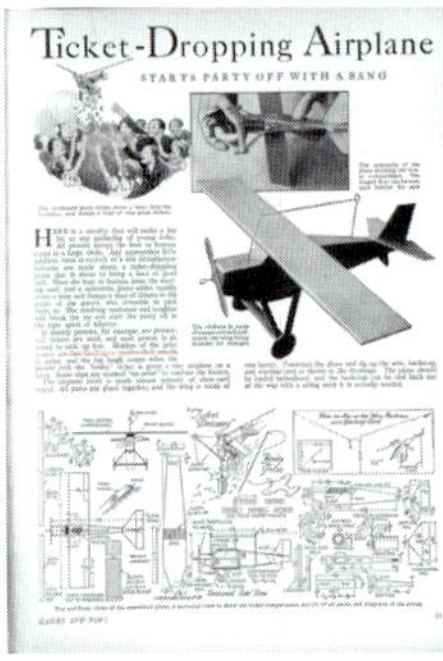

Ticket-Dropping Airplane

STARTS PARTY OFF WITH A BANG

H.B. Siegel is technology director for IMDb.com (part of Amazon.com). He has been CTO for ILM and in management at Pixar, Wavefront, and SGI.

Images courtesy of H.B. Siegel

RETROCOMPUTING Computers in the Machine

By Tom Owad

Emulation software gives you free, zero-footprint, vintage computers.

I've had a spate of flakey hardware recently. A wire-wrap board that only works when it's flexed, an SE/30 that randomly refuses to boot, and an Apple IIgs that only occasionally recognizes its hard drive have all prompted me to take a closer look at emulation. Space, too, has played a role. While classic computers overflow the garage, the emulators merely occupy a directory on my hard drive. The manuals and disks for them, instead of filling shelves, reside in PDFs and disk images.

Thanks to Mini vMac, my Mac Plus rarely comes off the shelf. I no longer worry about wear and tear on 20-year-old floppy disks. They've been imaged to my PowerBook and safely locked away. The emulator makes it possible to quickly inspect old programs for my website, without the painstaking task of physically transferring them from a PowerBook G4 to a Mac Plus.

All of my classic Mac games are now just a few clicks away, but as anybody knows, playing in an emulator just isn't the same as having the real thing. Granted, it's much more convenient when there's serious work to be done, but for just playing around, the hardware is half the fun. There are some systems, though, where the hardware is either too expensive or too large to make ownership practical. The DEC PDP-8, a classic mini-computer from the 60s and 70s, is a good example. Simply finding a PDP-8 is a significant undertaking, let alone amassing the necessary peripherals (paper tape reader, RK05 disks, etc.) and getting them all running.

PDP-8/E Simulator, by Bernhard Baehr, does a fantastic job of bringing the PDP-8 to life on your desktop, offering more functionality than the original. While the 8/E has 4K words of memory by default, the simulator comes with a plugin for a KM8-E Memory Extension, bringing the memory up to 32K words (each word is 12 bits). There are additional plugins for an ASR 33 Teletype, PC8-E High-Speed Paper Tape Reader, RK8-E Disk Cartridge System, LP8-E Line Printer, three different clocks, and a TSC8-75 board (for running the ETOS time-sharing operating system). There's even a KC8-EA Programmer's Console (i.e., "front panel") so you can toggle switches and watch the blinkenlights.

The API for plugins is well documented and included with the source code. With it you can write emulators for additional I/O devices or even create new devices of your own. The simulator is also well suited for development. I hear PDP-8 development isn't quite the burgeoning industry it used to be, but its educational value remains. Each device (plugin) has a window displaying its internal state. Through the PDP-8/E CPU window, it is possible to view and edit each register and the contents of memory. The CPU supports breakpoints, a trace mode, and many other features you'd expect to find in an advanced simulator. The front panel plugin is fully functional, so if you prefer the classic switches-and-lights view, it is entirely possible to edit memory, load programs, and single-step the processor using it.

> "PDP-8/E Simulator does a fantastic job of bringing the PDP-8 to life on your desktop, offering more functionality than the original."

In fact, Baehr's website contains a tutorial explaining how to operate a PDP-8 with no system software, by using the front panel. The process involves toggling in an 18-line program for reading paper tape (a paper tape reader and punch is built into the ASR 33 Teletype you have attached). This program is then used to load a more sophisticated paper tape reader, which you can finally use to load your own program.

For more sophisticated use, attach an RK8-E Disk System, and you can run Focal-8, OS/8, Pascal-S, and ETOS. Mount the RK05 DECpack that contains your desired system, toggle in the RK8-E boot code, and run. OS/8 was the most common operating system for the PDP-8 and is a good place to start. Included on the OS/8 DECpack is the classic text editor TECO, from which Emacs eventually developed. TECO documents are created by the OS/8 command "MAKE filename." For a glimpse at 1970s hacker humor, try entering the command "MAKE LOVE."

Tom Owad (owad@applefritter.com) is a Macintosh consultant in York, Pa., and editor of Applefritter (applefritter.com). He is the author of *Apple I Replica Creation* (Syngress, 2005).

MAKER'S CORNER

Save the Dates: Maker Faire

Saturday, April 22nd and Sunday, April 23rd

Maker Faire

Meet the Makers

Join the MAKE team and thousands

of other makers at MAKE magazine's first ever Maker Faire — inspiration, know-how, networking, and just plain fun for makers and aspiring makers of all ages and backgrounds. It's the first ever public gathering of tech DIY enthusiasts, crafters, educators, tinkerers, hobbyists, science clubs, students, authors, and exhibitors.

It's two days with six exposition and workshop pavilions, a 5-acre outdoor midway, over 100 exhibiting makers, hands-on workshops, demonstrations, DIY competitions, and the latest in tools, kits, and DIY resources.

It's Weird Science, Ultimate Garage, Robotics, Digital Entertainment/Gaming, Green Tech & Electronics Recycling, Ultimate Workshop, and the MAKE: Remix video film festival.

Bring your family, friends, and a student or three to the San Mateo Fairgrounds (centrally located in the San Francisco Bay Area) for a weekend of hands-on exploration, recipe-sharing, creative mischief-making, and wholesome play. All-day admission is a paltry $12 for adults, $5 for teens, and not a cent for kids 12 and under accompanied by an adult. Maker-friendly family tickets and full weekend passes are available as well!

Information and registration for the following can be found at makezine.com/faire:

Tickets: **50% off weekend passes for MAKE readers when purchased by March 31. Register and enter "MFWINTER" as your access code.**

MAKE: Remix Video Film Festival: **Contest details available online.**

To Exhibit as a Maker: **Further details and proposal information for makers interested in exhibiting opportunities are available online.**

To Sponsor or Exhibit at Maker Faire: **Please contact Larry Ecklund, the director of event sales at lecklund@makezine.com.**

The Silly Subscription Renewal Game

We're mad as hell, and we're not going to take it anymore! The magazine industry bombards us with renewal reminders, but by the time our subscription actually runs out, we get an incredible offer and a gift tossed in to boot. They've trained us to hold out, cluttering our mailboxes and wasting paper and money.

So here's our promise to you: Your best renewal offer from MAKE will always come up front. If you play the hold-out game, chances are you'll pay more. Sorry, but we think subscribers who cost us less ought to pay less.

We've also put together an auto-renew program. When it's time to renew, we'll send an email letting you know that we're renewing your subscription and charging your card. You'll always receive at least $5 off the standard subscription price, you'll receive just one email, and you'll have access to the entire MAKE archive when you access the MAKE Digital Edition, available only to MAKE subscribers.

To register for auto-renewal, please visit makezine.com/premier.

The Gift of MAKE

Buy a gift subscription for someone (yes, even if you are that special someone) and get a free MAKE T-shirt for yourself. Visit makezine.com/gift.

MAKE: The First Year Boxed Set Collector's Edition

For the most dedicated maker who wants to keep the complete first year of MAKE in pristine condition, we are offering a special 4-volume gift-box collector's edition. Just need the cool collector's box? We have that too (makezine.com/go/coolstuff).

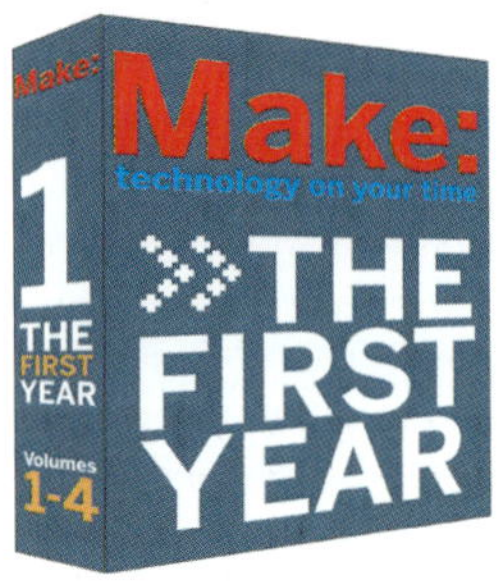

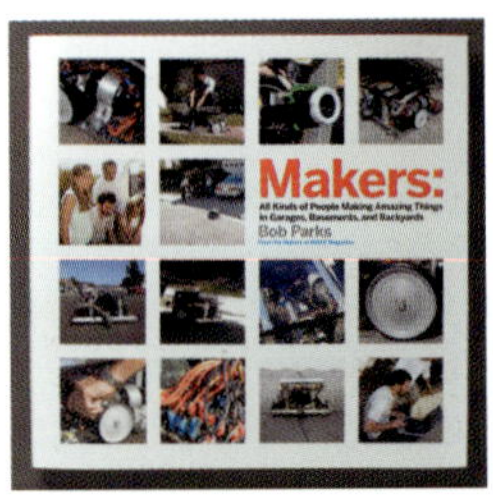

Makers: The Book

Steal away, kick your feet up, and lose yourself in the creativity and resourcefulness of the maker movement with this beautiful hardbound book. Bob Parks and the editors of MAKE profile 100 makers and their homebrew projects, backyard inventions, and basement creations (also at makezine.com/go/coolstuff).

MAKER'S CALENDAR Compiled by William Gurstelle

Our favorite events from around the world — March - April - May 2006

March

>> Space Shuttle STS: 121 Launch, March 4, Cape Canaveral, Fla. The space shuttle *Discovery* is scheduled to launch into orbit at 3:21 p.m. EST. This mission will test new equipment and procedures to increase shuttle safety and will deliver more supplies and cargo for future space station expansion. kennedyspacecenter.com/launches/scheduleStatus.asp

>> EDITOR'S CHOICE — Nevada Test Site: Public Access Day, March 23, April 26, and May 24, Las Vegas. The U.S. Department of Energy provides tours of the Nevada Test Site only a few times a year. The 250-mile bus tour begins in Las Vegas and takes visitors to nuclear test locations such as Frenchman's Flat, Mercury, and Control Point 1. nv.doe.gov/nts/tours.htm

>> The Texas Mile: Land Speed Racing Event, March 19–20, Goliad, Texas. Land speed racers from around the country meet on the long, long, flat runway of the Goliad Industrial Air Park to test their fastest motorcycles and automobiles. Some machines attain speeds of more than 250 mph. texasmile.com

April

>> Edinburgh Science Festival, April 6–16, Edinburgh, Scotland. A large and diverse series of science-oriented lectures, tours, and demonstrations, this world-class festival brings together many of the top thinkers and great science entertainers of the day in the U.K.'s largest public celebration of science and technology. sciencefestival.co.uk

>> Hacker Con Notacon, April 7–9, Cleveland. This annual conference explores and showcases the technologies, philosophy, and creativity associated with hacker life. notacon.org

>> Trinity Site: Public Access Day, April 8, White Sands Missile Range, N.M. Trinity Site was the first "ground zero": the spot where the first atomic bomb was placed on a 100-foot steel tower and detonated on July 16, 1945. Open to the public just two days a year. www.wsmr.army.mil/pao/TrinitySite/trndir.htm

>> 13th Annual Trinity College Fire Fighting Home Robot Contest, April 8–9, Hartford, Conn. This event features a robot fire fighting contest, a Robotics Olympiad, and a Robotics Symposium. trincoll.edu/events/robot

>> 12th Annual Mobot Races, April 21, Carnegie Mellon University, Pittsburgh. Student teams compete for prize money as they race autonomous, self-powered miniature robots on a torturous, curving, 255-foot-long course. cs.cmu.edu/~mobot

>> Thunder Over Louisville, April 22, Louisville, Ky. Each spring, 500,000 people gather to watch what is said to be the country's largest regularly scheduled pyrotechnics display. TOL also includes an air show with over 100 aircraft. It takes place just before the Kentucky Derby. thunderoverlouisville.org/show/default.asp

>> FIRST 2006 Robotics Championship, April 27–29, Atlanta. The annual FIRST Robotics Championship is the final step in a multinational robotics competition that involves 25,000 people, nearly 1,000 teams, and 30 preliminary competitions. Open to the public, free of charge. usfirst.org/robotics

May

>> EDITOR'S CHOICE — NRO L-26 Delta 4-Heavy Launch, May 1, Cape Canaveral, Fla. The biggest lifter in the U.S. rocket arsenal, the Boeing Delta 4-Heavy rocket, launches a classified spy satellite cargo on May 1. This rocket features three booster cores mounted together to form a triple-body rocket. More power and flames than anywhere else on the planet. kennedyspacecenter.com/launches/scheduleStatus.asp

>> Astronomy Day, May 6. Astronomy Day events take place across the United States. Stargazing, lectures, and demonstrations. astroleague.org/al/astroday/astroday.html

>> The Intel International Science and Engineering Fair, May 7–13, Indianapolis. The world's largest pre-college science competition, with over $3 million in awards and scholarships, will attract 1,500 top science high school students from the 50 U.S. states and 40 countries. sciserv.org/isef

>> Andrews Air Force Base Joint Services Open House, May 20–21, Washington, D.C. This event pulls in crowds of nearly 400,000 people, mainly for the spectacular aerial demonstrations, including precision aerial maneuvers by the Air Force's Thunderbirds, fly-bys, parachute jumping, and ground displays. public.andrews.amc.af.mil/index.asp

>> Jet Propulsion Laboratory Open House, May 20–21, Pasadena, Calif. This event showcases JPL's accomplishments with exhibits and demonstrations about the laboratory's ongoing research and space exploration. Many of the lab's scientists and engineers will be on hand to answer questions. www.jpl.nasa.gov/pso/oh.cfm

>> EDITOR'S CHOICE — World Championship 38th Annual Kinetic Sculpture Race, May 27–29, Arcata to Ferndale, Calif. Art and engineering combine to produce imaginative human-powered racing machines. Thousands watch as vehicles resembling an 85-foot salmon, a giant chocolate éclair, and a yellow submarine race for the finish line during the long and grueling race. kineticsculpturerace.org/

IMPORTANT: All times, dates, locations and events are unconfirmed and subject to change. Verify all information before making plans to attend.

Do you know of an event that should be included? Please use our online form at makezine.com/events. Sorry, it is not possible to include listings for all events in the magazine, but they will be listed online.

Did you attend an interesting event? If so, please tell us about it in our Forums (forums.makezine.com).

READER INPUT

Where makers tell their tales and offer praise, brickbats, and swell ideas.

The MAKE blog is the only site I check 3-5 times a day and actually enjoy.

MAKE made me realize that I'm not just someone with the hacker mindset — I'm someone who wants to create things myself. I never had an interest in crocheting, but the fact that I want to make a tux scarf makes me realize that things are just better when you make them yourself. *—Troy Fletcher*

I saw MAKE this morning. I subscribed immediately after seeing the VCR cat feeder in Vol. 3. I proceeded to check out the site, and haven't done a lick of web work since.

In fact, I will probably eat a large amount of chemical enhancements, shave my head, and devote my existence to building things from MAKE and/or coming up with at least one project to submit.

A haiku to express my newfound obsession with MAKE:

Client site is down,
I am owned by the makezine.
Where's my flamethrower?

—Benjamin Jones

I've been subscribing to MAKE since Vol. 2. As you can imagine, my subscription renewal is coming due. I was considering letting my subscription lapse, but Vol. 4 changed my mind. I think it was your first Really Good issue. Why? Because it finally had articles and projects I'm interested in.

I liked the projects on music bending, high-speed photography, turning a cup into a speaker, programming microcontrollers (though I don't understand everyone's fascination with the PIC; in my opinion the AVR chip is just as inexpensive and has a much more orthogonal [i.e. more consistent] instruction set), and the article about Dean Kamen. My personal interest is in electronic projects and robotics. Taking apart existing products and repurposing them is cool.

Your recent online list of $100-and-under gift ideas for a maker was spot-on. Loved it.

Can we see more articles about embedded processors and Linux devices?

—Barry Brown

Mister Jalopy,

You can't imagine how happy (and a bit disappointed) I was to find your site ... especially the LP ripping Apple/iPod 1960s radio. As you can see [from the photo], I thought I was pretty smarty pants when I built a networked MP3 Touchscreen Jukebox out of a 1940 RCA chassis that I bought for Can$40 ... but yours blows mine completely out of the water. ;-)

I wanted to say how much I enjoy your site and attitude toward the garage life and taking pleasure in cleaning, restoring, and recycling items that most folks would throw out! I read your article on welding in MAKE magazine, and I have to say it was well done. I am actually finishing my Arc welding class as we speak, so I am really looking forward to buying my first welder and getting set up ... I have been more of a woodworker up to now!

Anyways. I am now a convert to your site and want you to keep up the great work ... if ever you want a guest gadgeteer from the Great White North, let me know. I would be very happy to oblige. ***—Patrice Collin***

I owe you a debt of gratitude. Your magazine saved Christmas. I was a Grinch beyond anything Theodore Geisel could have imagined. I hated Christmas; I was tired of the rampant consumerism and sterilized joy that comes with only giving DVDs and video games to friends and family. But thanks to you, Christmas is fun again. After finally getting my subscription to MAKE and then getting all the back issues. I could only think of one thing to give on Christmas — marshmallow guns! I tinkered with the design a little, and gave it an amusing name, the Mallow Mauser. I handed out the guns disassembled; ten in all, with instructions, and in the confusion of the moment sprang a surprise attack on the family. The guns were put together fast, and thanks to a controlled leak on my part, safety glasses were already on hand. My family is made up of mechanics, engineers, and contractors and they all loved it. The adults definitely had more fun than their kids. To explain where this idea came from, I brought my collection of MAKE magazines. Two out of three uncles now say they are getting a subscription to MAKE. *—Mike Ciarlone*

I just want you to know that I am deeply, completely, and hopelessly in love with your magazine. I didn't think I could be any more in love until I logged on to your digital version. Now I realize that was only puppy love. This is the real thing. Please never stop.

—Suzanne Portnoy
Mother, Wife, Yoga Student,
PTA Member, Closet Hacker

When I was about 11 years old, my grandparents gave me a gift subscription to *National Geographic World*. Though I'd already been pretty heavily into the grownup *NG*, I was delighted with that kids' version, not least because it included how-to instructions that got me making stuff. The one I remember most vividly was an issue that had me building, understanding, and using a pinhole camera (aluminum foil on sticky paper for the aperture). I wish I still had those images, though the memories of the experience, shared with my photographer father, are more valuable by far.

When each new issue of MAKE arrives, I react, if possible, with more excitement than an 11-year-old receiving a long-awaited gift. Probably 75% of the stories in any given issue cover projects I would never actually presume to undertake, but that's not the point. Just knowing that there are people out there building aerial-photography kites, light-seeking robotic mice, or VCR-powered cat feeders, hacking this, that, and everything in between — this knowledge, the sheer range of possibility that your magazine deals with, and the sense that I could do it all, too — fills me with joy, and I devour every issue with the enthusiasm of an 11-year-old waking up to the world.

I also appreciate the editorial stance on digital rights. Your information about how to tweak my European DVD player so it will play what I own, where I own it, has been worth the price of admission alone. I've renewed my subscription. I hope to see you grow and prosper in Year 2.

—J.D. Stephens

I love MAKE & war not? Anyway, I love the online edition and plan to subscribe as a Christmas gift to myself and get (not make) my wife Fran to order the back issues for me. It is very comforting to know there are other subversives out there like myself, although I can't hold a candle to most of the skills displayed in your wonderful articles.

—Kevin P. Wilkinson

I love your blog on makezine.com and you guys are doing a great job with the magazine. I was glad to get the digital copy of the last issue since mail service here in New Orleans leaves something to be desired since Katrina hit. *—Clay McGovern*

MAKE AMENDS

In Volume 04, page 21, through some inexplicable turn of events, an old and inaccurate version of the "Unroll Your Own" story ran. The 2004 "Stuck at Prom" winners featured in the photograph are Kris Murray and Caitlyn Waters, from Oklahoma City, Okla.

In Volume 04, page 47, a line of text was left out. The sentence should have read: "It's way more fun than the Porsche, and it carries far more failed experiments to the beach," he said. The best thing about meeting the weirdoes is that the conversation works to communicate information first, NDAs later, if ever."

In Volume 04, page 130, the schematic for wiring the anonymous megaphone is drawn incorrectly. The diode and trimpot connections are reversed, and the polarities were left off the capacitors. A revised schematic can be downloaded at makezine.com/go/revised.

In Volume 04, pages 158-165, the Primer incorrectly uses the term "BASIC Stamp" to generically describe the products of Parallax and netMedia. The article incorrectly tells the reader that the "BASIC Stamp" is any full-blown module with a micro, memory, power supply, etc., and also incorrectly suggests that PICs are available from Microchip, Atmel, etc. In reality, PICs are only available from Microchip, and AVRs are available from Atmel.

TALES FROM MAKE: BLOG

Pushing Electrons

By Phillip Torrone

Bots, how-tos, and virtual worlds all on the MAKE: Blog. Take a tour of what's new with Associate Editor Phillip Torrone.

MAKE, Volume 03, displayed in a virtual library inside Second Life.

We have been pushing a lot of atoms and electrons since we kicked off MAKE, and before we knew it, the MAKE: Blog hit 3,000 posts in less than a year — during the first week of 2006. That's about ten posts a day, all about the amazing things makers are doing, building, and creating. And we're just getting warmed up: 24/7, makers are sending us their stories, their works, and their ideas. You, too, can visit Makezine.com and just clickity-click the submit button (makezine.com/blog).

We've also added a few new features and ways to get what you need from MAKE and to publish your projects.

First up is the MAKEbot. The MAKEbot is an instant message buddy you add to your AOL/AIM/iChat buddy list. He can tell you the latest news, photos, bookmarks, and forum posts, as well as search all of MAKE from your buddy list or from your phone. We're constantly making the MAKEbot smarter, so give it a try. Just type "help" to get a list of all the commands!

Next, we're pleased to announce that we've partnered up with Instructables. What the heck is an Instructable? Instructables is a step-by-step collaboration system that helps you record and share your projects with a mixture of images, text, CAD files, ingredient lists, and more. Head on over and add your projects to the MAKE group (instructables.com/make).

Then there's our MAKE Flickr photo pool. Have a hack, mod, tweak, or just something plain cool you want to share? Join in, upload, and share your photos with hundreds of makers around the world (flickr.com/groups/make/pool).

Forums! You asked for 'em, we got 'em! We launched the forums on Makezine.com, so anyone can talk about all things MAKE-related — from getting help in projects to selling MAKE wares in the Maker Marketplace. If you've ever posted on MAKE in the comments, your account works for the forums too, or just sign up and start posting with other makers.

Lastly, visit MAKE in the virtual world. We've created the first magazine based on a real magazine inside the virtual world of Second Life (secondlife.com). It's free to download and use for PCs and Macs, and a Linux version is coming soon at secondlife://Tenera/64/192.

Phillip Torrone is associate editor of MAKE.

Image courtesy of Torrone Trumbo

Create Explosive Visuals

By Chris Kenworthy

Real movie explosions are dangerous, expensive, and usually unnecessary. You can create a convincing explosion with well-lit water, and add it to your footage later. You need to choose a good sound effect to make this work well; check the internet.

You will need: A large, clean mirror, camera, towels, black balloon.

1 The balloon used in this example is red, but you should use a black balloon. A black balloon won't show up in the finished shot, ensuring that only the brightly lit water is visible. To create the illusion of a powerful explosion, fill the balloon halfway with water, and then add enough air to fill it up until it is close to bursting. This combination of air and water yields a good explosion, but you can experiment with other ratios.

2 To create this effect, you need to work in a dark, enclosed space. Lean the mirror against a wall or over a box, at a 45-degree angle to the floor. Suspend the balloon over the mirror. Your movie light should be positioned off to the side, aimed at the balloon. Keep the light far enough away that it won't be splashed when the balloon explodes.

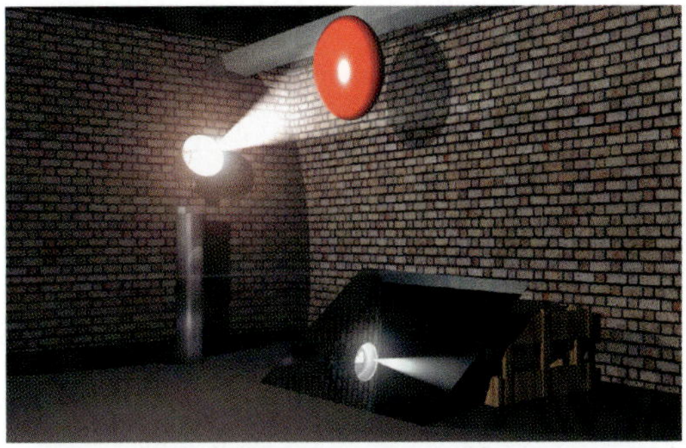

When you stand in front of the mirror, you should be able to see the balloon. (It shows up clearly in this example because the balloon is red.) You'll find it easier to line up your camera on the balloon if you tape a brightly colored piece of paper to it. Remove it when you're ready to shoot.

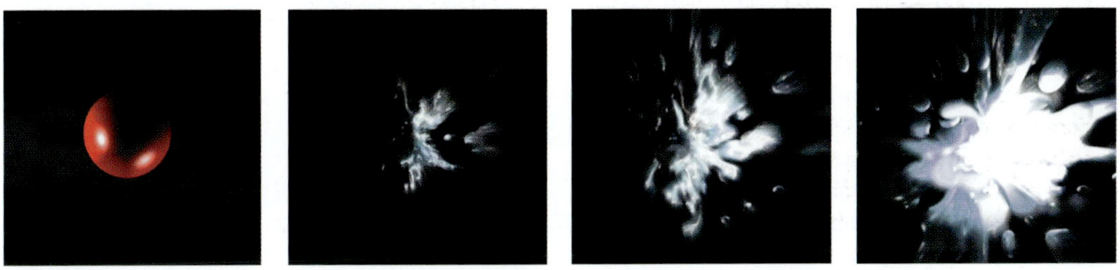

3 Zoom in on the mirror so that nothing is visible except the balloon. Use the zoom rather than moving the camera forward to avoid any risk of getting water on the camera.

4 Now pop the balloon. Use black tape to attach a sharp nail or tack to a black stick, which won't be visible when you move it into frame. Start shooting and pop your balloon. When it pops, brightly lit water will explode toward the camera.

5 In your editing software, layer the explosion footage over the object you want to blow up. Use a Color Correction filter to shift the colors toward yellow. If the object is moving, click the Keyframe button and drag the explosion layer so that it matches the movement of the object. Keep explosion shots brief, and cut them in with crowd reactions or people running away.

Excerpted from *Digital Video Production Cookbook*, ISBN 0596100310 (O'Reilly Media, Inc.).

Photography courtesy of Chris Kenworthy

A Treehouse Grows in British Columbia

GEORGE DYSON

Three years, 95 feet above the Earth.

David Brower taught me the laws of mountaineering. First: "Climbing is safer than staying home." Second: "Never step on anything you can step over, and never step over anything you can step around." When I left the mountains of California to become a boat builder in British Columbia, I recast this wisdom, in honor of one of the legendary scavengers of the Vancouver waterfront, as Jim Land's Law: "Never buy anything you can make, and never make anything you can find."

In 1972, at age 19 and facing my third winter in Canada, I built a small treehouse 95 feet up in a Douglas fir, and lived there for three years. This adventure began by accident, when the boat I was

"Never buy anything you can make, and never make anything you can find."

deckhanding on hit a large cedar log in Georgia Strait. We towed the log back with us to our anchorage within Vancouver's inner harbor, at the mouth of Indian Arm. As I began to split the log into shakes, I started to think of building a small cabin for the winter — like Malcolm Lowry, who had written his masterpiece *Under the Volcano* while squatting in a shack on the other side of the inlet, now Cates Park.

In the 1970s, the line between legitimate and illegitimate tenancy remained indistinct. Many coastal residents lived in houses built on floats, and when someone slid a house onto the shoreline (almost entirely "Crown land") no one paid much notice — or any tax. The Queen of England (also the Queen of Canada) rarely came by to check. One more cabin in the woods would probably not attract attention, but the rainforest was dark and damp. How about going up?

There happened to be one large, landmark Douglas fir right at the water's edge. The first set of branches were 30 feet up, and that would have been the logical place to build a house. But I kept climbing, and the view just got better and better, until, from near the top of the tree I had a panoramic view from Second Narrows to the 6,565-foot summit of Mount Meslillooet, directly north. There was a clear path through the branches from there to the ground, so I installed a pulley, found a 200-foot length of nylon

Photograph by Ann E. Yow

George Dyson's treehouse, constructed 95 feet above the ground in the branches of a Douglas fir on the shores of Burrard Inlet at Belcarra Park, British Columbia, in 1972. Oliver Thomas, George Dyson, and Claudia Thomas in the doorway of the treehouse. Too short to climb from branch to branch, the children ascended by rope.

rope, and started hauling things up.

For the initial framing, I used alder poles, lashed with two pounds of #15 tarred nylon net-mending twine. This and a length of stovepipe were the only materials I bought to build the house. Fourteen living branches were incorporated into the structure, with all the panels triangulated so that the house could flex in the wind but not twist itself apart.

Once the structure was framed, I installed a floor of salvaged boards, and then began covering the roof with cedar shakes and shingling the outer walls (this requiring some precarious, self-belayed rope work). The house had four windows — two made from tempered glass salvaged from junked console televisions and two octagonal windows salvaged from a remodeled house — and a "Trout" wood-burning stove. Along the other side, supported by one of the internal branches, I built a bunk with firewood storage underneath.

The November gales whistled through the shake walls and roof, so I insulated them with slabs of styrofoam packaging, and then paneled the entire house with split cedar boards on the inside. The cedar grain was so fine (almost 100 growth rings per inch) that some of the boards above my bunk spanned 700 years. During the winters I spent in the treehouse, I read the journals of Cook, Bering, La Perouse, Galiano, Vancouver, and the other early visitors to the Northwest Coast, and pondered how the entire recorded history of British Columbia and Southeast Alaska had only penetrated the outer few inches of the log I had collided with.

I climbed the tree in all conditions, day or night, and never had any misadventure going up, and only one misadventure coming down. I usually rappelled, using the doubled firewood-hauling line, one end clipped into a climber's descending harness and the other end through an aluminum brake. I would leave the house in near free-fall, and then after about 60 feet, start applying the brake. One day, my long, scraggly beard caught in the brake on the way down. *"Merde!"* as they say in Quebec.

I never imagined the treehouse would survive as well as it did. It withstood torrential rains, snow, ice, and winds gusting to 60 knots. During storms, the house gyrated eight or ten feet. When people think of the rainforest, they think damp moss, luxuriant fungi, and rotting wood. But that's on the ground; in the forest canopy, the rain is just passing by. Even though I would leave the house for weeks at a time, I never returned to find dampness, mildew, or mold. The sun hit my windows long before those of anyone else living on the slopes of Indian Arm, and I enjoyed a microclimate completely different from that on the ground.

The three winters I spent in the treehouse were the most memorable of my life. I had no political agenda, nor was I a recluse: I needed a place to live, I liked climbing, and I even entertained visitors now and then. More people should build more treehouses, and we need more Crown land. Sure, at 95 feet, it may have been a little dangerous, but you can get killed walking across the street. As Brower would say, it was probably safer than staying on the ground.

George Dyson, a kayak designer and historian of technology, lives in Bellingham, Wash., and is the author of *Baidarka, Project Orion,* and *Darwin Among the Machines.*

Photograph by Peter Thomas

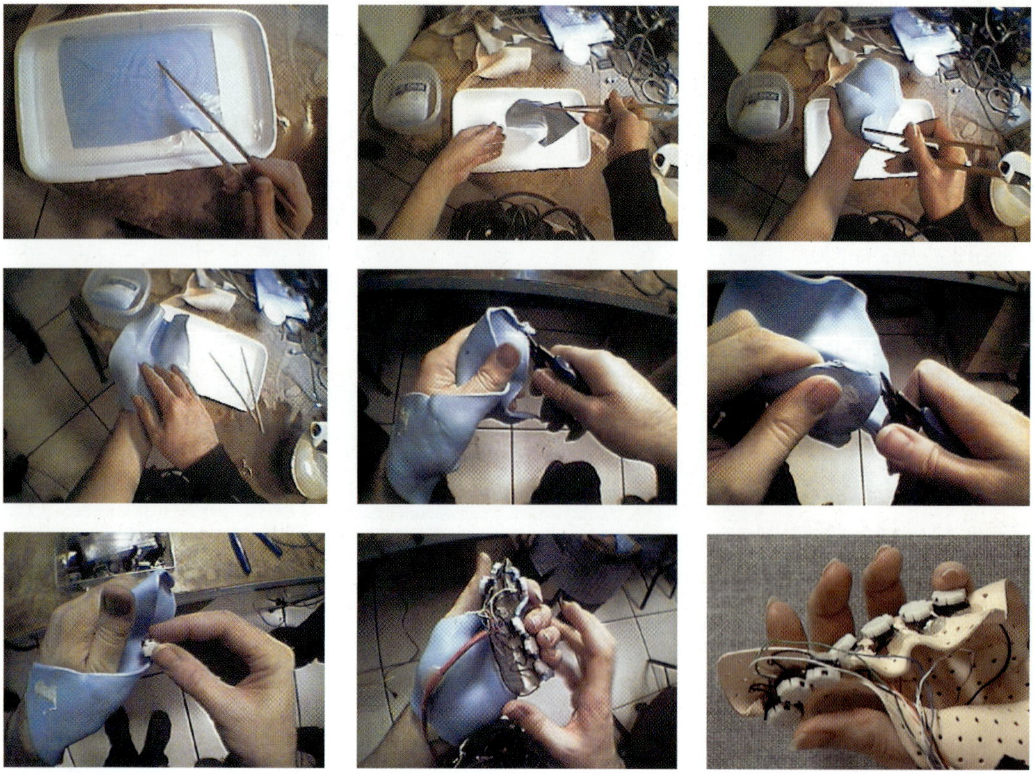

Photography courtesy of Steve Mann

HOMEBREW

My Multiambic Keyer

By Steve Mann

Back in the 1970s, I wanted to be able to type while walking around, or compose music while jogging, so I came up with various devices that were like a keyboard but without the board.

Of these, one of my favorites was a cluster of keys mounted to a piece of plastic that I'd molded to exactly fit my hand.

One of the simplest designs is the pentambic keyer, having 5 switches, one for each finger and thumb, resulting in 17,685 different combinations.

The septambic keyer provides a slight improvement by using 3 thumb switches, resulting in more than 29 million possible combinations, making it possible to map one English word to each possible combination.

In addition to merely typing text, I also use the keyer to compose music, so that I can generate notes in time to my footsteps. This results in the ability to produce music at a nice constant tempo, in time with a steady walk or jog. (I tend to keep jogging in place when a stoplight turns red.)

Periodic structures, like the granite slabs in a public square, or stairs, result in an enhanced sense of timing. For example, the stairs in my building have 9 steps per flight (half floor), so if I always take 3 steps per landing and I get 6 bars of music (in 4/4 time) per floor, going up 3 floors I can play exactly the first verse of a Gershwin lullaby ("Summertime, and the livin' is easy ...").

If I miss a beat, I know right away because I'll end up with an "off-by-one" error when I get to the top of the stairs. Instead of merely using switches, I use pressure sensors, so that the harder I squeeze, the louder the note or chord that is sounded, and thus, the keyer gives rise to a very expressive musical instrument.

If you would like to build a keyer for typing text, or a "musikeyer" for composing music while you jog, you can learn more from wearcam.org/septambi/index.html and from my textbook, *Intelligent Image Processing*, published by John Wiley and Sons.

Email Steve Mann at mann@eecg.toronto.edu.